# Lecture Notes in Artificial Intelligence 11714

Subseries of Lecture Notes in Computer Science

Serenella Cerrito · Andrei Popescu (Eds.)

# Automated Reasoning with Analytic Tableaux and Related Methods

28th International Conference, TABLEAUX 2019
London, UK, September 3–5, 2019
Proceedings

Springer

*Editors*
Serenella Cerrito
IBISC, Univ. Evry,
Université Paris-Saclay
Evry, France

Andrei Popescu
Middlesex University London
London, UK

ISSN 0302-9743 ISSN 1611-3349 (electronic)
Lecture Notes in Artificial Intelligence
ISBN 978-3-030-29025-2 ISBN 978-3-030-29026-9 (eBook)
https://doi.org/10.1007/978-3-030-29026-9

LNCS Sublibrary: SL7 – Artificial Intelligence



This Springer imprint is published by the registered company Springer Nature Switzerland AG
The registered company address is: Gewerbestrasse 11, 6330 Cham, Switzerland

# Preface

*In August 2018 Roy Dyckhoff left us. Roy was a prominent, very active member of the TABLEAUX community since its early period, in the nineties. He worked mainly in logic and proof theory, but his open mind also made him interested in other fields of computer science and mathematics and, beyond, in several other aspects of human activity and intellectual endeavor. Those of us who had the privilege of interacting with him have experienced his exceptional kindness and attention to other people's needs, supplemented by a wonderful sense of humor. Many people in the TABLEAUX community feel that they have lost not only an excellent researcher but also a dear friend, a mentor and, altogether, a beautiful human being. This volume is warmly dedicated to him, on behalf of all the authors and Program Committee members. Farewell, dear Roy!*

These proceedings contain the papers selected for presentation at the 28th International Conference on Automated Reasoning with Analytic Tableaux and Related Methods (TABLEAUX 2019). The conference was held during September 3–5, 2019 in London, UK, at Middlesex University. It was co-located with the 12th International Symposium on Frontiers of Combining Systems (FroCoS 2019).

TABLEAUX is the main international forum for presenting research on all aspects of the mechanization of tableaux-based reasoning and related methods, including theoretical foundations, implementation techniques, systems development, and applications. The first TABLEAUX conference was held in Lautenbach near Karlsruhe, Germany, in 1992. Since then it has been organized on an annual basis. In 2001, 2004, 2006, 2008, 2010, 2012, 2014, 2016, and 2018 it was a constituent of IJCAR.

TABLEAUX 2019 received 43 paper submissions, among which 37 regular research papers, 4 system descriptions, and 2 work-in-progress papers. The submissions were evaluated by the Program Committee on the basis of their significance, novelty, technical soundness, and appropriateness for the TABLEAUX audience. Reviewing was single-blind and each paper was subjected to at least three reviews, followed by a discussion within the Program Committee. In the end, 26 papers were selected for presentation at the conference: 24 regular papers, 1 system description, and 1 work-in-progress. This volume contains the accepted regular and system description papers, which have been grouped according to the following topic classification: (1) tableau calculi, (2) sequent calculi, (3) semantics and combinatorial proofs, (4) non-wellfounded proof systems, (5) automated theorem provers, and (6) logics for program or system verification.

This edition had two invited talks by leading experts in logic and mechanized reasoning:

- "Automated Reasoning for the Working Mathematician," by Jeremy Avigad
- "Remembering Roy Dyckhoff," by Stéphane Graham-Lengrand and Sara Negri

Sara and Stéphane were two of Roy's closest collaborators, and Stéphane had also been his PhD student. Jeremy Avigad's invited talk was shared with FroCoS 2019.

Conversely, one of the FroCoS invited talks, "Modularity and Automated Reasoning in Description Logics," by Uli Sattler, was shared with TABLEAUX.

The joint FroCoS/TABLEAUX event had two affiliated workshops:

- The 25th Workshop on Automated Reasoning (ARW 2019), organized by Alexander Bolotov and Florian Kammueller
- Journeys in Computational Logic: Tributes to Roy Dyckhoff, organized by Stéphane Graham-Lengrand, Ekaterina Komendantskaya, and Mehrnoosh Sadrzadeh

It also had two affiliated tutorials:

- Formalising Concurrent Computation: CLF, Celf, and Applications, by Sonia Marin
- How to Build an Automated Theorem Prover – An Introductory Tutorial (invited TABLEAUX tutorial), by Jens Otten

The program committee offered the TABLEAUX 2019 Best Paper Award to Björn Lellmann for his paper "Combining Monotone and Normal Modal Logic in Nested Sequents – with Countermodels". In addition, this year TABLEAUX also had the Best Paper by a Junior Researcher Award, supported financially by Springer; the award was offered to Timo Lang for his co-authored paper "A Game Model for Proofs with Costs".

We would like to thank all the people who contributed to making TABLEAUX 2019 a success. In particular, we thank the invited speakers for their inspiring talks, the authors for providing many high-quality submissions, the workshop and tutorial organizers for the interesting and engaging events, and all the attendees for contributing to the conference discussion. We thank the Program Committee members and the external reviewers for their careful, competent reviewing and discussion of the submissions on quite a tight schedule. We thank the Steering Committee members for their very helpful advice. We extend our thanks to the local Organizing Committee and to the Middlesex University staff, especially to Nicola Skinner, for offering their enthusiastic support to this event.

We gratefully acknowledge financial support from Amazon, Springer, and Middlesex University. The Association for Symbolic Logic (ASL) has kindly included TABLEAUX among the events for which students can apply to them for travel funding. Finally, we are grateful to EasyChair for allowing us to use their excellent conference management system.

September 2019

Serenella Cerrito
Andrei Popescu

# Organization

## Program Chairs

| | |
|---|---|
| Serenella Cerrito | Univ. Evry, Université Paris-Saclay, France |
| Andrei Popescu | Middlesex University London, UK |

## Local Organization

| | |
|---|---|
| Kelly Androutsopoulos | Middlesex University London, UK |
| Jaap Boender | Middlesex University London, UK |
| Michele Bottone | Middlesex University London, UK |
| Florian Kammueller | Middlesex University London, UK |
| Rajagopal Nagarajan | Middlesex University London, UK |
| Andrei Popescu (Conference Chair) | Middlesex University London, UK |
| Franco Raimondi | Middlesex University London, UK |

## Steering Committee

| | |
|---|---|
| Agata Ciabattoni | Vienna University of Technology, Austria |
| Didier Galmiche (Ex-officio) | LORIA – Université de Lorraine, France |
| Neil Murray | SUNY at Albany, USA |
| Cláudia Nalon (Ex-officio) | University of Brasília, Brazil |
| Hans de Nivelle | Nazarbayev University, Kazakhstan |
| Jens Otten (President) | University of Oslo, Norway |
| Andrei Popescu | Middlesex University London, UK |
| Renate Schmidt | University of Manchester, UK |

## Program Committee

| | |
|---|---|
| Peter Baumgartner | Data61/CSIRO, Australia |
| Maria Paola Bonacina | Universit degli Studi di Verona, Italy |
| James Brotherston | University College London, UK |
| Serenella Cerrito | Univ. Evry, Université Paris-Saclay, France |
| Agata Ciabattoni | Vienna University of Technology, Austria |
| Anupam Das | University of Copenhagen, Denmark |
| Clare Dixon | University of Liverpool, UK |
| Camillo Fiorentini | University of Milano, Italy |
| Pascal Fontaine | LORIA, Inria, University of Lorraine, France |

| | |
|---|---|
| Didier Galmiche | LORIA – Université de Lorraine, France |
| Martin Giese | University of Oslo, Norway |
| Laura Giordano | DISIT, Universit del Piemonte Orientale, Italy |
| Rajeev Goré | The Australian National University, Australia |
| Stéphane Graham-Lengrand | SRI International, USA |
| Reiner Hähnle | TU Darmstadt, Germany |
| Ori Lahav | Tel Aviv University, Israel |
| Tomer Libal | American University of Paris, France |
| George Metcalfe | University of Bern, Switzerland |
| Dale Miller | Inria, LIX/Ecole Polytechnique, France |
| Leonardo de Moura | Microsoft Research, USA |
| Neil Murray | SUNY at Albany, USA |
| Cláudia Nalon | University of Brasília, Brazil |
| Sara Negri | University of Helsinki, Finland |
| Hans de Nivelle | Nazarbayev University, Kazakhstan |
| Nicola Olivetti | LSIS, Aix-Marseille Universit, France |
| Jens Otten | University of Oslo, Norway |
| Valeria de Paiva | Nuance Communications, USA |
| Nicolas Peltier | CNRS, Laboratoire d'Informatique de Grenoble, France |
| Elaine Pimentel | Universidade Federal do Rio Grande do Norte, Brazil |
| Francesca Poggiolesi | CNRS, IHST Paris, France |
| Andrei Popescu | Middlesex University London, UK |
| Gian Luca Pozzato | University of Turin, Italy |
| Giles Reger | University of Manchester, UK |
| Giselle Reis | Carnegie Mellon University, Qatar |
| Renate Schmidt | University of Manchester, UK |
| Viorica Sofronie-Stokkermans | Universität Koblenz-Landau, Germany |
| Alwen Tiu | The Australian National University, Australia |
| Sophie Tourret | Max-Planck-Institut für Informatik, Saarbrücken, Germany |
| Dmitriy Traytel | ETH Zurich, Switzerland |
| Josef Urban | CIIRC, Czech Republic |
| Luca Viganö | King's College London, UK |
| Uwe Waldmann | Max-Planck-Institut für Informatik, Saarbrücken, Germany |
| Bruno Woltzenlogel Paleo | Vienna University of Technology, Austria |

## External Reviewers

Ruba Alassaf
Michael Benedikt
Julien Braine
Chad Brown
Karel Chvalovský
Tiziano Dalmonte
Simon Docherty
Mauro Ferrari
Marianna Girlando
Eduard Kamburjan

Max Kanovich
Stepan Kuznetsov
Peter Lammich
Dominique Larchey-Wendling
Sonia Marin
Eugenio Orlandelli
Edi Pavlovic
Benjamin Ralph
Christian Retoré
Reuben Rowe
Traian Florin Serbănută
Carsten Sinz
Dominic Steinhöfel
Lutz Straßburger
Yoni Zohar

# Abstracts of Invited Talks

# Automated Reasoning for the Working Mathematician

Jeremy Avigad

Carnegie Mellon University, Pittsburgh, Pennsylvania, USA
avigad@cmu.edu
http://www.andrew.cmu.edu/user/avigad/

**Abstract.** The mathematical literature is filled with minor errors and imprecision, and interactive proof assistants offer hope for making mathematics more reliable and exact. Given the gap between an informal proof and a formal derivation, one would expect automated reasoning tools to play a key role in formally verified mathematics. But this expectation has not been borne out in practice. Despite technological advances, automated reasoning is far from central to the field, and many of the most impressive accomplishments to date have used surprisingly little automation. The use of automated reasoning tools in mathematical discovery has been even more limited. In this talk, I will do my best to make sense of this state of affairs and offer guidance towards developing useful mathematical tools.

**Keywords:** Automated reasoning · Interactive theorem proving

# Remembering Roy Dyckhoff

Stéphane Graham-Lengrand[1] and Sara Negri[2]

[1] SRI International
stephane.graham-lengrand@csl.sri.com
[2] Department of Philosophy, University of Helsinki, Finland
sara.negri@helsinki.fi

**Abstract.** Roy Dyckhoff left us after a long illness in August 2018. Many of us have known him as a teacher, colleague, mentor, friend, collaborator, and coauthor. He is much missed in the academic world, and especially in the Tableaux community, community, of which he was a founding father and an extremely active member. We shall remember Roy as a scientist with a broad range of interests, care for rigour, passionate approach to new ideas, and enthusiasm for new projects. He showed a human approach to scientific endeavour, had great care for acknowledging priorities, was generous in helping others and did not spare his personal involvement in easing conflicts and striving for justice.

In his early years as a researcher, those of doctoral and postdoctoral study, Roy Dyckhoff gave substantial contributions to topology and category theory [2–8, 25], the latter also studied in relation to Martin-Löf's type theory [9]. Moving from mathematics to computer science he became interested in computational logic. In the early 1990s he started a systematic study of the use of sequent calculus as a basis for automated deduction, his most influential discovery being a terminating sequent calculus for intuitionistic propositional logic, known as G4 and published in 1992 [10]. He was not content with just any solution, but was always looking for the most elegant one. So he returned recently to this issue, improving the proof of the main result [15].

Intuitionistic logic was a main thread of his research; to use his words, he surveyed "the wide range of proof systems proposed for intuitionistic logic, emphasising the differences and their design for different purposes, ranging from ease of philosophical or other semantic justification through programming language semantics to automated reasoning" [11] as well as decision procedures and implementations thereof [12]. In investigating the relationship between natural deduction and sequent calculus he settled an old problem on the relationship between cut elimination, substitution and normalisation [13]. Furthermore, he studied the translations from intermediate logics to their modal companions as well as to the provability logic of Grzegorczyk logic, thus offering a fresh proof theoretic perspective on earlier semantical results [20, 22].

By his contributions relating sequent calculus and natural deduction he shed light on the connections between logic programming and functional programming [33], for instance regarding the concept of *uniform proof* [32]. Roy Dyckhoff was appealed by the use of term rewriting techniques in proof theory, and explored innovative extensions of the Curry-Howard-De Bruijn correspondence, which relates formulae to types and proofs to functional programs. He contributed to the development of proof-term grammars and typing systems corresponding to various sequent calculi, with the notion of cut giving a natural typing rule for explicit substitutions, and with cut-elimination being expressed as terminating proof-term normalisation procedures. His contributions to this approach involve for instance the focussed sequent calculi LJT [24] and LJQ [17], the sequent calculus G4 [18], and Pure Type Systems [29].

Roy Dyckhoff gave important contributions to the method of "axioms as rules"; in particular he proposed a view of rules as rewrite conditions and applied it to obtain a simple decision method, based on terminating proof search in a suitable sequent calculus, for the fragment of positively quantified formulas of the first-order theory of linearly ordered Heyting algebras [19]. Recent work [21] broadens the range of applications of the methodology of "axioms-as-rules." Not only many interesting mathematical theories can be expressed by means of coherent/geometric implications, classes of axioms that can be easily turned into rules, but any first order theory is amenable to such a treatment insofar it has a coherent conservative extension. Often classical conversion steps, such as those based on conjunctive and disjunctive normal form can (and should) be avoided. For this purpose, he devised a new algorithm of "coherentization" that preserves as much as possible of the formula structure.

Roy Dyckhoff investigated proof theory also from the more general perspective of proof-theoretic semantics, in particular various notions of harmony [26], the question of what it is to be a logical constant, favouring the view that leads to strong normalisation results, and the relationship between general and "flattened" elimination rules [14]. He also developed a proof-theoretic semantics for a fragment of natural language as an alternative to the traditional model-theoretic semantics [27, 28].

His scientific interests included systems of multimodal logics for encoding and reasoning about information and misinformation in multi-agent systems [23, 34]. For such logics he employed nested sequent calculi, a formalism beyond traditional Gentzen sequents and gave a Prolog implementation of a decision procedure [30].

Roy Dyckhoff has always been fascinated by the challenge of understanding the classics by modern means, as he did for Frege's Begriffsschrift notation [31]. More recently, he was engaged with Stoic logic. In [1] he showed that the extension of the Hertz-Gentzen Systems of 1933 (without thinning) by a rule and certain Stoic axioms preserves analyticity, which in turn yields decidability of propositional Stoic logic. His latest publication [16] shows how the rule of indirect proof, in the form with no multiple or vacuous discharges used by Aristotle, may be dispensed with, in a system comprising four basic rules of subalternation or conversion and six basic syllogisms.

As is clear from this necessarily incomplete summary, Roy Dyckhoff had a proactive attitude that fostered collaboration. He always took genuine interest in the work of others. Many researchers across computer science, mathematics, and philosophy profited from his wide knowledge, deep insights and open-minded approach. He was exemplary in giving credit to others rather than claiming it for himself, and in setting high standards, while at the same time being gracious to those who did not meet them. His humility and his approach to academic life will continue to be an inspiration to all.

## References


1. Bobzien, S., Dyckhoff, R.: Analyticity, balance and non-admissibility of cut in Stoic logic. Studia Logica **107**, 375–397 (2018)
2. Collins, P., Dyckhoff, R.: Connexion properties and factorisation theorems. Quaestiones Mathematicae **2**, 103–112 (1977)
3. Dyckhoff, R.: Factorisation theorems and projective spaces in topology. Mathematische Zeitschrift **127**, 256–264 (1972)
4. Dyckhoff, R.: Perfect light maps as inverse limits. Q. J. Math. **25**, 441–449 (1974)
5. Dyckhoff, R.: Simple classes of complete spaces. J. Lond. Math. Soci. **2**, 142–144 (1974)
6. Dyckhoff, R.: Projective resolutions of topological spaces. J. Pure Appl. Algebra **7**, 115–119 (1976)
7. Dyckhoff, R.: Categorical cuts. Gen. Topol. Appl. **6**, 291–295 (1976)
8. Dyckhoff, R.: Total reflections, partial products, and hereditary factorizations. Topol. Appl. **17**, 101–113 (1984)
9. Dyckhoff, R.: Category theory as an extension of Martin-Löf type theory. Technical report, Department of Computer Science, University of St. Andrews (1985)
10. Dyckhoff, R.: Contraction-free sequent calculi for intuitionistic logic. J. Symb. Log. **57**, 795–807 (1992)
11. Dyckhoff, R.: Proof Systems for Intuitionistic Logic. Abstract of a talk at the University of Oxford, 24 October 2008
12. Dyckhoff, R.: Intuitionistic decision procedures since Gentzen. In: Kahle, R., Strahm, T., Studer, T. (eds.) Advances in Proof Theory. Progress in Computer Science and Applied Logic, vol. 28, pp. 245–267. Birkhäuser Basel, Cham (2016)
13. Dyckhoff, R.: Cut elimination, substitution and normalisation. In: Wansing, H. (ed.) Dag Prawitz on Proofs and Meaning. Outstanding Contributions to Logic, vol. 7, pp. 163–187. Springer, Cham (2015). https://doi.org/10.1007/978-3-319-11041-7_7
14. Dyckhoff, R.: Some remarks on proof-theoretic semantics. In: Piecha, T., Schroeder-Heister, P. (eds.) Advances in Proof-Theoretic, Semantics. Trends in Logic (Studia Logica Library), vol. 43, pp. 79–93. Springer, Cham (2016). https://doi.org/10.1007/978-3-319-22686-6_5
15. Dyckhoff, R.: Contraction-free sequent calculi for intuitionistic logic: a correction. J. Symb. Log. **83**, 1680–1682 (2018)
16. Dyckhoff, R.: Indefinite proof and inversions of syllogisms. Bull. Symb. Log. (in press)
17. Dyckhoff, R., Graham-Lengrand, S.: Call-by-value $\lambda$-calculus and LJQ. J. Log. Comput. **17**, 1109–1134 (2007)
18. Dyckhoff, R., Kesner, D., Lengrand, S.: Strong cut-elimination systems for Hudelmaier's depth-bounded sequent calculus for implicational logic. In: Furbach, U., Shankar, N. (eds.) Automated Reasoning. IJCAR 2006. LNCS, vol. 4130, pp. 347–361. Springer, Heidelberg (2006). https://doi.org/10.1007/11814771_31
19. Dyckhoff, R., Negri, S.: Decision methods for linearly ordered Heyting algebras. Arch. Math. Log. **45**, 411–422 (2006)
20. Dyckhoff, R., Negri, S.: Proof analysis in intermediate logics. Arch. Math. Log. **51**, 71–92 (2012)
21. Dyckhoff, R., Negri, S.: Geometrisation of first-order logic. Bull. Symb. Log. **21**, 123–163 (2015)
22. Dyckhoff, R., Negri, S.: A cut-free sequent system for Grzegorczyk logic, with an application to the Godel-McKinsey-Tarski embedding. J. Log. Comput. **26**, 169–187 (2016)

23. Dyckhoff, R., Sadrzadeh, M., Truffaut, J.: Algebra, proof theory and applications for an intuitionistic logic of propositions, actions and adjoint modal operators. ACM Trans. Comput. Log. **14**, 1–37 (2013)
24. Dyckhoff, R., Urban, C.: Strong normalization of Herbelin's explicit substitution calculus with substitution propagation. J. Log. Comput. **13**, 689–706 (2003)
25. Dyckhoff, R., Tholen, W.: Exponentiable morphisms, partial products and pullback complements. J. Pure App. Algebra **49**, 103–116 (1987)
26. Francez, N., Dyckhoff, R.: A note on harmony. J. Philos. Log. **41**, 1–16 (2007)
27. Francez, N., Dyckhoff, R.: Proof-theoretic semantics for a natural language fragment. Linguist. Philos. **33**, 56–71 (2009)
28. Francez, N., Dyckhoff, R., Ben-Avi, G.: Proof-theoretic semantics for subsentential phrases. Studia Logica **94**, 381–401 (2010)
29. Graham-Lengrand, S., Dyckhoff, R., McKinna, J.: A focused sequent calculus framework for proof search in pure type systems. Log. Methods Comput. Sci. **7** (2010)
30. Kriener, J., Sadrzadeh, M., Dyckhoff, R.: Implementation of a cut-free sequent calculus for logics with adjoint modalities (2009)
31. Macinnis, R., McKinna, J., Parsons, J., Dyckhoff, R.: A mechanised environment for Frege's Begriffsschrift notation. In: Proceedings of Mathematical User-Interfaces Workshop 2004, Bialowieza, Poland (2004)
32. Pinto, L., Dyckhoff, R.: Uniform proofs and natural deductions. In: Galmiche, D., Wallen, L. (eds.) Proof Search in Type-Theoretic Languages; Proceedings of the 12th International Conference on Automated Deduction (CADE-12), pp. 17–23, Nancy, France (1994)
33. Pinto, L., Dyckhoff, R.: A constructive type system to integrate logic and functional programming. In: Galmiche, D., Wallen, L. (eds.) Proof Search in Type-Theoretic Languages; Proceedings of the 12th International Conference on Automated Deduction (CADE-12), pp. 70–81, Nancy, France (1994)
34. Sadrzadeh, M., Dyckhoff, R.: Positive logic with adjoint modalities: proof theory, semantics and reasoning about information. Electron. Notes Theoret. Comput. Sci. **3**, 451–470 (2009)

# Contents

## Semantics and Combinatorial Proofs



## Non-Wellfounded Proof Systems



## Automated Theorem Provers



## Logics for Program or System Verification

# Tableau Calculi

# A SAT-Based Encoding of the One-Pass and Tree-Shaped Tableau System for LTL

Luca Geatti[1,2], Nicola Gigante[1(✉)], and Angelo Montanari[1]

[1] University of Udine, Udine, Italy
{nicola.gigante,angelo.montanari}@uniud.it
[2] Fondazione Bruno Kessler, Trento, Italy
lgeatti@fbk.eu

**Abstract.** A new one-pass and tree-shaped tableau system for LTL satisfiability checking has been recently proposed, where each branch can be explored independently from others and, furthermore, directly corresponds to a potential model of the formula. Despite its simplicity, it proved itself to be effective in practice. In this paper, we provide a SAT-based encoding of such a tableau system, based on the technique of *bounded satisfiability checking*. Starting with a single-node tableau, *i.e.*, depth $k$ of the tree-shaped tableau equal to zero, we proceed in an incremental fashion. At each iteration, the tableau rules are encoded in a Boolean formula, representing all branches of the tableau up to the current depth $k$. A typical downside of such bounded techniques is the effort needed to understand when to stop incrementing the bound, to guarantee the completeness of the procedure. In contrast, termination and completeness of the proposed algorithm is guaranteed without computing any upper bound to the length of candidate models, thanks to the Boolean encoding of the PRUNE rule of the original tableau system. We conclude the paper by describing a tool that implements our procedure, and comparing its performance with other state-of-the-art LTL solvers.

**Keywords:** Tableau system · Temporal logic · Satisfiability · SAT

## 1 Introduction

*Linear Temporal Logic* (LTL) is one of the most used temporal logics in formal verification. In this context, the main problem is *model checking* [9], *i.e.*, deciding whether a given specification is satisfied by a given system. However, since testing a system against a valid or unsatisfiable formula can be useless at best, and dangerous at worst, *sanity checking* of specifications is another important step in model-based design [27]. For this reason, the *satisfiability problem*, *i.e.*, establishing whether a formula admits any model in the first place, has been given an important amount of research effort. In addition to its applications to formal verification, it also plays a role in AI systems [16,20], *e.g.*, in planning problems.


S. Cerrito and A. Popescu (Eds.): TABLEAUX 2019, LNAI 11714, pp. 3–20, 2019.
https://doi.org/10.1007/978-3-030-29026-9_1

Besides its relevant applications, the LTL satisfiability problem is theoretically by itself. Since the first computational complexity results [25], many techniques have been devised over the last decades, with *tableau methods* being among the first to be developed [18,19,24]. In contrast to earlier tableau methods for classical logic [4,10], that work by building a suitable derivation tree, most of these methods build a *graph* structure, whose paths represent possible evolutions of the computation, and then look for those ones that satisfy all the properties required by the formula. Recently, a novel one-pass tree-shaped tableau for LTL has been proposed by Reynolds [22]. In contrast to other tree-shaped systems [24], its novel termination condition allows each branch to be independently explored and accepted or rejected. Moreover, there is a direct relationship between the tableau branches and the models of the formula. These features led to an efficient implementation [3], a simple and fruitful parallelisation [21], and modular extensions to more expressive logics [13,14].

In this paper, we propose a satisfiability checking procedure for LTL formulae based on a *SAT encoding* of the one-pass and tree-shaped tableau by Reynolds [22]. The tableau tree is (symbolically) built in a breadth-first way, by means of Boolean formulae that encode all the tableau branches up to a given depth $k$, which is increased at every step. The expansion rules of the tableau system are encoded in the formulae in such a way that a successful assignment represents a branch of the tree of length $k$, which *directly* corresponds to a model of the original LTL formula. This breadth-first iterative deepening approach has been exploited in the past by *bounded satisfiability checking* and *bounded model checking* algorithms [7,15], which share with us the advantage of leveraging the great progress of SAT solvers in the last decades, and the *incrementality* of such solvers.

A common drawback of existing bounded satisfiability checking methods is the difficulty in identifying when to stop the search in the case of *unsatisfiable* formulae. In order to ensure termination, either a global upper bound has to be computed in advance, which is not always possible or feasible, or some other techniques are needed to identify where the search can be stopped. In our system, termination is guaranteed by a suitable encoding of the tableau's PRUNE rule. This rule was the main novelty of Reynolds' one-pass and tree-shaped system when it was originally proposed [22], has a clean model-theoretic interpretation [13], and the important role it plays in our encoding adds up to its interesting properties. The result is a simple and complete bounded satisfiability checking procedure based on a small and much simpler SAT encoding.

We implemented the proposed procedure and encoding in a tool, called BLACK for (Bounded LTL sAtisfiability ChecKer), and we report the outcomes of an initial experimental evaluation, comparing it with state-of-the-art tools. The results are promising, consistently improving over the tableau explicit construction.

The paper proceeds as follows. Section 2 includes a brief account of LTL and of Reynolds' one-pass and tree-shaped tableau system. Section 3 shows the base encoding of the tableau rules, excepting the PRUNE rule, building a system that terminates correctly on satisfiable instances. Later, Sect. 4 describes and discusses the encoding of the PRUNE rule, completing the procedure. Section 5 describes the BLACK tool, together with the results of the experimental evaluation. Section 6 concludes and highlights possible future developments.

## 2 Preliminaries

### 2.1 Linear Temporal Logic

*Linear Temporal Logic* (LTL) is a propositional modal logic interpreted over infinite (discrete) linear orders. Syntactically, LTL can be viewed as an extension of propositional logic with the *tomorrow* ($\mathsf{X}\,\phi$), *until* ($\alpha\,\mathcal{U}\,\beta$), and *release* ($\alpha\,\mathcal{R}\,\beta$) operators. Given a set $\Sigma = \{p, q, r, \ldots\}$ of atomic propositions, LTL formulae are inductively defined as follows:

$$\begin{aligned} \phi := p \mid \neg\phi_1 \mid \phi_1 \vee \phi_2 \mid \phi_1 \wedge \phi_2 & \qquad \text{Boolean operators} \\ \mid \mathsf{X}\,\phi_1 \mid \phi_1\,\mathcal{U}\,\phi_2 \mid \phi_1\,\mathcal{R}\,\phi_2 & \qquad \text{temporal operators} \end{aligned}$$

Note that, given disjunctions and the *until* operator, conjunctions and the *release* operator are not necessary (in particular, $\phi_1\,\mathcal{R}\,\phi_2 \equiv \neg(\neg\phi_1\,\mathcal{U}\,\neg\phi_2)$). However, it is useful to consider them as primitive, in order to allow any LTL formula $\phi$ to be put into *negated normal form*, producing a linear-size equivalent formula, noted as $\mathrm{nnf}(\phi)$, such that negations appear only applied to proposition letters. Moreover, common shorthands can be defined, such as the *eventually* ($\mathsf{F}\,\phi_1 \equiv \top\,\mathcal{U}\,\phi_1$) and *always* ($\mathsf{G}\,\phi_1 \equiv \neg\,\mathsf{F}(\neg\phi_1)$) operators.

LTL formulae are interpreted over infinite *state sequences* $\overline{\sigma} = \langle\sigma_0, \sigma_1, \ldots\rangle$, with $\sigma_i \subseteq \Sigma$ for each $i \geq 0$. Given a state sequence $\overline{\sigma}$, a position $i \geq 0$, and an LTL formula $\phi$, the satisfaction of $\phi$ by $\overline{\sigma}$ at position $i$, written $\overline{\sigma} \models_i \phi$, is inductively defined as follows:

1. $\overline{\sigma} \models_i p$ iff $p \in \sigma_i$
2. $\overline{\sigma} \models_i \neg\phi$ iff $\overline{\sigma} \not\models_i \phi$
3. $\overline{\sigma} \models_i \phi_1 \vee \phi_2$ iff either $\overline{\sigma} \models_i \phi_1$ or $\overline{\sigma} \models_i \phi_2$
4. $\overline{\sigma} \models_i \phi_1 \wedge \phi_2$ iff $\overline{\sigma} \models_i \phi_1$ and $\overline{\sigma} \models_i \phi_2$
5. $\overline{\sigma} \models_i \mathsf{X}\,\phi$ iff $\overline{\sigma} \models_{i+1} \phi$
6. $\overline{\sigma} \models_i \phi_1\,\mathcal{U}\,\phi_2$ iff there exists $j \geq i$ such that $\overline{\sigma} \models_j \phi_2$ and $\overline{\sigma} \models_k \phi_1$ for all $i \leq k < j$
7. $\overline{\sigma} \models_i \phi_1\,\mathcal{R}\,\phi_2$ iff for all $j \geq i$, either $\overline{\sigma} \models_j \phi_2$ or there exists $i \leq k < j$ such that $\overline{\sigma} \models_k \phi_1$.

We say that $\overline{\sigma}$ *satisfies* $\phi$, written $\overline{\sigma} \models \phi$, if and only if the state sequence $\overline{\sigma}$ satisfies $\phi$ at its first state, *i.e.*, $\overline{\sigma} \models_0 \phi$. In this case, we say that $\overline{\sigma}$ is a *model* of $\phi$.

### 2.2 The One-Pass and Tree-Shaped Tableau System

We now describe Reynolds' tableau system for LTL. After its original formulation in [22], the system was extended to support past operators [14] and more expressive real-time logics [13]. Here, we briefly recall its original future-only version, which is the one considered for the SAT encoding described in the next section.

**Table 1.** Tableau expansion rules. For each formula $\phi$ found in the label $\Gamma$ of a node $u$, one or two children $u'$ and $u''$, according to its type, are created with the same label as $u$ excepting for $\phi$, which is replaced, respectively, by the formulae from $\Gamma_1(\phi)$ and $\Gamma_2(\phi)$.

| Rule | $\phi$ | $\Gamma_1(\phi)$ | $\Gamma_2(\phi)$ |
|---|---|---|---|
| DISJUNCTION | $\alpha \vee \beta$ | $\{\alpha\}$ | $\{\beta\}$ |
| UNTIL | $\alpha \mathcal{U} \beta$ | $\{\beta\}$ | $\{\alpha, \mathsf{X}(\alpha \mathcal{U} \beta)\}$ |
| RELEASE | $\alpha \mathcal{R} \beta$ | $\{\alpha, \beta\}$ | $\{\beta, \mathsf{X}(\alpha \mathcal{R} \beta)\}$ |
| EVENTUALLY | $\mathsf{F}\,\beta$ | $\{\beta\}$ | $\{\mathsf{X}\,\mathsf{F}\,\beta\}$ |
| CONJUCTION | $\alpha \wedge \beta$ | $\{\alpha, \beta\}$ | |
| ALWAYS | $\mathsf{G}\,\alpha$ | $\{\alpha, \mathsf{X}\,\mathsf{G}\,\alpha\}$ | |

The tableau for a formula $\phi$ is a tree where each node $u$ is labelled by a set $\Gamma(u)$ of formulae from the closure $\mathcal{C}(\phi)$ of $\phi$. At each step of the construction, a set of rules is applied to each leaf node. Each rule can possibly append one or more children to the node, or either *accept* (✓) or *reject* (✗) the node. The construction continues until all leaves are either accepted or rejected, resulting into at least one accepted leaf if and only if the formula is satisfiable, with the corresponding branch representing a satisfying model for the formula. A node whose label contains only *elementary* formulae, *i.e.*, propositions or *tomorrow* operators, is called a *poised* node. At each step, the *expansion rules* are applied to any non-poised leaf node. The rules are given in Table 1. For each non-elementary formula $\psi \in \mathcal{C}(\phi)$, the corresponding expansion rule defines two sets of expanded formulae $\Gamma_1(\psi)$ and $\Gamma_2(\psi)$, with the latter possibly empty. The application of the rule to a node $u$ adds a child $u'$ to $u$ such that $\Gamma(u') = \Gamma(u) \setminus \{\psi\} \cup \Gamma_1(\psi)$, and, if $\Gamma_2(\psi) \neq \varnothing$, a second child $u''$ such that $\Gamma(u'') = \Gamma(u) \setminus \{\psi\} \cup \Gamma_2(\psi)$.

Expansion rules are applied to non-poised nodes until a poised node is produced. Then, a number of *termination rules* are applied, to decide whether the node can be accepted, rejected, or the construction can proceed. In what follows, a formula of the type $\mathsf{X}(\alpha \mathcal{U} \beta)$ is called $\mathsf{X}$-*eventuality*. Given a branch $\overline{u} = \langle u_0, \ldots, u_n \rangle$, an $\mathsf{X}$-eventuality $\psi$ is said to be *requested* in some node $u_i$ if $\psi \in \Gamma(u_i)$, and *fulfilled* in some node $u_j$, with $j \geq i$, if $\beta \in \Gamma(u_j)$.

Let $\overline{u} = \langle u_0, \ldots, u_n \rangle$ be a branch with poised leaf $u_n$. The termination rules are the following, to be applied in the given order:

EMPTY If $\Gamma(u_n) = \varnothing$, then $u_n$ is *accepted.*

CONTRADICTION If $\{p, \neg p\} \subseteq \Gamma(u_n)$, for some $p \in \Sigma$, then $u_n$ is *rejected.*

LOOP If there is a poised node $u_i < u_n$ such that $\Gamma(u_n) = \Gamma(u_i)$, and all the X-eventualities requested in $u_i$ are fulfilled in the nodes between $u_{i+1}$ and $u_n$, then $u_n$ is *accepted.*

PRUNE If there are three positions $i < j < n$, such that $\Gamma(u_i) = \Gamma(u_j) = \Gamma(u_n)$, and among the X-eventualities requested in these nodes, all those fulfilled between $u_{j+1}$ and $u_n$ are fulfilled between $u_{i+1}$ and $u_j$ as well, then $u_n$ is *rejected.*

If the branch is neither accepted nor rejected, the construction of the branch proceeds to the next temporal step by applying the STEP rule.

STEP A child $u_{n+1}$ is added to $u_n$ such that $\Gamma(u_{n+1}) = \{\psi \mid \mathsf{X}\,\psi \in \Gamma(u_n)\}$.

Intuitively, given an accepted branch of the complete tableau for $\phi$, the poised nodes are labelled by the formulae that hold in the states of the corresponding model for the formula. Depending on whether the branch is accepted by the EMPTY or the LOOP rule, it either corresponds to a finite (also called *loop-free*) model or to a periodic one (also called *lasso-shaped*), whose period corresponds to the segment in between the nodes that trigger the LOOP rule. If a branch is rejected, it happens either because of a logical contradiction, that triggers the CONTRADICTION rule, or because of the PRUNE rule, which avoids the tableau to infinitely postpone a request that is impossible to fulfil. From a model-theoretic point of view [13], the PRUNE rule allows one not to consider models that contain *redundant segments*, *i.e.*, segments that just repeat some previously done piece of work without contributing further to the satisfaction of all the pending requests. Recent work [13] studied this model-theoretic interpretation of the rule, showing a characterisation of the discarded models.

## 3 SAT-Based Encoding of the Tableau

This section describes the SAT-based encoding of Reynolds' tableau. We first describe the base encoding, leaving the PRUNE rule to the next section, which shows the complete satisfiability checking procedure.

As already pointed out, the overall structure of our procedure is similar to other *bounded satisfiability checking* approaches. At each step $k$, ranging from zero upwards, we produce a Boolean formula $|\phi|^k$, which represents all the accepted branches of the tableau of depth at most $k$. The satisfaction of such a formula witnesses the existence of an accepted branch of the tableau, which in turn proves the existence of a model for the formula. If the formula is unsatisfiable, we can proceed to the next depth level. Note that this corresponds to a symbolic breadth-first traversal of the complete tableau for $\phi$.

Such a procedure would be incomplete, possibly running forever on some unsatisfiable instances, without some halting criterion, which in our case is provided by the encoding of the PRUNE rule as described in Sect. 4. Let us now proceed with the description of the base encoding. In what follows, any LTL formula is assumed to be in *negated normal form.*

### 3.1 Notation

We now define some notation, useful for what follows. Let $\phi$ be an LTL formula (in negated normal form) over the alphabet $\Sigma$. The *closure* of $\phi$ is the set of formulae $\mathcal{C}(\phi)$ defined as follows:

1. $\phi \in \mathcal{C}(\phi)$;
2. if $\mathsf{X}\,\psi \in \mathcal{C}(\phi)$, then $\psi \in \mathcal{C}(\phi)$;
3. if $\psi \in \mathcal{C}(\phi)$, then $\Gamma_1(\phi) \subseteq \mathcal{C}(\phi)$ and $\Gamma_2(\phi) \subseteq \mathcal{C}(\phi)$ (as defined in Table 1).

Then, let $\mathsf{XR}(\phi) \subseteq \mathcal{C}(\phi)$ be the set of all the *tomorrow* formulae (X-*requests*) in $\mathcal{C}(\phi)$, *i.e.*, all the formulae $\mathsf{X}\,\psi \in \mathcal{C}(\phi)$, and let $\mathsf{XEV} \subseteq \mathsf{XR}(\phi)$ be the set of all the X-eventualities in $\mathcal{C}(\phi)$, *i.e.*, all the formulae $\mathsf{X}(\alpha\,\mathcal{U}\,\beta) \in \mathcal{C}(\phi)$.

The propositional encoding of the formula $\phi$ is defined over an extended alphabet $\Sigma_+$, which includes:

1. any proposition from the original alphabet $\Sigma$;
2. the *grounded* X-*requests*, *i.e.*, a proposition noted as $\psi_G$ for all $\psi \in \mathsf{XR}(\phi)$;
3. a *stepped* version $p^k$, for any $k \in \mathbb{N}$, of all the propositions $p$ above, with $p^0$ identified as $p$.

Some notation complements the above extended propositions. In particular, for all $\psi \in \mathcal{C}(\phi)$, we denote by $\psi_G$ the formula obtained by replacing $\rho$ with $\rho_G$ for any $\rho \in \mathsf{XR}(\phi)$ appearing in $\psi$. Similarly, for all $\psi \in \mathcal{C}(\phi)$, we denote as $\psi^k$, with $k \in \mathbb{N}$, the formula obtained from $\psi$ by replacing any proposition $p$ with $p^k$. Intuitively, different stepped versions of the same proposition $p$ are used to represent the value of $p$ at different states. From now on, for any formula $\psi \in \mathcal{C}(\phi)$, we will write $\psi_G^k$ as a shorthand for the formula $((\psi)_G)^k$.

Finally, we recall the definition of a simple transformation of LTL formulae which is heavily used in our encoding.

**Definition 1** (Next Normal Form). *An LTL formula $\phi$ is in* next normal form *iff every* until *or* release *subformula appears in the operand of a* tomorrow.

An LTL formula $\phi$ can be turned into its *next normal form* equivalent formula $\mathrm{xnf}(\phi)$ as follows:

1. $\mathrm{xnf}(p) \equiv p$ and $\mathrm{xnf}(\neg p) = \neg p$ for all $p \in \Sigma$;
2. $\mathrm{xnf}(\mathsf{X}\,\psi_1) \equiv \mathsf{X}\,\psi_1$ for all $\mathsf{X}\,\psi_1 \in \mathcal{C}(\phi)$;
3. $\mathrm{xnf}(\psi_1 \wedge \psi_2) \equiv \mathrm{xnf}(\psi_1) \wedge \mathrm{xnf}(\psi_2)$ for all $\psi_1$ and $\psi_2$;
4. $\mathrm{xnf}(\psi_1 \vee \psi_2) \equiv \mathrm{xnf}(\psi_1) \vee \mathrm{xnf}(\psi_2)$ for all $\psi_1$ and $\psi_2$;
5. $\mathrm{xnf}(\psi_1\,\mathcal{U}\,\psi_2) \equiv \mathrm{xnf}(\psi_2) \vee (\mathrm{xnf}(\psi_1) \wedge \mathsf{X}(\psi_1\,\mathcal{U}\,\psi_2))$ for all $\psi_1$ and $\psi_2$;
6. $\mathrm{xnf}(\psi_1\,\mathcal{R}\,\psi_2) \equiv \mathrm{xnf}(\psi_2) \wedge (\mathrm{xnf}(\psi_1) \vee \mathsf{X}(\psi_1\,\mathcal{R}\,\psi_2))$ for all $\psi_1$ and $\psi_2$.

The above definition has been recalled by other authors as well [17], but it follows the same structure of the expansion rules defined in Table 1, which is not surprising, since these rules trace back to earlier graph-shaped tableaux [18, 19]. This connection allows us to check that the above definition produces an equivalent formula, as $\psi \equiv \Gamma_1(\psi) \vee \Gamma_2(\psi)$ for all the cases covered by Table 1.

### 3.2 Expansion of the Tree

We can now define the first building block of our encoding. The *k-unravelling* of $\phi$, denoted as $[\![\phi]\!]^k$, is a propositional formula that encodes the expansion of all the branches of the tableau tree up to at most $k+1$ poised nodes per branch.

**Definition 2** ($k$-unraveling). *Let $\phi$ be an LTL formula over $\Sigma$ and some $k \in \mathbb{N}$. The* $k$-unravelling *of $\phi$ is a propositional formula $[\![\phi]\!]^k$ over $\Sigma_+$ defined as follows:*

$$\begin{aligned} [\![\phi]\!]^0 &= \mathrm{xnf}(\phi)_G \\ [\![\phi]\!]^{k+1} &= [\![\phi]\!]^k \wedge \textstyle\bigwedge_{\mathsf{X}\,\alpha \in \mathsf{XR}} \left( (\mathsf{X}\,\alpha)^k_G \leftrightarrow \mathrm{xnf}(\alpha)^{k+1}_G \right) \end{aligned}$$

Although such branches may in general have different length, they can be regarded as having the same depth as far as the corresponding model is concerned, since each state corresponds to a poised node. Thus, we may regard the $k$-unravelling as a symbolic encoding of a breadth-first traversal of the tree. The formula encodes the expansion rules by means of the next normal form transformation, and the STEP rule by tying the grounded X-requests at step $k$ with the grounding of the requested formulae at step $k+1$, ensuring temporal consistency between two adjacent states in the model (*i.e.*, $\sigma \models_i \mathsf{X}\,\psi$ iff $\sigma \models_{i+1} \psi$). Moreover, the CONTRADICTION rule is implicitly encoded as well, since satisfying assignments to the formula cannot represent branches containing propositional contradictions. Hence, the following holds.

**Proposition 1 (Soundness of the $k$-unraveling).** *Let $\phi$ be an LTL formula. Then, $[\![\phi]\!]^k$ is unsatisfiable if and only if the complete tableau for $\phi$ contains only branches with at most $k+1$ poised nodes crossed by contradiction.* □

Note that $[\![\phi]\!]^{k+1}$ can be computed incrementally from $[\![\phi]\!]^k$, by adding only the second conjunct of the definition. This speeds up the construction of the formula itself as well as the solution process of modern incremental SAT-solvers.

### 3.3 Encoding of Accepted Branches

Once all non-contradictory branches of a given depth have been identified with the $k$-unravelling, the *accepted* branches of such a depth can be represented by the conjunction of the propositional encoding of the EMPTY and LOOP rules of the tableau. This allows the unravelling process to be stopped in the case of satisfiable formulae.

The EMPTY rule, which is the simplest rule to encode, accepts *loop-free* models of the formula, that are identified by poised nodes lacking X-requests. In what follows, let $\mathsf{XR}_k \subseteq \mathsf{XR}$ be the set of X-requests that appear (grounded) in the $k$-th conjunct of the $k$-unravelling for $\phi$. Similarly, let $\mathsf{XEV}_k \subseteq \mathsf{XR}_k$ be the X-eventualities (*i.e.*, formulae of the form $\mathsf{X}(\psi_1\,\mathcal{U}\,\psi_2)$) found in $\mathsf{XR}_k$. The EMPTY rule can be encoded as follows:

$$E_k := \bigwedge_{\varphi \in \mathsf{XR}_k} \neg\varphi^k_G$$

Then, each satisfying assignment of the formula $[\![\phi]\!]^k \wedge E_k$ corresponds to a branch of the tableau for $\phi$, with exactly $k+1$ poised nodes, accepted by the EMPTY rule. Note that it would still be sound to use the full XR instead of $\mathsf{XR}_k$ in the definition above, but, in general, the latter is likely to be a smaller set, thus making the formula smaller.

The encoding of the LOOP rule, which accepts branches corresponding to *lasso-shaped* (periodic) models, is built on top of two pieces. For each $0 \le l < k$, let ${}_lR_k$ and ${}_lF_k$ be defined as follows:

$$ {}_lR_k := \bigwedge_{\psi \in \mathsf{XR}_k} \psi_G^l \leftrightarrow \psi_G^k $$

$$ {}_lF_k := \bigwedge_{\substack{\psi \in \mathsf{XEV}_k \\ \psi \equiv \mathsf{X}(\psi_1 \mathcal{U} \psi_2)}} \left( \psi_G^k \rightarrow \bigvee_{i=l+1}^{k} \mathrm{xnf}(\psi_2)_G^i \right) $$

Given a branch $\overline{u} = \langle u_0, \ldots, u_k \rangle$ identified by $[\![\phi]\!]^k$, ${}_lR_k$ states that the nodes $u_l$ and $u_k$ have the same set of X-requests, and ${}_lF_k$ states that all such X-requests are fulfilled between nodes $u_l$ and $u_k$. Together, they can be used to express the whole triggering condition of the LOOP rule:

$$ L_k := \bigvee_{l=0}^{k-1} ({}_lR_k \wedge {}_lF_k) $$

Then, each satisfying assignment of $[\![\phi]\!]^k \wedge L_k$ corresponds to a branch of the tableau for $\phi$, with exactly $k+1$ poised nodes, accepted by the LOOP rule, *i.e.*, with a satisfying loop between position $k$ and some previous position. Together, $[\![\phi]\!]^k$, $E_k$, and $L_k$ can represent any accepted branch of the tableau of the given depth.

**Definition 3** (Base encoding). *Let $\phi$ be an LTL formula over $\Sigma$ and $k \in \mathbb{N}$. The* base encoding *of $\phi$ at step $k$ is the formula $|\phi|^k$ over $\Sigma_+$ defined as follows:*

$$ |\phi|^k := \underbrace{[\![\phi]\!]^k}_{\substack{\textit{exp. rules} \\ \mathsf{STEP}\ \textit{rule}}} \wedge \Big( \underbrace{E_k}_{\mathsf{EMPTY}\ \textit{rule}} \vee \underbrace{L_k}_{\mathsf{LOOP}\ \textit{rule}} \Big) $$

Again, note that the base encoding can be built incrementally, allowing us to exploit the features of modern SAT solvers. Indeed, $|\phi|^k$ consists of the conjunction of $[\![\phi]\!]^k$, built from the already computed $[\![\phi]\!]^{k-1}$, and $E_k \vee L_k$.

The construction of $L_k$ gives us the following result.

**Proposition 2 (Soundness of the base encoding).** *Let $\phi$ be an LTL formula. Then, $|\phi|^k$ is satisfiable if and only if the complete tableau for $\phi$ contains at least an accepted branch with exactly $k+1$ poised nodes.* □

Propositions 1 and 2, together with the soundness result for Reynolds' tableau given in [22], lead us to the following result.

**Theorem 1 (Soundness).** *Given an* LTL *formula* $\phi$, *if* $|\phi|^k$ *is satisfiable, for some* $k \in \mathbb{N}$, *then* $\phi$ *is satisfiable.* □

Figure 1 shows a basic procedure that can be built on top of the encoding of Definition 3. The procedure starts with $k = 0$, and increments it at each step, looking for models of increasing size, stopping when a step $k$ is found with a satisfiable base encoding. The procedure is incomplete, as it may not terminate on unsatisfiable instances, similarly to early *bounded model checking* techniques.

```
procedure BSC(φ)
    k ← 0
    while True do
        generate |φ|^k
        if |φ|^k is SAT then
            φ is SAT
            stop
        k ← k + 1
```

**Fig. 1.** Incomplete satisfiability checking procedure built on top of the base encoding.

If the procedure terminates, then the satisfying assignment for $|\phi|^k$ can be used to build a model $\sigma \subseteq \Sigma^\omega$ of $\phi$ of minimal length, where, in the case of periodic models, the length is considered as the sum of the prefix and the period lengths. This breadth-first traversal, with the guarantee of finding a minimal model, would not be feasible if carried out explicitly, and it is a distinguishing feature of bounded satisfiability checking of this kind. Explicit implementations of Reynolds' tableau system [3] proceed instead in a depth-first way, and the models they find are *not* guaranteed to be minimal in length.

The next section adds to the picture the encoding of the PRUNE rule, showing how to integrate the above procedure in order to guarantee the termination for any unsatisfiable instance as well.

## 4 Completeness

In order to ensure termination of the algorithm in Fig. 1 also on unsatisfiable formulae, it is useful to look at the possible reasons why the base encoding $|\phi|^k$ of a formula $\phi$ may be unsatisfiable. We can distinguish two cases:

1. if the formula $[\![\phi]\!]^k$ is unsatisfiable, it means that all the branches of the tableau for $\phi$ are crossed by the CONTRADICTION rule at or before depth $k$ (see Proposition 1);
2. if both $[\![\phi]\!]^k \wedge E_k$ and $[\![\phi]\!]^k \wedge L_k$ are unsatisfiable, then there are no branches of depth $k$ accepted by the EMPTY rule or by the LOOP rule (see Proposition 2).

As an example of the first case, consider the formula $\mathsf{X}\,p \wedge \mathsf{X}\,\neg p$, whose 1-unravelling is $[\![\mathsf{X}\,p \wedge \mathsf{X}\,\neg p]\!]^1 \equiv (\mathsf{X}\,p)^0_G \wedge (\mathsf{X}\,\neg p)^0_G \wedge p^1 \wedge \neg p^1$. At step $k = 1$, the formula is found to be unsatisfiable because of a propositional contradiction between $p^1$ and $\neg p^1$. At this point there is no reason to continue looking further: we can stop incrementing $k$ and answer UNSAT.

The second case, instead, does *not* exclude that longer accepted branches exist, and require looking further. One interesting example is the (unsatisfiable) formula $\mathsf{G}\,\neg p \wedge q\,\mathcal{U}\,p$: it holds that $|\mathsf{G}\,\neg p \wedge q\,\mathcal{U}\,p|^k$ is unsatisfiable for all $k \geq 0$, since any branch can be accepted neither by the LOOP rule (because $\mathsf{G}\,\neg p$ forces $p^i$ to be false for each $0 \leq i \leq k$) nor by the EMPTY rule (because the failed fulfilment of $q\,\mathcal{U}\,p$ forces $\mathsf{X}(q\,\mathcal{U}\,p)^i$ to be true for each $0 \leq i \leq k$). Nevertheless, $[\![\mathsf{G}\,\neg p \wedge q\,\mathcal{U}\,p]\!]^k$ is satisfiable for all $k \geq 0$, because the branch of the tableau that indefinitely postpones the satisfaction of $q\,\mathcal{U}\,p$ is never closed by contradiction. Hence, the procedure in Fig. 1 can never be able to stop in this case.

In the tableau, such a branch is, instead, rejected by the PRUNE rule, whose role is exactly that of rejecting these potentially infinite branches. We can similarly recover termination and completeness of our procedure by introducing a propositional encoding of the rule.

Recall that the PRUNE rule rejects any branch of length $k$ that presents two positions $l < j < k$, with the same set of X-requests, such that all the X-eventualities fulfilled between $j+1$ and $k$ are fulfilled between $l+1$ and $j$ as well. Let $i$ and $j$ be one such pair of positions. We can encode the condition of the PRUNE rule by means of the following formula:

$$ {}_lP^k_j := \bigwedge_{\substack{\psi \in \mathsf{XEV}_k \\ \psi \equiv \mathsf{X}(\psi_1 \mathcal{U} \psi_2)}} \left( \psi^k_G \wedge \bigvee_{i=j+1}^{k} \mathrm{xnf}(\psi_2)^i_G \rightarrow \bigvee_{i=l+1}^{j} \mathrm{xnf}(\psi_2)^i_G \right) $$

Then, the above formula can be combined with the ${}_lR_k$ formula defined in the previous section to obtain the following encoding of the PRUNE rule:

$$ P^k := \bigvee_{l=0}^{k-2} \bigvee_{j=l+1}^{k-1} \left( {}_lR_j \wedge {}_jR_k \wedge {}_lP^k_j \right) $$

It is worth to note that the $P^k$ formula is of cubic size with respect to $k$ *and* the number of X-eventualities. With this formula, in case of an unsatisfiable base encoding, we can check whether there exists at least one branch of depth at most $k$ which does *not* satisfy the prune condition: if this is the case, then it makes sense to continue the search; otherwise, the procedure can stop reporting the unsatisfiability of the formula. This is done by testing the satisfiability of the *termination encoding* of $\phi$, defined as the following formula:

$$ |\phi|^k_T := \underbrace{[\![\phi]\!]^k}_{\substack{\text{exp. rules} \\ \text{STEP rule}}} \wedge \bigwedge_{i=0}^{k} \underbrace{\neg P^i}_{\text{PRUNE } rule} $$

The complete procedure is shown in Fig. 2, where the first step $k$ such that $|\phi|^k_T$ is unsatisfiable stops the search. Based on the soundness and completeness result for the encoded tableau system [22], we can state the following result.

**Theorem 2 (Soundness and completeness).** *For every LTL formula $\phi$, the procedure of Fig. 2 always terminates, and it answers SAT iff $\phi$ is satisfiable.*

Notably, the procedure guarantees termination and completeness without establishing *a priori* a bound to the depth of the tree, at the cost of a slightly bigger formula and three calls to the underlying solver.

```
1: procedure LTL-SAT-PRUNE(φ)
2:    k ← 0
3:    while True do
4:       generate ⟦φ⟧^k
5:       if ⟦φ⟧^k is UNSAT then
6:          φ is UNSAT
7:          stop
8:       generate |φ|^k
9:       if |φ|^k is SAT then
10:         φ is SAT
11:         stop
12:      generate |φ|^k_T
13:      if |φ|^k_T is UNSAT then
14:         φ is UNSAT
15:         stop
16:      k ← k + 1
```

**Fig. 2.** Complete and terminating satisfiability checking procedure based on the tableau encoding.

It is worth to spend some words on how the above procedure can exploit the *incrementality* of modern SAT solvers to speed up its execution. Many modern solvers have a push/pop interface that allows the client to push some conjuncts to a stack, solve them, then pop some of them while pushing others, maintaining all the information about the untouched conjuncts. In our case, the construction of $[\![\phi]\!]^k$ only requires the addition of a conjunct to $[\![\phi]\!]^{k-1}$, and $|\phi|^k$ only requires to join $E_k \vee L_k$ to $[\![\phi]\!]^k$. This means that such a conjunct can be pushed temporarily, while maintaining all the solver state about $[\![\phi]\!]^k$ for the next step. Moreover, the formula $[\![\phi]\!]^k$ generated and solved at Sect. 4 of Fig. 2 can be replaced by one built on top of the whole $|\phi|^{k-1}_T$ from the previous step, instead of only from $[\![\phi]\!]^{k-1}$. This allows us to avoid to backtrack the additional conjuncts of $|\phi|^k_T$. Since the PRUNE rule cuts redundant branches, maintaining the corresponding formulae from step to step helps guiding the solver through relevant branches.

## 5 Experimental Evaluation

The above-described procedure has been implemented in a tool called BLACK (Bounded LTL sAtisfiability ChecKer)[1]. This section presents some relevant aspects of the tool and shows the results of our preliminary experimental evaluation, where it has been compared with other state-of-the-art LTL solvers.

BLACK has been implemented from scratch in the C++17 language with the goals of efficiency, portability, and reusability. Most of the tool is implemented as a shared library with a well-defined API, that can be linked to other client applications as needed. The library provides basic formula handling facilities, and an interface to the main solving algorithm. The tool itself is as well a client of such a library, providing a simple command-line user interface.

The tool is currently implemented on top of MathSAT [5], used as its back-end SAT solver, which is actually a full-blown SMT solver. This choice was driven by the fact that, contrary to most pure-SAT solvers, MathSAT supports formulae with a general syntax, without the need of a preliminary conversion to CNF. This feature greatly simplified the initial development cycle of the project. Future plans include the support to multiple different SAT solvers, including those with simple CNF-based APIs, to find the most performant candidate.

The above-described satisfiability checking procedure is implemented on top of a formula handling layer, which eases the development of the solver by decoupling the logical encoding from low-level details. In particular, the lower layer transparently implements *subterm sharing*, *i.e.*, formulae are internally represented as *circuits*, by identifying repeated subformulae. Besides the positive effects on memory usage, this mechanism matches well with the term-based API of the MathSAT library. Most importantly, *syntactic equality* of two formulae reduces to a single pointer comparison, since building any two equal formulae results into two pointers to the same object. A peculiar feature of BLACK's formulae handling layer is that atomic propositions can be labelled by values of almost any data type, in contrast to being restricted to strings, integers, or similar identifiers. In this way, the *grounding* operation ($\psi_G$) performed on X-requests by our encoding (such as in $[\![\phi]\!]^k$) is effectively a *no-op*: the grounding of an X-request formula is just an atomic proposition labelled by the formula's representing object, with no need for any translation table between the formulae and their corresponding grounded symbols. Since formulae are uniquely identified by just the pointer to their object, this is implementable in such a way that the common cases of propositions labelled by short strings, formulae, and formula/integer pairs (for the *stepped* versions $\psi_G^k$) do not cause any memory allocation.

In our experiments, we compared BLACK with four competitors: *Aalta* v2.0 [17], *nuXmv* [6], *Leviathan* [3], and *PLTL* [1,24]. The *nuXmv* model checker is tested in two modes, which implement, respectively, the *Simple Bounded Model Checking* (SBMC) [15] and the *K-Liveness* [8] techniques. The SBMC mode

---

[1] BLACK can be downloaded from https://github.com/black-sat/black, together with the whole benchmarking suite and the raw results data.

is the most similar to ours among the tested solvers. The *PLTL* tool implements both a graph-shaped [1] tableau, and the one-pass tree-shaped tableau by Schwendimann [24]. Finally, *Leviathan* is an explicit implementation of Reynolds' tableau [3]. Because of technical issues, we could not include the LS4 [26] tool in our test. Future experiments will include this and other competitors as well.

We considered the comprehensive set of formulae collected by Schuppan and Darmawan [23], which contains a total of 3723 LTL formulae, grouped in seven families, `acacia`, `alaska`, `anzu`, `forobots`, `rozier`, `schuppan`, `trp`, named after their original source. We set a timeout of five minutes for each formula in the set.

We ran our tests on a Quad Core i5–2500k 3.30 GHz processor, with 8 GB of main memory. Processes were assigned a single CPU core each, with a memory limit of 2 GB per core (and the five minutes timeout). Figures 4 and 5 show six scatter plots comparing the execution times, while Fig. 3 shows the number of *timeouts* and *out of memory* interruptions for the tools on each class of formulae.

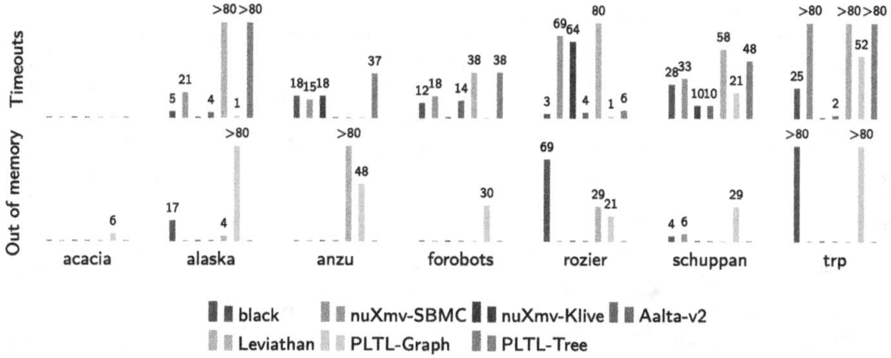


**Fig. 3.** Total number of *timeouts* and *out of memory* interruptions of the solvers on the different class of benchmark formulae.

Overall, the results are promising. Although *Aalta* remains the most performant tool in the majority of cases, the picture is mixed. In particular, BLACK is competitive with regards to *nuXmv*. With regards to the SBMC mode, the advantage is consistent but constant, showing similar trends both on satisfiable and unsatisfiable instances. The `rozier` set comes as an exception. Apart from the *counter* formulae, which are hard for both solvers, all these formulae have very short models, which is an advantage for iterative deepening approaches like ours. SBMC shares the same principle, but the large difference between BLACK and *nuXmv* on most of this set may be explained by (i) the simpler base encoding employed by BLACK, whose asymptotically larger size does not bite at lower values of the bound $k$, and/or (ii) differences between the SAT solvers underlying the two tools (the distributed binary of *nuXmv* is linked to *minisat* [12]).

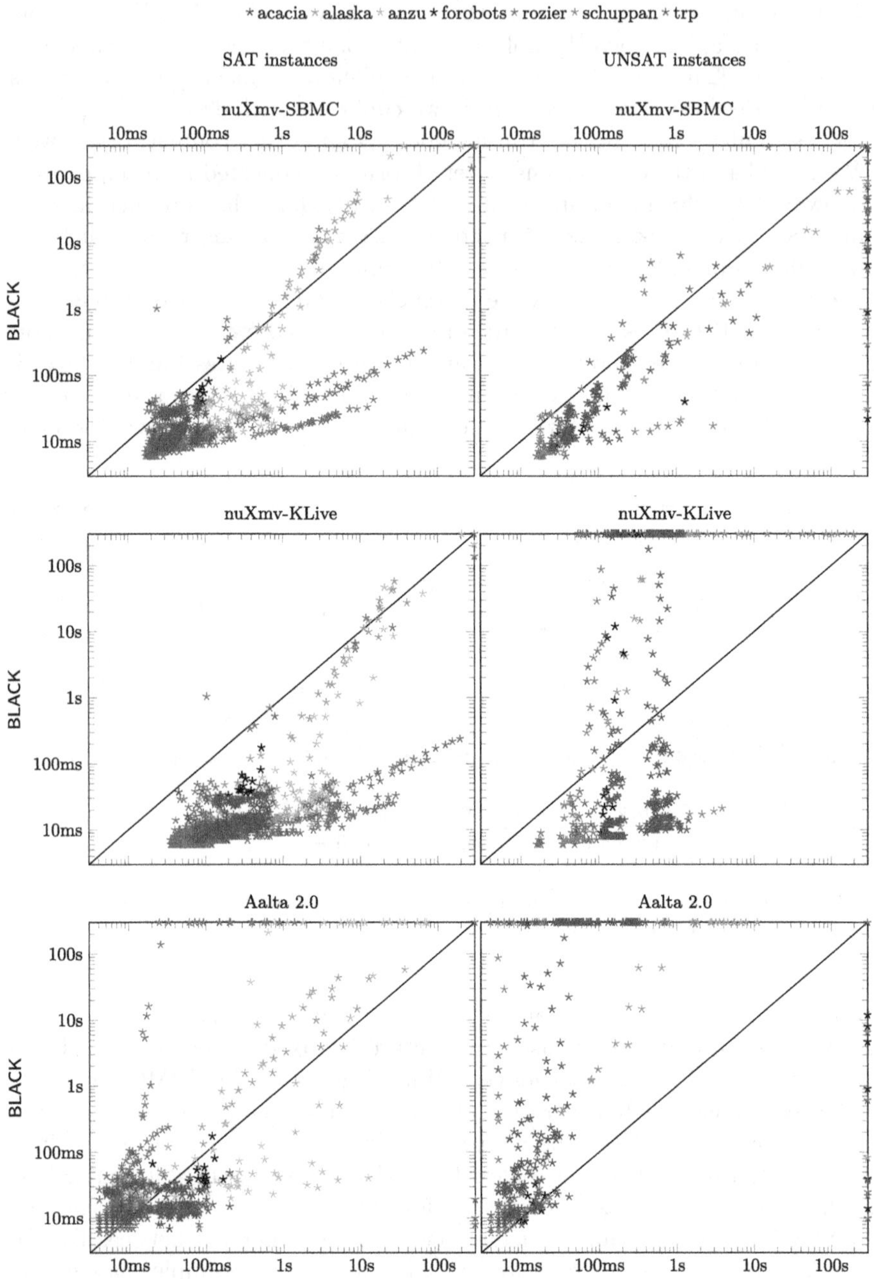


**Fig. 4.** Experimental comparison with *nuXmv* and *Aalta*.

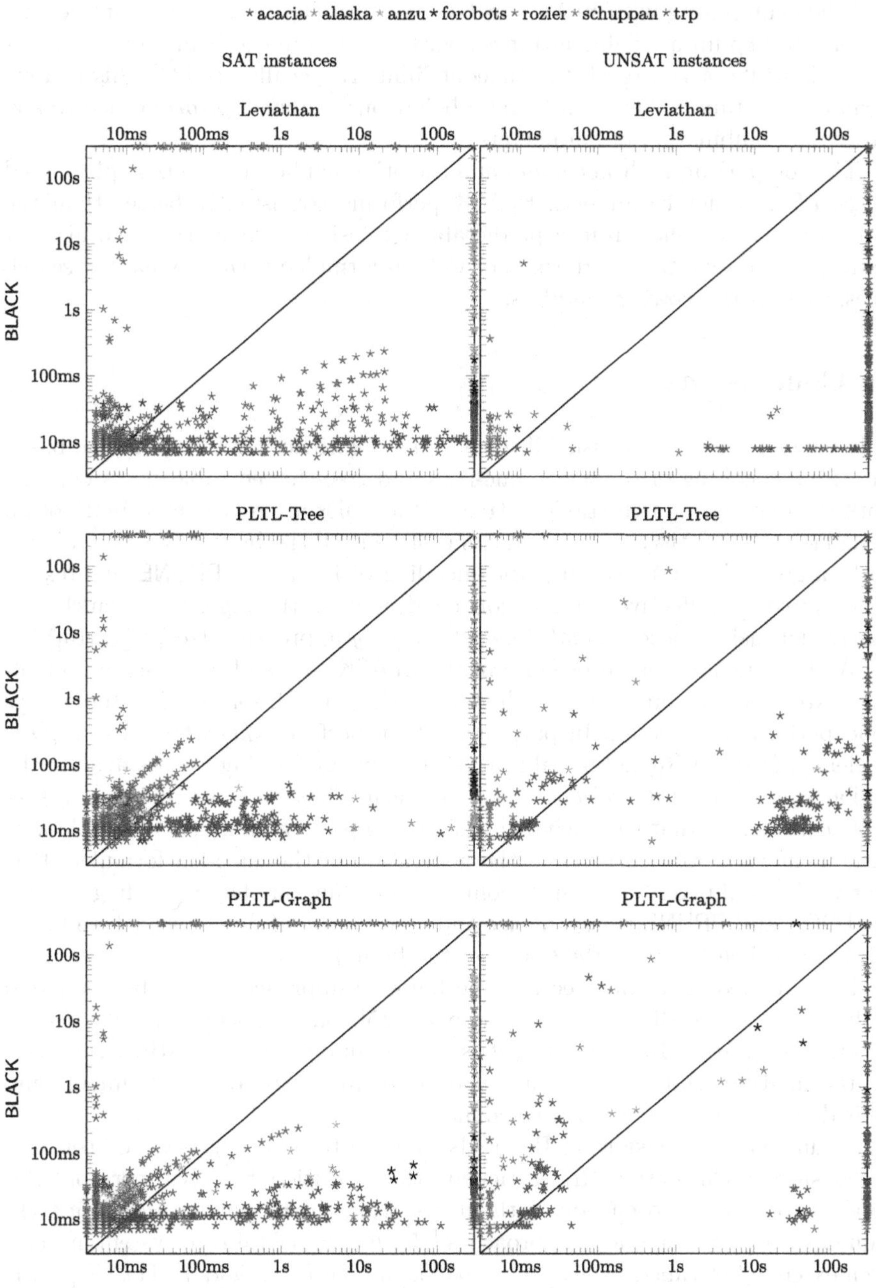


**Fig. 5.** Experimental comparison with *Leviathan* and *PLTL*.

When comparing with *nuXmv* in *k-liveness* mode, we can see an interesting pattern on `trp` unsatisfiable instances, with some formulae being solved in milliseconds while others reach the timeout limit. As recalled in [23], this is a set of *random* instances, hence the erratic behaviour cannot *a priori* be tied to any particular combination of parameters.

The comparison with *Leviathan* and the other explicit tableaux implemented by *PLTL* is easier to analyse. BLACK performs consistently better than the two tools, which suffer from a predictable explosion in memory usage in most instances. Notably, they perform very well on formulae with very narrow search trees, such as the `rozier` *counters*.

## 6 Conclusions

This paper proposed a satisfiability checking algorithm for LTL formulae based on a SAT encoding of Reynolds' one-pass and tree-shaped tableau system [22]. Both the expansion of the tableau tree and its rules are represented by Boolean formulae, whose satisfying assignments represent all the branches of the tableau up to a given depth $k$. Notably, the encoding of Reynolds' PRUNE rule results in a simple yet effective termination condition for the algorithm, which is a non-trivial task in other *bounded model checking* approaches (see, e.g., [15]).

We implemented our procedure in the BLACK tool and made some preliminary experimental comparison with state-of-the-art LTL solvers. The tool shows good performance overall. In particular, it outperforms *Leviathan*, the *explicit* implementation of Reynolds' tableau, and shows interesting results against the similar *simple bounded model checking* approach. The results are promising, especially considering that the encoding has been implemented in a very simple way, without any sort of heuristics in the generation of the encoded formulae. Further work should consider a more compact encoding for the unravelling and for the LOOP and PRUNE rules, the use and comparison of different back-end SAT solvers, and heuristics for the search of the bound.

From a theoretical perspective, the followed approach has to be compared with others, especially with *bounded* ones [15], on a conceptual, rather than experimental, level. In particular, it is worth comparing the PRUNE rule with the terminating conditions exploited in other bounded approaches, to understand their difference and draw possible connections.

A number of extensions of Reynolds' tableau to other logics have been proposed since its inception. In particular, the extension to past operators [14] appears to be easy to encode, without resorting to the *virtual unrolling* technique used in other bounded approaches [15]. Reynolds' tableau system has also been extended to timed logics [13], in particular TPTL [2] and $\mathsf{TPTL}_b$+P [11]. It is natural to ask whether the approach used here to encode the LTL tableau to SAT can be adapted to encode the timed extensions of the tableau to SMT.

**Acknowledgements.** This work has been supported by the PRID project *ENCASE - Efforts in the* uNderstanding of Complex interActing SystEms, and by the INdAM GNCS project *Formal Methods for Combined Verification*. The authors would like to thank *Alessandro Cimatti* and *Stefano Tonetta* for the helpful discussions about bounded satisfiability checking, *Valentino Picotti* for providing the benchmarking hardware, and *Nikhil Babu* for pointing out a bug in Leviathan that could have introduced a bias in the experimental evaluation. Thanks also to the anonymous reviewers for their helpful remarks.

## References

1. Abate, P., Goré, R., Widmann, F.: An on-the-fly tableau-based decision procedure for PDL-satisfiability. Electron. Notes Theor. Comput. Sci. **231**, 191–209 (2009). https://doi.org/10.1016/j.entcs.2009.02.036
2. Alur, R., Henzinger, T.A.: A really temporal logic. J. ACM **41**(1), 181–204 (1994)
3. Bertello, M., Gigante, N., Montanari, A., Reynolds, M.: Leviathan: a new LTL satisfiability checking tool based on a one-pass tree-shaped tableau. In: Proceedings of the 25th International Joint Conference on Artificial Intelligence, pp. 950–956. IJCAI/AAAI Press (2016)
4. Beth, E.W.: Semantic entailment and formal derivability. Sapientia **14**(54), 311 (1959)
5. Bruttomesso, R., Cimatti, A., Franzén, A., Griggio, A., Sebastiani, R.: The MATHSAT 4 SMT solver. In: Gupta, A., Malik, S. (eds.) CAV 2008. LNCS, vol. 5123, pp. 299–303. Springer, Heidelberg (2008). https://doi.org/10.1007/978-3-540-70545-1_28
6. Cavada, R., et al.: The NUXMV symbolic model checker. In: Biere, A., Bloem, R. (eds.) CAV 2014. LNCS, vol. 8559, pp. 334–342. Springer, Cham (2014). https://doi.org/10.1007/978-3-319-08867-9_22
7. Cimatti, A., Roveri, M., Sheridan, D.: Bounded verification of past LTL. In: Hu, A.J., Martin, A.K. (eds.) FMCAD 2004. LNCS, vol. 3312, pp. 245–259. Springer, Heidelberg (2004). https://doi.org/10.1007/978-3-540-30494-4_18
8. Claessen, K., Sörensson, N.: A liveness checking algorithm that counts. In: Proceedings of the 12th Formal Methods in Computer-Aided Design, pp. 52–59. IEEE (2012)
9. Clarke, E.M., Grumberg, O., Peled, D.A.: Model Checking. MIT Press, Cambridge (2001)
10. D'Agostino, M., Gabbay, D., Hähnle, R., Posegga, J. (eds.): Handbook of Tableau Methods. Springer, Dordrecht (1999). https://doi.org/10.1007/978-94-017-1754-0
11. Della Monica, D., Gigante, N., Montanari, A., Sala, P., Sciavicco, G.: Bounded timed propositional temporal logic with past captures timeline-based planning with bounded constraints. In: Proceedings of the 26th International Joint Conference on Artificial Intelligence, pp. 1008–1014 (2017). https://doi.org/10.24963/ijcai.2017/140
12. Eén, N., Sörensson, N.: An extensible sat-solver. In: Selected Revised Papers of the 6th International Conference on Theory and Applications of Satisfiability Testing, pp. 502–518 (2003). https://doi.org/10.1007/978-3-540-24605-3_37

13. Geatti, L., Gigante, N., Montanari, A., Reynolds, M.: One-pass and tree-shaped tableau systems for TPTL and TPTLb+Past. In: Orlandini, A., Zimmermann, M. (eds.) Proceedings 9th International Symposium on Games, Automata, Logics, and Formal Verification. EPTCS, vol. 277, pp. 176–190 (2018). https://doi.org/10.4204/EPTCS.277.13
14. Gigante, N., Montanari, A., Reynolds, M.: A one-pass tree-shaped tableau for LTL+Past. In: Proceedings of 21st International Conference on Logic for Programming, Artificial Intelligence and Reasoning. EPiC Series in Computing, vol. 46, pp. 456–473 (2017)
15. Heljanko, K., Junttila, T., Latvala, T.: Incremental and complete bounded model checking for full PLTL. In: Etessami, K., Rajamani, S.K. (eds.) CAV 2005. LNCS, vol. 3576, pp. 98–111. Springer, Heidelberg (2005). https://doi.org/10.1007/11513988_10
16. Kress-Gazit, H., Fainekos, G.E., Pappas, G.J.: Temporal-logic-based reactive mission and motion planning. IEEE Trans. Robot. **25**(6), 1370–1381 (2009)
17. Li, J., Zhu, S., Pu, G., Vardi, M.Y.: SAT-based explicit LTL reasoning. In: Piterman, N. (ed.) HVC 2015. LNCS, vol. 9434, pp. 209–224. Springer, Cham (2015). https://doi.org/10.1007/978-3-319-26287-1_13
18. Lichtenstein, O., Pnueli, A.: Propositional temporal logics: decidability and completeness. Logic J. IGPL **8**(1), 55–85 (2000). https://doi.org/10.1093/jigpal/8.1.55
19. Manna, Z., Pnueli, A.: Temporal Verification of Reactive Systems - Safety. Springer, New York (1995). https://doi.org/10.1007/978-1-4612-4222-2
20. Mayer, M.C., Limongelli, C., Orlandini, A., Poggioni, V.: Linear temporal logic as an executable semantics for planning languages. J. Logic, Lang. Inf. **16**(1), 63–89 (2007). https://doi.org/10.1007/s10849-006-9022-1
21. McCabe-Dansted, J.C., Reynolds, M.: A parallel linear temporal logic tableau. In: Bouyer, P., Orlandini, A., Pietro, P.S. (eds.) Proceedings of the 8th International Symposium on Games, Automata, Logics and Formal Verification. EPTCS, vol. 256, pp. 166–179 (2017)
22. Reynolds, M.: A new rule for LTL tableaux. In: Proceedings of the 7th International Symposium on Games, Automata, Logics and Formal Verification. EPTCS, vol. 226, pp. 287–301 (2016). https://doi.org/10.4204/EPTCS.226.20
23. Schuppan, V., Darmawan, L.: Evaluating LTL satisfiability solvers. In: Proceedings of the 9th International Symposium on Automated Technology for Verification and Analysis, pp. 397–413 (2011)
24. Schwendimann, S.: A new one-pass tableau calculus for **PLTL**. In: de Swart, H. (ed.) TABLEAUX 1998. LNCS (LNAI), vol. 1397, pp. 277–291. Springer, Heidelberg (1998). https://doi.org/10.1007/3-540-69778-0_28
25. Sistla, A.P., Clarke, E.M.: The complexity of propositional linear temporal logics. J. ACM **32**(3), 733–749 (1985). https://doi.org/10.1145/3828.3837
26. Suda, M., Weidenbach, C.: A PLTL-prover based on labelled superposition with partial model guidance. In: Gramlich, B., Miller, D., Sattler, U. (eds.) IJCAR 2012. LNCS (LNAI), vol. 7364, pp. 537–543. Springer, Heidelberg (2012). https://doi.org/10.1007/978-3-642-31365-3_42
27. Van Lamsweerde, A.: Goal-oriented requirements engineering: a guided tour. In: Proceedings of the 5th International Symposium on Requirements Engineering, pp. 249–262. IEEE (2001)

# Certification of Nonclausal Connection Tableaux Proofs

Michael Färber(✉) and Cezary Kaliszyk

Universität Innsbruck, Innsbruck, Austria
michael.faerber@gedenkt.at, cezary.kaliszyk@uibk.ac.at

**Abstract.** Nonclausal connection tableaux calculi enable proof search without performing clausification. We give a translation of nonclausal connection proofs to Gentzen's sequent calculus LK and compare it to an existing translation of clausal connection proofs. Furthermore, we implement the translation in the interactive theorem prover HOL Light, enabling certification of nonclausal connection proofs as well as a new, complementary automation technique in HOL Light.

## 1 Introduction

Most automated theorem provers (ATPs) output only limited proof traces for performance reasons. This is in contrast to the LCF approach, which hinges on the correctness of a small, trusted kernel [13]. One way to certify the correctness of proofs produced by ATPs is to translate them to interactive theorem provers (ITPs) [15,17]. Certification of proofs given by ATPs is also important for the integration of ATPs into interactive theorem provers, providing automation in the form of proof tactics [5].

Most ATPs convert their input problems to clausal normal form as preprocessing step [23]. To reconstruct the resulting clausal proofs in an ITP, it is necessary to verify in the ITP the conversion to clausal normal form. The ATP nanoCoP has demonstrated that a connection prover not requiring clausification can be effectively implemented [27]. The reconstruction of nonclausal proofs eliminates the necessity of proving the correctness of the clausification, but on the other hand, translating the proofs is more involved.

In this paper, we describe the translation of clausal and nonclausal connection proofs to Gentzen's LK. To ease the translation, we introduce slightly modified versions of the clausal and nonclausal connection calculus in Sect. 3. Using these calculi, we describe a translation method from clausal and nonclausal connection proofs to LK in Sect. 4. Based on this translation, we develop in Sect. 5 an automatic proof certification of clausal proofs from leanCoP as well as of nonclausal proofs from nanoCoP in the ITP HOL Light. We evaluate the performance of our implementations on HOL Light problem sets in Sect. 6.

This paper generalises work co-authored by the second author of this paper about the certification of clausal connection tableaux proofs [19]. Whereas [19] is concerned more with technical questions of implementing a clausal prover and


S. Cerrito and A. Popescu (Eds.): TABLEAUX 2019, LNAI 11714, pp. 21–38, 2019.
https://doi.org/10.1007/978-3-030-29026-9_2

a corresponding proof translation in a functional language, this paper abstracts more from technical details in order to treat the more involved nonclausal proof translation. This paper extends section 6.4 of the first author's PhD thesis [11], where a preliminary version of the nonclausal proof translation described in this paper was introduced.

## 2 Connection Calculi

In this section, we will give a brief overview of the clausal and the nonclausal connection tableaux calculus. For more details and examples, see [26,27].[1]

Let us start by fixing some notation. The transitive closure of a relation $R$ is denoted by $R^+$, and the transitive reflexive closure by $R^*$. A term $t$ is either a variable $x$, a constant $a$, or $f(t_1, \dots, t_n)$, where $f$ is a function symbol of arity $n$ and $t_1, \dots, t_n$ are terms. An atom $A$ is $P(t_1, \dots, t_n)$, where $P$ is a predicate of arity $n$ and $t_1, \dots, t_n$ are terms. A (first-order) formula $F$ is $(A)$, $(F_1 \lor F_2)$, $(F_1 \land F_2)$, $(F_1 \implies F_2)$, $(\neg F_1)$, $(\forall x.F_1)$, or $(\exists x.F_1)$, where $F_1$ and $F_2$ are formulas, $A$ is an atom, and $x$ is a variable. We write a sequence of quantifiers $\forall x_1 \dots x_n.F$ as $\forall \boldsymbol{x}.F$. The formula $F[t/x]$ denotes the formula $F$ with all unbound occurrences of $x$ replaced by $t$. A literal $L$ is either $\neg A$ or $A$, where $A$ is an atom. The complement $\overline{L}$ of a literal is $A$ if $L$ is of the shape $\neg A$, and $\neg A$ otherwise. A substitution $\sigma$ is a function from variables to terms.

In the clausal calculus, a clause $C$ is $\forall \boldsymbol{x}.(L_1 \lor \dots \lor L_n)$ and a matrix $M$ is $C_1 \land \dots \land C_n$. In the nonclausal calculus, a clause $C$ is $\forall \boldsymbol{x}.(X_1 \lor \dots \lor X_n)$, where $X$ is either a literal or a matrix, and a matrix $M$ is $C_1 \land \dots \land C_n$.[2] We refer to matrices in the clausal calculus as clausal matrices and to matrices in the nonclausal calculus as nonclausal matrices.

We can write a clause $\forall \boldsymbol{x}.(L_1 \lor \dots \lor L_n)$ as a set $\{L_1, \dots, L_n\}$ and we can write a matrix $C_1 \land \dots \land C_n$ as a set $\{C_1, \dots, C_n\}$. Alternatively, we write matrices as row vectors and clauses as column vectors.

For any formula $F$, there are equisatisfiable closed formulas $M(F)$ and $\bar{M}(F)$, where $M(F)$ is a nonclausal matrix and $\bar{M}(F)$ is a clausal matrix. We can convert any formula to a nonclausal matrix by conversion to negation normal form, Skolemisation (eliminating existential quantifiers), and pushing universal quantifiers inwards via $\forall \boldsymbol{x}.(F_1 \land F_2) \equiv (\forall \boldsymbol{x}.F_1) \land (\forall \boldsymbol{x}.F_2)$.

*Example 1.* Consider the following equivalent formulas $F$ and $\bar{F}$.

$$F = Q \land P(a) \land \forall x.(\neg P(x) \lor (\neg P(s^2 x) \land (P(sx) \lor \neg Q)))$$
$$\bar{F} = Q \land P(a) \land (\forall x.\neg P(x) \lor \neg P(s^2 x)) \land (\forall x.\neg P(x) \lor P(sx) \lor \neg Q)$$

[1] We diverge from [26] by using a refutational point of view; that is, instead of proving formulas directly, we refute their negations. This shows up for example when we interpret clauses and matrices: In this paper, a clause (of a negated formula) represents a disjunction, whereas in [26], a clause (of an unnegated formula) represents a conjunction. Our refutational view is historically motivated by other proof certification methods, namely those for MESON [15] and leanCoP [19].

[2] We represent clauses with quantifiers to reduce the size of the translated proofs.

For brevity, we write $sx$ for $s(x)$ and $s^2x$ for $s(s(x))$. The nonclausal matrix $M$ corresponds to $F$ and the clausal matrix $\bar{M}$ to $\bar{F}$:

$$M = \begin{bmatrix} [Q]\ [P(a)] \begin{bmatrix} \begin{bmatrix} & \neg P(x) & \\ [\neg P(s^2x)] & \begin{bmatrix} P(sx) \\ \neg Q \end{bmatrix} \end{bmatrix} \end{bmatrix} \end{bmatrix}$$

$$\bar{M} = \begin{bmatrix} [Q]\ [P(a)] \begin{bmatrix} \neg P(x) \\ \neg P(s^2x) \end{bmatrix} \begin{bmatrix} \neg P(x) \\ P(sx) \\ \neg Q \end{bmatrix} \end{bmatrix}$$

The words of the connection calculi treated in this paper are tuples $\langle C, M, Path \rangle$, where $C$ is a clause, $M$ is a matrix, and $Path$ is a set of literals and matrices called the active path.[3] In the calculus rules, $\sigma$ is a global (or rigid) term substitution, i.e. it is applied to the whole derivation. We say that a (non)clausal connection proof of $M$ is a derivation of $\langle \emptyset, M, \emptyset \rangle$ in the (non)clausal connection calculus.

The rules of the clausal connection calculus are shown in Fig. 1 [30]. For any closed formula $F$, we have that $F$ is unsatisfiable iff there is a clausal connection proof of $\bar{M}(F)$ [3]. A clausal connection proof of $\bar{M}$ from Example 1 is given in Fig. 2.

Axiom $\dfrac{}{\{\}, M, Path}$ A

Start $\dfrac{C_2, M, \{\}}{\varepsilon, M, \varepsilon}$ S where $C_2$ is copy of $C_1 \in M$

Reduction $\dfrac{C, M, Path \cup \{L'\}}{C \cup \{L\}, M, Path \cup \{L'\}}$ R where $\sigma(L) = \sigma(\overline{L'})$

Extension $\dfrac{C_2 \setminus \{L'\}, M, Path \cup \{L\} \quad C, M, Path}{C \cup \{L\}, M, Path}$ E

where $C_2$ is copy of $C_1 \in M$ and $L' \in C_2$ with $\sigma(L) = \sigma(\overline{L'})$

**Fig. 1.** Clausal connection calculus rules.

We now proceed to introduce definitions related to the nonclausal connection calculus.

**Definition 1 (Clause Predicates).** *A clause $C$* recursively contains *a literal or a matrix $X$ iff $X \in^+ C$.*[4] *A clause $C \in^+ M$ is* $\alpha$-related *to $X$ iff there is some*

[3] In the original description of the calculi, $Path$ denotes a set of literals. Our generalisation to literals and matrices is motivated by the correctness proof of our translation, in particular Theorem 1. It does, however, not alter the actual proof search with the calculi, as all active paths in a connection proof tree will only contain literals.

[4] We use the term "recursively contains" instead of "contains" as employed in [27] to clearly distinguish it from regular set membership.

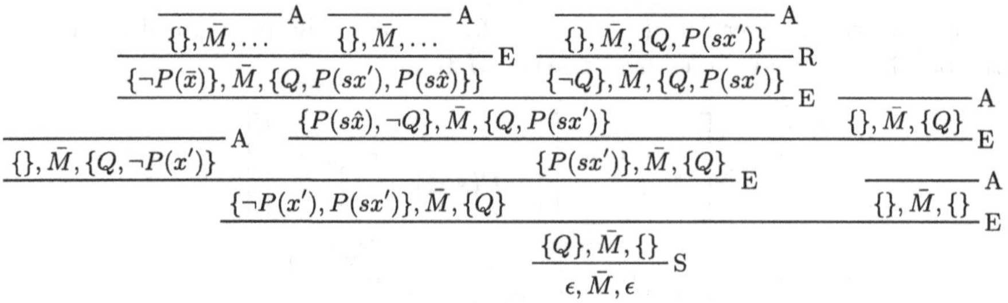


**Fig. 2.** Clausal connection proof with $\sigma = \{x' \mapsto a, \hat{x} \mapsto sx', \overline{x} \mapsto x'\}$.

*$M' \in^* M$ with $\{C_X, C_C\} \subseteq M'$ such that $C_X \neq C_C$, $X \in^+ C_X$, and $C \in^* C_C$. A variable is* free *in $C \in^+ M$ if it occurs only in literals recursively contained in $C$ and (possibly) in literals to which $C$ is $\alpha$-related. A clause $C'$ is a* parent clause *of $C$ iff $M' \in C'$ and $C \in M'$ for some matrix $M'$.*

**Definition 2 (Clause Functions).** *A* copy of the clause *$C \in^+ M$ is created by replacing all free variables in $C$ with fresh variables. $M[C_1 \backslash C_2]$ denotes the matrix $M$ in which the clause $C_1$ is replaced by the clause $C_2$.*

In a clausal matrix $\bar{M}$, all clauses in $\bar{M}$ can potentially give rise to an extension step. In a nonclausal matrix $M$, however, we have clauses $C$ for which $C \in^+ M$, but $C \notin M$. It depends on the active path which of these clauses may give rise to an extension step. Those clauses which do are called extension clauses.

**Definition 3 (Extension Clause).** *The clause $C \in^+ M$ is an* extension clause *(*e-clause*) of the matrix $M$ with respect to a set Path iff either (a) $C$ recursively contains an element of Path, or (b) $C$ is $\alpha$-related to all elements of Path recursively contained in $M$ and if $C$ has a parent clause, that parent clause recursively contains an element of Path.*

Given an extension clause, its $\beta$-clause removes from the clause those parts that are irrelevant to the current subgoal.

**Definition 4 ($\beta$-clause).** *The* $\beta$-clause *of $C$ with respect to $L$ is $C$ with $L$ and all clauses that are $\alpha$-related to $L$ removed.*

*Example 2.* Consider the nonclausal matrix

$$M = \left[ [Q][P(a)] \overbrace{\left[ \begin{bmatrix} \neg P(x) \\ \left[ \underbrace{[\neg P(s^2x)]}_{C_4} \underbrace{\begin{bmatrix} P(sx) \\ \neg Q \end{bmatrix}}_{C_5} \right] \end{bmatrix} \right]}^{C_3} \right]$$

from Example 1. The extension clauses with respect to $\{Q\}$ are all clauses $C \in M$. In particular, the first clause in $M$, $\{Q\}$, is an extension clause due to condition (a) of Definition 3, because it contains $Q$, and the other clauses in

$M$ are extension clauses due to condition (b), because they are $\alpha$-related to $Q$ and do not have parent clauses. Only one of the clauses in $M$ recursively contains $\neg Q$, namely $C_3$. The $\beta$-clause of $C_3$ with respect to $\neg Q$ is

$$\begin{bmatrix} \neg P(x) \\ [[P(sx)]] \end{bmatrix}$$

Let us now assume that $\sigma(x) = a$. The extension clauses with respect to $\{Q, P(sx)\} \cup \{P(s^2a)\}$ are all clauses in $M$, plus $C_4$ due to condition (b) and $C_5$ due to condition (a). Two of these extension clauses recursively contain the literal $\neg P(s^2x)$ that can be unified with $\neg P(s^2a)$, namely $C_3$ and $C_4$. The $\beta$-clause of $C_4$ with respect to $\neg P(s^2x)$ is $\{\}$, and the $\beta$-clause of $C_3$ with respect to $\neg P(s^2x)$ is

$$\begin{bmatrix} \neg P(x) \\ [[]] \end{bmatrix}$$

Some $\beta$-clauses in this example will be used in a nonclausal proof in Fig. 7.

The rules of the nonclausal calculus are shown in Fig. 3. The difference in the calculus rules to the clausal variant is the addition of a decomposition rule, and the adaptation of the extension rule to the nonclausal setting. For any closed formula $F$, we have that $F$ is unsatisfiable iff there is a nonclausal connection proof of $M(F)$ [26]. A nonclausal proof of $M$ from Example 1 as well as a shorter clausal proof of $\bar{M}$ from the same example will be given using slightly modified versions of the calculi in Sect. 3.

Axiom $\dfrac{}{\{\}, M, \mathit{Path}}\,\mathrm{A}$

Start $\dfrac{C_2, M, \{\}}{\varepsilon, M, \varepsilon}\,\mathrm{S}$ where $C_2$ is copy of $C_1 \in M$

Reduction $\dfrac{C, M, \mathit{Path} \cup \{L'\}}{C \cup \{L\}, M, \mathit{Path} \cup \{L'\}}\,\mathrm{R}$ where $\sigma(L) = \sigma(\overline{L'})$

Extension $\dfrac{C_3, M[C_1 \backslash C_2], \mathit{Path} \cup \{L\} \quad C, M, \mathit{Path}}{C \cup \{L\}, M, \mathit{Path}}\,\mathrm{E}$

where $C_3$ is the $\beta$-clause of $C_2$ with respect to $L'$, $C_2$ is copy of $C_1$, $C_1$ is e-clause of $M$ with respect to $\mathit{Path} \cup \{L\}$, $L' \in^+ C_2$ with $\sigma(L) = \sigma(\overline{L'})$

Decomposition $\dfrac{C \cup C', M, \mathit{Path}}{C \cup \{M'\}, M, \mathit{Path}}\,\mathrm{D}$ where $C' \in M'$

**Fig. 3.** Nonclausal connection calculus rules.

## 3 Compressed Connection Calculi

In Otten's presentation of connection calculi [26], all proof rules have a fixed number of premises. To ease the presentation of proofs in this paper, we present slightly reformulated versions of Otten's calculi. We call these calculi *compressed*, because proofs in these calculi usually consist of fewer proof steps and take up less space. The compressed calculi can be considered a mixture between Otten's and Letz's presentation of connection tableaux [21].

We introduce the following notation for rules with an arbitrary number of premises:

$$\frac{\bigwedge_i P_i}{C} \equiv \frac{P_1 \quad \dots \quad P_n}{C}$$

The compressed connection calculi are shown in Figs. 4 and 5. In the original calculi, the words are $\langle C, M, Path \rangle$. In the compressed calculi, the words are $\langle X, M, Path \rangle$, where $X$ denotes an arbitrary clause element, i.e. a matrix or a literal. In the compressed calculi, the axiom rule becomes obsolete.

Start $\dfrac{\bigwedge_i \langle X_i, M, \{\} \rangle}{\varepsilon, M, \varepsilon}$ S where $\{X_1, \dots, X_n\}$ is copy of $C \in M$

Reduction $\dfrac{}{L, M, Path \cup \{L'\}}$ R where $\sigma(L) = \sigma(\overline{L'})$

Extension $\dfrac{\bigwedge_i \langle L_i, M, Path \cup \{L\} \rangle}{L, M, Path}$ E

where $\{L_1, \dots, L_n\} \cup \{L'\}$ is copy of $C \in M$ and $\sigma(L) = \sigma(\overline{L'})$

**Fig. 4.** Compressed clausal connection calculus.

We will now show how proofs can be translated between the compressed calculi in this section and the original calculi in Sect. 2.

**Lemma 1.** *The sequent $\langle \{X_1, \dots, X_n\}, M, Path \rangle$ has a proof in a connection calculus iff all sequents $\langle X_1, M, Path \rangle, \dots, \langle X_n, M, Path \rangle$ have proofs in the corresponding compressed connection calculus.*

*Proof.* Any connection proof of $\langle \{X_1, \dots, X_n\}, M, Path \rangle$ has the following shape:

$$R_1 \frac{P_1 \qquad R_2 \dfrac{P_2 \qquad \dfrac{\vdots}{R_n \dfrac{P_n \qquad \dfrac{}{\{\}, M, Path} A}{}}}{\{X_2, \dots, X_n\}, M, Path}}{\{X_1, \dots, X_n\}, M, Path}$$

Start $\dfrac{\bigwedge_i \langle X_i, M, \{\}\rangle}{\varepsilon, M, \varepsilon}\,\text{S}$ where $\{X_1, \ldots, X_n\}$ is copy of $C \in M$

Reduction $\dfrac{}{L, M, Path \cup \{L'\}}\,\text{R}$ where $\sigma(L) = \sigma(\overline{L'})$

Extension $\dfrac{\bigwedge_i \langle X_i, M[C_1 \backslash C_2], Path \cup \{L\}\rangle}{L, M, Path}\,\text{E}$

where $\{X_1, \ldots, X_n\}$ is the $\beta$-clause of $C_2$ with respect to $L'$, $C_2$ is copy of $C_1$, $C_1$ is e-clause of $M$ with respect to $Path \cup \{L\}$, $L' \in^+ C_2$ with $\sigma(L) = \sigma(\overline{L'})$

Decomposition $\dfrac{\bigwedge_i \langle X_i, M, Path\rangle}{M', M, Path}\,\text{D}$ where $\{X_1, \ldots, X_n\} \in M'$

**Fig. 5.** Compressed nonclausal connection calculus.

From such a proof, we can recursively construct proofs of $\langle X_i, M, Path\rangle$ in the corresponding compressed calculus by

$$R_i\,\frac{P_i'}{X_i, M, Path}$$

where $P_i'$ is the translation of the proof $P_i$ to the compressed calculus. Similarly, we can translate proofs from the compressed to the original calculi. □

*Example 3.* For the matrices $M$ and $\bar{M}$ in Example 1, proofs in the compressed calculi are given in Figs. 6 and 7. The extension steps used to prove $\langle Q, M, \{\}\rangle$ and $\langle P(s\hat{x}), \hat{M}, \ldots\rangle$ in the nonclausal proof of $M$ are explained in Example 2.

$$\dfrac{\dfrac{\dfrac{}{\neg P(x'), \bar{M}, \{Q\}}\,\text{E} \qquad \dfrac{\dfrac{\dfrac{}{\neg P(\overline{x}), \bar{M}, \{Q, P(sx'), P(s\hat{x})\}}\,\text{E}}{P(s\hat{x}), \bar{M}, \{Q, P(sx')\}}\,\text{E} \qquad \dfrac{}{\neg Q, \bar{M}, \{Q, P(sx')\}}\,\text{R}}{P(sx'), \bar{M}, \{Q\}}\,\text{E}}{Q, \bar{M}, \{\}}\,\text{E}}{\epsilon, \bar{M}, \epsilon}\,\text{S}$$

**Fig. 6.** Proof in the compressed clausal calculus with $\sigma = \{x' \mapsto a, \hat{x} \mapsto sx', \overline{x} \mapsto x'\}$.

## 4 Connection Proof Translation

In this section, we propose a translation method from connection proofs to Gentzen's sequent calculus LK [12].

$$\dfrac{\dfrac{}{\neg P(x'), M', \{Q\}}\,\mathrm{E} \qquad \dfrac{\dfrac{\dfrac{}{P(s\hat{x}), \hat{M}, \{Q, P(sx')\}}\,\mathrm{E} \qquad \dfrac{}{\neg Q, \hat{M}, \{Q, P(sx')\}}\,\mathrm{R}}{\left[[\neg P(s^2\hat{x})] \begin{bmatrix} P(s\hat{x}) \\ \neg Q \end{bmatrix}\right], \hat{M}, \{Q, P(sx')\}}\,\mathrm{D}}{P(sx'), M', \{Q\}}\,\mathrm{E}}{\dfrac{Q, M, \{\}}{\epsilon, M, \epsilon}\,\mathrm{S}}\,\mathrm{E}$$

**Fig. 7.** Proof in the compressed nonclausal calculus with $\sigma = \{x' \mapsto a, \hat{x} \mapsto sx'\}$.

A connection proof for a first-order formula $F$ consists of a connection proof tree and a global substitution $\sigma$. Given this information, we want to construct a proof of $F \vdash \bot$, which is written in LK as $F \vdash$. To more concisely present the proof translation, we omit the substitution $\sigma$ in the LK translation; for example, instead of writing $\sigma(L), \sigma(M), \sigma(Path) \vdash$, we write $L, M, Path \vdash$.

We translate connection proof trees recursively by distinguishing the different rules of the calculus. We denote by $[\Gamma \vdash]$ the LK translation of the connection proof for $\Gamma$. We write that $C$ is in $M$ iff $M = C_1 \wedge \cdots \wedge C_n$ with $C = C_i$ for some $i$ with $1 \leq i \leq n$.

We use a rule $\wedge$L to extract a conjunct from a conjunction while keeping the conjunction in the context, as well as a rule $\bot$L to derive $\bot$ from two complementary literals in the context:[5]

$$\dfrac{\Gamma, C_i, C_1 \wedge \cdots \wedge C_n \vdash \Delta}{\Gamma, C_1 \wedge \cdots \wedge C_n \vdash \Delta}\,\wedge\mathrm{L} \qquad \dfrac{}{\Gamma, A, \overline{A} \vdash}\,\bot\mathrm{L}$$

We now describe the translation of connection proofs. Two rules of the connection calculi are translated the same way for clausal and nonclausal proofs, namely the start and the reduction rule. We show the translation of these rules in Fig. 8. For the start rule, the translation obtains the formula corresponding to the clause $C$ with the $\wedge$L rule, and instantiates it with the $\forall$L rule. The substitution $\sigma$ is used to determine the instantiations, where fresh names are invented when a variable is unbound in the substitution. As noted before, we omit $\sigma$ in the LK translation, writing $X_1 \vee \cdots \vee X_n, M \vdash$ to abbreviate $\sigma(X_1 \vee \cdots \vee X_n), \sigma(M) \vdash$. Then, the sequent is split into several proof trees $[X_i, M, \{\} \vdash]$, which represent the translations of the connection proofs for $\langle X_i, M, \{\}\rangle$.[6]

[5] These rules are not part of Gentzen's original LK calculus. However, translating them into Gentzen's LK is straightforward.

[6] In the clausal setting, $X_i$ could be written as $L_i$, but because the same rule is used in the nonclausal setting, where $X_i$ can represent either a literal or a matrix, we write $X_i$ for the common rules.

| Connection Calculus | LK |
|---|---|
| $\dfrac{\bigwedge_i \langle X_i, M, \{\}\rangle}{\varepsilon, M, \varepsilon}\,\mathrm{S}$ where $\{X_1, \dots, X_n\}$ is copy of $C \in M$ | $\dfrac{\dfrac{\dfrac{[X_1, M, \{\} \vdash] \quad \dots \quad [X_n, M, \{\} \vdash]}{X_1 \vee \dots \vee X_n, M \vdash}\,\vee\mathrm{L}}{\forall \boldsymbol{x}.(X_1 \vee \dots \vee X_n), M \vdash}\,\forall\mathrm{L}}{M \vdash}\,\wedge\mathrm{L}$ where $\forall \boldsymbol{x}.(X_1 \vee \dots \vee X_n)$ in $M$ |
| $\dfrac{}{L, M, \mathit{Path} \cup \{L'\}}\,\mathrm{R}$ where $\sigma(L) = \sigma(\overline{L'})$ | $\dfrac{}{L, M, \mathit{Path} \cup \{L'\} \vdash}\,\bot\mathrm{L}$ where $L = \overline{L'}$ |

**Fig. 8.** LK translation of common connection calculus rules.

### 4.1 Clausal Proof Translation

The translation of the clausal extension rule (shown in Fig. 4) is given in Fig. 9. First, $L, M, \mathit{Path} \vdash$ is transformed to the equivalent $M, P \vdash$, where $P = \mathit{Path} \cup \{L\}$. The remaining translation resembles that of the start rule, with the exception that it additionally closes a proof branch containing the negated literal $\overline{L}$.

$$\dfrac{\dfrac{\dfrac{\dfrac{[L_1, M, P \vdash] \quad \dots \quad \dfrac{}{\overline{L}, M, P \vdash}\,\bot\mathrm{L} \quad \dots \quad [L_n, M, P \vdash]}{L_1 \vee \dots \vee \overline{L} \vee \dots \vee L_n, M, P \vdash}\,\vee\mathrm{L}}{\forall \boldsymbol{x}.(L_1 \vee \dots \vee L_n), M, P \vdash}\,\forall\mathrm{L}}{M, P \vdash}\,\wedge\mathrm{L}}{L, M, \mathit{Path} \vdash}$$

where $\forall \boldsymbol{x}.(L_1 \vee \dots \vee L_n)$ in $M$ and $P = \mathit{Path} \cup \{L\}$

**Fig. 9.** LK translation of the clausal extension rule.

### 4.2 Nonclausal Proof Translation

We now proceed with the translation of nonclausal connection proofs, using the calculus introduced in Fig. 5. The LK context in the translation of nonclausal proofs now has the shape $X, \boldsymbol{M}, \mathit{Path}$, where $\boldsymbol{M}$ is a set of matrices instead of a single matrix $M$ as in the clausal case. During translation, $\boldsymbol{M}$ is extended such that for each word $\langle L, M, \mathit{Path}\rangle$ in the connection calculus and its corresponding sequent $L, \boldsymbol{M}, \mathit{Path} \vdash$ in LK, the e-clauses of $M$ with respect to $\mathit{Path} \cup \{L\}$ are the clauses $C$ for which $C$ in $M'$ and $M' \in \boldsymbol{M}$. We will see this in detail in the explanation for the extension rule.

The LK translation of nonclausal proofs reuses the translations of the start and the reduction rules given in Fig. 8. However, occurrences of $M$ in the LK translation are replaced by $\boldsymbol{M}$. The start rule uses $\boldsymbol{M} = \{M\}$, i.e. $\boldsymbol{M}$ contains only the initial problem matrix $M$.

The decomposition rule of the nonclausal calculus can be seen as a generalisation of the start rule. We give its translation to LK in Fig. 10.

| Connection Calculus | LK |
|---|---|
| $\dfrac{\bigwedge_i \langle X_i, M, Path\rangle}{M', M, Path}\,\text{D}$ where $\{X_1, \dots, X_n\} \in M'$ | $\dfrac{\dfrac{\dfrac{[X_1, \boldsymbol{M}', Path \vdash] \quad \dots \quad [X_n, \boldsymbol{M}', Path \vdash]}{X_1 \vee \dots \vee X_n, \boldsymbol{M}', Path \vdash}\,\vee\text{L}}{\forall \boldsymbol{x}.(X_1 \vee \dots \vee X_n), \boldsymbol{M}', Path \vdash}\,\forall\text{L}}{M', \boldsymbol{M}, Path \vdash}\,\wedge\text{L}$ where $\forall \boldsymbol{x}.(X_1 \vee \dots \vee X_n)$ in $M'$ and $\boldsymbol{M}' = \{M'\} \cup \boldsymbol{M}$ |

**Fig. 10.** LK translation of the decomposition rule.

Let us now consider a nonclausal extension step applied to $\langle L, M, Path\rangle$. Let $C_1$ denote the e-clause of $M$ with respect to $Path \cup \{L\}$ that was used for the extension step. By construction of $\boldsymbol{M}$ mentioned above, $C_1$ is some clause in $M_1 \in \boldsymbol{M}$. Furthermore, let $\beta_1$ be the $\beta$-clause of $C_1$ with respect to $\overline{L}$. Then we can find some $m$ such that $M_1$, $C_1$ and $\beta_1$ can be written as in Fig. 11.

$$M_i = \begin{cases} \left[\ \cdots\ \overbrace{\begin{bmatrix} X_{i,1} \\ \vdots \\ M_{i+1} \\ \vdots \\ X_{i,n_i} \end{bmatrix}}^{C_i}\ \cdots\ \right] & \text{if } i \le m \\ \overline{L} & \text{otherwise} \end{cases} \qquad \beta_i = \begin{cases} \begin{bmatrix} X_{i,1} \\ \vdots \\ [\beta_{i+1}] \\ \vdots \\ X_{i,n_i} \end{bmatrix} & \text{if } i \le m \\ [] & \text{otherwise} \end{cases}$$

**Fig. 11.** Definition of matrix $M_i$, clause $C_i$, and $\beta$-clause $\beta_i$.

The translation of the nonclausal extension rule is shown in Fig. 12. We first transform $L, \boldsymbol{M}, Path \vdash$ to $\boldsymbol{M}^0, P \vdash$ which is equivalent due to $\boldsymbol{M}^0 = \boldsymbol{M}$. We then determine $M_1 \in \boldsymbol{M}$ and put it into the context by contraction (CL).

Now we recursively prove the sequent $M_i, \boldsymbol{M}^{i-1}, P \vdash$ as follows: If $M_i$ is the literal $\overline{L}$, we prove the sequent $\overline{L}, \boldsymbol{M}^m, P \vdash$ with the $\bot$L rule. Otherwise, we proceed in the following way: First, we put the appropriate clause $C_i$ of $M_i$ that

$$
\dfrac{[X_{m,1}, \boldsymbol{M}^m, P \vdash] \quad \cdots \quad \dfrac{}{\overline{L}, \boldsymbol{M}^m, P \vdash}\,\bot\mathrm{L} \quad \cdots \quad [X_{m,n_m}, \boldsymbol{M}^m, P \vdash]}{}\,\vee\mathrm{L}
$$

$$
\dfrac{\dfrac{\dfrac{\dfrac{\dfrac{[X_{1,1}, \boldsymbol{M}^1, P \vdash] \quad \cdots \quad \dfrac{\vdots}{M_2, \boldsymbol{M}^1, P \vdash}\,\wedge\mathrm{L} \quad \cdots \quad [X_{1,n_1}, \boldsymbol{M}^1, P \vdash]}{X_{1,1} \vee \cdots \vee X_{1,n_1}, \boldsymbol{M}^1, P \vdash}\,\vee\mathrm{L}}{\forall \boldsymbol{x}.(X_{1,1} \vee \cdots \vee X_{1,n_1}), \boldsymbol{M}^1, P \vdash}\,\forall\mathrm{L}}{M_1, \boldsymbol{M}^0, P \vdash}\,\wedge\mathrm{L}}{\boldsymbol{M}^0, P \vdash}\,\mathrm{CL}}{L, \boldsymbol{M}, \mathit{Path} \vdash}
$$

where $\boldsymbol{M}^j = \boldsymbol{M} \cup \{M_i \mid 1 \leq i \leq j\}$ and $P = \mathit{Path} \cup \{L\}$

**Fig. 12.** LK translation of the nonclausal extension rule.

corresponds to $\beta_i$ into the context with the $\wedge$L rule. In the same step, we merge $M_i$ with $\boldsymbol{M}^{i-1}$, yielding $\boldsymbol{M}^i$. After the instantiation of $C_i$ with the $\forall$L rule, the clause elements $X_{i,1}$ to $X_{i,n_i}$ give rise to several proof branches where all but one are closed by translation of the proof branches of the connection proof. The one remaining clause element $M_{i+1}$ gives rise to a sequent $M_{i+1}, \boldsymbol{M}^i, P \vdash$, which we translate by recursion. This concludes the translation of the extension rule.

*Example 4.* Consider the nonclausal proof given in Fig. 7. We show its translation to LK in Fig. 13, where boxed sequents indicate words of the original proof. We use $F$ from Example 1 to define

$$
\begin{aligned}
\boldsymbol{M}_0 &= \{F\} \\
\boldsymbol{M}_1 &= \boldsymbol{M}_0 \cup \{\neg P(s^2 a) \wedge (P(sa) \vee \neg Q)\} \\
\boldsymbol{M}_2 &= \boldsymbol{M}_1 \cup \{\neg P(s^3 a) \wedge (P(s^2 a) \vee \neg Q)\}
\end{aligned}
$$

The question might arise whether the proof translation necessarily needs to keep a set of matrices $\boldsymbol{M}$ containing potential extension clauses. Could one instead reconstruct extension clauses from the initial $M$ and *Path*? The next example shows that extending $\boldsymbol{M}$ with extension clauses is indeed necessary.

*Example 5.* Consider the extension step that closes $\langle P(s\hat{x}), \hat{M}, \{Q, P(sx')\}\rangle$ in Fig. 7. The extension clause used in this extension step is $C_4$ from Example 2. However, the closest to $C_4$ we can obtain from $M$ and $\{Q, P(sx')\} \cup \{P(s\hat{x})\}$ is

$$
\begin{bmatrix} \neg P(x') \\ [[[\neg P(s^2 x')]]] \end{bmatrix}
$$

As performed by our translation, extending $\boldsymbol{M}$ in the translation of the extension step for $\langle P(sx'), M', \{Q\}\rangle$ with the $\alpha$-related clause $[\neg P(s^2 x')]$ corresponding to $C_4$ allows us to translate the extension step for $P(s\hat{x})$ with precisely that clause.

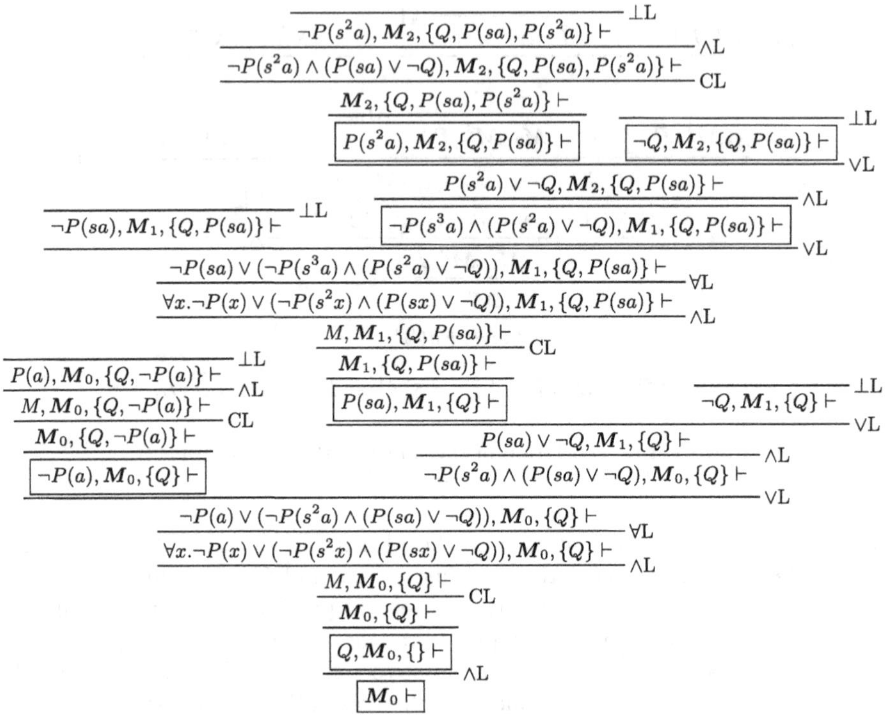

**Fig. 13.** Translation of the nonclausal proof in Fig. 7 to LK.

The LK translation uses *Path* only for reduction steps and $\boldsymbol{M}$ for extension steps, whereas the original calculus uses *Path* for both. Future work might explore whether a calculus closer to the translation yields more efficient proof search.

**Theorem 1.** *Let* $\langle X, M, Path\rangle$ *be a word in the nonclausal connection proof* $\Gamma$ *and let* $\boldsymbol{M}$ *contain the extension clauses of* $M$ *with respect to* $Path \cup \{X\}$. *For every premise* $\langle X', M', Path'\rangle$ *of the proof step in* $\Gamma$ *with the conclusion* $\langle X, M, Path\rangle$, *the translation* $[X, \boldsymbol{M}, Path \vdash]$ *has a sub-proof tree* $[X', \boldsymbol{M}', Path' \vdash]$ *such that* $\boldsymbol{M}'$ *contains the extension clauses of* $M'$ *with respect to* $Path' \cup \{X'\}$.

*Proof.* We distinguish the calculus rule to close $\langle X, M, Path\rangle$. The reduction rule is trivial because it has no premises.

Let us first consider the start rule in Fig. 8. The translation of the start rule yields several proof trees of the shape $[X_i, \boldsymbol{M}, \{\} \vdash]$, where $\boldsymbol{M} = \{M\}$. For every $i$, the extension clauses of $M$ with respect to $X_i$ are all the clauses in $M$, as was illustrated in Example 2. Because all clauses in $M$ are also contained in $\boldsymbol{M}$, the start rule satisfies the property.

Now for the decomposition rule shown in Fig. 10. By hypothesis, $\boldsymbol{M}$ contains the extension clauses of $M$ with respect to $Path \cup \{M'\}$. This implies that $\boldsymbol{M}$ contains all clauses that recursively contain $M'$. For every $i$, $X_i$ is contained in $M'$, therefore $\boldsymbol{M}'$ contains all clauses that recursively contain $X_i$, satisfying condition (a) of Definition 3. Furthermore, those clauses $\alpha$-related to $X_i$ that are required by condition (b) and that are not contained in $\boldsymbol{M}$ are $M' \setminus \{\forall \boldsymbol{x}.(X_1 \vee \cdots \vee X_n)\}$ and thus in $\boldsymbol{M}'$.

Finally we treat the extension rule shown in Fig. 12. By hypothesis, $\boldsymbol{M}$ contains the extension clauses of $M$ with respect to $Path \cup \{L\}$. We have to show that for each $i$ and $j$, the extension clauses of $M$ with respect to $P \cup \{X_{i,j}\}$ correspond to the clauses in $\boldsymbol{M}^i$. For every $i$ and $j$, we have that $\boldsymbol{M}^i$ contains all clauses that recursively contain $X_{i,j}$, which in addition to some clauses in $\boldsymbol{M}$ are the clauses $C_k$ (see Fig. 11) with $k \leq i$. This covers condition (a) of Definition 3. Furthermore, those clauses $\alpha$-related to $X_{i,j}$ that are required by condition (b) and that are not contained in $\boldsymbol{M}$ are the clauses $M_k \setminus \{C_k\}$ with $k \leq i$, which are contained in $\boldsymbol{M}^i$. □

**Corollary 1.** *For every formula $F$, if $\Gamma$ is a nonclausal connection proof of $M(F)$, then the translation $[\Gamma \vdash]$ is an LK proof of $M(F) \vdash$.*

*Proof.* By induction on $\Gamma$ and Theorem 1.

In four large test sets of nonclausal and clausal connection proofs, all translated proofs yielded by our implementations of the proof translations in this section are successfully verified by an interactive theorem prover, see Sect. 6.

## 5 Implementation

HOL Light is an interactive theorem prover developed by Harrison in OCaml [16]. leanCoP and nanoCoP are clausal and nonclausal connection provers developed by Otten in Prolog [27,30]. We developed proof search tactics for HOL Light based on leanCoP/nanoCoP and the proof translation shown in Sect. 4.[7] To ease integration with HOL Light, all parts of the tactics are written in OCaml, including functional implementations of leanCoP and nanoCoP using the compressed calculi in Sect. 3.

The structure of the proof search tactics is shown in Fig. 14: First, we convert given proof goals from higher-order logic to first-order logic. For this, we reuse a large part of the MESON [15] infrastructure, such as instantiation of higher-order axioms. This leaves us with first-order problems of the shape $(A_1 \wedge \cdots \wedge A_n) \implies C$, on which we run leanCoP and nanoCoP in the same interpreter as HOL Light [11]. Finally, we translate the resulting connection proofs to HOL Light proofs: We implemented the proof translation shown in Sect. 4 such that it directly yields HOL Light instead of LK proofs.

[7] The source code can be retrieved at http://cl-informatik.uibk.ac.at/users/mfaerber/tactics.html.

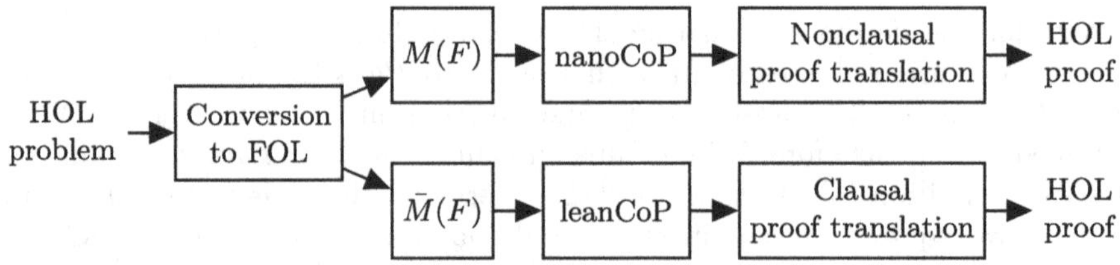


**Fig. 14.** Structure of the proof search tactics in HOL Light.

## 6 Evaluation

We compare the performance of our proof search tactics based on leanCoP 2.1 and nanoCoP 1.0 with the Metis [10] and MESON [15] tactics. Similarly to [19], we disable splitting for MESON. We evaluate the tactics on two kinds of problems derived from HOL Light: toplevel and MESON problems.

A toplevel problem results from any HOL Light theorem that is given a name on the OCaml toplevel. It consists of the conclusion of the theorem and the premises used to prove it. A MESON problem results from any call to the MESON tactic. It consists of the statement proven by MESON as well as the premises given to the MESON tactic. Note that toplevel problems are not necessarily solvable by first-order tactics, whereas MESON problems are, because the (first-order) tactic MESON is able to prove them.

We evaluate both toplevel and MESON problems with some tactic by letting the tactic find a proof of the problem conclusion using the problem premises. The problem counts as proven if the tactic finds a proof within a given time limit. We consider toplevel ("top") and MESON ("msn") problems from core HOL Light ("HL") and the Flyspeck project ("FS"), which finished in 2014 a formal proof of the Kepler conjecture [14]. We use the Git version 08f4461 of HOL Light from March 2017 (https://github.com/jrh13/hol-light/commit/08f4461), running every tactic with a timeout of 10 s on each problem. We use a 48-core server with AMD Opteron 6174 2.2 GHz CPUs, 320 GB RAM, and 0.5 MB L2 cache per CPU. Each problem is always assigned one CPU. We run all provers with a timeout of 10 s per problem.

**Listing 1.1.** Flyspeck problem WLOG_LINEAR_INJECTIVE_IMAGE_ALT.

```
!P. (!f s. P s /\ linear f ==> P (IMAGE f s))
    ==> (!f. linear f /\ (!x y. f x = f y ==> x = y)
         ==> (!s. P (IMAGE f s) <=> P s))
==>
!P f s. (!g t. P t /\ linear g ==> P (IMAGE g t)) /\
        linear f /\ (!x y. f x = f y ==> x = y)
        ==> (P (IMAGE f s) <=> P s)
```

The results are shown in Table 1: Metis solves the largest number of problems among all considered datasets. The comparatively low performance of leanCoP/nanoCoP inside HOL Light is due to their heavy use of array operations for unification: Array access is more than 30 times faster in native OCaml programs compared to programs compiled in OCaml's toplevel (as used in HOL Light). When compiled as native OCaml programs, we have shown that leanCoP/nanoCoP solve more problems than Metis on four out of six datasets that we evaluated [11]. Running leanCoP/nanoCoP outside HOL Light and translating the resulting proofs inside HOL Light would thus very likely increase the performance of the corresponding tactics.

**Table 1.** Number of problems solved by various HOL Light tactics.

| Prover | HL-top | HL-msn | FS-top | FS-msn |
|---|---|---|---|---|
| Problems in dataset | 2499 | 1119 | 27112 | 44468 |
| Metis | 807 | 1029 | 4626 | 42829 |
| MESON | 736 | 900 | 4221 | 39227 |
| leanCoP+cut | 724 | 948 | 3714 | 39922 |
| leanCoP−cut | 717 | 844 | 3800 | 38528 |
| nanoCoP+cut | 538 | 802 | 2743 | 34213 |
| nanoCoP−cut | 550 | 811 | 2351 | 34769 |

*Example 6.* Listing 1.1 shows a Flyspeck toplevel problem which among the evaluated tactics, only nanoCoP can solve in the given time limit of 10 s. It is proven by nanoCoP in 2.27 s.

## 7 Related Work

Certification of ATP found proofs has been especially important for the integration of ATPs into interactive proof assistants. Such components provide automation in the form of proof tactics for smaller steps. HOL Light includes the certified proof producing model elimination prover MESON [15]. The paramodulation-based prover Metis [17] was designed with a small certified proof core to simplify its integration with interactive theorem provers [10]. There exists a proof-certifying version of the intuitionistic first-order automated theorem prover JProver for Coq and Nuprl [20,33] as well as a proof certifying version of an ordered paramodulation prover for Matita [1]. Proofs from several SAT/SMT solvers can be certified in Coq [9] and Isabelle [4]. The logical framework Dedukti allows for the import of superposition proofs from iProver [6] as well as of tableaux proofs from Zenon [7]. The GAPT framework provides translations for a multitude of calculi and automated theorem provers, such as Vampire, E, Prover9, and leanCoP [8,31].

Among all provers whose proof certification is described in the cited work above, the only nonclausal one is JProver. However, its performance is far behind nanoCoP and the intuitionistic version of nanoCoP, nanoCoP-i, with nanoCoP and nanoCoP-i solving about three times as many problems as JProver on the TPTP and the ILTP benchmarks, respectively [27,28]. On the other hand, unlike for nanoCoP-i, there already exists a proof certification method for JProver in an intuitionistic proof assistant, namely in Coq. This leaves as future work the extension of the proof certification in this paper to an intuitionistic setting, in order to enable stronger automated proof search via nanoCoP-i in proof assistants like Coq.

## 8 Conclusion

We proposed a translation from clausal and nonclausal connection proofs to LK, yielding a sound proof certification and a proof search tactic for HOL Light. The tactic certifies every nanoCoP and leanCoP proof output in our evaluation.

Future work includes the improvement of the proof search tactics, for example by calling external instances of nanoCoP/leanCoP, but also by improved preprocessing of the tactics, for example by reordering the clauses in the ITP before proof search [29]. The proof search tactic could also be integrated into other ITPs, such as Isabelle [34] and Coq [2]. The latter being an intuitionistic system motivates the translation of nonclassical connection proofs, such as given by ileanCoP and nanoCoP-i [25,28]. Finally, we hope that the present article helps to prepare the ground for ITP-checked proofs of soundness/completeness of connection calculi as well as of their implementations.

**Acknowledgements.** We thank the reviewers of JAR and TABLEAUX for their valuable comments as well as Jens Otten for clarifying details of his work on nonclausal connection proving. This work has been supported by a doctoral scholarship of the University of Innsbruck and the European Research Council (ERC) grant no. 714034 *SMART*.

## References

1. Asperti, A., Tassi, E.: Higher order proof reconstruction from paramodulation-based refutations: the unit equality case. In: Kauers, M., Kerber, M., Miner, R., Windsteiger, W. (eds.) Calculemus/MKM -2007. LNCS (LNAI), vol. 4573, pp. 146–160. Springer, Heidelberg (2007). https://doi.org/10.1007/978-3-540-73086-6_14
2. Bertot, Y.: A short presentation of Coq. In: Mohamed et al. [22], pp. 12–16
3. Bibel, W.: Automated Theorem Proving. Artificial Intelligence, 2nd edn. Vieweg, Wiesbaden (1987)
4. Blanchette, J.C., Böhme, S., Paulson, L.C.: Extending Sledgehammer with SMT solvers. In: Bjørner, N., Sofronie-Stokkermans, V. (eds.) CADE 2011. LNCS (LNAI), vol. 6803, pp. 116–130. Springer, Heidelberg (2011). https://doi.org/10.1007/978-3-642-22438-6_11

5. Blanchette, J.C., Kaliszyk, C., Paulson, L.C., Urban, J.: Hammering towards QED. J. Formaliz. Reason. **9**(1), 101–148 (2016)
6. Burel, G.: A shallow embedding of resolution and superposition proofs into the $\lambda\Pi$-calculus modulo. In: Blanchette, J.C., Urban, J. (eds.) PxTP. EPiC Series in Computing, vol. 14, pp. 43–57. EasyChair (2013)
7. Cauderlier, R., Halmagrand, P.: Checking Zenon modulo proofs in Dedukti. In: Kaliszyk and Paskevich [18], pp. 57–73
8. Ebner, G., Hetzl, S., Reis, G., Riener, M., Wolfsteiner, S., Zivota, S.: System description: GAPT 2.0. In: Olivetti and Tiwari [24], pp. 293–301
9. Ekici, B., et al.: SMTCoq: a plug-in for integrating SMT solvers into Coq. In: Majumdar, R., Kunčak, V. (eds.) CAV 2017. LNCS, vol. 10427, pp. 126–133. Springer, Cham (2017). https://doi.org/10.1007/978-3-319-63390-9_7
10. Färber, M., Kaliszyk, C.: Metis-based paramodulation tactic for HOL Light. In: Gottlob, G., Sutcliffe, G., Voronkov, A., (eds.) GCAI. EPiC Series in Computing, vol. 36, pp. 127–136. EasyChair (2015)
11. Färber, M.: Learning proof search in proof assistants. Ph.D. thesis, Universität Innsbruck (2018)
12. Gentzen, G.: Untersuchungen über das logische Schließen. I. Mathematische Zeitschrift **39**(1), 176–210 (1935)
13. Gordon, M.: From LCF to HOL: a short history. In: Plotkin, G.D., Stirling, C., Tofte, M. (eds.) Proof, Language, and Interaction, Essays in Honour of Robin Milner, pp. 169–186. The MIT Press, Cambridge (2000)
14. Hales, T.C., et al.: A formal proof of the Kepler conjecture. Forum Math. Pi **5** (2017)
15. Harrison, J.: Optimizing proof search in model elimination. In: McRobbie, M.A., Slaney, J.K. (eds.) CADE 1996. LNCS, vol. 1104, pp. 313–327. Springer, Heidelberg (1996). https://doi.org/10.1007/3-540-61511-3_97
16. Harrison, J.: HOL Light: an overview. In: Berghofer, S., Nipkow, T., Urban, C., Wenzel, M. (eds.) TPHOLs 2009. LNCS, vol. 5674, pp. 60–66. Springer, Heidelberg (2009). https://doi.org/10.1007/978-3-642-03359-9_4
17. Hurd, J.: First-order proof tactics in higher-order logic theorem provers. In: Archer, M., Vito, B.D., Muñoz, C. (eds.) Design and Application of Strategies/Tactics in Higher Order Logics (STRATA). NASA Technical reports, no. NASA/CP-2003-212448, pp. 56–68, September 2003
18. Kaliszyk, C., Paskevich, A. (eds.): PxTP. EPTCS, vol. 186 (2015)
19. Kaliszyk, C., Urban, J., Vyskočil, J.: Certified connection tableaux proofs for HOL Light and TPTP. In: Leroy, X., Tiu, A. (eds.) CPP, pp. 59–66. ACM (2015)
20. Kreitz, C., Schmitt, S.: A uniform procedure for converting matrix proofs into sequent-style systems. Inf. Comput. **162**(1–2), 226–254 (2000)
21. Letz, R., Stenz, G.: Model elimination and connection tableau procedures. In: Robinson and Voronkov [32], pp. 2015–2114
22. Mohamed, O.A., Muñoz, C., Tahar, S. (eds.): TPHOLs 2008. LNCS, vol. 5170. Springer, Heidelberg (2008). https://doi.org/10.1007/978-3-540-71067-7
23. Nonnengart, A., Weidenbach, C.: Computing small clause normal forms. In: Robinson and Voronkov [32], pp. 335–367
24. Olivetti, N., Tiwari, A. (eds.): IJCAR 2016. LNCS (LNAI), vol. 9706. Springer, Cham (2016). https://doi.org/10.1007/978-3-319-40229-1
25. Otten, J.: Clausal connection-based theorem proving in intuitionistic first-order logic. In: Beckert, B. (ed.) TABLEAUX 2005. LNCS (LNAI), vol. 3702, pp. 245–261. Springer, Heidelberg (2005). https://doi.org/10.1007/11554554_19

26. Otten, J.: A non-clausal connection calculus. In: Brünnler, K., Metcalfe, G. (eds.) TABLEAUX 2011. LNCS (LNAI), vol. 6793, pp. 226–241. Springer, Heidelberg (2011). https://doi.org/10.1007/978-3-642-22119-4_18
27. Otten, J.: nanoCoP: a non-clausal connection prover. In: Olivetti and Tiwari [24], pp. 302–312
28. Otten, J.: Non-clausal connection calculi for non-classical logics. In: Schmidt, R.A., Nalon, C. (eds.) TABLEAUX 2017. LNCS (LNAI), vol. 10501, pp. 209–227. Springer, Cham (2017). https://doi.org/10.1007/978-3-319-66902-1_13
29. Otten, J.: Proof search optimizations for non-clausal connection calculi. In: Konev, B., Urban, J., Rümmer, P. (eds.) PAAR. CEUR Workshop Proceedings, vol. 2162, pp. 49–57. CEUR-WS.org (2018)
30. Otten, J., Bibel, W.: leanCoP: lean connection-based theorem proving. J. Symb. Comput. **36**(1–2), 139–161 (2003)
31. Reis, G.: Importing SMT and connection proofs as expansion trees. In: Kaliszyk and Paskevich [18], pp. 3–10
32. Robinson, J.A., Voronkov, A., (eds.): Handbook of Automated Reasoning (in 2 volumes). Elsevier and MIT Press (2001)
33. Schmitt, S., Lorigo, L., Kreitz, C., Nogin, A.: JProver: integrating connection-based theorem proving into interactive proof assistants. In: Goré, R., Leitsch, A., Nipkow, T. (eds.) IJCAR 2001. LNCS, vol. 2083, pp. 421–426. Springer, Heidelberg (2001). https://doi.org/10.1007/3-540-45744-5_34
34. Wenzel, M., Paulson, L.C., Nipkow, T.: The Isabelle framework. In: Mohamed et al. [22], pp. 33–38

# Preferential Tableaux for Contextual Defeasible $\mathcal{ALC}$

Katarina Britz[1] and Ivan Varzinczak[1,2(✉)]

[1] CAIR, Stellenbosch University, Stellenbosch, South Africa
abritz@sun.ac.za
[2] CRIL, Université d'Artois & CNRS, Lens, France
varzinczak@cril.fr

**Abstract.** In recent work, we addressed an important limitation in previous extensions of description logics to represent defeasible knowledge, namely the restriction in the semantics of defeasible concept inclusion to a single preference order on objects of the domain. Syntactically, this limitation translates to a context-agnostic notion of defeasible subsumption, which is quite restrictive when it comes to modelling different nuances of defeasibility. Our point of departure in our recent proposal allows for different orderings on the interpretation of roles. This yields a notion of contextual defeasible subsumption, where the context is informed by a role. In the present paper, we extend this work to also provide a proof-theoretic counterpart and associated results. We define a (naïve) tableau-based algorithm for checking preferential consistency of contextual defeasible knowledge bases, a central piece in the definition of other forms of contextual defeasible reasoning over ontologies, notably contextual rational closure.

**Keywords:** Description logics · Defeasible reasoning · Contexts · Tableaux

## 1 Introduction

Description logics (DLs) [1] are central to many modern AI and database applications since they provide the logical foundation of formal ontologies. Yet, as classical formalisms, DLs do not allow for the proper representation of and reasoning with defeasible information, as shown up in the following example from the access-control domain: employees have access to classified information; interns (who are also employees) do not; but graduate interns do. From a naïve (classical) formalisation of this scenario, one concludes that the class of interns is empty (just as that of graduate interns). But while concept unsatisfiability has been investigated extensively in ontology debugging and repair, our research problem here goes beyond that.

The past 25 years have witnessed many attempts to introduce defeasible-reasoning capabilities in a DL setting, usually drawing on a well-established body


S. Cerrito and A. Popescu (Eds.): TABLEAUX 2019, LNAI 11714, pp. 39–57, 2019.
https://doi.org/10.1007/978-3-030-29026-9_3

of research on non-monotonic reasoning (NMR). These comprise the so-called preferential approaches [13–15,25,26,29,30,34,35,47,48], circumscription-based ones [6,7,49], as well as others [2,3,5,8,27,37–39,45,46,51].

Preferential extensions of DLs [14,29] turn out to be particularly promising. There a notion of *defeasible subsumption* $\sqsubsetsim$ is introduced, the intuition of a statement of the form $C \sqsubsetsim D$ being that "usually, $C$ is subsumed by $D$" or "the normal $C$s are $D$s". The semantics is in terms of an ordering on the set of objects allowing us to identify the most normal elements in $C$ with the *minimal* $C$-instances w.r.t. the ordering.

The assumption of a single ordering on the domain of interpretation does not allow for different, possibly incompatible, notions of defeasibility in subsumption resulting from the fact that a given object may be more exceptional than another in some context but less exceptional in another. Defeasibility therefore introduces a new facet of contextual reasoning not present in *deductive* reasoning. In recent work [20] we addressed this limitation by allowing different orderings on objects, using preference relations on role interpretations [17]. Here we complete the picture by also providing a proof-theoretic counterpart in the form of a tableau algorithm for satisfiability checking of a defeasible $\mathcal{ALC}$ knowledge base. Even though the notion of entailment considered here is monotonic, it is required in order to compute a stronger non-monotonic version of entailment as, for example, used in the computation of rational closure [20].

The remainder of the present paper is organised as follows: In Sect. 2 we provide a summary of the DL $\mathcal{ALC}$ and set up the notation we shall follow. In Sect. 3, we recall our context-based defeasible DL, its properties, and in particular we show its fruitfulness in modelling context-based defeasibility. In Sect. 4, we define a naïve (i.e., doubly-exponential) tableau-based algorithm for checking consistency of contextual defeasible knowledge bases. After a discussion of and a comparison with related work (Sect. 5), we conclude with a note on future directions of investigation. (A preliminary version of this work was presented at the International Workshop on Description Logics [22].)

## 2 Logical Preliminaries

The (concept) language of $\mathcal{ALC}$ is built upon a finite set of atomic *concept names* $\mathsf{C}$, a finite set of *role names* $\mathsf{R}$ (a.k.a. *attributes*) and a finite set of *individual names* $\mathsf{I}$ such that $\mathsf{C}$, $\mathsf{R}$ and $\mathsf{I}$ are pairwise disjoint. In our scenario example, we can have for instance $\mathsf{C} = \{\mathsf{Classified}, \mathsf{Employee}, \mathsf{Graduate}, \mathsf{Intern}, \mathsf{ResAssoc}\}$, $\mathsf{R} = \{\mathsf{hasAcc}, \mathsf{hasJob}, \mathsf{hasQual}\}$, and $\mathsf{I} = \{\mathsf{anne}, \mathsf{bill}, \mathsf{chris}, \mathsf{doc123}\}$, with the obvious intuitions, and where ResAssoc, hasAcc and hasQual stand for 'research associate', 'has access' and 'has qualification', respectively. With $A, B, \ldots$ we denote atomic concepts, with $r, s, \ldots$ role names, and with $a, b, \ldots$ individual names. Complex concepts are denoted with $C, D, \ldots$ and are built using the constructors $\neg$ (complement), $\sqcap$ (concept conjunction), $\sqcup$ (concept disjunction), $\forall$ (value restriction) and $\exists$ (existential restriction) according to the following grammar rules:

$$C ::= \top \mid \bot \mid \mathsf{C} \mid (\neg C) \mid (C \sqcap C) \mid (C \sqcup C) \mid (\exists r.C) \mid (\forall r.C)$$

With $\mathcal{L}_{\mathcal{ALC}}$ we denote the *language* of all $\mathcal{ALC}$ concepts. Examples of $\mathcal{ALC}$ concepts in our scenario are $\mathsf{Employee} \sqcap \neg\mathsf{ResAssoc}$ and $\exists\mathsf{hasAcc}.\mathsf{Classified}$.

The semantics of $\mathcal{ALC}$ is the standard set-theoretic Tarskian semantics. An *interpretation* is a structure $\mathcal{I} =_{\text{def}} \langle \Delta^{\mathcal{I}}, \cdot^{\mathcal{I}} \rangle$, where $\Delta^{\mathcal{I}}$ is a non-empty set called the *domain*, and $\cdot^{\mathcal{I}}$ is an *interpretation function* mapping concept names $A$ to subsets $A^{\mathcal{I}}$ of $\Delta^{\mathcal{I}}$, role names $r$ to binary relations $r^{\mathcal{I}}$ over $\Delta^{\mathcal{I}}$, and individual names $a$ to elements of the domain $\Delta^{\mathcal{I}}$, i.e., $A^{\mathcal{I}} \subseteq \Delta^{\mathcal{I}}$, $r^{\mathcal{I}} \subseteq \Delta^{\mathcal{I}} \times \Delta^{\mathcal{I}}$, and $a^{\mathcal{I}} \in \Delta^{\mathcal{I}}$.

Figure 1 depicts an interpretation for our access-control example with domain $\Delta^{\mathcal{I}} = \{x_i \mid 0 \leq i \leq 11\}$, and interpreting the elements of the vocabulary as follows: $\mathsf{Classified}^{\mathcal{I}} = \{x_{10}\}$, $\mathsf{Employee}^{\mathcal{I}} = \{x_0, x_4, x_5, x_9\}$, $\mathsf{Graduate}^{\mathcal{I}} = \{x_4, x_5, x_6, x_9\}$, $\mathsf{Intern}^{\mathcal{I}} = \{x_0, x_4\}$, $\mathsf{ResAssoc}^{\mathcal{I}} = \{x_5, x_6, x_7\}$, $\mathsf{hasAcc}^{\mathcal{I}} = \{(x_4, x_{10}), (x_9, x_{10}), (x_6, x_{10}), (x_6, x_{11})\}$, $\mathsf{hasJob}^{\mathcal{I}} = \{(x_0, x_3), (x_4, x_3), (x_9, x_3), (x_5, x_1), (x_6, x_1)\}$, and $\mathsf{hasQual}^{\mathcal{I}} = \{(x_4, x_8), (x_9, x_8), (x_5, x_2), (x_6, x_2), (x_7, x_2)\}$. Further, $\mathsf{anne}^{\mathcal{I}} = x_5$, $\mathsf{bill}^{\mathcal{I}} = x_0$, $\mathsf{chris}^{\mathcal{I}} = x_6$, and $\mathsf{doc123}^{\mathcal{I}} = x_{10}$.

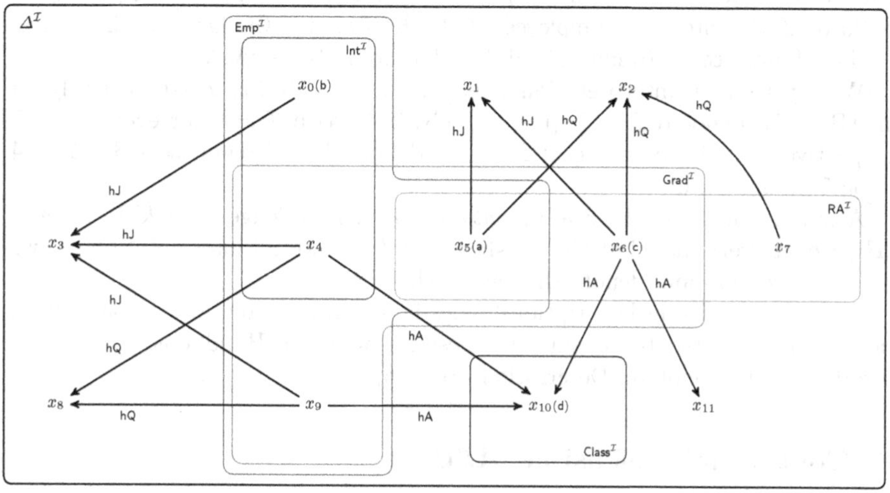


**Fig. 1.** An $\mathcal{ALC}$ interpretation for $\mathsf{C}$, $\mathsf{R}$ and $\mathsf{I}$ as above. For the sake of presentation, concept, role and individual names have been abbreviated.

Let $\mathcal{I} = \langle \Delta^{\mathcal{I}}, \cdot^{\mathcal{I}} \rangle$ be an interpretation and define $r^{\mathcal{I}}(x) =_{\text{def}} \{y \in \Delta^{\mathcal{I}} \mid (x, y) \in r^{\mathcal{I}}\}$, for $r \in \mathsf{R}$. We extend the interpretation function $\cdot^{\mathcal{I}}$ to interpret complex concepts of $\mathcal{L}_{\mathcal{ALC}}$ as follows:

$$\begin{aligned}
&\top^{\mathcal{I}} =_{\text{def}} \Delta^{\mathcal{I}}; \quad \bot^{\mathcal{I}} =_{\text{def}} \emptyset; \quad (\neg C)^{\mathcal{I}} =_{\text{def}} \Delta^{\mathcal{I}} \setminus C^{\mathcal{I}};\\
&(C \sqcap D)^{\mathcal{I}} =_{\text{def}} C^{\mathcal{I}} \cap D^{\mathcal{I}}; \quad (C \sqcup D)^{\mathcal{I}} =_{\text{def}} C^{\mathcal{I}} \cup D^{\mathcal{I}};\\
&(\exists r.C)^{\mathcal{I}} =_{\text{def}} \{x \in \Delta^{\mathcal{I}} \mid r^{\mathcal{I}}(x) \cap C^{\mathcal{I}} \neq \emptyset\};\\
&(\forall r.C)^{\mathcal{I}} =_{\text{def}} \{x \in \Delta^{\mathcal{I}} \mid r^{\mathcal{I}}(x) \subseteq C^{\mathcal{I}}\}.
\end{aligned}$$

For the interpretation $\mathcal{I}$ in Fig. 1, we have $(\mathsf{Employee} \sqcap \neg\mathsf{ResAssoc})^{\mathcal{I}} = \{x_0, x_4, x_9\}$ and $(\exists\mathsf{hasAcc.Classified})^{\mathcal{I}} = \{x_4, x_6, x_9\}$.

Given $C, D \in \mathcal{L}_{\mathcal{ALC}}$, a statement of the form $C \sqsubseteq D$ is called a *subsumption statement*, or *general concept inclusion* (GCI), read "$C$ is subsumed by $D$". Concrete examples of GCIs are $\mathsf{Intern} \sqsubseteq \mathsf{Employee}$ and $\mathsf{Intern} \sqcap \mathsf{Graduate} \sqsubseteq \exists\mathsf{hasAcc.Classified}$. $C \equiv D$ is an abbreviation for both $C \sqsubseteq D$ and $D \sqsubseteq C$. An $\mathcal{ALC}$ *TBox* $\mathcal{T}$ is a finite set of GCIs. Given $C \in \mathcal{L}_{\mathcal{ALC}}$, $r \in \mathsf{R}$ and $a, b \in \mathsf{I}$, an *assertional statement* (*assertion*, for short) is an expression of the form $a : C$ or $(a, b) : r$, read, respectively, "$a$ is an instance of $C$" and "$a$ is related to $b$ via $r$". Examples of assertions are $\mathsf{anne} : \mathsf{Employee}$ and $(\mathsf{chris}, \mathsf{doc123}) : \mathsf{hasAcc}$. An $\mathcal{ALC}$ *ABox* $\mathcal{A}$ is a finite set of assertional statements. We shall denote statements with $\alpha, \beta, \ldots$. Given $\mathcal{T}$ and $\mathcal{A}$, with $\mathcal{KB} =_{\mathrm{def}} \mathcal{T} \cup \mathcal{A}$ we denote an $\mathcal{ALC}$ *knowledge base*, a.k.a. an *ontology*.

An interpretation $\mathcal{I}$ *satisfies* a GCI $C \sqsubseteq D$ (denoted $\mathcal{I} \Vdash C \sqsubseteq D$) if $C^{\mathcal{I}} \subseteq D^{\mathcal{I}}$. (And then $\mathcal{I} \Vdash C \equiv D$ if $C^{\mathcal{I}} = D^{\mathcal{I}}$.) $\mathcal{I}$ *satisfies* an assertion $a : C$ (respectively, $(a, b) : r$), denoted $\mathcal{I} \Vdash a : C$ (respectively, $\mathcal{I} \Vdash (a, b) : r$), if $a^{\mathcal{I}} \in C^{\mathcal{I}}$ (respectively, $(a^{\mathcal{I}}, b^{\mathcal{I}}) \in r^{\mathcal{I}}$). In the interpretation $\mathcal{I}$ in Fig. 1, we have $\mathcal{I} \Vdash \mathsf{Intern} \sqsubseteq \mathsf{Employee}$, $\mathcal{I} \not\Vdash \mathsf{ResAssoc} \sqcap \mathsf{Graduate} \sqsubseteq \mathsf{Employee}$, $\mathcal{I} \Vdash \mathsf{bill} : \mathsf{Employee} \sqcap \neg\mathsf{Graduate}$ and $\mathcal{I} \not\Vdash (\mathsf{bill}, \mathsf{doc123}) : \mathsf{hasAcc}$.

We say that an interpretation $\mathcal{I}$ is a *model* of a TBox $\mathcal{T}$ (respectively, of an ABox $\mathcal{A}$), denoted $\mathcal{I} \Vdash \mathcal{T}$ (respectively, $\mathcal{I} \Vdash \mathcal{A}$) if $\mathcal{I} \Vdash \alpha$ for every $\alpha$ in $\mathcal{T}$ (respectively, in $\mathcal{A}$). We say that $\mathcal{I}$ is a model of a knowledge base $\mathcal{KB} = \mathcal{T} \cup \mathcal{A}$ if $\mathcal{I} \Vdash \mathcal{T}$ and $\mathcal{I} \Vdash \mathcal{A}$.

A statement $\alpha$ is (classically) *entailed* by a knowledge base $\mathcal{KB}$, denoted $\mathcal{KB} \models \alpha$, if every model of $\mathcal{KB}$ satisfies $\alpha$. If $\mathcal{I} \Vdash \alpha$ for all interpretations $\mathcal{I}$, we say $\alpha$ is a *validity* and denote this fact with $\models \alpha$.

For more details on Description Logics in general and on $\mathcal{ALC}$ in particular, the reader is invited to consult the Description Logic Handbook [1] and the introductory textbook on Description Logic [4].

## 3 Contextual Defeasible $\mathcal{ALC}$

The knowledge base $\mathcal{KB} = \mathcal{T} \cup \mathcal{A}$, with $\mathcal{T}$ and $\mathcal{A}$ as below, is a first stab at formalising our access-control example:

$$\mathcal{T} = \left\{\begin{array}{c} \mathsf{Intern} \sqsubseteq \mathsf{Employee}, \\ \mathsf{Employee} \sqsubseteq \exists\mathsf{hasJob}.\top, \\ \mathsf{Graduate} \sqsubseteq \mathsf{hasQual}.\top, \\ \mathsf{Employee} \sqsubseteq \exists\mathsf{hasAcc.Classified}, \\ \mathsf{Intern} \sqsubseteq \neg\exists\mathsf{hasAcc.Classified}, \\ \mathsf{Intern} \sqcap \mathsf{Graduate} \sqsubseteq \exists\mathsf{hasAcc.Classified}, \\ \mathsf{ResAssoc} \sqsubseteq \neg\mathsf{Employee}, \\ \mathsf{ResAssoc} \sqsubseteq \mathsf{Graduate} \end{array}\right\} \mathcal{A} = \left\{\begin{array}{c} \mathsf{anne} : \mathsf{ResAssoc}, \\ \mathsf{chris} : \mathsf{ResAssoc}, \\ \mathsf{doc123} : \mathsf{Classified}, \\ (\mathsf{chris}, \mathsf{doc123}) : \mathsf{hasAcc} \end{array}\right\}$$

It is not hard to see that this knowledge base is satisfiable and to check that $\mathcal{KB} \models \mathsf{Intern} \sqsubseteq \bot$, i.e., the ontology, although consistent, is *incoherent.* Incoherence of the knowledge base is but one of the (many) reasons to go defeasible. Armed with a notion of *defeasible subsumption* of the form $C \mathrel{\sqsubset\!\!\!\sim} D$ [15], of which the intuition is "normally, $C$ is subsumed by $D$", formalised by the adoption of a preferential semantics *à la* Shoham [50], we can give a more refined formalisation of our scenario example with $\mathcal{KB} = \mathcal{T} \cup \mathcal{D} \cup \mathcal{A}$, where $\mathcal{T}$ and $\mathcal{D}$ are given below ($\mathcal{D}$ standing for a *defeasible TBox*) and $\mathcal{A}$ is as above:

$$\mathcal{T} = \left\{ \begin{array}{c} \mathsf{Intern} \sqsubseteq \mathsf{Employee}, \\ \mathsf{Employee} \sqsubseteq \exists \mathsf{hasJob}.\top, \\ \mathsf{Graduate} \sqsubseteq \mathsf{hasQual}.\top \end{array} \right\} \quad \mathcal{D} = \left\{ \begin{array}{c} \mathsf{Employee} \mathrel{\sqsubset\!\!\!\sim} \exists \mathsf{hasAcc.Classified}, \\ \mathsf{Intern} \mathrel{\sqsubset\!\!\!\sim} \neg\exists \mathsf{hasAcc.Classified}, \\ \mathsf{Intern} \sqcap \mathsf{Graduate} \mathrel{\sqsubset\!\!\!\sim} \exists \mathsf{hasAcc.Classified}, \\ \mathsf{ResAssoc} \mathrel{\sqsubset\!\!\!\sim} \neg\mathsf{Employee}, \\ \mathsf{ResAssoc} \mathrel{\sqsubset\!\!\!\sim} \mathsf{Graduate} \end{array} \right\}$$

From such a defeasible knowledge base, one cannot conclude $\mathsf{Intern} \sqsubseteq \bot$, which is in line with the intuition. Pushing defeasible reasoning further, one could also ask whether intern research associates are usually graduates, and whether they should usually have access to classified information. It soon becomes clear that modelling defeasible information is more challenging than modelling classical information, and that it becomes problematic when defeasible information relating to different contexts are not modelled independently.

Suppose, for example, that Chris is a graduate research associate who is also an employee, and Anne is a research associate who is neither a graduate nor an employee. In any preferential model of the defeasible $\mathcal{KB}$, both Chris and Anne are exceptional in the class of research associates. This follows because Chris is an exceptional research associate w.r.t. employment status, and Anne is an exceptional research associate w.r.t. qualification. Also, in any preferential model of $\mathcal{KB}$ Chris and Anne are either incomparable, or one of them is more normal than the other. Since context has not been taken into account, there is no model in which Anne is more normal than Chris w.r.t. employment, but Chris is more normal than Anne w.r.t. qualification.

Contextual defeasible $\mathcal{ALC}$ ($d\mathcal{ALC}$) smoothly combines in a single logical framework the following features: all classical $\mathcal{ALC}$ constructs; defeasible value and existential restrictions [12,17]; defeasible concept inclusions [15], and context [20].

Let $\mathsf{C}$, $\mathsf{R}$ and $\mathsf{I}$ be as before. Complex $d\mathcal{ALC}$ concepts are denoted $C, D, \ldots$, and are built according to the rules:

$$C ::= \top \mid \bot \mid \mathsf{C} \mid (\neg C) \mid (C \sqcap C) \mid (C \sqcup C) \mid (\exists r.C) \mid (\forall r.C) \mid (\underset{\sim}{\exists} r.C) \mid (\underset{\sim}{\forall} r.C)$$

With $\mathcal{L}_{d\mathcal{ALC}}$ we denote the language of all $d\mathcal{ALC}$ concepts (including all $\mathcal{ALC}$ concepts). An example of $d\mathcal{ALC}$ concept in our access-control scenario is $\mathsf{ResAssoc} \sqcap (\underset{\sim}{\forall} \mathsf{hasAcc}.\neg\mathsf{Classified}) \sqcap (\exists \mathsf{hasAcc.Classified})$, denoting those research associates whose normal access is only to non-classified info but who also turn out to have some (exceptional) access to a classified document.

The semantics of $d\mathcal{ALC}$ is anchored in the well-known preferential approach to non-monotonic reasoning [42,43,50] and its extensions [9–11,16,18,19], especially those in DLs [15,17,32,47,52].

Let $X$ be a set. With $\#X$ we denote the *cardinality* of $X$. A binary relation is a *strict partial order* if it is irreflexive and transitive. If $<$ is a strict partial order on $X$, with $\min_< X =_{\text{def}} \{x \in X \mid \text{there is no } y \in X \text{ s.t. } y < x\}$ we denote the *minimal elements* of $X$ w.r.t. $<$. A strict partial order on a set $X$ is *well-founded* if for every $\emptyset \neq X' \subseteq X$, $\min_< X' \neq \emptyset$.

**Definition 1 (Ordered interpretation).** *An* ***ordered interpretation*** *is a tuple* $\mathcal{O} =_{\text{def}} \langle \Delta^{\mathcal{O}}, \cdot^{\mathcal{O}}, \ll^{\mathcal{O}} \rangle$ *such that:*

- $\langle \Delta^{\mathcal{O}}, \cdot^{\mathcal{O}} \rangle$ *is an* $\mathcal{ALC}$ *interpretation, with* $A^{\mathcal{O}} \subseteq \Delta^{\mathcal{O}}$, *for each* $A \in \mathsf{C}$, $r^{\mathcal{O}} \subseteq \Delta^{\mathcal{O}} \times \Delta^{\mathcal{O}}$, *for each* $r \in \mathsf{R}$, *and* $a^{\mathcal{O}} \in \Delta^{\mathcal{O}}$, *for each* $a \in \mathsf{I}$, *and*
- $\ll^{\mathcal{O}} =_{\text{def}} \langle \ll^{\mathcal{O}}_{r_1}, \ldots, \ll^{\mathcal{O}}_{r_{\#\mathsf{R}}} \rangle$, *where* $\ll^{\mathcal{O}}_{r_i} \subseteq r_i^{\mathcal{O}} \times r_i^{\mathcal{O}}$, *for* $i = 1, \ldots, \#\mathsf{R}$, *and such that each* $\ll^{\mathcal{O}}_{r_i}$ *is a well-founded strict partial order.*

Given $\mathcal{O} = \langle \Delta^{\mathcal{O}}, \cdot^{\mathcal{O}}, \ll^{\mathcal{O}} \rangle$, the intuition of $\Delta^{\mathcal{O}}$ and $\cdot^{\mathcal{O}}$ is the same as in a standard $\mathcal{ALC}$ interpretation. The intuition underlying each of the orderings in $\ll^{\mathcal{O}}$ is that they play the role of *preference relations* (or *normality orderings*), in a sense similar to the preference orders introduced by Shoham [50] in a propositional setting, and investigated by Kraus et al. [42,43] and others [10,11,14,29]: The pairs $(x, y)$ that are lower down in the ordering $\ll^{\mathcal{O}}_{r_i}$ are deemed as most normal (or typical, or expected, or conventional) in the context of (the interpretation of) $r_i$.

Figure 2 depicts an ordered interpretation in our example, where $\Delta^{\mathcal{O}}$ and $\cdot^{\mathcal{O}}$ are as in the interpretation $\mathcal{I}$ shown in Fig. 1, and $\ll^{\mathcal{O}} = \langle \ll^{\mathcal{O}}_{\mathsf{hasAcc}},$

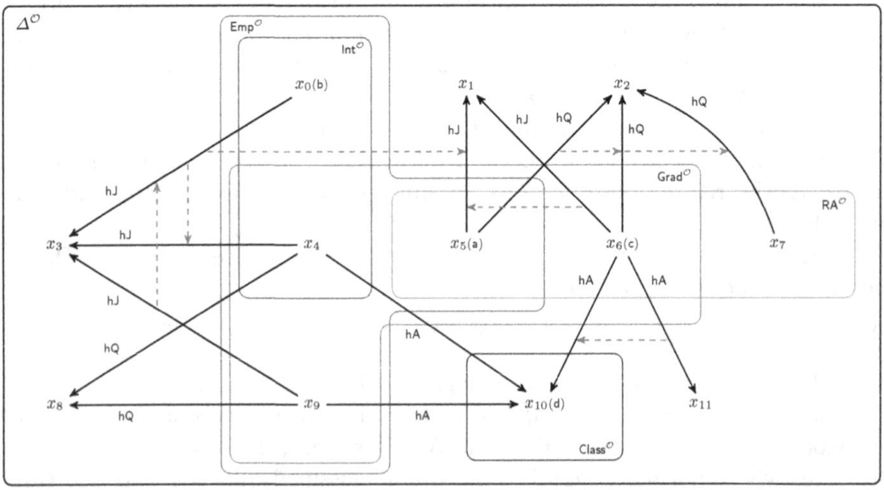


**Fig. 2.** An ordered interpretation. For the sake of presentation, we omit the transitive $\ll^{\mathcal{O}}_r$-arrows.

$\ll^{\mathcal{O}}_{\mathsf{hasJob}}, \ll^{\mathcal{O}}_{\mathsf{hasQual}}\rangle$, where $\ll^{\mathcal{O}}_{\mathsf{hasAcc}} = \{(x_6x_{11}, x_6x_{10})\}$, $\ll^{\mathcal{O}}_{\mathsf{hasJob}} = \{(x_9x_3, x_0x_3), (x_0x_3, x_4x_3), (x_9x_3, x_4x_3), (x_0x_3, x_5x_1), (x_9x_3, x_5x_1), (x_6x_1, x_5x_1)\}$, and $\ll^{\mathcal{O}}_{\mathsf{hasQual}} = \{(x_5x_2, x_6x_2), (x_6x_2, x_7x_2), (x_5x_2, x_7x_2)\}$.

For the sake of readability, we shall henceforth sometimes write $r$-tuples of the form $(x, y)$ as $xy$, as in the above example.

In the following definition we extend ordered interpretations to complex concepts of the language.

**Definition 2 (Interpretation of concepts).** *Let* $\mathcal{O} = \langle \Delta^{\mathcal{O}}, \cdot^{\mathcal{O}}, \ll^{\mathcal{O}}\rangle$, *let* $r \in \mathsf{R}$ *and, for each* $x \in \Delta^{\mathcal{O}}$, *let* $r^{\mathcal{O}|x} =_{\mathrm{def}} r^{\mathcal{O}} \cap (\{x\} \times \Delta^{\mathcal{O}})$ *(i.e., the restriction of the domain of* $r^{\mathcal{O}}$ *to* $\{x\}$*). The interpretation function* $\cdot^{\mathcal{O}}$ *interprets d*$\mathcal{ALC}$ *concepts as follows:*

$$\top^{\mathcal{O}} =_{\mathrm{def}} \Delta^{\mathcal{O}};\quad \bot^{\mathcal{O}} =_{\mathrm{def}} \emptyset;\quad (\neg C)^{\mathcal{O}} =_{\mathrm{def}} \Delta^{\mathcal{O}} \setminus C^{\mathcal{O}};$$
$$(C \sqcap D)^{\mathcal{O}} =_{\mathrm{def}} C^{\mathcal{O}} \cap D^{\mathcal{O}};\quad (C \sqcup D)^{\mathcal{O}} =_{\mathrm{def}} C^{\mathcal{O}} \cup D^{\mathcal{O}};$$
$$(\exists r.C)^{\mathcal{O}} =_{\mathrm{def}} \{x \in \Delta^{\mathcal{O}} \mid r^{\mathcal{O}}(x) \cap C^{\mathcal{O}} \neq \emptyset\};\ (\forall r.C)^{\mathcal{O}} =_{\mathrm{def}} \{x \in \Delta^{\mathcal{O}} \mid r^{\mathcal{O}}(x) \subseteq C^{\mathcal{O}}\};$$
$$(\underset{\sim}{\exists} r.C)^{\mathcal{O}} =_{\mathrm{def}} \{x \in \Delta^{\mathcal{O}} \mid \min_{\ll^{\mathcal{O}}_r}(r^{\mathcal{O}|x})(x) \cap C^{\mathcal{O}} \neq \emptyset\};$$
$$(\underset{\sim}{\forall} r.C)^{\mathcal{O}} =_{\mathrm{def}} \{x \in \Delta^{\mathcal{O}} \mid \min_{\ll^{\mathcal{O}}_r}(r^{\mathcal{O}|x})(x) \subseteq C^{\mathcal{O}}\}.$$

As an example, in the ordered interpretation $\mathcal{O}$ of Fig. 2, we have $((\underset{\sim}{\forall}\mathsf{hasAcc}.\neg\mathsf{Classified}) \sqcap (\exists\mathsf{hasAcc}.\mathsf{Classified}))^{\mathcal{O}} = \{x_6\}$.

Notice that, analogously to the classical case, $\underset{\sim}{\forall}$ and $\underset{\sim}{\exists}$ are dual to each other. As an example, for $\mathcal{O}$ as in Fig. 2, we have $(\underset{\sim}{\exists}\mathsf{hasAcc}.\mathsf{Classified})^{\mathcal{O}} = \{x_4, x_9\} = (\neg\underset{\sim}{\forall}\mathsf{hasAcc}.\neg\mathsf{Classified})^{\mathcal{O}}$.

Defeasible $\mathcal{ALC}$ also adds *contextual* defeasible subsumption statements to knowledge bases. Given $C, D \in \mathcal{L}_{d\mathcal{ALC}}$ and $r \in \mathsf{R}$, a statement of the form $C \sqsubseteq\!\!\!\sim_r D$ is a (contextual) *defeasible concept inclusion* (DCI), read "$C$ is usually subsumed by $D$ in the context $r$". A *d*$\mathcal{ALC}$ *defeasible TBox* $\mathcal{D}$ (or dTBox $\mathcal{D}$ for short) is a finite set of DCIs. A *d*$\mathcal{ALC}$ *classical TBox* $\mathcal{T}$ (or TBox $\mathcal{T}$ for short) is a finite set of (classical) subsumption statements $C \sqsubseteq D$ (i.e., $\mathcal{T}$ may contain defeasible concept constructs, but not defeasible concept inclusions). Given $\mathcal{T}$, $\mathcal{D}$ and $\mathcal{A}$, with $\mathcal{KB} =_{\mathrm{def}} \mathcal{T} \cup \mathcal{D} \cup \mathcal{A}$ we denote a *d*$\mathcal{ALC}$ *knowledge base*, a.k.a. a *defeasible ontology*, an example of which is given below:

$$\mathcal{T} = \left\{\begin{array}{c} \mathsf{Intern} \sqsubseteq \mathsf{Employee}, \\ \mathsf{Employee} \sqsubseteq \exists\mathsf{hasJob}.\top, \\ \mathsf{Graduate} \sqsubseteq \mathsf{hasQual}.\top, \\ \mathsf{ResAssoc} \sqsubseteq \underset{\sim}{\forall}\mathsf{hasAcc}.\neg\mathsf{Classified} \end{array}\right\} \quad \mathcal{A} = \left\{\begin{array}{c} \mathsf{anne} : \mathsf{Employee}, \\ \mathsf{anne} : \mathsf{ResAssoc}, \\ \mathsf{bill} : \mathsf{Intern}, \\ \mathsf{chris} : \mathsf{ResAssoc}, \\ \mathsf{doc123} : \mathsf{Classified}, \\ (\mathsf{chris}, \mathsf{doc123}) : \mathsf{hasAcc} \end{array}\right\}$$

$$\mathcal{D} = \left\{ \begin{array}{c} \mathsf{Employee} \precsim_{\mathsf{hasJob}} \exists \mathsf{hasAcc.Classified}, \\ \mathsf{Intern} \precsim_{\mathsf{hasJob}} \neg\exists \mathsf{hasAcc.Classified}, \\ \mathsf{Intern} \sqcap \mathsf{Graduate} \precsim_{\mathsf{hasJob}} \exists \mathsf{hasAcc.Classified}, \\ \mathsf{ResAssoc} \precsim_{\mathsf{hasJob}} \neg \mathsf{Employee}, \\ \mathsf{ResAssoc} \precsim_{\mathsf{hasQual}} \mathsf{Graduate} \end{array} \right\}$$

**Definition 3 (Satisfaction).** *Let $\mathcal{O} = \langle \Delta^{\mathcal{O}}, \cdot^{\mathcal{O}}, \ll^{\mathcal{O}} \rangle$, $r \in \mathsf{R}$, $C, D \in \mathcal{L}_{d\mathcal{ALC}}$, and $a, b \in \mathsf{I}$. Define $\prec_r^{\mathcal{O}} \subseteq \Delta^{\mathcal{O}} \times \Delta^{\mathcal{O}}$ as follows:*

$$\prec_r^{\mathcal{O}} =_{\text{def}} \{(x, y) \mid \textit{there is } (x, z) \in r^{\mathcal{O}} \textit{ s.t. for all } (y, v) \in r^{\mathcal{O}}, ((x, z), (y, v)) \in \ll_r^{\mathcal{O}}\}.$$

*The **satisfaction relation** $\Vdash$ is defined as follows:*

$$\begin{array}{llllll} \mathcal{O} \Vdash C \sqsubseteq D & \textit{if} & C^{\mathcal{O}} \subseteq D^{\mathcal{O}}; & \mathcal{O} \Vdash C \precsim_r D & \textit{if} & \min_{\prec_r^{\mathcal{O}}} C^{\mathcal{O}} \subseteq D^{\mathcal{O}}; \\ \mathcal{O} \Vdash a : C & \textit{if} & a^{\mathcal{O}} \in C^{\mathcal{O}}; & \mathcal{O} \Vdash (a, b) : r & \textit{if} & (a^{\mathcal{O}}, b^{\mathcal{O}}) \in r^{\mathcal{O}}. \end{array}$$

*If $\mathcal{O} \Vdash \alpha$, then we say $\mathcal{O}$ **satisfies** $\alpha$. $\mathcal{O}$ satisfies a $d\mathcal{ALC}$ knowledge base $\mathcal{KB}$, written $\mathcal{O} \Vdash \mathcal{KB}$, if $\mathcal{O} \Vdash \alpha$ for every $\alpha \in \mathcal{KB}$, in which case we say $\mathcal{O}$ is a **model** of $\mathcal{KB}$. We say $\mathcal{KB}$ is **preferentially consistent** if it admits a model. We say $C \in \mathcal{L}_{d\mathcal{ALC}}$ (resp. $r \in \mathsf{R}$) is **satisfiable** w.r.t. $\mathcal{KB}$ if there is a model $\mathcal{O}$ of $\mathcal{KB}$ s.t. $C^{\mathcal{O}} \neq \emptyset$ (resp. $r^{\mathcal{O}} \neq \emptyset$).*

One can check that the interpretation $\mathcal{O}$ in Fig. 2 satisfies the above knowledge base. To help in seeing why, Fig. 3 depicts the contextual orderings on objects (represented with dotted arrows) induced from those on roles in $\mathcal{O}$ as specified in Definition 3.

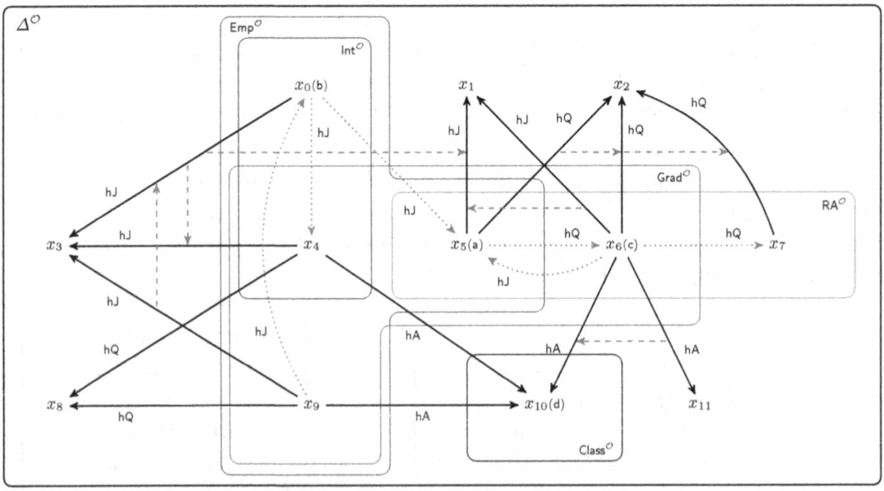


**Fig. 3.** Induced orderings on objects from the role orderings in Fig. 2. For the sake of presentation, we omit the transitive $\prec_r^{\mathcal{O}}$-arrows.

It follows from Definition 3 that, if $\ll_r^{\mathcal{O}} = \emptyset$, i.e., if no $r$-tuple is preferred to another, then $\sqsubseteq\!\!\!\sim_r$ reverts to a context-agnostic classical $\sqsubseteq$. A similar observation holds for individual concept inclusions: if $(C \sqcap \exists r.\top)^{\mathcal{O}} = \emptyset$, then $C \sqsubseteq\!\!\!\sim_r D$ reverts to $C \sqsubseteq D$. This reflects the intuition that the context $r$ is taken into account through the preference order on $r^{\mathcal{O}}$. In the absence of any preference, the context becomes irrelevant. This also shows why the classical counterpart of $\sqsubseteq\!\!\!\sim_r$ is independent of $r$ — context is taken into account in the form of a preference order, but preference has no bearing on the semantics of $\sqsubseteq$.

Contextual defeasible subsumption $\sqsubseteq\!\!\!\sim_r$ can also be viewed as defeasible subsumption based on a preference order on objects in the domain of $r^{\mathcal{O}}$ obtained from $\ll_r^{\mathcal{O}}$. Non-contextual defeasible subsumption can then be obtained as a special case by introducing a new role name $r$ and axiom $\top \sqsubseteq \exists r.\top$. More details can be found in our related work on contextual rational closure [21].

Given a $d\mathcal{ALC}$ knowledge base $\mathcal{KB}$, a fundamental task from the standpoint of knowledge representation and reasoning is that of deciding which statements follow from $\mathcal{KB}$ and which do not.

**Definition 4 (Preferential entailment).** *A statement $\alpha$ is* ***preferentially entailed*** *by a $d\mathcal{ALC}$ knowledge base $\mathcal{KB}$, written $\mathcal{KB} \models_{\mathsf{pref}} \alpha$, if $\mathcal{O} \Vdash \alpha$ for every $\mathcal{O}$ s.t. $\mathcal{O} \Vdash \mathcal{KB}$.*

The following lemma shows that deciding preferential entailment of GCIs and assertions can be reduced to $d\mathcal{ALC}$ knowledge base satisfiability, a result that will be used in the definition of a tableau system in Sect. 4. Its proof is analogous to that of its classical counterpart in the DL literature and we shall omit it here:

**Lemma 1.** *Let $\mathcal{KB}$ be a $d\mathcal{ALC}$ knowledge base and let $a$ be an individual name not occurring in $\mathcal{KB}$. For every $C, D \in \mathcal{L}_{d\mathcal{ALC}}$, $\mathcal{KB} \models C \sqsubseteq D$ iff $\mathcal{KB} \models C \sqcap \neg D \sqsubseteq \bot$ iff $\mathcal{KB} \cup \{a : C \sqcap \neg D\}$ is unsatisfiable. Moreover, for every $b \in \mathsf{I}$ and every $C \in \mathcal{L}_{d\mathcal{ALC}}$, $\mathcal{KB} \models b : C$ iff $\mathcal{KB} \cup \{b : \neg C\}$ is unsatisfiable.*

It turns out that deciding preferential entailment of DCIs too can be reduced to $d\mathcal{ALC}$ knowledge base satisfiability, but first, we introduce the tableau-based algorithm for deciding preferential consistency.

## 4 Tableau for Preferential Reasoning in $d\mathcal{ALC}$

In this section, we define a tableau method for deciding preferential consistency of a $d\mathcal{ALC}$ knowledge base. Our algorithm is based on that by Baader et al. [4] for the classical case; it therefore follows that it is doubly-exponential.

We start by observing that we can assume w.l.o.g. that all concepts appearing in a knowledge base are in negated normal form (NNF), i.e., concept complement $\neg$ occurs only in front of concept names.

Next, notice that for every ordered interpretation $\mathcal{O}$ and every $C, D \in \mathcal{L}_{d\mathcal{ALC}}$, $\mathcal{O} \Vdash C \sqsubseteq D$ if and only if $\mathcal{O} \Vdash \top \sqsubseteq \neg C \sqcup D$. In that respect, we can assume w.l.o.g. that all GCIs in a TBox are of the form $\top \sqsubseteq E$, for some $E \in \mathcal{L}_{d\mathcal{ALC}}$.

Notice also that we can assume w.l.o.g. that the ABox is not empty, for if it is, one can add to it the trivial assertion $a : \top$, for some new individual name $a$. It is easy to see that the resulting (non-empty) ABox is preferentially equivalent to the original one.

**Definition 5 (Subconcepts).** *Let $C \in \mathcal{L}_{d\mathcal{ALC}}$. The set of* **subconcepts** *of $C$, denoted $\mathsf{sub}(C)$, is defined inductively as follows:*

- *If $C = A$, for $A \in \mathsf{C} \cup \{\top, \bot\}$, then $\mathsf{sub}(C) =_{\text{def}} \{A\}$;*
- *If $C = C_1 \sqcap C_2$ or $C = C_1 \sqcup C_2$, then $\mathsf{sub}(C) =_{\text{def}} \{C\} \cup \mathsf{sub}(C_1) \cup \mathsf{sub}(C_2)$;*
- *If $C = \neg D$ or $C = \exists r.D$ or $C = \forall r.D$ or $C = \boldsymbol{\exists} r.D$ or $C = \boldsymbol{\forall} r.D$, then $\mathsf{sub}(C) =_{\text{def}} \{C\} \cup \mathsf{sub}(D)$.*

*Given a knowledge base $\mathcal{KB} = \mathcal{T} \cup \mathcal{D} \cup \mathcal{A}$, the set of subconcepts of $\mathcal{KB}$ is defined as $\mathsf{sub}(\mathcal{KB}) =_{\text{def}} \mathsf{sub}(\mathcal{T}) \cup \mathsf{sub}(\mathcal{D}) \cup \mathsf{sub}(\mathcal{A})$, where*

$$\mathsf{sub}(\mathcal{T}) =_{\text{def}} \bigcup_{C \sqsubseteq D \in \mathcal{T}} (\mathsf{sub}(C) \cup \mathsf{sub}(D)) \quad \mathsf{sub}(\mathcal{A}) =_{\text{def}} \bigcup_{a:C \in \mathcal{A}} \mathsf{sub}(C)$$

$$\mathsf{sub}(\mathcal{D}) =_{\text{def}} \bigcup_{C \sqsubset\!\!\!\sim_r D \in \mathcal{D}} (\mathsf{sub}(C) \cup \mathsf{sub}(D))$$

We say that an individual name $a$ *appears* in an ABox $\mathcal{A}$ if $\mathcal{A}$ contains an assertion of the form $a : C$, $(a, b) : r$ or $(b, a) : r$, for some $C \in \mathcal{L}_{d\mathcal{ALC}}$, $r \in \mathsf{R}$ and $b \in \mathsf{I}$.

**Definition 6 ($a$-concepts).** *Let $\mathcal{A}$ be an ABox and let $a$ be an individual name appearing in $\mathcal{A}$. With $\mathsf{con}_{\mathcal{A}}(a) =_{\text{def}} \{C \mid a : C \in \mathcal{A}\}$ we denote the* **set of concepts that $a$ is an instance of** *w.r.t. $\mathcal{A}$.*

We are now ready for the definition of the expansion rules for $d\mathcal{ALC}$-concepts. They are shown in Fig. 4. The $\sqcap$-, $\sqcup$-, $\forall$-, and $\mathcal{T}$-rules work as in the classical case [4], whereas the remaining rules handle the additional $d\mathcal{ALC}$ constructs according to our preferential semantics. We shall explain them in more detail below. Before doing so, we need a few more definitions, in particular of what it means for an individual to be *blocked*, as tested by the $\exists$-, $\boldsymbol{\exists}$-, and $\sqsubset\!\!\!\sim$-rules and needed to ensure termination of the algorithm we shall present.

As can be seen in the expansion rules, our tableau method makes use of a few auxiliary structures, which are built incrementally during the search for a model of the input knowledge base. The first one is a partial order on pairs of individuals $\rho^r_{\mathcal{A}}$, for each $r \in \mathsf{R}$. Its purpose is to build the skeleton of an $r$-preference relation on pairs of individual names appearing in an ABox $\mathcal{A}$. In the unravelling of the complete clash-free ABox (see below), if there is any, $\rho^r_{\mathcal{A}}$ is used to define a preference relation on the interpretation of role $r$ in the constructed ordered interpretation.

The second auxiliary structure is a pre-order $\sigma^r_{\mathcal{A}}$ on individual names, for each $r \in \mathsf{R}$. It fits the purpose of keeping track of which individuals are to be seen as more normal (or typical) relative to others in the application of the $\sqsubset\!\!\!\sim$-rule (see Fig. 4) so that the associated $\rho^r_{\mathcal{A}}$-ordering can be completed (by the $\ll$-rule) and, in the unravelling of the model, deliver an induced $\prec_r$ that is faithful to $\sigma^r_{\mathcal{A}}$.

(This point will be clarified in the explanation of the relevant rules.) Intuitively, $\sigma^r_{\mathcal{A}}$ corresponds to the converse of the preference order introduced in Definition 3.

Finally, the third structure used in the expansion rules is a labelling function $\tau^r_{\mathcal{A}}(a)$ mapping an individual name $a$ to the set of concepts $a$ ought to be a minimal instance of in the context $r$ w.r.t. the ABox $\mathcal{A}$. The purpose of $\tau^r_{\mathcal{A}}(a)$ is twofold: ($i$) whenever $C \in \tau^r_{\mathcal{A}}(a)$, it flags that every individual more preferred than $a$ should be marked as $\neg C$, as performed by the min-rule, and ($ii$) it plays a role in the blocking condition (see below) to prevent the generation of an infinite chain of increasingly more normal elements in $\sigma^r_{\mathcal{A}}$. Note that $\rho^r_{\mathcal{A}}$, $\sigma^r_{\mathcal{A}}$ and $\tau^r_{\mathcal{A}}(a)$ are only used in the inner workings of the tableau and are not accessible to the user.

**Definition 7 ($r$-ancestor).** *Let $\mathcal{A}$ be an ABox, $a, b \in \mathsf{I}$, and $r \in \mathsf{R}$. If $(a, b) : r \in \mathcal{A}$, we say $b$ is an* ***$r$-successor*** *of $a$ and $a$ is an* ***$r$-predecessor*** *of $b$. The transitive closure of the $r$-predecessor (resp. $r$-successor) relation is called* ***$r$-ancestor*** *(resp.* ***$r$-descendant****).*

**Definition 8 ($\sigma^r_{\mathcal{A}}$-ancestor).** *Let $\mathcal{A}$ be an ABox, $a, b \in \mathsf{I}$, and $r \in \mathsf{R}$. If $(a, b) \in \sigma^r_{\mathcal{A}}$, we say $b$ is a* ***$\sigma^r_{\mathcal{A}}$-successor*** *of $a$ and $a$ is an* ***$\sigma^r_{\mathcal{A}}$-predecessor*** *of $b$. The transitive closure of the $\sigma^r_{\mathcal{A}}$-predecessor (resp. $\sigma^r_{\mathcal{A}}$-successor) relation is called* ***$\sigma^r_{\mathcal{A}}$-ancestor*** *(resp.* ***$\sigma^r_{\mathcal{A}}$-descendant****).*

An individual is called a **root** if it has neither an $r$-ancestor nor a $\sigma^r_{\mathcal{A}}$-ancestor.
The following definition is used in the expansion rules of Fig. 4 to ensure termination:

**Definition 9 (Blocking).** *Let $\mathcal{A}$ be an ABox, $a, b \in \mathsf{I}$, and let $\sigma^r_{\mathcal{A}}$ and $\tau^r_{\mathcal{A}}$ be as above. We say that $b$ is* ***blocked*** *by $a$ in $\mathcal{A}$ in the context $r$ if (1) $a$ is either an $r$-ancestor or a $\sigma^r_{\mathcal{A}}$-ancestor of $b$, (2) $\mathsf{con}_{\mathcal{A}}(b) \subseteq \mathsf{con}_{\mathcal{A}}(a)$, and (3) $\tau^r_{\mathcal{A}}(b) \subseteq \tau^r_{\mathcal{A}}(a)$. We say $b$ is blocked in $\mathcal{A}$ if itself or some $r$-ancestor or $\sigma^r_{\mathcal{A}}$-ancestor of $b$ is blocked by some individual.*

The $\sqcap$-, $\sqcup$-, $\forall$-, and $\mathcal{T}$-rules in Fig. 4 are as in the classical case and need no further explanation.

The $\exists\!\!\!\sim$-rule creates a most preferred (relative to individual $a$) $r$-link to a new individual falling under concept $C$. Notice that this is achieved by just adding an assertion $(a, d) : r$ to $\mathcal{A}$, for $d$ new in $\mathcal{A}$, since there shall never be $(a, e)$ with $(ae, ad) \in \rho^r_{\mathcal{A}}$.

The $\forall\!\!\!\sim$-rule is analogous to the $\forall$-rule, but propagates a concept $C$ only to those individuals across preferred $r$-links (i.e., $r$-links that are minimal in $\rho^r_{\mathcal{A}}$).

The $\exists$-rule handles the creation of an $r$-successor without the information whether such an $r$-link is relatively preferred or not. In this case, both possibilities have to be explored, which is formalised by the or-branching in the rule. In one case, a preferred $r$-link is created just as in the $\exists\!\!\!\sim$- rule; in the other, an $r$-link is created along with an extra one which is then set as more preferred to it (in $\rho^r_{\mathcal{A}}$).

| | | |
|---|---|---|
| **$\sqcap$-rule:** | **if** | 1. $a : C \sqcap D \in \mathcal{A}$, and |
| | | 2. $\{a : C, a : D\} \not\subseteq \mathcal{A}$ |
| | **then** | $\mathcal{A} := \mathcal{A} \cup \{a : C, a : D\}$ |
| **$\sqcup$-rule:** | **if** | 1. $a : C \sqcup D \in \mathcal{A}$, and |
| | | 2. $\{a : C, a : D\} \cap \mathcal{A} = \emptyset$ |
| | **then** | $\mathcal{A} := \mathcal{A} \cup \{a : E\}$, for some $E \in \{C, D\}$ |
| **$\exists$-rule:** | **if** | 1. $a : \exists r.C \in \mathcal{A}$, and |
| | | 2. there is no $b$ s.t. $\{(a, b) : r, b : C\} \subseteq \mathcal{A}$, and |
| | | 3. $a$ is not blocked |
| | **then** | (a) $\mathcal{A} := \mathcal{A} \cup \{(a, c) : r, c : C\}$, for $c$ new in $\mathcal{A}$, **or** |
| | | (b) $\mathcal{A} := \mathcal{A} \cup \{(a, c) : r, c : C, (a, d) : r\}$, for $c, d$ new in $\mathcal{A}$, and $\rho^r_{\mathcal{A}} := \rho^r_{\mathcal{A}} \cup \{(ad, ac)\}$ |
| **$\forall$-rule:** | **if** | 1. $\{a : \forall r.C, (a, b) : r\} \subseteq \mathcal{A}$, and |
| | | 2. $b : C \notin \mathcal{A}$ |
| | **then** | $\mathcal{A} := \mathcal{A} \cup \{b : C\}$ |
| **$\tilde{\exists}$-rule:** | **if** | 1. $a : \tilde{\exists} r.C \in \mathcal{A}$, and |
| | | 2. there is no $b$ s.t. (*i*) $\{(a, b) : r, b : C\} \subseteq \mathcal{A}$, and (*ii*) there is no $c$ s.t. $(ac, ab) \in \rho^r_{\mathcal{A}}$, and |
| | | 3. $a$ is not blocked |
| | **then** | $\mathcal{A} := \mathcal{A} \cup \{(a, d) : r, d : C\}$, for $d$ new in $\mathcal{A}$ |
| **$\tilde{\forall}$-rule:** | **if** | 1. $\{a : \tilde{\forall} r.C, (a, b) : r\} \subseteq \mathcal{A}$, and |
| | | 2. there is no $c$ s.t. $(ac, ab) \in \rho^r_{\mathcal{A}}$, and |
| | | 3. $b : C \notin \mathcal{A}$ |
| | **then** | $\mathcal{A} := \mathcal{A} \cup \{b : C\}$ |
| **$\mathcal{T}$-rule:** | **if** | 1. $a$ appears in $\mathcal{A}$, $\top \sqsubseteq D \in \mathcal{T}$, and |
| | | 2. $a : D \notin \mathcal{A}$ |
| | **then** | $\mathcal{A} := \mathcal{A} \cup \{a : D\}$ |
| **$\sqsubset\!\!\!\sim$-rule:** | **if** | 1. $a$ appears in $\mathcal{A}$, $C \sqsubset\!\!\!\sim_r D \in \mathcal{D}$, and |
| | | 2. $\{a : \neg C, a : D\} \cap \mathcal{A} = \emptyset$, and |
| | | 3. either $a : C \notin \mathcal{A}$ or there is no $b$ s.t. $b : C \in \mathcal{A}$ and $(a, b) \in \sigma^r_{\mathcal{A}}$, and |
| | | 4. $a$ is not blocked |
| | **then** | (a) $\mathcal{A} := \mathcal{A} \cup \{a : \neg C\}$, **or** |
| | | (b) $\mathcal{A} := \mathcal{A} \cup \{a : C, c : C, c : D\}$, for $c$ new in $\mathcal{A}$, $\sigma^r_{\mathcal{A}} := \sigma^r_{\mathcal{A}} \cup \{(a, c)\}$, and $\tau^r_{\mathcal{A}}(c) := \{C\}$ **or** |
| | | (c) $\mathcal{A} := \mathcal{A} \cup \{a : D\}$ |
| min-**rule:** | **if** | 1. $C \in \tau^r_{\mathcal{A}}(a)$, and |
| | | 2. $b : \neg C \notin \mathcal{A}$, for some $b$ s.t. $(a, b) \in (\sigma^r_{\mathcal{A}})^+$ |
| | **then** | $\mathcal{A} := \mathcal{A} \cup \{b : \neg C\}$ |
| **$\ll$-rule:** | **if** | 1. $(b, a) \in \sigma^r_{\mathcal{A}}$, and |
| | | 2. there is no $c$ s.t. $(ac, bd) \in \rho^r_{\mathcal{A}}$ for every $(b, d) : r \in \mathcal{A}$, and |
| | | 3. $a$ is not blocked |
| | **then** | $\mathcal{A} := \mathcal{A} \cup \{(a, e) : r\}$, for $e$ new in $\mathcal{A}$, and $\rho^r_{\mathcal{A}} := \rho^r_{\mathcal{A}} \cup \{(ae, bf) \mid (b, f) : r \in \mathcal{A}\}$ |

**Fig. 4.** Expansion rules for the $d\mathcal{ALC}$ tableau.

The $\sqsubset\!\!\!\sim$-rule handles the presence of DCIs in the knowledge base, which have a global behaviour just as the GCIs in $\mathcal{T}$. Given an individual name $a$, it abides by a DCI $C \sqsubset\!\!\!\sim_r D$ if at least one of the following three possibilities holds: (*i*) $a$ is not in $C$; or (*ii*) $a$ falls under $C$ but there is another instance of $C$ that is more preferred than $a$, or (*iii*) $a$ is in $D$. This is captured by the or-like branch in the rule. Moreover, we need to check whether the node is not blocked in order to prevent the creation of an infinitely descending chain of increasingly more preferred objects. (This is needed to ensure termination of the algorithm and also that the preference relation on pairs of objects created when unraveling an open tableau is well-founded.)

The min-rule ensures that every individual that is more preferred than a typical instance of $C$ is marked as an instance of $\neg C$.

Finally, the $\ll$-rule takes care of completing $\rho^r_{\mathcal{A}}$ based on the information in $\sigma^r_{\mathcal{A}}$ so that the ordering on objects induced by that on pairs that $\rho^r_{\mathcal{A}}$ gives rise to coincides with the ordering on objects given by the strict version of $\sigma^r_{\mathcal{A}}$. (See also Definition 3.) This is needed because at the end of the tableau execution, $\sigma^r_{\mathcal{A}}$ is discarded and only $\rho^r_{\mathcal{A}}$ is used to define an ordering on objects against which to check satisfiability of DCIs.

**Definition 10 (Complete and clash-free ABox).** *Let $\mathcal{A}$ be an ABox. We say $\mathcal{A}$ contains a* ***clash*** *if there is some $a \in \mathsf{I}$ and $C \in \mathcal{L}_{d\mathcal{ALC}}$ such that $\{a : C, a : \neg C\} \subseteq \mathcal{A}$. We say $\mathcal{A}$ is* ***clash-free*** *if it does not contain a clash. $\mathcal{A}$ is* ***complete*** *if it contains a clash or if none of the expansion rules in Fig. 4 is applicable to $\mathcal{A}$.*

Let $\mathsf{ndexp}(\cdot)$ denote a function taking as input a clash-free ABox $\mathcal{A}$, a nondeterministic rule $\mathbf{R}$ from Fig. 4, and an assertion $\alpha \in \mathcal{A}$ such that $\mathbf{R}$ is applicable to $\alpha$ in $\mathcal{A}$. In our case, the nondeterministic rules are the $\sqcup$-, $\exists$- and $\precsim$-rules. The function returns a set $\mathsf{ndexp}(\mathcal{A}, \mathbf{R}, \alpha)$ containing each of the possible ABoxes resulting from the application of $\mathbf{R}$ to $\alpha$ in $\mathcal{A}$.

The tableau-based procedure for checking consistency of a $d\mathcal{ALC}$ knowledge base $\mathcal{KB} = \mathcal{T} \cup \mathcal{D} \cup \mathcal{A}$ is given in Algorithm 1 below. It uses Function Expand to apply the rules in Fig. 4 to $\mathcal{A}$ w.r.t. $\mathcal{T}$ and $\mathcal{D}$. Given an ABox $\mathcal{A}$, with $\rho_{\mathcal{A}}$, $\sigma_{\mathcal{A}}$ and $\tau_{\mathcal{A}}$ we denote, respectively, the sequences $\langle \rho^{r_1}_{\mathcal{A}}, \ldots, \rho^{r_{\#\mathsf{R}}}_{\mathcal{A}} \rangle$, $\langle \sigma^{r_1}_{\mathcal{A}}, \ldots, \sigma^{r_{\#\mathsf{R}}}_{\mathcal{A}} \rangle$ and $\langle \tau^{r_1}_{\mathcal{A}}, \ldots, \tau^{r_{\#\mathsf{R}}}_{\mathcal{A}} \rangle$.

**Lemma 2 (Termination).** *For every knowledge base $\mathcal{KB}$,* Consistent($\mathcal{KB}$) *terminates.*

The proof of Lemma 2 is similar to that showing termination of the classical $\mathcal{ALC}$ tableau for checking consistency of general knowledge bases [4, Lemma 4.10].

**Algorithm 1.** Consistent($\mathcal{KB}$)

```
Input: A dALC knowledge base KB = T ∪ D ∪ A
if Expand(KB) ≠ ∅ then
    return "Consistent"
else
    return "Inconsistent"
```

**Function** Expand($\mathcal{KB}$)

**Input:** A $d\mathcal{ALC}$ knowledge base $\mathcal{KB} = \mathcal{T} \cup \mathcal{D} \cup \mathcal{A}$

**if** *$\mathcal{A}$ is not complete* **then**
    Select a rule $\mathbf{R}$ that is applicable to $\mathcal{A}$;
    **if** *$\mathbf{R}$ is a nondeterministic rule* **then**
        Select an assertion $\alpha \in \mathcal{A}$ to which $\mathbf{R}$ is applicable;
        **if** *there is* $\mathcal{A}' \in \mathsf{ndexp}(\mathcal{A}, \mathbf{R}, \alpha)$ *with* Expand($\mathcal{T} \cup \mathcal{D} \cup \mathcal{A}'$) $\neq \emptyset$ **then**
            **return** Expand($\mathcal{T} \cup \mathcal{D} \cup \mathcal{A}'$)
        **else**
            **return** $\emptyset$
    **else**
        Apply $\mathbf{R}$ to $\mathcal{A}$
**if** *$\mathcal{A}$ contains a clash* **then**
    **return** $\emptyset$
**else**
    **return** $\langle \mathcal{A}, \rho_{\mathcal{A}}, \sigma_{\mathcal{A}}, \tau_{\mathcal{A}} \rangle$

**Theorem 1.** *Algorithm 1 is sound and complete w.r.t. preferential consistency of $d\mathcal{ALC}$ knowledge bases.*

**Corollary 1.** *Our tableau-based algorithm is a decision procedure for satisfiability of $d\mathcal{ALC}$ knowledge bases.*

## 5 Related Work

To the best of our knowledge, the first tableau system for preferential description logics was the one introduced by Giordano et al. [29,32]. They extend $\mathcal{ALC}$ with a typicality operator $\mathbf{T}(\cdot)$, which is applicable to concepts and for which they define a preferential semantics that is a special case of ours, in the sense that they place a preference relation only on objects of the domain. In their setting, a concept of the form $\mathbf{T}(C)$, understood as referring to the typical objects falling under $C$, serves as a macro for the sentence $C \sqcap \Box\neg C$ in a description language extended with a modality capturing the behaviour of a preference relation on objects. Hence, the intuition of $x \in (\mathbf{T}(C))^{\mathcal{I}} = (C \sqcap \Box\neg C)^{\mathcal{I}}$ is that $x$ is an instance of $C$ and any other object that is more preferred than $x$ falls under $\neg C$.

There are some similarities between Giordano et al.'s tableau system and the one we introduced here, but there are important differences as well. First, our method assumes an underlying language that is more expressive than $\mathcal{ALC}$ extended with $\mathbf{T}(\cdot)$. Second, our calculus does not have to explicitly handle an extra modality in the object language, since our preference relations are not part of the syntax and materialise only in the inner workings of the tableau. And finally, our tableau method allows for reasoning with several preference relations, in particular with possibly incompatible ones, which is not the case in frameworks that assume a single objective ordering on the domain.

Giordano et al.'s tableau system has been extended in a series of papers [30, 31,34,35], in particular also to deal with the computation of non-monotonic entailment from defeasible knowledge bases. In the latter case, the authors define a hyper-tableau calculus to compute the rational closure of a (context-less) defeasible ontology via a minimal model construction [33,35]. In recent work [20] we have shown how to compute context-based rational closure of $d\mathcal{ALC}$ knowledge bases, but instead of defining a hyper-tableau for that we rather rely on the use of a context-based version of Casini and Straccia's [25] algorithm, which is based on a polynomial number of calls to the preferential tableau we have described here and that can seamlessly be implemented as an extension of our Protégé plugin [23,24].

Although broadly similar in aim, our approach differs from that of Giordano and Gliozzi in their consideration of reasoning about multiple aspects in description logics [28]. Their aspects are linked to concept names, rather than to role names. Semantically equivalent concepts may therefore act as aspects, yet have unrelated associated preference orders. Also, only a single typicality operator is allowed in the language.

## 6 Concluding Remarks

In this paper, we have strengthened the case for a parameterised notion of defeasible concept inclusion in description logics introduced recently [20]. We have shown that preferential roles can be used to take context into account, and to deliver a simple, yet powerful, notion of contextual defeasible subsumption. Technically, this addresses an important limitation in previous defeasible extensions of description logics, namely the restriction in the semantics of defeasible concept inclusion to a single preference order on objects. Semantically, it answers the question of the meaning of multiple preference orders, namely that they reflect different contexts.

We have presented context as an explanation of the intuition underlying the introduction of multiple preference orders on objects, with defeasibility introducing a new facet of contextual reasoning not present in deductive reasoning. This offers a semantic treatment of contextual defeasible subsumption, requiring no extended vocabulary or further extension of the concept language. In contrast, an account of *deductive* reasoning with contexts in knowledge representation is not intrinsically linked to defeasible reasoning. The integration of defeasible description logics with such an account of contextual knowledge representation in description logics, for example, contextualised knowledge repositories [40] or two-sorted description logics of context [41], is orthogonal to our work, and has not yet been attempted.

The tableau procedure presented here can be implemented as a proof procedure for checking consistency of contextual defeasible knowledge bases. It can also be used to perform preferential (and modular) entailment checking, and hence can also be used as part of an algorithm to determine contextual rational closure [20]. In its current form the complexity of the naïve procedure here introduced is doubly-exponential. An optimal proof procedure along the lines of those by Nguyen and Szalas [44] and Goré and Nguyen [36] is currently under investigation. Given our previous results for similarly structured logics [18,19], we conjecture the satisfiability problem for contextual defeasible $\mathcal{ALC}$ is EXPTIME-complete, i.e., the same as that of reasoning with general TBoxes in classical $\mathcal{ALC}$.

## References


1. Baader, F., Calvanese, D., McGuinness, D., Nardi, D., Patel-Schneider, P. (eds.): The Description Logic Handbook: Theory, Implementation and Applications, 2nd edn. Cambridge University Press, Cambridge (2007)
2. Baader, F., Hollunder, B.: How to prefer more specific defaults in terminological default logic. In: Bajcsy, R. (ed.) Proceedings of the 13th International Joint Conference on Artificial Intelligence (IJCAI), pp. 669–675. Morgan Kaufmann Publishers (1993)
3. Baader, F., Hollunder, B.: Embedding defaults into terminological knowledge representation formalisms. J. Autom. Reasoning **14**(1), 149–180 (1995)
4. Baader, F., Horrocks, I., Lutz, C., Sattler, U.: An Introduction to Description Logic. Cambridge University Press, Cambridge (2017)
5. Bonatti, P., Faella, M., Petrova, I., Sauro, L.: A new semantics for overriding in description logics. Artif. Intell. **222**, 1–48 (2015)
6. Bonatti, P., Faella, M., Sauro, L.: Defeasible inclusions in low-complexity DLs. J. Artif. Intell. Res. **42**, 719–764 (2011)
7. Bonatti, P., Lutz, C., Wolter, F.: The complexity of circumscription in description logic. J. Artif. Intell. Res. **35**, 717–773 (2009)
8. Bonatti, P., Sauro, L.: On the logical properties of the nonmonotonic description logic $\mathrm{DL}^N$. Artif. Intell. **248**, 85–111 (2017)
9. Booth, R., Casini, G., Meyer, T., Varzinczak, I.: On the entailment problem for a logic of typicality. In: Proceedings of the 24th International Joint Conference on Artificial Intelligence (IJCAI), pp. 2805–2811 (2015)
10. Booth, R., Meyer, T., Varzinczak, I.: A propositional typicality logic for extending rational consequence. In: Fermé, E., Gabbay, D., Simari, G. (eds.) Trends in Belief Revision and Argumentation Dynamics, Studies in Logic - Logic and Cognitive Systems, vol. 48, pp. 123–154. King's College Publications, London (2013)
11. Boutilier, C.: Conditional logics of normality: a modal approach. Artif. Intell. **68**(1), 87–154 (1994)
12. Britz, K., Casini, G., Meyer, T., Varzinczak, I.: Preferential role restrictions. In: Proceedings of the 26th International Workshop on Description Logics, pp. 93–106 (2013)
13. Britz, K., Casini, G., Meyer, T., Varzinczak, I.: A KLM perspective on defeasible reasoning for description logics. In: Lutz, C., Sattler, U., Tinelli, A.Y., Wolter, F. (eds.) Description Logic, Theory Combination, and All That Essays Dedicated to Franz Baader on the Occasion of His 60th Birthday. LNCS, vol. 11560. Springer, Cham (2019). https://doi.org/10.1007/978-3-030-22102-7_7

14. Britz, K., Heidema, J., Meyer, T.: Semantic preferential subsumption. In: Lang, J., Brewka, G. (eds.) Proceedings of the 11th International Conference on Principles of Knowledge Representation and Reasoning (KR), pp. 476–484. AAAI Press/MIT Press (2008)
15. Britz, K., Meyer, T., Varzinczak, I.: Semantic foundation for preferential description logics. In: Wang, D., Reynolds, M. (eds.) AI 2011. LNCS (LNAI), vol. 7106, pp. 491–500. Springer, Heidelberg (2011). https://doi.org/10.1007/978-3-642-25832-9_50
16. Britz, K., Varzinczak, I.: Defeasible modalities. In: Proceedings of the 14th Conference on Theoretical Aspects of Rationality and Knowledge (TARK), pp. 49–60 (2013)
17. Britz, K., Varzinczak, I.: Introducing role defeasibility in description logics. In: Michael, L., Kakas, A. (eds.) JELIA 2016. LNCS (LNAI), vol. 10021, pp. 174–189. Springer, Cham (2016). https://doi.org/10.1007/978-3-319-48758-8_12
18. Britz, K., Varzinczak, I.: From KLM-style conditionals to defeasible modalities, and back. J. Appl. Non-Class. Logics (JANCL) **28**(1), 92–121 (2018)
19. Britz, K., Varzinczak, I.: Preferential accessibility and preferred worlds. J. Logic Lang. Inf. (JoLLI) **27**(2), 133–155 (2018)
20. Britz, K., Varzinczak, I.: Rationality and context in defeasible subsumption. In: Ferrarotti, F., Woltran, S. (eds.) FoIKS 2018. LNCS, vol. 10833, pp. 114–132. Springer, Cham (2018). https://doi.org/10.1007/978-3-319-90050-6_7
21. Britz, K., Varzinczak, I.: Contextual rational closure for defeasible $\mathcal{ALC}$. Annals of Mathematics and Artificial Intelligence, to appear (2019)
22. Britz, K., Varzinczak, I.: Reasoning with contextual defeasible $\mathcal{ALC}$. In: Proceedings of the 32nd International Workshop on Description Logics (2019)
23. Casini, G., Meyer, T., Moodley, K., Sattler, U., Varzinczak, I.: Introducing defeasibility into OWL ontologies. In: Arenas, M., et al. (eds.) ISWC 2015. LNCS, vol. 9367, pp. 409–426. Springer, Cham (2015). https://doi.org/10.1007/978-3-319-25010-6_27
24. Casini, G., Meyer, T., Moodley, K., Varzinczak, I.: Towards practical defeasible reasoning for description logics. In: Proceedings of the 26th International Workshop on Description Logics, pp. 587–599 (2013)
25. Casini, G., Straccia, U.: Rational closure for defeasible description logics. In: Janhunen, T., Niemelä, I. (eds.) JELIA 2010. LNCS (LNAI), vol. 6341, pp. 77–90. Springer, Heidelberg (2010). https://doi.org/10.1007/978-3-642-15675-5_9
26. Casini, G., Straccia, U.: Defeasible inheritance-based description logics. J. Artif. Intell. Res. (JAIR) **48**, 415–473 (2013)
27. Donini, F., Nardi, D., Rosati, R.: Description logics of minimal knowledge and negation as failure. ACM Trans. Comput. Logic **3**(2), 177–225 (2002)
28. Giordano, L., Gliozzi, V.: Reasoning about multiple aspects in DLs: semantics and closure construction. CoRR abs/1801.07161 (2018), http://arxiv.org/abs/1801.07161
29. Giordano, L., Gliozzi, V., Olivetti, N., Pozzato, G.L.: Preferential description logics. In: Dershowitz, N., Voronkov, A. (eds.) LPAR 2007. LNCS (LNAI), vol. 4790, pp. 257–272. Springer, Heidelberg (2007). https://doi.org/10.1007/978-3-540-75560-9_20
30. Giordano, L., Gliozzi, V., Olivetti, N., Pozzato, G.L.: Reasoning about typicality in preferential description logics. In: Hölldobler, S., Lutz, C., Wansing, H. (eds.) JELIA 2008. LNCS (LNAI), vol. 5293, pp. 192–205. Springer, Heidelberg (2008). https://doi.org/10.1007/978-3-540-87803-2_17

31. Giordano, L., Gliozzi, V., Olivetti, N., Pozzato, G.: Analytic tableaux calculi for KLM logics of nonmonotonic reasoning. ACM Trans. Comput. Logic **10**(3), 18:1–18:47 (2009)
32. Giordano, L., Gliozzi, V., Olivetti, N., Pozzato, G.: $\mathcal{ALC}+T$: a preferential extension of description logics. Fundam. Informaticae **96**(3), 341–372 (2009)
33. Giordano, L., Gliozzi, V., Olivetti, N., Pozzato, G.L.: A minimal model semantics for nonmonotonic reasoning. In: del Cerro, L.F., Herzig, A., Mengin, J. (eds.) JELIA 2012. LNCS (LNAI), vol. 7519, pp. 228–241. Springer, Heidelberg (2012). https://doi.org/10.1007/978-3-642-33353-8_18
34. Giordano, L., Gliozzi, V., Olivetti, N., Pozzato, G.: A non-monotonic description logic for reasoning about typicality. Artif. Intell. **195**, 165–202 (2013)
35. Giordano, L., Gliozzi, V., Olivetti, N., Pozzato, G.: Semantic characterization of rational closure: from propositional logic to description logics. Artif. Intell. **226**, 1–33 (2015)
36. Goré, R., Nguyen, L.: ExpTime tableaux for $\mathcal{ALC}$ using sound global caching. J. Autom. Reasoning **50**(4), 355–381 (2013)
37. Governatori, G.: Defeasible description logics. In: Antoniou, G., Boley, H. (eds.) RuleML 2004. LNCS, vol. 3323, pp. 98–112. Springer, Heidelberg (2004). https://doi.org/10.1007/978-3-540-30504-0_8
38. Grosof, B., Horrocks, I., Volz, R., Decker, S.: Description logic programs: Combining logic programs with description logic. In: Proceedings of the 12th International Conference on World Wide Web (WWW), pp. 48–57. ACM (2003)
39. Heymans, S., Vermeir, D.: A defeasible ontology language. In: Meersman, R., Tari, Z. (eds.) OTM 2002. LNCS, vol. 2519, pp. 1033–1046. Springer, Heidelberg (2002). https://doi.org/10.1007/3-540-36124-3_66
40. Homola, M., Serafini, L.: Contextualized knowledge repositories for the semantic web. Web Semant.: Sci., Serv. Agents World Wide Web **12**, 64–87 (2012)
41. Klarman, S., Gutiérrez-Basulto, V.: Description logics of context. J. Logic and Comput. **26**(3), 817–854 (2013)
42. Kraus, S., Lehmann, D., Magidor, M.: Nonmonotonic reasoning, preferential models and cumulative logics. Artif. Intell. **44**, 167–207 (1990)
43. Lehmann, D., Magidor, M.: What does a conditional knowledge base entail? Artif. Intell. **55**, 1–60 (1992)
44. Nguyen, L.A., Szałas, A.: ExpTime tableaux for checking satisfiability of a knowledge base in the description logic $\mathcal{ALC}$. In: Nguyen, N.T., Kowalczyk, R., Chen, S.-M. (eds.) ICCCI 2009. LNCS (LNAI), vol. 5796, pp. 437–448. Springer, Heidelberg (2009). https://doi.org/10.1007/978-3-642-04441-0_38
45. Padgham, L., Zhang, T.: A terminological logic with defaults: a definition and an application. In: Bajcsy, R. (ed.) Proceedings of the 13th International Joint Conference on Artificial Intelligence (IJCAI). pp. 662–668. Morgan Kaufmann Publishers (1993)
46. Qi, G., Pan, J.Z., Ji, Q.: Extending description logics with uncertainty reasoning in possibilistic logic. In: Mellouli, K. (ed.) ECSQARU 2007. LNCS (LNAI), vol. 4724, pp. 828–839. Springer, Heidelberg (2007). https://doi.org/10.1007/978-3-540-75256-1_72
47. Quantz, J., Royer, V.: A preference semantics for defaults in terminological logics. In: Proceedings of the 3rd International Conference on Principles of Knowledge Representation and Reasoning (KR), pp. 294–305 (1992)
48. Quantz, J., Ryan, M.: Preferential default description logics. Technical report, TU Berlin (1993). www.tu-berlin.de/fileadmin/fg53/KIT-Reports/r110.pdf

49. Sengupta, K., Krisnadhi, A.A., Hitzler, P.: Local closed world semantics: grounded circumscription for OWL. In: Aroyo, L., et al. (eds.) ISWC 2011. LNCS, vol. 7031, pp. 617–632. Springer, Heidelberg (2011). https://doi.org/10.1007/978-3-642-25073-6_39
50. Shoham, Y.: Reasoning about Change: Time and Causation from the Standpoint of Artificial Intelligence. MIT Press, Cambridge (1988)
51. Straccia, U.: Default inheritance reasoning in hybrid KL-ONE-style logics. In: Bajcsy, R. (ed.) Proceedings of the 13th International Joint Conference on Artificial Intelligence (IJCAI). pp. 676–681. Morgan Kaufmann Publishers (1993)
52. Varzinczak, I.: A note on a description logic of concept and role typicality for defeasible reasoning over ontologies. Logica Universalis **12**(3–4), 297–325 (2018)

# A Tableau Calculus for Non-clausal Maximum Satisfiability

Chu Min Li[1], Felip Manyà[2(✉)], and Joan Ramon Soler[2]

[1] MIS, Université de Picardie Jules Verne, Amiens, France
[2] Artificial Intelligence Research Institute (IIIA, CSIC), Bellaterra, Spain
felip@iiia.csic.es

**Abstract.** We define a non-clausal MaxSAT tableau calculus. Given a multiset of propositional formulas $\phi$, we prove that the calculus is sound in the sense that if the minimum number of contradictions derived among the branches of a completed tableau for $\phi$ is $m$, then the minimum number of unsatisfied formulas in $\phi$ is $m$. We also prove that it is complete in the sense that if the minimum number of unsatisfied formulas in $\phi$ is $m$, then the minimum number of contradictions among the branches of any completed tableau for $\phi$ is $m$. Moreover, we describe how to extend the proposed calculus to deal with hard and weighted soft formulas.

## 1 Introduction

We can distinguish between clausal MaxSAT and non-clausal MaxSAT. Clausal MaxSAT, usually known simply as MaxSAT, is to find an assignment that minimizes the number of unsatisfied clauses in a given multiset of clauses, and non-clausal MaxSAT is to find an assignment that minimizes the number of unsatisfied formulas in a given multiset of propositional formulas that are not necessarily in clausal form.

Inference systems for SAT are unsound for MaxSAT, because they preserve satisfiability but not the minimum number of unsatisfied formulas. Thus, we need to define logical calculi meeting that condition and show that they allow one to derive as many contradictions as the minimum number of unsatisfied formulas in the input multiset.

We count with complete resolution, natural deduction and tableau calculi for clausal MaxSAT [4,7,8,18]. Restrictions of MaxSAT resolution are routinely used to propagate information in branch-and-bound MaxSAT solvers [1,2,13,16,17]; and MaxSAT resolution was used to show that there exist polynomial-size MaxSAT resolution proofs of the pigeon hole principle (PHP) if PHP is encoded as a Partial MaxSAT instance using the dual rail encoding [14]. Indeed, the combination of the dual rail encoding and MaxSAT resolution is a stronger proof system than general resolution [6].

This work was supported by Project LOGISTAR from the EU H2020 Research and Innovation Programme under Grant Agreement No. 769142, MINECO-FEDER projects RASO (TIN2015-71799-C2-1-P) and Generalitat de Catalunya SGR 172.


S. Cerrito and A. Popescu (Eds.): TABLEAUX 2019, LNAI 11714, pp. 58–73, 2019.
https://doi.org/10.1007/978-3-030-29026-9_4

In this paper we address the problem of defining a complete calculus for non-clausal MaxSAT. As far as we know, non-clausal MaxSAT has not yet been considered in the community. Thus, inspired by the work on clausal MaxSAT tableaux [4,18], we define the first sound and complete non-clausal MaxSAT tableau calculus, and describe how it can be extended to deal with hard and weighted soft formulas.

The paper is mainly theoretical, but it is important to highlight that MaxSAT solving has been applied to solve problems in a range of real-world domains as diverse as bioinformatics [11,22], circuit design and debugging [23], community detection in complex networks [15], diagnosis [10], FPGA routing [25], planning [26], scheduling [5] and team formation [21], among many others.

The paper is structured as follows. Section 2 reviews how tableaux solve non-clausal SAT and clausal MaxSAT. Section 3 defines a complete non-clausal MaxSAT tableau calculus. Section 4 describes how to extend the proposed calculus to deal with hard and weighted soft formulas. Section 5 gives the conclusions.

## 2 Background

A propositional formula is an expression constructed from propositional variables by means of the propositional connectives $\land, \lor, \rightarrow$ and $\neg$ in accordance with the following rules: (i) each propositional variable is a propositional formula; and (ii) if $A$ and $B$ are propositional formulas, then so are $(A \land B)$, $(A \lor B)$, $(A \rightarrow B)$, and $(\neg A)$. A non-clausal MaxSAT instance is a multiset of propositional formulas.[1] A truth assignment is a mapping that assigns 0 (false) or 1 (true) to each propositional variable. A propositional formula is satisfied by an assignment if it is true under the usual truth-functional interpretation of the connectives and the truth values assigned to the variables.

Given a non-clausal MaxSAT instance $\phi$, non-clausal MaxSAT is the problem of finding an assignment of $\phi$ that minimizes the number of unsatisfied formulas.

Clauses are a particular type of propositional formulas defined as follows. A clause is a disjunction of literals, where a literal $l_i$ is a variable $x_i$ or its negation $\neg x_i$. A clausal MaxSAT instance is a multiset of clauses. Given a clausal MaxSAT instance $\phi$, clausal MaxSAT is the problem of finding an assignment of $\phi$ that minimizes the number of unsatisfied clauses.

A weighted formula is a pair $(A, w)$, where $A$ is a propositional formula and $w$, its weight, is a positive number. A non-clausal weighted MaxSAT instance is a multiset of weighted formulas. Given a non-clausal weighted MaxSAT instance $\phi$, non-clausal weighted MaxSAT is the problem of finding an assignment of $\phi$ that minimizes the sum of weights of unsatisfied formulas.

A weighted clause is a pair $(C, w)$, where $C$ is a clause and $w$, its weight, is a positive number. A clausal weighted MaxSAT instance is a multiset of weighted clauses. Given a clausal weighted MaxSAT instance $\phi$, clausal weighted MaxSAT

[1] We use multisets of formulas instead of sets of formulas because duplicated formulas cannot be collapsed into one formula because then the minimum number of unsatisfied formulas might not be preserved.

is the problem of finding an assignment of $\phi$ that minimizes the sum of weights of unsatisfied clauses.

A non-clausal partial MaxSAT instance is a multiset of formulas in which some formulas are declared to be relaxable or soft and the rest are declared to be non-relaxable or hard. Given a non-clausal partial MaxSAT instance $\phi$, non-clausal partial MaxSAT is the problem of finding an assignment of $\phi$ that satisfies all the hard formulas and minimizes the number of unsatisfied soft formulas.

A clausal partial MaxSAT instance is a multiset of clauses in which some clauses are declared to be relaxable or soft and the rest are declared to be non-relaxable or hard. Given a clausal partial MaxSAT instance $\phi$, clausal partial MaxSAT is the problem of finding an assignment of $\phi$ that satisfies all the hard clauses and minimizes the number of unsatisfied soft clauses.

The weighted partial MaxSAT problem is the combination of partial MaxSAT and weighted MaxSAT. Given a multiset $\phi$ composed of hard formulas (clauses) and soft weighted formulas (clauses), non-clausal (clausal) weighted partial MaxSAT is the problem of finding an assignment of $\phi$ that satisfies all the hard formulas (clauses) and minimizes the sum of weights of unsatisfied soft formulas (clauses).

We can group all propositional formulas of the form $(A \circ B)$ and $\neg(A \circ B)$, where $A$ and $B$ denote propositional formulas and $\circ \in \{\vee, \wedge, \rightarrow\}$, into two categories so that the presentation and proofs are simplified. Those that act conjunctively, which are called $\alpha$-formulas, and those that act disjunctively, which are called $\beta$-formulas. The different formulas in each category are displayed in Table 1. To complete a taxonomy of propositional formulas, excluding literals, we also need the propositional formulas of the form $\neg\neg A$. This notation is known as uniform notation.

**Table 1.** $\alpha$-formulas and $\beta$-formulas.

| $\alpha$ | $\alpha_1$ | $\alpha_2$ |
|---|---|---|
| $A \wedge B$ | $A$ | $B$ |
| $\neg(A \vee B)$ | $\neg A$ | $\neg B$ |
| $\neg(A \rightarrow B)$ | $A$ | $\neg B$ |

| $\beta$ | $\beta_1$ | $\beta_2$ |
|---|---|---|
| $A \vee B$ | $A$ | $B$ |
| $\neg(A \wedge B)$ | $\neg A$ | $\neg B$ |
| $A \rightarrow B$ | $\neg A$ | $B$ |

Note that $\alpha$ is logically equivalent to $\alpha_1 \wedge \alpha_2$, $\beta$ is logically equivalent to $\beta_1 \vee \beta_2$ and $\neg\neg A$ is logically equivalent to $A$. In SAT tableaux, these equivalences are used to reduce the problem of finding a satisfying assignment of $\alpha$ to that of finding a satisfying assignment of both $\alpha_1$ and $\alpha_2$, of $\beta$ to that of finding a satisfying assignment of $\beta_1$ or $\beta_2$ and of $\neg\neg A$ to that of finding a satisfying assignment of $A$. Thus, using the expansion rules of Table 2 we obtain a complete tableau calculus for non-clausal SAT. We introduced the contradiction rule ($\Box$-rule), where $l$ denotes a literal, because it will be necessary in MaxSAT; in the literature, applying this rule is usually referred to as closing the branch. Note that uniform notation allows to define tableau rules for arbitrary propositional formulas in a concise way.

The tableau method is used to determine the satisfiability of a given set of propositional formulas [9,12,24]. It starts creating an initial tableau composed of a single branch that has a node for each formula in the input set of formulas. Then, it applies the expansion rules of Table 2 until a contradiction is derived in each branch (in this case, the input set of formulas is unsatisfiable) or a branch is saturated without deriving a contradiction (in this case, the input set of formula is satisfiable). A branch is saturated in a SAT tableau when all the possible applications of the expansion rules have been applied in that branch.

**Table 2.** Tableau expansion rules for SAT

| | | | |
|---|---|---|---|
| $\dfrac{\alpha}{\alpha_1 \quad \alpha_2}$ | $\dfrac{\beta}{\beta_1 \mid \beta_2}$ | $\dfrac{\neg\neg A}{A}$ | $\dfrac{l \quad \neg l}{\Box}$ |
| $\alpha$-rule | $\beta$-rule | $\neg$-rule | $\Box$-rule |

The single tableau calculus for MaxSAT [18] defined in the literature limits the input to multisets of clauses; i.e., it is a clausal tableau calculus that cannot solve non-clausal MaxSAT. This calculus does not contain the $\alpha$- and $\neg$-rule. It consists of the $\beta$- and $\Box$-rule. In fact, as all the formulas in the tableau are clauses and the formulas of type $\beta$ are always disjunctions of literals of the form $l_1 \vee l_2 \vee \cdots \vee l_n$, the previous $\beta$-rule is replaced with the following $n$-ary $\beta$-rule:

$$\frac{l_1 \vee l_2 \vee \cdots \vee l_n}{l_1 \mid l_2 \mid \cdots \mid l_n}$$

$n$-ary $\beta$-rule

Note that the $n$-ary $\beta$-rule collapses $n-1$ applications of the $\beta$-rule over the clause $l_1 \vee l_2 \vee \cdots \vee l_n$.

In clausal MaxSAT tableaux, all the clauses in the initial tableau are declared to be active. Clauses become inactive in a branch once they have been used as premises of the $\beta$- or $\Box$-rule, and then the added conclusions become active. The application of expansion rules is restricted to active clauses. In this way, the preservation of the minimum number of unsatisfied clauses is guaranteed. Active and inactive clauses are not needed in SAT because the goal is to preserve satisfiability and the application of rules in a branch stops once a contradiction is detected. The application of rules in MaxSAT continues until no more tableau rules can be applied to the formulas in the branch, because the aim is to derive all the possible contradictions. Thus, the saturation of branches is also different in SAT and MaxSAT.

Figure 1 shows the differences between clausal SAT and clausal MaxSAT tableaux using the multiset of clauses is $\phi = \{\neg x_1, \neg x_2, \neg x_3, x_1 \vee x_2, \neg x_1 \vee x_3\}$. In the SAT case, it is enough with applying the $\beta$-rule to $x_1 \vee x_2$. Since a contradiction is detected in each branch, the input multiset of formulas is declared unsatisfiable. However, in the MaxSAT case, the $\beta$-rule must also applied to $\neg x_1 \vee x_3$ and all the possible contradictions must be detected to complete the tableau. Note that in the leftmost branch of the clausal MaxSAT tableau there is just one contradiction because we have just one occurrence of $x_1$, which became inactive after detecting the first contradiction.

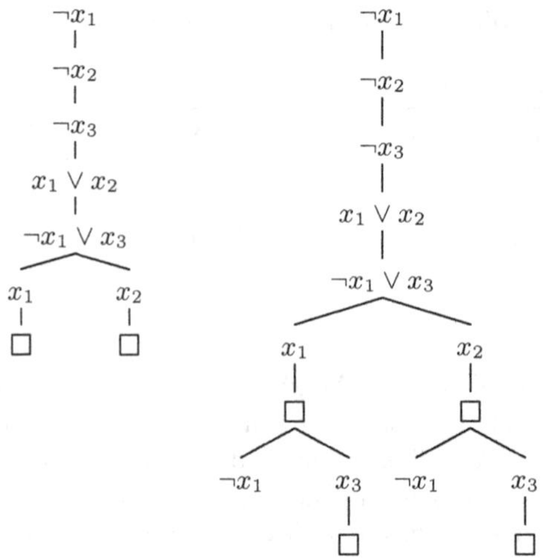

**Fig. 1.** Completed clausal SAT tableau (left) and completed clausal MaxSAT tableau (right) when the input multiset of clauses is $\phi = \{\neg x_1, \neg x_2, \neg x_3, x_1 \vee x_2, \neg x_1 \vee x_3\}$. The left tableau proves that $\phi$ is unsatisfiable and the right tableau proves that the minimum number of unsatisfied clauses in $\phi$ is 1.

The soundness of the previous clausal MaxSAT tableau calculus states that the $\beta$- and $\Box$-rule preserve the minimum number of unsatisfied clauses between a tableau and its extension; in particular, the $\beta$-rule preserves that number in at least one branch and does not decrease it in the rest of branches. So, once all branches have been saturated, the minimum number of contradictions derived among the branches of a completed tableau is the minimum number of unsatisfied clauses in the input multiset of clauses. The completeness states that any completed tableau for a multiset of clauses $\phi$, whose minimum number of clauses that can be unsatisfied in it is $k$, has a branch with $k$ contradictions and the rest of branches contain at least $k$ contradictions [18].

If we move to deal with arbitrary propositional formulas (i.e., non-clausal MaxSAT), the first problem we encounter is that the $\alpha$-rule does not preserve the minimum number of unsatisfied formulas as the $\beta$-rule does for clauses. Assume that we want to solve the non-clausal MaxSAT instance $\{x_1, x_2, \neg x_1 \wedge \neg x_2\}$, whose single optimal assignment is the one that sets $x_1$ and $x_2$ to true, and only falsifies $\neg x_1 \wedge \neg x_2$. If we apply the $\alpha$-rule to $\neg x_1 \wedge \neg x_2$, we add two nodes, labelled with $\neg x_1$ and $\neg x_2$, to the initial tableau. Then, we can derive two contradictions by applying the $\Box$-rule to $\{x_1, \neg x_1\}$ and $\{x_2, \neg x_2\}$, but the minimum number of formulas unsatisfied by the optimal assignment is just one. Figure 2 displays the resulting tableau. This counterexample shows that the $\alpha$-rule is unsound in MaxSAT. So, we need to define a new and sound $\alpha$-rule as a first step towards getting a sound and complete non-clausal MaxSAT calculus.

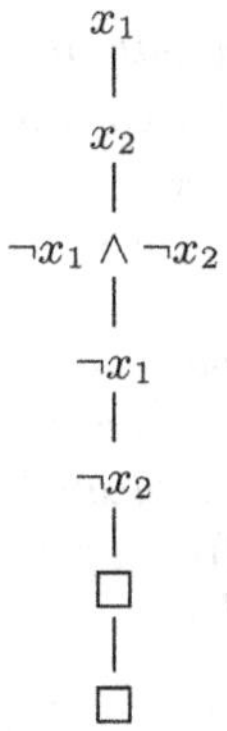


**Fig. 2.** Counterexample that shows that the $\alpha$-rule is unsound for non-clausal MaxSAT. The input multiset is $\phi = \{x_1, x_2, \neg x_1 \wedge \neg x_2\}$.

The previous example also illustrates that the standard conversion to clausal form is not valid in MaxSAT because it does not preserve the number of unsatisfied clauses. The clausal form of $\{x_1, x_2, \neg x_1 \wedge \neg x_2\}$ is $\{x_1, x_2, \neg x_1, \neg x_2\}$. However, the MaxSAT solution of $\{x_1, x_2, \neg x_1 \wedge \neg x_2\}$ is 1 and the MaxSAT solution of $\{x_1, x_2, \neg x_1, \neg x_2\}$ is 2. Thus, it is not possible to solve non-clausal MaxSAT by first translating to clausal form and then using clausal MaxSAT tableaux. We refer the reader to [19] for a recent paper on clausal form transformations for MaxSAT.

## 3 A Non-clausal MaxSAT Tableau Calculus

We formally define a non-clausal MaxSAT tableau calculus and prove its soundness and completeness. In the rest of the section, unless otherwise stated, when we say tableau we refer to a non-clausal MaxSAT tableau.

**Definition 1.** *A tableau is a tree with a finite number of branches whose nodes are labelled by either a propositional formula or a box ($\Box$). A box in a tableau denotes a contradiction. A branch is a maximal path in a tree, and we assume that branches have a finite number of nodes.*

**Table 3.** Tableau expansion rules for non-clausal MaxSAT

| $\alpha$ | | $\beta$ | | $\neg\neg A$ | $l$ |
|---|---|---|---|---|---|
| $\square$ | $\alpha_1$ | $\beta_1$ | $\beta_2$ | $A$ | $\neg l$ |
| | $\alpha_2$ | | | | $\square$ |
| $\alpha$-rule | | $\beta$-rule | | $\neg$-rule | $\square$-rule |

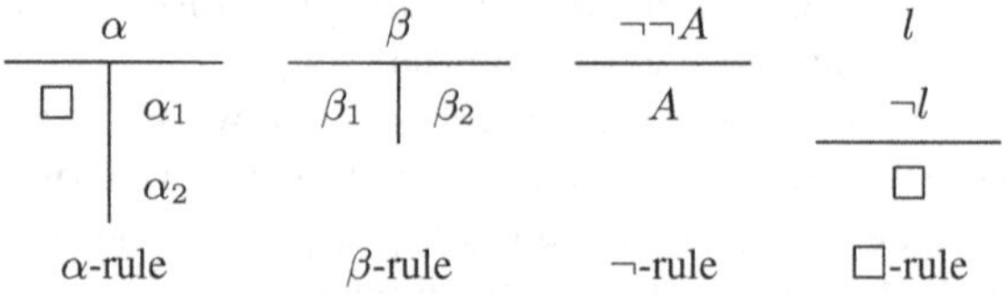

**Definition 2.** *Let $\phi = \{\phi_1, \ldots, \phi_m\}$ be a multiset of propositional formulas. A tableau for $\phi$ is constructed by a sequence of applications of the following rules:*

**Initialize** *A tree with a single branch with m nodes such that each node is labelled with a formula of $\phi$ is a tableau for $\phi$. Such a tableau is called initial tableau and its formulas are declared active.*

*Given a tableau $T$ for $\phi$, a branch $b$ of $T$, and a node of $b$ labelled with an active formula $F$,*

$\alpha$**-rule** *If $F$ is of type $\alpha$, the tableau obtained by appending a new left node below $b$ labelled with $\square$ and a new right branch with two nodes below $b$ labelled with $\alpha_1$ and $\alpha_2$ is a tableau for $\phi$. Formula $F$ becomes inactive in $b$ and $\alpha_1$ and $\alpha_2$ are declared active.*

$\beta$**-rule** *If $F$ is of type $\beta$, the tableau obtained by appending a new left node below $b$ labelled with $\beta_1$ and a new right node below $b$ labelled with $\beta_2$ is a tableau for $\phi$. Formula $F$ becomes inactive in $b$ and $\beta_1$ and $\beta_2$ are declared active.*

$\neg$**-rule** *If $F$ is of type $\neg\neg A$, the tableau obtained by appending a new node below $b$ labelled with $A$ is a tableau for $\phi$. Formula $\neg\neg A$ becomes inactive in $b$ and $A$ is declared active.*

$\square$**-rule** *Given a tableau $T$ for $\phi$, a branch $b$ of $T$, and two nodes of $b$ labelled with two active complementary literals $l$ and $\neg l$, the tableau obtained by appending a node below $b$ labelled with $\square$ is a tableau for $\phi$. Literals $l$ and $\neg l$ become inactive in $b$.*

The expansion rules of the previous definition are summarized in Table 3. Note that all the rules preserve the number of premises falsified by an assignment $I$ in at least one branch and do not decrease that number in the other branch (if any). In particular, in the $\alpha$-rule, we have that if $I$ falsifies $\alpha$, the left branch contains one contradiction and $\alpha_1$ and $\alpha_2$ cannot be used to derive any other contradiction in that branch because they are not expanded; moreover, $I$ falsifies $\alpha_1$ or $\alpha_2$ (or both) on the right branch. On the other hand, if $I$ satisfies $\alpha$, then $I$ also satisfies $\alpha_1$ and $\alpha_2$ on the right branch.

**Definition 3.** *Let $T$ be a tableau for a multiset of propositional formulas $\phi$. A branch $b$ of $T$ is saturated when no further expansion rules can be applied on $b$, and $T$ is completed when all its branches are saturated. The cost of a saturated branch is the number of boxes on the branch. The cost of a completed tableau is the minimum cost among all its branches.*

As we show below, the minimum number of formulas that can be unsatisfied in a multiset of propositional formulas $\phi$ is $k$ iff the cost of a completed tableau for $\phi$ is $k$. Thus, the systematic construction of a completed tableau for $\phi$ provides an exact method for non-clausal MaxSAT.

*Example 1.* We can determine the minimum number of unsatisfied formulas in the multiset $\phi = \{x_1, x_2, \neg x_1 \wedge \neg x_2\}$ using the previous tableau calculus. Figure 3 displays how the tableau is constructed. We start by constructing the initial tableau (the leftmost tableau) and then apply the $\alpha$-rule to $\neg x_1 \wedge \neg x_2$, getting as a result the second tableau in the figure. The leftmost branch is saturated and we apply the $\Box$-rule to $\{x_1, \neg x_1\}$ on the rightmost branch, getting as a result the third tableau. Finally, we apply the $\Box$-rule to $\{x_2, \neg x_2\}$ on the same branch and get the rightmost tableau in the figure. Since the minimum number of boxes among the branches of the last tableau is 1, the minimum number of formulas that can be unsatisfied in $\phi$ is 1.

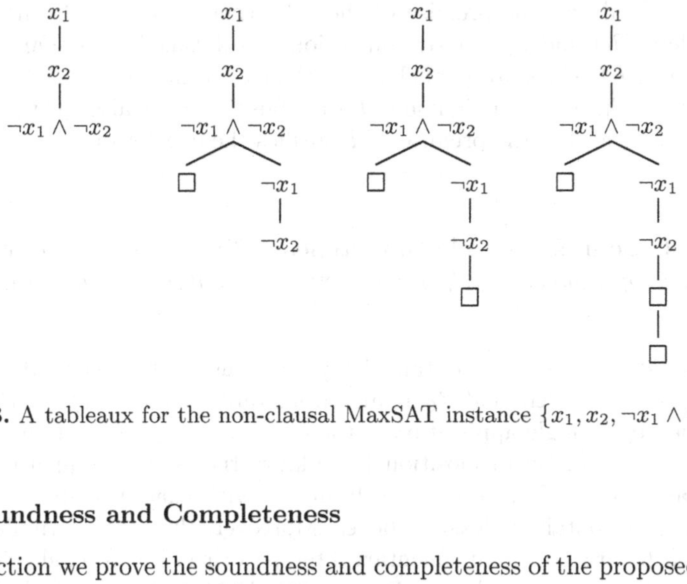

**Fig. 3.** A tableaux for the non-clausal MaxSAT instance $\{x_1, x_2, \neg x_1 \wedge \neg x_2\}$.

### 3.1 Soundness and Completeness

In this section we prove the soundness and completeness of the proposed tableau calculus for non-clausal MaxSAT. We start by proving two propositions needed later.

**Proposition 1.** *A tableau for a multiset of propositional formulas $\phi$ is completed in a finite number of steps.*

*Proof.* We start by creating an initial tableau and then apply rules in the newly created branches until they are saturated. The $\alpha$-, $\beta$- and $\neg$-rule reduce the number of connectives. Since we began with a finite number of connectives, these rules can only be applied a finite number of times. The $\Box$-rule inactivates two literals and adds a box. Since we began with a finite number of literals and boxes cannot be premises of any expansion rule, this rule can only be applied a finite number of times. Hence, the construction of any completed tableau terminates in a finite number of steps. □

**Proposition 2.** *An assignment $I$ falsifies $k$ premises of a $\alpha$-, $\beta$-, $\neg$- and $\Box$-rule iff assignment $I$ falsifies $k$ conclusions in one branch of the conclusions of the rule and at least $k$ conclusions in the other branch (if any).*

*Proof.* We prove the result for each rule:

- $\Box$-rule: Any assignment $I$ always falsifies one premise and satisfies the other. Since the single conclusion is a box and denotes a contradiction, $I$ falsifies the same number of formulas in the premises and the conclusion.
- $\alpha$-rule: If $I$ falsifies the premise of the rule, then $I$ falsifies at least one conclusion in each branch. The left conclusion is a box and is falsified by any assignment, and $I$ falsifies $\alpha_1$ or $\alpha_2$ (o both) of the right conclusion. On the other direction, if $I$ falsifies at least one conclusion in each branch, then $I$ falsifies $\alpha_1$ or $\alpha_2$ (o both) and therefore $I$ falsifies the premise $\alpha_1 \wedge \alpha_2$.
- $\beta$-rule: If $I$ falsifies the premise of the rule, then $I$ falsifies $\beta_1$ and $\beta_2$, and so the left ($\beta_1$) and right ($\beta_2$) conclusions are falsified by $I$. On the other direction, if $I$ falsifies both conclusions, then $I$ falsifies $\beta_1 \vee \beta_2$.
- The $\neg$-rule: Since any assignment $I$ that falsifies $\neg\neg A$ also falsifies $A$, and vice versa, $I$ falsifies the premise iff $I$ falsifies the conclusion.

□

**Theorem 1. Soundness and completeness.** *The cost of a completed tableau for a multiset of formulas $\phi$ is $k$ iff the minimum number of unsatisfied formulas in $\phi$ is $k$.*

*Proof.* (*Soundness:*) $T$ was obtained by creating a sequence of tableaux $T_0, \ldots, T_n$ ($n \geq 0$) such that $T_0$ is an initial tableau for $\phi$, $T_n = T$, and $T_i$ was obtained by a single application of the $\alpha$-, $\beta$-, $\neg$- or $\Box$-rule on an branch of $T_{i-1}$ for $i = 1, \ldots, n$. By Proposition 1, we know that such a sequence is finite. Since $T$ has cost $m$, $T_n$ contains one branch $b$ with exactly $m$ boxes and the rest of branches contain at least $m$ boxes. Moreover, the active formulas in the branches of $T_n$ are non-complementary literals; otherwise we could yet apply expansion rules and $T_n$ could not be completed. The assignment that sets to true each active literal in $b$, only falsifies the $m$ boxes and there cannot be any assignment satisfying less than $m$ formulas in a branch of $T_n$ because each branch contains at least $m$ boxes. Therefore, the minimum number of active formulas than can be unsatisfied among the branches of $T_n$ is $m$.

Proposition 2 guarantees that the minimum number of unsatisfied active formulas is preserved in the sequence of tableaux $T_0, \ldots, T_n$. Thus, the minimum number of unsatisfied active formulas in $T_0$ is also $m$. Since $T_0$ is formed by a single branch that only contains the formulas in $\phi$ and all these formulas are active, the minimum number of formulas that can be unsatisfied in $\phi$ is $m$.

(*Completeness:*) Assume that there is a completed tableau $T$ for $\phi$ that does not have cost $m$. We distinguish two cases:

(i) $T$ has a branch $b$ of cost $k$, where $k < m$. Then, $T$ has a branch with $k$ boxes and a satisfiable multiset of non-complementary literals because $T$ is completed. This implies that the minimum number of unsatisfied active formulas among the branches of $T$ is at most $k$. By Proposition 2, this also holds for $T_0$, but this is in contradiction with $m$ being the minimum number of unsatisfied formulas in $\phi$ because $k < m$. Thus, any branch of $T$ has at least cost $m$.

(ii) $T$ has no branch of cost $m$. This is in contradiction with $m$ being the minimum number of unsatisfied formulas in $\phi$. Since the tableau rules preserve the minimum number of unsatisfied formulas and the branches of any completed tableau only contain active formulas that are boxes or non-complementary literals, $T$ must have a saturated branch with $m$ boxes. Thus, $T$ has a branch of cost $m$.

Hence, each completed tableau $T$ for a multiset of formulas $\phi$ has cost $m$ if the minimum number of formulas that can be unsatisfied in $\phi$ is $m$. $\square$

## 4 Extension to Hard and Weighted Formulas

We presented the tableau calculus for non-clausal unweighted MaxSAT (i.e,; non-clausal MaxSAT) for ease of presentation but tableaux can be extended to deal with hard and soft formulas, and soft formulas can be weighted as well.

In the case of non-clausal partial MaxSAT, there are three basic observations:

- The hard literals of the initial tableau, as well as any other literal derived by the application of an expansion rule to an input hard formula or a subformula derived from a hard formula, remain always active. In the rest of the section, we will refer to such literals as hard literals and to the subformulas derived from a hard formula as hard subformulas.
- If the $\square$-rule is applied to two hard literals, then the current branch is pruned. This means that we have found a contradiction among hard clauses. This corresponds to an unfeasible solution.
- When the premise of the $\alpha$-rule is a hard formula or subformula, the $\alpha$-rule of Table 2 can be used instead of the $\alpha$-rule of Table 3. The calculus remains sound and complete but branching is reduced. This is so because hard formulas must be satisfied by any optimal assignment.

*Example 2.* Let $\phi = \mathcal{H} \cup \mathcal{S}$ be a non-clausal partial MaxSAT instance, where $\mathcal{H}$ is the multiset of hard formulas and $\mathcal{S}$ is the multiset of soft formulas. Given the multiset of propositional formulas $\{x_1 \wedge x_2 \wedge x_3, \neg x_1, \neg x_2, \neg x_3\}$, we analyze the different tableaux obtained when we vary the formulas declared as hard and soft.

The first tableau of Fig. 4 displays a completed tableau when all the formulas are soft; in this case $\phi = \mathcal{H} \cup \mathcal{S} = \emptyset \cup \{x_1 \wedge x_2 \wedge x_3, \neg x_1, \neg x_2, \neg x_3\}$.

The second tableau displays a completed tableau when $x_1 \wedge x_2 \wedge x_3$ is hard and the rest of formulas are soft; in this case $\phi = \mathcal{H} \cup \mathcal{S} = \{x_1 \wedge x_2 \wedge x_3\} \cup \{\neg x_1, \neg x_2, \neg x_3\}$. Notice that the input hard formulas and derived hard subformulas are in bold. We applied the $\alpha$-rule of Table 2 because the premise is hard.

The third tableau displays a completed tableau when $\neg x_1$, $\neg x_2$ and $\neg x_3$ are hard, and $x_1 \wedge x_2 \wedge x_3$ is soft; in this case $\phi = \mathcal{H} \cup \mathcal{S} = \{\neg x_1, \neg x_2, \neg x_3\} \cup \{x_1 \wedge x_2 \wedge x_3\}$. We applied the $\alpha$-rule of Table 3 because the premise is soft.

The fourth tableau displays a completed tableau when $x_1 \wedge x_2 \wedge x_3$ and $\neg x_1$ are hard, and $\neg x_2$ and $\neg x_3$ are soft; in this case $\phi = \mathcal{H} \cup \mathcal{S} = \{x_1 \wedge x_2 \wedge x_3, \neg x_1\} \cup \{\neg x_2, \neg x_3\}$. Notice that the single branch of the tableau is pruned as soon as the $\Box$-rule has two hard premises ($\neg x_1$ and $x_1$). We use a filled box to denote that there is no feasible solution.

In the first case, the minimum number of unsatisfied soft formulas is 1. In the second case, the minimum number of unsatisfied soft formulas among the assignments that satisfy the hard formulas is 3. In the third case, the minimum number of unsatisfied soft formulas among the assignments that satisfy the hard formulas is 1. In the fourth case, there is no optimal solution because the subset of hard formulas is unsatisfiable.

Table 4 displays the expansion rules for weighted formulas. The $\alpha$-, $\beta$- and $\neg$-rule have just one premise and the weight associated to the premise is transferred to the conclusions. The $\Box$-rule has two premises and so the contradiction takes as weight the minimum of the weights associated to the premises

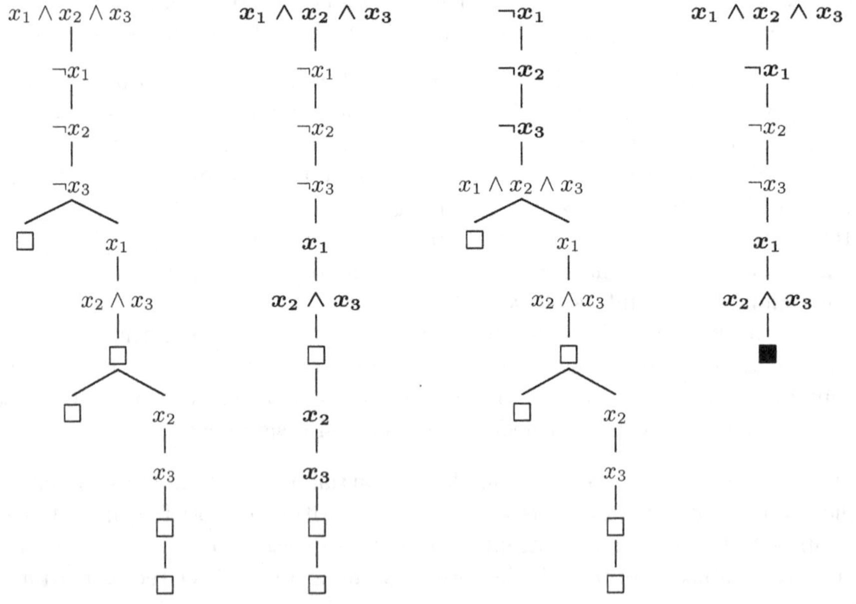

**Fig. 4.** Examples of non-clausal partial MaxSAT tableaux. Input hard formulas and derived hard subformulas are in bold.

**Table 4.** Tableau expansion rules for non-clausal weighted MaxSAT

| α-rule | β-rule | ¬-rule | □-rule |
|---|---|---|---|
| $\dfrac{(\alpha, w)}{(\Box, w) \mid \begin{array}{l}(\alpha_1, w)\\(\alpha_2, w)\end{array}}$ | $\dfrac{(\beta, w)}{(\beta_1, w) \mid (\beta_2, w)}$ | $\dfrac{(\neg\neg A, w)}{(A, w)}$ | $\dfrac{\begin{array}{c}(l, w_1)\\(\neg l, w_2)\end{array}}{\begin{array}{c}(\Box, \min(w_1, w_2))\\(l, w_1 - \min(w_1, w_2))\\(\neg l, w_2 - \min(w_1, w_2))\end{array}}$ |

$(\min(w_1, w_2))$. If the premises have different weights, the remaining weight in the premise with the greatest weight can be used to detect further contradictions. The compensation weight of the other premise is 0, and formulas with weight 0 are removed. In the weighted case, when a branch has repeated occurrences of a formula A, say $(A, w_1), \ldots, (A, w_s)$, such occurrences can be replaced with the single formula $(A, w_1 + \cdots + w_s)$. Moreover, the cost of a saturated weighted branch is the sum of weights of the boxes that appear in the branch, and the cost of a completed weighted tableau is the minimum cost among all its branches.

The expansion rules of Table 4 provide a sound and complete calculus for non-clausal weighted MaxSAT. The correctness of such rules follows from the correctness of the unweighted tableau rules and the fact that having a weighted formula $(A, w)$ is equivalent to having $w$ copies of the unweighted formula $A$.

*Example 3.* Let $\phi = \{(\neg x_1 \rightarrow x_2, 3), (x_1 \wedge x_3, 2), (\neg x_1, 5), (\neg x_1, 5), (\neg x_3, 2)\}$ be a non-clausal weighted MaxSAT instance. Figure 5 displays a completed tableau for $\phi$. This tableau has been obtained by applying the expansion rules of Table 4. The costs of the branches, from left to right, are 5, 7, 5 and 7. So, the minimum sum of weights of unsatisfied formulas is 5.

Finally, we show how to solve non-clausal weighted partial MaxSAT instances with tableaux. The first observation is that hard formulas can be considered as weighted formulas with infinity weight, and this observation is important to understand the □-rule in weighted partial MaxSAT. Notice that the □-rule is the only rule with two premises; in the rest os cases, if the premise is hard, we proceed as in partial MaxSAT, and if it is soft, we proceed as in weighted MaxSAT. If the two premises of the □-rule are hard, then the branch is pruned because we are in front of an unfeasible solution. If the two premises are soft, then the □-rule of Table 4 is applied. If there is a hard premise $l$ and a soft premise $(\neg l, w)$, then $(\Box, w)$ is derived, $(\neg l, w)$ becomes inactive and $l$ remains active.

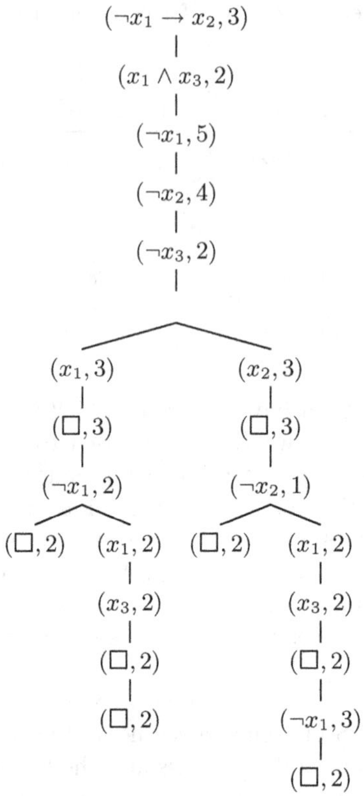

**Fig. 5.** Examples of non-clausal weighted MaxSAT tableaux.

*Example 4.* Let $\phi = \{(x_1 \wedge x_3, (\neg x_1 \rightarrow x_2, 3), (\neg x_1, 5), (\neg x_2, 1), (\neg x_3, 2)\}$ be a non-clausal weighted partial MaxSAT instance, where the first formula is hard and the rest of formulas are soft. Figure 6 displays a completed tableau for $\phi$. This tableau has been obtained by applying the expansion rules for non-clausal weighted partial MaxSAT explained above. The cost of the left branch is 10 and the cost of the right branch is 8. Thus, the minimum sum of weights of unsatisfied soft formulas among the assignments that satisfy the hard formula is 8.

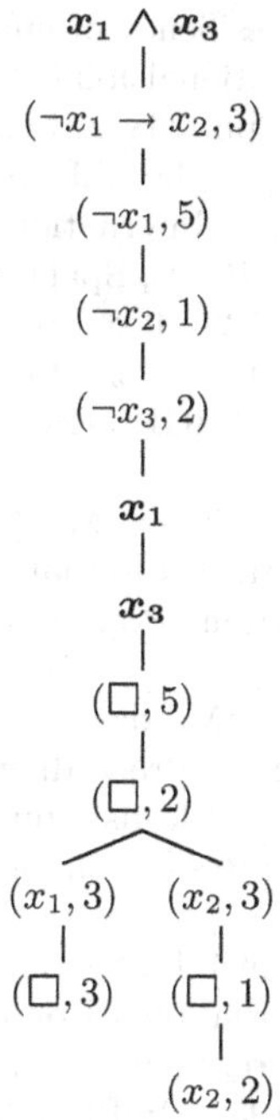


**Fig. 6.** Example of non-clausal weighted partial MaxSAT tableau. Input hard formulas and derived hard subformulas are in bold.

## 5 Conclusions

The main contributions of this paper are a non-clausal MaxSAT tableau calculus, the corresponding proofs of soundness and completeness, and its extension to deal with hard and weighted soft formulas. We claim that improvements defined for SAT, like detection of contradictory subformulas instead of contradictory literals, are also valid in our framework or can be easily adapted.

Tableaux have played a central role in automated deduction in first-order logic, as well as in other non-classical logics [9,12], and this work might be a first step towards dealing with optimization problems in those logics. From the propositional perspective, tableaux might be used to find new proof complexity results as the ones found for MaxSAT resolution [6,14], as well as to better understand MaxSAT and the logic behind. An interesting open problem is to find out how to define a complete tableau calculus for non-clausal MinSAT [3,20].

## References

1. Abramé, A., Habet, D.: Efficient application of max-sat resolution on inconsistent subsets. In: Proceedings of the 20th International Conference on Principles and Practice of Constraint Programming, CP, Lyon, France, pp. 92–107 (2014)

2. Abramé, A., Habet, D.: On the resiliency of unit propagation to Max-Resolution. In: Proceedings of the 24th International Joint Conference on Artificial Intelligence, IJCAI, Buenos Aires, Argentina, pp. 268–274 (2015)
3. Argelich, J., Li, C.M., Manyà, F., Soler, J.R.: Clause branching in MaxSAT and MinSAT. In: Proceedings of the 21st International Conference of the Catalan Association for Artificial Intelligence, Roses, Spain. Frontiers in Artificial Intelligence and Applications, vol. 308, pp. 17–26. IOS Press (2018)
4. Argelich, J., Li, C.M., Manyà, F., Soler, J.R.: Clause tableaux for maximum and minimum satisfiability. Logic J. IGPL (2019). https://doi.org/10.1093/jigpal/jzz025
5. Bofill, M., Garcia, M., Suy, J., Villaret, M.: MaxSAT-based scheduling of B2B meetings. In: Proceedings of the12th International Conference on Integration of AI and OR Techniques in Constraint Programming, CPAIOR, Barcelona, Spain, pp. 65–73 (2015)
6. Bonet, M.L., Buss, S., Ignatiev, A., Marques-Silva, J., Morgado, A.: MaxSAT resolution with the dual rail encoding. In: Proceedings of the 32nd AAAI Conference on Artificial Intelligence, AAAI, New Orleans, Louisiana, USA, pp. 6565–6572 (2018)
7. Bonet, M.L., Levy, J., Manyà, F.: Resolution for Max-SAT. Artif. Intell. **171**(8–9), 240–251 (2007)
8. Casas-Roma, J., Huertas, A., Manyà, F.: Solving MaxSAT with natural deduction. In: Proceedings of the 20th International Conference of the Catalan Association for Artificial Intelligence, Deltebre, Spain. Frontiers in Artificial Intelligence and Applications, vol. 300, pp. 186–195. IOS Press (2017)
9. D'Agostino, M.: Tableaux methods for classical propositional logic. In: D'Agostino, M., Gabbay, D., Hähnle, R., Posegga, J. (eds.) Handbook of Tableau Methods, pp. 45–123. Springer, Dordrecht (1999). https://doi.org/10.1007/978-94-017-1754-0_2
10. D'Almeida, D., Grégoire, É.: Model-based diagnosis with default information implemented through MAX-SAT technology. In: Proceedings of the IEEE 13th International Conference on Information Reuse & Integration, IRI, Las Vegas, NV, USA, pp. 33–36 (2012)
11. Guerra, J., Lynce, I.: Reasoning over biological networks using maximum satisfiability. In: Proceedings of the 18th International Conference on Principles and Practice of Constraint Programming, CP, Québec City, QC, Canada, pp. 941–956 (2012)
12. Hähnle, R.: Tableaux and related methods. In: Robinson, J.A., Voronkov, A. (eds.) Handbook of Automated Reasoning, pp. 100–178. Elsevier and MIT Press (2001)
13. Heras, F., Larrosa, J.: New inference rules for efficient Max-SAT solving. In: Proceedings of the National Conference on Artificial Intelligence, AAAI-2006, Boston/MA, USA, pp. 68–73 (2006)
14. Ignatiev, A., Morgado, A., Marques-Silva, J.: On tackling the limits of resolution in SAT solving. In: Gaspers, S., Walsh, T. (eds.) SAT 2017. LNCS, vol. 10491, pp. 164–183. Springer, Cham (2017). https://doi.org/10.1007/978-3-319-66263-3_11
15. Jabbour, S., Mhadhbi, N., Raddaoui, B., Sais, L.: A SAT-based framework for overlapping community detection in networks. In: Proceedings of the 21st Pacific-Asia Conference on Advances in Knowledge Discovery and Data Mining, Part II, PAKDD, Jeju, South Korea, pp. 786–798 (2017)
16. Larrosa, J., Heras, F.: Resolution in Max-SAT and its relation to local consistency in weighted CSPs. In: Proceedings of the International Joint Conference on Artificial Intelligence, IJCAI-2005, Edinburgh, Scotland, pp. 193–198. Morgan Kaufmann (2005)

17. Li, C.M., Manyà, F., Planes, J.: New inference rules for Max-SAT. J. Artif. Intell. Res. **30**, 321–359 (2007)
18. Li, C.M., Manyà, F., Soler, J.R.: A clause tableaux calculus for MaxSAT. In: Proceedings of the 25th International Joint Conference on Artificial Intelligence, IJCAI-2016, New York, USA, pp. 766–772 (2016)
19. Li, C.M., Manyà, F., Soler, J.R.: Clausal form transformation in MaxSAT. In: Proceedings of the 49th IEEE International Symposium on Multiple-Valued Logic, ISMVL, Fredericton, Canada, pp. 132–137 (2019)
20. Li, C.M., Zhu, Z., Manyà, F., Simon, L.: Optimizing with minimum satisfiability. Artif. Intell. **190**, 32–44 (2012)
21. Manyà, F., Negrete, S., Roig, C., Soler, J.R.: A MaxSAT-based approach to the team composition problem in a classroom. In: Sukthankar, G., Rodriguez-Aguilar, J.A. (eds.) AAMAS 2017. LNCS (LNAI), vol. 10643, pp. 164–173. Springer, Cham (2017). https://doi.org/10.1007/978-3-319-71679-4_11
22. Marques-Silva, J., Argelich, J., Graça, A., Lynce, I.: Boolean lexicographic optimization: algorithms & applications. Ann. Math. Artif. Intell. **62**(3–4), 317–343 (2011)
23. Safarpour, S., Mangassarian, H., Veneris, A.G., Liffiton, M.H., Sakallah, K.A.: Improved design debugging using maximum satisfiability. In: Proceedings of 7th International Conference on Formal Methods in Computer-Aided Design, FMCAD, Austin, Texas, USA, pp. 13–19 (2007)
24. Smullyan, R.: First-Order Logic. Dover Publications, New York (1995). Second corrected edition, First published 1968 by Springer-Verlag
25. Xu, H., Rutenbar, R.A., Sakallah, K.A.: Sub-sat: a formulation for relaxed boolean satisfiability with applications in routing. IEEE Trans. CAD Integr. Circuits Syst. **22**(6), 814–820 (2003)
26. Zhang, L., Bacchus, F.: MAXSAT heuristics for cost optimal planning. In: Proceedings of the 26th AAAI Conference on Artificial Intelligence, Toronto, Ontario, Canada, pp. 1846–1852 (2012)

# Sequent Calculi

# First-Order Quasi-canonical Proof Systems

Yotam Dvir(✉) and Arnon Avron(✉)

Tel-Aviv University, Tel Aviv, Israel
yotamdvir@mail.tau.ac.il, aa@cs.tau.ac.il

**Abstract.** Quasi-canonical Gentzen-type systems with dual-arity quantifiers is a wide class of proof systems. Using four-valued non-deterministic semantics, we show that every system from this class admits strong cut-elimination iff it satisfies a certain syntactic criterion of coherence. As a specific application, this result is applied to the framework of Existential Information Processing (EIP), in order to extend it from its current propositional level to the first-order one—a step which is crucial for its usefulness for handling information that comes from different sources (that might provide contradictory or incomplete information).

**Keywords:** Gentzen-type proof systems · Cut-elimination · Coherence · Non-deterministic matrices · Information processing · Knowledge bases

## 1 Introduction

Proving the cut-elimination theorem for a given Gentzen-type system **G** is usually a difficult and detail intensive task, especially if **G** involves quantifiers that bind variables. In [3] this problem was solved for the wide class of *canonical* Gentzen-type proof systems. These are the systems in which the language features dual-arity quantifiers (i.e. quantifiers that may bind several variables and at the same time connect several formulas), and in which all the logical rules are of the ideal type which was used by Gentzen in [12]. The solution was achieved by formulating an easily checkable syntactic criterion of coherence, and showing that for canonical systems coherence is equivalent both to strong cut-elimination and to strong soundness and completeness with respect to some two-valued non-deterministic matrix. Based on results in [1], we extend this theory here to *quasi-canonical* systems, i.e. systems which are canonical 'up to negation'. (See Definitions 18 and 20 below). Our main theorem is fairly similar to that in [3], but it is more general, and has the significant difference that the semantics we use is based on *four-valued* (rather than two-valued) non-deterministic matrices.

This research was supported by The Israel Science Foundation (Grant No. 817-15).


S. Cerrito and A. Popescu (Eds.): TABLEAUX 2019, LNAI 11714, pp. 77–93, 2019.
https://doi.org/10.1007/978-3-030-29026-9_5

As a very important specific application of the general theory described above, we take the problem of gathering and processing information from a set of sources. In [6,7], Belnap proposed a propositional framework to this end, based on Dunn's four-valued matrix [11]. In his model, sources of information are only allowed to provide information on *atomic* formulas. However, this model is inadequate for dealing with knowledge bases in which information about complex formulas may not originate from information about atomic formulas. Therefore Belnap's framework is generalized in [2] to the Existential Information Processing (EIP) framework, where sources may provide information on complex formulas too. For example, a source which does not state that $\varphi$ is true, nor that $\psi$ is true, may still state that their disjunction is true. For reasoning under those circumstances, a corresponding strongly sound and complete Gentzen-type proof system, that admits strong cut-elimination, is provided.

The EIP framework of [2] is still confined to the propositional level. However, a knowledge base should permit queries in a first-order language in order to really be useful. Using our extension to quasi-canonical systems we are able to extend the EIP framework to the first-order level, carrying over its induced semantics and proof system, and prove that the latter admits strong cut-elimination.

**Note.** Due to space constraints, some of the proofs are omitted here. They will be given in a future publication.

## 2 Preliminaries

The following conventions are used throughout this paper.

- $\mathbb{N}$ is the set of natural numbers (which includes 0).
- A prefix of $\mathbb{N}$ is a set $\{n \in \mathbb{N} \mid n < k\}$, where $k \in \mathbb{N} \cup \{\infty\}$.
- A function $f : X \to Y$ where $X \cap \mathcal{P}[X] = \emptyset$ is implicitly extended to $f : X \cup \mathcal{P}[X] \to Y \cup \mathcal{P}[Y]$ by acting point-wise, i.e.

$$f[\zeta] = \begin{cases} f[\zeta] & \zeta \in X \\ \{f[z] \mid z \in \zeta\} & \zeta \subseteq X \end{cases}$$

This paper considers first-order languages with dual-arity quantifiers, i.e. $\langle n, k\rangle$-quantifiers for some $n, k \in \mathbb{N}$. Such a quantifier connects $n$ formulas and binds $k$ variables. Connectives of arity $n$ are seen as $\langle n, 0\rangle$-quantifiers.

*Example 1.* The language of first-order logic is usually defined to have the $\langle 1, 0\rangle$-quantifier $\neg$, the $\langle 2, 0\rangle$-quantifiers $\vee$, $\wedge$, $\to$, and the $\langle 1, 1\rangle$-quantifiers $\exists$, $\forall$.

For the rest of this paper $L$ is a fixed first-order language with dual-arity quantifiers. Constants of $L$ are taken as 0-ary function symbols.

Construction of $L$-terms and atomic $L$-formulas is standard, and that of $L$-formulas is a simple generalization of the usual construction: If $\mathsf{Q}$ is an $\langle n, k\rangle$-ary quantifier in $L$, $z_1, \ldots z_k$ are distinct variables, and $A_1, \ldots A_n$ are

$L$-formulas, then $\mathsf{Q}\, z_1 \dots z_k\, (A_1, \dots A_n)$ is an $L$-formula where the free occurrences of $z_1, \dots z_k$ in each of the formulas $A_1, \dots A_n$ become bound. Here $\mathsf{Q}$ is said to connect $A_1, \dots A_n$ and bind $z_1, \dots z_k$.

If $A$ and $A'$ are $L$-formulas that are equal up to renaming bound variables, we write $A \overset{\alpha}{\sim} A'$. If $A$ is an $L$-formula, $t_1, \dots t_k$ are $L$-terms, and $z_1, \dots z_k$ are distinct variables, then $A\{t_1/z_1, \dots t_k/z_k\}$ is obtained from $A$ by simultaneously replacing free occurrences of $z_i$ by $t_i$ for all $i \in \{1, \dots k\}$. The accompanying concept of $t_1, \dots t_k$ being substitutable for $z_1, \dots z_k$ in $A$ is defined as usual.

An $L$-sequent is a construct of the form $\Gamma \Rightarrow \Delta$, where $\Gamma$ and $\Delta$ are finite sets of $L$-formulas. We make use of the list-for-union shorthand, e.g. $\Gamma, A, \Delta, B \Rightarrow$ stands for $\{A, B\} \cup \Gamma \cup \Delta \Rightarrow \emptyset$.

**Definition 1.** *Let $V \subseteq \mathsf{Var}$ (the set of all variables). An $L$-formula (-term) is* $V$**-open** *if it has no free variables outside of $V$; it is* **closed** *if it is $\emptyset$-open.*

Non-deterministic matrices [4] provide a rich and modular semantic framework. First defined for propositional logic, the concept was later generalized to predicate logic with dual-arity quantifiers [3]. In what follows, we restrict the domain of our structure to at most countable, so without loss of generality the domain may be taken to be a prefix of $\mathbb{N}$.

**Definition 2.** *A* **generalized non-deterministic matrix (GNmatrix) for** *$L$ is a triple $\langle \mathcal{V}, \mathcal{D}, \mathcal{O} \rangle$ such that:*

- *$\mathcal{V}$ is a set (of truth values).*
- *$\mathcal{D}$ is a non-empty proper subset of $\mathcal{V}$ (of designated truth values).*
- *$\mathcal{O}$ associates with every non-empty prefix $X$ of $\mathbb{N}$ and every $\langle n, k \rangle$-quantifier $\mathsf{Q}$ a function $\tilde{\mathsf{Q}}_X : (X^k \to \mathcal{V}^n) \to \mathcal{P}^+[\mathcal{V}]$ (truth table).*

Note that the quantifiers' interpretations return *sets* of truth values. This will give rise to the semantics' non-determinism, specifically in Definition 12 below.

For the rest of this section $\mathcal{M} = \langle \mathcal{V}, \mathcal{D}, \mathcal{O} \rangle$ is a fixed GNmatrix.

**Notation.** If $\mathsf{Q}$ is a connective, $\tilde{\mathsf{Q}}_X$ may be abbreviated to $\tilde{\mathsf{Q}}$.

**Definition 3.** *An $L$-***algebra** *(in the sense of [5]) $\mathcal{A}$ consists of:*

- *A non-empty prefix $\mathsf{Dom}\,\mathcal{A}$ of $\mathbb{N}$ called the* **domain of** $\mathcal{A}$.
- *For each $m$-ary func. symbol $f$ in $L$, a function $f^{\mathcal{A}} : (\mathsf{Dom}\,\mathcal{A})^m \to \mathsf{Dom}\,\mathcal{A}$ called the* **interpretation of** $f$ **in** $\mathcal{A}$.

For the rest of this section $\mathcal{A}$ is a fixed $L$-algebra.

**Notation.** If $t$ is a closed $L$-term, then $t^{\mathcal{A}}$ denotes its interpretation in the $L$-algebra $\mathcal{A}$, defined inductively: $(f(t_1, \dots t_m))^{\mathcal{A}} = f^{\mathcal{A}}[t_1{}^{\mathcal{A}}, \dots t_m{}^{\mathcal{A}}]$.

**Definition 4.** *An $\mathcal{A}$-***based** *$L$-***informer for** *$\mathcal{M}$, $\mathcal{I}$, consists of the following:*

- *For every m-ary predicate symbol p in L, a predicate* $p^{\mathcal{I}} : (\mathsf{Dom}\,\mathcal{A})^m \to \mathcal{V}$ *called the* **interpretation of** $p$ **in** $\mathcal{I}$.

▌ For the rest of this section $\mathcal{I}$ is a fixed $\mathcal{A}$-based $L$-informer for $\mathcal{M}$.

**Definition 5.** *A pair* $\langle \mathcal{A}, \mathcal{I} \rangle$*, where* $\mathcal{A}$ *and* $\mathcal{I}$ *are as above, is called an* **L-structure for** $\mathcal{M}$ *(which is* **based at** $\mathcal{A}$ *and* **informed by** $\mathcal{I}$*).*[1]

▌ For the rest of this section $\mathcal{S} = \langle \mathcal{A}, \mathcal{I} \rangle$ is a fixed $L$-structure for $\mathcal{M}$.

Substitutional semantics [14] is used to handle assignment of elements of the domain to free variables when evaluating a formula. This contrasts with the prevailing denotational semantics which is inadequate in the non-deterministic context. What follows is a condensed and slightly adapted presentation of notions that appear in [4] (for more see references there).

**Definition 6.** *The set* $\{\overline{a} \mid a \in \mathsf{Dom}\,\mathcal{A}\}$ *of the* **individual constants of** $\mathcal{A}$ *is obtained by associating a constant with every member of* $\mathsf{Dom}\,\mathcal{A}$.

**Notation.** $L(\mathcal{A})$ is obtained by extending $L$ with $\{\overline{a} \mid a \in \mathsf{Dom}\,\mathcal{A}\}$.

**Definition 7.** *The extension of* $\mathcal{A}$ *to an* $L(\mathcal{A})$*-algebra is obtained by letting* $\overline{a}^{\mathcal{A}} = a$ *for every* $a \in \mathsf{Dom}\,\mathcal{A}$.

**Definition 8.** *An* **$\mathcal{A}$-substitution** *is a* $\mathsf{Var} \to \{\overline{a} \mid a \in \mathsf{Dom}\,\mathcal{A}\}$ *function.*

**Definition 9.** *Let* $t$ *be an* $L(\mathcal{A})$*-term. The* **normal form of** $t$*, denoted* $|t|$*, is defined inductively as follows:*

- *If* $t = f(t_1, \dots t_m)$*, then* $|t| = \overline{t^{\mathcal{A}}}$ *if* $t$ *is closed, otherwise* $|t| = f(|t_1|, \dots |t_m|)$.
- *Otherwise (i.e.* $t$ *is a variable),* $|t| = t$.

*For an* $L(\mathcal{A})$*-term* $t'$*, we write* $t \overset{\mathcal{A}}{\sim} t'$ *if* $|t| = |t'|$.

**Definition 10.** *Let* $\varphi$*,* $\varphi'$ *be* $L(\mathcal{A})$*-formulas. We write* $\varphi \overset{\mathcal{A}}{\sim} \varphi'$ *if* $|\varphi| \overset{\alpha}{\sim} |\varphi'|$*, where* $|\varphi|$*, the* **normal form of** $\varphi$*, is defined inductively as follows:*

- *If* $\varphi = p(t_1, \dots t_m)$ *is atomic, then* $|\varphi| = p(|t_1|, \dots |t_m|)$.
- *If* $\varphi = \mathsf{Q}\, z_1 \dots z_k (\psi_1, \dots \psi_n)$*, then* $|\varphi| = \mathsf{Q}\, z_1 \dots z_k (|\psi_1|, \dots |\psi_n|)$.

Valuations are functions that assign truth values to all formulas in a way that is compatible with a particular GNmatrix and structure. In many cases, and in Sect. 4 specifically, it is desirable to define valuations only on some of the formulas.

[1] This is equivalent to the usual definition of a structure. However, it is more convenient for our purposes. See e.g. the independence of Definitions 9 and 10 below from the informer, and the statement of Proposition 2. The convenience is further evident in Sect. 4, where the base algebra remains fixed while the informer varies.

**Definition 11.** *A set $\Phi$ of $L(\mathcal{A})$-formulas is* **closed under subsentences (Sclosed)** *if every formula in $\Phi$ is closed, and $\mathsf{Q}\, z_1 \ldots z_k (\psi_1, \ldots \psi_n) \in \Phi$ implies that for all $i \in \{1, \ldots n\}$ and $a_1, \ldots a_k \in \mathsf{Dom}\,\mathcal{A}$, $\psi_i \{\overline{a_1}/z_1, \ldots \overline{a_k}/z_k\} \in \Phi$.*

**Notation.** $\tilde{\mathsf{Q}}_{\mathsf{Dom}\,\mathcal{A}}$ may be abbreviated to $\tilde{\mathsf{Q}}_{\mathcal{A}}$.

**Definition 12.** *Let $\Phi$ be an Sclosed set of $L(\mathcal{A})$-formulas, and let $v : \Phi \to \mathcal{V}$. Consider the following conditions:*

*A. If $\varphi \overset{\mathcal{A}}{\sim} \varphi'$, then $v[\varphi] = v[\varphi']$.*
*I. $v[p(t_1, \ldots t_m)] = p^{\mathcal{I}}(t_1{}^{\mathcal{A}}, \ldots t_m{}^{\mathcal{A}})$.*
*Q. $v[\mathsf{Q}\, z_1 \ldots z_k (\psi_1, \ldots \psi_n)] \in \tilde{\mathsf{Q}}_{\mathcal{A}}[h]$, where $h$ is $\lambda a_1, \ldots a_k \in \mathsf{Dom}\,\mathcal{A}\,.\, \langle v[\psi_1 \{\overline{a_1}/z_1, \ldots \overline{a_k}/z_k\}], \ldots v[\psi_n \{\overline{a_1}/z_1, \ldots \overline{a_k}/z_k\}]\rangle$.*

- *$v$ is a* **partial $\mathcal{M}$-legal $\mathcal{A}$-valuation** *if conditions A and Q hold.*
- *$v$ is a* **partial $\mathcal{M}$-legal $\mathcal{S}$-valuation** *if conditions A, I and Q hold.*
- *The word 'partial' may be omitted if $\Phi$ includes all closed $L(\mathcal{A})$-formulas.*

**Proposition 1** [4]. *Every partial $\mathcal{M}$-legal $\mathcal{S}$-valuation $v$ is extendable to an $\mathcal{M}$-legal $\mathcal{S}$-valuation (and similarly for partial $\mathcal{M}$-legal $\mathcal{A}$-valuations).*

**Proposition 2.** *For every partial $\mathcal{M}$-legal $\mathcal{A}$-valuation $v$ there exists an $\mathcal{A}$-based $L$-informer $\tilde{\mathcal{I}}$ for $\mathcal{M}$ such that $v$ is a partial $\mathcal{M}$-legal $\langle \mathcal{A}, \tilde{\mathcal{I}} \rangle$-valuation.*

**Definition 13.** *Let $C$ be an $L$-formula, $\Theta \cup \{\Gamma \Rightarrow \Delta\}$ be a set of $L$-sequents, $v$ be an $\mathcal{M}$-legal $\mathcal{S}$-valuation, and $\sigma$ be an $\mathcal{S}$-substitution. Define:*

- *$\mathcal{S}, v, \sigma \models C$ if $v[\sigma[C]] \in \mathcal{D}$.*
- *$\mathcal{S}, v, \sigma \models \Gamma \Rightarrow \Delta$ if there exists $A \in \Gamma$ such that $\mathcal{S}, v, \sigma \not\models A$ or $B \in \Delta$ such that $\mathcal{S}, v, \sigma \models B$.*
- *$\mathcal{S}, v, \sigma \models \Theta$ if $\mathcal{S}, v, \sigma \models \Gamma' \Rightarrow \Delta'$ for every $\Gamma' \Rightarrow \Delta' \in \Theta$.*
- *$\mathcal{S}, v \models \star$ if $\mathcal{S}, v, \sigma' \models \star$ for every $\mathcal{S}$-substitution $\sigma'$ ($\star$ is a formula, sequent, or set of sequents).*
- *$\Theta \vdash_{\mathcal{M}} \Gamma \Rightarrow \Delta$ if the following holds for every $L$-structure $\mathcal{S}'$ for $\mathcal{M}$ and $\mathcal{M}$-legal $\mathcal{S}'$-valuation $v'$: if $\mathcal{S}', v' \models \Theta$, then $\mathcal{S}', v' \models \Gamma \Rightarrow \Delta$.*[2]

**Definition 14.** *Let $\Theta \cup \{\Gamma \Rightarrow \Delta\}$ be a set of $L$-sequents.*

- *$\Theta \vdash_G \Gamma \Rightarrow \Delta$ if $\Gamma \Rightarrow \Delta$ is derivable from $\Theta$ in $G$.*
- *$G$ is* **strongly sound** *for $\mathcal{M}$ if $\vdash_G \subseteq \vdash_{\mathcal{M}}$.*
- *$G$ is* **strongly complete** *for $\mathcal{M}$ if $\vdash_{\mathcal{M}} \subseteq \vdash_G$.*
- *$\mathcal{M}$ is* **strongly characteristic** *for $G$ if $G$ is strongly sound and strongly complete for $\mathcal{M}$.*

**Notation.** It will often be convenient to use a structure instead of its base algebra or informer:

- $\mathsf{Dom}\,\mathcal{S} = \mathsf{Dom}\,\mathcal{A}$; $L(\mathcal{S}) = L(\mathcal{A})$; $\overset{\mathcal{S}}{\sim} = \overset{\mathcal{A}}{\sim}$.
- For a function symbol $f$: $f^{\mathcal{S}} = f^{\mathcal{A}}$; and for a closed term $t$: $t^{\mathcal{S}} = t^{\mathcal{A}}$.
- For a predicate symbol $p$: $p^{\mathcal{S}} = p^{\mathcal{I}}$.

[2] Two consequence relations for formulas $\Gamma \vdash_{\mathcal{M}} \varphi$ are definable using this consequence relation for sequents: 'truth' $\vdash_{\mathcal{M}} \Gamma \Rightarrow \varphi$ and 'validity' $\{\Rightarrow \psi \mid \psi \in \Gamma\} \vdash_{\mathcal{M}} \Rightarrow \varphi$.

## 3 Quasi-canonical Proof Systems and Their Semantics

As we said in the introduction, a characterization for strong cut-elimination was given in [3] for *canonical* Gentzen-type systems. Specifically, the following properties of a canonical system $G$ with dual-arity quantifiers are shown to be equivalent: (i) $G$ is coherent, (ii) $G$ admits strong cut-elimination, and (iii) $G$ has a strongly characteristic GNmatix of a particular kind. In [1] it is shown that for a *quasi*-canonical system $G$ of *proposition* logic (i) entails (ii) and (iii). This section combines these two results, thus generalizing both (Theorem 1): (i), (ii) and (iii) are found to be equivalent for a quasi-canonical system $G$ with dual-arity quantifiers.

For the rest of this paper assume $L$ includes the 1-ary connective $\neg$.

### 3.1 Introducing Quasi-canonical Proof Systems

The family of highly simplified representation languages defined below suffices for expressing the logical rules of a quasi-canonical system.

**Definition 15.** *The language $L_k^n$ is the language that consists – aside from the mandatory variables and auxiliary symbols – of enumerably many constants $\mathsf{Con} = \{c_i \mid i \in \mathbb{N}\}$, predicate symbols $p_1, \ldots p_n$ of arity $k$, and the connective $\neg$.*

**Notation.** Let $\overset{a}{\neg}$ denote $\neg$ if $a = 1$, and the empty string if $a = 0$.

**Definition 16.** *An* **$\langle n, k\rangle$-literal** *is an $L_k^n$-formula of the form $\overset{a}{\neg}p_i(t_1, \ldots t_k)$, where $a \in \{0, 1\}$, $i \in \{1, \ldots n\}$, and for every $j \in \{1, \ldots k\}$, $t_j \in \mathsf{Con} \cup \mathsf{Var}$. An* **$\langle n, k\rangle$-gc** *(generalized clause) is a sequent of $\langle n, k\rangle$-literals.*

**Definition 17.** *Let $\mathsf{Q}$ be an $\langle n, k\rangle$-ary quantifier. A* **quasi-canonical rule for** *$\mathsf{Q}$ is a construct of the form $\Lambda/T$, where $\Lambda$ is a set of $\langle n, k\rangle$-gcs, and $T$ is the rule's type—one of the following: $(\mathsf{Q} \Rightarrow)$, $(\Rightarrow \mathsf{Q})$, $(\neg\mathsf{Q} \Rightarrow)$, $(\Rightarrow \neg\mathsf{Q})$. An* **$\langle n, k\rangle$-rule** *is a quasi-canoncial rule for an $\langle n, k\rangle$-ary quantifier.*

To apply an $\langle n, k\rangle$-rule as an inference in a proof one must first instantiate the schematic constituents of $L_k^n$ by constituents of $L$.

**Definition 18.** *Let $r = \Lambda/T$ be an $\langle n, k\rangle$-rule. Let $\Phi$ be a set of $L$-formulas and $z_1, \ldots z_k$ be distinct variables. An* **$\langle L, r, \Phi, z_1, \ldots z_k\rangle$-mapping** *is any function $\chi$ from the terms and predicate symbols of $L_k^n$ to terms and formulas of $L$, satisfying the following conditions:*

- *For every $y \in \mathsf{Var}$, $\chi[y] \in \mathsf{Var}$, and for every $x \in \mathsf{Var}$ such that $x \neq y$, $\chi[x] \neq \chi[y]$.*
- *For every $c \in \mathsf{Con}$, $\chi[c]$ is an $L$-term, such that for every $x \in \mathsf{Var}$ occurring in $\Lambda$, $\chi[x]$ does not occur in $\chi[c]$.*
- *For every $i \in \{1, \ldots n\}$, $\chi[p_i]$ is an $L$-formula. If $\overset{a}{\neg}p_i(t_1, \ldots t_k)$ occurs in $\Lambda$, then for every $j \in \{1, \ldots k\}$: $\chi[t_j]$ is substitutable for $z_j$ in $\chi[p_i]$, and if $t_j \in \mathsf{Var}$, then $\chi[t_j]$ does not occur free in $\Phi \cup \{\mathsf{Q}\, z_1 \ldots z_k (\chi[p_1], \ldots \chi[p_n])\}$.*

*$\chi$ extends to $\langle n,k\rangle$-literals by* $\chi\left[\overset{a}{\neg} p_i(t_1,\dots t_n)\right]=\overset{a}{\neg}\chi[p_i]\{\chi[t_1]/z_1,\dots\chi[t_k]/z_k\}$.

**Definition 19.** *Let* Q *be an $\langle n,k\rangle$-ary quantifier, and* $r=\{\Pi_\ell\Rightarrow\Sigma_\ell\}_{\ell=1}^m/(\mathsf{Q}\Rightarrow)$ *be a quasi-canonical rule for* Q. *An* **application of** $r$ *is any inference step of the form:*

$$\frac{\{\Gamma,\chi[\Pi_\ell]\Rightarrow\chi[\Sigma_\ell],\Delta\}_{\ell=1}^m}{\Gamma,\mathsf{Q}\,z_1\dots z_k\,(\chi[p_i],\dots\chi[p_n])\Rightarrow\Delta}\;(\mathsf{Q}\Rightarrow)$$

*where $\chi$ is some $\langle L,r,\Gamma\cup\Delta,z_1,\dots z_k\rangle$-mapping.*

*Applications of the other types of quasi-canonical rules are defined similarly.*

*Example 2.* Consider the following quasi-canonical rules for $\exists$:

$$\{\Rightarrow\neg p_1(v_1)\}/(\Rightarrow\neg\exists)\qquad\{\neg p_1(c_1)\Rightarrow\}/(\neg\exists\Rightarrow)$$

Application of these rules has the forms:

$$\frac{\Gamma\Rightarrow\neg A\{x/z\},\Delta}{\Gamma\Rightarrow\neg\exists zA,\Delta}\;(\Rightarrow\neg\exists)\qquad\frac{\Gamma,\neg A\{t/z\}\Rightarrow\Delta}{\Gamma,\neg\exists zA\Rightarrow\Delta}\;(\neg\exists\Rightarrow)$$

where $x$ is not free in $\Gamma\cup\Delta\cup\{\neg\exists zA\}$, and $x$ and $t$ are substitutable for $z$ in $A$.

**Definition 20.** *A* **full quasi-canonical calculus for** $L$ *is a Gentzen-type system that consists of rules of the following types:*

- *Logical rules: a finite number of quasi-canonical inference rules.*
- *Structural rules: the $\alpha$-axiom scheme* (A)*, the weakening rule* (W)*, the cut rule* (C)*, and the substitution rule* (S)*, with application forms*

$$\frac{}{A\Rightarrow A'}\;(\mathsf{A})\qquad\frac{\Gamma\Rightarrow\Delta}{\Gamma',\Gamma\Rightarrow\Delta,\Delta'}\;(\mathsf{W})\qquad\frac{\Gamma'\Rightarrow\Delta,A\quad A,\Gamma\Rightarrow\Delta'}{\Gamma',\Gamma\Rightarrow\Delta,\Delta'}\;(\mathsf{C})$$

$$\frac{\Gamma\Rightarrow\Delta}{\Gamma\{t_1/x_1,\dots t_m/x_m\}\Rightarrow\Delta\{t_1/x_1,\dots t_m/x_m\}}\;(\mathsf{S})$$

*where $\Gamma,\Gamma',\Delta,\Delta',\{A,A'\}$ are sets of L-formulas such that $A\overset{\alpha}{\sim}A'$; $x_1,\dots x_m$ are distinct variables; $t_1,\dots t_m$ are L-terms substitutable for $x_1,\dots x_m$ in every formula in $\Gamma\cup\Delta$.*

*A* **full 4-quasi-canonical calculus for** $L$ *is a full quasi-canonical calculus in which there are no rules of the types* $(\neg\Rightarrow)$ *and* $(\Rightarrow\neg)$.

The structural rules are sound in the following sense:

**Proposition 3.** *Let $\Theta\cup\{\Gamma\Rightarrow\Delta\}$ be a set of L-sequents such that $\Theta/\Gamma\Rightarrow\Delta$ is an application of a structural rule $r$. Let $\mathcal{M}$ be an GNmatix. Let $\mathcal{S}$ be an L-structure for $\mathcal{M}$, and $v$ be a $\mathcal{M}$-legal $\mathcal{S}$-valuation, such that $\mathcal{S},v\models\Theta$. Let $\sigma$ be an $\mathcal{S}$-substitution. Then $\mathcal{S},v,\sigma\models\Gamma\Rightarrow\Delta$.*

*Proof.* By case analysis on the rule $r$:

- If $r$ is the $\alpha$-axiom (A) this follows from the definition of valuations (and $\overset{\mathcal{S}}{\sim}$).
- If $r$ is the weakening rule (W) or cut rule (C) this follows trivially as usual.
- If $r$ is the substitution rule (S), then for every variable $x$, denote by $t_x$ its $L$-term replacement (or simply $t_x = x$ if $x$ was not replaced). Let $\sigma'$ be the $\mathcal{S}$-substitution such that $\sigma'[x] = \overline{(\sigma[t_x])}^{\mathcal{S}}$. In particular, $\sigma'[x] \overset{\mathcal{S}}{\sim} \sigma[t_x]$. By assumption, $\mathcal{S}, v, \sigma' \models \Theta$. Consequently, $\mathcal{S}, v, \sigma \models \Gamma \Rightarrow \Delta$.

Coherence [3] is a syntactic property of quasi-canonical systems that will later be used to determine whether the system admits strong cut-elimination.

**Definition 21.** *A set $\Lambda$ of $\langle n, k\rangle$-gcs is* **inconsistent** *if there is a proof of $\Rightarrow$ from $\Lambda$ using only* (C) *and* (S)*; otherwise it is* **consistent**.

**Definition 22.** *Let $\Lambda_1$ and $\Lambda_2$ be sets of $\langle n, k\rangle$-gcs. $\Lambda_1 \Cup \Lambda_2$ is $\Lambda_1 \cup \Lambda_2'$, where $\Lambda_2'$ is obtained from $\Lambda_2$ by fresh renaming of constants and variables in $\Lambda_1$.*

**Definition 23.** *Rules $\Lambda_1/T_1$ and $\Lambda_2/T_2$ are* **conflicting** *if for some quantifier* Q *either $T_1 = (\mathsf{Q} \Rightarrow)$ and $T_2 = (\Rightarrow \mathsf{Q})$, or $T_1 = (\neg\mathsf{Q} \Rightarrow)$ and $T_2 = (\Rightarrow \neg\mathsf{Q})$.*

**Definition 24.** *A full quasi-canonical calculus for $L$ is* **coherent** *if for every pair of conflicting rules $\Lambda_1/T_1$ and $\Lambda_2/T_2$, the set $\Lambda_1 \Cup \Lambda_2$ is inconsistent.*

*Example 3.* Consider the full quasi-canonical calculus in which the inference rules are those from Example 2. These rules are conflicting. However, the set $\{\neg p_1(c_1) \Rightarrow, \Rightarrow \neg p_1(v_1)\}$ is clearly inconsistent, so the calculus is coherent.

**Proposition 4.** *Let $\Lambda \cup \{\Pi \Rightarrow \Sigma\}$ be a set of $\langle n, k\rangle$-gcs. If there is a proof of $\Pi \Rightarrow \Sigma$ from $\Lambda$ using only* (C) *and* (S)*, then there is such a proof in which* (S) *is used only as the first inference step on leaves of the proof tree, and only for substituting by constants that appear in $\Lambda \cup \{\Pi \Rightarrow \Sigma\}$.*

*Proof.* Note that an application of (C) followed by an application of (S) can be replaced with an a pair of applications of (S) followed by an application of (C); and two consecutive applications of (S) can be replaced with one. Using induction on the given proof's height, applications of (S) can thus be pushed to the leaves. Next, using induction on the given proof's height, the obtained proof remains valid after replacing all variables and constants that do not appear in $\Lambda \cup \{\Pi \Rightarrow \Sigma\}$ with a variable or constant that does appear in $\Lambda \cup \{\Pi \Rightarrow \Sigma\}$.

**Corollary 1.** *The coherence of a full 4-quasi-canonical calculus is decidable.*

### 3.2 The Semantics of Quasi-canonical Proof Systems

The semantics of quasi-canonical proof systems is based on Dunn's four truth values [11,13], where each truth value is a different subset of $\{0,1\}$, and the presence of 1 (0) indicates evidence supporting (opposing) the truth of a formula.

**Notation.** $\bot = \{\}$; $\mathrm{f} = \{0\}$; $\mathrm{t} = \{1\}$; $\top = \{0,1\}$.

A statement is considered true iff it has supporting evidence, and its negation true iff the statement has opposing evidence.

**Definition 25.** *A GNmatix $\mathcal{M} = \langle \mathcal{V}, \mathcal{D}, \mathcal{O} \rangle$ for $L$ is a* **¬-GNmatix** *if:*

- $\mathcal{V} \subseteq \{\mathrm{t}, \mathrm{f}, \top, \bot\}$, *and* $\mathcal{D} = \mathcal{V} \cap \{\mathrm{t}, \top\}$.
- *The following hold for the operation* $\tilde{\neg}$ *of* $\mathcal{O}$:
  - *If* $\mathrm{t} \in \mathcal{V}$, *then* $\tilde{\neg}\mathrm{t} \subseteq \{\mathrm{f}, \bot\}$.
  - *If* $\mathrm{f} \in \mathcal{V}$, *then* $\tilde{\neg}\mathrm{f} \subseteq \{\mathrm{t}, \top\}$.
  - *If* $\top \in \mathcal{V}$, *then* $\tilde{\neg}\top \subseteq \{\mathrm{t}, \top\}$.
  - *If* $\bot \in \mathcal{V}$, *then* $\tilde{\neg}\bot \subseteq \{\mathrm{f}, \bot\}$.
- *All operations* $\tilde{\mathsf{Q}}$ *of* $\mathcal{O}$ *return members of*

$$\{\mathcal{V}, \{\mathrm{t}, \top\}, \{\mathrm{t}, \bot\}, \{\mathrm{f}, \top\}, \{\mathrm{f}, \bot\}, \{\mathrm{t}\}, \{\mathrm{f}\}, \{\top\}, \{\bot\}\}$$

**Definition 26.** $\mathcal{M}_4 = \langle \{\mathrm{t}, \mathrm{f}, \top, \bot\}, \{\mathrm{t}, \top\}, \{\tilde{\neg}^4\} \rangle$ *with* $\tilde{\neg}^4\mathrm{t} = \tilde{\neg}^4\bot = \{\mathrm{f}, \bot\}$, $\tilde{\neg}^4\mathrm{f} = \tilde{\neg}^4\top = \{\mathrm{t}, \top\}$.

The next couple of definitions are adapted from [1]. First, a function is defined to take a quasi-canonical rule for some quantifier Q and return a set of truth values. Intuitively, the set returned consists of those truth values Q can take for the rule's conclusion to hold.

**Definition 27.** *The function $F$ on quasi-canonical rules is defined as follows:*

$$F[r] = \begin{cases} \{\mathrm{t}, \top\} & r \textit{ is of type } (\Rightarrow \mathsf{Q}) \\ \{\mathrm{f}, \bot\} & r \textit{ is of type } (\mathsf{Q} \Rightarrow) \\ \{\mathrm{f}, \top\} & r \textit{ is of type } (\Rightarrow \neg\mathsf{Q}) \\ \{\mathrm{t}, \bot\} & r \textit{ is of type } (\neg\mathsf{Q} \Rightarrow) \end{cases}$$

Next, the function is used to provide an interpretation to quantifiers that corresponds to a given Gentzen-type proof system.

**Definition 28.** *Let $G$ be a coherent full 4-quasi-canonical calculus for $L$. The* **¬-GNmatrix induced by** *$G$, denoted $\mathcal{M}_G$, is the ¬-GNmatrix $\langle \mathcal{V}_4, \{\mathrm{t}, \top\}, \mathcal{O}_G \rangle$ in which, for every non-empty prefix $X$ of $\mathbb{N}$, the interpretation $\tilde{\mathsf{Q}}_X$ in $\mathcal{O}_G$ of an $\langle n, k \rangle$-quantifier* Q *in $L$ is defined as follows:*

$$\tilde{\mathsf{Q}}_X[h] = \begin{cases} \bigcap\{F[r] \mid r \in R_G[\mathsf{Q}, X, h]\} & \mathsf{Q} \neq \neg \\ \tilde{\neg}^4[h] \cap \bigcap\{F[r] \mid r \in R_G[\mathsf{Q}, X, h]\} & \mathsf{Q} = \neg \end{cases}$$

*where $R_G[\mathsf{Q}, X, h]$ is the set of rules $\Lambda/T$ for* Q *in $G$ that satisfy the following: an $L_k^n$-structure $\mathcal{N}$ for $\mathcal{M}_4$ exists such that* $\mathsf{Dom}\,\mathcal{N} = X$, $p_i^{\mathcal{N}} = h_i$, *and* $\mathcal{N} \models \Lambda$.

Examples where Definitions 27 and 28 are employed can be found in the proof of Theorem 2 below.

**Proposition 5.** *Let $G$ be a coherent full 4-quasi-canonical calculus for $L$. Then $\mathcal{M}_G$ is a well-defined four-valued ¬-GNmatix.*

### 3.3 Soundness, Completeness, and Cut-Elimination

**Proposition 6.** *Let $G$ be a coherent full 4-quasi-canonical calculus for $L$. Then $G$ is strongly sound for $\mathcal{M}_G$.*

**Definition 29.** *Let $G$ be a full quasi-canonical calculus.*

- *Let $\Theta \cup \{\Gamma \Rightarrow \Delta\}$ be some set of $L$-sequents. A proof in $G$ of $\Gamma \Rightarrow \Delta$ from $\Theta$ is* ***$\Theta$-cut-free*** *if all cuts in the proof are on substitution instances of formulas from $\Theta$.*
- *$G$ admits* **strong cut-elimination** *if for every set of $L$-sequents $\Theta \cup \{\Gamma \Rightarrow \Delta\}$ that satisfies the free-variable condition (no variable occurs both free and bound): if there is a proof in $G$ of $\Gamma \Rightarrow \Delta$ from $\Theta$, there is also such a proof which is $\Theta$-cut-free.*

*Example 4.* Consider the following proofs of $\Rightarrow$ from $\{\Rightarrow \neg p(x), \neg p(c) \Rightarrow\}$ in the system from Example 3:

$$\dfrac{\dfrac{\Rightarrow \neg p(x)}{\Rightarrow \neg\exists x p(x)}\,(\Rightarrow\neg\exists) \qquad \dfrac{\neg p(c) \Rightarrow}{\neg\exists x p(x) \Rightarrow}\,(\neg\exists\Rightarrow)}{\Rightarrow}\,(\mathsf{C}) \quad \rightsquigarrow \quad \dfrac{\dfrac{\Rightarrow \neg p(x)}{\Rightarrow \neg p(c)}\,(\mathsf{S}) \qquad \neg p(c) \Rightarrow}{\Rightarrow}\,(\mathsf{C})$$

The cut in the proof on the left was eliminated by using the substitution rule, resulting in the proof on the right which is $\{\Rightarrow \neg p(x), \neg p(c) \Rightarrow\}$-cut-free.

**Proposition 7.** *Let $G$ be a coherent full 4-quasi-canonical calculus. Let $\Theta \cup \{\Gamma \Rightarrow \Delta\}$ be a set of $L$-sequents that satisfies the free-variable condition. If $\Gamma \Rightarrow \Delta$ has no $\Theta$-cut-free proof from $\Theta$ in $G$, then $\Theta \nvdash_{\mathcal{M}_G} \Gamma \Rightarrow \Delta$.*

**Proposition 8.** *Let $\Lambda \cup \{\Pi \Rightarrow \Sigma\}$ be a set of $\langle n, k\rangle$-gcs.*

1. *If there is a proof of $\Pi \Rightarrow \Sigma$ from $\Lambda$ using only* (A), (W), (C), *and* (S), *then there are $\Pi' \subseteq \Pi$ and $\Sigma' \subseteq \Sigma$ such that there is a proof of $\Pi' \Rightarrow \Sigma'$ from $\Lambda$ using only* (A), (C), *and* (S).
2. *If there is a proof of $\Pi \Rightarrow \Sigma$ from $\Lambda$ using only* (A), (C), *and* (S), *and $\Pi \Rightarrow \Sigma$ is not an instance of* (A), *then there is a proof of $\Pi \Rightarrow \Sigma$ from $\Lambda$ using only* (C) *and* (S).

**Corollary 2.** *If a set $\Lambda$ of $\langle n, k\rangle$-gcs is consistent, then there is an $L^n_k$-structure $\mathcal{N}$ for $\mathcal{M}_4$ such that $\mathcal{N} \models \Lambda$.*

**Theorem 1.** *Let $G$ be a full 4-quasi-canonical calculus for $L$. The following are equivalent:*

1. *$G$ is coherent.*
2. *$G$ is coherent and $\mathcal{M}_G$ is strongly characteristic for $G$.*
3. *$G$ has a strongly characteristic $\neg$-GNmatix.*
4. *$G$ admits strong cut-elimination.*

*Proof.* We prove $1 \Longrightarrow 2 \Longrightarrow 3 \Longrightarrow 1$ and $1 \Longrightarrow 4 \Longrightarrow 1$:

1 $\Longrightarrow$ 2. Assume $G$ is coherent. Then by Proposition 6, $\mathcal{M}_G$ is strongly sound for $G$. It remains to show that $\mathcal{M}_G$ is strongly complete for $G$. Let $\Theta \cup \{\Gamma \Rightarrow \Delta\}$ be a set of $L$-sequents such that $\Gamma \Rightarrow \Delta$ has no proof from $\Theta$ in $G$. Rename variables in $\Theta \cup \{\Gamma \Rightarrow \Delta\}$ as necessary to obtain $\Theta' \cup \{\Gamma' \Rightarrow \Delta'\}$ satisfying the free-variable condition. Then $\Gamma' \Rightarrow \Delta'$ has no proof from $\Theta'$ in $G$, otherwise a proof of $\Gamma \Rightarrow \Delta$ from $\Theta$ in $G$ could be obtained by using (A) and (C). By Proposition 7, $\Theta' \nvdash_{\mathcal{M}_G} \Gamma' \Rightarrow \Delta'$. Since valuations respect $\alpha$-equivalence, $\Theta \nvdash_{\mathcal{M}_G} \Gamma \Rightarrow \Delta$. Therefore, if $\Theta \vdash_{\mathcal{M}_G} \Gamma \Rightarrow \Delta$, then $\Theta \vdash_G \Gamma \Rightarrow \Delta$, the required strong completeness.

2 $\Longrightarrow$ 3. $\mathcal{M}_G$ is a $\neg$-GNmatix by Proposition 5.

3 $\Longrightarrow$ 1. Assume $G$ has a strongly characteristic $\neg$-GNmatix $\mathcal{M}$. Suppose for contradiction that $G$ is not coherent. Then there must exist two $\langle n, k\rangle$-rules $r_1 = \Lambda_1 / \left(\overset{a}{\neg}\mathsf{Q} \Rightarrow\right)$ and $r_2 = \Lambda_2 / \left(\Rightarrow \overset{a}{\neg}\mathsf{Q}\right)$ in $G$ such that $\Lambda_1 \uplus \Lambda_2$ is consistent. By Corollary 2, there exist an $L_k^n$-structure $\mathcal{N}$ for $\mathcal{M}_4$ and an $\mathcal{M}_4$-legal $\mathcal{N}$-valuation $u$ such that $\mathcal{N}, u \models \Lambda_1 \uplus \Lambda_2$. Pick an $L$-structure $\mathcal{S}$ that extends[3] $\mathcal{N}$ and an $\mathcal{M}$-legal $\mathcal{S}$-valuation $v$ such that for every closed $L(\mathcal{S})$-literal $l$ it holds that $v[l] \in \{\mathsf{t}, \top\}$ iff $u[l] \in \{\mathsf{t}, \top\}$. Such $v$ exists since $\mathcal{M}$ is a $\neg$-GNmatix. Thus $\mathcal{S}, v \models \Lambda_1 \uplus \Lambda_2$. However, $\Lambda_1 \uplus \Lambda_2 \vdash_G \Rightarrow$, so by strong soundness $\mathcal{S}, v \models$ $\Rightarrow$ which is impossible.

1 $\Longrightarrow$ 4. Let $\Theta \cup \{\Gamma \Rightarrow \Delta\}$ be a set of $L$-sequents that satisfies the free-variable condition such that $\Theta \vdash_G \Gamma \Rightarrow \Delta$. We have already shown that $\mathcal{M}_G$ is strongly sound for $G$, and therefore $\Theta \vdash_{\mathcal{M}_G} \Gamma \Rightarrow \Delta$. By Proposition 7, $\Gamma \Rightarrow \Delta$ has a $\Theta$-cut-free proof from $\Theta$ in $G$. Thus $G$ admits strong cut-elimination.

4 $\Longrightarrow$ 1. Assume that $G$ admits strong cut-elimination. Suppose $G$ is not coherent. Then there exist two rules $\Lambda_1 / \left(\overset{a}{\neg}\mathsf{Q} \Rightarrow\right)$ and $\Lambda_2 / \left(\Rightarrow \overset{a}{\neg}\mathsf{Q}\right)$ in $G$ such that $\Lambda_1 \uplus \Lambda_2$ is consistent. Obtain $\Lambda_1 \uplus \Lambda_2 \vdash_G \Rightarrow$ by applying each rule once and following with an application of (C). The set $(\Lambda_1 \uplus \Lambda_2) \cup \{\Rightarrow\}$ clearly satisfies the free-variable condition as there are no bound variable occurrences there at all. Since $G$ admits strong cut-elimination, there must be a $\Lambda_1 \uplus \Lambda_2$-cut-free proof in $G$ of $\Rightarrow$ from $\Lambda_1 \uplus \Lambda_2$.

Suppose there was an application of a logical rule in the proof. Since the rule is neither of type $(\neg \Rightarrow)$ nor of type $(\Rightarrow \neg)$, such an application must introduce a non-literal formula. It is easy to show that the existence of a non-literal formula must be retained throughout a proof in which applications of (C) eliminate only literals, in contradiction to the conclusion being $\Rightarrow$.

Therefore, the only rules applied in the proof are (A), (W), (C), and (S). By Proposition 8, the proof can be reduced to one using only (C) and (S). Yet this is a contradiction to the fact that $\Lambda_1 \uplus \Lambda_2$ is consistent.

## 4 Existential Information Processing

In [2] a propositional framework of Existential Information Processing (EIP) is suggested as a means to handle inconsistent information in knowledge bases.[4]

[3] Without loss of generality, $L_k^n \subseteq L$.

[4] See [8] for a different approach that uses logics of formal inconsistency.

This involves indiscriminately gathering information from a set of sources and then processing it in order to discern further logical conclusions, while keeping inconsistencies to a minimum. In this section the framework is extended to predicate logic using the tools developed above.

> For the rest of this paper assume the quantifiers of $L$ are the 1-ary connective $\neg$, the 2-ary connectives $\vee$ and $\wedge$, and the $\langle 1,1\rangle$-ary quantifiers $\exists$ and $\forall$; and assume $\mathcal{A}$ is a fixed $L$-algebra.

## 4.1 Sources of Information

In the EIP framework, sources provide information on *arbitrary* formulas, in the form of truth values from $\{\mathrm{i},0,1\}$, where i means that the source doesn't know. This fact enables them to possess disjunctive information: a source may know that $\varphi \vee \psi$ holds without knowing which of $\varphi$ and $\psi$ holds; and dually, a source may know that $\varphi \wedge \psi$ does not hold without knowing which of $\varphi$ and $\psi$ does not hold. To extend this framework to predicate logic, sources must provide information on formulas with the classical quantifiers. This will be done here by following the classical intuition that $\exists x\varphi \equiv \bigvee_a \varphi\{\overline{a}/x\}$ and $\forall x\varphi \equiv \bigwedge_a \varphi\{\overline{a}/x\}$, where $a$ ranges over the domain (which may be infinite).

**Definition 30.** *Let* $\mathcal{QM}_r^3 = \langle\{\mathrm{i},0,1\},\{1\},\mathcal{QO}_r^3\rangle$, *where* $\mathcal{QO}_r^3$ *is detailed below:*

| $a$ | $\tilde{\neg}a$ |
|---|---|
| i | $\{\mathrm{i}\}$ |
| 0 | $\{1\}$ |
| 1 | $\{0\}$ |

| $\tilde{\vee}$ | i | 0 | 1 |
|---|---|---|---|
| i | $\{\mathrm{i},1\}$ | $\{\mathrm{i}\}$ | $\{1\}$ |
| 0 | $\{\mathrm{i}\}$ | $\{0\}$ | $\{1\}$ |
| 1 | $\{1\}$ | $\{1\}$ | $\{1\}$ |

| $\tilde{\wedge}$ | i | 0 | 1 |
|---|---|---|---|
| i | $\{\mathrm{i},0\}$ | $\{0\}$ | $\{\mathrm{i}\}$ |
| 0 | $\{0\}$ | $\{0\}$ | $\{0\}$ |
| 1 | $\{\mathrm{i}\}$ | $\{0\}$ | $\{1\}$ |

| $h[X]$ | $\tilde{\exists}_X[h]$ |
|---|---|
| $\{\mathrm{i}\}$ | $\{\mathrm{i},1\}$ |
| $\{\mathrm{i},0\}$ | $\{\mathrm{i},1\}$ |
| $\{0\}$ | $\{0\}$ |
| else | $\{1\}$ |

| $h[X]$ | $\tilde{\forall}_X[h]$ |
|---|---|
| $\{\mathrm{i}\}$ | $\{\mathrm{i},0\}$ |
| $\{\mathrm{i},1\}$ | $\{\mathrm{i},0\}$ |
| $\{1\}$ | $\{1\}$ |
| else | $\{0\}$ |

**Definition 31.** *An* $\mathcal{A}$**-source** *is a partial* $\mathcal{QM}_r^3$*-legal* $\mathcal{A}$*-valuation.*
*An* $\mathcal{A}$**-reservoir** *is a set of* $\mathcal{A}$*-sources.*[5]

Sources in a reservoir share an algebra, thus agreeing on the objects under discussion. This means that disagreement is limited to *properties* of said objects.

> For the rest of this section $R$ is a fixed $\mathcal{A}$-reservoir.

## 4.2 Gathering and Processing the Information

The next step is to gather the information from the reservoir for processing.

**Definition 32.** *The* **existential gathering function of** $R$ *is the function* $g_R$ *from the closed* $L(\mathcal{A})$*-formulas to* $\mathcal{V}_4$ *defined as follows:*

$$g_R = \lambda\varphi\,.\,\{b \in \{0,1\} \mid \exists u \in R\,.\,b \in u[\varphi]\}$$

[5] Note how dividing structures into an algebra and an informer is convenient here.

There may be knowledge that can only be learned by processing the information in the reservoir. For example, if source $a$ says $\varphi$ holds and source $b$ says $\psi$ holds, then the reservoir $\{a, b\}$ provides evidence supporting $\varphi \wedge \psi$. The gatherer will not observe this fact if neither $a$ nor $b$ say $\varphi \wedge \psi$ holds.

**Definition 33.** *Let $g$ be a function from the closed $L(\mathcal{A})$-formulas to $\mathcal{V}_4$. The* **information processing valuation induced by** *$g$ is the function $d$ from the closed $L(\mathcal{A})$-formulas to $\mathcal{V}_4$ inductively defined as follows (for any $b \in \{0,1\}$, $x \in \mathsf{Var}$, $\theta$ an $\{x\}$-open $L(\mathcal{A})$-formulas, and $\varphi, \varphi', \varphi_l, \varphi_r$ closed $L(\mathcal{A})$-formulas such that $\varphi \overset{\mathcal{A}}{\sim} \varphi'$):*

*(d0)* $b \in g[\varphi'] \implies b \in d[\varphi]$.
*(d1)* $b \in d[\varphi] \implies 1 - b \in d[\neg\varphi]$.
*(d2)* $1 \in d[\varphi_l] \cup d[\varphi_r] \implies 1 \in d[\varphi_l \vee \varphi_r]$.
*(d3)* $0 \in d[\varphi_l] \cap d[\varphi_r] \implies 0 \in d[\varphi_l \vee \varphi_r]$.
*(d4)* $1 \in \bigcup_{a \in \mathsf{Dom}\,\mathcal{A}} d[\theta\{\overline{a}/x\}] \implies 1 \in d[\exists x\theta]$.
*(d5)* $0 \in \bigcap_{a \in \mathsf{Dom}\,\mathcal{A}} d[\theta\{\overline{a}/x\}] \implies 0 \in d[\exists x\theta]$.

*The dual items for $\wedge$ and $\forall$ are omitted.*

**Proposition 9.** *Let $\theta, \varphi$ be closed $L(\mathcal{A})$-formulas. If $\theta \overset{\mathcal{A}}{\sim} \varphi$, then $d[\theta] = d[\varphi]$.*

**Definition 34.** *The* **existential information processing valuation induced by** *$R$, $d_R$, is the information processing valuation induced by $g_R$.*

**Proposition 10.** *For existential information processing, (d1), (d3) and (d5) hold in the other direction ($\Longleftarrow$) as well (likewise for their duals).*

These facts permit capturing the semantics of processors using a $\neg$-GNmatix.

**Definition 35.** *Let $\mathcal{QM}^4_E = \langle \mathcal{V}_4, \{\mathrm{t}, \top\}, \mathcal{QO}^4_E \rangle$, where $\mathcal{QO}^4_E$ is detailed below:*

| $a$ | $\tilde{\neg} a$ |
|---|---|
| $\bot$ | $\{\bot\}$ |
| f | $\{\mathrm{t}\}$ |
| t | $\{\mathrm{f}\}$ |
| $\top$ | $\{\top\}$ |

| $\tilde{\vee}$ | $\bot$ | f | t | $\top$ |
|---|---|---|---|---|
| $\bot$ | $\{\bot, \mathrm{t}\}$ | $\{\bot, \mathrm{t}\}$ | $\{\mathrm{t}\}$ | $\{\mathrm{t}\}$ |
| f | $\{\bot, \mathrm{t}\}$ | $\{\mathrm{f}, \top\}$ | $\{\mathrm{t}\}$ | $\{\top\}$ |
| t | $\{\mathrm{t}\}$ | $\{\mathrm{t}\}$ | $\{\mathrm{t}\}$ | $\{\mathrm{t}\}$ |
| $\top$ | $\{\mathrm{t}\}$ | $\{\top\}$ | $\{\mathrm{t}\}$ | $\{\top\}$ |

| $\tilde{\wedge}$ | $\bot$ | f | t | $\top$ |
|---|---|---|---|---|
| $\bot$ | $\{\bot, \mathrm{f}\}$ | $\{\mathrm{f}\}$ | $\{\bot, \mathrm{f}\}$ | $\{\mathrm{f}\}$ |
| f | $\{\mathrm{f}\}$ | $\{\mathrm{f}\}$ | $\{\mathrm{f}\}$ | $\{\mathrm{f}\}$ |
| t | $\{\bot, \mathrm{f}\}$ | $\{\mathrm{f}\}$ | $\{\mathrm{t}, \top\}$ | $\{\top\}$ |
| $\top$ | $\{\mathrm{f}\}$ | $\{\mathrm{f}\}$ | $\{\top\}$ | $\{\top\}$ |

| $h[X]$ | $\tilde{\exists}_X[h]$ |
|---|---|
| $\{\bot\}$ | $\{\bot, \mathrm{t}\}$ |
| $\{\bot, \mathrm{f}\}$ | $\{\bot, \mathrm{t}\}$ |
| $\{\mathrm{f}\}$ | $\{\mathrm{f}, \top\}$ |
| $\{\mathrm{f}, \top\}$ | $\{\top\}$ |
| $\{\top\}$ | $\{\top\}$ |
| else | $\{\mathrm{t}\}$ |

| $h[X]$ | $\tilde{\forall}_X[h]$ |
|---|---|
| $\{\bot\}$ | $\{\bot, \mathrm{f}\}$ |
| $\{\bot, \mathrm{t}\}$ | $\{\bot, \mathrm{f}\}$ |
| $\{\mathrm{t}\}$ | $\{\mathrm{t}, \top\}$ |
| $\{\mathrm{t}, \top\}$ | $\{\top\}$ |
| $\{\top\}$ | $\{\top\}$ |
| else | $\{\mathrm{f}\}$ |

**Corollary 3.** *$d_R$ is a $\mathcal{QM}^4_E$-legal $\mathcal{A}$-valuation.*

**Proposition 11.** *For every $\mathcal{QM}^4_E$-legal $\mathcal{A}$-valuation $v$ there is an $\mathcal{A}$-reservoir $R_v$ such that $v = d_{R_v}$.*

**Corollary 4.** *The set of all $\mathcal{QM}^4_E$-legal $\mathcal{A}$-valuations is identical to the set of all existential information processing valuations induced by $\mathcal{A}$-reservoirs.*

### 4.3 Proof System for the Logic Induced by Processors

A Gentzen-type system for processors is defined based on the propositional one from [2] using the same intuition for the quantifiers that was used for $\mathcal{QM}_E^4$.

**Definition 36.** $\mathbf{QG}^4_{\mathbf{EIP}}$ *is the full 4-quasi-canonical calculus for L with the following logical rules:*

$\neg$. $\{\neg\neg p_1 \Rightarrow\}/(\neg\neg\Rightarrow)$, $\{\Rightarrow \neg\neg p_1\}/(\Rightarrow\neg\neg)$.
$\vee$. $\{\Rightarrow p_1, p_2\}/(\Rightarrow\vee)$, $\{\neg p_1, \neg p_2 \Rightarrow\}/(\neg\vee\Rightarrow)$, $\{\Rightarrow \neg p_1, \Rightarrow \neg p_2\}/(\Rightarrow\neg\vee)$.
$\wedge$. $\{p_1, p_2 \Rightarrow\}/(\wedge\Rightarrow)$, $\{\Rightarrow p_1, \Rightarrow p_2\}/(\Rightarrow\wedge)$, $\{\Rightarrow \neg p_1, \neg p_2\}/(\Rightarrow\neg\wedge)$.
$\exists$. $\{\Rightarrow p_1(c_1)\}/(\Rightarrow\exists)$, $\{\neg p_1(c_1) \Rightarrow\}/(\neg\exists\Rightarrow)$, $\{\Rightarrow \neg p_1(v_1)\}/(\Rightarrow\neg\exists)$.
$\forall$. $\{p_1(c_1) \Rightarrow\}/(\forall\Rightarrow)$, $\{\Rightarrow p_1(v_1)\}/(\Rightarrow\forall)$, $\{\Rightarrow \neg p_1(c_1)\}/(\Rightarrow\neg\forall)$.

Figure 1 below presents the application forms of the logical rules of $\mathbf{QG}^4_{\mathbf{EIP}}$, where the usual restrictions on variables apply.

$$\frac{\Gamma, \varphi \Rightarrow \Delta}{\Gamma, \neg\neg\varphi \Rightarrow \Delta}\ (\neg\neg\Rightarrow) \qquad \frac{\Gamma \Rightarrow \varphi, \Delta}{\Gamma \Rightarrow \neg\neg\varphi, \Delta}\ (\Rightarrow\neg\neg)$$

$$\frac{\Gamma \Rightarrow \varphi, \psi, \Delta}{\Gamma \Rightarrow \varphi\vee\psi, \Delta}\ (\Rightarrow\vee) \qquad \frac{\Gamma \Rightarrow \varphi\{t/x\}, \Delta}{\Gamma \Rightarrow \exists x\varphi, \Delta}\ (\Rightarrow\exists)$$

$$\frac{\Gamma, \neg\varphi, \neg\psi \Rightarrow \Delta}{\Gamma, \neg(\varphi\vee\psi) \Rightarrow \Delta}\ (\neg\vee\Rightarrow) \qquad \frac{\Gamma, \neg\varphi\{t/x\} \Rightarrow \Delta}{\Gamma, \neg\exists x\varphi \Rightarrow \Delta}\ (\neg\exists\Rightarrow)$$

$$\frac{\Gamma \Rightarrow \neg\varphi, \Delta \quad \Gamma \Rightarrow \neg\psi, \Delta}{\Gamma \Rightarrow \neg(\varphi\vee\psi), \Delta}\ (\Rightarrow\neg\vee) \qquad \frac{\Gamma \Rightarrow \neg\varphi\{y/x\}, \Delta}{\Gamma \Rightarrow \neg\exists x\varphi, \Delta}\ (\Rightarrow\neg\exists)$$

$$\frac{\Gamma, \varphi, \psi \Rightarrow \Delta}{\Gamma, \varphi\wedge\psi \Rightarrow \Delta}\ (\wedge\Rightarrow) \qquad \frac{\Gamma, \varphi\{t/x\} \Rightarrow \Delta}{\Gamma, \forall x\varphi \Rightarrow \Delta}\ (\forall\Rightarrow)$$

$$\frac{\Gamma \Rightarrow \neg\varphi, \neg\psi, \Delta}{\Gamma \Rightarrow \neg(\varphi\wedge\psi), \Delta}\ (\Rightarrow\neg\wedge) \qquad \frac{\Gamma \Rightarrow \neg\varphi\{t/x\}, \Delta}{\Gamma \Rightarrow \neg\forall x\varphi, \Delta}\ (\Rightarrow\neg\forall)$$

$$\frac{\Gamma \Rightarrow \varphi, \Delta \quad \Gamma \Rightarrow \psi, \Delta}{\Gamma \Rightarrow \varphi\wedge\psi, \Delta}\ (\Rightarrow\wedge) \qquad \frac{\Gamma \Rightarrow \varphi\{y/x\}, \Delta}{\Gamma \Rightarrow \forall x\varphi, \Delta}\ (\Rightarrow\forall)$$

**Fig. 1.** The system $\mathbf{QG}^4_{\mathbf{EIP}}$ in standard form

**Theorem 2.** $\mathbf{QG}^4_{\mathbf{EIP}}$ *admits strong cut-elimination, and* $\mathcal{QM}_E^4$ *is strongly characteristic for it.*

*Proof.* One can mechanically check that $\mathbf{QG}^4_{\mathbf{EIP}}$ is coherent (e.g. see Example 3). It follows from Theorem 1 that $\mathbf{QG}^4_{\mathbf{EIP}}$ admits strong cut-elimination and that $\mathcal{M}_{\mathbf{QG}^4_{\mathbf{EIP}}}$ is characteristic for it. It remains to show that $\mathcal{QM}_E^4 = \mathcal{M}_{\mathbf{QG}^4_{\mathbf{EIP}}}$.

As an example, consider a non-empty prefix $X$ of $\mathbb{N}$ and a function $h : X \to \mathcal{V}$ with image $\{\bot, \mathrm{t}\}$. In $\mathcal{QM}_E^4$ one has $\tilde{\forall}_X [h] = \{\bot, \mathrm{f}\}$. For $\mathcal{M}_{\mathbf{QG}^4_{\mathbf{EIP}}}$ one must find which $\forall$-rules of $\mathbf{QG}^4_{\mathbf{EIP}}$ are members of $R_{\mathbf{QG}^4_{\mathbf{EIP}}} [\forall, X, h]$. Let $\mathcal{N}$ be a $L^1_1$-structure for $\mathcal{M}_4$ such that $\mathsf{Dom}\,\mathcal{N} = X$ and $p_1{}^{\mathcal{N}} = h$. Pick $\xi_\bot \in h^{-1}[\bot]$ and $\xi_\mathrm{t} \in h^{-1}[\mathrm{t}]$. Consider each $\forall$-rule of $\mathbf{QG}^4_{\mathbf{EIP}}$:

- If $c_1{}^{\mathcal{N}} = \xi_\mathrm{t}$, then $p_1{}^{\mathcal{N}} \left[c_1{}^{\mathcal{N}}\right] = \mathrm{t}$, and so $\mathcal{N} \models \{p_1(c_1) \Rightarrow\}$.
  Thus $\{p_1(c_1) \Rightarrow\} / (\forall \Rightarrow) \in R_{\mathbf{QG}^4_{\mathbf{EIP}}} [\forall, X, h]$.
- There exists an $\mathcal{N}$-substitution $\tau$ such that $(\tau[v_1])^{\mathcal{N}} = \xi_\bot$, so $\mathcal{N} \not\models \{\Rightarrow p_1(v_1)\}$.
  Thus $\{\Rightarrow p_1(v_1)\} / (\Rightarrow \forall) \notin R_{\mathbf{QG}^4_{\mathbf{EIP}}} [\forall, X, h]$.
- Note that $p_1{}^{\mathcal{N}} \left[c_1{}^{\mathcal{N}}\right] \in \{\mathrm{t}, \bot\}$, so $\tilde{\neg}^4 p_1{}^{\mathcal{N}} \left[c_1{}^{\mathcal{N}}\right] \in \{\mathrm{f}, \bot\}$, and so $\mathcal{N} \not\models \{\Rightarrow \neg p_1(c_1)\}$.
  Thus $\{\Rightarrow \neg p_1(c_1)\} / (\Rightarrow \neg\forall) \notin R_{\mathbf{QG}^4_{\mathbf{EIP}}} [\forall, X, h]$.

Therefore, in $\mathcal{M}_{\mathbf{QG}^4_{\mathbf{EIP}}}$, $\tilde{\forall}_X [h] = \bigcap \{F [\{p_1(c_1) \Rightarrow\} / (\forall \Rightarrow)]\} = \{\mathrm{f}, \bot\}$. The other cases are similar.

## 5 Conclusion and Future Research

We have shown that for a very wide class of quasi-canonical Gentzen-type proof systems, our syntactic criterion of coherence is equivalent to both strong cut-elimination and to strong soundness and completeness. Hence the task of proving cut-elimination (which is often rather difficult) now becomes very easy for systems in this class, since it involves only the trivial matter of verifying the coherence criterion. Using this result we extended the framework of Existential Information Processing to predicate logic with dual-arity quantifiers. Parallelizing the propositional case, non-deterministic semantics and a strongly sound and complete proof system were given for this extension, and the admissibility of strong cut-elimination for that system was shown.

There are several directions of further research following this paper.

- Including function symbols in the schematic representation language(s) from Definition 15 (not just constants) to express explicit dependencies between variables and terms in the application forms of canonical rules.
- Definition 20 only addresses systems in which there are no rules of type $(\neg \Rightarrow)$ or $(\Rightarrow \neg)$, however in [1] systems with one such rule (of a specific shape) are also considered, yielding systems for three-valued logics.[6] These systems require a bit more care in their analysis (c.f. [1, Definition 5.5] of $\bar{x}$-inconsistency where $x \in \{\mathrm{f}, \mathrm{t}, \top, \bot\}$). Still, we expect such 3-quasi-canonical systems could similarly be extended to first-order logic.

[6] The addition of more than one such rule is uninteresting as it result in a system that is either trivial or equivalent to a (non-quasi) canonical one.

- Theorem 1 may be seen as evidence that canonicity is a flexible concept, and so similar theorems may be provable for other kinds of Gentzen-type proof systems. The systems dealt with in [9] and [10] are natural candidates.
- The existential strategy is just one possible information gathering strategy. A more interesting one involves a reservoir equipped with an order indicating authority. This enables the *authoritative* strategy, in which information is gathered only from sources that have not been overruled by a superior one.
- Sources in a reservoir share the same algebra. This means they are all aware of the same individuals, and agree about the meaning of all function symbols. A generalization which captures situations where this is not the case would be interesting, and increase the usefulness of this framework.
- Formulas that are classically equivalent are not equivalent in this framework. For example, a source may assign 1 to $\varphi \vee (\psi \wedge \theta)$ yet assign 0 to $(\varphi \vee \psi) \wedge (\varphi \vee \theta)$. The issue is in mitigating this with minimal complications.

## References

1. Avron, A.: Quasi-canonical systems and their semantics. Synthese, pp. 1–19, December 2018. https://doi.org/10.1007/s11229-018-02045-0
2. Avron, A., Konikowska, B.: Finite-valued logics for information processing. Fund. Inform. **114**(1), 1–30 (2012)
3. Avron, A., Zamansky, A.: A triple correspondence in canonical calculi: strong cut-elimination, coherence, and non-deterministic semantics. In: Hirsch, E.A., Razborov, A.A., Semenov, A., Slissenko, A. (eds.) CSR 2008. LNCS, vol. 5010, pp. 52–63. Springer, Heidelberg (2008). https://doi.org/10.1007/978-3-540-79709-8_9
4. Avron, A., Zamansky, A.: Non-deterministic semantics for logical systems. In: Gabbay, D., Guenthner, F. (eds.) Handbook of philosophical logic, pp. 227–304. Springer, Dordrecht (2011). https://doi.org/10.1007/978-94-007-0479-4_4
5. Baader, F., Nipkow, T.: Term Rewriting and All That. Cambridge University Press, Cambridge (1999)
6. Belnap, N.D.: How a computer should think. In: Ryle, G. (ed.) Contemporary Aspects of Philosophy. Oriel Press, Stockfield (1977)
7. Belnap, N.D.: A useful four-valued logic. In: Dunn, J.M., Epstein, G. (eds.) Modern Uses of Multiple-Valued Logic, pp. 5–37. Springer, Dordrecht (1977). https://doi.org/10.1007/978-94-010-1161-7_2
8. Carnielli, W., Marcos, J., de Amo, S.: Formal inconsistency and evolutionary databases. Logic Logical Philos. **8**(8), 115–152 (2004). https://apcz.umk.pl/czasopisma/index.php/LLP/article/view/LLP.2000.008
9. Ciabattoni, A., Lahav, O., Zamansky, A.: Basic constructive connectives, determinism and matrix-based semantics. In: Brünnler, K., Metcalfe, G. (eds.) Automated Reasoning with Analytic Tableaux and Related Methods, pp. 119–133. Springer, Heidelberg (2011)
10. Ciabattoni, A., Terui, K.: Towards a semantic characterization of cut-elimination. Stud. Logica **82**(1), 95–119 (2006). https://doi.org/10.1007/s11225-006-6607-2
11. Dunn, J.M.: The Algebra of Intensional Logics. Ph.D. thesis, University of Pittsburgh (1966)

12. Gentzen, G.: Investigations into logical deduction. Am. Philos. Q. **1**(4), 288–306 (1964). http://www.jstor.org/stable/20009142
13. Schöter, A.: Evidential bilattice logic and lexical inference. J. Logic Lang. Inform. **5**(1), 65–105 (1996). https://doi.org/10.1007/BF00215627
14. Shoenfield, J.R.: Mathematical Logic. Addison-Wesley Pub. Co., Reading (1967)

# Bounded Sequent Calculi for Non-classical Logics via Hypersequents

Agata Ciabattoni(✉), Timo Lang, and Revantha Ramanayake

Technische Universität Wien (TU Wien), Vienna, Austria
{agata,timo,revantha}@logic.at

**Abstract.** Using substructural and modal logics as case studies, a uniform method is presented for transforming cut-free hypersequent proofs into sequent calculus proofs satisfying relaxations of the subformula property. As a corollary we prove decidability for a large class of commutative substructural logics with contraction and mingle, and get a simple syntactic proof of a well known result: the sequent calculus for **S5** is analytic.

## 1 Introduction

In 1935, Gentzen introduced the sequent calculi **LJ** and **LK** for intuitionistic and classical logic as alternatives to the prevailing axiomatic systems. For this purpose, he replaced the rule of *modus ponens* in the latter with the more general *cut rule*. His motivation was to obtain the *subformula property* (also called *analyticity*) which asserts that a proof need only contain subformulas of the end formula. This was achieved by exploiting the additional structure in the sequent calculus formalism to show the redundancy of the cut rule. Analyticity yields a strong restriction on the proof search space and it is this that is the key for using a proof calculus to prove metalogical results (e.g. decidability, complexity, interpolation, disjunction properties) and for automated reasoning.

Unfortunately, the sequent calculus is not expressive enough to support analyticity for most logics of interest. The structural proof theoretic response has been the development of numerous *exotic proof formalisms* (e.g. hypersequent, nested sequent, display, labelled calculi, tree-hypersequent)—typically extending the syntax of the sequent calculus—with the aim of regaining analyticity via cut-elimination. The *hypersequent calculus*, introduced independently by Mints [19], Pottinger [22] and Avron [1], is one of the most successful such formalisms. Cut-free hypersequent calculi have been presented for many non-classical logics that resist an analytic sequent calculus formulation. Especially noteworthy are the uniform constructions of cut-free hypersequent calculi via structural/modal rule extensions for commutative substructural [6] and modal [13,14,16] logics.

Many non-classical logics possess a cut-free calculus in some exotic formalism but such calculi tend to be less useful than cut-free sequent calculi because the presence of the extended structure is a hinderance to proving metalogical results.

Here we propose an alternative: *retain the sequent calculus and seek systematic relaxations of analyticity.* Of course, most logics will have a sequent calculus with arbitrary cuts that is complete for it, but this does not meaningfully


S. Cerrito and A. Popescu (Eds.): TABLEAUX 2019, LNAI 11714, pp. 94–110, 2019.
https://doi.org/10.1007/978-3-030-29026-9_6

restrict the proof search space. Therefore what we seek here a restriction on the 'quality' of the cut-formula in terms of shape, complexity and composition. Such a cut-restricted sequent calculus will be called a *bounded sequent calculus.*

In this work we obtain bounded sequent calculi by transforming cut-free hypersequent calculi. This is a natural starting point: hypersequents are simple extension of sequents (in fact, just one step further); the existing uniform constructions of cut-free hypersequent calculi can be exploited to obtain a uniform method for constructing bounded sequent calculi; and, given the novelty and inherent technicalities in our proposal, there is an advantage in simplifying one aspect of the problem by starting from proofs that already possess a nice structure (i.e. cut-free hypersequent proofs). The bounded sequent calculi that we obtain in this way are novel: a consideration of the quality of cut-formulas has never been attempted for logics lacking an analytic sequent calculus.

Specifically, we present a methodology to uniformly transform cut-free hypersequent calculi for a large class of propositional non-classical logics (substructural, intermediate and modal logics) into bounded sequent calculi. As a corollary we obtain the decidability of all acyclic $\mathcal{P}_3'$-axiomatic extensions (c.f. the substructural hierarchy [6]) of the commutative Full Lambek calculus with contraction and mingle [11] (including, e.g., **UML** [18]). This implies the decidability of the equational theory of the corresponding classes of residuated lattices [8]. We also obtain a simple and new syntactic proof of a well-known result [7]: analyticity of the sequent calculus for the modal logic **S5**. We note that the syntactic proof from the literature due to Takano [23] is highly intricate.

**Related Work.** Using algebraic methods, Bezhanishvili and Ghilardi [4] show that several modal logics satisfy the *bounded proof property*, a restriction on the modal complexity of formulas that need appear in a Hilbert-style proof. However all those logics already have well-known analytic sequent calculi. Bezhanishvili *et al.* [5] extend these methods to cut-free hypersequent calculi for intermediate logics. In particular, it is shown that is it possible to restrict hypersequent calculus proofs (with cuts) to proofs consisting of formulas whose implicational depth is bounded by the implicational depth of the endsequent. This is in the spirit of this work (systematic relaxations of analyticity), although here our aim is not only to restrict the formulas in the proof but to eliminate the hypersequent structure as well. Moreover, our methods apply also to substructural logics. Lahav and Zohar [15] establish syntactic criteria for determining if a pure sequent calculus has analyticity. They introduce a subformula property modulo leading negation symbols and provide a method for constructing analytic calculi for sub-logics of a base logic from simple derivable rules in the base calculus. In contrast, for us, relaxations of analyticity are *the* parameter for capturing extensions of the base logic. In this sense, analyticity is the lower-limit of our investigation: we are willing to give up analyticity in a carefully considered way, to preserve the sequent calculus formalism. Fitting [7] proved analyticity of the sequent calculus for several modal logics by logic-specific semantic argument and asked if the "theorems could be established by a more uniform approach".

Our methodology suggests that it may indeed be possible to obtain analyticity (and its relaxations) for modal logics in a uniform manner.

$$\frac{}{A \Rightarrow A}\ (id) \qquad \frac{A, \Delta \Rightarrow \Pi \quad \Gamma \Rightarrow A}{\Gamma, \Delta \Rightarrow \Pi}\ (cut) \qquad \frac{\Gamma \Rightarrow A_i}{\Gamma \Rightarrow A_1 \vee A_2}\ (\vee_R)_{i \in \{1,2\}} \qquad \frac{\Gamma, A \Rightarrow B}{\Gamma \Rightarrow A \rightarrow B}\ (\rightarrow_R)$$

$$\frac{}{\Rightarrow 1}\ (1_R) \qquad \frac{\Gamma \Rightarrow \Pi}{\Gamma, 1 \Rightarrow \Pi}\ (1_L) \qquad \frac{\Gamma \Rightarrow}{\Gamma \Rightarrow 0}\ (0_R) \qquad \frac{}{0 \Rightarrow}\ (0_L) \qquad \frac{}{\Gamma \Rightarrow \top}\ (\top) \qquad \frac{}{\Gamma, \bot \Rightarrow \Pi}\ (\bot)$$

$$\frac{\Gamma \Rightarrow A \quad \Delta \Rightarrow B}{\Gamma, \Delta \Rightarrow A \cdot B}\ (\cdot_R) \qquad \frac{\Gamma, A \Rightarrow \Pi \quad \Gamma, B \Rightarrow \Pi}{\Gamma, A \vee B \Rightarrow \Pi}\ (\vee_L) \qquad \frac{\Gamma, A, B \Rightarrow \Pi}{\Gamma, A \cdot B \Rightarrow \Pi}\ (\cdot_L)$$

$$\frac{\Gamma \Rightarrow A \quad \Delta, B \Rightarrow \Pi}{\Gamma, \Delta, A \rightarrow B \Rightarrow \Pi}\ (\rightarrow_L) \qquad \frac{\Gamma \Rightarrow A \quad \Gamma \Rightarrow B}{\Gamma \Rightarrow A \wedge B}\ (\wedge_R) \qquad \frac{\Gamma, A_i \Rightarrow \Pi}{\Gamma, A_1 \wedge A_2 \Rightarrow \Pi}\ (\wedge_L)_{i \in \{1,2\}}$$

**Fig. 1.** The single-conclusioned sequent calculus $\mathbf{FL_e}$

## 2 Preliminaries

In this paper we consider extensions of the commutative Full Lambek calculus $\mathbf{FL_e}$ (see Fig. 1), including intermediate and normal modal logics. The language of these logics may be inferred from their calculi. The connective $\cdot$ is called *fusion* (or multiplicative conjunction), e.g. [6,8]. A *sequent* is a tuple $(\Gamma, \Delta)$ of formula multisets (written as $\Gamma \Rightarrow \Delta$). It is *single-conclusioned* if $\Delta$ contains at most one formula, and *multi-conclusioned* otherwise. Throughout, $\neg A$ will abbreviate $A \rightarrow \bot$. $A, B, C, \ldots$ will be used for formulas/formula variables, $\Gamma, \Delta, \Pi, \ldots$ for formula multisets/formula multiset variables. $\Pi$ is taken to contain at most one formula. A $\Omega$-instantiation of a formula $A$ is a uniform substitution of the propositional variables of $A$ by elements from the set $\Omega$.

**Rules and Rule Instances.** An explicit distinction between a rule and a rule instance will be made only where required. An *instance of a rule* $(r)$ is denoted $\sigma(r)$, where $\sigma$ is a function mapping the structure variables in $(r)$ to concrete elements of the corresponding type. E.g. in an instance $\sigma(cut)$ of $(cut)$ (Fig. 1), $\sigma$ maps the multiset variables $\Gamma$ and $\Delta$ to (possibly empty) multisets of formulas, the formula variable $A$ to a formula, and the structure variable $\Pi$ to a multiset of formulas of size $\leq 1$.

**Axiomatic Extensions.** Let $\mathbf{S}$ be a sequent calculus and $\mathcal{F}$ a set of formulas. $\mathbf{S} + \mathcal{F}$ denotes the extension of $\mathbf{S}$ with initial sequents $\{\Rightarrow A | A \in \mathcal{F}\}$. Initial sequents are rules with no premises. Except in special cases, it is easily seen that $\mathbf{S} + \mathcal{F}$ fails cut-elimination even if $\mathbf{S}$ has cut-elimination.

**Derivability.** For a set $\mathcal{F} \cup \{S\}$ of sequents, $\mathcal{F} \vdash_{\mathbf{S}} S$ (resp. $\mathcal{F} \vdash^{cf}_{\mathbf{S}} S$) denotes that $S$ is derivable (resp. cut-free derivable) from $\mathcal{F}$ using the rule instances in $\mathbf{S}$. If $\mathcal{F} = \emptyset$, then we say that $S$ is derivable (cut-free derivable) and write $\vdash_{\mathbf{S}} S$ ($\vdash^{cf}_{\mathbf{S}} S$). Note: $\mathcal{F} \vdash_{\mathbf{S}} S$ denotes a derivation from a fixed set $\mathcal{F}$. In contrast, substitution instances of $\mathcal{F}$ *can* be used in $\vdash_{\mathbf{S}+\mathcal{F}} S$.

Let $\text{subf}(S)$ denote the set of subformulas in a formula/sequent $S$. For a multiset $\mathcal{F}$ of formulas, let $\odot\mathcal{F}$ be the fusion of all formulas in $\mathcal{F}$ (1 if $\mathcal{F}$ is empty). For sequent calculi where conjunction $\wedge$ and fusion $\cdot$ conflate (i.e. in the presence of contraction **c** and weakening **w**), we use just the single connective $\wedge$. Then $\odot\mathcal{F}$ is defined as a conjunction of all formulas in $\mathcal{F}$ ($\top$ if $\mathcal{F}$ is empty).

A *bounding function* is a map from a sequent to a set of formulas. In the following two definitions, **S** is a sequent calculus, $g$ is a bounding function, $S$ is an arbitrary sequent and $\mathcal{F}$ is a set of initial sequents of **S**.

**Definition 1 ($g$-, $(g, \mathcal{F})$-bounded derivation).** *A derivation of $S$ in* **S** *is* $g$-bounded *if every formula in the derivation is a subformula of an instantiation of an initial sequent of* **S** *by formulas in $g(S)$.*

$(g, \mathcal{F})$-bounded *if it is $g$-bounded and additionally every cut rule instance and every initial sequent instance $\Rightarrow A$ from $\mathcal{F}$ occurs together as shown below, where $A$ is a $g(\Gamma \Rightarrow \Pi)$-instantiation of a formula in $\mathcal{F}$.*

$$\frac{\Rightarrow A \qquad \Gamma, A \Rightarrow \Pi}{\Gamma \Rightarrow \Pi}\ (cut)$$

Intuitively, $g$-boundedness is a global relaxed-analyticity property on the derivation. Meanwhile, $(g, \mathcal{F})$-boundedness specifies also that cuts and initial sequent instances of $\mathcal{F}$ occur together and only together, and that the cut-formula satisfies a local relaxed-analyticity property.

The particular relaxation of analyticity is determined by the bounding function $g$. In particular, a $g_a$-bounded derivation of $S$ with $g_a(S) = \{A | A \in \text{subf}(S)\}$ is essentially an analytic derivation (but not quite, since subformulas of the initial sequents may also occur).

The global/local relaxed analyticity properties are analogous to the global/local subformula properties considered in Kowalski and Ono [12].

**Definition 2 ($g$-, $(g, \mathcal{F})$-bounded sequent calculus).** *A sequent calculus* **S** *is* $g$-bounded *($(g, \mathcal{F})$-bounded) if every sequent derivable in* **S** *has a $g$-bounded (resp. $(g, \mathcal{F})$-bounded) derivation.*

A $g$ - or $(g, \mathcal{F})$-bounded derivation/sequent calculus for some $g$ and $\mathcal{F}$ is referred to as a *bounded derivation/sequent calculus.*

For an associative binary connective $\heartsuit$, define the bounding functions:

$$g^{\heartsuit}(S) := \{A_1 \heartsuit \ldots \heartsuit A_n | A_i \in \text{subf}(S)\}$$
$$g^{1\heartsuit}(S) := \{A_1 \heartsuit \ldots \heartsuit A_n | A_i \in \text{subf}(S), \text{ and } A_i = A_j \text{ iff } i = j\}$$

Note that the set $g^{1\heartsuit}(S)$ is always finite, whereas $g^{\heartsuit}(S)$ is not. As an example, $g^{1\cdot}(p \Rightarrow q) = \{p, q, p \cdot q, q \cdot p\}$. A $g^{\heartsuit}$-bounded derivation of $S$ would only contain subformulas of instantiations of the initial sequents by formulas of the form $A_1 \heartsuit \ldots \heartsuit A_n$ where $A_i \in \text{subf}(S)$. A $g^{1\heartsuit}$-bounded derivation additionally requires that there is no repetition in $A_1, \ldots, A_n$.

**Hypersequent Calculi** are a generalisation of sequent calculi. Each proof rule is built from *hypersequents* i.e. finite multisets of sequents $S_1 \mid \ldots \mid S_n$. Each $S_i$ is said to be a *component* of the hypersequent.

Every sequent calculus **S** can be embedded into a hypersequent calculus **HS**; replace each rule $(r)$ in **S** with $(Hr)$ (see below) where the new structure variable $G$ can be instantiated with a hypersequent (possibly empty). In addition to the rules $(Hr)$, **HS** contains the structural rules of *external weakening* $(ew)$ and *external contraction* $(ec)$.

$$\frac{S_1 \quad \ldots \quad S_n}{S'}\ (r) \qquad \frac{G \mid S_1 \quad \ldots \quad G \mid S_n}{G \mid S'}\ (Hr) \qquad \frac{G}{G \mid S}\ (ew) \qquad \frac{G \mid S \mid S}{G \mid S}\ (ec)$$

The embedding is conservative, i.e. no new sequents are provable in **HS**.

Some axiomatic extensions of **S** cannot be captured analytically by extending **S** with sequent rules, but they can be captured analytically by extending **HS** with "proper" hypersequent rules that act on many sequents simultaneously.

*Example 3.* Let **lin** denote $(p \to q) \vee (q \to p)$. A sequent calculus for propositional *Gödel logic* is obtained by adding $\Rightarrow$ **lin** to Full Lambek calculus with exchange, contraction and weakening (denoted $\mathbf{FL_{ecw}}$, or **LJ**). Cut is ineliminable. A cut-free hypersequent calculus is obtained by adding (*com*) [2] to **HLJ**.

$$\frac{G \mid \Sigma_1, \Gamma_1 \Rightarrow \Pi_1 \quad G \mid \Sigma_2, \Gamma_2 \Rightarrow \Pi_2}{G \mid \Sigma_1, \Gamma_2 \Rightarrow \Pi_1 \mid \Sigma_2, \Gamma_1 \Rightarrow \Pi_2}\ (com)$$

Several [3,13,21] cut-free hypersequent calculi for **S5** have been given. E.g. in [3], a cut-free hypersequent calculus for **S5** is obtained by adding $(MS_{Av})$ to **HS4**.

$$\frac{G \mid \Gamma_1, \Box\Gamma_2 \Rightarrow \Box\Delta_2, \Delta_1}{G \mid \Gamma_1 \Rightarrow \Delta_1 \mid \Box\Gamma_2 \Rightarrow \Box\Delta_2}\ (MS_{Av})$$

In the above examples, the structure variable $G$ is called the *context*. The remaining components in the rule are called the *active components*.

## 3 A Guided Example Demonstrating the Methodology

We demonstrate by transforming a cut-free hypersequent derivation $d_h$ of $\Rightarrow F$ in **HLJ** + (*com*) into a bounded sequent derivation $d_s$ of $\Rightarrow F$ in **LJ** + **lin** through an example. Assume that $d_h$ contains a single instance of (com) above an instance of (ec). Then it has the following form:

$$\dfrac{\dfrac{\dfrac{\begin{matrix}\pi_1 \\ \Sigma_1, \Gamma_1 \Rightarrow \Pi_1\end{matrix} \qquad \begin{matrix}\pi_2 \\ \Sigma_2, \Gamma_2 \Rightarrow \Pi_2\end{matrix}}{\Sigma_1, \Gamma_2 \Rightarrow \Pi_1 | \Sigma_2, \Gamma_1 \Rightarrow \Pi_2}\ (\text{com})}{\begin{matrix}\cdots \\ \Gamma' \Rightarrow \Pi' | \Gamma' \Rightarrow \Pi'\end{matrix}}}{\begin{matrix}\Gamma' \Rightarrow \Pi' \\ \cdots \\ \Rightarrow F\end{matrix}}\ (\text{ec})$$

Construct the sequent derivation $d_s$ as follows, utilising portions of $d_h$:

$$
\begin{array}{c}
\begin{array}{cc}
\dfrac{\Gamma_2 \Rightarrow \wedge\Gamma_2 \qquad \begin{array}{c}\pi_1\\ \Sigma_1, \Gamma_1 \Rightarrow \Pi_1\\ \cdots\\ \Sigma_1, \wedge\Gamma_1 \Rightarrow \Pi_1\end{array}}{\wedge\Gamma_2 \rightarrow \wedge\Gamma_1, \Sigma_1, \Gamma_2 \Rightarrow \Pi_1}\rightarrow_L
&
\dfrac{\Gamma_1 \Rightarrow \wedge\Gamma_1 \qquad \begin{array}{c}\pi_2\\ \Sigma_2, \Gamma_2 \Rightarrow \Pi_2\\ \cdots\\ \Sigma_2, \wedge\Gamma_2 \Rightarrow \Pi_2\end{array}}{\wedge\Gamma_1 \rightarrow \wedge\Gamma_2, \Sigma_2, \Gamma_1 \Rightarrow \Pi_2}\rightarrow_L
\\
\cdots & \cdots
\end{array}
\\
\dfrac{\Rightarrow \alpha \qquad \dfrac{\wedge\Gamma_2 \rightarrow \wedge\Gamma_1, \Gamma' \Rightarrow \Pi' \qquad \wedge\Gamma_1 \rightarrow \wedge\Gamma_2, \Gamma' \Rightarrow \Pi'}{(\wedge\Gamma_2 \rightarrow \wedge\Gamma_1) \vee (\wedge\Gamma_1 \rightarrow \wedge\Gamma_2), \Gamma' \Rightarrow \Pi'}\vee_L}{\Gamma' \Rightarrow \Pi'}\text{cut}(\alpha)
\\
\cdots
\\
\Rightarrow F
\end{array}
$$

The cut formula $\alpha$ is $\sigma(\mathbf{lin})$ where $\sigma(p) = \wedge\Gamma_2$ and $\sigma(q) = \wedge\Gamma_1$. Since $d_h$ is cut-free: $\Gamma_1 \cup \Gamma_2 \subseteq \text{subf}(F)$. So $d_s$ is a $g^\wedge$-bounded derivation. By construction, the cut-rule occurs together with and only with the **lin** initial sequent instance. Furthermore, again because $d_h$ is cut-free: $\Gamma_1 \cup \Gamma_2 \subseteq \text{subf}(\Gamma' \cup \Pi')$. Thus a stronger result holds: $d_s$ is a $(g^\wedge, \{\mathbf{lin}\})$-bounded derivation.

## 4 The Disjunction Form of a Rule: A Formal Definition

Let us summarise the idea in the previous section. Given a cut-free hypersequent derivation of $\Rightarrow F$, we aim to obtain a sequent calculus derivation of each component of each hypersequent in it. If the hypersequent derivation contains an instance of a "proper" structural rule $(r)$, the sequent calculus derivation is forced to append to the LHS of its $i^{\text{th}}$ active conclusion component a suitable formula $D_i$. The formula $\vee_i D_i$ can be defined explicitly (Definition 9) from the form of $(r)$ such that it satisfies (Theorem 12) the properties of a *disjunction form* (Definition 5), which is formally defined in this section; these conditions make the transformation work. E.g. (provability) guarantees that $\vee_i D_i$ is no stronger than the axiom corresponding to $(r)$. Thus $\Rightarrow \vee_i D_i$ is used as an initial sequent without extending the logic. The disjunction form formulas are eliminated at the bottom of the sequent calculus derivation of $\Rightarrow F$ via bounded cuts on these initial sequents.

**Definition 4.** *For a multiset $\Delta$ of formulas, define $\Delta\#(\Gamma \Rightarrow \Pi)$ as $\Delta, \Gamma \Rightarrow \Pi$.*

Let $\Gamma_1, \ldots, \Gamma_m$ be the structure variables in a hypersequent rule $(r)$; associate with each $\Gamma_i$ a propositional variable $\widehat{\Gamma_i}$. Given an instantiation $\sigma$ on $(r)$, define the extended instantiation $\widehat{\sigma}$ which maps each $\widehat{\Gamma_i}$ to the formula $\odot\sigma(\Gamma_i)$.

**Definition 5 (disjunction form of a rule).** *Let* $\mathbf{H}$ *be a hypersequent calculus and $(r)$ a hypersequent rule with set $\mathcal{H}$ of premises and conclusion $G|S_1|\ldots|S_n$ built from the structure variables $\Gamma_1, \ldots, \Gamma_m$. A formula $A_1 \vee \ldots \vee A_n$ built from the propositional variables $\widehat{\Gamma_1}, \ldots, \widehat{\Gamma_m}$ is a* disjunction form *of $(r)$ if:*

*(splitting) For every rule instance $\sigma(r)$ and every $i \leq n$:*

$$\sigma(\mathcal{H}) \vdash^{cf}_{\mathbf{H}} \sigma(G \mid \widehat{\sigma}(A_i)\#S_i)$$

*(provability)* $\vdash_{\mathbf{H}+(r)} A_1 \vee \ldots \vee A_n$

We use the term "splitting" because the condition asserts that we can split the active components of a structural rule instance: the $i^{\text{th}}$ active component $\sigma(S_i)$ appended with the disjunct $\widehat{\sigma}(A_i)$ in the antecedent is cut-free derivable from the premises of the rule without using $(r)$. In effect:

$$\frac{\sigma(\mathcal{H})}{\sigma(G \mid S_1 \mid \ldots \mid S_n)}\,(r) \quad \rightsquigarrow \quad \left\{ \begin{array}{ccccc} \sigma(\mathcal{H}) & & & & \sigma(\mathcal{H}) \\ \vdots\ \mathbf{H} & & & & \vdots\ \mathbf{H} \\ \sigma(G \mid \widehat{\sigma}(A_1)\#S_1) & , & \ldots & , & \sigma(G \mid \widehat{\sigma}(A_n)\#S_n) \end{array} \right\}$$

There are pathological ways to obtain (*splitting*), for example by setting each $A_i$ as $\bot$. Such formulas are ruled out by the (*provability*) condition.

*Example 6.* $(\widehat{\Gamma}_2 \to \widehat{\Gamma}_1) \vee (\widehat{\Gamma}_1 \to \widehat{\Gamma}_2)$ is a disjunction form of (com) in Eg. 3.

## 5 Disjunction Forms for Commutative Substructural Logics

We show how to compute a disjunction form of analytic rules for substructural logics. The logics we consider are extensions of $\mathbf{FL_e}$ by axioms in the class $\mathcal{P}_3$ ($\mathcal{P}_3'$) of the *substructural hierarchy* [6]. Recall that the class $\mathcal{P}_3'$ is a modification of $\mathcal{P}_3$ used in absence of weakening. Let us write $B_{\wedge 1}$ to denote $B \wedge 1$. For $A = A_1 \vee \cdots \vee A_n$ (head connective of $A_i$ is not disjunction), set $A^{\vee} := (A_1)_{\wedge 1} \vee \cdots \vee (A_n)_{\wedge 1}$ and let $\mathcal{P}_3' := \{A^{\vee} | A \in \mathcal{P}_3\}$. Let $\mathcal{F}^{\vee}$ denote $\{A^{\vee} | A \in \mathcal{F}\}$.

**Definition 7 (amenable).** *A set $\mathcal{F}$ of formulas is* amenable *if (i) $\mathcal{F} \subseteq \mathcal{P}_3$ and contains weakening $p \cdot q \to p$, or (ii) $\mathcal{F} \subseteq \mathcal{P}_3'$ consists of acyclic formulas.*

The interest in amenable axiomatic extensions is that they admit a cut-free hypersequent calculus. This result is established in [6] and summarised below.

**Theorem 8 ([6]).** *From every finite set $\mathcal{F}$ of amenable formulas, a finite set $\mathcal{R}_{\mathcal{F}}$ of ('analytic') structural hypersequent rules can be computed such that*

$$\textit{for every sequent } S: \vdash_{\mathbf{FL_e}+\mathcal{F}} S \textit{ if and only if } \vdash^{cf}_{\mathbf{HFL_e}+\mathcal{R}_{\mathcal{F}}} S$$

E.g., $\mathcal{F} = \{p \cdot q \to p, \mathbf{lin}\}$ is an amenable set of formulas and $\mathcal{R}_{\mathcal{F}}$ is the set containing the rules of weakening and (*com*) (Example 3); hence $\mathbf{HFL_{ew}} + (com)$ is a cut-free hypersequent calculus for $\mathbf{FL_{ew}} + \mathbf{lin}$. Likewise, the set $\mathcal{F}' = \{\mathbf{lin}^{\vee}\}$ is amenable (where $\mathbf{lin}^{\vee} = (p \to q)_{\wedge 1} \vee (q \to p)_{\wedge 1}$), $\mathcal{R}_{\mathcal{F}'}$ is the rule (*com*), and so $\mathbf{HFL_e} + (com)$ is a cut-free hypersequent calculus for $\mathbf{FL_e} + \mathbf{lin}$.

Analytic structural hypersequent rules have one active component in each premise and additionally satisfy the following properties.

**(linear conclusion)** All structure variables in the conclusion are distinct.

**(separation)** No structure variable occurs both on the left hand side (LHS) and the right hand side (RHS) of a sequent.

**(coupling)** For each conclusion component with variable $\Pi$ on the RHS there is a variable $\Sigma$ on the LHS such that the pair $(\Sigma, \Pi)$ always occur together in the premises.

**(subformula property)** Each variable in the premise occurs in the conclusion.

**Computing the Disjunction Form of an Analytic Rule.** Select exactly one structure variable occurrence in the active component of each premise ('distinguished variable occurrence'). This induces an association of the distinguished variable (and the premise it is contained in) to the unique conclusion component containing this variable. We furthermore stipulate that every variable $\Sigma$ that is coupled (i.e. as $(\Sigma, \Pi)$ for some $\Pi$) is chosen as distinguished.

$$\frac{\{G \mid \mathcal{S}_{ij}, \mathbf{\Sigma}_i \Rightarrow \Pi_i\}_{i \in I, j \in J_i} \quad \{G \mid \mathcal{T}_{ijl}, \mathbf{\Gamma}_{ij} \Rightarrow\}_{i \in I, j \leq r_i, l \in M_{ij}} \quad \{G \mid \mathcal{U}_{ijl}, \mathbf{\Delta}_{ij} \Rightarrow\}_{i \in L, j \leq s_i, l \in N_{ij}}}{G \mid [\mathcal{V}_i, \mathbf{\Gamma}_{i1}, \dots, \mathbf{\Gamma}_{ir_i}, \mathbf{\Sigma}_i \Rightarrow \Pi_i]_{i \in I} \mid [\mathcal{W}_i, \mathbf{\Delta}_{i1}, \dots, \mathbf{\Delta}_{is_i} \Rightarrow]_{i \in L}}$$

**Fig. 2.** Association form. $\mathcal{S}, \mathcal{T}, \mathcal{U}, \mathcal{V}, \mathcal{W}$ denote multisets of structure variables. The distinguished variable occurrences in the premises and their associated occurrences in the components of the conclusion are indicated in boldface. The index sets $I, L, J_i, M_{ij}$ and $N_{ij}$ are assumed to be pairwise disjoint.

The analytic rule together with the choice of distinguished variables can be pictured in *association form* (see Fig. 2). Observe that:

- A structure variable declared as distinguished in a premise with empty RHS may appear in a conclusion component with or without empty RHS.
- Distinct premises may be associated to the same conclusion component, although not necessarily due to the same distinguished variable.
- Some conclusion components with empty RHS might not be associated to any premise (captured by the possibility that $s_i = 0$).
- The multisets $\mathcal{S}, \mathcal{T}$ and $\mathcal{U}$ may contain further (non-distinguished) occurences of the distinguished variables $\Gamma$ and $\Delta$, but no further occurences of $\Sigma$ due to the coupling property. The multisets $\mathcal{V}$ and $\mathcal{W}$ do not contain any further occurences of distinguished variables due to the linear conclusion property.

For a multiset $\mathcal{S} = \{\Gamma_1, \dots, \Gamma_n\}$ of structure variables, let $\widehat{\mathcal{S}}$ denote the multiset $\{\widehat{\Gamma}_1, \dots, \widehat{\Gamma}_n\}$ of propositional variables.

**Definition 9 (Form$(r, i)$).** *For a rule $(r)$ in association form (Fig. 2), let*

$$Form(r,i) := \left( \odot\widehat{\mathcal{V}}_i \cdot \odot \left\{ \widehat{\Gamma}_{ij} \wedge (\neg \bigvee_{l \in M_{ij}} \odot\widehat{\mathcal{T}}_{ijl}) \mid j \leq r_i \right\} \to \bigvee_{j \in J_i} \odot\widehat{\mathcal{S}}_{ij} \right)_{\wedge 1} \qquad (i \in I)$$

$$Form(r,i) := \left( \neg \left( \odot\widehat{\mathcal{W}}_i \cdot \odot \left\{ \widehat{\Delta}_{ij} \wedge (\neg \bigvee_{l \in N_{ij}} \odot\widehat{\mathcal{U}}_{ijl}) \mid j \leq s_i \right\} \right) \right)_{\wedge 1} \qquad (i \in L)$$

*Finally, let $Form(r) := \bigvee_{i \in I \cup L} Form(r, i)$.*

*Example 10.* Here are association forms of three well-known structural rules. In (com), the choice of distinguished variables $\Sigma_1$ and $\Sigma_2$ is determined by the coupling property. In (lq) and (wc), we could also have chosen $\Gamma$ resp. the second occurence of $\Delta$ as distinguished.

$$\frac{G \mid \Gamma_1, \mathbf{\Sigma_1} \Rightarrow \Pi_1 \quad G \mid \Gamma_2, \mathbf{\Sigma_2} \Rightarrow \Pi_2}{G \mid \Gamma_2, \mathbf{\Sigma_1} \Rightarrow \Pi_1 \mid \Gamma_1, \mathbf{\Sigma_2} \Rightarrow \Pi_2}\ (com) \qquad \frac{G \mid \mathbf{\Delta}, \Gamma \Rightarrow}{G \mid \mathbf{\Delta} \Rightarrow \mid \Gamma \Rightarrow}\ (lq) \qquad \frac{G \mid \mathbf{\Delta}, \Delta \Rightarrow}{G \mid \mathbf{\Delta} \Rightarrow}\ (wc)$$

*Example 11.* Consider the $(com)$ rule from Example 10. Pattern-matching the rule with Fig. 2 we obtain: $I = \{1, 2\}$, $L = \emptyset$, $\mathcal{V}_1 = \{\Gamma_2\}$, $\mathcal{V}_2 = \{\Gamma_1\}$, $J_1 = J_2 = \{1\}$, $\mathcal{S}_{11} = \{\Gamma_1\}$, $\mathcal{S}_{21} = \{\Gamma_2\}$, $r_1 = r_2 = 0$:

$$\text{Form}(com, 1) := \left( \odot\widehat{\mathcal{V}}_1 \cdot \odot\cancel{\{\widehat{\Gamma}_{1j} \wedge (\neg \textstyle\bigvee_{l \in M_{1j}} \odot\widehat{\mathcal{T}}_{1jl}) \mid j \leq r_1\}} \rightarrow \bigvee_{j \in \{1\}} \odot\widehat{\mathcal{S}}_{1j} \right)_{\wedge 1}$$

$$\text{Form}(com, 2) := \left( \odot\widehat{\mathcal{V}}_2 \cdot \odot\cancel{\{\widehat{\Gamma}_{2j} \wedge (\neg \textstyle\bigvee_{l \in M_{2j}} \odot\widehat{\mathcal{T}}_{2jl}) \mid j \leq r_2\}} \rightarrow \bigvee_{j \in \{1\}} \odot\widehat{\mathcal{S}}_{2j} \right)_{\wedge 1}$$

So $\text{Form}(com) = (\widehat{\Gamma}_2 \cdot 1 \rightarrow \widehat{\Gamma}_1)_{\wedge 1} \vee (\widehat{\Gamma}_1 \cdot 1 \rightarrow \widehat{\Gamma}_2)_{\wedge 1}$. Also:

$$\text{Form}(lq) = (\neg(1 \cdot (\widehat{\Delta} \wedge \neg\widehat{\Gamma})))_{\wedge 1} \vee (\neg(\widehat{\Gamma} \cdot 1))_{\wedge 1} \qquad \text{Form}(wc) = (\neg(1 \cdot (\widehat{\Delta} \wedge \neg\widehat{\Delta})))_{\wedge 1}$$

**Theorem 12.** *Form(r) is a disjunction form of the analytic rule (r).*

*Proof.* Given an analytic rule $(r)$, obtain Form$(r)$ from its association form. We require (c.f. Definition 5) (i) *provability*, i.e. $\vdash_{\mathbf{HFL_e}+(r)} \Rightarrow \text{Form}(r)$, and (ii) *splitting*.

(i) Apply the invertible rules $(ec)$, $(\vee_L)$, $(\rightarrow_R)$, $(\cdot_L)$ backwards from $\Rightarrow \text{Form}(r)$ to obtain the hypersequent below. The substitution $\sigma$ that makes it the conclusion of an instance $\sigma(r)$ of $(r)$ in association form (cf. Fig. 2) is obtained by pattern-matching (refer to variables shown above hypersequent below),

$$\left[ \widehat{\mathcal{V}}_i, \overbrace{\{\widehat{\Gamma}_{ij} \wedge (\neg \bigvee_{l \in M_{ij}} \odot\widehat{\mathcal{T}}_{ijl}) \mid j \leq r_i\}}^{\Gamma_{ij}} \overset{\Sigma_i}{\Rightarrow} \overbrace{\bigvee_{j \in I_j} \odot\widehat{\mathcal{S}}_{ij}}^{\Pi_i} \right]_{i \in I} \mid \left[ \widehat{\mathcal{W}}_i, \overbrace{\{\widehat{\Delta}_{ij} \wedge (\neg \bigvee_{l \in N_{ij}} \odot\widehat{\mathcal{U}}_{ijl}) \mid j \leq s_i\}}^{\Delta_{ij}} \Rightarrow \right]_{i \in L}$$

$$\sigma(G) := \emptyset \qquad \sigma(\Sigma_i) := \emptyset \qquad \sigma(\Pi_i) := \bigvee_{j \in J_i} \odot(\widehat{\mathcal{S}}_{ij}) \qquad (i \in I)$$

For $\mathcal{V}_i = \{Q_1, \ldots, Q_n\}$, set $\sigma(Q_s) := \widehat{Q}_s$

$$\sigma(\Gamma_{ij}) := \widehat{\Gamma}_{ij} \wedge \neg \bigvee_{l \in M_{ij}} \odot(\widehat{\mathcal{T}}_{ijl}) \qquad (i \in I, j \leq r_i)$$

For $\mathcal{W}_i = \{Q_1, \ldots, Q_n\}$, set $\sigma(Q_s) := \widehat{Q}_s$ $\qquad (i \in L)$

$$\sigma(\Delta_{ij}) := \widehat{\Delta}_{ij} \wedge \neg \bigvee_{l \in N_{ij}} \odot(\widehat{\mathcal{S}}_{ijl}) \qquad (i \in L, j \leq s_i)$$

Applying $\sigma(r)$ backwards to the hypersequent above, it remains to derive each premise of $\sigma(r)$ in $\mathbf{HFL_e}$.
We illustrate with the premise $G \mid \mathcal{S}_{ij}, \Sigma_i \Rightarrow \Pi_i$ of $(r)$ $(i \in I, j \in J_i)$. In $\sigma(r)$ this becomes $\sigma(\mathcal{S}_{ij}) \Rightarrow \bigvee_{j' \in J_i} \odot(\widehat{\mathcal{S}}_{ij'})$. Obtain $\sigma(\mathcal{S}_{ij}) \Rightarrow \odot(\widehat{\mathcal{S}}_{ij})$ using $(\vee_R)$. Let $\mathcal{S}_{ij} = \{P_1, \ldots, P_n\}$ (each $P_s$ is a structure variable). Applying $(\cdot_R)$ backwards to the latter sequent we obtain $\sigma(P_s) \Rightarrow \widehat{P}_s$ $(1 \le s \le n)$. It remains to verify derivability of the latter. Since $P_s$ occurs in the premise in the LHS, it must occur in the conclusion (subformula property) in the LHS (separation). Additionally it cannot be a $\Sigma$ variable (coupling). Therefore either $P_s \in \mathcal{V}_i$, $P_s \in \mathcal{W}_i$, $P_s = \Gamma_{uv}$ or $P_s = \Delta_{uv}$. In the first two cases, due to the definition of $\sigma(\mathcal{V}_i)$ and $\sigma(\mathcal{W}_i)$, we have the assignment $\sigma(P_s) := \widehat{P}_s$ and hence derivability. In the latter two cases we get $\widehat{\Gamma}_{uv} \wedge \neg \bigvee_{l \in M_{uv}} \odot(\widehat{\mathcal{T}}_{uvl}) \Rightarrow \widehat{\Gamma}_{uv}$ and $\widehat{\Delta}_{uv} \wedge \neg \bigvee_{l \in N_{uv}} \odot(\widehat{\mathcal{S}}_{uvl}) \Rightarrow \widehat{\Delta}_{uv}$ respectively. Applying $(\wedge_L)$ backwards we get $\widehat{\Gamma}_{uv} \Rightarrow \widehat{\Gamma}_{uv}$ and $\widehat{\Delta}_{uv} \Rightarrow \widehat{\Delta}_{uv}$.

(ii) Proving that Form$(r)$ satisfies (*splitting*) follows from a straightforward inspection so we simply set out what needs to be proved. Let $(r)$ be given as

$$\frac{\mathcal{H}}{G \mid [S_i]_{i \in I \cup L}}\ (r)$$

We have to show that for any instantiation $\sigma$ and for any $i \in I \cup L$, the hypersequent $\sigma(G \mid \widehat{\sigma}(\text{Form}(r,i)) \# S_i)$ is derivable from $\sigma(\mathcal{H})$ *without* invoking $(r)$ or $(cut)$. For $i \in I$, the hypersequent $\sigma(G \mid \widehat{\sigma}(\text{Form}(r,i)) \# S_i)$ is

$$\sigma(G) \mid \widehat{\sigma}(\text{Form}(r,i)), \sigma(\mathcal{V}_i), \sigma(\Gamma_{i1}), \ldots, \sigma(\Gamma_{ir_i}), \sigma(\Sigma_i) \Rightarrow \sigma(\Pi_i) \qquad (1)$$

From Definition 9 we have that $\widehat{\sigma}(\text{Form}(r,i))$ has the following form:

$$\left( \odot\sigma(\mathcal{V}_i) \cdot \odot \left\{ \sigma(\Gamma_{ij}) \wedge (\neg \bigvee_{l \in M_{ij}} \odot\sigma(\mathcal{T}_{ijl})) \mid j \le r_i \right\} \rightarrow \bigvee_{j \in I_j} \odot\sigma(\mathcal{S}_{ij}) \right)_{\wedge 1}$$

Now $\sigma(\mathcal{H}) \vdash^{cf}_{\mathbf{HFL_e}} (1)$ can be witnessed by decomposing $\widehat{\sigma}(\text{Form}(r,i))$. ∎

*Remark 13.* Form$(r)$ is not necessarily the $\mathcal{P}_3/\mathcal{P}'_3$ formula that generates $(r)$ and might not coincide with the formula obtained by suitably reversing the algorithm in [6]. E.g., in the guided example (Sect. 3) we used the formula $(\widehat{\Gamma}_2 \rightarrow \widehat{\Gamma}_1) \vee (\widehat{\Gamma}_1 \rightarrow \widehat{\Gamma}_2)$ as a disjunction form of (*com*), but our method computes a slightly different (though equivalent) form (Example 11). The advantage of the method given here is that it works uniformly for substructural and modal logics, and it does not require any familiarity with the algorithm in [6].

## 6 Bounded Calculi for Commutative Substructural Logics

Let $\mathcal{F}$ be a set of amenable axioms and $\mathcal{R}_{\mathcal{F}}$ the corresponding set of analytic structural hypersequent rules. In some cases (e.g. weakening, contraction axioms), the computed rule(s) may have just a single active component conclusion and hence they correspond to sequent structural rules. We call these *sequent axioms*, and they belong to the class $\mathcal{N}_2$ in the $(\mathcal{P}_i, \mathcal{N}_i)$ substructural hierarchy.

**Theorem 14.** *Let $\mathcal{F}_{seq} \cup \mathcal{F}$ be a finite set of amenable axioms such that $\mathcal{F}_{seq}$ is a set of sequent axioms with corresponding sequent rules $\mathcal{R}_{seq}$. Also set $\mathcal{F}' = \{Form(r) | r \in \mathcal{R}_{\mathcal{F}}\}$. For every sequent $S$, the following are equivalent:*

1. $\vdash_{\mathbf{FL_e}+\mathcal{F}_{seq}+\mathcal{F}} S$
2. $\vdash^{cf}_{\mathbf{FL_e}+\mathcal{R}_{seq}} \Gamma_S \# S$ *for a multiset $\Gamma_S$ of $g^{\cdot}(S)$-instantiations of elements in $\mathcal{F}'$.*
3. *$S$ has a $(g^{\cdot}, \mathcal{F}')$-bounded derivation in $\mathbf{FL_e} + \mathcal{R}_{seq} + \mathcal{F}'$.*

*Proof.* (1) $\Rightarrow$ (2). Suppose that $\vdash_{\mathbf{FL_e}+\mathcal{F}_{seq}+\mathcal{F}} S$. By Theorem 8: $\vdash^{cf}_{\mathbf{HFL_e}+\mathcal{R}_{\mathcal{F}_{seq}}+\mathcal{R}_{\mathcal{F}}} S$. Let $d_0$ be the hypersequent derivation witnessing the latter. Define the *rank* of a derivation in $\mathbf{HFL_e} + \mathcal{R}_{seq} + \mathcal{R}_{\mathcal{F}}$ as the maximum number of $\mathcal{R}_{\mathcal{F}}$-instances on a branch. We successively eliminate all bottommost occurrences of $\mathcal{R}_{\mathcal{F}}$, obtaining a hypersequent derivation of $\Gamma_S \# S$ where $\Gamma_S$ is an increasing (with each round of elimination/reduction of rank) multiset of $g^{\cdot}(S)$-instantiations of $\mathcal{F}'$.

First observe that since $d_0$ is cut-free, it has the following property:

($*$) every instance of a rule from $\mathcal{R}_{\mathcal{F}}$ instantiates its structure variables with a multiset of elements from $\mathrm{subf}(S)$

Identify the bottommost $\mathcal{R}_{\mathcal{F}}$-instances $\sigma_1(r_1), \ldots, \sigma_n(r_n)$ in $d_0$. Denote the conclusion of $\sigma_i(r_i)$ by $G|S_i^1|\ldots|S_i^{k_i}$. By Theorem 12, $\mathrm{Form}(r_i) = \bigvee_{j \leq k_i} \mathrm{Form}(r_i, j)$ is a disjunction form for $r_i$, i.e. a formula built from the structure variables in $r_i$ satisfying (*splitting*) and (*provability*) in Definition 5. From ($*$) we establish that each $\widehat{\sigma}_i(\mathrm{Form}(r_i, j))$ is an instantiation of $\mathrm{Form}(r_i, j)$ by formulas in $g^{\cdot}(S)$.

Set $\delta_1 := d_0$ and fix an $n$-tuple $(j_1, \ldots, j_n)$ satisfying $j_i \leq k_i$ $(i \leq n)$.

**for** $i = 1$ to $n$ **do**

Use (*splitting*) to obtain a derivation of $G|\widehat{\sigma}_i(\mathrm{Form}(r_i, j_i)) \# S_i^{j_i}$ using the derivations of the premises of $\widehat{\sigma}_i(r_i)$ in $d_0$. Now use (*ew*) to derive the following.

$$G|S_i^1|\ldots|\widehat{\sigma}_i(\mathrm{Form}(r_i, j_i)) \# S_i^{j_i}|\ldots|S_i^{k_i} \tag{2}$$

Replace the subderivation (in $\delta_i$) of the conclusion $G|S_i^1|\ldots|S_i^{k_i}$ of $\sigma_i(r_i)$ with the above derivation of (2). The result object is not yet a derivation. The following changes are required: when an additive rule or (*ec*) occurs below (2) (left column below), proceed as in the right column to add the missing formula. Here we are making use of the fact that every $\mathrm{Form}(r_i, j_i)$ has the form $B \wedge 1$ and hence can be inserted in the LHS using $(1_L)$ and $(\wedge_L)$.

$$\frac{G|\Gamma' \Rightarrow \Pi'|\Gamma' \Rightarrow \Pi'}{G|\Gamma' \Rightarrow \Pi'} \qquad \dfrac{\dfrac{G|\widehat{\sigma}_i(\mathrm{Form}(r_i, j_i)), \Gamma' \Rightarrow \Pi'|\Gamma' \Rightarrow \Pi'}{G|\widehat{\sigma}_i(\mathrm{Form}(r_i, j_i)), \Gamma' \Rightarrow \Pi'|\widehat{\sigma}_i(\mathrm{Form}(r_i, j_i)), \Gamma' \Rightarrow \Pi'}\,(1_L), (\wedge_L)}{G|\widehat{\sigma}_i(\mathrm{Form}(r_i, j_i)), \Gamma' \Rightarrow \Pi'}$$

$$\frac{G|\Gamma' \Rightarrow A \qquad G|\Gamma' \Rightarrow B}{G|\Gamma' \Rightarrow A \wedge B} \qquad \dfrac{G|\widehat{\sigma}_i(\mathrm{Form}(r_i, j_i)), \Gamma' \Rightarrow A \qquad \dfrac{G|\Gamma' \Rightarrow B}{G|\widehat{\sigma}_i(\mathrm{Form}(r_i, j_i)), \Gamma' \Rightarrow B}\,(1_L), (\wedge_L)}{G|\widehat{\sigma}_i(\mathrm{Form}(r_i, j_i)), \Gamma' \Rightarrow A \wedge B}$$

A derivation of $\widehat{\sigma}_1(\mathrm{Form}(r_1, j_1)), \ldots, \widehat{\sigma}_{i-1}(\mathrm{Form}(r_{i-1}, j_{i-1})), \widehat{\sigma}_i(\mathrm{Form}(r_i, j_i))\#S$ is obtained. Call this derivation $\delta_i'$ and set $\delta_{i+1} := \delta_i'$. **end for**

The output is a derivation of the hypersequent below left for every $(j_1, \ldots, j_n)$. Since $\widehat{\sigma}_i(\mathrm{Form}(r_i)) = \bigvee_{j \leq k_i} \widehat{\sigma}_i(\mathrm{Form}(r_i, j))$, repeatedly apply $(\vee_L)$ to this family of derivations to obtain ultimately the derivation $d_1$ of below right.

$$\widehat{\sigma}_1(\mathrm{Form}(r_1, j_1)), \ldots, \widehat{\sigma}_n(\mathrm{Form}(r_n, j_n))\#S \quad \widehat{\sigma}_1(\mathrm{Form}(r_1)), \ldots, \widehat{\sigma}_n(\mathrm{Form}(r_n))\#S$$

By construction, each $\widehat{\sigma}_i(\mathrm{Form}(r_i))$ is a $g^{\cdot}(S)$-instantiation of $\mathrm{Form}(r_i)$. Derivation $d_1$ was obtained from $d_0$ (without adding any cuts) by eliminating all bottommost $\mathcal{R}_{\mathcal{F}}$-instances, without modifying any non-bottommost $\mathcal{R}_{\mathcal{F}}$-instances. Thus $d_1$ is cut-free, has lesser rank than $d_0$ and satisfies $(*)$.

Identify the bottommost $\mathcal{R}_{\mathcal{F}}$ instances in $d_1$ and repeat the above argument, to obtain ultimately a cut-free derivation $d_N$ of $\Gamma_S\#S$ in $\mathbf{HFL_e} + \mathcal{R}_{seq} + \mathcal{R}_{\mathcal{F}}$ with rank 0, and hence also in $\mathbf{HFL_e} + \mathcal{R}_{seq}$. As the derivation contains no rules which act on more than one component in the premise or conclusion, we obtain the cut-free sequent derivation of $\Gamma_S\#S$ in $\mathbf{FL_e} + \mathcal{R}_{seq}$.

(2) $\Rightarrow$ (3). Given a cut-free derivation of $\{A_1, \ldots, A_n\}\#S$ in $\mathbf{FL_e} + \mathcal{R}_{seq}$ where each $A_i$ is a $g^{\cdot}(S)$-instantiations of some element in $\mathcal{F}'$, perform cuts on $A_1, \ldots, A_n$ to obtain a derivation of $S$. This derivation is $(g^{\cdot}, \mathcal{F}')$-bounded.

(3) $\Rightarrow$ (1). Theorem 12 states that each $\mathrm{Form}(r) \in \mathcal{F}'$ is derivable in $\mathbf{HFL_e} + \mathcal{R}_{\mathcal{F}}$. So it follows from (3) that $\vdash^{cf}_{\mathbf{HFL_e}+\mathcal{R}_{seq}+\mathcal{R}_{\mathcal{F}}} S$. Then Theorem 8 implies (1). ∎

**Corollary 15.** *$\mathbf{FL_e} + \mathcal{F}$ has a $(g^{\cdot}, \mathcal{F}')$-bounded sequent calculus for every finite set $\mathcal{F}$ of amenable axioms and $\mathcal{F}' = \{\mathrm{Form}(r) | r \in \mathcal{R}_{\mathcal{F}}\}$.*

### 6.1 Application: Decidability and Complexity Of $\mathbf{FL_{ecm}}$ Extensions

The bounded sequent calculi obtained above can be used to give a simple and uniform proof of decidability for every amenable axiomatic extension of $\mathbf{FL_{ecm}}$. Here $\mathbf{m}$ is the *mingle rule* corresponding to the sequent axiom $p \to p \cdot p$.

$$\frac{\Delta, \Gamma_1 \Rightarrow \Pi \quad \Delta, \Gamma_2 \Rightarrow \Pi}{\Delta, \Gamma_1, \Gamma_2 \Rightarrow \Pi}\,(m)$$

**Theorem 16.** *$\mathbf{FL_{ecm}} + \mathcal{F}$ is decidable for each finite set $\mathcal{F}$ of amenable axioms.*

*Proof.* Let $\mathcal{F}' := \{\mathrm{Form}(r) | r \in \mathcal{R}_{\mathcal{F}}\}$. Given a sequent $S$, let $\Gamma^*$ be the (finite) multiset of all $g^{1\cdot}(S)$-instantiations of $\mathcal{F}'$ without repeats. We claim that

$$\vdash_{\mathbf{FL_{ecm}}+\mathcal{F}} S \qquad \text{iff} \qquad \vdash_{\mathbf{FL_{ecm}}} \Gamma^*\#S$$

The result follows since $\mathbf{FL_{ecm}}$ is decidable [10, 11] and $\Gamma^*$ is computable from $S$. The direction right to left follows from (2) $\Longrightarrow$ (1) in Theorem 14.

For the other direction, (1) $\Longrightarrow$ (2) in Theorem 14 guarantees the existence of a multiset $\Gamma_S$ of $g^{\cdot}(S)$-instantiations of $\mathcal{F}'$ such that $\vdash_{\mathbf{FL_{ecm}}} \Gamma_S\#S$. Due to mingle and contraction in $\mathbf{FL_{ecm}}$, we have that $\vdash_{\mathbf{FL_{ecm}}} B \leftrightarrow B^n$ for every formula $B$ and $B^n = B \cdot \ldots \cdot B$ ($n \geq 1$ occurrences). It follows that for every $g^{\cdot}(S)$-instantiation $A$ of a formula in $\mathcal{F}'$, there is a $g^{1\cdot}(S)$-instantiation $A^1$ of the same formula such that $\vdash_{\mathbf{FL_{ecm}}} A^1 \Rightarrow A$. By applying cuts with such sequents $A^1 \Rightarrow A$ to the proof of $\vdash_{\mathbf{FL_{ecm}}} \Gamma_S \# S$, we obtain a derivation of $\Gamma'_S \# S$. Applying contractions to this sequent to remove repeated elements in $\Gamma'_S$, we obtain a derivation of $\Gamma''_S \# S$ such that $\Gamma''_S \subseteq \Gamma^*$. Now obtain $\Gamma^* \# S$ by introducing the elements in $\Gamma^* \setminus \Gamma''_S$ by $(1_L)$, $(\wedge_L)$ and $(\vee_L)$ (each $A^1 \in \Gamma^*$ has the form $(A_1)_{\wedge 1} \vee \ldots \vee (A_n)_{\wedge 1}$). ■

From the above proof we can also obtain a complexity upper bound. The size of $\Gamma^*$ in the above proof is $O(2^{|S|})$ and this multiset can be computed from $S$ in exponential time. It follows that the decision problem for each amenable extension of $\mathbf{FL_{ecm}}$ is at most exponentially greater than $\mathbf{FL_{ecm}}$.

Deciding if a formula is derivable in $\mathbf{FL_{ecm}}$ is known to be PSPACE-hard [9] but as far as we are aware, no upperbound has been presented in the literature. Let us sketch how to obtain an EXPTIME upperbound using forward proof search. In the presence of contraction and mingle, we can treat the antecedent of a sequent as a *set* instead of a multiset. In an analytic proof of a sequent $S$, there are at most $2^{|S|} \cdot |S|$ different sequents (with sets as antecedents) that could appear in the proof. Compute in successive steps which of these sequents is derivable in a proof with depth at most $1, 2, 3, \ldots, 2^{|S|} \cdot |S|$, terminating if a step does not derive any new sequents. Since each step except perhaps the last derives at least one new sequent, and since no more than $2^{|S|} \cdot |S|$ sequents may be derived, it follows that $S$ is derivable iff $S$ is encountered in one of these steps. Each step takes $O(2^{|S|})$ time so the entire procedure takes $O(2^{|S|}) \cdot 2^{|S|} \cdot |S|$ time and the EXPTIME upperbound follows.

In terms of the algebraic semantics [8], Theorem 16 establishes the decidability of the equational theory for the corresponding classes of residuated lattices.

*Example 17.* Our decidability result applies to a large class of logics including Uninorm Mingle Logic **UML** [18] (see [17] for an alternative proof of decidability) axiomatized as $\mathbf{FL_{ecm}} + (p \to q)_{\wedge 1} \vee (q \to p)_{\wedge 1}$, as well as $\mathbf{FL_{ecm}} + (p \cdot \neg p) \to p$ ($\subset$ **LJ**), and $\mathbf{FL_{ecm}} + (Bwk)$ ($k \geq 2$) where $(Bwk)$ is $\vee_{i=0}^{k}(p_i \to \vee_{j \neq i} p_j)_{\wedge 1}$.

The proof of Theorem 16 also yields the following refinement of Corollary 15.

**Lemma 18.** $\mathbf{FL_{ecm}} + \mathcal{F}$ *has a* $(g^{1\cdot}, \mathcal{F}')$*-bounded calculus for every finite set* $\mathcal{F}$ *of amenable axioms and* $\mathcal{F}' = \{Form(r) | r \in \mathcal{R}_{\mathcal{F}}\}$.

## 7 The Methodology Applied to Modal Logics

We shall extract bounded sequent calculi from cut-free hypersequent calculi for three normal modal logics. First, observe that the sequent calculus **S4** is obtained by the addition of the rules (T) and (4) to the multi-conclusioned sequent calculus **LK** for classical propositional logic. While **S4** has cut-elimination, the sequent calculus **S5** = **S4** + (5) famously fails cut-elimination [20] and, despite much effort, no natural cut-free sequent calculus for the logic has been found.

$$\frac{A, \Gamma \Rightarrow \Delta}{\Box A, \Gamma \Rightarrow \Delta}\,(T) \qquad \frac{\Box\Gamma \Rightarrow A}{\Box\Gamma \Rightarrow \Box A}\,(4) \qquad \frac{\Box\Gamma \Rightarrow A, \Box\Delta}{\Box\Gamma \Rightarrow \Box A, \Box\Delta}\,(5)$$

Let **HS4** denote the hypersequent version of the sequent calculus **S4**. Kurokawa [13] has shown that the hypersequent calculi in the first column below satisfy cut-elimination, and are sound and complete for the corresponding axiomatisations.

$$\begin{array}{ll}
\mathbf{HS4} + (RMS) & \mathbf{S4.2_{sc}} = \mathbf{S4} + \neg\Box\neg\Box A \to \Box\neg\Box\neg A \\
\mathbf{HS4} + (MC) & \mathbf{S4.3_{sc}} = \mathbf{S4} + \Box(\Box A \to B) \vee \Box(\Box B \to A) \\
\mathbf{HS4} + (MS) & \mathbf{S5_{sc}} = \mathbf{S4} + \neg\Box A \to \Box\neg\Box A
\end{array}$$

The rules (RMS), (MC) and (MS) are given below. The methodology in Sect. 5 has been used to identify the distinguished variables (highlighted in bold). Note: for this purpose we consider a term of the form $\Box\Gamma$ to be a single structure variable. Also let $\widehat{\Box\Gamma}$ denote a propositional variable.

$$\frac{G|\Box\boldsymbol{\Gamma}, \Box\Delta \Rightarrow}{G|\Box\boldsymbol{\Gamma} \Rightarrow |\Box\Delta \Rightarrow}\,(RMS) \quad \frac{G|\Gamma_1, \Box\boldsymbol{\Sigma_2} \Rightarrow \Pi_1 \quad G|\Gamma_2, \Box\boldsymbol{\Sigma_1} \Rightarrow \Pi_2}{G|\Gamma_1, \Box\boldsymbol{\Sigma_1} \Rightarrow \Pi_1|\Gamma_2, \Box\boldsymbol{\Sigma_2} \Rightarrow \Pi_2}\,(MC) \quad \frac{G|\Box\Gamma, \boldsymbol{\Delta} \Rightarrow \Pi}{G|\Box\Gamma \Rightarrow |\boldsymbol{\Delta} \Rightarrow \Pi}\,(MS)$$

Directly from Definition 9 we obtain:

$$\begin{aligned}
\mathrm{Form}(RMS) &= (\neg(1 \cdot (\widehat{\Box\Gamma} \wedge \neg\widehat{\Box\Delta})))_{\wedge 1} \vee (\neg(\widehat{\Box\Delta} \cdot 1))_{\wedge 1} \\
\mathrm{Form}(MC) &= (\widehat{\Box\Sigma}_1 \cdot 1 \to \widehat{\Box\Sigma}_2)_{\wedge 1} \vee (\widehat{\Box\Sigma}_2 \cdot 1 \to \widehat{\Box\Sigma}_1)_{\wedge 1} \\
\mathrm{Form}(MS) &= (\neg(\widehat{\Box\Gamma} \cdot 1))_{\wedge 1} \vee (\widehat{\Box\Gamma} \cdot 1)_{\wedge 1}
\end{aligned}$$

The modal case requires the following additional *uniform amendments*: (i) the $\wedge 1$ in every disjunct is replaced by a leading $\Box$, (ii) the 1 is omitted, (iii) a $\Box$ is introduced in front of every propositional variable $\widehat{\Box\Gamma}$, and (iv) every maximal subformula $\neg B$ with $B$ not boxed is substituted by $\neg\Box B$. Let $\mathrm{Form}^{\Box}(r)$ denote the image under these amendments. Then

$$\begin{aligned}
\mathrm{Form}^{\Box}(RMS) &= \Box\neg\Box(\Box\widehat{\Box\Gamma} \wedge \neg\Box\widehat{\Box\Delta}) \vee \Box\neg\Box(\Box\widehat{\Box\Delta}) \\
\mathrm{Form}^{\Box}(MC) &= \Box(\Box\widehat{\Box\Sigma}_1 \to \Box\widehat{\Box\Sigma}_2) \vee \Box(\Box\widehat{\Box\Sigma}_2 \to \Box\widehat{\Box\Sigma}_1) \\
\mathrm{Form}^{\Box}(MS) &= \Box\neg\Box(\Box\widehat{\Box\Gamma}) \vee \Box(\Box\widehat{\Box\Gamma})
\end{aligned}$$

The motivation for these amendments is that the analogue of Theorem 12 can now be verified by inspection for $r \in \{RMS, MC, MS\}$ i.e. *provability* of $\mathrm{Form}^{\Box}(r)$ in $\mathbf{HS4} + r$, and that $\mathrm{Form}^{\Box}(r)$ satisfies *splitting* in **HS4**.

For example, here is a derivation of $\text{Form}^{\Box}(RMS)$ in $\mathbf{HS4} + (RMS)$:

$$
\dfrac{\dfrac{\dfrac{\dfrac{\dfrac{\widehat{\Box\Delta} \Rightarrow \widehat{\Box\Delta}}{\neg\Box\widehat{\Box\Delta}, \Box\widehat{\Box\Delta} \Rightarrow}\;(\neg_L)}{\Box(\Box\widehat{\Box\Gamma} \wedge \neg\Box\widehat{\Box\Delta}), \Box(\Box\widehat{\Box\Delta}) \Rightarrow}\;(T),(\wedge_L)}{\Box(\Box\widehat{\Box\Gamma} \wedge \neg\Box\widehat{\Box\Delta}) \Rightarrow \;|\; \Box(\Box\widehat{\Box\Delta}) \Rightarrow}\;(RMS)}{\Rightarrow \Box\neg\Box(\Box\widehat{\Box\Gamma} \wedge \neg\Box\widehat{\Box\Delta}) \;|\; \Rightarrow \Box\neg\Box(\Box\widehat{\Box\Delta})}\;(4),(\neg_R)}{\Rightarrow \Box\neg\Box(\Box\widehat{\Box\Gamma} \wedge \neg\Box\widehat{\Box\Delta}) \vee \Box\neg\Box(\Box\widehat{\Box\Delta})}\;(ec),(\vee_L)
$$

For an instantiation $\sigma$ on $(r)$, define the extended instantiation $\hat{\sigma}$ which maps each $\widehat{\Box\Gamma}$ to the formula $\wedge\Box\sigma(\Gamma)$ (c.f. paragraph following Definition 4). Hence $\Box\sigma(\Gamma) \Rightarrow \hat{\sigma}(\widehat{\Box\Gamma})$ is derivable. The following witnesses the *splitting* of $\text{Form}^{\Box}(RMS)$ in **HS4**.

$$
\dfrac{\dfrac{\dfrac{\dfrac{\Box\sigma(\Gamma) \Rightarrow \hat{\sigma}(\widehat{\Box\Gamma})}{\Box\sigma(\Gamma) \Rightarrow \Box\hat{\sigma}(\widehat{\Box\Gamma})}\;(4) \qquad \dfrac{\dfrac{\dfrac{\Box\sigma(\Gamma), \Box\sigma(\Delta) \Rightarrow}{\Box\sigma(\Gamma), \hat{\sigma}(\widehat{\Box\Delta}) \Rightarrow}\;(\wedge_L)}{\Box\sigma(\Gamma), \Box\hat{\sigma}(\widehat{\Box\Delta}) \Rightarrow}\;(T)}{\Box\sigma(\Gamma) \Rightarrow \neg\Box\hat{\sigma}(\widehat{\Box\Delta})}\;(\neg_R)}{\dfrac{\Box\sigma(\Gamma) \Rightarrow \Box\hat{\sigma}(\widehat{\Box\Gamma}) \wedge \neg\Box\hat{\sigma}(\widehat{\Box\Delta})}{\Box\sigma(\Gamma) \Rightarrow \Box(\Box\hat{\sigma}(\widehat{\Box\Gamma}) \wedge \neg\Box\hat{\sigma}(\widehat{\Box\Delta}))}\;(4)}\;(\wedge_R)}{\Box\sigma(\Gamma), \neg\Box(\Box\hat{\sigma}(\widehat{\Box\Gamma}) \wedge \neg\Box\hat{\sigma}(\widehat{\Box\Delta})) \Rightarrow}\;(\neg_L)}{\Box\sigma(\Gamma), \Box\neg\Box(\Box\hat{\sigma}(\widehat{\Box\Gamma}) \wedge \neg\Box\hat{\sigma}(\widehat{\Box\Delta})) \Rightarrow}\;(T)
\qquad
\dfrac{\dfrac{\dfrac{\dfrac{\Box\sigma(\Delta) \Rightarrow \hat{\sigma}(\widehat{\Box\Delta})}{\Box\sigma(\Delta) \Rightarrow \Box\hat{\sigma}(\widehat{\Box\Delta})}\;(4)}{\Box\sigma(\Delta) \Rightarrow \Box(\Box\hat{\sigma}(\widehat{\Box\Delta}))}\;(4)}{\Box\sigma(\Delta), \neg\Box(\Box\hat{\sigma}(\widehat{\Box\Delta})) \Rightarrow}\;(\neg_L)}{\Box\sigma(\Delta), \Box\neg\Box(\Box\hat{\sigma}(\widehat{\Box\Delta})) \Rightarrow}\;(T)
$$

We can thus obtain the following (the proof is analogous to that for Theorem 14 and the strengthening in Lemma 18).

**Theorem 19.** *The calculus* $\mathbf{S4.2_{sc}}$ *(*$\mathbf{S4.3_{sc}}$*,* $\mathbf{S5_{sc}}$*) has a* $(g^{1\wedge}, \mathit{Form}^{\Box}(RMS))$*- (resp.* $(g^{1\wedge}, \mathit{Form}^{\Box}(MC))$*-,* $(g^{1\wedge}, \mathit{Form}^{\Box}(MS))$*-) bounded sequent calculus.*

### A New Syntactic Proof of Analyticity for S5

Although cut-elimination fails in **S5**, Takano [23] gave an intricate syntactic proof of analyticity by establishing that only cuts on subformulas are required. Prior to this, only a semantic argument was known, see Fitting [7].

Although the $(g^{1\wedge}, \text{Form}^{\Box}(MS))$-bounded sequent calculus we obtained in Theorem 19 has a finite proof search space and hence is suited for meta-theoretic argument, it is natural to ask if it is possible to modify the methodology of this paper to obtain Takano's (sharper) result for **S5**. We are able to answer this in the affirmative. Here is a simple and rather short proof of analyticity for **S5**.

**Theorem 20.** *The sequent calculus* **S5** *is analytic.*

*Proof.* Cut-free **HS4** + $(MS)$ derives exactly the same sequents as $\mathbf{S5_{sc}}$, and the latter is (of course) equivalent to **S5**. Moreover, any instance of the rule $(MS)$ can be replaced (without introducing cuts) using multiple instances of its single formula version $(MS_1)$ below, so cut-free **HS4** + $(MS_1)$ derives exactly the same sequents too.

$$\frac{G \mid \Box A, \Delta \Rightarrow \Pi}{G \mid \Box A \Rightarrow \mid \Delta \Rightarrow \Pi}\ (MS_1)$$

Consider a cut-free derivation $d$ in **HS4** + $(MS_1)$ of a sequent $S$. For simplicity, suppose that $d$ contains a single instance of $(MS_1)$ (in the general case, bottom-most instances of $(MS_1)$ are eliminated at each step, c.f. proof of Theorem 14). From this instance, we can obtain the following derivations in **HS4** + $(MS_1)$:

$$\frac{\Box A \Rightarrow \Box A}{G|\Box A \Rightarrow \Box A|\Delta \Rightarrow \Pi, \Box A}\ (\text{ew}) \qquad \frac{G|\Box A, \Delta \Rightarrow \Pi}{G|\Box A \Rightarrow |\Box A, \Delta \Rightarrow \Pi}\ (\text{ew})$$

Above left (right), the component $\Box A \Rightarrow \Box A$ (resp. $\Box A, \Delta \Rightarrow \Pi$) has a $\Box A$ in the RHS (resp. LHS) that was not present in the original derivation $d$. Proceed downward from each hypersequent following the rules in $d$, (propagating also these additional $\Box A$ formulas downwards from premise to conclusion).

This is not possible only if (4) is encountered as this rule permits only a single formula in the RHS and the additional $\Box A$ in the RHS would violate this. Solution: use (5) at this point instead of (4). In this way we obtain $(MS_1)$-free hypersequent derivations (hence in **HS5**) of $\Box A\#S$ and $S\#\Box A$ (the latter denotes that $\Box A$ is added to the RHS of $S$). Applying the cut-rule on $\Box A$ on these sequents, we obtain a derivation of $S$ in **HS5**. Since $\Box A$ occurred in the cut-free derivation $d$, it is a subformula of $S$. Finally: every rule in **HS5** has one active component, so we can extract an analytic derivation of $S$ in **S5**. ■

**Concluding Remark:** We have investigated what is required in terms of a relaxation of analyticity—represented via a bounding function—in order to trade the extended syntax of the hypersequent calculus for a sequent calculus. This paves the way for a new classification of logics based on their bounding functions. Identifying which functions are amenable to various meta-theoretic arguments (think decidability, interpolation and so on) could lead to the development of a common toolbox of methods applicable over a range of different logics. The immediate corollaries obtained in this paper demonstrate the potential of our approach. Finally, our transformations may provide a means of assessing the logical content of analyticity in the hypersequent calculus (by pegging it to its corresponding bounded sequent calculus), a problem hitherto unstudied.

**Acknowledgments.** Work supported by FWF projects: START 544-N2, W1255-N23 and I 2982. We thank B. Lellmann for bringing the forward proof search complexity argument to our attention.

## References


1. Avron, A.: A constructive analysis of RM. J. Symbol. Logic **52**(4), 939–951 (1987)
2. Avron, A.: Hypersequents, logical consequence and intermediate logics for concurrency. Ann. Math. Artif. Intell. **4**(3–4), 225–248 (1991)
3. Avron, A.: The method of hypersequents in the proof theory of propositional nonclassical logics. In: Logic: From Foundations to Applications (Staffordshire, 1993), pp. 1–32. Oxford Sci. Publ., Oxford Univ. Press, New York (1996)
4. Bezhanishvili, N., Ghilardi, S.: The bounded proof property via step algebras and step frames. Ann. Pure Appl. Logic **165**(12), 1832–1863 (2014)
5. Bezhanishvili, N., Ghilardi, S., Lauridsen, F.M.: One-step heyting algebras and hypersequent calculi with the bounded proof property. J. Logic Comput. **27**(7), 2135–2169 (2016)
6. Ciabattoni, A., Galatos, N., Terui, K.: From axioms to analytic rules in nonclassical logics. LICS **2008**, 229–240 (2008)
7. Fitting, M.: Subformula results in some propositional modal logics. Studia Logica **37**(4), 387–391 (1979), (1978)
8. Galatos, N., Jipsen, P., Kowalski, T., Ono, H.: Residuated lattices: analgebraic glimpse at substructural logics. Studies in Logic and the Foundations of Mathematics, vol. 151. Elsevier (2007)
9. Horcík, R., Terui, K.: Disjunction property and complexity of substructural logics. Theor. Comput. Sci. **412**(31), 3992–4006 (2011)
10. Hori, R., Ono, H., Schellinx, H.: Extending intuitionistic linear logic with knotted structural rules. Notre Dame J. Formal Logic‘ **35**(2), 219–242 (1994)
11. Kamide, N.: Substructural logics with mingle. J. Logic Lang. Inform. **11**(2), 227–249 (2002)
12. Kowalski, T., Ono, H.: Analytic cut and interpolation for bi-intuitionistic logic. Rev. Symbolic Logic **10**(2), 259–283 (2017)
13. Kurokawa, H.: Hypersequent calculi for modal logics extending S4. In: Nakano, Y., Satoh, K., Bekki, D. (eds.) JSAI-isAI 2013. LNCS (LNAI), vol. 8417, pp. 51–68. Springer, Cham (2014). https://doi.org/10.1007/978-3-319-10061-6_4
14. Lahav, O.: From frame properties to hypersequent rules in modal logics. In: LICS 2013, IEEE. pp. 408–417 (2013)
15. Lahav, O., Zohar, Y.: On the construction of analytic sequent calculi for subclassical logics. In: 21st International Workshop, WoLLIC 2014, Proceedings, pp. 206–220 (2014)
16. Lellmann, B.: Hypersequent rules with restricted contexts for propositional modal logics. Theor. Comput. Sci. **656**, 76–105 (2016)
17. Marchioni, E., Montagna, F.: On triangular norms and uninorms definable in $\L\pi\frac{1}{2}$. Int. J. Approx. Reason. **47**(2), 179–201 (2008)
18. Metcalfe, G., Montagna, F.: Substructural fuzzy logics. J. Symbolic Logic **72**(3), 834–864 (2007)
19. Minc, G.E.: Some calculi of modal logic. Trudy Mat. Inst. Steklov **98**, 88–111 (1968)
20. Ohnishi, M., Matsumoto, K.: Gentzen method in modal calculi. ii. Osaka Math. J. **11**(2), 115–120 (1959)
21. Poggiolesi, F.: A cut-free simple sequent calculus for modal logic S5. Rev. Symbolic Logic **1**(1), 3–15 (2008)
22. Pottinger, G.: Uniform, cut-free formulations of T, S4 and S5 (abstract). J. Symbolic Logic **48**(3), 900 (1983)
23. Takano, M.: Subformula property as a substitute for cut-elimination in modal propositional logics. Math. Jpn. **37**, 1129–1145 (1992)

# A Proof-Theoretic Perspective on SMT-Solving for Intuitionistic Propositional Logic

Camillo Fiorentini[1(✉)], Rajeev Goré[2], and Stéphane Graham-Lengrand[3]

[1] Department of Computer Science, University of Milan, Milan, Italy
fiorentini@di.unimi.it
[2] Research School of Computer Science, Australian National University, Canberra, Australia
[3] SRI International, Menlo Park, USA

**Abstract.** Claessen and Rósen have recently presented an automated theorem prover, `intuit`, for intuitionistic propositional logic which utilises a SAT-solver. We present a sequent calculus perspective of the theory underpinning `intuit` by showing that it implements a generalisation of the implication-left rule from the sequent calculus LJT, also known as G4ip and popularised by Roy Dyckhoff.

## 1 Introduction

Intuitionistic propositional logic IPL is one of the most important "non-classical" logics due to its constructive reading of implication. There is a long history of automated reasoning techniques for deciding validity of IPL-formulae, but most of them are based on either sequent or tableaux calculi. One of the simplest procedures for IPL is root-first *(a.k.a. backward or goal-directed)* proof search in the LJT (a.k.a. G4ip) sequent calculus [2], as it is guaranteed to terminate without implementing loop-checking.

Claessen and Rósen [1] have recently presented an automated theorem prover, `intuit`, for intuitionistic propositional logic, based on a Satisfiability-Modulo-Theories (SMT) approach. Their procedure also terminates without requiring any loop-detection machinery. As of 2015, the `intuit` prover was the best performing IPL prover, at least when evaluated on about 1200 standard benchmarks [1], which include for instance the ILTP library [12].

The SMT approach embraced by `intuit`, organised around the top-level loop of the DPLL($\mathcal{T}$) procedure [11], and the proof-theoretic approach based on root-first proof search, appear as radically different methodologies; the potential connections between them was left as an open question. In this paper we reconcile the two approaches, formalising an explicit connection. In particular, we reformulate a variant of `intuit` using (a suitable generalisation of) one of the rules of LJT. The procedure builds an explicit proof when the input formula is valid, and builds an explicit Kripke counter-model when the formula is not valid.


S. Cerrito and A. Popescu (Eds.): TABLEAUX 2019, LNAI 11714, pp. 111–129, 2019.
https://doi.org/10.1007/978-3-030-29026-9_7

## 2 Syntax and Kripke Semantics of IPL

In this paper, formulae of IPL, denoted by lowercase Greek letters, are built from an infinite set of propositional variables $V$, the "falsum" constant $\bot$ and the connectives $\wedge$, $\vee$, $\rightarrow$ with negation defined as $\neg\alpha := (\alpha \rightarrow \bot)$. We use $\mathrm{Atm} := V \cup \{\bot\}$ for the set of "atoms", denoted by lowercase Roman letters.

A rooted Kripke model for IPL is a quadruple $\langle W, \leq, r, \vartheta\rangle$ where $W$ is a non-empty set of "worlds" containing $r$, and $\leq$ is a reflexive and transitive binary relation over $W$, and the root world $r$ is minimal wrt $\leq$, and $\vartheta : W \mapsto 2^{\mathrm{Atm}}$ is a "valuation" mapping each world to a set of propositional variables which obeys the "persistence" condition: $\forall w, v \in W$, if $w \leq v$ and $p \in \vartheta(w)$ then $p \in \vartheta(v)$. Given a Kripke model $\langle W, \leq, r, \vartheta\rangle$, the valuation $\vartheta$ can be extended into a "forcing" relation between worlds and formulae as shown below:

$w \Vdash p$ iff $p \in \vartheta(w)$ $\qquad$ $w \Vdash \alpha \rightarrow \beta$ iff $\forall v \geq w$, $v \Vdash \alpha$ implies $v \Vdash \beta$
$w \Vdash \alpha \wedge \beta$ iff $w \Vdash \alpha$ and $w \Vdash \beta$ $\qquad$ $w \Vdash \bot$ never holds
$w \Vdash \alpha \vee \beta$ iff $w \Vdash \alpha$ or $w \Vdash \beta$

A formula $\alpha$ is IPL-valid if, for all Kripke models $\langle W, \leq, r, \vartheta\rangle$, we have $r \Vdash \alpha$. The problem of deciding whether a formula is IPL-valid is known to be PSPACE-complete [13]. For formula set or multiset $\Gamma$, we write $w \Vdash \Gamma$ for $\forall\gamma \in \Gamma . w \Vdash \gamma$.

A model $M$ for classical propositional logic (CPL), or "classical model", is just a set of propositional variables (assigned true). By $M \models \alpha$ we mean that $\alpha$ is true in model $M$ (following the Boolean truth tables). We write $M \models \Gamma$ iff $\forall\gamma \in \Gamma.\ M \models \gamma$. We write $\Gamma \vdash_{\mathrm{ipl}} \delta$ when the formula $\bigwedge \Gamma \rightarrow \delta$ is IPL-valid and $\Gamma \vdash_{\mathrm{cpl}} \delta$ when it is CPL-valid, that is, when $M \models \gamma$ for all classical models $M$.

## 3 The Theorem Prover `intuit`

The `intuit` theorem prover is an intuitionistic prover built on top of a SAT-solver, following a Satisfiability-Modulo-Theories (SMT) approach. Despite the fact that SMT-solving works primarily in classical logic and with first-order theories, Claessen and Rosén [1] show this approach to be relevant to the problem of deciding IPL-validity. They use a variant of the SMT scheme known as DPLL($\mathcal{T}$) [11], where DPLL is the well-known procedure for SAT-solving and $\mathcal{T}$ is here the "theory of intuitionistic implications". The main loop of DPLL($\mathcal{T}$) can be seen as a particular case of Counter-Example Guided Abstraction Refinement (CEGAR), as we describe next, for the particular case of IPL. A formula $\alpha$ whose IPL-validity is to be determined is transformed into a set $R$ of classical "flat clauses", a set $X$ of intuitionistic "implication clauses", and an atomic formula $q$, such that $\vdash_{\mathrm{ipl}} \alpha$ iff $R, X \vdash_{\mathrm{ipl}} q$ (the definitions are in Sect. 3.1). The *sequent* $R \Rightarrow q$ constitutes an "abstraction" of the input formula $\alpha$. A SAT-solver tries to find a classical counter-model $M$ for it, in that $M \models R$ but $M \not\models q$. If no such counter-model exists then $\alpha$ is not only CPL-valid but also IPL-valid. Otherwise the SAT-solver returns such a counter-model $M$, although the existence of $M$ does not necessarily mean that $R, X \nvdash_{\mathrm{ipl}} q$, as the implication clauses

```
1 procedure prove(R, X, q)
2 |   s = newSolver()
3 |   for φ ∈ R do addClause(s, φ)
4 |   for (a → b) → c ∈ X do addClause(s, b → c)
5 |   return intuitProve(s, X, ∅, q)
6 end
7 procedure intuitProve(s, X, A, q)
8 |   loop // CEGAR LOOP
9 |   |   τ₀ ← satProve(s, A, q)
10 |   |   if τ₀ = Yes(A') then return Yes(A')
11 |   |   else // τ₀ = No(M), with M a putative counter-model
12 |   |   |   if intuitCheck(s, X, M) then return No(M)
13 |   |   end
14 |   end
15 end
16 procedure intuitCheck(s, X, M)
17 |   for ι = (a → b) → c ∈ X such that a ∉ M and b ∉ M and c ∉ M do
18 |   |   τ₁ ← intuitProve(s, Xι, M ∪ {a}, b)  // Xι = X \ {ι}
19 |   |   if τ₁ = Yes(A₁) then
20 |   |   |   addClause(s, ⋀(A₁ \ {a}) → c)
21 |   |   |   return False
22 |   |   end
23 |   end
24 |   return True
25 end
```

**Fig. 1.** Main algorithms of intuit [1]

$X$ have so far been ignored. Therefore another procedure "checks" model $M$, in that it tries to produce a new abstraction $R' \Rightarrow q$ of $\alpha$ that refines $R \Rightarrow q$ (technically, $R \subseteq R'$) and *defeats* model $M$, meaning that $M \not\models R'$ while still ensuring $\vdash_{ipl} \alpha$ iff $R', X \vdash_{ipl} q$. If it fails, then indeed $\alpha$ is not IPL-valid. If it does produce a refinement, then the procedure loops with $R'$ instead of $R$. Eventually, it either finds a counter-model for $\alpha$, or exhausts the set of putative counter-models and conclude that $\alpha$ is IPL-valid.

A key element of the approach is that $R \vdash_{cpl} q$ iff $R \vdash_{ipl} q$, as the clauses in $R$ are "flat". A twist of the approach, compared to the standard DPLL($\mathcal{T}$) loop, is that the procedure that checks model $M$ has to solve a new IPL-validity problem (for a different $R, X, q$), so that it recursively calls a new DPLL($\mathcal{T}$) loop. In other words intuit implements a recursive version of DPLL($\mathcal{T}$), although a single SAT-solver is used, incrementally, for all recursive calls.

```
1  procedure intuitProve1(s, X, A, q)
2      loop
3          τ0 ← satProve(s, A, q)
4          if τ0 = Yes(A') then return Yes(A')
5          else  // τ0 = No(M)
6              for ι = (a → b) → c ∈ X such that a ∉ M and b ∉ M and c ∉ M do
7                  τ1 ← intuitProve1(s, X_ι, M ∪ {a}, b)  // X_ι = X \ {ι}
8                  if τ1 = Yes(A1) then
9                      addClause(s, ⋀(A1 \ {a}) → c)
10                     go to line 2 (outer loop)
11                 end
12             end
13             return No(M)
14         end
15     end
16 end
```

**Fig. 2.** The function `intuitProve1`

### 3.1 intuit in Detail

Firstly, the formula $\alpha$ is transformed into a formula $\bigwedge \Gamma \to q$, where $q$ is an atom and $\Gamma$ is a set of *flat clauses* $\varphi$ and *implication clauses* $\iota$, where:

$$\varphi \;::=\; \bigwedge A_1 \to \bigvee A_2 \qquad A_1 \cup A_2 \subseteq \text{Atm}$$
$$\iota \;::=\; (a \to b) \to c \qquad \{a,\, b,\, c\} \subseteq \text{Atm}$$

$\bigwedge A_1$ is the conjunction of the atoms in $A_1$

$\bigvee A_2$ is the disjunction of the atoms in $A_2$ if $A_1 = \emptyset$ then $\varphi = \bigvee A_2$

A flat clause where $A_1$ is empty is simply a disjunction $\bigvee A_2$ of atoms, and simply an atom $a$ when $A_2$ is the singleton $\{a\}$. Henceforth, we write $R$, $R_1$, $R'$ etc. to denote sets of flat clauses; $X$, $X_1$, $X'$ etc. for sets of implication clauses; $A$, $A_1$, $A'$ etc. for sets of atoms; and $X_\iota$, $X_{\iota'}$ for the sets $X \setminus \{\iota\}$ and $X \setminus \{\iota'\}$, respectively. A clausification procedure is presented in [1], similar to Tseitin's [15], where clauses are created by naming subformulae with new propositional atoms. Technically, it transforms any IPL formula $\alpha$ into a triple $\mathsf{clausal}(\alpha) = (R, X, q)$ whose cumulative size is linear in the size of $\alpha$ and that is equiprovable to $\alpha$:

**Lemma 1.** *For every $\alpha$ with $\mathsf{clausal}(\alpha) = (R, X, q)$, $\vdash_{\text{ipl}} \alpha$ iff $R, X \vdash_{\text{ipl}} q$ [1].*

From now on, we focus on deciding $R, X \vdash_{\text{ipl}} q$. The `intuit` algorithm [1], outlined in Fig. 1 with only slight modifications, consists of three procedures. It exploits a single SAT-solver $s$ that is *incremental*: clauses can be added to $s$ but not removed, and problems can be solved with varying atomic assumptions. So the clauses in $s$ are "global clauses" which must hold at any point of proof-search. Technically, the SAT-solver has the following API, i.e., it supports the following operations, where $\mathrm{R}(s)$ denotes the set of clauses stored in $s$:

```
procedure intuitPR(s, X, A, q)
    τ0 ← satProve(s, A, q)
    if τ0 = Yes(A') then return Yes(A')
    else  // τ0 = No(M)
        for ι = (a → b) → c ∈ X such that a ∉ M and b ∉ M and c ∉ M do
            τ1 ← intuitPR(s, Xι, M ∪ {a}, b)   // Xι = X \ {ι}
            if τ1 = Yes(A1) then
                addClause(s, ⋀(A1 \ {a}) → c)
                return intuitPR(s, X, A, q)
            end
        end
        return No(M)
    end
end
```

**Fig. 3.** The procedure `intuitPR` (recursive variant of `intuitProve`)

`newSolver()`: create a new SAT-solver;
`addClause(`$s$, $\varphi$`)`: add the flat clause $\varphi$ to the SAT-solver $s$;
`satProve(`$s$, $A$, $q$`)`: call the SAT-solver $s$ to decide whether $\mathrm{R}(s), A \vdash_{\mathrm{cpl}} q$, where $A$ is a local set of assumptions and $q$ is an atom.

The call `satProve(`$s$, $A$, $q$`)` yields one of the following answers:

Yes($A'$): thus $A' \subseteq A$ and $\mathrm{R}(s), A' \vdash_{\mathrm{cpl}} q$;
No($M$): thus $M$ is a classical model such that $M \models \mathrm{R}(s) \cup A$ and $M \not\models q$.

The main function `prove(`$R$, $X$, $q$`)` of `intuit` yields:

Yes($\emptyset$) if $R, X \vdash_{\mathrm{ipl}} q$;
No($M$) if there is a Kripke model $\mathcal{K} = \langle W, \leq, r, \vartheta \rangle$ such that $\vartheta(r) = M$ and $r \Vdash R \cup X$ and $r \nVdash q$; this implies $R, X \nvdash_{\mathrm{ipl}} q$.

As sketched by Claessen and Rósen [1], if `prove(`$R$, $X$, $q$`)` returns No($M$), then one can actually build the mentioned model $\mathcal{K}$ by tracking the sets $M'$ returned by `intuitProve`.

To reason about `intuit`, it is convenient to merge the functions `intuitProve` and `intuitCheck` into one recursive function. Firstly, we plug `intuitCheck` into `intuitProve` and obtain the function `intuitProve1` in Fig. 2. Then, we remove the outer loop by replacing the "go to" statement at line 10 with a recursive call; we get the recursive procedure `intuitPR` in Fig. 3. We henceforth consider the `intuit` algorithm as implemented by the main function `prove` in Fig. 1, with function `intuitProve` at line 5 replaced by function `intuitPR` in Fig. 3.

## 4 Adapting the Sequent Calculus LJT to Clausal Forms

The sequent calculus LJT is a variant of Gentzen's sequent calculus LJ for intuitionistic logic [6,7] that was discovered many times, as outlined by Roy Dyckhoff [2]. Its rules are given in Fig. 4, where $\Gamma \Rightarrow \alpha$ denotes a sequent whose *antecedent* $\Gamma$ is a multiset of assumptions, and whose provability in LJT is denoted $\vdash_{\mathrm{LJT}} \Gamma \Rightarrow \alpha$. The key difference from LJ lies in the left introduction rules for implication. In order to introduce an implication $\eta \to \gamma$ on the left, LJT offers four rules, depending on the form of $\eta$, namely: either $\eta = p$, with $p \in V$; or $\eta = \alpha \wedge \beta$; or $\eta = \alpha \vee \beta$; or $\eta = \alpha \to \beta$[1].

$$\frac{}{\Gamma, p \Rightarrow p}\,\mathrm{Ax} \qquad \frac{}{\Gamma, \bot \Rightarrow \delta}\,L\bot \qquad \frac{\Gamma, \alpha, \beta \Rightarrow \delta}{\Gamma, \alpha \wedge \beta \Rightarrow \delta}\,L\wedge \qquad \frac{\Gamma \Rightarrow \alpha \qquad \Gamma \Rightarrow \beta}{\Gamma \Rightarrow \alpha \wedge \beta}\,R\wedge$$

$$\frac{\Gamma, \alpha \Rightarrow \delta \qquad \Gamma, \beta \Rightarrow \delta}{\Gamma, \alpha \vee \beta \Rightarrow \delta}\,L\vee \qquad \frac{\Gamma \Rightarrow \alpha_k}{\Gamma \Rightarrow \alpha_1 \vee \alpha_2}\,R\vee_k \qquad \frac{\Gamma, \alpha \Rightarrow \beta}{\Gamma \Rightarrow \alpha \to \beta}\,R\to$$

$$\frac{\Gamma, p, \beta \Rightarrow \delta}{\Gamma, p, p \to \beta \Rightarrow \delta}\,L0\to \qquad \frac{\Gamma, \alpha \to (\beta \to \gamma) \Rightarrow \delta}{\Gamma, (\alpha \wedge \beta) \to \gamma \Rightarrow \delta}\,L\wedge\to \qquad \frac{\Gamma, \alpha \to \gamma, \beta \to \gamma \Rightarrow \delta}{\Gamma, (\alpha \vee \beta) \to \gamma \Rightarrow \delta}\,L\vee\to$$

$$\frac{\Gamma, \beta \to \gamma, \alpha \Rightarrow \beta \qquad \Gamma, \gamma \Rightarrow \delta}{\Gamma, (\alpha \to \beta) \to \gamma \Rightarrow \delta}\,L\to\to \qquad p \in V,\ k \in \{1, 2\}$$

**Fig. 4.** The calculus LJT (a.k.a. G4ip)

### 4.1 Root-First Proof Search, Invertibility and Recursivity

The purpose of replacing Gentzen's left-introduction of implication by those four rules is to ensure that root-first proof search terminates. When given a sequent to prove, root-first proof search matches it against the conclusion of one of the rules, and recursively tries to prove each of its premises. In every rule of LJT, the multiset $\Gamma_i, \alpha_i$ corresponding to the $i$th premise $\Gamma_i \Rightarrow \alpha_i$ is strictly smaller than the multiset $\Gamma, \alpha$ corresponding to the conclusion $\Gamma \Rightarrow \alpha$, according to the well-founded multiset ordering based on formula size.[2] Hence, the recursions of root-first proof search terminate. Note that keeping several copies of an assumption is never useful for proof search, so from now on, the antecedents of sequents will be considered sets. If, following the application of a rule, the recursive call that attempts to prove any one of its premises fails, then proof search attempts to apply another rule or another instance of the rule. Conceptually, a backtrack point was set when the original rule was applied. However, no such backtrack is needed if the rule is *invertible*, in the sense that, whenever the rule's conclusion is provable (in LJT), so is each of its premises. In that case indeed, if any one of the premises is not provable, then neither is the conclusion, so there is no point in trying out another rule. In LJT, all rules are invertible except $R\vee_k$ and $(L\to\to)$, and therefore backtrack points need to be set only when applying those two rules. However $(L\to\to)$ is *right-invertible* in that, if the conclusion is provable (in LJT), then so is the right premise (while the left premise may or

[1] We follow Troelstra and Schwichtenberg [14] where the calculus is called G4ip; the original $(L\to\to)$ rule by Dyckhoff [2] has $\Gamma, \beta \to \gamma \Rightarrow \alpha \to \beta$ as the left premise.

[2] This is the number of connectives, each conjunction counting for two [2].

may not be provable). This means that $(L\rightarrow\rightarrow)$ can be seen as a one-premise invertible rule, guarded by a side-condition:

$$(\Gamma,\beta\rightarrow\gamma,\alpha\Rightarrow\beta)\ \frac{\Gamma,\gamma\Rightarrow\delta}{\Gamma,(\alpha\rightarrow\beta)\rightarrow\gamma\Rightarrow\delta}.$$

When applying this rule, root-first proof search makes a first recursive call that checks that the side-condition holds and, if successful, makes a second one on the premise. Because of invertibility, that second call is *tail-recursive*, as no backtrack point is needed: the output of proof search is the output of the recursive call. The next sections describe how these two recursive calls, for a generalisation of rule $(L\rightarrow\rightarrow)$ satisfying the same invertibility properties, correspond to the two recursive calls of `intuitPR` in Fig. 3 (lines 6 and 9). The (right-)invertibility of that generalised rule means that the recursive call on the (right) premise is tail-recursive and proof-search can thus be implemented by a while loop: namely the DPLL($\mathcal{T}$) loop of `intuit`. To see this we specialise LJT to clausal forms.

### 4.2 LJT Specialised to Clausal Forms

We now consider sequents in clausal form, namely sequents of the form $R, X \Rightarrow q$. The only LJT rule manipulating $X$ is then $(L\rightarrow\rightarrow)$, which becomes:

$$\frac{R, b\rightarrow c, X_\iota, a\Rightarrow b \qquad R, X_\iota, c\Rightarrow q}{R, X\Rightarrow q}\qquad \begin{array}{l}\iota=(a\rightarrow b)\rightarrow c \text{ in } X\\ X_\iota = X\setminus\{\iota\}\end{array}$$

All other rules concern $R$ and $q$, or do not apply because the sequent to prove is already in clausal form. Hence, these rules can be replaced by the use of a SAT-solver, remembering that $R\vdash_{\text{cpl}} q$ iff $R\vdash_{\text{ipl}} q$.

To prove the left premiss of $(L\rightarrow\rightarrow)$, atom $a$ is added as an assumption, and will be taken into account by the SAT-solver. As $a$ is not present in the right premise, it must be removed from the assumptions if the same SAT-solver is used for the right premise. Hence, it is useful to refine the notion of sequent in clausal form into sequents of the form $R, X, A\Rightarrow q$, where $R$ is the set of clauses present in the SAT solver, which can only grow bigger, and $A$ is a set of atomic assumptions that can vary from one call to the next, relying on the SAT-solver's API presented in Sect. 3.1. On such sequents, rule $(L\rightarrow\rightarrow)$ becomes:

$$\frac{R, b\rightarrow c, X_\iota, A, a\Rightarrow b \qquad R, X_\iota, A, c\Rightarrow q}{R, X, A\Rightarrow q}\qquad \begin{array}{l}\iota=(a\rightarrow b)\rightarrow c \text{ in } X\\ X_\iota = X\setminus\{\iota\}\end{array}$$

We can start relating root-first proof search to `intuit` by relating the application of the above rule to a call to function `intuitPR`($s$,$X$,$A$,$q$), described in Fig. 3, considering $R = \mathrm{R}(s)$.

Indeed the left premise of the above rule is very similar to the first recursive call `intuitPR`($s$, $X_\iota$, $M\cup\{a\}$, $b$) on line 6 of Fig. 3, except that $b\rightarrow c$ is added to $R$ in the premise, and model $M$ may differ from $A$: According to its definition on line 4 of Fig. 3, $M$ must satisfy (and therefore contain) all atoms in $A$, but other atoms could be assigned true in $M$ that are not in $A$. Note however that

R($s$) does contain $b \to c$ if $(a \to b) \to c$ is in $X$: it has been added to $s$ at the beginning of the computation of function `prove` from Fig. 1.

Likewise, the right premise of the above rule is very similar to the second recursive call `intuitPR(`$s$, $X$, $A$, $q$`)` on line 9 of Fig. 3, except that the right premise contains the atomic formula $c$ and the recursive call keeps the implication $(a \to b) \to c$.

Furthermore, the use of the SAT-solver in procedure `intuitPR` has side-effects: let $R^0$ denote R($s$) at the time when `intuitPR` is called, $R^1$ denote R($s$) at the time of the first recursive call (line 6), and $R^2$ denote R($s$) at the time of the second recursive call (line 9). We have $R^0 \subseteq R^1 \subseteq R^1 \cup \{\bigwedge(A_1 \setminus \{a\}) \to c\} \subseteq R^2$, for a subset $A_1$ of $M \cup \{a\}$ such that $R^1, X_\iota, A_1 \vdash_{\mathrm{ipl}} b$. This incremental use of the SAT-solver is not reflected in root-first proof search using rule $(L{\to}{\to})$. Hence, rule $(L{\to}{\to})$ has to be generalised to account for these differences.

### 4.3 A Generalised Version of $(L{\to}{\to})$

The first generalisation consists in allowing the addition, in the left premise of $(L{\to}{\to})$, of extra atomic assumptions that were not assumptions in the conclusion. By allowing this, the rule can model either the extra atoms that are in $M$ but not in $A$ for the first recursive call of `intuitPR` (line 6 of Fig. 3), or the atoms $A_1$ that are returned by the call if successful. This ambiguity is systemic to the description of root-first proof search as presented in Sect. 4.1, which makes a double usage of sequent calculus rules. Consider a sequent calculus rule.

- Firstly, if the conclusion describes the arguments of a proof search, then the premises describe the arguments of the recursive calls; the rule describes the descent into the recursions.
- Secondly, if the recursive calls succeed, then a proof of each premise has been completed, either explicitly or implicitly, and a proof of the conclusion can be constructed; the rule describes the ascent back from the recursions.

It is useful to enhance proof search by considering, for each rule, a variant used for the first purpose and a variant used for the second purpose. Typically for the first purpose, it is convenient to integrate the weakening rule of the sequent calculus (say in LJ) to the axiom rule, and use context-sharing rules, as in Fig. 4. For the second purpose, it is useful to push all weakenings down towards the proof-tree root, using context-splitting rules. This strengthens the proved sequents by pruning the input sequent of all assumptions that were not used in the proof. This was described for instance in [8,9], which also connects the said pruning to the notion of *conflict analysis* used in SAT and SMT-solving. This is relevant for the connection between LJT and `intuit`, as a call to function `intuitPR(`$s$, $X$, $A$, $q$`)`, if successful, precisely performs this pruning by outputting a subset $A'$ of $A$ that is sufficient for provability. In that spirit, we present the generalisation of $(L{\to}{\to})$ in a context-splitting style, emphasising what happens upon the completion of the recursive calls.

The addition of extra atomic assumptions in the left premise of $(L{\to}{\to})$ makes the premise *easier* to prove than with the original rule (when seen as a one-premise rule with a side-condition, the rule applies more often). The price to pay for this is that the new assumption that is learnt and that helps proving the right premise, has to be weakened from $c$ to the flat clause $\varphi = \bigwedge(A_1 \setminus \{a\}) \rightarrow c$, which in `intuitPR` is added to the SAT-solver on line 8 in Fig. 3. Note that clause $\varphi$ is a consequence of the original problem. Weakening this newly available assumption in turn means that it no longer subsumes the original implication clause $\iota$, which has to stay in the right premise. The resulting rule is rule (ljt) in Fig. 5.

$$\frac{R \vdash_{\text{cpl}} q}{R, X \Rightarrow q}\ (\text{cpl}) \qquad \frac{R_1, X_1 \vdash_{\text{ipl}} \varphi \qquad \varphi, R_2, X_2 \Rightarrow q}{R_1, R_2, X_1, X_2 \Rightarrow q}\ (\text{cut}_{\text{ipl}})$$

$$\frac{R_1,\ b \rightarrow c,\ X,\ A_1 \Rightarrow b \qquad R_2,\ \bigwedge(A_1 \setminus \{a\}) \rightarrow c,\ X,\ (a \rightarrow b) \rightarrow c \Rightarrow q}{R_1,\ R_2,\ X,\ (a \rightarrow b) \rightarrow c \Rightarrow q}\ (\text{ljt})$$

**Fig. 5.** The calculus $\text{LJT}_{\text{SAT}}$

When emulating SAT or SMT-solving, the sequent calculus has to deal with the effectful aspect of these solvers, which learn clauses that are consequences of the input problem, such as clause $\varphi$ above.[3] This effect was described in terms of *memoisation* of root-first proof search in [8,9], and in terms of *cuts* in [4,9]. Here again we use cuts to model the phenomenon: the added clauses can be deleted at the end of the proof search computation by applying rule $(\text{cut}_{\text{ipl}})$ of Fig. 5.

## 5 The Calculus $\text{LJT}_{\text{SAT}}$

We introduce the calculus $\text{LJT}_{\text{SAT}}$ (LJT with SAT-solver) for sequents of the form $R, X \Rightarrow q$, whose provability in $\text{LJT}_{\text{SAT}}$ is denoted $\vdash_{\text{LJT}_{\text{SAT}}} R, X \Rightarrow q$. It consists of the three rules of Fig. 5. Rule (cpl) has the premise judgment $R \vdash_{\text{cpl}} q$ and the sequent $R, X \Rightarrow q$ as conclusion, with $X$ any set of implicational clauses. The rule can be applied if $R \vdash_{\text{cpl}} q$ holds, as checked by a SAT-solver (hence $\text{LJT}_{\text{SAT}}$). The rule $(\text{cut}_{\text{ipl}})$ is a cut rule having the judgment $R_1, X_1 \vdash_{\text{ipl}} \varphi$ as left premise. In the proof-search procedure, whenever we apply $(\text{cut}_{\text{ipl}})$, the left-premise is a judgment of the kind $R_0, X_0 \vdash_{\text{ipl}} \varphi$, where the cut formula $\varphi$ is a clause already stored in the SAT-solver, and we can take for granted that the assertion holds (we do not have to invoke an external prover to check it). The rule (ljt) is a sort of context-splitting generalisation of $(L{\rightarrow}{\rightarrow})$ needed to capture the `intuit` procedure. We point out that the sets $R_1$ and $R_2$ may overlap, thus the common part $R_1 \cap R_2$ is kept in both the premises. The formula $\bigwedge(A_1 \setminus \{a\}) \rightarrow c$ in the right premise is needed to guarantee the soundness of the rule, since $A_1$ is any set of atoms. To get completeness, we have to keep the main formula $(a \rightarrow b) \rightarrow c$ in the right premise; as a side-effect, the termination of proof-search is now trickier

[3] These effectful additions account for the distinction between $R^0$, $R^1$ and $R^2$ above.

to prove. If the sets $R_1$ and $R_2$ coincide and $A_1 = \{a\}$ (thus $\bigwedge(A_1 \setminus \{a\}) \to c$ is the atom $c$), we get the instance

$$\frac{R,\ b \to c,\ X,\ a \Rightarrow b \qquad R,\ c,\ X,\ (a \to b) \to c \Rightarrow q}{R,\ X,\ (a \to b) \to c \Rightarrow q}\ (\text{ljt})$$

The formula $(a \to b) \to c$ in the right premise is now redundant since it is implied by the occurrence of $c$, thus we recover Dyckhoff's rule $(L{\to}{\to})$.

To prove the soundness of $\text{LJT}_{\text{SAT}}$, we show that an $\text{LJT}_{\text{SAT}}$-derivation can be translated into the calculus LJT. In derivations, a double line marks the application of more than one rule. Firstly, we prove the soundness of rule (cpl).

**Lemma 2 (Soundness of rule** (cpl)**).** *If* $R \vdash_{\text{cpl}} q$ *then* $\vdash_{\text{LJT}} R \Rightarrow q$.

*Proof.* We proceed via contraposition, so suppose $\nvdash_{\text{LJT}} R \Rightarrow q$. By completeness, there is a Kripke model $\langle W, \leq, r, \vartheta \rangle$ containing a world $w \in W$ such that $w \Vdash \bigwedge R$ and $w \nVdash q$. Now consider any flat clause $\varphi = \bigwedge A_1 \to \bigvee A_2 \in R$. The valuation $\vartheta(w)$ either has $A_2 \cap \vartheta(w) \neq \emptyset$ or $A_2 \cap \vartheta(w) = \emptyset$. If $A_2 \cap \vartheta(w) \neq \emptyset$ then $\varphi$ is classically true at $w$. If $A_2 \cap \vartheta(w) = \emptyset$ then reflexivity demands $A_1 \not\subseteq \vartheta(w)$, as otherwise $w \nVdash \varphi$, contradicting our assumption. Again, $\varphi$ is classically true at $w$. That is, $w$ by itself is a classical model that also makes $\bigwedge R$ true and $q$ false, so $R \nvdash_{\text{cpl}} q$. By contraposition, if $R \vdash_{\text{cpl}} q$ then $\vdash_{\text{LJT}} R \Rightarrow q$. Notice that this proof only works because $R$ contains flat clauses. □

A syntactic way of proving Lemma 2 is to consider the proof returned by a (proof-producing) SAT-solver, justifying the unsatisfiability of $R, \overline{q}$ (where $\overline{q}$ is the negation of $q$) with a resolution proof concluding the empty clause $\bot$ from the flat clauses $R$, and $\overline{q}$. Indeed, the resolution rule is perfectly valid in intuitionistic logic if a clause $\overline{a_1} \vee \cdots \vee \overline{a_n} \vee b_1 \vee \cdots \vee b_m$ is read as the flat clause $a_1 \wedge \cdots \wedge a_n \to b_1 \vee \cdots \vee b_m$, as pointed out by Claessen and Rósen [1]. Removing $q \to \bot$ from the leaves of the resolution tree leaves $q$ at its root, yielding an intuitionistic proof of $R \vdash_{\text{ipl}} q$. Completeness of LJT [2] concludes $\vdash_{\text{LJT}} R \Rightarrow q$.

We prove the main lemma for the soundness of $\text{LJT}_{\text{SAT}}$.

**Lemma 3.** *If* $\vdash_{\text{LJT}_{\text{SAT}}} R, X \Rightarrow q$ *then* $\vdash_{\text{LJT}} R, X \Rightarrow q$.

*Proof.* Let $\mathcal{D}$ be an $\text{LJT}_{\text{SAT}}$-derivation of $R, X \Rightarrow q$; we prove the lemma by induction on the depth of $\mathcal{D}$. If the root rule of $\mathcal{D}$ is (cpl), the assertion follows by Lemma 2. Let us assume that $\mathcal{D}$ is

$$\frac{R_1,\ X_1 \vdash_{\text{ipl}} \varphi \qquad \overset{\mathcal{D}_2}{\varphi,\ R_2,\ X_2 \Rightarrow q}}{R_1,\ R_2,\ X_1,\ X_2 \Rightarrow q}\ (\text{cut}_{\text{ipl}})$$

By the completeness of LJT [2], there exists an LJT-derivation $\mathcal{E}_1$ of $R_1, X_1 \Rightarrow \varphi$. By the induction hypothesis, there exists an LJT-derivation $\mathcal{E}_2$ of $\varphi, R_2, X_2 \Rightarrow q$.

```
procedure LJTSatMain(R₀, X₀, q₀)
    // s, R₀, X₀ are global parameters
    s ← newSolver()
    for φ ∈ R₀ do addClause(s, φ)
    for (a → b) → c ∈ X₀ do addClause(s, b → c)
    τ ← LJTSat(R₀, X₀, ∅, q₀, [])
    if τ = 𝒦 then return 𝒦
    else // τ = (𝒟, R, ∅), with R = {φ₁,...,φₙ}
        return the LJT_SAT-derivation
```

$$\dfrac{R_0, X_0 \vdash_{\text{ipl}} \varphi_1 \quad \ldots \quad R_0, X_0 \vdash_{\text{ipl}} \varphi_n \qquad \begin{matrix}\mathcal{D}\\ R_0, R, X_0 \Rightarrow q_0\end{matrix}}{R_0, X_0 \Rightarrow q_0}\ (\text{cut}^*_{\text{ipl}})$$

```
    end
end
```

**Fig. 6.** The procedure `LJTSatMain`

Since (cut) is admissible in LJT [2], from $\mathcal{E}_1$ and $\mathcal{E}_2$ we get an LJT-derivation of $R_1, R_2, X_1, X_2 \Rightarrow q$. Otherwise, $\mathcal{D}$ has the form

$$\dfrac{\begin{matrix}\mathcal{D}_1\\ R_1,\ b \to c,\ X_\iota,\ A_1 \Rightarrow b\end{matrix} \qquad \begin{matrix}\mathcal{D}_2\\ R_2,\ \varphi,\ X \Rightarrow q\end{matrix}}{R_1,\ R_2,\ X \Rightarrow q}\ (\text{ljt}) \qquad \begin{aligned} \iota &= (a \to b) \to c\\ X_\iota &= X \setminus \{\iota\}\\ \varphi &= \textstyle\bigwedge(A_1 \setminus \{a\}) \to c\end{aligned}$$

Obtaining $\mathcal{E}_1$ (resp. $\mathcal{E}_2$) from $\mathcal{E}_1$ (resp. $\mathcal{D}_2$) by the induction hypothesis, we get the following LJT-derivation of $R, X \Rightarrow q$. We use here the fact that in LJT, weakenings are admissible (so we can assume $a \in A_1$), and so are cuts.

$$\dfrac{\dfrac{\dfrac{\begin{matrix}\mathcal{E}_1\\ R_1,\ b \to c,\ X_\iota,\ A_1 \Rightarrow b\end{matrix} \qquad \dfrac{}{R_1,\ X_\iota,\ c,\ A_1 \setminus \{a\} \Rightarrow c}\ \text{Ax}}{R_1,\ X,\ A_1 \setminus \{a\} \Rightarrow c}\ L{\to}{\to}}{R_1,\ X \Rightarrow \varphi}\ L\wedge,\ R\to \qquad \begin{matrix}\mathcal{E}_2\\ R_2,\ \varphi,\ X \Rightarrow q\end{matrix}}{R_1,\ R_2,\ X \Rightarrow q}\ (\text{cut})$$

□

By Lemma 3 and the soundness of LJT, we conclude:

**Theorem 1 (Soundness of $\text{LJT}_{\text{SAT}}$).** *$\vdash_{\text{LJT}_{\text{SAT}}} R, X \Rightarrow q$ implies $R, X \vdash_{\text{ipl}} q$.*

## 6 Proof-Search Using $\text{LJT}_{\text{SAT}}$

We present the proof-search procedure based on the calculus $\text{LJT}_{\text{SAT}}$, implemented by the main function `LJTSatMain` (Fig. 6), which exploits the auxiliary recursive function `LJTSat` (Fig. 7). They correspond to the functions `prove` and `intuitPR` respectively, enhanced with explicit proof/counter-model-construction. The worlds of the counter-models to be constructed are sequences of implication clauses, with [] denoting the empty sequence and $w{::}\iota$ denoting

**procedure** `LJTSat`($R$, $X$, $A$, $q$, $r$)
  $\tau_0 \leftarrow$ `satProve`($s$, $A$, $q$)
  **if** $\tau_0 = \text{Yes}(A')$ **then**
    **return** ($\mathcal{D}$, R($s$), $A'$) where $\mathcal{D}$ is

$$\frac{\mathrm{R}(s), A' \vdash_{\text{cpl}} q}{\mathrm{R}(s), X, A' \Rightarrow q}\ (\text{cpl})$$

  **else** // $\tau_0 = \text{No}(M)$
    $\Theta \leftarrow \emptyset$ // `Empty family of Kripke models`
    **for** $\iota = (a \to b) \to c \in X$ *such that* $a \notin M$ *and* $b \notin M$ *and* $c \notin M$ **do**
      $\tau_1 \leftarrow$ `LJTSat`($R \cup \{b \to c\}$, $X_\iota$, $M \cup \{a\}$, $b$, $r{::}\iota$) // `where` $X_\iota = X \setminus \{\iota\}$
      **if** $\tau_1 = \mathcal{K}_1$ **then** $\Theta \leftarrow \Theta \uplus (\iota \mapsto \mathcal{K}_1)$ // `adding` $\mathcal{K}_1$ `with index` $\iota$
      **else** // $\tau_1 = (\mathcal{D}_1, R_1, A_1)$
        $\tilde{\varphi} \leftarrow \bigwedge(A_1 \setminus \{a\}) \to c$
        `addClause`($s$, $\tilde{\varphi}$)
        $\tau_2 \leftarrow$ `LJTSat`($R \cup \{\tilde{\varphi}\}$, $X$, $A$, $q$, $r$)
        **if** $\tau_2 = \mathcal{K}_2$ **then return** $\mathcal{K}_2$
        **else** // $\tau_2 = (\mathcal{D}_2, R_2, A_2)$
          **return** ($\mathcal{D}$, $R_1 \cup R_2$, $A_2$) where $\mathcal{D}$ is

$$\frac{\begin{array}{c}\mathcal{D}_1\\ R, R_1, b \to c, X_\iota, A_1 \Rightarrow b\end{array} \qquad \begin{array}{c}\mathcal{D}_2\\ R, R_2, \tilde{\varphi}, X, A_2 \Rightarrow q\end{array}}{R,\ R_1,\ R_2,\ X,\ A_2 \Rightarrow q}\ (\text{ljt})$$

        **end**
      **end**
    **end**
    **return** $\text{Mod}(r, \Theta, M)$
  **end**
**end**

**Fig. 7.** The procedure `LJTSat`

the extension of sequence $w$ with clause $\iota$. Function `LJTSat` takes as its last argument the root world of the counter-model to be constructed, if the input is not IPL-valid. Initially, `LJTSatMain` calls `LJTSat` with the empty sequence as root world, as shown in Fig. 6, where $(\text{cut}^*_{\text{ipl}})$ also denotes a chain of $n$ successive applications of $(\text{cut}_{\text{ipl}})$.

A Kripke model $\mathcal{K} = \langle W, \leq, r, \vartheta \rangle$ is a *counter-model* for a sequent $\sigma = R, X \Rightarrow q$, written $\mathcal{K} \not\models \sigma$, if $r \Vdash R \cup X$ and $r \nVdash q$. In our incarnation of `intuit`, the counter-model is obtained by gluing Kripke models, as explained next.

Let $\Theta$ be a family $(\mathcal{K}_i)_{i \in I}$ of Kripke models and $M$ a classical model such that, for every $i \in I$, the root $r_i$ of $\mathcal{K}_i$ obeys $r_i \Vdash M$. We write $\text{Mod}(r, \Theta, M)$ for the model $\mathcal{K} = \langle W, \leq, r, \vartheta \rangle$ obtained by gluing all the models in $\Theta$ over $r$, where $r$ is a new world such that $\vartheta(r) = M$. More specifically, if $\Theta = (\langle W_i, \leq_i, r_i, \vartheta_i\rangle)_{i\in I}$, where the sets $W_i$ are pairwise disjoint and none of them contains $r$, then $\mathrm{Mod}(r, \Theta, M)$ is the model $\mathcal{K} = \langle W, \leq, r, \vartheta\rangle$ such that

$$W = \{r\} \uplus \biguplus_{i=1}^{n} W_i \quad \leq_0 = \{\,(r, r_1), \cdots, (r, r_n)\,\} \cup \bigcup_{i=1}^{n} \leq_i \quad \vartheta = \bigcup_{i=1}^{n} \vartheta_i \cup \{(r, M)\}$$

and $\leq$ is the reflexive-transitive closure of $\leq_0$ ($\uplus$ denotes disjoint union).

If $\Theta = \emptyset$, then $\mathcal{K} = \langle\{r\}, \{(r, r)\}, r, \vartheta\rangle$ only contains the reflexive world $r$ with $\vartheta(r) = M$. Given a set $X$ and a classical model $M$, we write $X_M$ for $\{((a \to b) \to c) \in X \mid a \notin M, b \notin M, c \notin M\}$.

**Lemma 4.** *Let $\sigma = R, X \Rightarrow q$ be a sequent, $r$ be a world, $M$ be a classical model and $\Theta$ be a family $(\mathcal{K}_\iota)_{\iota\in X_M}$ of Kripke models indexed by $X_M$ such that:*

*(i)* $(a \to b) \to c \in X$ *implies* $b \to c \in R$;
*(ii)* $M \models R$ *and* $M \not\models q$;
*(iii) For every* $\iota \in X_M$, *we have* $\mathcal{K}_\iota \not\models R, X_\iota, M, a \Rightarrow b$.

*Then,* $\mathrm{Mod}(r, \Theta, M) \not\models \sigma$.

*Proof.* Let $\mathcal{K}$ be $\mathrm{Mod}(r, \Theta, M) = \langle W, \leq, r, \vartheta\rangle$, whence $r \Vdash M$. We have to show that, at $r$, all the formulae in $R \cup X$ are forced and $q$ is not forced. By (ii), we immediately get $r \nVdash q$. We prove the cases for $R$ and $X$.

*Proof that $r \Vdash R$:* Suppose $\varphi \in R$ and assume $\varphi = \bigwedge A_1 \to \bigvee A_2$ with $A_1 \neq \emptyset$. Let $w \in W$ be any world such that $r \leq w$ and $w \Vdash \bigwedge A_1$; we prove $w \Vdash \bigvee A_2$. If $w = r$, we have $r \Vdash \bigwedge A_1$, which implies $A_1 \subseteq M$. Since $M \models \varphi$, we get $M \models \bigvee A_2$, which implies $r \Vdash \bigvee A_2$. If $w \neq r$, $w$ must be in some $W_\iota$ for some $\mathcal{K}_\iota = \langle W_\iota, \leq_\iota, r_\iota, \vartheta_\iota\rangle$ in $\Theta$, with $\iota \in X_M$. Thus $r_\iota \leq w$ in $\mathcal{K}$. By (iii), $r_\iota \Vdash R$ in $\mathcal{K}_\iota$ and $\varphi \in R$, so $r_\iota \Vdash \varphi$ in $\mathcal{K}_\iota$. Hence $r_\iota \Vdash \varphi$ in $\mathcal{K}$, and the persistence of $\Vdash$ gives $w \Vdash \varphi$ in $\mathcal{K}$. Since $w \Vdash \bigwedge A_1$, we obtain $w \Vdash \bigvee A_2$. The case $A_1 = \emptyset$ (namely, $\varphi = \bigvee A_2$) is similar.

*Proof that $r \Vdash X$:* First note that by (i), $r \Vdash b \to c$ for every $(a \to b) \to c \in X$. Choose any $\iota = (a \to b) \to c \in X$ and let $w$ be any world such that $r \leq w$ and $w \Vdash a \to b$; we prove $w \Vdash c$.

If $c \in M$, then $r \Vdash c$ by construction, hence $w \Vdash c$. If $b \in M$ then $r \Vdash b$ by construction, and we already have $r \Vdash b \to c$, so we get $r \Vdash c$, hence $w \Vdash c$. If $a \in M$ then $r \Vdash a$ and $r \Vdash b \to c$ and $w \Vdash a \to b$, giving $w \Vdash c$. The previous three cases are independent, thus $w \Vdash c$ if $a \in M$ or $b \in M$ or $c \in M$.

So suppose $a \notin M$ and $b \notin M$ and $c \notin M$. By (iii), $\Theta$ contains a model $\mathcal{K}_\iota = \langle W_\iota, \leq_\iota, r_\iota, \vartheta_\iota\rangle$ such that $\mathcal{K}_\iota \not\models R, X_\iota, M, a \Rightarrow b$. Thus $r_\iota \nVdash a \to b$ in $\mathcal{K}_\iota$, hence $r_\iota \nVdash a \to b$ in $\mathcal{K}$. By reverse persistence $r \nVdash a \to b$ (in $\mathcal{K}$), which implies $w \neq r$. There is also a model $\mathcal{K}_{\iota'} = \langle W', \leq', r', \vartheta'\rangle$ of $\Theta$ such that $w \in W'$, hence $r \leq r' \leq w$ (in $\mathcal{K}$). We need to cover the two cases $\iota = \iota'$ and $\iota \neq \iota'$:

1. If $\iota' = \iota$, by (iii) $r_\iota \Vdash a$ in $\mathcal{K}_\iota$, thus $w \Vdash a$ (in $\mathcal{K}$). Since $r \Vdash b \to c$, we have $w \Vdash b \to c$. Then $w \Vdash a \to b$ gives $w \Vdash c$.

2. If $\iota' \neq \iota$ then $\iota \in X_{\iota'}$, hence $r' \Vdash \iota$ in $\mathcal{K}_{\iota'}$, whence it follows that $w \Vdash \iota$ (in $\mathcal{K}$). Our initial assumption that $w \Vdash a \to b$ gives $w \Vdash c$. □

In Fig. 6 we define the proof-search function `LJTSatMain` such that `LJTSatMain(`$R_0$`, `$X_0$`, `$q_0$`)` returns either an $\mathrm{LJT}_{\mathtt{SAT}}$-derivation of $\sigma_0 = R_0, X_0 \Rightarrow q_0$ or a counter-model $\mathcal{K}$ for $\sigma_0$. Worlds of $\mathcal{K}$ are sequences of implication clauses, ordered by the prefix order on sequences, and the empty sequence [] is its root world. We set:

$$H_0 = \{ b \to c \mid (a \to b) \to c \in X_0 \} \qquad V_0 = \{ p \in V \mid p \text{ occurs in } \sigma_0 \}$$
$$\mathrm{M}_{/V_0}(R) = \{ M \subseteq V_0 \mid M \models R \} \quad \text{for any set } R \text{ of flat clauses}$$

The call `LJTSatMain(`$R_0$`, `$X_0$`, `$q_0$`)` defines a SAT-solver $s$ and initializes it by storing all the clauses in $R_0 \cup H_0$; we consider $s$, $R_0$ and $X_0$ as global parameters. It exploits the auxiliary recursive procedure `LJTSat` defined in Fig. 7. A call `LJTSat(`$R$`, `$X$`, `$A$`, `$q$`, `$r$`)` performed during the computation of the main call `LJTSatMain(`$R_0$`, `$X_0$`, `$q_0$`)` has the following specification.

Input Assumptions (IA):
- $R \subseteq \mathrm{R}(s)$ and $X \subseteq X_0$;
- for every $\varphi \in \mathrm{R}(s)$, we have $R_0, X_0 \vdash_{\mathrm{ipl}} \varphi$;
- $r$ is a sequence of implication clauses.

Output Properties (OP):
`LJTSat(`$R$`, `$X$`, `$A$`, `$q$`, `$r$`)` yields a triple $(\mathcal{D}, R', A')$ or else a model $\mathcal{K}$ with:
- $R' \subseteq \mathrm{R}(s)$ and $A' \subseteq A$;
- for every $\varphi \in \mathrm{R}(s)$, we have $R_0, X_0 \vdash_{\mathrm{ipl}} \varphi$;
- $\mathcal{D}$ is an $\mathrm{LJT}_{\mathtt{SAT}}$-derivation of $R, R', X, A' \Rightarrow q$;
- $\mathcal{K}$ has root $r$ and worlds are ordered by the prefix order on sequences;
- $\mathcal{K} \not\models R, H_0, X, A \Rightarrow q$.

In (IA), $\mathrm{R}(s)$ refers to the clauses in the SAT-solver $s$ at the beginning of the computation of `LJTSat(`$R$,$X$,$A$,$q$,$r$`)`; in (OP), $\mathrm{R}(s)$ is the set of clauses in $s$ at the end of the computation. Note that (OP) implies that the call `LJTSat(`$R$,$X$,$A$,$q$`)` terminates. To prove the correctness of `LJTSat`, we have to show that, if the assumptions (IA) are matched, then (OP) holds. We need the following property about derivability in IPL.

**Lemma 5.** *$R, X, A \vdash_{\mathrm{ipl}} b$ implies $R, X, (a \to b) \to c \vdash_{\mathrm{ipl}} \bigwedge(A \setminus \{a\}) \to c$.*

*Proof.* Let $A' = A \setminus \{a\}$. If $R, X, A \vdash_{\mathrm{ipl}} b$, then $R, X, A' \vdash_{\mathrm{ipl}} a \to b$, which implies $R, X, (a \to b) \to c, A' \vdash_{\mathrm{ipl}} c$, hence $R, X, (a \to b) \to c \vdash_{\mathrm{ipl}} \bigwedge A' \to c$. □

To prove correctness, put the following order relation on pairs $(R, X)$ such that $R$ is any set of flat clauses and $X \subseteq X_0$:

$$(R', X') \prec (R, X) \quad \text{iff} \quad (X' \subset X) \text{ or } \big(X' = X \text{ and } \mathrm{M}_{/V_0}(R') \subset \mathrm{M}_{/V_0}(R)\big)$$

Since the sets $X$ and $\mathrm{M}_{/V_0}(R)$ are finite, the relation $\prec$ is well-founded, hence we can prove correctness of `LJTSat` (Lemma 6) by induction on $\prec$.

**Lemma 6.** *If a call* `LJTSat(`$R$, $X$, $A$, $q$, $r$`)` *satisfies (IA) then (OP) holds.*

*Proof.* We use the main induction hypothesis (MIH) below and show that the invariant (Inv) holds at any point of the computation described in Fig. 7:

(MIH): if $(R', X') \prec (R, X)$, then the lemma holds for `LJTSat(`$R'$,$X'$,$A'$,$q'$,$r'$`)`;
(Inv): for every $\varphi \in \mathrm{R}(s)$, we have $R_0, X_0 \vdash_{\mathrm{ipl}} \varphi$.

At the start of the computation (Inv) holds by (IA). Let $\tau_0$ be the value computed at line 2. If $\tau_0 = \mathrm{Yes}(A')$, then the triple $(\mathcal{D}, \mathrm{R}(s), A')$ is returned at line 4, with $\mathcal{D}$ defined at line 5; by definition of `satProve`, it holds that $\mathrm{R}(s), A' \vdash_{\mathrm{cpl}} q$ and $A' \subseteq A$, hence (OP) holds. Otherwise, since $R \cup H_0 \subseteq \mathrm{R}(s)$, we have:

(P0) $\tau_0 = \mathrm{No}(M)$ and $M \models R \cup H_0 \cup A$ and $M \not\models q$.

Without loss of generality, we can assume $M \subseteq V_0$, namely $M \in \mathrm{M}_{/V_0}(R)$. If, for every $(a \to b) \to c \in X$, the loop condition at line 8 does not hold, then the loop at lines 8–21 is skipped and the model $\mathrm{Mod}(r, \emptyset, M)$ is returned at line 22, which is a counter-model for $R, H_0, A \Rightarrow q$ by Lemma 4; thus (OP) holds. Let us assume that the loop at lines 8–21 is entered. We prove that at every iteration of the loop the following properties hold, where $\tau_1$ and $\tau_2$ are the values computed at lines 9 and 14 respectively, and $\tilde{\varphi} = \bigwedge(A_1 \setminus \{a\}) \to c$ is defined at line 12:

(P1) $\tau_1 = (\mathcal{D}_1, R_1, A_1)$ or $\tau_1 = \mathcal{K}_1$ where:
- $R_1 \subseteq \mathrm{R}(s)$ and $A_1 \subseteq M \cup \{a\}$;
- $\mathcal{D}_1$ is an $\mathrm{LJT}_{\mathrm{SAT}}$-derivation of $R, R_1, b \to c, X_\iota, A_1 \Rightarrow b$;
- $\mathcal{K}_1$ has root $r{::}\iota$ and worlds are ordered by the prefix order on sequences;
- $\mathcal{K}_1 \not\models R, H_0, X_\iota, M, a \Rightarrow b$.

(P2) $\tau_2 = (\mathcal{D}_2, R_2, A_2)$ or $\tau_2 = \mathcal{K}_2$ where:
- $R_2 \subseteq \mathrm{R}(s)$ and $A_2 \subseteq A$;
- $\mathcal{D}_2$ is an $\mathrm{LJT}_{\mathrm{SAT}}$-derivation of $R, R_2, \tilde{\varphi}, X, A_2 \Rightarrow q$;
- $\mathcal{K}_2$ has root $r$ and worlds are ordered by the prefix order on sequences;
- $\mathcal{K}_2 \not\models R, \tilde{\varphi}, H_0, X, A \Rightarrow q$.

Let us consider the first iteration of the loop and let $\iota = (a \to b) \to c \in X$ be the selected clause (hence, $a \notin M$, $b \notin M$ and $c \notin M$). The call to `LJTSat` at line 9 satisfies (IA). Since $X_\iota \subset X$, we have $(R \cup \{b \to c\}, X_\iota) \prec (R, X)$. By (MIH) $\tau_1$ satisfies (OP); this proves (P1). Note that (OP) guarantees that (Inv) holds after the computation of $\tau_1$. At line 13, $\tilde{\varphi}$ is added to $s$; we check that (Inv) is preserved, namely $R_0, X_0 \vdash_{\mathrm{ipl}} \tilde{\varphi}$. By (P1) and Soundness of $\mathrm{LJT}_{\mathrm{SAT}}$ (Theorem 1), $R, R_1, b \to c, X_\iota, A_1 \vdash_{\mathrm{ipl}} b$ and, by Lemma 5, we get $R, R_1, b \to c, X \vdash_{\mathrm{ipl}} \tilde{\varphi}$, hence $R, R_1, b \to c, X_0 \vdash_{\mathrm{ipl}} \tilde{\varphi}$. Since $R \cup R_1 \cup \{b \to c\} \subseteq \mathrm{R}(s)$, from (Inv) it follows that $R_0, X_0 \vdash_{\mathrm{ipl}} \tilde{\varphi}$. The call to `LJTSat` at line 14 matches (IA). To apply (MIH), we have to check that:

(P3) $(R \cup \{\tilde{\varphi}\}, X) \prec (R, X)$.

Clearly, $\mathrm{M}_{/V_0}(R \cup \{\tilde{\varphi}\}) \subseteq \mathrm{M}_{/V_0}(R)$; to conclude the proof of (P3), we show that the inclusion is strict. Note that $M \in \mathrm{M}_{/V_0}(R)$ see (P0) and the subsequent remark). For a contradiction, assume $M \models \tilde{\varphi}$. Since $A_1 \subseteq M \cup \{a\}$, we get $A_1 \setminus \{a\} \subseteq M$. By definition of $\tilde{\varphi}$, it follows that $M \models c$, namely $c \in M$, a contradiction. Thus, $M \not\models \tilde{\varphi}$, which implies $M \notin \mathrm{M}_{/V_0}(R \cup \{\tilde{\varphi}\})$; this proves (P3). We can apply (MIH) to the recursive call at line 14 and we get (P2) and the preservation of (Inv). Let us consider the iteration $k+1$ of the loop ($k \geq 1$). We can repeat the above reasoning to prove that (P1) and (P2) hold at iteration $k+1$; the invariant property (Inv) is crucial to guarantee that the recursive calls at lines 9 and 14 satisfy (IA). We conclude that (P1) and (P2) hold at every iteration of the loop.

Let us assume that, at some iteration of the loop, $\tau_1$ is not a model, namely $\tau_1 = (\mathcal{D}_1, R_1, A_1)$. If $\tau_2$ is a model $\mathcal{K}_2$, then $\mathcal{K}_2$ is returned at line 15 and (OP) follows from (P2). Otherwise, $\tau_2 = (\mathcal{D}_2, R_2, A_2)$ and $(\mathcal{D}, R_1 \cup R_2, A_2)$ is returned at line 17, where $\mathcal{D}$ is the $\mathrm{LJT}_{\mathsf{SAT}}$-derivation displayed at line 18; accordingly (OP) holds. Finally, let us assume that, at every iteration, $\tau_1$ is a model. Since $X$ is finite, the loop eventually ends and the model $\mathrm{Mod}(r, \Theta, M)$ is returned at line 22. At that point, $\Theta$ has been completed into a family $(\mathcal{K}_\iota)_{\iota \in X_M}$ such that for every $\iota$ in $X_M$, $\mathcal{K}_\iota$ has root $r::\iota$ and its worlds are ordered by the prefix order on sequences, by (P1). $\mathrm{Mod}(r, \Theta, M)$ has root $r$ and also uses the prefix order on sequences. So in order to prove (OP), we only have to check that $\mathrm{Mod}(r, \Theta, M)$ is a counter-model for $R, H_0, X, A \Rightarrow q$. Let $R' = R \cup H_0 \cup M$ and $\sigma' = R', X \Rightarrow q$; by (P0) and (P1), it follows that $\sigma'$, $M$ and $\Theta$ satisfy the assumptions of Lemma 4. Since $A \subseteq M$, $\mathrm{Mod}(r, \Theta, M) \not\models \sigma'$ and (OP) holds. □

By Lemma 6, we get:

**Theorem 2 (Correctness of `LJTSatMain`).** `LJTSatMain(`$R_0$, $X_0$, $q_0$`)` *returns either an* $\mathrm{LJT}_{\mathsf{SAT}}$*-derivation of* $\sigma_0 = R_0, X_0 \Rightarrow q_0$ *or a counter-model for* $\sigma_0$.

*Proof.* Consider the call `LJTSat(`$R_0$,$X_0$,$\emptyset$,$q_0$,`[])` at line 6. When `LJTSat` is called, $\mathrm{R}(s) = R_0 \cup H_0$. Let $b \to c \in H_0$; since $X_0$ contains a formula of the kind $(a \to b) \to c$ and $(a \to b) \to c \vdash_{\mathrm{ipl}} b \to c$, it follows that $R_0, X_0 \vdash_{\mathrm{ipl}} b \to c$. Thus, the call to `LJTSat` satisfies (IA); by Lemma 6, the returned value $\tau$ satisfies (OP). If $\tau$ is a counter-model $\mathcal{K}$, then $\mathcal{K}$ is returned at line 7; since $\mathcal{K} \not\models R_0, H_0, X_0 \Rightarrow q$, we get $\mathcal{K} \not\models \sigma_0$. Otherwise $\tau = (\mathcal{D}, R, \emptyset)$, where $\mathcal{D}$ is an $\mathrm{LJT}_{\mathsf{SAT}}$-derivation $\mathcal{D}$ of $R_0, R, X_0 \Rightarrow q_0$ and $R_0, X_0 \vdash_{\mathrm{ipl}} \varphi$, for every $\varphi \in R$. Accordingly, the returned derivation, displayed at line 10, is an $\mathrm{LJT}_{\mathsf{SAT}}$-derivation of $\sigma_0$. □

As a consequence, we get:

**Theorem 3.** ($\mathrm{LJT}_{\mathsf{SAT}}$**completeness**). *$R, X \vdash_{\mathrm{ipl}} q$ implies $\vdash_{\mathrm{LJT}_{\mathsf{SAT}}} R, X \Rightarrow q$.*

Let us consider the call `LJTSatMain(`$R_0$,$X_0$,$q_0$`)`; we show that we can build an LJT-derivation $\mathcal{D}_\varphi$ of $R_0, X_0 \Rightarrow \varphi$, for every clause $\varphi$ stored in the SAT-solver $s$ during the computation. Let $\varphi = b \to c$ be a clause introduced at the beginning of `LJTSatMain` (line 5 in Fig. 6). Then, $(a \to b) \to c \in X_0$ and, setting $X' = X_0 \setminus \{(a \to b) \to c\}$, $\mathcal{D}_\varphi$ is:

$$\dfrac{\dfrac{\dfrac{}{R_0,\, X',\, b,\, b \to c,\, a \Rightarrow b} \qquad \dfrac{}{R_0,\, X',\, c,\, b \Rightarrow c}}{R_0,\, X_0,\, b \Rightarrow c}\; L{\to}{\to}}{R_0,\, X_0 \Rightarrow b \to c}\; R\to$$

Let us consider the clause $\tilde{\varphi}$ added in the loop of `LJTSat` when the clause $\iota = (a \to b) \to c$ is considered (line 13 in Fig. 7). By Point (P1) in the proof of Lemma 6, there exists an $\mathrm{LJT}_{\mathrm{SAT}}$-derivation $\mathcal{D}_1$ of $R, R_1, b \to c, X_\iota, A_1 \Rightarrow b$. Note that $R \cup R_1 \cup \{b \to c\}$ has the form $R_0 \cup \{\varphi_1, \ldots, \varphi_n\}$, where the clauses $\varphi_1, \ldots, \varphi_n$ are in $s$ and $X_\iota \subseteq X_0$. Let $\mathcal{D}_1'$ be the $\mathrm{LJT}_{\mathrm{SAT}}$-derivation:

$$\dfrac{R_0, X_0 \vdash_{\mathrm{ipl}} \varphi_1 \quad \ldots \quad R_0, X_0 \vdash_{\mathrm{ipl}} \varphi_n \qquad \begin{array}{c}\mathcal{D}_1\\ R_0,\, \varphi_1, \ldots, \varphi_n,\, X_\iota,\, A_1 \Rightarrow b\end{array}}{R_0,\, X_0,\, A_1 \Rightarrow b}\; (\mathrm{cut}^*_{\mathrm{ipl}})$$

By construction, for every judgment $R_0, X_0 \vdash_{\mathrm{ipl}} \varphi'$ occurring in $\mathcal{D}_1'$, the clause $\varphi'$ is in $s$, thus we can assume that the LJT-derivation $\mathcal{D}_{\varphi'}$ has already been defined. We can turn $\mathcal{D}_1'$ into an LJT-derivation $\mathcal{E}_1$ of $R_0, X_0, A_1 \Rightarrow b$.

If $a \notin A_1$, then $\tilde{\varphi} = \bigwedge A_1 \to c$ and $\mathcal{D}_{\tilde{\varphi}}$ is the LJT-derivation

$$\dfrac{\dfrac{\begin{array}{c}\mathcal{E}_1\\ R_0, X_0, A_1 \Rightarrow b\end{array} \qquad \dfrac{\dfrac{}{b,\, b \to c,\, a \Rightarrow b}\;\mathrm{Ax} \qquad \dfrac{}{b,\, c \Rightarrow c}\;\mathrm{Ax}}{b,\, (a \to b) \to c \Rightarrow c}\; L{\to}{\to}}{R_0, X_0, A_1 \Rightarrow c}\; (\mathrm{cut})}{R_0, X_0 \Rightarrow \bigwedge A_1 \to c}\; L\wedge,\, R\to$$

If $A_1 = \emptyset$ (hence $\tilde{\varphi} = c$) the bottom applications of $L\wedge$ and $R \to$ are crossed out. Let $a \in A_1$ and $A_1' = A_1 \setminus \{a\}$. Thus, $\tilde{\varphi} = \bigwedge A' \to c$ and $\mathcal{D}_{\tilde{\varphi}}$ is the derivation

$$\dfrac{\dfrac{\dfrac{\begin{array}{c}\mathcal{E}_1\\ R_0, X_0, A_1', a \Rightarrow b\end{array}}{R_0, X_0, A_1' \Rightarrow a \to b}\; R\to \qquad \dfrac{\dfrac{\dfrac{}{b,\, b \to c,\, a \Rightarrow b}\;\mathrm{Ax}}{a \to b,\, b \to c,\, a \Rightarrow b}\; L0\to \qquad \dfrac{}{a \to b,\, c \Rightarrow c}\;\mathrm{Ax}}{a \to b,\, (a \to b) \to c \Rightarrow c}\; L{\to}{\to}}{R_0, X_0, A_1' \Rightarrow c}\; (\mathrm{cut})}{R_0, X_0 \Rightarrow \bigwedge A_1' \to c}\; L\wedge,\, R\to$$

If $A' = \emptyset$ (hence $\tilde{\varphi} = c$) the bottom applications of $L\wedge$ and $R \to$ are skipped.

By the above discussion, we can enhance the procedures `LJTSatMain` and `LJTSat` so that, whenever a flat clause $\varphi$ is added to the SAT-solver, an LJT-derivation $\mathcal{D}_\varphi$ of $R_0, X_0 \Rightarrow \varphi$ is stored. Let us assume that `LJTSatMain(`$R_0$`,` $X_0$`,` $q_0$`)` returns an $\mathrm{LJT}_{\mathrm{SAT}}$-derivation $\mathcal{D}$ of $\sigma_0 = R_0, X_0 \Rightarrow q_0$ Proceeding as in the proof of Lemma 3, we can exploit the derivations $\mathcal{D}_\varphi$ to translate $\mathcal{D}$ into an LJT-derivation of $\sigma_0$.

## 7 Discussion, Further Work and Conclusions

The construction of Kripke countermodels from failed proof search in (a variant of) LJT was already explored by Dyckhoff and Pinto [3]. In this paper we have merged proof search and countermodel construction into one procedure based on LJT while benefiting at the same time from the insights from `intuit` regarding incremental SAT-solvers.

The rule $(L \rightarrow\rightarrow)$ from Fig. 4 can be interpreted semantically as follows by reading it from conclusion to premises. The antecedent of the conclusion requires the current world $w$ to make $(a \rightarrow b) \rightarrow c$ true, to make all members of $\Gamma$ true and to make $\delta$ false. If $w \Vdash c$ then we have the right premise. Else $w \nVdash c$ and therefore, $w \nVdash a \rightarrow b$. But that means that there exists a $v \geq w$ such that $v \Vdash \Gamma$ and $v \Vdash a$ and $v \nVdash b$, implying that $v \Vdash b \rightarrow c$, which is the left premise.

We have shown that our simpler recursive version of `intuit` can be reconciled with this semantic view if we generalise the rule $(L \rightarrow\rightarrow)$ into the rule (ljt). By doing so, we can utilise an incremental SAT-solver to implement the right premise of the rule (ljt) by "restarting" the SAT-solver with additional flat clauses learned during the process of finding the derivation of the left premise of (ljt).

There are many sequent and natural deduction calculi that contain rules which have this "here" or "at some successor" flavour. For example, the LSJ calculus of Ferrari et al. [5] and the traditional tableau calculus for linear temporal logic PLTL [16]. Can we extend our insights to such calculi to obtain **incremental** SAT-based decision procedures for these calculi too [10]?

Another direction for future work is the extent to which this approach relies on the clausification of the input formula. Indeed, LJT is able to natively treat any IPL formula. Technically, could the calculus $\text{LJT}_{\texttt{SAT}}$ be extended to any sequent, not necessarily in clausal form? If so how would the interaction with the SAT solver be organised?

**Acknowledgment.** This project has received funding from the European Union's Horizon 2020 research and innovation programme under the Marie Skłodowska-Curie grant agreement No 689176.

## References

1. Claessen, K., Rosén, D.: SAT modulo intuitionistic implications. In: Davis, M., Fehnker, A., McIver, A., Voronkov, A. (eds.) LPAR 2015. LNCS, vol. 9450, pp. 622–637. Springer, Heidelberg (2015). https://doi.org/10.1007/978-3-662-48899-7_43
2. Dyckhoff, R.: Contraction-free sequent calculi for intuitionistic logic. J. Symb. Log. **57**(3), 795–807 (1992)
3. Dyckhoff, R., Pinto, L.: Implementation of a loop-free method for construction of counter-models for intuitionistic propositional logic. University of St Andrews Report CS/96/8 (1996)
4. Farooque, M., Graham-Lengrand, S., Mahboubi, A., A bisimulation between DPLL(T) and a proof-search strategy for the focused sequent calculus. In: Momigliano, A., Pientka, B., Pollack, R. (eds.) Proceedings of the 2013 International Workshop on Logical Frameworks and Meta-Languages: Theory and Practice, LFMTP 2013. ACM Press, September 2013
5. Ferrari, M., Fiorentini, C., Fiorino, G.: Contraction-free linear depth sequent calculi for intuitionistic propositional logic with the subformula property and minimal depth counter-models. J. Autom. Reason. **51**(2), 129–149 (2013)

6. Gentzen, G.: Untersuchungen über das logische schließen. i. Mathematische Zeitschrift **39**, 176–210 (1934)
7. Gentzen, G.: Investigations into logical deduction i. In: Szabo, M. (ed.) The Collected Papers of Gerhard Gentzen. North-Holland, Amsterdam (1969)
8. Graham-Lengrand, S.: PSYCHE: a proof-search engine based on sequent calculus with an LCF-style architecture. In: Galmiche, D., Larchey-Wendling, D. (eds.) TABLEAUX 2013. LNCS (LNAI), vol. 8123, pp. 149–156. Springer, Heidelberg (2013). https://doi.org/10.1007/978-3-642-40537-2_14
9. Graham-Lengrand, S.: Polarities & Focussing: a journey from Realisability to Automated Reasoning. Habilitation thesis, Université Paris-Sud (2014)
10. Li, J., Zhu, S., Pu, G., Zhang, L., Vardi, M.Y.: Sat-based explicit LTL reasoning and its application to satisfiability checking. Form. Methods Syst. Des. (2019). https://doi.org/10.1007/s10703-018-00326-5
11. Nieuwenhuis, R., Oliveras, A., Tinelli, C.: Solving SAT and SAT modulo theories: from an abstract Davis-Putnam-Logemann-Loveland procedure to DPLL(T). J. ACM **53**(6), 937–977 (2006)
12. Raths, T., Otten, J., Kreitz, C.: The ILTP problem library for intuitionistic logic. J. Autom. Reason. **38**(1), 261–271 (2007)
13. Statman, R.: Intuitionistic propositional logic is polynomial-space complete. Theor. Comput. Sci. **9**, 67–72 (1979)
14. Troelstra, A.S., Schwichtenberg, H.: Basic Proof Theory, volume 43 of Cambridge Tracts in Theoretical Computer Science. Cambridge University Press, Cambridge (2000)
15. Tseitin, G.S.: On the Complexity of Derivation in Propositional Calculus, pp. 466–483. Springer, Heidelberg (1983). https://doi.org/10.1007/978-3-642-81955-1_28
16. Wolper, P.: Temporal logic can be more expressive. Inf. Control **56**(1/2), 72–99 (1983)

# Relating Labelled and Label-Free Bunched Calculi in BI Logic

Didier Galmiche, Michel Marti, and Daniel Méry(✉)

Université de Lorraine, CNRS, LORIA, 54506 Vandoeuvre-lès-Nancy, France
Daniel.Mery@loria.fr

**Abstract.** In this paper we study proof translations between labelled and label-free calculi for the logic of Bunched Implications (BI). We first consider the bunched sequent calculus LBI and define a labelled sequent calculus, called GBI, in which labels and constraints reflect the properties of a specifically tailored Kripke resource semantics of BI with two total resource composition operators and explicit internalization of inconsistency. After showing the soundness of GBI w.r.t. our specific Kripke frames, we show how to translate any LBI-proof into a GBI-proof. Building on the properties of that translation we devise a tree property that every LBI-translated GBI-proof enjoys. We finally show that any GBI-proof enjoying this tree property (and not only LBI-translated ones) can systematically be translated to an LBI-proof.

## 1 Introduction

The ubiquitous notion of resource is a basic one in many fields but has become more and more central in the design and validation of modern computer systems over the past twenty years. Resource management encompasses various kinds of behaviours and interactions including consumption and production, sharing and separation, spatial distribution and mobility, temporal evolution, sequentiality or non-determinism, ownership and access control, etc.

Dealing with the various aspects of resource management is mostly in the territory of substructural logics, and more precisely, resource-aware logics such as Linear Logic (LL) [10] with its resource consumption interpretation, the logic of Bunched Implications (BI) [17,18] with its resource sharing interpretation, or order-aware non-commutative logic (NL) [1]. As specification logics, they model features like resource distribution and mobility, non-determinism, sequentiality or coordination of entities [4]. Separation Logic, of which BI is the logical kernel, has proved itself very successful as an assertion language for verifying programs that handle mutable data structures via pointers [12,19].

From a semantic point of view, resource interactions such as production and consumption, or separation and sharing are handled in resource models at the

Work supported by the TICAMORE project (ANR grant 16-CE91-0002).


S. Cerrito and A. Popescu (Eds.): TABLEAUX 2019, LNAI 11714, pp. 130–146, 2019.
https://doi.org/10.1007/978-3-030-29026-9_8

level of resource composition. For example, various semantics have been proposed to capture the resource sharing interpretation of BI including categorical, topological, relational and monoidal models [9]. From a proof-theoretic and purely syntactic point of view, the subtleties of a particular resource composition usually lead to the definition of distinct sets of connectives (*e.g.*, additive vs multiplicative, commutative vs non-commutative).

Capturing the interaction between various kinds of connectives often results in label-free calculi that deal with structures more elaborated than sets or multisets of formulas. For example, the standard label-free sequent calculus for BI, which is called LBI, admits sequents the left-hand part of which are structured as binary trees called bunches [15,18]. Resource interaction is usually much simpler to handle in labelled calculi since labels and label constraints are allowed to reflect and mimic, inside the calculus, the fundamental properties of the resource models they are drawn from. Several labelled tableaux or sequent-style systems have been proposed for BI and its variants [9,11,14].

Categorical, relational, topological and monoidal resource models with a Beth interpretation of the additive disjunction have all been proven sound and complete w.r.t. both LBI and TBI in [9,17,18]. Unfortunately, although by far the most widely used models of BI in the literature, monoidal resource models with a more usual Kripke interpretation of the additive disjunction have only been proven complete w.r.t. TBI. Their status w.r.t. LBI is not known and still a difficult open problem as many attempts at solving it from a semantic point of view have failed over the past fifteen years. Therefore, a better understanding of how LBI relates to labelled calculi could be very helpful as a first step towards solving the problem from the more syntactic standpoint of proof translations.

Our work takes place in the general context of studying the relationships between labelled and label-free calculi. In this paper we more particularly focus on the relationships between GBI, a sequent-style reworking of the labelled tableaux calculus TBI [8], and the label-free bunched sequent calculus LBI [18].

In Sect. 2 we recall the basic notions about BI and its label-free bunch sequent calculus LBI. We also introduce a non-standard resource semantics for BI based on two total monoidal operators with an explicit treatment of inconsistency from which we derive a new sequent-style labelled calculus called GBI in Sect. 3. GBI can be seen as an intermediate calculus between TBI and LBI as both calculi share the idea of sets of labels and constraints arranged as a resource graph, but the resource graph in GBI is partially constructed on the fly using explicit structural rules rather than being obtained as the result of a closure operator [8].

Section 4 is devoted to our first contribution which is a translation of any LBI-proof into a GBI-proof. This translation is not a one-to-one correspondence sending each LBI-rule occurring in the original proof to its corresponding GBI counterpart in the translated proof. Indeed, most of the translations patterns require several additional structural steps to obtain an actual GBI-proof. However, these patterns are such that the rule-application strategy of the original proof will be contained in the translated proof, making our translation structure preserving in that particular sense.

Section 5 investigates how GBI-proofs relate to LBI-proofs. We first restrict GBI to have a single formula on the right-hand side. This is justified by the fact that, contrary to related works on translating labelled or prefixed calculi to label-free sequent calculi, mainly in modal and (bi-)intuitionistic logics [16, 20], we cannot rely on the existence of a multi-conclusioned variant of LBI. Such a variant would require bunches on the right-hand side of sequents, and thus the definition of an intuitionistic dual to multiplicative conjunction, which seems problematic, although there exists a multi-conclusioned display calculus for Boolean BI [3].

We define a tree property for single-conclusioned GBI labelled sequents which allows us to translate the left-hand side of a labelled sequent to a bunch according to the label of the formula on its right-hand side. Refining our analysis of the LBI-translation, we show that every sequent in a GBI-proof obtained by translation of an LBI-proof satisfies our tree property.

The second and main contribution finally follows the definition of a restricted variant of GBI the proofs of which always satisfy the tree property and can moreover systematically be translated into LBI-proofs. Let us remark that this result does not depend on a GBI-proof being some translated image of an LBI-proof. We thus observe that our tree property can serve as a criterion for defining a notion of normal GBI-proofs for which normality also means LBI-translatability.

## 2 The Logic BI

In this section, we give a short introduction to BI (see [18] for more details). We recall the bunched sequent calculus LBI and introduce a variant of the usual Kripke resource semantics.

### 2.1 Syntax and Sequent Calculus LBI

Let $\mathsf{Prop} = \{\mathsf{p}, \mathsf{q}, \dots\}$ be a countable set of propositional letters. The formulas of BI, the set of which is denoted $\mathsf{Fm}$, are given by the grammar:

$$\mathrm{A} ::= \mathsf{p} \mid \top_{\mathsf{m}} \mid \mathrm{A} * \mathrm{A} \mid \mathrm{A} \mathrel{-\!\!*} \mathrm{A} \mid \top_{\mathsf{a}} \mid \bot \mid \mathrm{A} \wedge \mathrm{A} \mid \mathrm{A} \vee \mathrm{A} \mid \mathrm{A} \rightarrow \mathrm{A}$$

Bunches are rooted trees given by the following grammar:

$$\Gamma ::= \mathrm{A} \mid \varnothing_{\mathsf{a}} \mid \Gamma \,;\, \Gamma \mid \varnothing_{\mathsf{m}} \mid \Gamma \,,\, \Gamma$$

Equivalence of bunches $\equiv$ is given by commutative monoid equations for “;” and “,” with units $\varnothing_{\mathsf{a}}$ and $\varnothing_{\mathsf{m}}$ respectively, together with the substitution congruence for subbunches.

The LBI sequent calculus is depicted in Fig. 1. LBI derives sequents of the form $\Gamma \vdash \mathrm{C}$, where $\Gamma$ is a bunch and $\mathrm{C}$ is a formula. The notation $\Gamma(\Delta)$ denotes a bunch $\Gamma$ that contains the bunch $\Delta$ as a subtree.

$$\frac{}{A \vdash A}\,\mathrm{id} \qquad \frac{}{\varnothing_m \vdash \top_m}\,\top_{mR} \qquad \frac{}{\varnothing_a \vdash \top_a}\,\top_{aR} \qquad \frac{}{\Gamma(\bot) \vdash A}\,\bot_L$$

$$\frac{\Gamma(\varnothing_m) \vdash A}{\Gamma(\top_m) \vdash A}\,\top_{mL} \qquad \frac{\Gamma(\varnothing_a) \vdash A}{\Gamma(\top_a) \vdash A}\,\top_{aL} \qquad \frac{\Gamma(B) \vdash A \quad \Gamma(C) \vdash A}{\Gamma(B \vee C) \vdash A}\,\vee_L \qquad \frac{\Gamma \vdash A_{i \in \{1,2\}}}{\Gamma \vdash A_1 \vee A_2}\,\vee_R^i$$

$$\frac{\Delta \vdash B \quad \Gamma(C) \vdash A}{\Gamma(B \mathrel{-\!\!*} C, \Delta) \vdash A}\,\mathrel{-\!\!*}_L \qquad \frac{\Gamma, A \vdash B}{\Gamma \vdash A \mathrel{-\!\!*} B}\,\mathrel{-\!\!*}_R \qquad \frac{\Gamma(B, C) \vdash A}{\Gamma(B * C) \vdash A}\,*_L \qquad \frac{\Gamma \vdash A \quad \Delta \vdash B}{\Gamma, \Delta \vdash A * B}\,*_R$$

$$\frac{\Delta \vdash B \quad \Gamma(C) \vdash A}{\Gamma(B \to C ; \Delta) \vdash A}\,\to_L \qquad \frac{\Gamma ; A \vdash B}{\Gamma \vdash A \to B}\,\to_R \qquad \frac{\Gamma(B ; C) \vdash A}{\Gamma(B \wedge C) \vdash A}\,\wedge_L \qquad \frac{\Gamma \vdash A \quad \Delta \vdash B}{\Gamma ; \Delta \vdash A \wedge B}\,\wedge_R$$

$$\frac{\Gamma(\Delta_1) \vdash A}{\Gamma(\Delta_1 ; \Delta_2) \vdash A}\,\mathrm{W} \qquad \frac{\Gamma(\Delta ; \Delta) \vdash A}{\Gamma(\Delta) \vdash A}\,\mathrm{C} \qquad \frac{\Gamma \vdash A}{\Delta \vdash A}\,\Gamma \equiv \Delta \qquad \frac{\Delta \vdash B \quad \Gamma(B) \vdash A}{\Gamma(\Delta) \vdash A}\,\mathrm{Cut}$$

Rules replacing $\equiv$:

$$\frac{\Gamma(\Delta_1 , \Delta_2) \vdash A}{\Gamma(\Delta_2 , \Delta_1) \vdash A}\,\mathrm{E}_m \qquad \frac{\Gamma((\Delta_1 , \Delta_2) , \Delta_3) \vdash A}{\Gamma(\Delta_1 ,(\Delta_2 , \Delta_3)) \vdash A}\,\mathrm{A}_m \qquad \frac{\Gamma(\Delta) \vdash A}{\Gamma(\varnothing_m , \Delta) \vdash A}\,\mathrm{U}_m$$

$$\frac{\Gamma(\Delta_1 ; \Delta_2) \vdash A}{\Gamma(\Delta_2 ; \Delta_1) \vdash A}\,\mathrm{E}_a \qquad \frac{\Gamma((\Delta_1 ; \Delta_2) ; \Delta_3) \vdash A}{\Gamma(\Delta_1 ;(\Delta_2 ; \Delta_3)) \vdash A}\,\mathrm{A}_a \qquad \frac{\Gamma(\Delta) \vdash A}{\Gamma(\varnothing_a ; \Delta) \vdash A}\,\mathrm{U}_a$$

**Fig. 1.** The sequent calculus LBI.

A formula C is a *theorem* of LBI iff $\varnothing_m \vdash C$ is provable in LBI. Let us remark that the cut rule is admissible in LBI [18]. In order to make LBI-proofs shorter, we often skip explicit uses of the exchange rules. We thus consider bunches up to commutativity of "," and ";". However, we do not consider associativity of bunches as implicit (*i.e.*, we do not consider "," and ";" as $n$-ary functors) since it easily leads to unexpected difficulties when adapting results from unassociative systems, *e.g.* in [13] where the decidability of BI is erroneously concluded from the decidability of the Lambek calculus using length and depth arguments on the representation of bunches from [7] that actually fail in the presence of associativity (and contraction).

The rule for equivalence of bunches can easily be replaced with the last six rules given in Fig. 1, where double lines indicate rules that work both ways (*i.e.*, rules for which the premiss and the conclusion can be swapped). We distinguish bottom-up and top-down uses of such rules in LBI-proofs with up and down arrows respectively. For technical reasons, the rules replacing $\equiv$ will be prefered in the proofs of the forthcoming translation theorems.

Figure 2 gives an example of a proof in LBI, which also shows that the set of derivable sequents in cut-free LBI gets strictly smaller if contraction is removed or restricted to a single formula.

**Lemma 1.** *The following semi-distributivity rule is derivable in* LBI*:*

$$\dfrac{\Gamma((\Delta_1\,,\Delta_2)\,;(\Delta_1\,,\Delta_3))\vdash A}{\Gamma(\Delta_1\,,(\Delta_2\,;\Delta_3))\vdash A}\ \mathrm{Sd}$$

*Proof.* Use contraction on $(\Delta_1\,,(\Delta_2\,;\Delta_3))$ followed by two weakenings.

$$\dfrac{\dfrac{\dfrac{\dfrac{\dfrac{}{p\vdash p}\,\mathrm{id}\qquad \dfrac{\dfrac{}{p\vdash p}\,\mathrm{id}\qquad \dfrac{\dfrac{}{q\vdash q}\,\mathrm{id}\qquad \dfrac{}{r\,;q\vdash r}\,\mathrm{id}}{q\to r\,;q\vdash r}\,\to_L}{q\to r\,;(p \mathrel{-\!\!*} q\,,p)\vdash r}\,\mathrel{-\!\!*}_L}{(p\mathrel{-\!\!*}(q\to r)\,,p)\,;(p\mathrel{-\!\!*}q\,,p)\vdash r}\,\mathrel{-\!\!*}_L}{(p\mathrel{-\!\!*}(q\to r)\,,p)\,;((p\mathrel{-\!\!*}(q\to r)\,;p\mathrel{-\!\!*}q)\,,p)\vdash r}\,\mathrm{W}}{((p\mathrel{-\!\!*}(q\to r)\,;p\mathrel{-\!\!*}q)\,,p)\,;((p\mathrel{-\!\!*}(q\to r)\,;p\mathrel{-\!\!*}q)\,,p)\vdash r}\,\mathrm{W}}{(p\mathrel{-\!\!*}(q\to r)\,;p\mathrel{-\!\!*}q)\,,p\vdash r}\,\mathrm{C}$$

**Fig. 2.** A proof in LBI.

**Lemma 2.** *Adding semi-distributivity to* LBI *while restricting contraction to* $\varnothing_m$ *(or* $\top_m$*) leads to the same set of derivable sequents.*

*Proof.* Contraction is derivable from contraction on $\varnothing_m$ and semi-distributivity:

$$\dfrac{\dfrac{\dfrac{\dfrac{\Gamma(\Delta\,;\Delta)\vdash A}{\Gamma((\Delta\,,\varnothing_m)\,;(\Delta\,,\varnothing_m))\vdash A}\,\mathrm{U_m}\!\uparrow}{\Gamma((\varnothing_m\,;\varnothing_m)\,,\Delta)\vdash A}\,\mathrm{E_m}+\mathrm{Sd}}{\Gamma(\varnothing_m\,,\Delta)\vdash A}\,\mathrm{C}}{\Gamma(\Delta)\vdash A}\,\mathrm{U_m}\!\downarrow$$

## 2.2 Semantics of BI

BI admits various semantics: monoidal, relational, topological, categorical, with or without explicit inconsistency [18]. We introduce a variant of the total (*i.e.*, with an explicit treatment of inconsistency) monoidal semantics [9] that makes use of two monoidal functors to better reflect the syntactic structure of bunches. Although the labelled tableau calculus TBI is known to be complete w.r.t. this semantics [9], whether it is also the case for LBI is still an open problem.

**Definition 1 (Resource Monoid).** *A* resource monoid (RM) *is a structure* $\mathcal{M}=(M,\otimes,1,\oplus,0,\infty,\sqsubseteq)$ *where* $(M,\otimes,1)$, $(M,\oplus,0)$ *are commutative monoids and* $\sqsubseteq$ *is a preordering relation on* $M$ *such that:*

- *for all* $m\in M$, $m\sqsubseteq\infty$ *and* $\infty\sqsubseteq\infty\otimes m$,
- *for all* $m,n\in M$, $m\sqsubseteq m\oplus n$ *and* $m\oplus m\sqsubseteq m$,

– *if* $m \sqsubseteq n$ *and* $m' \sqsubseteq n'$, *then* $m \otimes m' \sqsubseteq n \otimes n'$ *and* $m \oplus m' \sqsubseteq n \oplus n'$.

Let us remark that the conditions of Definition 1 imply that $\infty$ and 0 respectively are greatest and least elements and that $\oplus$ is idempotent.

**Definition 2 (Resource Interpretation).** *Given a resource monoid* $\mathcal{M}$, *a* resource interpretation (RI) for $\mathcal{M}$, *is a function* $[\![-]\!] : \mathsf{Fm} \longrightarrow \mathcal{P}(M)$ *satisfying* $\forall \mathrm{p} \in \mathsf{Prop}$, $\infty \in [\![\mathrm{p}]\!]$ *and* $\forall m, n \in M$ *such that* $m \sqsubseteq n$, $m \in [\![\mathrm{p}]\!] \Rightarrow n \in [\![\mathrm{p}]\!]$.

**Definition 3 (Kripke Resource Model).** *A* Kripke resource model (KRM) *is a structure* $\mathcal{K} = (\mathcal{M}, \models, [\![-]\!])$ *where* $\mathcal{M}$ *is a resource monoid,* $[\![-]\!]$ *is a resource interpretation and* $\models$ *is a forcing relation such that:*

- $m \models \mathrm{p}$ *iff* $m \in [\![\mathrm{p}]\!]$,
- $m \models \bot$ *iff* $\infty \sqsubseteq m$, $m \models \top_{a}$ *iff* $0 \sqsubseteq m$, $m \models \top_{m}$ *iff* $1 \sqsubseteq m$,
- $m \models \mathrm{A} * \mathrm{B}$ *iff for some* $n, n'$ *in* $M$ *such that* $n \otimes n' \sqsubseteq m$, $n \models \mathrm{A}$ *and* $n' \models \mathrm{B}$,
- $m \models \mathrm{A} \wedge \mathrm{B}$ *iff for some* $n, n'$ *in* $M$ *such that* $n \oplus n' \sqsubseteq m$, $n \models \mathrm{A}$ *and* $n' \models \mathrm{B}$,
- $m \models \mathrm{A} \mathrel{-\!\!*} \mathrm{B}$ *iff for all* $n, n'$ *in* $M$ *such that* $n \models \mathrm{A}$ *and* $m \otimes n \sqsubseteq n'$, $n' \models \mathrm{B}$,
- $m \models \mathrm{A} \rightarrow \mathrm{B}$ *iff for all* $n, n'$ *in* $M$ *such that* $n \models \mathrm{A}$ *and* $m \oplus n \sqsubseteq n'$, $n' \models \mathrm{B}$,
- $m \models \mathrm{A} \vee \mathrm{B}$ *iff* $m \models \mathrm{A}$ *or* $m \models \mathrm{B}$.

The semantic clauses for the additive connectives are stated so as to be perfectly symmetric with their multiplicative counterparts (as is the case of their corresponding syntactic rules in LBI). Although such clauses might seem strange at first sight, they are easily proven equivalent to their more usual definitions.

A formula A is valid in the Kripke resource semantics iff $1 \models \mathrm{A}$ in all Kripke resource models.

## 3 The Labelled Calculus GBI

In this section we define a new labelled calculus for BI in the spirit of [2,5,6] and prove its soundness w.r.t. the resource semantics given in Sect. 2.

A countable set $L$ of symbols is a set of *label letters* if it is disjoint from the set $U = \{\,\mathrm{m}, \mathrm{a}, \varpi\,\}$ of *label units*. $\mathcal{L}^0_L = L \cup U$ is the set of *atomic labels over* $L$. The set $\mathcal{L}_L$ of *labels over* $L$ is defined as $\bigcup_{n \in \mathbb{N}} \mathcal{L}^n_L$ where

$$\mathcal{L}^{n+1}_L = \mathcal{L}^n_L \cup \{\, \mathfrak{r}(\ell, \ell') \mid \ell, \ell' \in \mathcal{L}^n_L \text{ and } \mathfrak{r} \in \{\, \mathfrak{m}, \mathfrak{a}\,\}\,\}.$$

For readability, we often drop the subscript $L$ when $L$ is clear from the context. A *label constraint* is an expression $\ell \leqslant \ell'$, where $\ell$ and $\ell'$ are labels. A *labelled formula* is an expression $\mathrm{A} : \ell$, where A is a formula and $\ell$ is a label.

In full generality, the labelled sequent calculus GBI deals with sequents of the form $\Gamma \vdash \Delta$, where $\Gamma$ is a multiset mixing both labelled formulas and label constraints and $\Delta$ is a multiset of labelled formulas. From now on, we only deal with the single-conclusioned variant of GBI where $\Delta$ is restricted to exactly one labelled formula. This restriction is justified by the fact that this paper is a first step at understanding how purely syntactic LBI-proofs relate to GBI-proofs and

$$\frac{\ell \leqslant \ell, \Gamma \vdash \Delta}{\Gamma \vdash \Delta} \text{ R} \qquad \frac{\ell_0 \leqslant \ell, \ell_0 \leqslant \ell_1, \ell_1 \leqslant \ell, \Gamma \vdash \Delta}{\ell_0 \leqslant \ell_1, \ell_1 \leqslant \ell, \Gamma \vdash \Delta} \text{ T}$$

$$\frac{\mathfrak{r}(\ell, r) \leqslant \ell, \Gamma \vdash \Delta}{\Gamma \vdash \Delta} \text{ U}^1_\mathfrak{r} \qquad \frac{\mathfrak{r}(r, \ell) \leqslant \ell, \Gamma \vdash \Delta}{\Gamma \vdash \Delta} \text{ U}^2_\mathfrak{r} \qquad \frac{\mathfrak{r}(\ell_2, \ell_1) \leqslant \ell, \Gamma \vdash \Delta}{\mathfrak{r}(\ell_1, \ell_2) \leqslant \ell, \Gamma \vdash \Delta} \text{ E}_\mathfrak{r}$$

$$\frac{\mathfrak{r}(\ell_3, \ell_2) \leqslant \ell_0, \mathfrak{r}(\ell_4, \ell_0) \leqslant \ell, \Gamma \vdash \Delta}{\mathfrak{r}(\ell_4, \ell_3) \leqslant \ell_1, \mathfrak{r}(\ell_1, \ell_2) \leqslant \ell, \Gamma \vdash \Delta} \text{ A}^1_\mathfrak{r} \qquad \frac{\mathfrak{r}(\ell_1, \ell_4) \leqslant \ell_0, \mathfrak{r}(\ell_0, \ell_3) \leqslant \ell, \Gamma \vdash \Delta}{\mathfrak{r}(\ell_4, \ell_3) \leqslant \ell_2, \mathfrak{r}(\ell_1, \ell_2) \leqslant \ell, \Gamma \vdash \Delta} \text{ A}^2_\mathfrak{r}$$

$$\frac{\mathfrak{a}(\ell, \ell) \leqslant \ell, \Gamma \vdash \Delta}{\Gamma \vdash \Delta} \text{ I}_\mathfrak{a} \qquad \frac{\ell_i \leqslant \ell, \mathfrak{a}(\ell_1, \ell_2) \leqslant \ell, \Gamma \vdash \Delta}{\mathfrak{a}(\ell_1, \ell_2) \leqslant \ell, \Gamma \vdash \Delta} \text{ P}^i_\mathfrak{a}$$

$$\frac{\ell_i \leqslant \ell, \mathfrak{m}(\ell_1, \ell_2) \leqslant \ell, \Gamma \vdash \Delta}{\mathfrak{m}(\ell_1, \ell_2) \leqslant \ell, \Gamma \vdash \Delta} \text{ P}^i_\mathfrak{m}$$

$$\frac{\mathfrak{r}(\ell_0, \ell_2) \leqslant \ell, \ell_0 \leqslant \ell_1, \mathfrak{r}(\ell_1, \ell_2) \leqslant \ell, \Gamma \vdash \Delta}{\ell_0 \leqslant \ell_1, \mathfrak{r}(\ell_1, \ell_2) \leqslant \ell, \Gamma \vdash \Delta} \text{ C}^1_\mathfrak{r} \qquad \frac{\ell \leqslant \ell_1, \Gamma, A : \ell_1 \vdash \Delta}{\ell \leqslant \ell_1, \Gamma, A : \ell \vdash \Delta} \text{ K}_L$$

$$\frac{\mathfrak{r}(\ell_1, \ell_0) \leqslant \ell, \ell_0 \leqslant \ell_2, \mathfrak{r}(\ell_1, \ell_2) \leqslant \ell, \Gamma \vdash \Delta}{\ell_0 \leqslant \ell_2, \mathfrak{r}(\ell_1, \ell_2) \leqslant \ell, \Gamma \vdash \Delta} \text{ C}^2_\mathfrak{r} \qquad \frac{\ell_1 \leqslant \ell, \Gamma \vdash A : \ell_1, \Delta}{\ell_1 \leqslant \ell, \Gamma \vdash A : \ell, \Delta} \text{ K}_R$$

$$\frac{\Gamma_0 \vdash \Delta}{\Gamma_0, \Gamma_1 \vdash \Delta} \text{ W}_L \qquad \frac{\Gamma \vdash \Delta_0}{\Gamma \vdash \Delta_0, \Delta_1} \text{ W}_R \qquad \frac{\Gamma_0, \Gamma_1, \Gamma_1 \vdash \Delta}{\Gamma_0, \Gamma_1 \vdash \Delta} \text{ C}_L \qquad \frac{\Gamma \vdash \Delta_0, \Delta_1, \Delta_1}{\Gamma \vdash \Delta_0, \Delta_1} \text{ C}_R$$

**Side conditions:**

$i \in \{1, 2\}$ and $\mathfrak{r} \in \{\mathfrak{m}, \mathfrak{a}\}$.

$\ell_0$ is a fresh label letter in $\text{A}^i_\mathfrak{r}$. $\ell_i$ in $\text{P}^i_\mathfrak{m}$ must be in $\{\mathfrak{m}, \varpi\}$.

$\ell$ in R and $\text{I}_\mathfrak{a}$, $\ell_1, \ell_2$ in $\text{P}^i_\mathfrak{a}$ and $\ell_{3-i}$ in $\text{P}^i_\mathfrak{m}$ must occur in $\Gamma$, $\Delta$ or $\{\mathfrak{m}, \mathfrak{a}, \varpi\}$.

**Fig. 3.** Structural rules of GBI.

LBI is a single-conclusioned calculus. Similarly to bunches, we use the notation $\Gamma(\Delta)$ for a multiset $\Gamma$ which contains $\Delta$ as a sub-multiset.

The structural rules of GBI are given in Fig. 3. They syntactically reflect the semantic properties of the binary operators $\otimes$, $\oplus$ and the binary relation $\sqsubseteq$ into the binary functors $\mathfrak{m}$, $\mathfrak{a}$ and the binary relation $\leqslant$. The units 1, 0 and $\infty$ are reflected into the labels units $\mathfrak{m}$, $\mathfrak{a}$ and $\varpi$. We generically write $\mathfrak{r}$ (resp. r) to denote either $\mathfrak{m}$ or $\mathfrak{a}$ (resp. m and a) in contexts where the multiplicative or additive nature of the functor (resp. unit) is not important (*e.g.*, for properties that hold in both cases).

We begin with rules R and T to capture the reflexivity and transitivity of the accessibility relation. Then we continue with rules $\text{U}^i_\mathfrak{r}$ that capture the identity of the functors $\mathfrak{m}$ and $\mathfrak{a}$ w.r.t. m and a. The superscript $i \in \{1, 2\}$ in GBI-rule names denotes which argument of an underlying $\mathfrak{r}$-functor is treated by the rule. We then proceed with rules $\text{A}^i_\mathfrak{r}$ and $\text{E}_\mathfrak{r}$ for associativity and commutativity of the $\mathfrak{r}$-functors. In the presence of explicit exchange rules $\text{E}_\mathfrak{r}$, or if we implicitly consider the $\mathfrak{r}$-functors as commutative (which we do not), the superscript variants of the rules are not needed. We nevertheless keep them as they help drastically reduce explicit uses of $E_{\mathfrak{r}}$. The rule $I_{\mathfrak{a}}$ reflects the idempotency of $\oplus$ into the $\mathfrak{a}$-functor. The projection rules $P^i_{\mathfrak{a}}$ reflect into the $\mathfrak{a}$-functor the fact that $\oplus$ is increasing, *i.e.*, $m \sqsubseteq m \oplus n$. The projection rules $P^i_{\mathfrak{m}}$ capture the fact that $m \sqsubseteq m \otimes n$ generally only holds if $n$ is $\infty$ or 1. The compatibility rules $C^i_{\mathfrak{r}}$ reflect that $\oplus$ and $\otimes$ are both order preserving. Finally the last six rules simply express Kripke monotonicity, weakening and contraction.

$$\frac{}{\Gamma, \varpi \leqslant \ell \vdash A : \ell, \Delta}\ \bot_R \qquad \frac{}{\Gamma, A : \ell \vdash A : \ell, \Delta}\ \mathrm{id} \qquad \frac{}{\Gamma, m \leqslant \ell \vdash \top_m : \ell, \Delta}\ \top_{mR}$$

$$\frac{\Gamma, \varpi \leqslant \ell \vdash \Delta}{\Gamma, \bot : \ell \vdash \Delta}\ \bot_L \qquad \frac{\Gamma, m \leqslant \ell \vdash \Delta}{\Gamma, \top_m : \ell \vdash \Delta}\ \top_{mL} \qquad \frac{\Gamma, a \leqslant \ell \vdash \Delta}{\Gamma, \top_a : \ell \vdash \Delta}\ \top_{aL} \qquad \frac{}{\Gamma, a \leqslant \ell \vdash \top_a : \ell, \Delta}\ \top_{aR}$$

$$\frac{\mathfrak{a}(\ell, \ell_1) \leqslant \ell_2, \Gamma \vdash A : \ell_1, \Delta \qquad \mathfrak{a}(\ell, \ell_1) \leqslant \ell_2, \Gamma, B : \ell_2 \vdash \Delta}{\mathfrak{a}(\ell, \ell_1) \leqslant \ell_2, \Gamma, A \to B : \ell \vdash \Delta}\ \to_L$$

$$\frac{\mathfrak{m}(\ell, \ell_1) \leqslant \ell_2, \Gamma \vdash A : \ell_1, \Delta \qquad \mathfrak{m}(\ell, \ell_1) \leqslant \ell_2, \Gamma, B : \ell_2 \vdash \Delta}{\mathfrak{m}(\ell, \ell_1) \leqslant \ell_2, \Gamma, A \mathrel{-\!\!*} B : \ell \vdash \Delta}\ \mathrel{-\!\!*}_L$$

$$\frac{\mathfrak{a}(\ell, \ell_1) \leqslant \ell_2, \Gamma, A : \ell_1 \vdash B : \ell_2, \Delta}{\Gamma \vdash A \to B : \ell, \Delta}\ \to_R \qquad \frac{\mathfrak{m}(\ell, \ell_1) \leqslant \ell_2, \Gamma, A : \ell_1 \vdash B : \ell_2, \Delta}{\Gamma \vdash A \mathrel{-\!\!*} B : \ell, \Delta}\ \mathrel{-\!\!*}_R$$

$$\frac{\mathfrak{a}(\ell_1, \ell_2) \leqslant \ell, \Gamma, A : \ell_1, B : \ell_2 \vdash \Delta}{\Gamma, A \wedge B : \ell \vdash \Delta}\ \wedge_L \qquad \frac{\mathfrak{m}(\ell_1, \ell_2) \leqslant \ell, \Gamma, A : \ell_1, B : \ell_2 \vdash \Delta}{\Gamma, A * B : \ell \vdash \Delta}\ *_L$$

$$\frac{\mathfrak{a}(\ell_1, \ell_2) \leqslant \ell, \Gamma \vdash A : \ell_1, \Delta \qquad \mathfrak{a}(\ell_1, \ell_2) \leqslant \ell, \Gamma \vdash B : \ell_2, \Delta}{\mathfrak{a}(\ell_1, \ell_2) \leqslant \ell, \Gamma \vdash A \wedge B : \ell, \Delta}\ \wedge_R$$

$$\frac{\mathfrak{m}(\ell_1, \ell_2) \leqslant \ell, \Gamma \vdash A : \ell_1, \Delta \qquad \mathfrak{m}(\ell_1, \ell_2) \leqslant \ell, \Gamma \vdash B : \ell_2, \Delta}{\mathfrak{m}(\ell_1, \ell_2) \leqslant \ell, \Gamma \vdash A * B : \ell, \Delta}\ *_R$$

$$\frac{\Gamma, A : \ell \vdash \Delta \qquad \Gamma, B : \ell \vdash \Delta}{\Gamma, A \vee B : \ell \vdash \Delta}\ \vee_L \qquad \frac{\Gamma \vdash A_{i \in \{1,2\}} : \ell, \Delta}{\Gamma \vdash A_1 \vee A_2 : \ell, \Delta}\ \vee^i_R$$

**Side conditions:** $\ell_1$ and $\ell_2$ must be fresh label letters in $*_L$, $\wedge_L$, $\mathrel{-\!\!*}_R$, and $\to_R$.

**Fig. 4.** Logical rules of GBI.

The logical rules of GBI are given in Fig. 4. and are direct translations of their semantic clauses. Figure 7 gives an example of a proof in GBI, where the notation "–" subsumes all the elements we omit to keep proofs more compact.

**Definition 4.** *A formula* A *is a theorem of* GBI *if the sequent* $m \leqslant \ell \vdash A : \ell$ *is provable in* GBI *for some label letter* $\ell$.

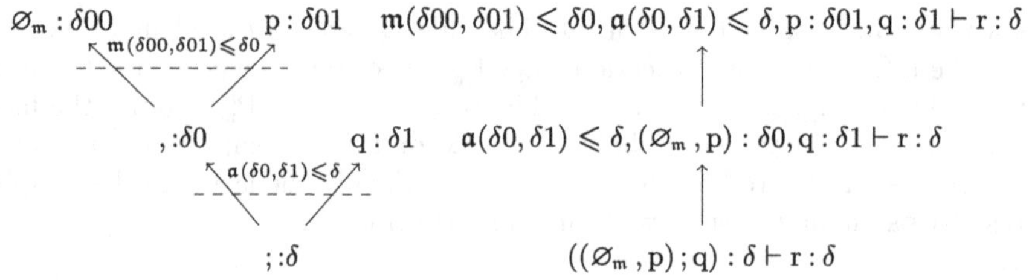

$\varnothing_m : \delta 00$
$p : \delta 01$
$m(\delta 00, \delta 01) \leqslant \delta 0$
$, : \delta 0$
$q : \delta 1$
$a(\delta 0, \delta 1) \leqslant \delta$
$; : \delta$
$m(\delta 00, \delta 01) \leqslant \delta 0, a(\delta 0, \delta 1) \leqslant \delta, p : \delta 01, q : \delta 1 \vdash r : \delta$
$a(\delta 0, \delta 1) \leqslant \delta, (\varnothing_m , p) : \delta 0, q : \delta 1 \vdash r : \delta$
$((\varnothing_m , p) ; q) : \delta \vdash r : \delta$


**Fig. 5.** Translation of the LBI-sequent $(\varnothing_m , p) \; ; \; q \vdash r$

$$\frac{\Gamma(a(\delta s0, \delta s1) \leqslant \delta s, \Theta : \delta s0, \Theta : \delta s1) : \delta \vdash A : \delta}{\Gamma(\Theta : \delta s) : \delta \vdash A : \delta} \; C_T$$

$$\frac{\Gamma(\mathfrak{r}(\delta s0, \delta s1) \leqslant \delta s, r \leqslant \delta s\bar{x}, \Theta : \delta s x) : \delta \vdash A : \delta}{\Gamma(\Theta : \delta s) : \delta \vdash A : \delta} \; Z_{\mathfrak{r}}^{x+1}$$

**Fig. 6.** Tree-like structural rules of GBI.

$$\dfrac{\dfrac{}{-, p : \ell_4 \vdash p : \ell_4}\,\text{id} \qquad \dfrac{\dfrac{\dfrac{}{-, q : \ell_1 \vdash q : \ell_1}\,\text{id}}{\ell_1 \leqslant \ell_2, -, q : \ell_1, p : \ell_4 \vdash q : \ell_2}\,K_R}{-, m(\ell_0, \ell_1) \leqslant \ell_2, -, q : \ell_1, p : \ell_4 \vdash q : \ell_2}\,P^2_m}{\dfrac{m(\ell_3, \ell_4) \leqslant \ell_1, -, p \mathrel{-\!\!*} q : \ell_3, p : \ell_4 \vdash q : \ell_2}{\dfrac{m(\ell_0, \ell_1) \leqslant \ell_2, -, (p \mathrel{-\!\!*} q) * p : \ell_1 \vdash q : \ell_2}{m \leqslant \ell_0 \vdash ((p \mathrel{-\!\!*} q) * p) \mathrel{-\!\!*} q : \ell_0}\,{-\!\!*}_R}\,{*}_L}\,{-\!\!*}_L$$

**Fig. 7.** A proof in GBI.

### 3.1 Soundness of GBI

**Definition 5 (Realization).** *Let $\mathcal{K} = (\mathcal{M}, \models, [\![-]\!])$ be a Kripke resource model with $\mathcal{M} = (M, \otimes, 1, \oplus, 0, \infty, \sqsubseteq)$. Let $s = \Gamma \vdash \Delta$ be a labelled sequent. A* realization *of $s$ in $\mathcal{K}$ is a total function $\rho$ from the labels of $s$ to $M$ such that:*

- $\rho(\mathrm{m}) = 1$, $\rho(m(\ell_1, \ell_2)) = \rho(\ell_1) \otimes \rho(\ell_2)$,
- $\rho(\mathrm{a}) = 0$, $\rho(\varpi) = \infty$, $\rho(a(\ell_1, \ell_2)) = \rho(\ell_1) \oplus \rho(\ell_2)$,
- *for all $\ell_1 \leqslant \ell_2$ in $\Gamma$, $\rho(\ell_1) \sqsubseteq \rho(\ell_2)$ in $\mathcal{M}$,*
- *for all $A : \ell$ in $\Gamma$, $\rho(\ell) \models A$ and for all $A : \ell$ in $\Delta$, $\rho(\ell) \not\models A$.*

*We say that $s$ is* realizable in $\mathcal{K}$ *if there exists a realization of $s$ in $\mathcal{K}$ and that $s$ is* realizable *if it is realizable in some Kripke resource model $\mathcal{K}$.*

**Lemma 3.** *If in a* GBI*-proof the sequent $s = \Gamma \vdash \Delta$ is an initial sequent,* i.e., *a leaf sequent that is the conclusion of a zero-premiss rule, then $s$ is not realizable.*

*Proof.* Suppose that $s$ is realizable, then we have a realization $\rho$ of $s$ in some Kripke resource model $\mathcal{K} = (\mathcal{M}, \models, [-])$. We proceed by case analysis on the zero-premiss rule $r$ of which $s$ is the conclusion. If $r$ is id then $s$ has the form $\Gamma, \mathrm{A} : \ell \vdash \mathrm{A} : \ell, \Delta$, which implies the contradiction $\rho(\ell) \models \mathrm{A}$ and $\rho(\ell) \not\models \mathrm{A}$. If $r$ is $\top_{\mathrm{m}\mathrm{L}}$ then $s$ has the form $\mathrm{m} \leqslant \ell, \Gamma \vdash \top_{\mathrm{m}} : \ell, \Delta$ so that both $\rho(\ell) \not\models \top_{\mathrm{m}}$ and $1 \sqsubseteq \rho(\ell)$, which is a contradiction since $1 \sqsubseteq \rho(\ell)$ implies $\rho(\ell) \models \top_{\mathrm{m}}$. Similarly for the case when $r$ is $\top_{\mathrm{a}\mathrm{L}}$. Finally, if $r$ is $\perp_{\mathrm{R}}$ then $s$ has the form $\Gamma, \varpi \leqslant \ell \vdash \mathrm{A} : \ell, \Delta$ so that $\infty \sqsubseteq \rho(\ell)$ and $\rho(\ell) \not\models \mathrm{A}$, which is a contradiction because by Kripke monotonicity, $\rho(\ell) \models \mathrm{A}$.

**Lemma 4.** *Every proof-rule in* GBI *preserves realizability.*

*Proof.* By case analysis of the proof rules of GBI.

**Theorem 1 (Soundness).** *If a formula* A *is provable in* GBI, *then it is valid in the Kripke resource semantics of* BI.

*Proof.* Suppose that A is provable in GBI but not valid in the Kripke resource semantics of BI. Then, the sequent $\vdash \mathrm{A} : \mathrm{m}$ is trivially realizable and we have a GBI-proof $\mathcal{P}$ of A. It follows from Lemma 4 that $\mathcal{P}$ contains a branch the sequents of which are all realizable. Since $\mathcal{P}$ is a proof, the branch ends with an initial (axiom) sequent and Lemma 3 implies that this initial sequent is not realizable, which is a contradiction. Therefore, A is valid.

## 4 From LBI-Proofs to GBI-Proofs

In this section, we introduce the concepts for translating sequents of LBI to sequents of GBI. In order to highlight the relationships between the labels and the tree structure of bunches more easily we use label letters of the form $xs$ where $x$ is a non-greek letter and $s \in \{0,1\}^*$ is a binary string that encodes the path of the node $xs$ in a tree structure the root of which is $x$. We thus call $x$ the root of a label letter $xs$. We use greek letters to range over label letters with the convention that distinct greek letters denote label letters with distinct roots.

**Definition 6.** *Given a bunch* $\Gamma$ *and a label letter* $\delta$, *we define* $\mathfrak{L}(\Gamma, \delta)$, *the* translation of $\Gamma$ *according to* $\delta$, *by induction on the structure of* $\Gamma$ *as follows:*

- $\mathfrak{L}(\mathrm{A}, \delta) = \{\mathrm{A} : \delta\}$, $\mathfrak{L}(\varnothing_{\mathrm{a}}, \delta) = \{\mathrm{a} \leqslant \delta\}$, $\mathfrak{L}(\varnothing_{\mathrm{m}}, \delta) = \{\mathrm{m} \leqslant \delta\}$,
- $\mathfrak{L}((\Delta_0 , \Delta_1), \delta) = \mathfrak{L}(\Delta_0, \delta 0) \cup \mathfrak{L}(\Delta_1, \delta 1) \cup \{\mathfrak{m}(\delta_0, \delta_1) \leqslant \delta\}$,
- $\mathfrak{L}((\Delta_0 ; \Delta_1), \delta) = \mathfrak{L}(\Delta_0, \delta 0) \cup \mathfrak{L}(\Delta_1, \delta 1) \cup \{\mathfrak{a}(\delta_0, \delta_1) \leqslant \delta\}$.

*The definition extends to* LBI*-sequents as follows:* $\mathfrak{L}(\Gamma \vdash \mathrm{A}, \delta) = \mathfrak{L}(\Gamma, \delta) \vdash \mathrm{A} : \delta$.

We write $\Gamma : \delta$ as a shorthand for $\mathfrak{L}(\Gamma, \delta)$ so that $\mathfrak{L}(\Gamma \vdash \mathrm{A}, \delta) = \Gamma : \delta \vdash \mathrm{A} : \delta$. An illustration of Definition 6 is given in Fig. 5. Let $\Delta$ be a sub-bunch of $\Gamma$, then for any label letter $\delta$, $\Gamma : \delta$ will contain the multiset $\Delta : \delta s$ for some (possibly empty) binary suffix $s$, in which case we write $\Gamma(\Delta : \delta s) : \delta$.

Before translating LBI-proofs into GBI-proofs we introduce the notion of *label substitution*, which is a mapping from label letters to atomic labels, written $[\alpha_1 \mapsto \ell_1, \ldots, \alpha_n \mapsto \ell_n]$. Since label letters have the form $xs$ where $s$ is a binary string, we write $\alpha \hookrightarrow \ell$ as a shorthand for $\forall s.\, \alpha s \mapsto \ell s$.

**Theorem 2.** *If a sequent* $\Gamma \vdash \mathrm{A}$ *is provable in* LBI*, then for any label letter* $\delta$*, the labelled sequent* $\Gamma : \delta \vdash \mathrm{A} : \delta$ *is provable in* GBI.

*Proof.* The proof is by induction on the height of LBI-proofs, using a case distinction on the last rule R applied. We show that for an arbitrary label letter $\delta$, we can build a GBI-proof of the translation of the conclusion of R from translations of its premises. Several LBI-rules that operate on a bunch $\Delta$ that can be nested inside a bunch $\Gamma(\Delta)$ require a careful distinction between their shallow (no actual $\Gamma$ around $\Delta$) and deep variants. We only consider a few cases, the others being similar.

- Axiom id: $\dfrac{}{\mathrm{A} \vdash \mathrm{A}}\,\mathrm{id}$ is translated to $\dfrac{}{\mathrm{A} : \alpha \vdash \mathrm{A} : \alpha}\,\mathrm{id}$
- Axiom $\top_{\mathrm{m}}\mathrm{R}$: $\dfrac{}{\varnothing_{\mathrm{m}} \vdash \top_{\mathrm{m}}}\,\top_{\mathrm{m}}\mathrm{R}$ is translated to $\dfrac{}{\mathrm{m} \leqslant \alpha \vdash \top_{\mathrm{m}} : \alpha}\,\top_{\mathrm{m}}\mathrm{R}$
- Axiom $\top_{\mathrm{a}}\mathrm{R}$: This case is similar to $\top_{\mathrm{m}}\mathrm{R}$
- Case $-\!\!*\mathrm{R}$: Consider the LBI-proof depicted below on the left-hand side where $\mathcal{D}$ is a proof of $\Gamma\,;\mathrm{A} \vdash \mathrm{B}$, the premiss of $-\!\!*\mathrm{R}$

$$\dfrac{\dfrac{\mathcal{D}}{\Gamma\,, \mathrm{A} \vdash \mathrm{B}}}{\Gamma \vdash \mathrm{A} -\!\!* \mathrm{B}}\,-\!\!*\mathrm{R} \qquad\qquad \dfrac{\mathcal{P}}{\mathfrak{m}(\alpha 0, \alpha 1) \leqslant \alpha, \Gamma : \alpha 0, \mathrm{A} : \alpha 1 \vdash \mathrm{B} : \alpha}$$

Given an arbitrary label letter $\delta$, we are required to build a GBI-proof of $\Gamma : \delta \vdash \mathrm{A} -\!\!* \mathrm{B} : \delta$. By I.H. on $\mathcal{D}$ for some label letter $\alpha$, we have a proof $\mathcal{P}$ of $(\Gamma\,, \mathrm{A}) : \alpha \vdash \mathrm{B} : \alpha$ depicted above on the right-hand side from which we get

$$\dfrac{\dfrac{\mathcal{P}[\alpha 0 \hookrightarrow \delta]}{\mathfrak{m}(\delta, \alpha 1) \leqslant \alpha, \Gamma : \delta, \mathrm{A} : \alpha 1 \vdash \mathrm{B} : \alpha}}{\Gamma : \delta \vdash \mathrm{A} -\!\!* \mathrm{B} : \delta}\,-\!\!*\mathrm{R}$$

Let us note that $\alpha 1$ and $\alpha$ are indeed fresh labels in the premiss of $-\!\!*\mathrm{R}$ since by convention $\alpha$ and $\delta$ have distinct roots.
- Case W (Shallow): By I.H. suppose we have for some $\alpha$

$$\dfrac{\dfrac{\mathcal{D}}{\Delta_0 \vdash \mathrm{A}}}{\Delta_0\,; \Delta_1 \vdash \mathrm{A}}\,\mathrm{W} \qquad\qquad \dfrac{\mathcal{P}}{\Delta_0 : \alpha \vdash \mathrm{A} : \alpha}$$

We then construct the following proof

$$\dfrac{\dfrac{\dfrac{\dfrac{\mathcal{P}[\alpha \hookrightarrow \delta 0]}{\Delta_0 : \delta 0 \vdash \mathrm{A} : \delta 0}}{\delta 0 \leqslant \delta, \mathfrak{a}(\delta 0, \delta 1) \leqslant \delta, \Delta_0 : \delta 0, \Delta_1 : \delta 1 \vdash \mathrm{A} : \delta 0}\,\mathrm{W_L}}{\delta 0 \leqslant \delta, \mathfrak{a}(\delta 0, \delta 1) \leqslant \delta, \Delta_0 : \delta 0, \Delta_1 : \delta 1 \vdash \mathrm{A} : \delta}\,\mathrm{K_R}}{\mathfrak{a}(\delta 0, \delta 1) \leqslant \delta, \Delta_0 : \delta 0, \Delta_1 : \delta 1 \vdash \mathrm{A} : \delta}\,\mathrm{P}^1_{\mathfrak{a}}$$

Let us note that we used $W_L$ to make the premiss of $K_R$ exactly match $\mathcal{P}[\alpha \hookrightarrow \delta 0]$. We can get rid of $W_L$ in all translation patterns by pasting the missing material to every sequent in the proofs obtained by I.H.

- Case $U_m\!\downarrow$ (Deep): Suppose we have a proof

$$\dfrac{\dfrac{\mathcal{D}}{\Gamma(\varnothing_m, \Delta) \vdash A}}{\Gamma(\Delta) \vdash A}\ U_m\!\downarrow$$

By I.H., for some $\alpha$, $s \in \{0,1\}^*$ and $x \in \{0,1\}$, we have a proof

$$\dfrac{\mathcal{P}}{\Gamma(m(\alpha s x 0, \alpha s x 1) \leqslant \alpha s x, m \leqslant \alpha s x 0, \Delta : \alpha s x 1) : \alpha \vdash A : \alpha}$$

We then construct the following proof

$$\dfrac{\dfrac{\dfrac{\mathcal{P}[\alpha s x 0 \mapsto m][\alpha s x 1 \hookrightarrow \delta s x][\alpha \hookrightarrow \delta]}{\Gamma(m(m, \delta s x) \leqslant \delta s x, m \leqslant m, \Delta : \delta s x) : \delta \vdash A : \delta}}{\Gamma(\delta s x \leqslant \delta s x, m \leqslant m, \Delta : \delta s x) : \delta \vdash A : \delta}\ U_m^1}{\Gamma(\Delta : \delta s x) : \delta \vdash A : \delta}\ R$$

Using tree-like identity we get an alternative proof

$$\dfrac{\dfrac{\mathcal{P}[\alpha \hookrightarrow \delta]}{\Gamma(m(\delta s x 0, \delta s x 1) \leqslant \delta s x, m \leqslant \delta s 0, \Delta : \delta s x 1) : \delta \vdash A : \delta}}{\Gamma(\Delta : \delta s x) : \delta \vdash A : \delta}\ Z_m^1$$

Figure 8 summarizes the translation patterns from LBI to GBI, where left-to-right reading of the rules means bottom-up application in a proof. We write ~~$W_L$~~ to indicate the patterns for which explicit uses of weakening in GBI can be discarded as explained in the proof of Theorem 2. In LBI, the rules $-\!\!*_L$, $\to_L$, $*_R$ and $\wedge_R$ require context splitting, which is problematic for bottom-up proof-search. Removing weakening from GBI is desirable as context splitting is no longer needed, which also makes the labelled calculus more interesting as its sequents become more than just an isomorphic term-like transcription of bunches. Besides, removing $W_L$ allows the translation to send all logical rules in LBI directly to their counterpart in GBI. Finally, we also learn from the patterns that $K_R$ instead of $C_t^i$ is what distinguishes the shallow cases from the deep ones, that $I_a$ identifies contraction in LBI while R identifies upward identity $U_t\!\uparrow$ and that T and $K_L$ are never used and can thus be removed from GBI without harming its ability to prove any LBI-provable formula.

| LBI | GBI | LBI | GBI | LBI | GBI | LBI | GBI |
|---|---|---|---|---|---|---|---|
| id | id | $-\!\!*_{\mathrm{R}}$ | $-\!\!*_{\mathrm{R}}$ | $\wedge_{\mathrm{L}}$ | $\wedge_{\mathrm{L}}$ | W(S) | $\mathrm{P}^1_{\mathfrak{a}}\ \mathrm{K_R}\ \cancel{\mathrm{W_L}}$ |
| $\top_{\mathfrak{m}\mathrm{R}}$ | $\top_{\mathfrak{m}\mathrm{R}}$ | $\rightarrow_{\mathrm{R}}$ | $\rightarrow_{\mathrm{R}}$ | $*_{\mathrm{R}}$ | $*_{\mathrm{R}}\ \cancel{\mathrm{W_L}}$ | W(D) | $\mathrm{P}^1_{\mathfrak{a}}\ \mathrm{C}^{\mathrm{i}}_{\mathrm{t}}\ \cancel{\mathrm{W_L}}$ |
| $\top_{\mathfrak{a}\mathrm{R}}$ | $\top_{\mathfrak{a}\mathrm{R}}$ | $-\!\!*_{\mathrm{L}}$ | $-\!\!*_{\mathrm{L}}\ \cancel{\mathrm{W_L}}$ | $\wedge_{\mathrm{R}}$ | $\wedge_{\mathrm{R}}\ \cancel{\mathrm{W_L}}$ | C | $(\mathrm{C_L}\ \mathrm{I}_{\mathfrak{a}})$ or $\mathrm{C_T}$ |
| $\bot_{\mathrm{L}}$(S) | $\bot_{\mathrm{L}}\ \bot_{\mathrm{R}}$ | $\rightarrow_{\mathrm{L}}$ | $\rightarrow_{\mathrm{L}}\ \cancel{\mathrm{W_L}}$ | $\mathrm{E}_{\mathfrak{r}}$ | $\mathrm{E}_{\mathfrak{r}}$ | $\mathrm{U}_{\mathfrak{r}}\!\uparrow$(S) | $\mathrm{C}^1_{\mathfrak{r}}\ \mathrm{P}^2_{\mathfrak{r}}\ \mathrm{K_R}\ \cancel{\mathrm{W_L}}$ |
| $\bot_{\mathrm{L}}$(D) | $\bot_{\mathrm{L}}\ (\mathrm{C}^{\mathrm{i}}_{\mathrm{t}}\ \ \mathrm{P}^{\mathrm{i}}_{\mathrm{t}})^+\ \bot_{\mathrm{R}}$ | $\vee_{\mathrm{L}}$ | $\vee_{\mathrm{L}}$ | $\mathrm{A}_{\mathfrak{r}}\!\uparrow$ | $\mathrm{A}^1_{\mathfrak{r}}\ \cancel{\mathrm{W_L}}$ | $\mathrm{U}_{\mathfrak{r}}\!\uparrow$(D) | $\mathrm{C}^1_{\mathfrak{r}}\ \mathrm{P}^2_{\mathfrak{r}}\ \mathrm{C}^{\mathrm{i}}_{\mathrm{t}}\ \cancel{\mathrm{W_L}}$ |
| $\top_{\mathfrak{m}\mathrm{L}}$ | $\top_{\mathfrak{m}\mathrm{L}}$ | $\vee^{\mathrm{i}}_{\mathrm{R}}$ | $\vee^{\mathrm{i}}_{\mathrm{R}}$ | $\mathrm{A}_{\mathfrak{r}}\!\downarrow$ | $\mathrm{A}^2_{\mathfrak{r}}\ \cancel{\mathrm{W_L}}$ | $\mathrm{U}_{\mathfrak{r}}\!\downarrow$ | $(\mathrm{R}\ \mathrm{U}^1_{\mathfrak{r}})$ or $\mathrm{Z}^1_{\mathfrak{m}}$ |
| $\top_{\mathfrak{a}\mathrm{L}}$ | $\top_{\mathfrak{a}\mathrm{L}}$ | $*_{\mathrm{L}}$ | $*_{\mathrm{L}}$ | | | | |

**Fig. 8.** Translation patterns with $\mathrm{t}, \mathfrak{r} \in \{\mathfrak{m}, \mathfrak{a}\}$, $\mathrm{i} \in \{1, 2\}$, S = Shallow, D = Deep.

## 5 Back from GBI-Proofs to LBI-Proofs

In this section we define the notion of normal GBI-proofs and show how to translate them into LBI-proofs. The main problem is that bunches are binary trees, while label constraints describe graphs that capture the accessibility relations between the worlds of a resource model. We observe that translating a bunch as of Definition 6 results in label constraints encoding a binary tree, which might only be destroyed by the rules $\mathrm{W_L}$, $\mathrm{I}_{\mathfrak{a}}$ and $\mathrm{U}^{\mathrm{i}}_{\mathfrak{r}}$. Using label letters of the form $xs$, we can formulate (without requiring explicit substitutions) two tree preserving rules $\mathrm{C_T}$ and $\mathrm{Z}^{\mathrm{i}}_{\mathfrak{r}}$ described in Fig. 6. $\mathrm{C_T}$ duplicates the whole subtree $\Theta$ rooted at $\delta s$ into two subtrees rooted at $\delta s0$ and $\delta s1$ (thus renaming all labels in the new subtrees) and inserts a new node $\delta s$ as the parent of the duplicated subtrees. $\mathrm{Z}^{\mathrm{i}}_{\mathfrak{r}}$ behaves similarly except that one of the new subtrees is linked with the unit r.

From now on, without harming completeness w.r.t. LBI, we restrict GBI to the rules that are actually used in the patterns of Fig. 8 (discarding $\mathrm{W_L}$) and replace $\mathrm{C_L}$ and $\mathrm{U}^{\mathrm{i}}_{\mathfrak{r}}$ with $\mathrm{C_T}$ and $\mathrm{Z}^{\mathrm{i}}_{\mathfrak{r}}$ of Fig. 6. We also slightly modify LBI: we discard the surrounding $\Gamma(-)$ in the axiom $\bot_{\mathrm{L}}$ and extend the weakening rule to "," whenever the bunch to weaken is $\bot$.

Let $\Gamma \vdash A$ be labelled sequent with label letters in a set of label letters $L$. For $\mathfrak{r} \in \{\mathfrak{a}, \mathfrak{m}\}$, $\Gamma$ induces a *subterm relation* $\twoheadrightarrow = (\twoheadrightarrow_{\mathfrak{a}} \cup \twoheadrightarrow_{\mathfrak{m}})$ defined as follows:

$$\ell_0 \twoheadrightarrow_{\mathfrak{r}} \ell_1 \text{ iff } \ell_1 \in L \text{ and } \exists \ell_2(\mathfrak{r}(\ell_1, \ell_2) \leqslant \ell_0 \in \Gamma \text{ or } \mathfrak{r}(\ell_2, \ell_1) \leqslant \ell_0 \in \Gamma).$$

Intuitively, the subterm relation is intended to characterize the links from parent to children nodes when the relation represents a tree.

$\Gamma$ also induces a *reduction relation* $\rightsquigarrow$ defined as follows:

$$\ell_0 \rightsquigarrow \ell_1 \text{ iff } \ell_1 \leqslant \ell_0 \in \Gamma, \ell_1 \in \mathcal{L}^0_L, \ell_0 \in L \text{ and } \ell_1 \neq \ell_0.$$

Intuitively, the reduction relation will help us track steps that trigger weakenings in LBI. A label $\ell_0$ is irreducible in $\Gamma$ if $\Gamma$ has no redex $\ell_0 \rightsquigarrow \ell_1$. A redex $\ell_0 \rightsquigarrow \ell_1$ is minimal if $\ell_1$ is irreducible. A reduction of $\ell_0$ to $\ell_n$ in $\Gamma$ is a path $\ell_0 \rightsquigarrow \ell_1 \ldots \rightsquigarrow \ell_n$ such that for all $0 \leqslant i < n$, $\ell_i \rightsquigarrow \ell_{i+1}$ in $\Gamma$. A reduction of $\ell_0$ to $\ell_n$ is minimal if $\ell_n$ is irreducible. If all minimal reductions of $\ell_0$ terminate with the same irreducible label $\ell_n$, then $\ell_n$ is called the normal form of $\ell_0$ (in $\Gamma$).

A label $\ell'$ is reachable from a label $\ell$ in $\Gamma$, written $\ell \rightarrowtail \ell'$, if $\ell = \ell'$ or there is a path $P$ from $\ell$ to $\ell'$ with no redex pointing outside $P$, more formally, $P$ is a sequence $\ell_0 \twoheadrightarrow \ell_1 \ldots \twoheadrightarrow \ell_n$ such that $\ell_0 = \ell$, $\ell_n = \ell'$ and for all $0 \leqslant i < n$ and all $\ell''$ such that $\ell_i \rightsquigarrow \ell''$, $\ell'' \in P$. If $\mathrm{A} : \ell \in \Gamma$ then A is an $\ell$-leaf in $\Gamma$. A label constraint $\ell_2 \leqslant \ell_1$ is reachable from $\ell_0$ in $\Gamma$ if $\ell_1$ is reachable from $\ell_0$ and there is no formula A and no irreducible $\ell'$ on the path from $\ell_0$ to $\ell_1$ such $\mathrm{A} : \ell' \in \Gamma$.

**Definition 7 (Tree Property).** *A labelled sequent* $\Gamma \vdash \Delta$ *has the* tree property *if it satisfies all of the following conditions:*

($T_1$) $\Delta = \{\mathrm{A} : \ell\}$ *and* $\mathrm{A} : \ell$ *is called the* root formula *with* root label $\ell$,
($T_2$) *for all* $\mathrm{C} : \ell_0 \in \Gamma \cup \Delta$, $\ell_0$ *is a label letter,*
($T_3$) *for all* $\ell_1 \leqslant \ell_0 \in \Gamma$, $\ell_0$ *is a label letter and if so is* $\ell_1$ *then* $\ell_0 \twoheadrightarrow \ell_1$,
($T_4$) *for all* $\mathfrak{r}(\ell_1, \ell_2) \leqslant \ell_0 \in \Gamma$, $\ell_1$ *and* $\ell_2$ *are atomic,*
($T_5$) *if* $\ell \rightarrowtail \ell_0$ *and* $\ell_0$ *is reducible then* $\ell_0$ *has a normal form and* $\Gamma$ *has no* $\ell_0$*-leaf,*
($T_6$) *if* $\ell \rightarrowtail \ell_0$ *and* $\ell_0$ *is irreducible,* $\Gamma$ *has exactly one* $\ell_0$*-leaf,*
($T_7$) *the set* $\{\ell_1 \twoheadrightarrow \ell_0 \mid \ell \rightarrowtail \ell_0\}$ *is a tree with root* $\ell$ *in which all internal nodes have exactly two children linked with* $\twoheadrightarrow_{\mathfrak{r}}$ *arrows of the same* $\mathfrak{r}$ *type.*

*A* GBI*-proof has the* tree property *iff all of its sequents have the tree property.*

A careful analysis of the translation patterns shows that all LBI-translated GBI-proofs satisfy conditions ($T_1$) to ($T_6$). ($T_7$) might seem very restrictive as it implies that for all sequents $s$ in a GBI-proof and all labels $\ell$ in $s$, $s$ contains at most one corresponding label constraint of the form $\mathfrak{r}(\ell_1, \ell_2) \leqslant \ell$. Actually, we can allow sequents in a proof to have more than one label constraint with the same label on its right-hand side as long as we can decide which one has to be used for the subterm relation to represent a tree structure. This can be achieved either by managing label constraints with a stack strategy, always picking the one which has been introduced into the sequent the most recently, or by using a notion of rank corresponding to the depth at which the label constraint has been introduced in (a bottom up reading of) the proof.

A sequent $\Gamma \vdash \mathrm{A} : \ell$ is *terminal* if it admits a proof of height 0. For any GBI proof-rule $R$, the *principal label* and *principal label constraints* of $R$ are the labels and label constraints explicitly mentioned in the conclusion of $R$ as written in Figs. 3, 4 or 6.

**Definition 8.** *A* GBI*-proof is* normal *if it satisfies the tree property, all of its terminal sequents are initial sequents and in all sequents s that are the conclusion of an instance of a proof-rule R, the principal label and principal label constraints of R in s can be reached from the root label of s.*

Given a finite set $B$ of bunches we define (up to associativity and commutativity of bunches) $\mathcal{B}_{\mathfrak{a}}(B)$ as $\varnothing_{\mathfrak{a}}$ if $B$ is empty and $B_1 ; \ldots ; B_n$ with $B_i \in B$ otherwise. Similarly for $\mathcal{B}_{\mathfrak{m}}(B)$ w.r.t. $\varnothing_{\mathfrak{m}}$ and ",".

**Definition 9.** *Given a labelled sequent* $\Gamma \vdash \mathrm{A} : \ell$ *in a normal* GBI*-proof, its translation to an* LBI*-sequent is defined as* $\mathfrak{B}(\Gamma \vdash \mathrm{A} : \ell) = \Gamma@\ell \vdash \mathrm{A}$ *where* $\Gamma@\ell$ *is defined by induction as follows:*

- $\Gamma@\mathrm{m} = \varnothing_{\mathrm{m}}$, $\Gamma@\mathrm{a} = \varnothing_{\mathrm{a}}$, $\Gamma@\varpi = \bot$,
- $\Gamma@\ell = \Gamma@\ell'$ *if for some* $\ell'$, $\ell \rightsquigarrow \ell'$ *in* $\Gamma$,
- *let* $\mathrm{L} = \{\, \mathrm{A}_i \mid \mathrm{A}_i : \ell \in \Gamma \,\}$ *and* $\mathrm{S}_{\mathfrak{r}} = \{\, \ell_i \mid \ell \twoheadrightarrow_{\mathfrak{r}} \ell_i \text{ in } \Gamma \,\}$,

$$\Gamma@\ell = \begin{cases} \mathcal{B}_{\mathrm{a}}(\mathrm{L}) & \text{if } \mathrm{L} \neq \varnothing. \\ \mathcal{B}_{\mathrm{a}}(\mathrm{S_a}) & \text{if } \mathrm{L} = \varnothing, \mathrm{S_m} = \varnothing, \mathrm{S_a} \neq \varnothing. \\ \mathcal{B}_{\mathrm{m}}(\mathrm{S_m}) & \text{if } \mathrm{L} = \varnothing, \mathrm{S_m} \neq \varnothing, \mathrm{S_a} = \varnothing. \end{cases}$$

**Theorem 3.** *Any normal* GBI*-proof of a formula* A *can be translated into an* LBI*-proof of* $\varnothing_{\mathrm{m}} \vdash \mathrm{A}$.

*Proof.* The proof is by induction on the height of normal GBI-proofs. We only give a few illustrative cases, the others being similar.

- Base Case id: We show that the normal GBI-proof

$$\dfrac{}{\Gamma(\mathrm{A} : \ell) \vdash \mathrm{A} : \ell}\ \mathrm{id} \qquad \text{translates to} \qquad \dfrac{}{\Delta\, ; \mathrm{A} \vdash \mathrm{A}}\ \mathrm{id}$$

  Since A is a $\ell$-leaf in $\Gamma$, $\Gamma$ has no redex for $\ell$. Therefore, $\Gamma@\ell$ is by definition a bunch of the form $\mathrm{A}_1 ; \ldots ; \mathrm{A}_n$ where $\mathrm{A} = \mathrm{A}_i$ for some $1 \leq i \leq n$ and $\mathrm{A}_i : \ell \in \Gamma$ for all $1 \leq i \leq n$. Up to associativity and commutativity, $\Gamma@\ell$ can therefore be rewritten as a bunch $\Delta ; \mathrm{A}$.
- Base Case $\top_{\mathrm{m}}\mathrm{R}$: We show that the normal GBI-proof

$$\dfrac{}{\Gamma(\mathrm{m} \leqslant \ell) \vdash \top_{\mathrm{m}} : \ell}\ \top_{\mathrm{m}}\mathrm{R} \qquad \text{translates to} \qquad \dfrac{}{\varnothing_{\mathrm{m}} \vdash \top_{\mathrm{m}}}\ \top_{\mathrm{m}}\mathrm{R}$$

  Since $\ell$ is a label letter, $\Gamma$ has a redex $\ell \rightsquigarrow \mathrm{m}$. Therefore, $\Gamma$ cannot have any $\ell$-leaf, so that $\Gamma(\mathrm{m} \leqslant \ell)@\ell = \Gamma(\mathrm{m} \leqslant \ell)@\mathrm{m} = \varnothing_{\mathrm{m}}$.
- Case $\top_{\mathrm{m}}\mathrm{L}$: We show that the normal GBI-proof (below)

$$\dfrac{\begin{array}{c}\mathcal{D} \\ \hdashline s_1 = \Gamma(\mathrm{m} \leqslant \ell) \vdash \mathrm{A} : \ell_0\end{array}}{s_0 = \Gamma(\top_{\mathrm{m}} : \ell) \vdash \mathrm{A} : \ell_0}\ \top_{\mathrm{m}}\mathrm{L} \qquad \text{translates to} \qquad \dfrac{\begin{array}{c}\mathcal{P} \\ \hdashline \Delta(\varnothing_{\mathrm{m}}) \vdash \mathrm{A}\end{array}}{\Delta(\top_{\mathrm{m}}) \vdash \mathrm{A}}\ \top_{\mathrm{m}}\mathrm{L}$$

  By I.H., we have an LBI-proof $\mathcal{P}$ of the sequent $\Gamma(\mathrm{m} \leqslant \ell)@\ell_0 \vdash \mathrm{A}$. Since $\mathcal{D}$ is normal, we have $\ell_0 \rightarrowtail \ell$ in the last two sequents $s_0$, $s_1$ so that $\ell$ is actually treated by the translation of $s_1$. Then $\Gamma(\mathrm{m} \leqslant \ell)@\ell_0$ is of the form $\Delta(\varnothing_{\mathrm{m}})$. Since $\top_{\mathrm{m}}$ is the only $\ell$-leaf in $s_0$, $\Gamma(\top_{\mathrm{m}} : \ell)@\ell_0$ is $\Delta(\top_{\mathrm{m}})$.
- Case $\mathrm{K_R}$: Suppose we have a normal GBI-proof

$$\dfrac{\begin{array}{c}\mathcal{D} \\ \hdashline s_1 = \Gamma(\ell_1 \leqslant \ell) \vdash \mathrm{A} : \ell_1\end{array}}{s_0 = \Gamma(\ell_1 \leqslant \ell) \vdash \mathrm{A} : \ell}\ \mathrm{K_R}$$

  Since $\ell$ is a label letter, $\Gamma$ has a redex $\ell \rightsquigarrow \ell_1$. Therefore, $\Gamma$ has no $\ell$-leaf so that $\Gamma@\ell = \Gamma@\ell_1$ by definition. By I.H. we have an LBI-proof of $s_1@\ell_1$, which is also an LBI-proof of $s_0@\ell$.

– Case $-\!\!*_{\mathrm{L}}$: Suppose we have a normal GBI-proof

$$\dfrac{\overset{\mathcal{D}_1}{s_1 = \Gamma(\mathfrak{m}(\ell_1, \ell_2) \leqslant \ell) \vdash \mathrm{B} : \ell_2} \qquad \overset{\mathcal{D}_2}{s_2 = \Gamma(\mathfrak{m}(\ell_1, \ell_2) \leqslant \ell, \mathrm{C} : \ell) \vdash \mathrm{A} : \ell_0}}{s_0 = \Gamma(\mathfrak{m}(\ell_1, \ell_2) \leqslant \ell, \mathrm{B} -\!\!* \mathrm{C} : \ell_1) \vdash \mathrm{A} : \ell_0}\ -\!\!*_{\mathrm{L}}$$

Since $\mathfrak{m}(\ell_1, \ell_2) \leqslant \ell$ is reachable from $\ell_0$ in $s_0$, $s_0$ contains no $\ell$-leaf. $s_0@\ell_0$ then has the form $\Delta(\mathrm{B} -\!\!* \mathrm{C}, \Gamma@\ell_2) \vdash \mathrm{A}$ and $s_2@\ell_0$ has the form $\Delta(\mathrm{C}) \vdash \mathrm{A}$ since $s_2$ has a $\ell$-leaf C making $\mathfrak{m}(\ell_1, \ell_2) \leqslant \ell$ unreachable from $\ell_0$. By I.H., we have LBI-proofs $\mathcal{P}_1$, $\mathcal{P}_2$ of $s_1$, $s_2$ respectively, leading to the LBI-proof

$$\dfrac{\overset{\mathcal{P}_1}{\Gamma@\ell_2 \vdash \mathrm{B}} \qquad \overset{\mathcal{P}_2}{\Delta(\mathrm{C}) \vdash \mathrm{A}}}{\Delta(\mathrm{B} -\!\!* \mathrm{C}, \Gamma@\ell_2) \vdash \mathrm{A}}\ -\!\!*_{\mathrm{L}}$$

## 6 Conclusion and Future Work

In this paper we have shown how to translate any LBI-proof into a GBI-proof. We also showed how to translate (normal) GBI-proofs satisfying the tree property back into an LBI-proof. A first perspective is to investigate whether any GBI-proof can be normalized so as to satisfy the tree property. We conjecture that it is indeed the case. A second interesting perspective would be to find an effective (algorithmic) procedure translating TBI-proofs into GBI-proofs since TBI is known to be sound and complete w.r.t. total KRMs. Finally, a third perspective relies on the construction of counter-models in the KRM semantics of BI directly from failed GBI-proof attempts. This direction requires building countermodels from a single-conclusioned calculus in which backtracking is allowed. Those perspectives would help us to show that total Kripke monoidal models with explicit inconsistency are complete w.r.t. the label-free sequent calculus LBI, thus solving a long-lasting open problem.

## References


1. Abrusci, M., Ruet, P.: Non-commutative logic I: the multiplicative fragment. Ann. Pure Appl. Log. **101**, 29–64 (2000)
2. Basin, D., D'Agostino, M., Gabbay, D.M., Matthews, S., Viganó, L. (eds.): Labelled Deduction, Volume 17 of Applied Logic Series. Kluwer Academic Publishers, Dordrecht (2000)
3. Brotherston, J., Villard, J.: Sub-classical Boolean bunched logics and the meaning of par. In: International Conference on Computer Science Logic, CSL 2014, LIPIcs, Dagsthul, Vienna, Austria, pp. 325–342 (2014)
4. Collinson, M., Pym, D.: Algebra and logic for resource-based systems modelling. Math. Struct. Comput. Sci. **19**(5), 959–1027 (2009)
5. Fitting, M.: Proof Methods for Modal and Intuitionistic Logics. Reidel Publishing Company, Dordrecht (1983)

6. Gabbay, D.M.: Labelled Deductive Systems, Volume I - Foundations. Oxford University Press, Oxford (1996)
7. Galatos, N., Jipsen, P.: Distributive residuated frames and generalized bunched implication algebras. Algebra Universalis **78**(3), 303–336 (2017)
8. Galmiche, D., Méry, D.: Semantic labelled tableaux for propositional BI without bottom. J. Log. Comput. **13**(5), 707–753 (2003)
9. Galmiche, D., Méry, D., Pym, D.: The semantics of BI and Resource Tableaux. Math. Struct. Comput. Science **15**(6), 1033–1088 (2005)
10. Girard, J.Y.: Linear logic. Theor. Comput. Sci. **50**(1), 1–102 (1987)
11. Hóu, Z., Tiu, A., Goré, R.: A labelled sequent calculus for BBI: proof theory and proof search. In: Galmiche, D., Larchey-Wendling, D. (eds.) TABLEAUX 2013. LNCS (LNAI), vol. 8123, pp. 172–187. Springer, Heidelberg (2013). https://doi.org/10.1007/978-3-642-40537-2_16
12. Ishtiaq, S., O'Hearn, P.: BI as an assertion language for mutable data structures. In: 28th ACM Symposium on Principles of Programming Languages, POPL 2001, London, UK, pp. 14–26 (2001)
13. Kaminski, M., Francez, N.: The Lambek calculus extended with intuitionistic propositional logic. Studia Logica **104**(5), 1051–1082 (2016)
14. Larchey-Wendling, D.: The formal strong completeness of partial monoidal Boolean BI. J. Log. Comput. **26**(2), 605–640 (2014)
15. O'Hearn, P.W., Pym, D.: The logic of bunched implications. Bull. Symb. Log. **5**(2), 215–244 (1999)
16. Pinto, L., Uustalu, T.: Proof search and counter-model construction for bi-intuitionistic propositional logic with labelled sequents. In: Giese, M., Waaler, A. (eds.) TABLEAUX 2009. LNCS (LNAI), vol. 5607, pp. 295–309. Springer, Heidelberg (2009). https://doi.org/10.1007/978-3-642-02716-1_22
17. Pym, D.: On bunched predicate logic. In: 14th Symposium on Logic in Computer Science, Trento, Italy, pp. 183–192. IEEE Computer Society Press, July 1999
18. Pym, D.: The Semantics and Proof Theory of the Logic of Bunched Implications, volume 26 of Applied Logic Series. Kluwer Academic Publishers, Dordrecht (2002)
19. Reynolds, J.: Separation logic: a logic for shared mutable data structures. In: IEEE Symposium on Logic in Computer Science, Copenhagen, Danemark, pp. 55–74, July 2002
20. Schmitt, S., Kreitz, C.: Converting non-classical matrix proofs into sequent-style systems. In: McRobbie, M.A., Slaney, J.K. (eds.) CADE 1996. LNCS, vol. 1104, pp. 418–432. Springer, Heidelberg (1996). https://doi.org/10.1007/3-540-61511-3_104

# Sequentialising Nested Systems

Elaine Pimentel[1(✉)], Revantha Ramanayake[2], and Björn Lellmann[2]

[1] Department of Mathematics, UFRN, Natal, Brazil
elaine.pimentel@gmail.com
[2] Institute of Logic and Computation, TU Wien, Vienna, Austria

**Abstract.** In this work, we investigate the proof theoretic connections between sequent and nested proof calculi. Specifically, we identify general conditions under which a nested calculus can be transformed into a sequent calculus by restructuring the nested sequent derivation (proof) and shedding extraneous information to obtain a derivation of the same formula in the sequent calculus. These results are formulated generally so that they apply to calculi for intuitionistic, normal modal logics and negative modalities.

## 1 Introduction

Contemporary proof theory can be traced to Gentzen's seminal work [8] where analytic proof calculi for classical and intuitionistic logic were presented. Proof calculi consist of formal rules of inference which describe the logic under consideration; in an analytic calculus, every formula that occurs in a proof generated by the calculus is a subformula of the end formula being proved. Analyticity is crucial because it induces a structure on the proofs (in terms of the end formula). This proof structure can be exploited to formalise reasoning, investigate metalogical properties of the logic e.g. decidability, complexity and interpolation, and develop automated deduction procedures.

The wide applicability of logical methods and their use in new subject areas has resulted in an explosion of new logics different from classical logic; their usefulness depends on the availability of an analytic proof calculus. The sequent calculus is the simplest and best-known formalism for constructing analytic proof calculi. Unfortunately, there are many natural non-classical logics—for example, most extensions of intuitionistic and modal logic—for which the sequent calculus formalism is unable to provide an analytic calculus (the precise reasons for this inability are still not well understood). In response, many more new formalisms have been proposed, such as *hypersequents* [2,21], *labelled sequents* [6,18], *nested sequents* [3,10] and *linear nested sequents* [14,16]. This work is primarily concerned with the nested sequent formalism which is obtained by replacing the sequent in the sequent calculus with a tree of sequents.

Funded by the project Reasoning Tools for Deontic Logic and Applications to Indian Sacred Texts, CAPES, CNPq, START, FWF-ANR and TICAMORE (I 2982).


S. Cerrito and A. Popescu (Eds.): TABLEAUX 2019, LNAI 11714, pp. 147–165, 2019.
https://doi.org/10.1007/978-3-030-29026-9_9

While the trend has been towards developing formalisms with greater sophistication in order to provide non-classical logics with analytic calculi, in this work we look in the reverse direction by investigating which aspects of this sophistication are extraneous. More specifically, we identify general conditions under which a nested calculus can be transformed into a sequent calculus by restructuring the nested sequent derivation (proof) and shedding information to obtain a derivation of the same formula in the sequent calculus. Our approach identifies a class of nested systems, called *basic nested systems*, suitable for such transformations. In these systems, nested rules either create new nestings (creation rules), or manage sequent *contexts* (update rules), moving formulae to *deeper nestings*, with nesting depth difference *exactly equal to one*. This builds an interesting connection with Avron and Lahav's *basic sequent systems* [1,11], since the systematic separation of the behaviour of principal-auxiliary/context formulae in basic sequent systems and creation/update rules allows for a neater way of relating sequent and nested frameworks.

We exploit this separation of rules as follows: after creating a new nesting, upgrade rules control the flow of formulas from the surrounding context to nestings. We show that, if this flow is restricted to stepwise, (bottom-up) outside-in moves, then the whole block of applications of nested rules can be seen as a single sequent rule, with the principal and auxiliary formulae determined by the creation rule, and the context restrictions determined by the upgrade rules. Observe that the proof strategy described above is only possible since basic nested systems allow for a general form of the *disjunction property.* We apply this method to intuitionistic, normal modal logics and negative modalities.

We believe that the material presented here is not only a mere technicality for establishing connections between proof formalisms: on pinpointing the key differences between sequent and nested systems, we are in fact shedding some light on the discussion of to what extent sequent calculus is an adequate meta-language for producing analytic systems. We thus finish the paper by showing how our ideas can be used in order to better understand the bounds for analyticity in sequent systems.

*Organisation and Contributions.* Section 2 presents the notation for basic sequent systems; Sect. 3 introduces the notion of basic nested systems and shows a normalisation procedure for nested sequent derivations; Sect. 4 explains how to recover sequent systems from nested ones; Sect. 5 applies our results to some example logics and brings a discussion about nestings and cut-elimination; Sect. 6 concludes the paper.

## 2 Sequent Systems

In [1] a family of sequent systems (called *basic systems*) was uniformly presented by explicitly differentiating the *context* and *non-context* portions of a rule. The former is defined using binary *context relations* and the latter via a specified rigid structure. The advantage of such a presentation is that it allows us to relate the properties of the rule with the formal specification of its content and non-context. We will greatly explore this separation when relating sequent rule applications to blocks of nested derivations in Sect. 4. In this work, we will adopt the presentation for basic systems given in [11]. Where convenient, we will also present sequent systems using the traditional *rule schemas* built from meta-variables for formulae and sets of formulae.

Let $\mathcal{L}$ denote a propositional language and $wff_{\mathcal{L}}$ the set of its well formed formulae, built using a countable set $\mathsf{Var} = \{p, p_1, p_2, \ldots\}$ of propositional variable symbols.

**Definition 1.** *A* signed formula *is an expression of the form* $\mathsf{T} : A$ *or* $\mathsf{F} : A$ *where* $A \in wff_{\mathcal{L}}$. *A* sequent *is a finite set of signed formulae. As in [11], we will adopt the usual sequent notation* $\Gamma \vdash \Delta$, *where* $\Gamma, \Delta$ *are (possibly empty) finite sets of formulae and* $\Gamma \vdash \Delta$ *is interpreted as the sequent* $\{\mathsf{F} : A \mid A \in \Gamma\} \cup \{\mathsf{T} : A \mid A \in \Delta\}$.

*A* substitution *is a function* $\sigma : wff_{\mathcal{L}} \to wff_{\mathcal{L}}$ *such that*

$$\sigma(\heartsuit(A_1, \ldots, A_k)) = \heartsuit(\sigma(A_1), \ldots, \sigma(A_k))$$

*for every* $k$*-ary connective* $\heartsuit$ *of* $\mathcal{L}$. *Substitutions extend to signed formulae (preserving sign), sequents and (later) to nested sequents in the standard way.*

*A* context relation *is a finite binary relation on the set of signed formulae. Given a context relation* $C$, *we denote by* $\overline{C}$ *the binary relation between signed formulae* $\overline{C} = \{\langle \sigma(\alpha), \sigma(\beta) \rangle \mid \sigma$ *is a substitution, and* $\langle \alpha, \beta \rangle \in C\}$.

*A* $C$-instance *is a pair of sequents* $\langle S_1, S_2 \rangle$ *such that for some enumeration* $S_1 = \{\alpha_1, \ldots, \alpha_k\}$ *and* $S_2 = \{\beta_1, \ldots, \beta_k\}$ *and every* $1 \leq i \leq k$, *it is the case that* $\alpha_i \overline{C} \beta_i$.

*Example 2.* From the *trivial relation* $C_{\mathtt{id}} := \{\langle \mathsf{F} : p, \mathsf{F} : p \rangle, \langle \mathsf{T} : p, \mathsf{T} : p \rangle\}$ it follows that a signed formula is $\overline{C}_{\mathtt{id}}$-related to another iff the two signed formulae are identical. It follows that the $C_{\mathtt{id}}$-instances are precisely the sets $\{\langle S, S \rangle \mid S$ is a sequent$\}$ of pairs.

From the relation $C_{\mathsf{int}} := \{\langle \mathsf{F} : p, \mathsf{F} : p \rangle\}$ it follows that while $(\mathsf{F} : A)\overline{C}_{\mathsf{int}}(\mathsf{F} : A)$ for every formula $A$, it is *not* the case that $(\mathsf{T} : A)\overline{C}_{\mathsf{int}}(\mathsf{T} : A)$. In particular, the $C_{\mathsf{int}}$-instances are precisely those sequent pairs of the form $\langle \Gamma \vdash, \Gamma \vdash \rangle$. Informally, $C_{\mathsf{int}}$-instances are identical sequents with empty right hand side.

Let $\Box\Gamma$ denote the set $\{\Box A \mid A \in \Gamma\}$. Then from $C_{\mathsf{K}} := \{\langle \mathsf{F} : p, \mathsf{F} : \Box p \rangle\}$ it follows that $C_{\mathsf{K}}$-instances are precisely those sequent pairs of the form $\langle \Gamma \vdash, \Box\Gamma \vdash \rangle$.

Finally, from the relation $C_4 := \{\langle \mathsf{F} : \Box p, \mathsf{F} : \Box p \rangle\}$ it follows that $C_4$-instances are precisely those sequent pairs of the form $\langle \Box\Gamma \vdash, \Box\Gamma \vdash \rangle$.

We define the concatenation $(\Gamma_1 \vdash \Delta_1) \otimes (\Gamma_2 \vdash \Delta_2)$ as $\Gamma_1, \Gamma_2 \vdash \Delta_1, \Delta_2$, and $\emptyset \otimes \Pi$ as $\emptyset$.

**Definition 3.** *A* basic premise *is a pair* $\langle S; C \rangle$ *where* $S$ *is a sequent and* $C$ *is a context relation. A* basic rule *is a pair* $s/S$ *where* $s = \{\langle S_i; C_i \rangle\}_{1 \leq i \leq k}$ *is a finite* *set of basic premises and* $S$ *is the* conclusion *sequent of the rule. A basic rule is represented explicitly as:*

$$\frac{\langle S_1; C_1\rangle \quad \cdots \quad \langle S_k; C_k\rangle}{S}$$

*The formulae in* $S_i$ $(1 \leq i \leq k)$ *are called* auxiliary formulae *and the formulae in* $S$ *are called the* principal formulae.

*A rule with an empty set of basic premises is called an* axiom. *A* basic sequent system (SC) *consists of a set of basic rules.*

*An* application of a basic rule *has the following form, where* $\sigma$ *is a substitution,* $\Pi_1, \Pi_1', \ldots, \Pi_k, \Pi_k'$ *are sequents and* $\langle \Pi_i, \Pi_i'\rangle$ *is a* $C_i$*-instance for each* $i$ $(1 \leq i \leq k)$.

$$\frac{\sigma(S_1) \otimes \Pi_1 \quad \cdots \quad \sigma(S_k) \otimes \Pi_k}{\sigma(S) \otimes \Pi_1' \otimes \cdots \otimes \Pi_k'}\ r$$

*The notion of premise, conclusion, principal and auxiliary formulae extends to applications of rules in the standard way.*

*A* derivation *in a* SC *is defined in the usual way as a finite labelled rooted tree: the root is labelled by the* end-sequent, *the labels of each node and its children correspond to the conclusion and premises of a rule application, and axioms label the leaves.*

$$\frac{\Gamma \vdash A, \Delta \quad \Gamma, B \vdash \Delta}{\Gamma, A \to B \vdash \Delta} \to_L \qquad \frac{\Gamma, A \vdash B}{\Gamma \vdash A \to B} \to_R \qquad \frac{\Gamma, A, B \vdash \Delta}{\Gamma, A \wedge B \vdash \Delta} \wedge_L$$

$$\frac{\Gamma \vdash A, \Delta \quad \Gamma \vdash B, \Delta}{\Gamma \vdash A \wedge B, \Delta} \wedge_R \qquad \frac{\Gamma, A, \vdash \Delta \quad \Gamma, B \vdash \Delta}{\Gamma, A \vee B \vdash \Delta} \vee_L \qquad \frac{\Gamma \vdash A, B, \Delta}{\Gamma \vdash A \vee B, \Delta} \vee_R \qquad \frac{}{\Gamma, \bot \vdash \Delta} \bot_L$$

**Fig. 1.** Multi-conclusion intuitionistic calculus $\mathsf{SC}_{\mathsf{mLJ}}$.

*Example 4.* The rule below on the left has principal formula $p_1 \to p_2$, auxiliary formulae $p_1, p_2$, and application depicted on the right

$$\frac{\langle p_1 \vdash p_2; C_{\mathsf{int}}\rangle}{\vdash p_1 \to p_2} \qquad \frac{\Gamma, A \vdash B}{\Gamma \vdash A \to B} \to_R$$

*Example 5.* The axiom init, and the right and left weakening rules are defined as follows:

$$\emptyset / p \vdash p \qquad \langle \emptyset; C_{\mathtt{id}}\rangle / \vdash p \qquad \langle \emptyset; C_{\mathtt{id}}\rangle / p \vdash$$

In the presence of the above rules, the following rules can be seen to be derivable:

$$\frac{}{\Gamma, A \vdash A, \Delta}\ \mathsf{init} \qquad \frac{\Gamma \vdash \Delta}{\Gamma, \Gamma' \vdash \Delta, \Delta'}\ \mathsf{W}$$

*Remark 6.* The systems that we consider will be *fully structural* in the sense that free application of the schemas init and W is permitted (as originally in the definition of basic sequent systems in [11]). Observe that, since sequents are *sets*, we do not the copy formulae in the contexts.

Figure 1 presents (the schema representation of) $\mathsf{SC}_{\mathsf{mLJ}}$ [17], a multiple conclusion sequent system for propositional intuitionistic logic. Observe that all rules, except $\rightarrow_R$, have the trivial relation in the basic premises. On the other hand, the relation $C_{\mathsf{int}}$ in the implication right rule enforces that the only formula in the succedent of the conclusion is the principal formula.

In what follows, for readability, we may omit the word *basic* when referring to rules, applications and systems.

## 3 Nested Systems

Nested systems [4,20] are extensions of the sequent framework where each sequent is replaced by a tree of sequents. In this work, we will identify a family of *basic* nested systems, inspired by [1,13].

**Definition 7.** *A* nested sequent *is defined inductively as follows:*

*(i) if $S$ is a sequent, then it is a nested sequent;*
*(ii) if $\Gamma \vdash \Delta$ is a sequent and $G_1, \ldots, G_k$ are nested sequents, then $\Gamma \vdash \Delta, [G_1], \ldots, [G_k]$ is a nested sequent.*

*A* nested rule *is a pair $\upsilon/\Upsilon$ represented as follows, where $\upsilon = \{\Upsilon_1, \ldots, \Upsilon_k\}$ is a finite set of nested sequents (the premises) and $\Upsilon$ is the conclusion nested sequent of the rule.*

$$\frac{\Upsilon_1 \quad \cdots \quad \Upsilon_k}{\Upsilon}$$

*The non-context formulae in the premises are called* auxiliary formulae *and the non-context formulae in the conclusion are called* principal formulae.

*For a sequent $S = \Gamma \vdash \Delta$, define $S \otimes (\vdash [\vdash])$ to be the nested sequent $\Gamma \vdash \Delta, [\vdash]$.*

*Let $S, S_1, \ldots, S_k$ be sequents. A* basic nested rule *has one of the following forms:*

*i.* sequent-like *rules*

$$\frac{S_1 \quad \cdots \quad S_k}{S}$$

*ii.* nested-like *rules*

*(a)* creation rules

$$\frac{\vdash [S_1] \quad \cdots \quad \vdash [S_k]}{S}$$

*(b)* upgrade rules*

$$\frac{\vdash [S_1]}{S \otimes (\vdash [\vdash])}$$

**Upgrade rules must have exactly one principal and auxiliary formulae.*

*We will call the nestings in the premises of a creation rule its* auxiliary nestings. *Nestings containing principal or auxiliary formulae are called* active.

*Example 8.* Consider the following nested-like rules

$$\frac{\vdash [p_1 \vdash p_2]}{\vdash p_1 \to p_2} \qquad\qquad \frac{\vdash [p \vdash]}{p \vdash [\vdash]}$$

The first is a creation rule, with auxiliary nesting $[p_1 \vdash p_2]$; the second an upgrade rule.

*Remark 9.* Observe that our definition of nested-like rules restricts the rule form in three ways. First, nested-like rules must have *exactly* one nesting in the premises or conclusion. Second, information in nested rules always moves *deeper* inside nestings, when reading rules bottom-up. Finally, upgrade rules move *only one* piece of information at a time. The first restriction is crucial for avoiding non-determinism when defining of applications of rules; the second one will be key for stating sufficient conditions for the linearisation of nested systems; the third restriction is natural but not necessary. In fact, nested rules usually are *local*, acting in one formula at a time. Also, upgrade rules naturally have this shape, which will allow for the identification of upgrade nested rules as basic sequent context relations later in Sect. 4.

We will present, in Sect. 5, some examples of basic nested systems. It is worth mentioning that every nested calculus we know that has a correspondence (in the sense of this paper) with sequent systems is equivalent to a basic nested system. On the other hand, the restrictions above exclude, *e.g.*, the representation of the rules for modal axioms 5 and B [4]. But then, there are no known simple, cut-free sequent systems for logics K5 and KB. We will discuss some cases that fall outside our scope also in Sect. 5.

For readability, we will denote by $\Gamma, \Delta$ sequent contexts and by $\Lambda$ sets of nestings. In this way, every nested sequent has the shape $\Gamma \vdash \Delta, \Lambda$ where elements of $\Lambda$ have the shape $[\Gamma' \vdash \Delta', \Lambda']$ and so on. We will denote by $\Upsilon$ arbitrary nested sequents. Application of rules and schemas in nested systems will be represented using *holed contexts.*[1]

**Definition 10.** *A* nested-holed context *is a nested sequent that contains a hole of the form* $\{\ \}$ *in place of nestings. We represent such a context as* $\mathcal{S}\{\ \}$. *Given a holed context and a nested sequent* $\Upsilon$, *we write* $\mathcal{S}\{\Upsilon\}$ *to stand for the nested sequent where the hole* $\{\ \}$ *has been replaced by* $[\Upsilon]$, *assuming that the hole is removed if* $\Upsilon$ *is empty and if* $\mathcal{S}$ *is empty then* $\mathcal{S}\{\Upsilon\} = \Upsilon$.

For example, $(\Gamma \vdash \Delta, \{\ \})\{\Gamma' \vdash \Delta'\} = \Gamma \vdash \Delta, [\Gamma' \vdash \Delta']$ while $\{\ \}\{\Gamma' \vdash \Delta'\} = \Gamma' \vdash \Delta'$ and $(\Gamma \vdash \Delta, \{\ \})\{\ \} = \Gamma \vdash \Delta$.

[1] Observe that, since in basic nested systems nested-like rules must have exactly one nesting in the premises or conclusion, only one hole is enough for describing both schemas and applications of rules. Compare with, *e.g.*, the schematic nested rule for 5 in [5].

**Definition 11.** *An* application of a basic nested rule *is given by*[2]

$$\frac{\mathcal{S}\{\sigma(\Upsilon_1) \otimes \mathcal{G}\} \quad \cdots \quad \mathcal{S}\{\sigma(\Upsilon_k) \otimes \mathcal{G}\}}{\mathcal{S}\{\sigma(\Upsilon) \otimes \mathcal{G}\}}\ r^n$$

*where $\sigma$ is a substitution, $\mathcal{G}$ is the* nested sequent context. *The definition of* derivations *in a* NS *is a natural extension of the one for* SC, *only replacing sequents by nested sequents. The notion of principal and auxiliary formulae is extended to applications of rules in the standard way.*

*Remark 12.* In this work we will assume that nested systems are *fully structural*, *i.e.*, including the following nested versions for the initial axiom and weakening

$$\frac{}{\mathcal{S}\{\Gamma, A \vdash \Delta, A, \Lambda\}}\ \mathsf{init}^n \qquad \frac{\mathcal{S}\{\Gamma \vdash \Delta, \Lambda\}}{\mathcal{S}\{\Gamma, \Gamma' \vdash \Delta, \Delta', \Lambda, \Lambda'\}}\ \mathsf{W}^n$$

Also, we only consider *cut-free* nested systems (we will discuss cut-freeness in Sect. 5).

By treating nested contexts as sets, we are setting the context relations to be the identity function. In this way, every basic rule having only $C_{\mathtt{id}}$ as contexts relations in its premises is a sequent-like basic nested rule (and vice-versa). Note that this also implies that basic nested rules are *invertible*.

*Example 13.* Applications of the basic rules in Example 8 have, respectively, the form

$$\frac{\mathcal{S}\{\Gamma \vdash \Delta, \Lambda, [A \vdash B]\}}{\mathcal{S}\{\Gamma \vdash A \to B, \Delta, \Lambda\}}\ \to_R^n \qquad \frac{\mathcal{S}\{\Gamma \vdash \Delta, \Lambda, [\Gamma', A \vdash \Delta', \Lambda']\}}{\mathcal{S}\{\Gamma, A \vdash \Delta, \Lambda, [\Gamma' \vdash \Delta', \Lambda']\}}\ \mathtt{lift}^n$$

Figure 2 presents (the schema representation of) $\mathsf{NS}_{\mathsf{mLJ}}$ [7], a basic nested system for mLJ.

$$\frac{\mathcal{S}\{\Gamma \vdash \Delta, A, \Lambda\} \quad \mathcal{S}\{\Gamma, B \vdash \Delta, \Lambda\}}{\mathcal{S}\{\Gamma, A \to B \vdash \Delta, \Lambda\}}\ \to_L^n \quad \frac{\mathcal{S}\{\Gamma \vdash \Delta, \Lambda, [A \vdash B]\}}{\mathcal{S}\{\Gamma \vdash A \to B, \Delta, \Lambda\}}\ \to_R^n \quad \frac{\mathcal{S}\{\Gamma, A, B \vdash \Delta, \Lambda\}}{\mathcal{S}\{\Gamma, A \wedge B \vdash \Delta, \Lambda\}}\ \wedge_L^n$$

$$\frac{\mathcal{S}\{\Gamma \vdash A, \Delta, \Lambda\} \quad \mathcal{S}\{\Gamma \vdash B, \Delta, \Lambda\}}{\mathcal{S}\{\Gamma \vdash A \wedge B, \Delta, \Lambda\}}\ \wedge_R^n \quad \frac{\mathcal{S}\{\Gamma, A \vdash \Delta, \Lambda\} \quad \mathcal{S}\{\Gamma, B \vdash \Delta, \Lambda\}}{\mathcal{S}\{\Gamma, A \vee B \vdash \Delta, \Lambda\}}\ \vee_L^n \quad \frac{\mathcal{S}\{\Gamma \vdash A, B, \Delta, \Lambda\}}{\mathcal{S}\{\Gamma \vdash A \vee B, \Delta, \Lambda\}}\ \vee_R^n$$

$$\frac{\mathcal{S}\{\Gamma \vdash \Delta, \Lambda, [\Gamma', A \vdash \Delta', \Lambda']\}}{\mathcal{S}\{\Gamma, A \vdash \Delta, \Lambda, [\Gamma' \vdash \Delta', \Lambda']\}}\ \mathtt{lift}^n \quad \frac{}{\mathcal{S}\{\Gamma, \bot \vdash \Delta, \Lambda\}}\ \bot_L^n$$

**Fig. 2.** Nested system $\mathsf{NS}_{\mathsf{mLJ}}$.

[2] Throughout, we will use $n$ as a superscript, etc for indicating "nested". Hence *e.g.*, $\to_R^n$ will be the designation of the implication right rule in the nesting framework.

### 3.1 Normal Forms in NS

While adding a tree structure to sequents enhances the expressiveness of the nesting framework when compared with the sequent one, the price to pay is that the obvious proof search procedure may be of suboptimal complexity, since there can be an exponential blow-up due to the nestings [4]. Hence the importance of proposing *normal form derivations* and/or *proof search strategies* for taming the proof search space.

In this section, we will propose a *normalisation procedure* for basic nested systems, which will be crucial for transforming a nested sequent derivation into a sequent derivation.

The first result states that the *disjunction property* holds for basic nested systems.

**Theorem 14.** *Let* NS *be a basic nested system and let* $\Lambda_i$ *be nestings in* NS. *Then* $\vdash \Lambda_1, \ldots, \Lambda_k$ *is derivable iff* $\vdash \Lambda_i$ *is derivable for some* $i \in \{1, \ldots, k\}$.

*Proof.* ($\Leftarrow$) Trivial due to $\mathsf{W}^n$.

($\Rightarrow$) Due to the shape of basic nested rules, it is immediate to see that any derivation $\pi$ of the nested sequent $\vdash \Lambda_1, \ldots, \Lambda_k$ has the form

$$\dfrac{\overset{\pi_1}{\vdash \Lambda_1, \ldots, \Lambda_i^1, \ldots, \Lambda_k} \quad \cdots \quad \overset{\pi_h}{\vdash \Lambda_1, \ldots, \Lambda_i^h, \ldots, \Lambda_k}}{\vdash \Lambda_1, \ldots, \Lambda_i, \ldots, \Lambda_k}\ r^n$$

By inductive hypothesis, for each premise $j$, either $\Lambda_i^j$ is provable for all $1 \leq j \leq h$ or there is a $m \neq i$ such that $\Lambda_m$ is provable. In both cases, the result follows trivially.

This result generalises not only the disjunction property for mLJ [23] but also for Horn relational sequent theories for modal logics (see [24], Prop. 8.2.9).

The next definition explains how to determine the exact position of nestings and formulae occurring in a nested sequent, as well as the nesting-size of a sequent. Intuitively, the depth of a hole/formula is the number of nodes on the branch of its nesting tree (inside-out measure). The depth of a sequent, however, measures the number of nodes on a branch of the nesting tree of maximal length (outside-in measure). We will overload the function symbol dp in order to keep the notation light.

**Definition 15.** *The* depth *of* $\mathcal{S}\{\ \}$, *denoted by* $\mathsf{dp}(\mathcal{S}\{\ \})$, *is defined inductively by* $\mathsf{dp}(\{\ \}) = 0$, $\mathsf{dp}(\Gamma \vdash \Delta, \Lambda, \{\ \}) = 1$ *and* $\mathsf{dp}(\Gamma \vdash \Delta, \Lambda, [\mathcal{S}\{\ \}]) = \mathsf{dp}(\mathcal{S}\{\ \}) + 1$. *If a formula* $A \in \Gamma, \Delta$, *then the* depth *of* $A$ *in* $\mathcal{S}\{\Gamma \vdash \Delta, \Lambda\}$ *is defined as* $\mathsf{dp}(\mathcal{S}\{\ \})$. *Finally, the* depth *of a nested sequent* $\Upsilon$, *written* $\mathsf{dp}(\Upsilon)$, *is defined as the maximum depth of formulae in* $\Upsilon$.

For example, if $\mathcal{S}\{\ \} = \Gamma \vdash \Delta, \{\ \}$ and $\Upsilon = \mathcal{S}\{\Gamma' \vdash \Delta', A, [\vdash B]\}$, then $\mathsf{dp}(\mathcal{S}\{\ \}) = 1$, the depth of $A$ in $\Upsilon$ is 1 and $\mathsf{dp}(\Upsilon) = 2$.

**Definition 16.** *Let* NS *be a nested system. The* depth of an application *of a rule* $r^n$ *in a derivation* ($\mathsf{dp}\,(r^n)$) *is the depth of the principal formula in the conclusion of* $r^n$.

*Example 17.* In the following derivation

$$\dfrac{\dfrac{\dfrac{\mathcal{S}\{\Gamma, B, C \vdash \Delta, \Lambda, [A, \Gamma' \vdash \Delta', D, E]\}}{\mathcal{S}\{\Gamma, B, C \vdash \Delta, \Lambda, [A, \Gamma' \vdash \Delta', D \vee E]\}}\,\vee_R^n}{\mathcal{S}\{\Gamma, B \wedge C \vdash \Delta, \Lambda, [A, \Gamma' \vdash \Delta', D \vee E]\}}\,\wedge_L^n}{\mathcal{S}\{\Gamma, A, B \wedge C \vdash \Delta, \Lambda, [\Gamma' \vdash \Delta', D \vee E]\}}\,\mathtt{lift}^n$$

$\mathsf{dp}\,(\mathtt{lift}^n) = \mathsf{dp}\,(\wedge_L^n) = \mathsf{dp}\,(\mathcal{S}\{\,\})$, while $\mathsf{dp}\,(\vee_R^n) = \mathsf{dp}\,(\mathcal{S}\{\,\}) + 1$.

The next definition brings a variant of nested systems where rules can be applied only in the deep-most nestings of a sequent (this is an adaptation to nested systems of the similar definition for *linear nested systems* [16]).

**Definition 18.** *Let* $\Upsilon$ *be a nested sequent with* $\mathsf{dp}\,(\Upsilon) \leq 1$ *and* $m = \mathsf{dp}\,(\mathcal{S}\{\,\})$. *An application of a basic nested sequent rule* $r^n$ *over* $\mathcal{S}\{\Upsilon\}$ *is* end-active *if* $\mathsf{dp}\,(r^n) = m$ *and*

- $r^n$ *is sequent-like and* $\mathsf{dp}\,(\Upsilon) = 0$*; or*
- $r^n$ *is a creation rule; or*
- $r^n$ *is an upgrade rule and* $\mathsf{dp}\,(\Upsilon) = 1$.

*The* end-active variant *of a* NS *calculus is the calculus with the rules restricted to end-active applications.*

*Example 19.* Consider the following (open) derivations of $A \wedge B \vdash C \to D, E \to (F \to G)$ in $\mathsf{NS}_{\mathsf{mLJ}}$.

$$(a)\quad \dfrac{\dfrac{\dfrac{\dfrac{\dfrac{A \vdash [B, C \vdash D], [E \vdash [F \vdash G]]}{A \vdash [B, C \vdash D], [E \vdash (F \to G)]}\,\to_R^n}{A \vdash E \to (F \to G), [B, C \vdash D]}\,\to_R^n}{A, B \vdash E \to (F \to G), [C \vdash D]}\,\mathtt{lift}^n}{A \wedge B \vdash E \to (F \to G), [C \vdash D]}\,\wedge_L^n}{A \wedge B \vdash C \to D, E \to (F \to G)}\,\to_R^n$$

$$(b)\quad \dfrac{\dfrac{\dfrac{\dfrac{\dfrac{A \vdash [B, C \vdash D], [E \vdash [F \vdash G]]}{A, B \vdash [C \vdash D], [E \vdash [F \vdash G]]}\,\mathtt{lift}^n}{A, B \vdash C \to D, [E \vdash [F \vdash G]]}\,\to_R^n}{A, B \vdash C \to D, [E \vdash (F \to G)]}\,\to_R^n}{A, B \vdash C \to D, E \to (F \to G)}\,\to_R^n}{A \wedge B \vdash C \to D, E \to (F \to G)}\,\wedge_L^n$$

In $(a)$, the application of the rule $\wedge_L^n$ is not end-active since $\mathsf{dp}\,(A \wedge B \vdash E \to (F \to G), [C \vdash D]) = 1$. In $(b)$, the topmost application of the rule $\to_R^n$ and the application of rule $\mathtt{lift}^n$ are not end-active since $\mathsf{dp}\,(A, B \vdash C \to D, [E \vdash [F \vdash G]]) = 2$. All the other rule applications are end-active.

The schematic representation of end-active basic nested rules is as follows:

i. *sequent-like* rules

$$\frac{\mathcal{S}\{\Gamma_1 \vdash \Delta_1\} \quad \cdots \quad \mathcal{S}\{\Gamma_k \vdash \Delta_k\}}{\mathcal{S}\{\Gamma \vdash \Delta\}}$$

ii. *nested-like* rules

(a) *creation rules* (b) *upgrade rules*

$$\frac{\mathcal{S}\{\Gamma \vdash \Delta, \Lambda, [S_1]\} \quad \cdots \quad \mathcal{S}\{\Gamma \vdash \Delta, \Lambda, [S_k]\}}{\mathcal{S}\{\Gamma \vdash \Delta, \Lambda\}} \qquad \frac{\mathcal{S}\{\Gamma \vdash \Delta, \Lambda, [S \otimes S']\}}{\mathcal{S}\{\Gamma \vdash \Delta, \Lambda, [S]\}}$$

where $\Lambda = \{\Lambda_1, \ldots, \Lambda_l\}$ is such that $\mathsf{dp}\,(\Lambda_i) = 1$ for all $0 \leq i \leq l$.

It turns out that basic nested systems always admit end-active versions, since some applications of rules *permute*. The following extends the definition of permutability to the nested setting.

**Definition 20.** *Let* NS *be a nested system,* $r_1, r_2$ *be applications rules and* $\Upsilon$ *be a nested sequent. We say that* $r_2$ permutes down $r_1$ *(*$r_2 \downarrow r_1$*) if, for every derivation in which* $r_1$ *has as conclusion* $\Upsilon$ *and* $r_2$ *is applied over one or more of* $r_1$*'s premises (but not on auxiliary formulae/nestings of* $r_1$*), there exists another derivation of* $\Upsilon$ *in which* $r_2$ *has as conclusion* $\Upsilon$ *and* $r_1$ *is applied over zero or more of* $r_2$*'s premises (but not on auxiliary formulae/nestings of* $r_2$*).*

*Example 21.* In Example 19 the application of the rule $\wedge_L^n$ permutes down w.r.t. $\rightarrow_R^n$ in $(a)$, the same with the applications of $\rightarrow_R^n$ and $\mathtt{lift}^n$ in $(b)$.

$$(a)\quad \dfrac{\dfrac{\dfrac{\dfrac{\dfrac{A \vdash [B, C \vdash D], [E \vdash [F \vdash G]]}{A \vdash [B, C \vdash D], [E \vdash F \rightarrow G]}\rightarrow_R^n}{A \vdash E \rightarrow (F \rightarrow G), [B, C \vdash D]}\rightarrow_R^n}{A, B \vdash E \rightarrow (F \rightarrow G), [C \vdash D]}\mathtt{lift}^n}{A, B \vdash C \rightarrow D, E \rightarrow (F \rightarrow G)}\rightarrow_R^n}{A \wedge B \vdash C \rightarrow D, E \rightarrow (F \rightarrow G)}\wedge_L^n$$

$$(b)\quad \dfrac{\dfrac{\dfrac{\dfrac{\dfrac{A \vdash [B, C \vdash D], [E \vdash [F \vdash G]]}{A \vdash [B, C \vdash D], [E \vdash F \rightarrow G]}\rightarrow_R^n}{A, B \vdash [C \vdash D], [E \vdash F \rightarrow G]}\mathtt{lift}^n}{A, B \vdash C \rightarrow D, [E \vdash F \rightarrow G]}\rightarrow_R^n}{A, B \vdash C \rightarrow D, E \rightarrow (F \rightarrow G)}\rightarrow_R^n}{A \wedge B \vdash C \rightarrow D, E \rightarrow (F \rightarrow G)}\wedge_L^n$$

In the derivations above, all applications of rules are end-active. Observe that they are different, but equivalent up-to-permutation derivations. Note also that end-activeness implies that the deep-most implication can be unfolded *only after* the application of all shallower rules.

**Definition 22.** *Let* NS *be a basic nested system. In any derivation* $\pi$ *in* NS*, a* sequential block $\mathcal{B}^s$ *(*nested block $\mathcal{B}^n$*) is a maximal bottom-up sequence of applications of sequent-like (nested-like) rules in a branch of* $\pi$ *having the same depth* $d$*. We will define the* depth *of such sequential (nested) block as* $\mathsf{dp}\,(\mathcal{B}^s) = d$ *(*$\mathsf{dp}\,(\mathcal{B}^n) = d$*).*

**Theorem 23.** *Any basic nested system admits an end-active variant. Moreover, in any end-active derivation, if $\mathcal{B}^s$ is the immediate successor of $\mathcal{B}^n$ then* $\mathsf{dp}(\mathcal{B}^s) = \mathsf{dp}(\mathcal{B}^n) + 1$.

*Proof.* The proof is by permutation of rules, using the fact that nested-like rules do not modify outer sequents, hence not extruding information. For example, upgrade rules ($r_2$) permute down creation rules ($r_1$):

$$\dfrac{\dfrac{\overset{\pi_1}{\mathcal{S}\{\Gamma \vdash \Delta, \Lambda, \Lambda'', [\Omega_1 \vdash \Theta_1]\}}}{\mathcal{S}\{\Gamma \vdash \Delta, \Lambda, \Lambda', [\Omega_1 \vdash \Theta_1]\}}\, r_2 \quad \cdots \quad \overset{\pi_k}{\mathcal{S}\{\Gamma \vdash \Delta, \Lambda, \Lambda', [\Omega_k \vdash \Theta_k]\}}}{\mathcal{S}\{\Gamma \vdash \Delta, \Lambda, \Lambda'\}}\, r_1 \quad \rightsquigarrow$$

$$\dfrac{\dfrac{\overset{\pi_1}{\mathcal{S}\{\Gamma \vdash \Delta, \Lambda, \Lambda'', [\Omega_1 \vdash \Theta_1]\}} \quad \cdots \quad \overset{\pi_k'}{\mathcal{S}\{\Gamma \vdash \Delta, \Lambda, \Lambda'', [\Omega_k \vdash \Theta_k]\}}}{\mathcal{S}\{\Gamma \vdash \Delta, \Lambda, \Lambda''\}}\, r_1}{\mathcal{S}\{\Gamma \vdash \Delta, \Lambda, \Lambda'\}}\, r_2$$

Observe that if $\mathcal{S}\{\Gamma \vdash \Delta, \Lambda, \Lambda', [\Omega_i \vdash \Theta_i]\}$ is provable with proof $\pi_i$ then it is the case that $\mathcal{S}\{\Gamma \vdash \Delta, \Lambda, \Lambda'', [\Omega_i \vdash \Theta_i]\}$ is provable with proof $\pi_i'$, a weakened version of $\pi_i$.

Note that restricting systems to its end-active form is not enough for guaranteeing that derivations occur in alternating sequent and nested blocks. Next, we define a depth first normalisation procedure for basic nested systems.

**Definition 24.** *Let* NS *be an end-active basic nested system. We say that a derivation $\pi$ of $\Upsilon$ in* NS *is in* normal form *(or $\pi$ is a normal derivation) if, for each branch of $\pi$,*

*a. if $\mathcal{B}^n$ is the immediate successor of $\mathcal{B}^s$ then* $\mathsf{dp}(\mathcal{B}^n) = \mathsf{dp}(\mathcal{B}^s)$*;*
*b. axioms are applied eagerly (i.e. as soon as possible).*

*Example 25.* The following end-active derivations in $\mathsf{NS_{mLJ}}$ *are not* in normal form

$$\dfrac{\dfrac{\dfrac{}{\Gamma \vdash \Delta, [A, B \vdash B]}\,\mathsf{init}^n}{\Gamma, A \vdash [B \vdash B], \Delta}\,\mathtt{lift}^n}{\Gamma, A \vdash B \rightarrow B, \Delta}\,\rightarrow_R^n \qquad \dfrac{\dfrac{\dfrac{\dfrac{}{\Gamma \vdash \Delta, [B, C \vdash C, D]}\,\mathsf{init}^n}{\Gamma, C \vdash \Delta, [B \vdash C, D]}\,\mathtt{lift}^n}{\Gamma, C \vdash \Delta, [B \vdash C \vee D]}\,\vee_R^n}{\Gamma, C \vdash B \rightarrow (C \vee D), \Delta}\,\rightarrow_R^n$$

The first since the axiom was not applied eagerly; and the second since the sequential block of depth 1 is succeeded by a nested block of depth 0.

**Theorem 26.** *Let* NS *be a basic nested system. Then any provable nested sequent $\Upsilon$ in* NS *has a normal derivation.*

*Proof.* By Theorem 23, we may consider the end-active variant of NS. The result follows by observing that nested-like rules permute down sequent-like rules.

In Sect. 5, we will show representative examples of systems falling into the class of end-active basic nested systems.

# 4 Recovering Sequent Systems

The proof of Theorem 26 provides a way of *pruning* nested derivations so to reach normal forms. However, the normalisation procedure given by Definition 24 may produce several different normal forms (see Example 21). We will show next how to further polish the normalisation process, so to avoid useless creation steps and output a *unique* normal form, that will allow for sequential proofs.

*Example 27.* Consider the following normal-form derivation in $\mathsf{NS}_{\mathsf{mLJ}}$, where $\Gamma_i \subseteq \Gamma$ and the top-most premise marks the end of a nested block (see Definition 22).

$$\dfrac{\dfrac{\overset{\pi}{\Gamma \vdash [\Gamma_1, A_1 \vdash B_1], \ldots, [\Gamma_k, A_k \vdash B_k]}}{\Gamma \vdash [A_1 \vdash B_1], \ldots, [A_k \vdash B_k]}\ \mathtt{lift}^n}{\Gamma \vdash A_1 \to B_1, \ldots, A_k \to B_k}\ \to_R^n$$

Since $\pi$ is in normal form, no rules can be applied over $\Gamma$. Hence, by Theorem 14, $\Gamma_i, A_i \vdash B_i$ is provable for some $1 \leq i \leq k$. Let $\pi_i$ be a normal-form proof of such sequent. Thus the proof above can be replaced by

$$\dfrac{\dfrac{\overset{\pi_i}{\Gamma \vdash A_1 \to B_1, \ldots, A_{i-1} \to B_{i-1}, A_{i+1} \to B_{i+1}, \ldots, A_k \to B_k, [\Gamma_i, A_i \vdash B_i]}}{\Gamma \vdash A_1 \to B_1, \ldots, A_{i-1} \to B_{i-1}, A_{i+1} \to B_{i+1}, \ldots, A_k \to B_k, [A_i \vdash B_i]}\ \mathtt{lift}^n}{\Gamma \vdash A_1 \to B_1, \ldots, A_k \to B_k}\ \to_R^n$$

Note that, since $\pi_1$ is normal, no rule can be applied to the outer context, which will be erased in the leaves by the $\mathsf{init}^n$ rule.

This idea can be generalised (with a trivial proof) to basic nested systems.

**Lemma 28.** *Let* NS *be a basic nested system. Then every normal derivation of a nested sequent* $\Upsilon$ *can be restricted so that* exactly one *creation rule is applied in any nested block.*

That is, normal derivations have alternating sequential and nested blocks with non-decreasing depth, such that the nested blocks are restricted to *one* application of a creation rule followed by possible applications of upgrade rules.

The next result shows how nested blocks are transformed into basic sequent rules.

**Theorem 29.** *Let* $r_c$ *be the creation rule and* $r_{u_i}$ *be the upgrade rules*

$$\dfrac{\vdash [\Omega_1 \vdash \Theta_1] \quad \cdots \quad \vdash [\Omega_k \vdash \Theta_k]}{\Gamma \vdash \Delta}\ r_c \qquad\qquad \dfrac{\vdash [\Psi_i \vdash \Xi_i]}{\Sigma_i \vdash \Phi_i, [\vdash]}\ r_{u_i}$$

*where* $\{\Sigma_i, \Phi_i\} = \{F_i\}, \{\Psi_i, \Xi_i\} = \{G_i,\}$, $F_i, G_i$ *formulae,* $1 \leq i \leq l$. *Then a nested block consisting of the application of* $r_c$ *followed by applications of* $r_{u_i}$

*in a normal-form derivation corresponds to the application of the basic sequent rule*

$$\frac{\langle \Omega_1 \vdash \Theta_1; C_u\rangle \quad \dots \quad \langle \Omega_k \vdash \Theta_k; C_u\rangle}{\Gamma \vdash \Delta}$$

*where* $C_u = \{\langle \mathsf{I} : G_i, \mathsf{J} : F_i\rangle \mid 1 \leq i \leq l\}$, $\mathsf{I}, \mathsf{J} \in \{\mathsf{T}, \mathsf{F}\}$.

*Proof.* Any nested block consisting of the application of $r_c$ followed by (maximal blocks of) applications of $r_{u_i}$ has the shape

$$\cfrac{\cfrac{\overset{\pi_1}{\mathcal{S}\{\Gamma \vdash \Delta, [\Gamma', \Omega_1 \vdash \Delta', \Theta_1]\}}}{\mathcal{S}\{\Gamma \vdash \Delta, [\Omega_1 \vdash \Theta_1]\}}\, r_{u_i} \quad \dots \quad \cfrac{\overset{\pi_k}{\mathcal{S}\{\Gamma \vdash \Delta, [\Gamma', \Omega_k \vdash \Delta', \Theta_k]\}}}{\mathcal{S}\{\Gamma \vdash \Delta, [\Omega_k \vdash \Theta_k]\}}\, r_{u_i}}{\mathcal{S}\{\Gamma \vdash \Delta\}}\, r_c$$

$$\mathsf{K}\ \Box(A \to B) \to (\Box A \to \Box B) \qquad \frac{A}{\Box A}\ \mathsf{nec}$$

**Fig. 3.** Modal axiom K and necessitation rule nec.

$$\frac{\mathcal{S}\{\Gamma, A \vdash B, \Delta, \Lambda\}}{\mathcal{S}\{\Gamma \vdash A \to B, \Delta, \Lambda\}}\ \to_R^n \qquad \frac{\mathcal{S}\{\Gamma \vdash \Delta, [\Gamma', A \vdash \Delta'], \Lambda\}}{\mathcal{S}\{\Gamma, \Box A \vdash \Delta, [\Gamma' \vdash \Delta'], \Lambda\}}\ \Box_L^n \qquad \frac{\mathcal{S}\{\Gamma \vdash \Delta, \Lambda, [\ \vdash A]\}}{\mathcal{S}\{\Gamma \vdash \Delta, \Box A, \Lambda\}}\ \Box_R^n$$

**Fig. 4.** Nested system $\mathsf{NS_K}$. Rules $\to_L^n, \wedge_R^n, \wedge_L^n, \vee_R^n, \vee_L^n$ and $\bot_L^n$ are the same as in Fig. 2.

Considering that $\pi_j$ is in normal form, $1 \leq j \leq k$, the only active formulae in the leaves will be in $\Gamma', \Omega_j \vdash \Delta', \Theta_j$. Thus nested blocks transform sequents into sequents, and they can be seen as a macro sequent-like rule. With this thinking, an upgrade nested rule $r_{u_i}$ is actually a *context relation* of the form $C_{u_i} = \langle \mathsf{I} : G_i, \mathsf{J} : F_i\rangle$, $1 \leq i \leq l$. Hence the result follows.

**Corollary 30.** *Let* NS *be a basic nested system which sequentialises to the sequent system* SC. *Then the sequent* $\Gamma \vdash \Delta$ *is provable in* NS *iff it is provable in* SC.

*Example 31.* In $\mathsf{NS_{mLJ}}$, a nested block containing the creation rule $\to_R^n$ and the upgrade rule $\mathtt{lift}^n$ has the shape

$$\cfrac{\cfrac{\mathcal{S}\{\Gamma \vdash \Delta, [\Gamma', A \vdash B]\}}{\mathcal{S}\{\Gamma \vdash \Delta, [A \vdash B]\}}\ \mathtt{lift}^n}{\mathcal{S}\{\Gamma \vdash \Delta, A \to B\}}\ \to_R^n$$

with $\Gamma' \subseteq \Gamma$. Observe that $\texttt{lift}^n$ maps an F formula into itself and there are no context relations on T formulae. Hence $C_u = C_{\mathsf{int}}$, and the corresponding sequent rule is

$$\frac{\langle p_1 \vdash p_2; C_{\mathsf{int}} \rangle}{\vdash p_1 \rightarrow p_2}$$

which is the implication right rule for mLJ. That is, sequentialising the basic nested system $\mathsf{NS}_{\mathsf{mLJ}}$ (Fig. 2) results in the sequent system mLJ (Fig. 1).

## 5 Examples and Discussion

In the previous sections we used intuitionistic logic as a running example for illustrating our method approach. In this section we will apply the sequentialisation procedure to other well known logical systems.

**Normal Modalities.** The modal logic K is obtained from classical propositional logic by adding the unary modal connective $\Box$ to the set of classical connectives, together with the necessitation rule and the K axiom (see Fig. 3 for the Hilbert-style axiom schemata) to the set of axioms for propositional classical logic.

The nested framework provides an elegant way of formulating modal systems, since no context restriction is imposed on rules. Figure 4 presents the schemata of the modal rules for the nested sequent calculus $\mathsf{NS}_\mathsf{K}$ for the modal logic K [4,20]. Observe that there are two rules for handling the box operator ($\Box_L$ and $\Box_R$), which allows the treatment of one formula at a time. While this is one of the main features of nested sequent calculi and deep inference in general [9], being able to separate the left/right behaviour of the modal connectives is the key to modularity for nested calculi [14,22]. Indeed, K can be modularly extended by adding to $\mathsf{NS}_\mathsf{K}$ the nested rules corresponding to other modal axioms.

D $\Box A \rightarrow \Diamond A$ T $\Box A \rightarrow A$ 4 $\Box A \rightarrow \Box\Box A$

(a) Axioms D, T, 4.

B $A \rightarrow \Box\Diamond A$ 5 $\Diamond A \rightarrow \Box\Diamond A$

(b) Axioms B, 5.

**Fig. 5.** Axioms D, T, 4, B and 5, where $\Diamond A$ is a short for $\neg\Box\neg A$.

$$\frac{S\{\Gamma \vdash \Delta, [A \vdash\ ], \Lambda\}}{S\{\Gamma, \Box A \vdash \Delta, \Lambda\}}\ \mathsf{d}^n \qquad \frac{S\{\Gamma, A \vdash \Delta, \Lambda\}}{S\{\Gamma, \Box A \vdash \Delta, \Lambda\}}\ \mathsf{t}^n \qquad \frac{S\{\Gamma \vdash \Delta, [\Gamma', \Box A \vdash \Delta'], \Lambda\}}{S\{\Gamma, \Box A \vdash \Delta, [\Gamma' \vdash \Delta'], \Lambda\}}\ 4^n$$

$$\mathsf{NS}_{\mathsf{K}\mathcal{A}}: \quad \{\Box^n_R, \Box^n_L\} \cup \mathcal{A} \quad \text{for } \mathcal{A} \subseteq \{\mathsf{D}, \mathsf{T}, 4\}$$

**Fig. 6.** Nested sequent rules for $\{\mathsf{D}, \mathsf{T}, 4\}$ extensions of K.

Let us first consider the axioms D, T and 4 (Fig. 5a). Figure 6 shows the modal nested rules for such extensions: for a logic $\mathsf{K}\mathcal{A}$ with $\mathcal{A} \subseteq \{\mathsf{D}, \mathsf{T}, 4\}$ the calculus $\mathsf{NS}_{\mathsf{K}\mathcal{A}}$ extends $\mathsf{NS}_\mathsf{K}$ with the corresponding nested modal rules.

Note that $\mathsf{t}^n$ is actually a sequent-like rule. On the other hand, $\Box_R^n$ and $\mathsf{d}^n$ are creation rules while $\Box_L^n$ and $4^n$ are upgrade rules. It is straightforward to verify that $\mathsf{NS}_{\mathsf{K}\mathcal{A}}$ is basic. Observe that $4^n$ maps a boxed F formula into itself, $\Box_L^n$ maps F formulae into the boxed versions and there are no context relations on T formulae. Hence $C_u = C_{\mathsf{K}} \cup C_4$, and the basic sequent rules corresponding to T, K and D (with possibly 4) are, respectively

$$\frac{\langle p \vdash; C_{\mathtt{id}}\rangle}{\Box p \vdash} \qquad \frac{\langle \vdash p; C_{\mathsf{K}} \cup C_4\rangle}{\vdash \Box p} \qquad \frac{\langle p \vdash; C_{\mathsf{K}} \cup C_4\rangle}{\Box p \vdash}$$

understanding that if the axiom 4 is not present in the logic then the relation $C_4$ is dropped. Hence sequentialising the nested system $\mathsf{NS}_{\mathsf{K}\mathcal{A}}$ results in the sequent system $\mathsf{SC}_{\mathsf{K}\mathcal{A}}$ (shown as rule schemas) in Fig. 7a.

We now move our attention to the extension of K4 containing axiom 5 (Fig. 5b). The rule 45 presented in Fig. 8 is a local rule schema for capturing the behaviour of 5 in the presence of the nested rules for $\mathsf{NS}_{\mathsf{K4}}$ (see [4] for a discussion on the decomposition of $5^n$ in local rules). Observe that the rule $45^n$ is an upgrade rule, hence sequentialising the nested system $\mathsf{NS}_{\mathsf{KT45}}$ results in the sequent system $\mathsf{SC}_{\mathsf{KT45}}$ (Fig. 7b). Hence, since $\mathsf{SC}_{\mathsf{KT45}}$ is not cut-free, this implies that $\mathsf{NS}_{\mathsf{KT45}}$ is also not cut-free (see *e.g.* [25]).

The rule $\mathsf{b}^n$ (Fig. 8), corresponding to axiom B, is *not basic*. Hence systems $\mathsf{NS}_{\mathsf{K}\mathcal{A}}$ extended with this nested rule cannot be sequentialised. However, our approach gives a good insight on the relationship between the extruding information from nestings and cut-elimination in sequent systems (which will be discussed later in this Section).

**Negative Modalities.** While normal modal modalities satisfy the monotone property "if $A \vdash B$ then $\Box A \vdash \Box B$", negative modalities satisfy the antitone: "if $A \vdash B$ then $\Box B \vdash \Box A$". The logic $PK$ [12] has four 1-ary connectives $\smile, \frown, \ominus, \ominus$, interpreted non-locally in terms of a Kripke model $\mathcal{M} = \langle W, R, V\rangle$ as follows

- $\mathcal{M}, w \Vdash {\smile}A$ iff $\mathcal{M}, v \nVdash A$ for some $v \in W$ such that $wRv$;
- $\mathcal{M}, w \Vdash {\frown}A$ iff $\mathcal{M}, v \nVdash A$ for every $v \in W$ such that $wRv$;
- $\mathcal{M}, w \Vdash \ominus A$ iff $\mathcal{M}, w \nVdash A$ or $\mathcal{M}, w \nVdash -A$ for $- \in \{\smile, \frown\}$;

$$\frac{\Gamma \vdash A}{\Box\Gamma \vdash \Box A}\,\mathsf{k} \quad \frac{\Gamma, A \vdash \Delta}{\Gamma, \Box A \vdash \Delta}\,\mathsf{t} \quad \frac{\Gamma, A \vdash}{\Box\Gamma, \Box A \vdash}\,\mathsf{d} \quad \frac{\Box\Gamma_4, \Gamma_{\mathsf{K}} \vdash A}{\Box\Gamma_4, \Box\Gamma_{\mathsf{K}} \vdash \Box A}\,\mathsf{k4} \quad \frac{\Box\Gamma_4, \Gamma_{\mathsf{K}}, A \vdash}{\Box\Gamma_4, \Box\Gamma_{\mathsf{K}}, \Box A \vdash}\,\mathsf{d4} \quad \frac{\Gamma \vdash A, \Box\Delta}{\Box\Gamma \vdash \Box A, \Delta}\,\mathsf{b} \quad \frac{\Box\Gamma, \vdash A, \Box\Delta}{\Box\Gamma \vdash \Box A, \Box\Delta}\,45$$

$\mathsf{SC}_{\mathsf{K}}$ {k} $\mathsf{SC}_{\mathsf{KT}}$ {k, t} $\mathsf{SC}_{\mathsf{KD}}$ {k, d} $\mathsf{SC}_{\mathsf{K4}}$ {k4} $\mathsf{SC}_{\mathsf{KD4}}$ {d4} $\mathsf{SC}_{\mathsf{KB}}$ {B} $\mathsf{SC}_{\mathsf{K45}}$ {45}

**(a) Seq. rules for T, D, 4.** **(b) Seq. rules for B, 45.**

**Fig. 7.** Modal sequent rules for normal modal logics $\mathsf{SC}_{\mathsf{K}\mathcal{A}}$, for $\mathcal{A} \subseteq \{\mathsf{T}, \mathsf{D}, 4, \mathsf{B}, 45\}$.

$$\frac{\mathcal{S}\{\Gamma \vdash \Delta, \Lambda, [\Gamma' \vdash \Box A, \Delta', \Lambda']\}}{\mathcal{S}\{\Gamma \vdash \Delta, \Lambda, \Box A, [\Gamma' \vdash \Delta', \Lambda']\}}\,45^n \qquad \frac{\mathcal{S}\{\Gamma, A \vdash \Delta, \Lambda, [\Gamma' \vdash \Delta', \Lambda']\}}{\mathcal{S}\{\Gamma \vdash \Delta, \Lambda, [\Gamma', \Box A \vdash \Delta', \Lambda']\}}\,\mathsf{b}^n$$

**Fig. 8.** Basic nested sequent rules for axioms 45 and B.

We present in Fig. 9a proposal of a nested system for such negative modalities. Observe that rules for $\ominus, \circleddash$ are sequent-like, the lift rules are upgrade and the rules for $\smile, \frown$ are creation rules. Hence by sequentialising $\mathsf{NS}_{PK}$ we obtain the basic sequent rules

$$\frac{\langle \vdash p; C_{PK}\rangle}{\smile p \vdash} \qquad \frac{\langle p \vdash; C_{PK}\rangle}{\vdash \frown p} \qquad \frac{\langle p, \smile p \vdash; C_{\mathtt{id}}\rangle}{\vdash \ominus p}$$

$$\frac{\langle \vdash p; C_{\mathtt{id}}\rangle \quad \langle \vdash \smile p; C_{\mathtt{id}}\rangle}{\ominus p \vdash} \qquad \frac{\langle p \vdash; C_{\mathtt{id}}\rangle \quad \langle \frown p \vdash; C_{\mathtt{id}}\rangle}{\vdash \circleddash p} \qquad \frac{\langle \vdash p, \frown p; C_{\mathtt{id}}\rangle}{\circleddash p \vdash}$$

where $C_{PK} := \{\langle \mathsf{F} : p, \mathsf{T} : \smile p\rangle, \langle \mathsf{T} : p, \mathsf{F} : \frown p\rangle\}$, which are *exactly* the basic sequent rules for the system $\mathsf{SC}_{PK}$ presented in [12]. Hence $\mathsf{NS}_{PK}$ is sound and complete w.r.t. the Kripke semantics described above by Corollary 30.

**Learning from Failure: The Case of B.** The work in [15] suggests that it should be hard, if not impossible, to define simple, cut-free sequent systems for logics with no corresponding *basic nested* systems[3]. In this work, we advocate that the sole responsible for this impossibility are: (a) on allowing extruding formulae from nestings (when seen bottom-up), one could gather more information, adding an extra-advantage not allowed in the meta-language of sequents; and (b) on allowing information to "jump" over more than one nesting level, the stepwise nature of sequents forces this information to be lost. In fact, our sequentialisation method is *heavily* based on the fact that basic nested-like rules move formulae to *deeper nestings*, with depth difference *exactly equal to one*.

For stressing these points better, we will present a relation between analytic cuts and the lack of basic nested rules for KB (Fig. 8). The following definition shows how to interpret nestings as formulae in the modal framework in the S5 cube.

$$\frac{S\{\Gamma \vdash \Delta, \Lambda, [\vdash A]\}}{S\{\Gamma, \smile A \vdash \Delta, \Lambda\}} \smile^n \qquad \frac{S\{\Gamma \vdash \Delta, \Lambda, [A \vdash]\}}{S\{\Gamma \vdash \Delta, \frown A, \Lambda\}} \frown^n \qquad \frac{S\{\Gamma, A, \smile A \vdash \Delta, \Lambda\}}{S\{\Gamma \vdash \Delta, \ominus A, \Lambda\}} \ominus^n_R$$

$$\frac{S\{\Gamma \vdash \Delta, \Lambda\, [\Gamma' \vdash \Delta', A, \Lambda']\}}{S\{\Gamma, \frown A \vdash \Delta, \Lambda, [\Gamma' \vdash \Delta', \Lambda']\}} \mathtt{lift}^n_L \qquad \frac{S\{\Gamma \vdash \Delta, \Lambda\, [\Gamma', A \vdash \Delta', \Lambda']\}}{S\{\Gamma \vdash \Delta, \smile A, \Lambda, [\Gamma' \vdash \Delta', \Lambda']\}} \mathtt{lift}^n_R$$

$$\frac{S\{\Gamma \vdash \Delta, A, \Lambda\} \quad S\{\Gamma \vdash \Delta, \smile A, \Lambda\}}{S\{\Gamma, \ominus A \vdash \Delta, \Lambda\}} \ominus^n_L \qquad \frac{S\{\Gamma \vdash \Delta, A, \frown A, \Lambda\}}{S\{\Gamma, \circleddash A \vdash \Delta, \Lambda\}} \circleddash^n_L$$

$$\frac{S\{\Gamma, A \vdash \Delta, \Lambda\} \quad S\{\Gamma, \frown A \vdash \Delta, \Lambda\}}{S\{\Gamma \vdash \Delta, \circleddash A, \Lambda\}} \circleddash^n_R$$

**Fig. 9.** Nested system $\mathsf{NS}_{PK}$.

[3] We observe that, in [11] the basic *sequent* systems for KB and S5 were proved to be *analytic* (although not cut-free).

**Definition 32.** *The modal interpretation* $\iota_\Box$ *for modal nested sequents is given by*

- *if* $\Gamma \vdash \Delta$ *is a sequent, then* $\iota_\Box(\Gamma \vdash \Delta) = \bigwedge \Gamma \rightarrow \bigvee \Delta$.
- $\iota_\Box(\Gamma \vdash \Delta, [\Gamma_1 \vdash \Delta_1, \Lambda_1], \ldots, [\Gamma_n \vdash \Delta_n, \Lambda_n]) =$
  $\bigwedge \Gamma \rightarrow (\bigvee \Delta \vee \Box(\iota_\Box(\Gamma_1 \vdash \Delta_1, \Lambda_1)) \vee \ldots \vee \Box(\iota_\Box(\Gamma_n \vdash \Delta_n, \Lambda_n)))$.

That is, the structural connective $[\cdot]$ is interpreted by the logical connective $\Box$.

Consider a proof of the shape

$$\dfrac{\begin{matrix}\pi \\ \Gamma, A \vdash \Delta, [\Gamma', \Box A \vdash \Delta']\end{matrix}}{\Gamma \vdash \Delta, [\Gamma', \Box A \vdash \Delta']}\ \mathsf{b}^n$$

where $\pi$ has no occurrences of the $\mathsf{b}^n$ rule. We may assume that $A$ is principal in $\pi$, otherwise this instance application of $\mathsf{b}^n$ can be discarded and the results from Sect. 4 apply immediately. Thus, all rules applied in $\pi$ are basic and, by Theorem 26, we may assume that $\pi$ is in normal form. Hence, by Theorem 29, $\pi$ can be transformed into a derivation $\iota_\Box(\pi)$ of the sequent $A, \Gamma \vdash \Delta, \Box(\bigwedge \Gamma' \wedge \Box A \rightarrow \bigvee \Delta')$. Now, the following is derivable in $\mathsf{KB}$

$$\dfrac{\dfrac{\dfrac{\dfrac{}{\Gamma', \Box A \vdash \Delta', \Box A}\ \mathsf{init}}{\vdash \bigwedge \Gamma' \wedge \Box A \rightarrow \bigvee \Delta', \Box A}\ {\rightarrow_R}, {\wedge_L}, {\vee_R}}{\vdash \Box(\bigwedge \Gamma' \wedge \Box A \rightarrow \bigvee \Delta'), A}\ \mathsf{b} \qquad \begin{matrix}\iota_\Box(\pi) \\ A, \Gamma \vdash \Delta, \Box(\bigwedge \Gamma' \wedge \Box A \rightarrow \bigvee \Delta')\end{matrix}}{\Gamma \vdash \Delta, \Box(\bigwedge \Gamma' \wedge \Box A \rightarrow \bigvee \Delta')}\ \mathsf{cut}$$

That is, the *end-active* application of the $\mathsf{b}^n$ rule in $\mathsf{NS_{KB}}$ can be mimicked by a proof in $\mathsf{KB}$ with an *analytic cut* whose cut-formula is the auxiliary formula in $\mathsf{b}^n$. This establishes a (so far, weak) connection between nested derivations and analytic cuts in sequent calculi for $\mathsf{KB}$. Generalising this correlation is an ongoing work.

## 6 Conclusion and Future Work

A common theme in recent structural proof theory is the development of new proof formalisms generalising and extending the sequent calculus, in order to present analytic proof calculi for the ever growing number of logics of interest. This work considers the reverse direction: how can we transform an analytic nested calculus into an analytic sequent calculus? Given that the nested sequent calculi generalise the sequent calculus, and because the former has been used to present logics that have defined presentation in the latter, an underlying aim of this work is to identify general characteristics that make nested calculi "sequentialisable". In doing so, we open a new insight into the discussion of the bounds for analyticity in sequent systems.

There are many ways of continuing this research topic. First of all, since we showed how to transform nested into sequent systems it would be interesting to ask: how about the other way around? Or: when it is possible to transform basic sequent rules into basic nested rules? This would allow for the automatic generation of (analytic, possibly cut-free) nested systems from sequent systems. One possible attempt would be analysing if the Kripke-style semantic interpretation of basic sequent systems given in [11] can lifted to the nestings-as-worlds interpretation of nested systems [7,19].

Another path worth investigating is if our approach entails *negative results*, such as the impossibility of cut-free basic systems for KB, for example, as a generalisation of results in [15]. Not mentioning developing further the relationship between the introduction of cuts in sequent calculi and the need for nested-like rules in nested systems, discussion we have started here.

Finally, we would like to analyse to what extend our setting can handle other systems, such as the known ones for non-normal modal logics, $GL$ and $PDL$.

**Acknowledgments.** We would like to thank Agata Ciabattoni for our fruitful discussions and the anonymous reviewers for their valuable comments.

## References

1. Avron, A.: Simple consequence relations. Inf. Comput. **92**(1), 105–140 (1991). https://doi.org/10.1016/0890-5401(91)90023-U
2. Avron, A.: The method of hypersequents in the proof theory of propositional non-classical logics. In: Logic: From Foundations to Applications (Staffordshire, 1993), Oxford Science Publishing, pp. 1–32. Oxford University Press, New York (1996)
3. Brünnler, K.: Deep sequent systems for modal logic. In: Governatori, G., Hodkinson, I., Venema, Y. (eds.) Advances in Modal Logic, vol. 6, pp. 107–119. College Publications, London (2006)
4. Brünnler, K.: Deep sequent systems for modal logic. Arch. Math. Log. **48**, 551–577 (2009)
5. Chaudhuri, K., Marin, S., Straßburger, L.: Modular focused proof systems for intuitionistic modal logics. In: FSCD, pp. 16:1–16:18 (2016)
6. Fitting, M.: Proof methods for modal and intuitionistic logics. In: Synthese Library, vol. 169. Reidel Publishing Company, Dordrecht (1983)
7. Fitting, M.: Nested sequents for intuitionistic logics. Notre Dame J. Form. Log. **55**(1), 41–61 (2014)
8. Gentzen, G.: Investigations into logical deduction. In: The Collected Papers of Gerhard Gentzen, pp. 68–131 (1969)
9. Guglielmi, A., Straßburger, L.: Non-commutativity and MELL in the calculus of structures. In: Fribourg, L. (ed.) CSL 2001. LNCS, vol. 2142, pp. 54–68. Springer, Heidelberg (2001). https://doi.org/10.1007/3-540-44802-0_5
10. Kashima, R.: Cut-free sequent calculi for some tense logics. Studia Logica **53**(1), 119–135 (1994)
11. Lahav, O., Avron, A.: A unified semantic framework for fully structural propositional sequent systems. ACM Trans. Comput. Log. **14**(4), 27:1–27:33 (2013). http://doi.acm.org/10.1145/2528930
12. Lahav, O., Marcos, J., Zohar, Y.: Sequent systems for negative modalities. Logica Universalis **11**(3), 345–382 (2017). https://doi.org/10.1007/s11787-017-0175-2

13. Lellmann, B.: Sequent calculi with context restrictions and applications to conditional logic. Ph.D. thesis, Imperial College London (2013). http://hdl.handle.net/10044/1/18059
14. Lellmann, B.: Linear nested sequents, 2-sequents and hypersequents. In: De Nivelle, H. (ed.) TABLEAUX 2015. LNCS (LNAI), vol. 9323, pp. 135–150. Springer, Cham (2015). https://doi.org/10.1007/978-3-319-24312-2_10
15. Lellmann, B., Pattinson, D.: Correspondence between modal Hilbert axioms and sequent rules with an application to S5. In: Galmiche, D., Larchey-Wendling, D. (eds.) TABLEAUX 2013. LNCS (LNAI), vol. 8123, pp. 219–233. Springer, Heidelberg (2013). https://doi.org/10.1007/978-3-642-40537-2_19
16. Lellmann, B., Pimentel, E.: Modularisation of sequent calculi for normal and non-normal modalities. ACM Trans. Comput. Log. **29** (2019). Article No. 7
17. Maehara, S.: Eine Darstellung der intuitionistischen Logik in der klassischen. Nagoya Math. J. **7**, 45–64 (1954)
18. Mints, G.: Indexed systems of sequents and cut-elimination. J. Philos. Logic **26**(6), 671–696 (1997)
19. Pimentel, E.: A semantical view of proof systems. In: Moss, L.S., de Queiroz, R., Martinez, M. (eds.) WoLLIC 2018. LNCS, vol. 10944, pp. 61–76. Springer, Heidelberg (2018). https://doi.org/10.1007/978-3-662-57669-4_3
20. Poggiolesi, F.: The method of tree-hypersequents for modal propositional logic. In: Makinson, D., Malinowski, J., Wansing, H. (eds.) Towards Mathematical Philosophy. TL, vol. 28, pp. 31–51. Springer, Dordrecht (2009). https://doi.org/10.1007/978-1-4020-9084-4_3
21. Pottinger, G.: Uniform, cut-free formulations of $t$, $s4$ and $s5$ (abstract). J. Symbolic Logic **48**, 900–901 (1983)
22. Straßburger, L.: Cut elimination in nested sequents for intuitionistic modal logics. In: Pfenning, F. (ed.) FoSSaCS 2013. LNCS, vol. 7794, pp. 209–224. Springer, Heidelberg (2013). https://doi.org/10.1007/978-3-642-37075-5_14
23. Troelstra, A.S., Schwichtenberg, H.: and Helmut Schwichtenberg. Basic Proof Theory. Cambridge University Press, Cambridge (1996)
24. Luca, V.: Labelled Non-classical Logics. Kluwer, Dordrecht (2000)
25. Wansing, H.: Sequent systems for modal logics. In: Gabbay, D.M., Guenthner, F. (eds.) Handbook of Philosophical Logic, vol. 8. Springer, Heidelberg (2002). https://doi.org/10.1007/978-94-010-0387-2_2

# A Hypersequent Calculus with Clusters for Data Logic over Ordinals

Anthony Lick(✉)

LSV, CNRS & ENS Paris-Saclay, Université Paris-Saclay, Cachan, France
anthony.lick@lsv.fr

**Abstract.** We study freeze tense logic over well-founded data streams. The logic features past- and future-navigating modalities along with freeze quantifiers, which store the datum of the current position and test data (in)equality later in the formula. We introduce a decidable fragment of that logic, and present a proof system that is sound for the whole logic, and complete for this fragment. Technically, this is a hypersequent system enriched with an ordering, clusters, and annotations. The proof system is tailored for proof search, and yields an optimal coNP complexity for validity and a small model property for our fragment.

**Keywords:** Modal logic · Data ordinals · Freeze logic · Proof system · Hypersequent calculus

## 1 Introduction

*Data Streams.* Many applications can generate *data streams*, such as traces of a program's execution [22], system logs [7], XML streams [19], or intrusions detection [21], which motivates the study of *data words* and *data $\omega$-words* in order to be able to formally reason about such streams. They consist respectively of finite and infinite sequences in which each position carries a *label* from a finite alphabet and a *datum* from an infinite domain.

Consider for instance a system where multiple processes could be editing the same file on some server. The log of their execution can be represented as an infinite data word, the datum being an integer identifying the process, and the label representing their action: $b$ for the beginning of a process, $e$ for its ending, and $r$ (resp. $w$) when a process reads (resp. writes) the file. On such a data $\omega$-word, we could want to verify various properties:

1. Every process does not do anything after it stops or before it starts, i.e. for every datum, the corresponding subword belongs to $b(r+w)^*e$.
2. For every position labelled by $w$, there exists an earlier position labelled by $r$ with the same datum such that there is no position in-between labelled by $w$ and carrying a different datum.

Funded by ANR-14-CE28-0005 PRODAQ. I am grateful to David Baelde and Sylvain Schmitz for their valuable advice and for proofreading this paper.


S. Cerrito and A. Popescu (Eds.): TABLEAUX 2019, LNAI 11714, pp. 166–184, 2019.
https://doi.org/10.1007/978-3-030-29026-9_10

On the following infinite data word, only the first property is respected.

$$(b,1)(r,1)(b,2)(r,2)(w,1)(w,2)(e,2)(e,1)(b,3)(r,3)(e,3)\ \cdots$$

Moreover, working with ordinals instead of words can sometimes be useful to model some problems more easily. For instance, Demri and Novak [16] use ordinals to model Zeno behaviours in physical systems [16], and Godefroid and Wolper [20] to model-check $n$ concurrent executions while avoiding exploring their $n!$ interleavings. Hence, extending LTL to ordinal structures in the data-free case has already been investigated [15].

*Data Logics.* Among the many logics developed to reason about data words, freeze LTL [14,18,30] extends linear temporal logic [37] with *freeze quantifiers*: a formula $\downarrow_r \varphi$ stores the current datum in register $r$ and evaluates $\varphi$; in this scope, $\uparrow_r$ is satisfied if the current datum matches the one stored in $r$. As always with data logics, the satisfiability problem for freeze LTL is undecidable and its known decidable fragments are untractable [14,18].

*Contributions.* One of the main computational problems associated with a logic is the *satisfiability* problem. In this paper, we investigate the satisfiability problem of the *freeze tense logic* over data ordinals, which we call $\mathbf{K_t^\downarrow L_\ell.3}$, and which combines freeze quantifiers à la Demri and Lazić [14] with the tense logic over ordinals $\mathbf{K_t L_\ell.3}$. Our temporal core is thus Prior's tense logic [9,40], which only features the strict 'past' $\mathsf{H}$ and 'future' $\mathsf{G}$ temporal modalities (and their duals $\mathsf{P}$ and $\mathsf{F}$), but this is sufficient for many modelling tasks [42], and is known to lead to an NP-complete satisfiability problem over arbitrary linear time flows [36], over $\omega$-words [41], and over arbitrary ordinals [3]. For instance, Property 1 above can be expressed by

$$\begin{aligned}&\mathsf{G}\left(b \supset \downarrow_r\left(\mathsf{H}\neg\uparrow_r \wedge \mathsf{G}(b \supset \neg\uparrow_r) \wedge \mathsf{F}\left(\uparrow_r \wedge e \wedge \mathsf{G}\neg\uparrow_r \wedge \mathsf{H}\left(e \supset \neg\uparrow_r\right)\right)\right)\right)\\ &\wedge \mathsf{G}((e \vee r \vee w) \supset \downarrow_r \mathsf{P}\left(\uparrow_r \wedge b\right))\end{aligned}$$

The full freeze tense logic $\mathbf{K_t^\downarrow L_\ell.3}$ is already undecidable with a single register, just like freeze LTL over finite words [14,18]. We present a decidable fragment, dubbed $\mathbf{K_t^d L_\ell.3}$, in which the use of registers is further restricted, and which is exactly as expressive as the two-variable fragment of the first order logic on data words [6]. We show in particular that

1. the satisfiability problem for $\mathbf{K_t^d L_\ell.3}$ over the class of ordinals is NP-complete,
2. a formula $\varphi$ of $\mathbf{K_t^d L_\ell.3}$ has a well-founded linear model if and only if it has a model of order type $\alpha$ for some $\alpha < \omega \cdot (4 \cdot |\varphi|^2 + |\varphi| + 2)$; this should be contrasted with the corresponding $\omega \cdot (|\varphi| + 2)$ bound proven in [3, Prop. 4.1] for the underlying data-free logic $\mathbf{K_t L_\ell.3}$.

These results are however just by-products of our main contribution, which is a sound and complete proof system for $\mathbf{K_t^d L_\ell.3}$ in which proof search is in coNP. Moreover, our system allows to work not only with data $\omega$-words but with arbitrary data ordinals, which provides greater modelling flexibility.

*Algorithmic Approaches.* Algorithmic results for data logics are often obtained via *automata-theoretic* techniques [6,14,43], by building an enriched automaton recognising the models of a given formula and testing it for emptiness. However, this kind of approach might not be modular as the type of enriched automata is often tailored for each specific logic. Moreover, if one's interest is to check that a formula $\varphi$ is valid, the automata-theoretic approach does not yield a 'natural' certificate that could be checked by simple independent means.

All these considerations motivate our use of *proof-theoretic* techniques. The primary example of proof system amenable to automated reasoning is Gentzen's sequent calculus. However, it is often too limited for modal logics. Hence, it has been enriched in various ways, using e.g. labelled sequents [35], display calculus [5,28], nested sequents [8,27,31,38,39], or hypersequents [1,23,24,29]. These enriched formalisms remain quite modular and sustain extensions simply by adding a few rules. They can be exploited to provide optimal complexity solutions to the validity problem directly by proof search [2–4,12,26,34], which may sometimes avoid the worst-case complexity of the problem and rely in practice on various heuristics. Finally, this approach obviously yields a proof of validity as a certificate in case of success.

Specifically, we use the framework of ordered hypersequents *with clusters* introduced in [2] as an elaboration, with terminating proof search, of Indrzejczak's ordered hypersequent calculus for $\mathbf{K_t4.3}$ [24,25], and which we have generalised [3] to work over ordinals. Conceptually, re-using the framework required to adapt it to work with data ordinals, to use additional rules to deal with registers, and to develop a strategy to make sure that proof search always produces proof attempts of polynomial depth. Moreover, this framework uses *annotations* to bound the proof search, and we managed to handle them as a new type of formulæ rather than just as an artefact of the proof system.

Furthermore, as in [3], our proof system can be easily adapted to also address the more precise problems of validity over all the data ordinals of order type $\beta < \alpha + 1$ for a given $\alpha$ and of order type exactly $\alpha < \omega^2$—which is recalled in Sect. 6. Such a result seems out of reach for axiomatisations, and yields for instance a coNP decision procedure for validity over data $\omega$-words.

The detailed proof of every claim of this paper is available in the full version of the paper [33] (https://hal.archives-ouvertes.fr/hal-02165359).

## 2 Freeze Tense Logic over Ordinals

*Syntax.* Our logic, called $\mathbf{K_t^{\downarrow}L_{\ell}.3}$, features two unary temporal operators from the tense logic, countably many freeze operators, and a countable set $\Phi$ of propositional variables, with the following syntax:

$$\varphi ::= \bot \mid p \mid \varphi \supset \varphi \mid \mathsf{G}\varphi \mid \mathsf{H}\,\varphi \mid \downarrow_r\varphi \mid \uparrow_r \qquad \text{(where } p \in \Phi \text{ and } r \in \mathbb{N}\text{)}$$

Formulæ $\mathsf{G}\varphi$ and $\mathsf{H}\,\varphi$ are called *modal formulæ*. Intuitively, $\mathsf{G}\varphi$ expresses that $\varphi$ holds 'globally' in all future worlds, while $\mathsf{H}\,\varphi$ expresses that $\varphi$ holds 'historically' in all past worlds. Other Boolean connectives may be encoded from

$\perp$ and $\supset$, and as usual $\mathsf{F}\,\varphi = \neg\mathsf{G}\neg\varphi$ expresses that $\varphi$ will hold 'in the future' and $\mathsf{P}\,\varphi = \neg\mathsf{H}\neg\varphi$ that it held 'in the past'. Formulæ $\downarrow_r\varphi$ are called *freeze formulæ*, and atoms $\uparrow_r$ are called *thaw formulæ*. Intuitively, $\downarrow_r\varphi$ stores the datum of the current world in the register $r$, and evaluates $\varphi$, and $\uparrow_r$ tests if the current world has the same datum as the one stored in the register $r$. Any occurrence of a thaw $\uparrow_r$ within the scope of a freeze quantifier $\downarrow_r$ is bounded by it; otherwise, that thaw is *free*.

Furthermore, in order to guide the proof search, our calculus will have to manipulate a different kind of future formulæ called *annotations*: these formulæ will be written $(\mathsf{G}\varphi)$, where $\mathsf{G}\varphi$ is a future modal formula, and will express that $\mathsf{G}\varphi$ holds starting from a specific later position. Note that such formulæ cannot ever appear as a subformula.

*Data Ordinal Semantics.* Recall that an ordinal $\alpha$ is seen set-theoretically as $\{\beta \in \mathrm{Ord} \mid \beta < \alpha\}$. An ordinal is either 0 (the empty linear order), a *limit* ordinal $\lambda$ (such that for all $\beta < \lambda$ there exists $\gamma$ with $\beta < \gamma < \lambda$), or a *successor* ordinal $\alpha + 1$. In the case of $\mathbf{K}_\mathbf{t}^{\downarrow}\mathbf{L}_\ell\mathbf{.3}$, our formulæ shall be evaluated on *data ordinals*, which are tuples $(\alpha, \delta)$ with $\alpha$ an ordinal and $\delta$ a function mapping elements from $\alpha$ to a datum from an infinite[1] domain $\mathbb{D}$. Models of our logic are Kripke *structures* $\mathfrak{M} = (\mathfrak{F}, V)$, where the frame $\mathfrak{F} = (\alpha, \delta)$ is a data ordinal, and $V : \Phi \to \alpha$ is a valuation of the propositional variables. A *register valuation* is a finite partial map $\nu$ from $\mathbb{N}$ to $\mathbb{D}$. The domain of such a $\nu$ must contain all the free registers that appear in the formulæ it evaluates.

Given a structure $\mathfrak{M} = ((\alpha, \delta), V)$ and a register valuation $\nu$, we define the *satisfaction* relation $\mathfrak{M}, \beta \models_\nu^{(\theta)} \varphi$, where $\beta < \alpha$, $\theta < \alpha$ and $\varphi$ is a formula, by structural induction on $\varphi$. Notice that $\theta$ is only used for the annotations.

$$
\begin{array}{ll}
\mathfrak{M}, \beta \not\models_\nu^{(\theta)} \perp & \\
\mathfrak{M}, \beta \models_\nu^{(\theta)} p & \text{iff } \beta \in V(p) \\
\mathfrak{M}, \beta \models_\nu^{(\theta)} \varphi \supset \psi & \text{iff if } \mathfrak{M}, \beta \models_\nu^{(\theta)} \varphi \text{ then } \mathfrak{M}, \beta \models_\nu^{(\theta)} \psi \\
\mathfrak{M}, \beta \models_\nu^{(\theta)} \mathsf{G}\varphi & \text{iff } \mathfrak{M}, \gamma \models_\nu^{(\theta)} \varphi \text{ for all } \beta < \gamma < \alpha \\
\mathfrak{M}, \beta \models_\nu^{(\theta)} \mathsf{H}\,\varphi & \text{iff } \mathfrak{M}, \gamma \models_\nu^{(\theta)} \varphi \text{ for all } \gamma < \beta \\
\mathfrak{M}, \beta \models_\nu^{(\theta)} \downarrow_r\varphi & \text{iff } \mathfrak{M}, \beta \models_{\nu[r \mapsto \delta(\beta)]}^{(\theta)} \varphi \\
\mathfrak{M}, \beta \models_\nu^{(\theta)} \uparrow_r & \text{iff } \delta(\beta) = \nu(r) \\
\mathfrak{M}, \beta \models_\nu^{(\theta)} (\mathsf{G}\varphi) & \text{iff } \beta < \theta, \text{and } \mathfrak{M}, \gamma \models_\nu^{(\theta)} \varphi \\
& \text{for all } \gamma \text{ such that } \theta \leq \gamma < \alpha
\end{array}
$$

When $\mathfrak{M}, \beta \models_\nu^{(\theta)} \varphi$, we say that $(\mathfrak{M}, \nu, \beta, (\theta))$ is a *model* of $\varphi$. Note that, since annotations cannot appear as subformulæ, we have $\mathfrak{M}, \beta \models_\nu^{(\theta)} \varphi$ if and only if $\mathfrak{M}, \beta \models_\nu^{(\theta')} \varphi$ for any $\theta'$, when $\varphi$ is not an annotation.

[1] Since we will only be able to perform equality tests between data values, we can assume that $\mathbb{D}$ is countable.

*Substitutions.* We note $[x/y](\varphi)$ for the formula $\varphi$ where every free occurrence of the register $y$ is replaced by the register $x$. More formally, we define it by structural induction as follows:

$$
\begin{array}{rlrl}
[x/y](\bot) &= \bot & [x/y](p) &= p \\
[x/y](\mathsf{H}\,\varphi) &= \mathsf{H}\,[x/y](\varphi) & [x/y](\mathsf{G}\varphi) &= \mathsf{G}[x/y](\varphi) \\
[x/y](\varphi_1 \supset \varphi_2) &= [x/y](\varphi_1) \supset [x/y](\varphi_2) & [x/y]((\mathsf{G}\varphi)) &= (\mathsf{G}[x/y](\varphi)) \\
[x/y](\uparrow_r) &= \uparrow_r \text{ if } r \neq y & [x/y](\uparrow_y) &= \uparrow_x \\
[x/y](\downarrow_r\varphi) &= \downarrow_r[x/y](\varphi) \text{ if } r \neq y \text{ and } r \neq x & [x/y](\downarrow_y\varphi) &= \downarrow_y\varphi \\
[x/y](\downarrow_x\varphi) &= \downarrow_r[x/y]([r/x](\varphi)) \text{ where } r \text{ is fresh} & &
\end{array}
$$

*Example 1.* Even though the underlying data-free logic $\mathbf{K_tL_\ell.3}$ cannot express that a model is of order type at least $\omega^2$ [3], this can be done with $\mathbf{K_t^\downarrow L_\ell.3}$, even without using any propositional variable. Consider for this $\varphi_1 = \mathsf{G}(\downarrow_r \mathsf{F}\uparrow_r)$, $\varphi_2 = \mathsf{G}(\downarrow_r \mathsf{F}\neg\uparrow_r)$, and $\varphi_3 = \mathsf{G}(\downarrow_r \mathsf{G}(\mathsf{F}\uparrow_r \supset \uparrow_r))$. Then, $\varphi = \mathsf{F}\top \wedge \varphi_1 \wedge \varphi_2 \wedge \varphi_3$ is satisfiable, and any model of $\varphi$ is of order type at least $\omega^2$.

Thanks to the conjunct $\mathsf{F}\top$, the other formulæ do not quantify over an empty set of future positions: there exists at least a future $\beta_1$. The conjunct $\varphi_1$ forces that every datum appears infinitely many times, and $\varphi_3$ forces that every such infinite sequence of positions carrying the same datum is continuous (two such sequences for two different data cannot be interleaved). Hence, any model of $\varphi$ starts with at least $\omega$ positions carrying $d_1 = \delta(\beta_1)$. In turn, $\varphi_2$ forces the existence of $\beta_2$ carrying a datum $d_2$ such that $d_1 \neq d_2$, which due to $\varphi_3$ must be after the positions carrying $d_1$; and because of $\varphi_2$ we must have at least $\omega$ positions carrying the datum $d_2$. Again, $\varphi_2$ forces the existence of $\beta_3$ carrying $d_3$ different from $d_1$ and $d_2$, and thus $\omega$ positions carrying $d_3$ must exist, etc. By repeating this reasoning, any model of $\varphi$ must comprise at least $\omega$ positions carrying the datum $d$, for infinitely many $d \in \mathbb{D}$, so is of order type at least $\omega^2$.

Moreover, $\varphi$ is indeed satisfied by a model of order type $\omega^2$, where the $i$th $\omega$ carries $d_i$, for an enumeration $(d_i)_{i\in\mathbb{N}}$ of $\mathbb{D}$.

## 3 Hypersequents with Clusters

As is often the case with modal logics, Gentzen's sequent calculus does not provide a rich enough framework to obtain complete proof systems. The extension we consider is to use *hypersequents* [1], which are essentially sets of sequents logically interpreted as a disjunction. Indrzejczak has moved to *ordered* hypersequents [24,25] (which are lists of hypersequents) to obtain a sound and complete calculus for $\mathbf{K_t4.3}$. We have further enriched the structure of his ordered hypersequents with *clusters* and annotations to obtain calculi for $\mathbf{K_t4.3}$ [2] and $\mathbf{K_tL_\ell.3}$ [3] for which proof search terminates and, in fact, yields an optimal complexity decision procedure. We keep the same structure in the present work, but simplify the annotation mechanism, and add rules to handle freeze formulæ.

In this section, we recall some definitions introduced in [3], with some minor generalisations for working with data ordinals. In Sect. 4, we present our proof system, which extends the proof system from [3], and prove that it is sound for $\mathbf{K}_\mathrm{t}^{\downarrow}\mathbf{L}_\ell\mathbf{.3}$. In Sect. 5, we focus on a decidable fragment of $\mathbf{K}_\mathrm{t}^{\downarrow}\mathbf{L}_\ell\mathbf{.3}$, and prove that our calculus is complete for that fragment, and has a proof strategy of optimal complexity.

### 3.1 Syntax

*Sequents.* A *sequent* (denoted $S$) is a tuple consisting of two finite sets $\Gamma, \Delta$ of formulæ, written $\Gamma \vdash \Delta$. It is satisfied by worlds $\beta$ and $\theta$ of a structure $\mathfrak{M}$ if there exists a register valuation $\nu$ such that $\mathfrak{M}, \beta \models_\nu^{(\theta)} \bigwedge \Gamma \supset \bigvee \Delta$ (where $\bigwedge \Gamma$ and $\bigvee \Delta$ denote respectively the conjunction of the formulæ of $\Gamma$ and the disjunction of the formulæ of $\Delta$).

*Hypersequents.* A *hypersequent* is a list of *cells*, each cell being either a sequent or a non-empty list of sequents called a (syntactic) *cluster*. We shall use the following abstract syntax, where both operators ';' and '$\|$' are associative with unit '$\bullet$':

$$H ::= C \mid H \,;\, H \qquad \text{(hypersequents)}$$

$$C ::= \bullet \mid S \mid \{Cl\} \qquad \text{(cells)}$$

$$Cl ::= S \mid Cl \parallel Cl \qquad \text{(cluster contents)}$$

Note that this definition allows for empty cells and hypersequents '$\bullet$', but these notational conveniences will never arise in actual proofs—and should not be confused with the empty sequent '$\vdash$'. We will see that the order of cells in a hypersequent is semantically relevant, but the order of sequents inside a cluster is not. Nevertheless, assuming an ordering as part of the syntactic structure of clusters is useful in order to refer to specific sequents or positions.

### 3.2 Semantics

The semantics of an ordered hypersequent with clusters relies on a notion of embedding, building on a view of hypersequents as partially ordered structures.

*Partial Order of a Hypersequent.* Let $H$ be a hypersequent containing $n$ sequents, counting both the sequents found directly in its cells and those in its clusters. In this context, any $i \in [1;n]$ is called a *position* of $H$, and we write $H(i)$ for the $i$-th sequent of $H$. We define a partial order $\precsim$ on the positions of $H$ by setting $i \precsim j$ if and only if either the $i$-th and $j$-th sequents are in the same cluster, or the $i$-th sequent is in a cell that lies strictly to the left of the cell of the $j$-th sequent. We write $i \prec j$ when $i \precsim j$ but $j \not\precsim i$, i.e. $j$ lies strictly to the right of

$i$ in $H$. We write $i \sim j$ when $i \precsim j \precsim i$. Finally, the *domain* of $H$ is defined as $\text{dom}(H) = ([1;n], \precsim)$; note that empty cells are ignored in $\text{dom}(H)$.

While a hypersequent is syntactically a finite partial order, its semantics will refer to a linear well-founded order, obtained by 'bulldozing' its clusters into copies of $\omega$. This defines the *order type* $o(H)$ of $H$ by induction on its structure: for cells, $o(\bullet) = 0$, $o(S) = 1$, and $o(\{Cl\}) = \omega$, and for hypersequents, $o(H_1 \,;\, H_2) = o(H_1) + o(H_2)$. Thus, $o(H) = \omega \cdot k + m$ where $k$ is the number of clusters in $H$ and $m$ the number of non-empty cells to the right of the rightmost cluster.

*Embeddings.* Let $H$ be a hypersequent and $\alpha$ an ordinal. We say that $\mu : \text{dom}(H) \to \alpha + 1 \setminus \{0\}$ is an *embedding* of H into $\alpha$, written $H \hookrightarrow_\mu \alpha$, if

- for all $i, j \in \text{dom}(H)$, $i \prec j$ implies $\mu(i) < \mu(j)$ and $i \sim j$ implies $\mu(i) = \mu(j)$
- and for all $i \in \text{dom}(H)$, $i$ is in a cluster if and only if $\mu(i)$ is a limit ordinal.

Observe that, if $H \hookrightarrow_\mu \alpha$, then $o(H) < \alpha + 1$.

*Semantics.* A structure $\mathfrak{M}$ is a *model* of a hypersequent $H$ if there exists a register valuation $\nu$, an embedding $\mathfrak{M} \hookrightarrow_\mu H$, and a position $i$ of $H$ such that for all $d \in \mathbb{D}$ there exists an ordinal $\beta_d < \mu(i)$ such that for all $\gamma$ such that $\beta_d \leq \gamma < \mu(i)$ and $\delta(\gamma) = d$, we have $\mathfrak{M}, \gamma \models_\nu^{(\mu(i))} H(i)$. In that case, we write $\mathfrak{M}, \nu, \mu \models H$. Following this definition, we say that a hypersequent is *valid* if for any $\mathfrak{M} = ((\alpha, \delta), V)$, any embedding $H \hookrightarrow_\mu \mathfrak{M}$, and any register valuation $\nu$ we have $\mathfrak{M}, \nu, \mu \models H$. A formula $\varphi$ is valid in the usual sense (i.e., satisfied in every world of every ordinal structure) if and only if the hypersequent $\vdash \varphi$ is valid in our sense.

If a hypersequent $H$ is not valid, then it has a *counter-model*, that is a structure $\mathfrak{M} = ((\alpha, \delta), V)$, an embedding $H \hookrightarrow_\mu \mathfrak{M}$ and a register valuation $\nu$ such that for every $i \in \text{dom}(H)$ there exists $d_i \in \mathbb{D}$ such that for every $\beta < \mu(i)$, there exists $\gamma$ with $\beta \leq \gamma < \mu(i)$ and $\delta(\gamma) = d_i$ such that $\mathfrak{M}, \gamma \not\models_\nu^{(\mu(i))} H(i)$. For the positions $i \in \text{dom}(H)$ that are not in clusters, $\mu(i)$ is a successor ordinal $\gamma + 1$ and this amounts to asking that $\mathfrak{M}, \gamma \not\models_\nu^{(\gamma+1)} H(i)$. When $i$ is in a cluster, the condition implies the existence of an infinite increasing sequence $(\gamma_j)_j$ of ordinals carrying the same datum, and with limit $\mu(i) = \sup_j \gamma_j$, such that $\mathfrak{M}, \gamma_j \not\models_\nu^{(\mu(i))} H(i)$ for all $j$.

## 4 Proof System

We now present our proof system for $\mathbf{K_t^{\downarrow}L_\ell.3}$, called $\mathbf{HK_t^dL_\ell.3}$. The rules of $\mathbf{HK_t^dL_\ell.3}$ are given in Figs. 1, 2 and 3: the first group includes the usual propositional rules, the second deals with modalities. They are the same rules as in [3]. Figure 3 shows the new annotation rule—subsuming the annotation rules from [3]—, and additional rules to deal with freeze formulæ. The figures make use of some notations introduced in [3] which we recall next, before commenting on the rule definitions themselves.

$$\frac{}{H[\varphi, \Gamma \vdash \Delta, \varphi]}\ (\mathsf{ax}) \qquad \frac{H[\varphi \supset \psi, \Gamma \vdash \Delta, \varphi] \quad H[\varphi \supset \psi, \psi, \Gamma \vdash \Delta]}{H[\varphi \supset \psi, \Gamma \vdash \Delta]}\ (\supset\vdash)$$

$$\frac{}{H[\Gamma, \bot \vdash \Delta]}\ (\bot) \qquad \frac{H[\varphi, \Gamma \vdash \Delta, \psi, \varphi \supset \psi]}{H[\Gamma \vdash \Delta, \varphi \supset \psi]}\ (\vdash\supset)$$

**Fig. 1.** Propositional rules of $\mathbf{HK^d_t L_\ell.3}$.

### 4.1 Notations

First, we use hypersequents with *holes*. One-placeholder hypersequents, cells, and clusters are defined by the following syntax:

$$H[] ::= H\,; C[]\,; H \quad C[] ::= \star \mid \{\, Cl[] \,\} \quad Cl[] ::= Cl_\bullet \parallel \star \parallel Cl_\bullet \quad Cl_\bullet ::= \bullet \mid Cl$$

Two-placeholder cells and hypersequents have two holes identified by $\star_1$ and $\star_2$:

$$H[][] ::= H\,; C[][]\,; H \mid H[\star_1]\,; H[\star_2]$$
$$C[][] ::= \{\, Cl[\star_1] \parallel Cl[\star_2] \,\} \mid \{\, Cl[\star_2] \parallel Cl[\star_1] \,\}$$

As usual, $C[S]$ (resp. $C[Cl]$) denotes the same cell with $S$ (resp. $Cl$) substituted for $\star$; two-placeholder cells and hypersequents with holes behave similarly. In terms of the partial orders underlying hypersequents with two holes, observe that the positions $i$ and $j$ associated resp. to $\star_1$ and $\star_2$ are exactly such that $i \precsim j$.

Second, we use a convenient notation for *enriching* a sequent: if $S$ is a sequent $\Gamma \vdash \Delta$, then $S \ltimes (\Gamma' \vdash \Delta')$ is the sequent $\Gamma, \Gamma' \vdash \Delta, \Delta'$. Moreover, we sometimes need to enrich an arbitrary sequent of a cluster $\{Cl\}$ with a sequent $S$; then $\{Cl\} \ltimes S$ denotes the cluster with its leftmost sequent enriched.

Finally, we write $[x/y](H)$ for the hypersequent $H$ where the operator $[x/y]$ has been applied to every formula.

### 4.2 Rules

We now comment on our rules. The rules from Figs. 1 and 2 are the same as in [3]. The propositional rules of Fig. 1 are straightforward: they are the usual ones applied to an arbitrary sequent of the hypersequent.

The first four modal rules of Fig. 2 should not be surprising. For instance, in $(\mathsf{G}\vdash)$, if the conclusion has a counter-model, then $\mathsf{G}\varphi$ holds at some ordinal and thus both $\varphi$ and $\mathsf{G}\varphi$ must also hold at strictly greater ordinals. The rule also applies to two distinct sequents inside the same cluster. The $(\{\mathsf{G}\vdash\})$ rule allows to proceed in the same way inside a cluster when the sequent further to the right is the original sequent itself, something that our notations do not allow in $(\mathsf{G}\vdash)$. Finally, $(\mathsf{H}\vdash)$ and $(\{\mathsf{H}\vdash\})$ are symmetric to the two previous rules.

$$\frac{H\,[\mathsf{G}\,\varphi,\Gamma \vdash \Delta]\,[\varphi,\mathsf{G}\,\varphi,\Pi \vdash \Sigma]}{H\,[\mathsf{G}\,\varphi,\Gamma \vdash \Delta]\,[\Pi \vdash \Sigma]}\;(\mathsf{G}\vdash) \qquad \frac{H_1;\{Cl_\bullet \parallel \varphi,\mathsf{G}\,\varphi,\Gamma \vdash \Delta \parallel Cl'_\bullet\};H_2}{H_1;\{Cl_\bullet \parallel \mathsf{G}\,\varphi,\Gamma \vdash \Delta \parallel Cl'_\bullet\};H_2}\;(\{\mathsf{G}\vdash\})$$

$$\frac{H\,[\varphi,\mathsf{H}\,\varphi,\Pi \vdash \Sigma]\,[\mathsf{H}\,\varphi,\Gamma \vdash \Delta]}{H\,[\Pi \vdash \Sigma]\,[\mathsf{H}\,\varphi,\Gamma \vdash \Delta]}\;(\mathsf{H}\vdash) \qquad \frac{H_1;\{Cl_\bullet \parallel \varphi,\mathsf{H}\,\varphi,\Gamma \vdash \Delta \parallel Cl'_\bullet\};H_2}{H_1;\{Cl_\bullet \parallel \mathsf{H}\,\varphi,\Gamma \vdash \Delta \parallel Cl'_\bullet\};H_2}\;(\{\mathsf{H}\vdash\})$$

$$\frac{\begin{array}{ll} H_1\,;C\,[\Gamma \vdash \Delta,\mathsf{G}\,\varphi]\,;(\mathsf{G}\,\varphi) \vdash \varphi\,;C'\,;H_2 & \\ H_1\,;C\,[\Gamma \vdash \Delta,\mathsf{G}\,\varphi]\,;\{(\mathsf{G}\,\varphi) \vdash \varphi\}\,;C'\,;H_2 & \\ H_1\,;C\,[\Gamma \vdash \Delta,\mathsf{G}\,\varphi \parallel (\mathsf{G}\,\varphi) \vdash \varphi]\,;C'\,;H_2 & \text{if } C \neq \star \\ H_1\,;C\,[\Gamma \vdash \Delta,\mathsf{G}\,\varphi]\,;C' \ltimes (\vdash \mathsf{G}\,\varphi)\,;H_2 & \text{if } C' \neq \bullet \\ H_1\,;C\,[\Gamma \vdash \Delta,\mathsf{G}\,\varphi]\,;C' \ltimes ((\mathsf{G}\,\varphi) \vdash \varphi)\,;H_2 & \text{if } C' \neq \bullet \text{ and } C' \neq \{Cl\} \end{array}}{H_1\,;C\,[\Gamma \vdash \Delta,\mathsf{G}\,\varphi]\,;C'\,;H_2}\;(\vdash\mathsf{G})$$

$$\frac{\begin{array}{ll} H_2\,;C'\,;\mathsf{H}\,\varphi \vdash \varphi\,;C\,[\Gamma \vdash \Delta,\mathsf{H}\,\varphi]\,;H_1 & \\ H_2\,;C' \ltimes (\vdash \mathsf{H}\,\varphi)\,;C\,[\Gamma \vdash \Delta,\mathsf{H}\,\varphi]\,;H_1 & \text{if } C' \neq \bullet \\ H_2\,;C' \ltimes (\mathsf{H}\,\varphi \vdash \varphi)\,;C\,[\Gamma \vdash \Delta,\mathsf{H}\,\varphi]\,;H_1 & \text{if } C' \neq \bullet \text{ and } C' \neq \{Cl\} \end{array}}{H_2\,;C'\,;C\,[\Gamma \vdash \Delta,\mathsf{H}\,\varphi]\,;H_1}\;(\vdash\mathsf{H})$$

**Fig. 2.** Modal rules of $\mathbf{HK}^{\mathrm{d}}_{\mathrm{t}}\mathbf{L}_\ell\mathbf{.3}$. In $(\vdash\mathsf{G})$ and $(\vdash\mathsf{H})$, we allow $C' = \bullet$ only when $H_2 = \bullet$.

The rules $(\vdash\mathsf{G})$ and $(\vdash\mathsf{H})$ are the most complex ones. We shall not try to justify their soundness at this point, but simply make a few remarks that are important to understand their definition. First, these rules are the only ones that may introduce new cells in hypersequents. In $(\vdash\mathsf{G})$, new cells come with the annotation $(\mathsf{G}\varphi)$ of the principal formula $\mathsf{G}\varphi$, which will help bounding the proof search, as we will see in Lemma 1. In $(\vdash\mathsf{H})$, new sequents come instead with the principal formula $\mathsf{H}\,\varphi$ on their left hand-side, which will have the same effect of bounding the proof search. This difference comes from the well-foundedness of our models when navigating to the past. As a result, we do not need annotations for past operators. In the next section, one should notice the similar roles played in Lemma 1 by past formulæ of the form $\mathsf{H}\,\varphi$ and annotations of the form $(\mathsf{G}\varphi)$. Also, the principal cell $C\,[\Gamma \vdash \Delta, \mathsf{G}\varphi]$ in $(\vdash\mathsf{G})$ may be the rightmost cell of the conclusion hypersequent, in which case both $C'$ and $H_2$ are empty, and the rule has two or three premises depending on whether the principal cell is a cluster or not. When the principal cell is not rightmost, then $C'$ is not allowed to be empty, and the rule has one or two extra premises depending on whether $C'$ is a cluster or not. The situation is symmetric for $(\vdash\mathsf{H})$.

The annotation rules from [3] are now subsumed by the new rule $((\mathsf{G}))$ from Fig. 3 which is similar to $(\mathsf{G}\vdash)$, the difference being that an annotation $(\mathsf{G}\varphi)$ cannot affect the current cluster.

The other rules from Fig. 3 are new. The rule $(\uparrow\,\vdash)$ unifies two registers when they must contain the same datum, and is helpful to bound the number of registers appearing in the proof search. The rules $(\downarrow\vdash)$ and $(\vdash\downarrow)$ both handle the freeze quantifier $\downarrow_r$ by adding a version of $\varphi$ where $r$ has been replaced by either an already used register matching the current datum if any, or a fresh one otherwise.

Finally, we have designed our rules so that they are all *invertible*: by keeping in premises all the formulæ from the conclusion, we ensure that validity is never lost by applying a rule; this is proved in [33].

**Proposition 1 (invertibility).** *In any rule instance, if a premise has a counter-model, then so does its conclusion.*

In practice, keeping all formulæ can be unnecessarily heavy. Fortunately, the following weakening rules are admissible:

$$\frac{H\,[\Gamma \vdash \Delta]}{H\,[\Gamma, \varphi \vdash \Delta]}\ (\text{weak} \vdash) \qquad \frac{H\,[\Gamma \vdash \Delta]}{H\,[\Gamma \vdash \varphi, \Delta]}\ (\vdash \text{weak})$$

This may not seem obvious since the new rules $(\downarrow \vdash)$ and $(\vdash \downarrow)$ require some specific checks. To prove this claim, once the original proof system is proven complete, one could just prove that the weakening rules are sound. Nonetheless, we will sometimes omit formulæ when they do not play any role to lighten some examples. Every time we do so, the exact same proof could be derived without omitting any formulæ.

$$\frac{H_1\,[(\mathsf{G}\,\varphi), \Gamma \vdash \Delta]\,;H_2\,[\varphi, \mathsf{G}\,\varphi, \Pi \vdash \Sigma]}{H_1\,[(\mathsf{G}\,\varphi), \Gamma \vdash \Delta]\,;H_2\,[\Pi \vdash \Sigma]}\ ((\mathsf{G})) \qquad \frac{[x/y](H\,[\uparrow_x, \Gamma \vdash \Delta])}{H\,[\uparrow_x, \uparrow_y, \Gamma \vdash \Delta]}\ (\uparrow \vdash)$$

$$\frac{\begin{array}{ll} H\,[\Gamma, \downarrow_r\varphi, \uparrow_x, [x/r](\varphi) \vdash \Delta] & \text{if } \forall y \in \mathbb{N}, \uparrow_y \notin \Gamma, \text{ with } x \text{ fresh} \\ H\,[\Gamma, \downarrow_r\varphi, [x/r](\varphi) \vdash \Delta] & \text{if } \uparrow_x \text{ is the only thaw atom in } \Gamma \end{array}}{H\,[\Gamma, \downarrow_r\varphi \vdash \Delta]}\ (\downarrow \vdash)$$

$$\frac{\begin{array}{ll} H\,[\Gamma, \uparrow_x \vdash [x/r](\varphi), \downarrow_r\varphi, \Delta] & \text{if } \forall y \in \mathbb{N}, \uparrow_y \notin \Gamma, \text{ with } x \text{ fresh} \\ H\,[\Gamma \vdash [x/r](\varphi), \downarrow_r\varphi, \Delta] & \text{if } \uparrow_x \text{ is the only thaw atom in } \Gamma \end{array}}{H\,[\Gamma \vdash \downarrow_r\varphi, \Delta]}\ (\vdash \downarrow)$$

**Fig. 3.** Annotation, freeze, and thaw rules of $\mathbf{HK_t^dL_\ell.3}$. By fresh, we mean that $x$ does not appear as a free register anywhere in the conclusion.

### 4.3 Soundness

Our calculus is sound w.r.t $\mathbf{K_t^dL_\ell.3}$, which is proved in [33].

**Proposition 2.** *The rules of* $\mathbf{HK_t^dL_\ell.3}$ *are sound: if the premises of a rule instance are valid, then so is its conclusion.*

Invertibility is not enough to obtain completeness since proof search does not terminate, $\mathbf{K_t^\downarrow L_\ell.3}$ being undecidable. We now investigate a decidable fragment for which $\mathbf{HK_t^dL_\ell.3}$ is complete and has a proof strategy of optimal complexity.

# 5 Restricted Logic and Completeness

The $\mathbf{K}_\mathrm{t}^{\downarrow}\mathbf{L}_\ell\mathbf{.3}$ logic is known to be undecidable, even with only one register [14] and some restrictions regarding the use of that register [18]. Here, we consider another restriction of the logic, and prove that our calculus is complete for that fragment, with proof search in coNP.

## 5.1 Restricted Syntax

We consider the following fragment of $\mathbf{K}_\mathrm{t}^{\downarrow}\mathbf{L}_\ell\mathbf{.3}$, which we call $\mathbf{K}_\mathrm{t}^{\mathrm{d}}\mathbf{L}_\ell\mathbf{.3}$:

$$\begin{aligned}\varphi ::= \; & \bot \mid p \mid \varphi \supset \varphi \mid \mathsf{G}\varphi \mid \mathsf{H}\,\varphi \\ & \mid {\downarrow_r}\mathsf{G}(\uparrow_r \supset \varphi) \mid {\downarrow_r}\mathsf{G}(\neg\uparrow_r \supset \varphi) \\ & \mid {\downarrow_r}\mathsf{H}\,(\uparrow_r \supset \varphi) \mid {\downarrow_r}\mathsf{H}\,(\neg\uparrow_r \supset \varphi) \qquad \text{(where } p \in \Phi \text{ and } r \in \mathbb{N})\end{aligned}$$

Because the use of registers is restricted to specific formulæ, we define the following syntactic sugar:

$$\mathsf{G}_{=r}\,\varphi = \mathsf{G}(\uparrow_r \supset \varphi) \qquad \mathsf{G}_{\neq r}\,\varphi = \mathsf{G}(\neg\uparrow_r \supset \varphi)$$
$$\mathsf{H}_{=r}\,\varphi = \mathsf{H}\,(\uparrow_r \supset \varphi) \qquad \mathsf{H}_{\neq r}\,\varphi = \mathsf{H}\,(\neg\uparrow_r \supset \varphi)$$

Intuitively, $\mathsf{G}_{=r}\,\varphi$ (resp. $\mathsf{G}_{\neq r}\,\varphi$) expresses the fact that $\varphi$ holds in every future position with the same (resp. a different) datum as the one stored in the register $r$; and $\mathsf{H}_{=r}\,\varphi$, $\mathsf{H}_{\neq r}\,\varphi$ express the same for past positions. Moreover, since a negation before a freeze quantifier can be moved inside its scope, e.g. $\neg{\downarrow_r}\mathsf{G}_{\neq r}\,\neg\varphi \equiv {\downarrow_r}\neg\mathsf{G}_{\neq r}\,\neg\varphi$, we can also define their dual diamond modalities, e.g. $\mathsf{F}_{\neq r}\,\varphi = \neg\mathsf{G}_{\neq r}\,\neg\varphi$. Formulæ of the form ${\downarrow_r}\mathsf{G}_{=r}\,\varphi$ (resp. ${\downarrow_r}\mathsf{G}_{\neq r}\,\varphi$) corresponds to formulæ denoted $\Box_{=}\varphi$ (resp. $\Box_{\neq}\varphi$) from [4], which works over data trees. As explained in [33], this fragment is furthermore exactly as expressive as a two-variable fragment of first-order logic over data ordinals investigated in [6].

*Example 2.* The formula from Example 1, forcing its models to have order type at least $\omega^2$, does not belong to this fragment. Furthermore, there is no equivalent formula belonging to $\mathbf{K}_\mathrm{t}^{\mathrm{d}}\mathbf{L}_\ell\mathbf{.3}$, as we will show later that satisfiable formulæ from this fragment always have a model of order type strictly below $\omega^2$.

Property 2 from the introduction does not seem expressible either, since it would require to perform nested data tests. However, Property 1 can be expressed by the following formula:

$$\begin{aligned}&\mathsf{G}(b \supset (\neg{\downarrow_r}\mathsf{P}_{=r}\top \wedge \neg{\downarrow_r}\mathsf{F}_{=r}\,b \wedge {\downarrow_r}\mathsf{F}_{=r}\,(e \wedge \neg{\downarrow_r}\mathsf{F}_{=r}\top \wedge \neg{\downarrow_r}\mathsf{P}_{=r}\,e))) \\ &\wedge\mathsf{G}((e \vee r \vee w) \supset {\downarrow_r}\mathsf{P}_{=r}\,b)\end{aligned}$$

From now on, we only consider formulæ from $\mathbf{K}_\mathrm{t}^{\mathrm{d}}\mathbf{L}_\ell\mathbf{.3}$.

### 5.2 Completeness and Complexity

As in [2,3], completeness is a by-product of a rather simple proof-search strategy. As already stated in Proposition 1, all the rules are invertible; and as we shall see, our strategy only produces proof trees with branches that are polynomially bounded for the restricted logic, as it will avoid any pitfall that could happen. Thus it is unnecessary to backtrack during proof-search. Moreover, proof attempts result in finite (polynomial depth) partial proofs, whose unjustified leaves yield counter-models that amount (by invertibility) to counter-models of the conclusion. Hence the completeness of our calculus. We detail this argument below, and its corollary: proof-search yields an optimal coNP procedure for validity.

We characterise next the proof attempts that we consider for proof search, and show how to extract counter-models when such attempts fail.

**Lemma 1.** *If a hypersequent $H$ satisfies one of these conditions, then $H$ is provable (and we say that $H$ is* immediately provable*).*

*(a) There exists a formula $\varphi$, and two positions $i \prec j$ of $H$ such that $H(i)$ and $H(j)$ both contain $(\mathsf{G}\varphi) \vdash \varphi$.*
*(b) There exists a formula $\varphi$, and two positions $i \prec j$ of $H$ such that $H(i)$ and $H(j)$ both contain $\mathsf{H}\,\varphi \vdash \varphi$.*
*(c) There exists a formula $\varphi$, three positions $i \prec j \prec k$ of $H$, and three registers $x, y, z \in \mathbb{N}$ such that:*
  - *$H(i)$ contains $(\mathsf{G}_{\neq x}\,\varphi) \vdash \neg\uparrow_x \supset \varphi$.*
  - *$H(j)$ contains $(\mathsf{G}_{\neq y}\,\varphi) \vdash \neg\uparrow_y \supset \varphi$.*
  - *$H(k)$ contains $(\mathsf{G}_{\neq z}\,\varphi) \vdash \neg\uparrow_z \supset \varphi$.*

*(d) There exists a formula $\varphi$, three positions $i \prec j \prec k$ of $H$, and three registers $x, y, z \in \mathbb{N}$ such that:*
  - *$H(i)$ contains $\mathsf{H}_{\neq x}\,\varphi \vdash \neg\uparrow_x \supset \varphi$.*
  - *$H(j)$ contains $\mathsf{H}_{\neq y}\,\varphi \vdash \neg\uparrow_y \supset \varphi$.*
  - *$H(k)$ contains $\mathsf{H}_{\neq z}\,\varphi \vdash \neg\uparrow_z \supset \varphi$.*

We provide a proof tree for every case in [33]. The intuition behind (d) is the following: if there exists $\gamma$ where $\mathsf{H}_{\neq z}\,\varphi$ holds, and $\gamma' < \gamma$ where $\mathsf{H}_{\neq y}\,\varphi$ holds, and if $y$ and $z$ stores different data, then $\varphi$ holds in every past position of $\gamma'$ (at any position, either $\neg\uparrow_z$ or $\neg\uparrow_y$ holds) and thus any $\mathsf{H}_{\neq x}\,\varphi$ holds in the past. The intuition behind (c) is similar. This reasoning fails if $y$ and $z$ store the same datum, but this cannot be assumed during proof search only when $\uparrow_y$ and $\uparrow_z$ appear on the left-hand side of the same sequent, and in this case we should apply $(\uparrow\vdash)$ in priority.

*Partial Proofs.* We characterise now the proof attempts that we consider for proof search, and show how to extract counter-models when such attempts fail. We call *partial proof* a finite derivation tree whose internal nodes correspond to rule applications, but whose leaves may be unjustified hypersequents, and that satisfies three conditions:

(a) no rule application should be such that, if $H$ is the conclusion hypersequent,
  (i) one of the premises is also $H$, or
  (ii) the rule being applied is $(\vdash \mathsf{G})$ on a formula $\mathsf{G}\varphi$ at position $i$ such that there exists $j \sim i$ such that $H(j)$ contains $(\mathsf{G}\varphi) \vdash \varphi$, or
  (iii) the rule being applied is $(\vdash \mathsf{G})$ on a formula $\mathsf{G}(\neg\uparrow_x \supset \varphi)$ at position $i$ such that there exists $j \sim i$ and $y \neq x$ such that $H(j)$ contains $(\mathsf{G}(\neg\uparrow_y \supset \varphi)) \vdash \neg\uparrow_y \supset \varphi$ and does not contain $\uparrow_x$ on its left-hand side.

(b) If the rule $(\uparrow\vdash)$ is applicable, or if the rule $(\vdash\supset)$ is applicable on a formula of the form $\uparrow_x \supset \varphi$, then the other rules cannot be applied.

(c) immediately provable hypersequents must be proven immediately as sketched in the proof of Lemma 1.

Finally, we call *failure hypersequent* a hypersequent on which any rule application would not respect condition (a).

The second part of condition (b) is there to optimize the use of its first part, which in turn is there to bound the number of registers our calculus manipulates during a proof search. Conditions (a) and (c) amount to a simple proof search strategy that avoids loops, and addresses especially loops arising from repeated applications of $(\vdash\mathsf{H})$ or $(\vdash\mathsf{G})$, in branches where several new cells are created for the same modal formula (up to maybe a different register): this results either in immediately provable hypersequents from Lemma 1, or failure hypersequents on which the proof strategy is stuck and for which we prove next that we can always construct a counter-model.

*Example 3.* The annotation rules from [3] are subsumed by our new version of the rule $((\mathsf{G}))$. For instance, if $H$ has a position $i$ that is not in a cluster such that $H(i)$ contains $(\mathsf{G}\varphi) \vdash \mathsf{G}\varphi$, the branch can be immediately closed by some rule from [3]. Let us show that such an $H$ is provable by $\mathbf{HK_t^dL_\ell.3}$. First of all, since $(\mathsf{G}\varphi) \in H(i)$, then either $H(i)$ contains $(\mathsf{G}\varphi) \vdash \varphi$, or there exist $j \prec i$ such that $H(j)$ contains it. Then:

- If $(\vdash \mathsf{G})$ cannot be applied on $\mathsf{G}\varphi$, it is either because $H(i+1)$ contains $(\mathsf{G}\varphi) \vdash \varphi$, and then $H$ is immediately provable, or because $\mathsf{G}\varphi$ also appears on the right-hand side of $H(i+1)$, and the same formula can also be sent on its left-hand side (if not already present) by applying $((\mathsf{G}))$ on the annotation $(\mathsf{G}\varphi)$, and then $(\mathsf{ax})$ can be used.
- Else, we apply $(\vdash\mathsf{G})$ on $\mathsf{G}\varphi$. All premises are immediately provable, except for the premise sending $\mathsf{G}\varphi$ on the right-hand side of $H(i+1)$ which can be proved as in the previous case.

**Proposition 3.** *Any failure hypersequent $H$ has a counter-model.*

*Proof (sketch).* The detailed proof is provided in [33]. We first construct a counter-model $\mathfrak{M} = ((\alpha, \delta), V)$ of $H$ with $\alpha = o(H)$ and a straightforward embedding $\mathfrak{M} \hookrightarrow_\mu H$. We also describe a function $\mathrm{pos} : \alpha \to \mathrm{dom}(H)$ which will act as the reverse of $\mu$. Then, $\delta$ is chosen such that every pair of worlds from $\mathfrak{M}$ carry distinct data unless their corresponding sequent in $H$ carries the same atomic $\uparrow_r$ on their left-hand side; and $V$ is constructed similarly: every atom $p$ is false except in worlds for which the corresponding sequent carries $p$ on its left-hand side. We finally prove that $\mathfrak{M}$ is a counter-model of $H$ by structural induction on the subformulæ of $H$.

*Example 4.* Consider the following partial proof of the sequent $\vdash \downarrow_r \mathsf{G}\uparrow_r$:

$$\dfrac{\uparrow_x \vdash \downarrow_r \mathsf{G}\uparrow_r, \mathsf{G}\uparrow_x \,;\, (\mathsf{G}\uparrow_x) \vdash \uparrow_x \qquad \dfrac{\cdots \quad \uparrow_x \vdash \downarrow_r \mathsf{G}\uparrow_r, \mathsf{G}\uparrow_x \,;\, \{(\mathsf{G}\uparrow_x) \vdash \uparrow_x, \mathsf{G}\uparrow_x\}}{\uparrow_x \vdash \downarrow_r \mathsf{G}\uparrow_r, \mathsf{G}\uparrow_x \,;\, \{(\mathsf{G}\uparrow_x) \vdash \uparrow_x\}}\,(\vdash \mathsf{G})}{\dfrac{\uparrow_x \vdash \downarrow_r \mathsf{G}\uparrow_r, \mathsf{G}\uparrow_x}{\vdash \downarrow_r \mathsf{G}\uparrow_r}\,(\vdash \downarrow)}\,(\vdash \mathsf{G})$$

The first leaf is a failure hypersequent, since case (i) of condition (a) prevents any rule application. Its corresponding counter-model consists of to worlds with distinct data. The other branch will reach immediately provable hypersequent (not displayed on the figure), and another failure hypersequent, since cases (i) and (ii) of condition (a) prevent any rule application. Its corresponding counter-model consists of a first world with a datum $d_x$, followed by an infinite sequence of worlds all carrying another datum $d_y \neq d_x$.

We now turn to establishing that proof search terminates, and always produces branches of polynomial length. For a hypersequent $H$, let $\mathsf{len}(H)$ be its number of sequents (i.e., the size of $\mathrm{dom}(H)$), and $|H|$ the number of distinct subformulæ occurring in $H$.

**Lemma 2.** *For any partial proof of a hypersequent $H$, any branch of the proof is of length at most* $2|H|(4|H| + \mathsf{len}(H))((4|H| + \mathsf{len}(H))|H| + \mathsf{len}(H) + 1)$.

*Proof (sketch).* Each creation of a new position along a branch of the proof search could lead to the creation of a new register later in the branch, which in turn could create a new renamed copy of some subformula of $H$, which then could lead to the creation of another position. We must make sure that such a process cannot go ad infinitum: we first bound the numbers of free registers that can appear along a branch respecting our proof strategy, then the size of the hypersequents of such a branch, and finally the number of rules that can be applied. The details are presented in [33].

*Example 5.* If we did not follow our strategy, a bad case such as described at the beginning of the previous proof could happen on the hypersequent $H = \vdash ; \varphi \vdash$ with $\varphi = \mathsf{H}(\neg\downarrow_r \mathsf{H}_{=r} \bot)$. In practice, $\varphi$ can send the subformula $\downarrow_r \mathsf{H}_{=r} \bot$ on the right-hand side of any past position by using $(\mathsf{H} \vdash)$ and then handling the negation. A proof of $H$ will start as follows:

$$\dfrac{\dfrac{\dfrac{\mathsf{H}_{=x} \bot \vdash \uparrow_x \supset \bot \,;\, \varphi, \uparrow_x \vdash \mathsf{H}_{=x} \bot, \downarrow_r \mathsf{H}_{=r} \bot \,;\, \varphi \vdash}{\varphi, \uparrow_x \vdash \mathsf{H}_{=x} \bot, \downarrow_r \mathsf{H}_{=r} \bot \,;\, \varphi \vdash}\,(\vdash \mathsf{H})}{\varphi \vdash \downarrow_r \mathsf{H}_{=r} \bot \,;\, \varphi \vdash}\,(\vdash \downarrow)}{\vdash ; \varphi \vdash}$$

Our strategy would now force to handle the formula $\uparrow_x \supset \bot$. If we do not respect that, and instead send $\downarrow_r \mathsf{H}_{=r} \bot$ on the right-hand side of the leftmost position and apply $(\vdash\downarrow)$ again, a new register $y$ would be created, along with the formula $\mathsf{H}_{=y} \bot$, which in turn will create a new position more in the past when applying $(\vdash\mathsf{H})$. If we never deal with a formula of the form $\uparrow_x \supset \bot$, this process could go on ad infinitum, alternating between creating a new register and creating a new position. However, if we respect our strategy, the proof search will reach an immediately provable hypersequent after creating a fourth position. It is not surprising, since we can prove that a counter-model of $H$ should be such that every datum appearing in the past does so infinitely many times, which is impossible as our models are well-founded.

We now conclude that $\mathbf{HK^d_tL_\ell.3}$ is complete for $\mathbf{K^d_tL_\ell.3}$, and also enjoys $\mathsf{coNP}$ proof search. This complexity is optimal, since our logic contains the propositional logic [10].

**Theorem 1 (completeness).** *Our calculus is complete for* $\mathbf{K^d_tL_\ell.3}$*: every valid hypersequent* $H$ *has a proof in* $\mathbf{HK^d_tL_\ell.3}$.

*Proof.* Assume that $H$ is not provable. Consider a partial proof $\mathcal{P}$ of $H$ that cannot be expanded any more: its unjustified leaves are failure hypersequents. Such a partial proof exists by Lemma 2. Any unjustified leaf of that partial proof has a counter-model by Proposition 3, and by invertibility shown in Proposition 1 it is also a counter-model of $H$.

**Proposition 4.** *Proof search in* $\mathbf{HK^d_tL_\ell.3}$ *is in* $\mathsf{coNP}$.

*Proof.* Proof search can be implemented in an alternating Turing machine maintaining the current hypersequent on its tape, with only universal states (choosing a premise of the rule): by Proposition 1, we can choose an arbitrary order in which to apply rules; and the choice of a fresh $x$ by any application of $(\downarrow\vdash)$ or $(\vdash\downarrow)$ does not matter (e.g., $x$ can be taken as the next unused integer). Moreover, by Lemma 2, the computation branches are of length bounded by a polynomial, hence the Turing machine is in $\mathsf{coNP}$.

## 6 Restricted Logic on Given Ordinals

We have designed a proof system that is sound and complete for $\mathbf{K^d_tL_\ell.3}$, and enjoys optimal complexity proof search. Moreover, as in [3], we can derive a small model property from the proof of completeness: the logic $\mathbf{K^d_tL_\ell.3}$ can only distinguish ordinals up to $\omega^2$, as the underlying data-free logic [3].

**Proposition 5 (small model property).** *If a hypersequent* $H$ *has a counter-model, then it has one of order type* $\alpha < \omega \cdot ((4|H| + \mathsf{len}(H))|H| + \mathsf{len}(H) + 1)$.

*Proof.* This is a corollary of Theorem 1. By the proof of Lemma 2, the hypersequents in a failure branch—which are not immediately provable—have at most $(4|H| + \mathsf{len}(H))|H| + \mathsf{len}(H) + 1$ non-empty sequents. The counter-model extracted in Proposition 3 from a failure hypersequent $H'$ is over $o(H') < \omega \cdot ((4|H| + \mathsf{len}(H))|H| + \mathsf{len}(H) + 1)$. A counter-model for $H$ is then obtained by Proposition 1, with a different embedding but the same structure.

In particular, for a formula $\varphi$, the hypersequent $H = \vdash \varphi$ has $|H| = |\varphi|$ and $\mathsf{len}(H) = 1$, hence the $\omega \cdot (4 \cdot |\varphi|^2 + |\varphi| + 2)$ bound announced in the introduction.

Furthermore, as in [3], we can easily enrich our calculus by the following rule to obtain a sound and complete proof system for tense logic over data ordinals below a certain type $\alpha$.

$$\frac{}{H}\ (\mathsf{ord}_\alpha) \text{ if } o(H) > \alpha$$

We can also capture validity at a fixed ordinal $\alpha < \omega^2$, by padding the input with enough empty sequents to start with a hypersequent $H$ of order type $\alpha$, and enriching our calculus with rule $(\mathsf{ord}_\alpha)$ to forbid larger ordinals (as in [3]). The only catch is that we should check that the formula of interest in valid in all possible positions, i.e. considering all possible paddings leading to a hypersequent of order type $\alpha$. When checking validity of a formula $\varphi$ in all structures of order type exactly $\alpha$, we must prove in $\mathbf{HK_t^dL_\ell.3}$ extended with $(\mathsf{ord}_\alpha)$ all hypersequents of order type $\alpha$ containing one sequent $\vdash \varphi$ and otherwise only empty sequents. For instance, when $\alpha = \omega$ we must check $\vdash \varphi\,;\{\vdash\}$.

## 7 Related Work and Conclusion

We have investigated $\mathbf{K_t^{\downarrow}L_\ell.3}$—the freeze tense logic over ordinals—and proposed a decidable fragment, namely $\mathbf{K_t^dL_\ell.3}$, for which we designed a sound and complete proof system.

Thanks to Indrzejczak's ordered hypersequents [24], enriched with clusters and annotations as in [2,3], our system enjoys optimal coNP proof search, allows to derive small model properties, and can be extended into a proof system for variants of the logic over bounded or fixed data ordinals.

*First-Order Logic with Two Variables.* Bojańczyk et al. [6] have shown that validity in first-order logic with two variables over data words and data $\omega$-words is in coNEXP. The same statement can be derived from our results, since $\mathbf{K_t^dL_\ell.3}$ is exactly as expressive as $\mathbf{FO}^2(<, \sim)$. We detail this aspect in [33]: converting a first-order formulæ into an equivalent $\mathbf{K_t^dL_\ell.3}$ formulæ can be done by adapting the proof from [17]—which involves an exponential blow-up—, we can then apply Theorem 4 to get a coNEXP decision procedure.

*Other Logics.* Our fragment can be encoded in the Logic of Repeating Values with Past from [13] (PLRV), for which the satisfiability problem is equivalent to the problem of reachability in VASS, which is TOWER-hard [11], and with an ACKERMANN complexity upper bound [32]. $\mathbf{K_t^d L_\ell.3}$ can also be encoded in the fragment of XPath with data tests and navigation among siblings which has been proved undecidable [18].

In both cases, the main difference is that $\mathbf{K_t^d L_\ell.3}$ cannot perform nested data tests. However, this restriction allowed us to get a logic for which the satisfiability problem has a smaller complexity (NP), as established by our proof system. The complexities of various logics on data words and their inclusions is summed up in [33].

The systems most closely related to $\mathbf{HK_t^d L_\ell.3}$ are obviously the calculus for $\mathbf{K_t 4.3}$ [2] and $\mathbf{K_t L_\ell.3}$ [3] in which we respectively introduced the notions of clusters and annotations, and adapted them to work over ordinals. The main contribution of the paper is being able to maintain the small branch property of the calculus with the addition of data registers. Another contribution is the shift of perspective about the annotations. In [3], they were only considered as an artefact of the proof system being able to guide the proof search; but in this paper, they are treated as a new kind of formula, and generalising the notion of immediately provable hypersequents introduced in [3] allowed us to mimic the syntactic condition they were previously bound to.

## References


1. Avron, A.: Hypersequents, logical consequence and intermediate logics for concurrency. Ann. Math. Artif. Intell. **4**(3–4), 225–248 (1991). https://doi.org/10.1007/BF01531058
2. Baelde, D., Lick, A., Schmitz, S.: A hypersequent calculus with clusters for linear frames. In: AiML 2018: Advances in Modal Logic, vol. 12, pp. 36–55. College Publications (2018). https://hal.inria.fr/hal-01756126
3. Baelde, D., Lick, A., Schmitz, S.: A hypersequent calculus with clusters for tense logic over ordinals. In: FSTTCS 2018. Leibniz International Proceedings in Informatics (LZI), vol. 122, pp. 15:1–15:19 (2018). https://doi.org/10.4230/LIPIcs.FSTTCS.2018.15
4. Baelde, D., Lunel, S., Schmitz, S.: A sequent calculus for a modal logic on finite data trees. In: CSL 2016. Leibniz International Proceedings in Informatics (LZI), vol. 62, pp. 32:1–32:16 (2016). https://doi.org/10.4230/LIPIcs.CSL.2016.32
5. Belnap, N.D.: Display logic. J. Philos. Logic **11**(4), 375–417 (1982). https://doi.org/10.1007/BF00284976
6. Bojańczyk, M., David, C., Muscholl, A., Schwentick, T., Segoufin, L.: Two-variable logic on data words. ACM Trans. Comput. Log. **12**(4), 27:1–27:26 (2011). https://doi.org/10.1145/1970398.1970403
7. Bollig, B.: An automaton over data words that captures EMSO logic. In: Katoen, J.-P., König, B. (eds.) CONCUR 2011. LNCS, vol. 6901, pp. 171–186. Springer, Heidelberg (2011). https://doi.org/10.1007/978-3-642-23217-6_12
8. Brünnler, K.: Deep sequent systems for modal logic. Archiv für Mathematische Logik und Grundlagenforschung **48**(6), 551–577 (2009). https://doi.org/10.1007/s00153-009-0137-3

9. Cocchiarella, N.B.: Tense and modal logic: a study in the topology of temporal reference. Ph.D. thesis, University of California, Los Angeles (1965)
10. Cook, S.A.: The complexity of theorem-proving procedures. In: STOC 1971, pp. 151–158. ACM (1971). https://doi.org/10.1145/800157.805047
11. Czerwiński, W., Lasota, S., Lazić, R., Leroux, J., Mazowiecki, F.: The reachability problem for Petri nets is not elementary. In: STOC 2019. ACM (2019, to appear)
12. Das, A., Pous, D.: A cut-free cyclic proof system for kleene algebra. In: Schmidt, R.A., Nalon, C. (eds.) TABLEAUX 2017. LNCS (LNAI), vol. 10501, pp. 261–277. Springer, Cham (2017). https://doi.org/10.1007/978-3-319-66902-1_16
13. Demri, S., Figueira, D., Praveen, M.: Reasoning about data repetitions with counter systems. Log. Methods Comput. Sci. **12**(3), 1 (2016). https://doi.org/10.2168/LMCS-12(3:1)2016
14. Demri, S., Lazić, R.: LTL with the freeze quantifier and register automata. ACM Trans. Comput. Log. **10**(3), 16 (2009). https://doi.org/10.1145/1507244.1507246
15. Demri, S., Rabinovich, A.: The complexity of linear-time temporal logic over the class of ordinals. Log. Methods Comput. Sci. **6**(4), 9 (2010). https://doi.org/10.2168/LMCS-6(4:9)2010
16. Demri, S., Nowak, D.: Reasoning about transfinite sequences. Int. J. Fund. Comput. Sci. **18**(01), 87–112 (2007). https://doi.org/10.1142/S0129054107004589
17. Etessami, K., Vardi, M.Y., Wilke, T.: First-order logic with two variables and unary temporal logic. Inf. Comput. **179**(2), 279–295 (2002). https://doi.org/10.1006/inco.2001.2953
18. Figueira, D., Segoufin, L.: Future-looking logics on data words and trees. In: Královič, R., Niwiński, D. (eds.) MFCS 2009. LNCS, vol. 5734, pp. 331–343. Springer, Heidelberg (2009). https://doi.org/10.1007/978-3-642-03816-7_29
19. Gauwin, O., Niehren, J., Tison, S.: Queries on XML streams with bounded delay and concurrency. Inf. Comput. **209**(3), 409–442 (2011). https://doi.org/10.1016/j.ic.2010.08.003
20. Godefroid, P., Wolper, P.: A partial approach to model checking. Inf. Comput. **110**(2), 305–326 (1994). https://doi.org/10.1006/inco.1994.1035
21. Goubault-Larrecq, J., Olivain, J.: On the efficiency of mathematics in intrusion detection: the NetEntropy case. In: Danger, J.-L., Debbabi, M., Marion, J.-Y., Garcia-Alfaro, J., Zincir Heywood, N. (eds.) FPS -2013. LNCS, vol. 8352, pp. 3–16. Springer, Cham (2014). https://doi.org/10.1007/978-3-319-05302-8_1
22. Christ, J., Hoenicke, J., Nutz, A.: Proof tree preserving interpolation. In: Piterman, N., Smolka, S.A. (eds.) TACAS 2013. LNCS, vol. 7795, pp. 124–138. Springer, Heidelberg (2013). https://doi.org/10.1007/978-3-642-36742-7_9
23. Indrzejczak, A.: Cut-free hypersequent calculus for S4.3. Bull. Sec. Logic **41**(1/2), 89–104 (2012)
24. Indrzejczak, A.: Linear time in hypersequent framework. Bull. Symb. Log. **22**(1), 121–144 (2016). https://doi.org/10.1017/bsl.2016.2
25. Indrzejczak, A.: Cut elimination theorem for non-commutative hypersequent calculus. Bull. Sec. Logic **46**(1/2), 135–149 (2017). https://doi.org/10.18778/0138-0680.46.1.2.10
26. Kanovich, M.: The multiplicative fragment of linear logic is NP complete. Technical report X-91-13, Institute for Language, Logic, and Information (1991)
27. Kashima, R.: Cut-free sequent calculi for some tense logics. Studia Logica **53**(1), 119–135 (1994). https://doi.org/10.1007/BF01053026
28. Kracht, M.: Power and weakness of the modal display calculus. In: Wansing, H. (ed.) Proof Theory of Modal Logic, pp. 93–121. Springer, Dordrecht (1996). https://doi.org/10.1007/978-94-017-2798-3_7

29. Kurokawa, H.: Hypersequent calculi for modal logics extending S4. In: Nakano, Y., Satoh, K., Bekki, D. (eds.) JSAI-isAI 2013. LNCS (LNAI), vol. 8417, pp. 51–68. Springer, Cham (2014). https://doi.org/10.1007/978-3-319-10061-6_4
30. Lazić, R.: Safety alternating automata on data words. ACM Trans. Comput. Log. **12**(2), 10:1–10:24 (2011). https://doi.org/10.1145/1877714.1877716
31. Lellmann, B.: Linear nested sequents, 2-sequents and hypersequents. In: De Nivelle, H. (ed.) TABLEAUX 2015. LNCS (LNAI), vol. 9323, pp. 135–150. Springer, Cham (2015). https://doi.org/10.1007/978-3-319-24312-2_10
32. Leroux, J., Schmitz, S.: Reachability in vector addition systems is primitive-recursive in fixed dimension. In: LICS 2019. IEEE (2019, to appear). http://arxiv.org/abs/1903.08575
33. Lick, A.: A hypersequent calculus with clusters for data logic over ordinals. In: Cerrito, S., Popescu, A., (eds.) TABLEAUX 2019. LNAI, vol. 11714, pp. 166–184 (2019). https://hal.archives-ouvertes.fr/hal-02165359
34. Lincoln, P., Mitchell, J., Scedrov, A., Shankar, N.: Decision problems for propositional linear logic. Ann. Pure Appl. Log. **56**(1–3), 239–311 (1992). https://doi.org/10.1016/0168-0072(92)90075-B
35. Negri, S.: Proof analysis in modal logic. J. Philos. Logic **34**(5), 507–544 (2005). https://doi.org/10.1007/s10992-005-2267-3
36. Ono, H., Nakamura, A.: On the size of refutation Kripke models for some linear modal and tense logics. Studia Logica **39**(4), 325–333 (1980). https://doi.org/10.1007/BF00713542
37. Pnueli, A.: The temporal logic of programs. In: FOCS 1977, pp. 46–57. IEEE (1977). https://doi.org/10.1109/SFCS.1977.32
38. Poggiolesi, F.: The method of tree-hypersequents for modal propositional logic. In: Makinson, D., Wansing, H. (eds.) Towards Mathematical Philosophy. Trends in Logic, vol. 28, pp. 31–51. Springer, Dordrecht (2009). https://doi.org/10.1007/978-1-4020-9084-4_3
39. Poggiolesi, F.: A purely syntactic and cut-free sequent calculus for the modal logic of provability. Rev. Symb. Log. **2**(4), 593–611 (2009). https://doi.org/10.1017/S1755020309990244
40. Prior, A.N.: Time and Modality. Oxford University Press, Oxford (1957)
41. Sistla, A.P., Clarke, E.M.: The complexity of propositional linear temporal logics. J. ACM **32**(3), 733–749 (1985). https://doi.org/10.1145/3828.3837
42. Sistla, A.P., Zuck, L.D.: Reasoning in a restricted temporal logic. Inf. Comput. **102**(2), 167–195 (1993). https://doi.org/10.1006/inco.1993.1006
43. Vardi, M.Y., Wolper, P.: An automata-theoretic approach to automatic program verification. In: LICS 1986, pp. 322–331. IEEE (1986)

# Syntactic Cut-Elimination and Backward Proof-Search for Tense Logic via Linear Nested Sequents

Rajeev Goré[1] and Björn Lellmann[2(✉)]

[1] Research School of Computer Science, Australian National University, Canberra, Australia
[2] Faculty of Informatics, Technical University of Vienna, Vienna, Austria
lellmann@logic.at

**Abstract.** We give a linear nested sequent calculus for the basic normal tense logic Kt. We show that the calculus enables backwards proof-search, counter-model construction and syntactic cut-elimination. Linear nested sequents thus provide the minimal amount of nesting necessary to provide an adequate proof-theory for modal logics containing converse. As a bonus, this yields a cut-free calculus for symmetric modal logic KB.

## 1 Introduction

The two main proof-calculi for normal modal logics are sequent calculi and tableau calculi [4]. Tableau calculi are algorithmic, directly providing a decision procedure via cut-free completeness. Sequent calculi are proof-theoretic, requiring us to show completeness via cut-admissibility. Often, there is a direct relationship between these two formalisms, where one can be seen as the "upside down" variant of the other. However, this direct relationship breaks down for modal logics where the modalities are interpreted with respect to a Kripke reachability relation as well as its converse relation, as in modal tense logic Kt.

Modal sequent calculi go back to at least 1957 [16]. Sequent calculi for normal modal tense logics have proved more elusive, with some previous published attempts failing cut-elimination [19]: the counter-example is $p \to \Box\neg\blacksquare\neg p$. But we now have several extended sequent frameworks for tense logics: for example, display calculi [20]; nested sequents [6, 10] and labelled sequents [1]. The main disadvantage is the rather heavy machinery required to achieve cut-elimination. Tableau calculi for tense logics in contrast take a global view of proof-search, permitting to expand any node in the search space but requiring technical novelties such as dynamic blocking [8] and the use of a "restart" rule [7].

But there is a glaring disparity between the simplicity of tableau calculi for tense logics versus the mentioned extended sequent frameworks, giving rise to the question: What is the minimum extension over traditional sequents enabling a proof-theory for tense logics amenable to (algorithmic) backward proof-search?

Supported by WWTF project MA16-28.


S. Cerrito and A. Popescu (Eds.): TABLEAUX 2019, LNAI 11714, pp. 185–202, 2019.
https://doi.org/10.1007/978-3-030-29026-9_11

Here, we address this question by giving a sequent-style calculus for tense logic Kt which includes two "restart" rules. The calculus is given in the *linear nested sequent* framework. This framework, essentially a reformulation of *2-sequents* [15], lies between the original sequent framework and the nested sequent framework, in that it extends the sequent structure to *lists* of sequents. Apart from *op.cit.*, this framework yielded, e.g., cut-free calculi for a number of standard normal and non-normal modal logics [13,14,17] as well as temporal or intermediate logics of linear frames [9,11]. Yet, so far the only examples were logics which either have a cut-free sequent formulation, or where the underlying semantic structure exactly matches that of linear nested sequents. The calculus presented here thus is interesting for two reasons: First, it shows that not the full complexity of nested sequents is necessary to capture tense logic without cuts; second, it provides a non-trivial example showing that the linear nested sequent framework can handle interesting logics beyond the reach of standard sequents, with models not mirroring the linear structure.

In the following, we present the calculus, then show how to use it for backward proof-search and cut-free completeness. We also show that it is amenable to the usual proof-theoretic results such as the admissibility of the structural rules and cut. As a bonus, this yields a calculus for symmetric modal logic KB, suggesting that the linear nested sequent framework so far is the simplest purely syntactic extension of the standard sequent framework capturing KB in a cut-free way, since even hypersequent systems for KB, such as that of Lahav [12], seem to require an analytic cut rule and hence are not completely cut-free.

We thank Reviewer 2 for many helpful suggestions.

## 2 Preliminaries

We assume that the reader is familiar with normal modal tense logics and their associated Kripke semantics but give a very terse introduction below.

*Formulae* of normal modal tense logics are built from a given set $\mathtt{Atm}$ of atomic formula via the BNF grammar below where $p \in \mathtt{Atm}$:

$$A := p \mid \bot \mid A \to A \mid \Box A \mid \Diamond A \mid \blacksquare A \mid \blacklozenge A$$

We assume conjunction, disjunction and negation are defined as usual.

The Kripke semantics for Kt is given by a non-empty set (of worlds) $W$, a binary relation $R$ over $W$, and a valuation function $V$ mapping a world $w \in W$ and an atomic formula $p \in \mathtt{Atm}$ to either "true" or "false". Given a Kripke model $\langle W, R, V \rangle$, the forcing relation $w \Vdash A$ between a world $w \in W$ and a formula $A$ is defined as follows (omitting clauses for the propositional connectives):

$$w \Vdash p \quad \text{if} \quad V(w,p) = true$$

$$w \Vdash \Diamond A \text{ if } \exists v \in W.\ wRv \ \&\ v \Vdash A \qquad w \Vdash \blacklozenge A \text{ if } \exists v \in W.\ vRw \ \&\ v \Vdash A$$

$$w \Vdash \Box A \text{ if } \forall v \in W.\ wRv \Rightarrow v \Vdash A \qquad w \Vdash \blacksquare A \text{ if } \forall v \in W.\ vRw \Rightarrow v \Vdash A$$

$$\frac{\mathcal{G} \updownarrow \Gamma \Rightarrow \Delta, A \swarrow \Sigma \Rightarrow \Pi, \Box A \quad \mathcal{G} \updownarrow \Gamma \Rightarrow \Delta \swarrow \Sigma \Rightarrow \Pi, \Box A \nearrow \epsilon \Rightarrow A}{\mathcal{G} \updownarrow \Gamma \Rightarrow \Delta \swarrow \Sigma \Rightarrow \Pi, \Box A} \; \Box^1_R$$

$$\frac{\mathcal{G} \updownarrow \Gamma \Rightarrow \Delta, A \nearrow \Sigma \Rightarrow \Pi, \blacksquare A \quad \mathcal{G} \updownarrow \Gamma \Rightarrow \Delta \nearrow \Sigma \Rightarrow \Pi, \blacksquare A \swarrow \epsilon \Rightarrow A}{\mathcal{G} \updownarrow \Gamma \Rightarrow \Delta \nearrow \Sigma \Rightarrow \Pi, \blacksquare A} \; \blacksquare^1_R$$

$$\frac{\mathcal{G} \nearrow \Gamma \Rightarrow \Delta, \Box A \nearrow \epsilon \Rightarrow A}{\mathcal{G} \nearrow \Gamma \Rightarrow \Delta, \Box A} \; \Box^2_R \qquad \frac{\mathcal{G} \swarrow \Gamma \Rightarrow \Delta, \blacksquare A \swarrow \epsilon \Rightarrow A}{\mathcal{G} \swarrow \Gamma \Rightarrow \Delta, \blacksquare A} \; \blacksquare^2_R$$

$$\frac{\mathcal{G} \updownarrow \Gamma, \Box A \Rightarrow \Delta \nearrow \Sigma, A \Rightarrow \Pi}{\mathcal{G} \updownarrow \Gamma, \Box A \Rightarrow \Delta \nearrow \Sigma \Rightarrow \Pi} \; \Box^1_L \qquad \frac{\mathcal{G} \updownarrow \Gamma, \blacksquare A \Rightarrow \Delta \swarrow \Sigma, A \Rightarrow \Pi}{\mathcal{G} \updownarrow \Gamma, \blacksquare A \Rightarrow \Delta \swarrow \Sigma \Rightarrow \Pi} \; \blacksquare^1_L$$

$$\frac{\mathcal{G} \updownarrow \Gamma, A \Rightarrow \Delta}{\mathcal{G} \updownarrow \Gamma \Rightarrow \Delta \swarrow \Sigma, \Box A \Rightarrow \Pi} \; \Box^2_L \qquad \frac{\mathcal{G} \updownarrow \Gamma, A \Rightarrow \Delta}{\mathcal{G} \updownarrow \Gamma \Rightarrow \Delta \nearrow \Sigma, \blacksquare A \Rightarrow \Pi} \; \blacksquare^2_L$$

$$\frac{}{\mathcal{G} \updownarrow \Gamma, p \Rightarrow p, \Delta} \; (id) \qquad \frac{}{\mathcal{G} \updownarrow \Gamma, \bot \Rightarrow \Delta} \; \bot_L \qquad \frac{\mathcal{G}}{\mathcal{G} \updownarrow \Gamma \Rightarrow \Delta} \; \mathsf{EW}$$

$$\frac{\mathcal{G} \updownarrow \Gamma, A \Rightarrow \Delta, A \to B, B}{\mathcal{G} \updownarrow \Gamma \Rightarrow \Delta, A \to B} \; \to_R \qquad \frac{\mathcal{G} \updownarrow \Gamma, A \to B, B \Rightarrow \Delta \quad \mathcal{G} \updownarrow \Gamma, A \to B \Rightarrow \Delta, A}{\mathcal{G} \updownarrow \Gamma, A \to B \Rightarrow \Delta} \; \to_L$$

**Fig. 1.** The system $\mathsf{LNS_{Kt}}$ where $\updownarrow$ stands for either $\nearrow$ or $\swarrow$

As usual, a formula $A$ is *satisfiable* if there is some Kripke model $\langle W, R, V \rangle$, and some world $w \in W$ such that $w \Vdash A$. A formula $A$ is *valid* if $\neg A$ is unsatisfiable. Formally, the logic $\mathsf{Kt}$ is the set of all valid formulae.

The traditional Hilbert system $\mathsf{HKt}$ for tense logic $\mathsf{Kt}$ takes all classical propositional tautologies as axioms, adds the axioms $\Box(A \to B) \to (\Box A \to \Box B)$ and $\blacksquare(A \to B) \to (\blacksquare A \to \blacksquare B)$, the necessitation rules $\mathrm{Nec}_\Box : A/\Box A$ and $\mathrm{Nec}_\blacksquare : A/\blacksquare A$, and the two interaction axioms $\Diamond\blacksquare p \to p$ and $\blacklozenge\Box p \to p$. The system $\mathsf{HKt}$ is sound and complete w.r.t. the Kripke semantics.

## 3 A Linear Nested Sequent Calculus for Kt

Unlike standard Hilbert-calculi, our calculus operates on linear nested sequents instead of formulae, defined and adapted from Lellmann [13] as follows.

**Definition 1.** *A* component *is an expression $\Gamma \Rightarrow \Delta$ where the* antecedent *$\Gamma$ and the* succedent *$\Delta$ are finite, possibly empty, multisets of formulae. We write $\epsilon$ to stand for an empty antecedent or succedent to avoid confusion. A* linear nested sequent *is an expression obtained via the following BNF grammar:*

$$S := \Gamma \Rightarrow \Delta \mid \Gamma \Rightarrow \Delta \nearrow S \mid \Gamma \Rightarrow \Delta \swarrow S\,.$$

We often write $\mathcal{G}$ for a possibly empty *context*: e.g., $\mathcal{G} \nearrow \Gamma \Rightarrow \Delta$ stands for $\Gamma \Rightarrow \Delta$ if $\mathcal{G}$ is empty, and for $\Sigma \Rightarrow \Pi \swarrow \Omega \Rightarrow \Theta \nearrow \Gamma \Rightarrow \Delta$ if $\mathcal{G}$ is the linear nested sequent $\Sigma \Rightarrow \Pi \swarrow \Omega \Rightarrow \Theta$. Figure 1 shows the rules of our calculus $\mathsf{LNS_{Kt}}$. As usual, each rule has a number of *premisses* above the horizontal line and a single *conclusion* below it. The single formula in the conclusion is the *principal* formula and the formulae in the premisses are the *side-formulae.*

Every instance of the rule (id) is a *derivation* of height 0, and if $(\rho)$ is an $n$-ary rule and we are given $n$ premiss derivations $d_1, \cdots, d_n$, each of height $h_1, \cdots, h_n$, with respective conclusions $c_1, \cdots, c_n$, and $c_1, \cdots, c_n / d_0$ is an instance of $(\rho)$ then $d_1, \cdots, d_n / d_0$ is a derivation of height $1 + max\{h_1, \cdots, h_n\}$. We write $\mathcal{D} \vdash S$ if $\mathcal{D}$ is a derivation in $\mathsf{LNS_{Kt}}$ of the linear nested sequent $S$, and $\vdash S$ if there is a derivation $\mathcal{D}$ with $\mathcal{D} \vdash S$.

Note that our calculus is *end-active*, i.e., in every logical rule and every premiss, at least one active formula occurs in the last component.

*Example 2.* Consider the end-sequent $\Rightarrow \Box p, \Box q, r \to \blacksquare\neg\Box\neg r$ where $r \to \blacksquare\neg\Box\neg r$ is the axiom $r \to \blacksquare\Diamond r$ with the definition of $\Diamond$ expanded. Suppose we apply the rule $(\to_R)$ upward to obtain $r \Rightarrow \Box p, \Box q, \blacksquare\neg\Box\neg r$. Then there are two different instances of the rule $\Box^2_R$ using two different principal formulae, neither of which leads to a derivation, and one instance of the rule $\blacksquare^2_R$ which leads to a derivation:

$$\dfrac{r \Rightarrow \Box p, \Box q, \blacksquare\neg\Box\neg r \nearrow \epsilon \Rightarrow p}{r \Rightarrow \Box p, \Box q, \blacksquare\neg\Box\neg r}\,\Box^2_R \qquad \dfrac{r \Rightarrow \Box p, \Box q, \blacksquare\neg\Box\neg r \nearrow \epsilon \Rightarrow q}{r \Rightarrow \Box p, \Box q, \blacksquare\neg\Box\neg r}\,\Box^2_R$$

$$\dfrac{\dfrac{\dfrac{\dfrac{\dfrac{}{r, \neg r \Rightarrow r, \Box p, \Box q, \blacksquare\neg\Box\neg r}\,\text{id}}{r, \neg r \Rightarrow \Box p, \Box q, \blacksquare\neg\Box\neg r}\,\neg_L}{r \Rightarrow \Box p, \Box q, \blacksquare\neg\Box\neg r \swarrow \Box\neg r \Rightarrow \neg\Box\neg r}\,\Box^2_L}{r \Rightarrow \Box p, \Box q, \blacksquare\neg\Box\neg r \swarrow \epsilon \Rightarrow \neg\Box\neg r}\,\neg_R}{r \Rightarrow \Box p, \Box q, \blacksquare\neg\Box\neg r}\,\blacksquare^2_R$$

Intuitively, each component of a linear nested sequent corresponds to a world of a Kripke model, and the structural connectives $\nearrow$ and $\swarrow$ between components corresponds to the relations $R$ and $R^{-1}$ that connect these worlds.

These intuitions can be made formal since linear nested sequents have a natural interpretation as formulae given by taking $\nearrow$ and $\swarrow$ to be the structural connectives corresponding to $\Box$ and $\blacksquare$, respectively:

**Definition 3.** *If $\Gamma = \{A_1, \cdots, A_n\}$ then we write $\hat{\Gamma}$ for $A_1 \wedge \cdots \wedge A_n$ and $\check{\Gamma}$ for $A_1 \vee \cdots \vee A_n$. The* formula translation *of a linear nested sequent is given recursively by $\tau(\Gamma \Rightarrow \Delta) = \hat{\Gamma} \to \check{\Delta}$ and*

$$\tau(\Gamma \Rightarrow \Delta \nearrow \mathcal{G}) = \hat{\Gamma} \to (\check{\Delta} \vee \Box\, \tau(\mathcal{G})) \quad \tau(\Gamma \Rightarrow \Delta \swarrow \mathcal{G}) = \hat{\Gamma} \to (\check{\Delta} \vee \blacksquare\, \tau(\mathcal{G}))\,.$$

*A sequent $S$ is* falsifiable *if there exists a model $\langle W, R, v\rangle$ and a world $w \in W$ such that $w \not\Vdash \tau(S)$. A sequent $S$ is* valid *if it is not falsifiable.*

Soundness of the calculus then follows by induction on the depth of the derivation from the following theorem.

**Theorem 4 (Soundness).** *For every rule, if the conclusion is falsifiable then so is one of the premisses.*

*Proof.* We only give the interesting cases going beyond the standard calculi.

For rule $\Box^1_R$, suppose that for $\mathfrak{M} = \langle W, R, V\rangle$ and $w_1 \in W$ we have $\mathfrak{M}, w_1 \not\Vdash \tau(\Gamma_1 \Rightarrow \Delta_1 \updownarrow \ldots \updownarrow \Gamma_n \Rightarrow \Delta_n \updownarrow \Gamma \Rightarrow \Delta \swarrow \Sigma \Rightarrow \Pi, \Box A)$. Hence there are worlds $w_2, \ldots, w_n, x, y \in W$ with $w_1 R^{\epsilon_1} w_2 R^{\epsilon_2} \ldots R^{\epsilon_{n-1}} w_n R^{\epsilon_n} x R^{-1} y$, for $\epsilon_i$ empty or $-1$ as needed, such that $w_i \Vdash \hat{\Gamma}_i \wedge \neg\check{\Delta}_i$ for every $i \leq n$, as well as $x \Vdash \hat{\Gamma} \wedge \neg\check{\Delta}$

and $y \Vdash \hat{\Sigma} \wedge \neg\breve{\Pi} \wedge \neg\Box A$. Hence there is a world $z \in W$ with $yRz$ such that $z \not\Vdash A$. If $z = x$, then $\mathfrak{M}, w_1$ falsifies the interpretation of the first premiss. If $z \neq x$, we have a model falsifying the interpretation of the second premiss. The case of rule $\blacksquare^1_R$ is analogous.

For the "restart" rule ($\blacksquare^2_L$), suppose that the conclusion $\mathcal{G} \updownarrow \Gamma \Rightarrow \Delta \nearrow \Sigma, \blacksquare A \Rightarrow \Pi$ is falsifiable. Thus there is a world $w$ such that $w \not\Vdash \hat{\Gamma} \rightarrow \breve{\Delta} \vee \Box(\hat{\Sigma} \wedge \blacksquare A \rightarrow \breve{\Pi})$. So $w \Vdash \hat{\Gamma}$ and $w \not\Vdash \breve{\Delta}$ and $w$ must have an $R$-successor $v$ such that $v \Vdash \hat{\Sigma}$ and $v \not\Vdash \breve{\Pi}$ and $v \Vdash \blacksquare A$. But then $w \Vdash A$, exactly as desired to conclude that the premiss $\Gamma, A \Rightarrow \Delta$ is falsifiable. ⊣

**Corollary 5.** *For every linear nested sequent $S$, if $\vdash S$, then $\tau(S)$ is valid.* ⊣

Why does the premiss of the rule $\blacksquare^2_L$ not contain the sequent $\Sigma, \blacksquare A \Rightarrow \Pi$ ? Because there may be an incompatibility between $w$ and its $R$-successor $v$. The $\blacksquare^2_L$ rule removes this incompatibility by propagating $A$ to the $R$-predecessor $w$. But $A$ could be arbitrarily complex and we must again saturate the predecessor before re-creating $v$. The current $v$ must be deleted and we must "restart" $w$.

Before showing completeness of $\mathsf{LNS_{Kt}}$ we remark on a simplification of the calculus. Let $\mathsf{LNS^*_{Kt}}$ be the calculus obtained from $\mathsf{LNS_{Kt}}$ by replacing the modal right rules $\Box^1_R, \blacksquare^1_R, \Box^2_R$ and $\blacksquare^2_R$ with the following two rules:

$$\frac{\mathcal{G} \updownarrow \Gamma \Rightarrow \Delta, \Box A \nearrow \epsilon \Rightarrow A}{\mathcal{G} \updownarrow \Gamma \Rightarrow \Delta, \Box A}\ \Box_R \qquad \frac{\mathcal{G} \updownarrow \Gamma \Rightarrow \Delta, \blacksquare A \swarrow \epsilon \Rightarrow A}{\mathcal{G} \updownarrow \Gamma \Rightarrow \Delta, \blacksquare A}\ \blacksquare_R$$

Soundness of these rules can be shown exactly as in Theorem 4. Moreover, since derivations in the system $\mathsf{LNS_{Kt}}$ can be converted straightforwardly into derivations in the system $\mathsf{LNS^*_{Kt}}$ by simply omitting the subderivations of the left premisses of $\Box^1_R$ and $\blacksquare^1_R$ respectively, we immediately obtain:

**Proposition 6.** *If $\mathsf{LNS_{Kt}}$ is cut-free complete for $\mathsf{Kt}$, then so is $\mathsf{LNS^*_{Kt}}$.* ⊣

For technical reasons, in particular to facilitate a cut elimination proof when the cut formula is principal in the rules $\Box^2_L$ or $\blacksquare^2_L$, in the following we take $\mathsf{LNS_{Kt}}$ as the main system, but it is worth keeping in mind that the completeness results automatically extend to $\mathsf{LNS^*_{Kt}}$. Note also that, modulo the structural rules and deleting the last component in the rules $\Box^2_L$ and $\blacksquare^2_L$, $\mathsf{LNS^*_{Kt}}$ is essentially a two-sided linear end-active reformulation of the cut-free nested sequent calculus $S2K_t$ for $\mathsf{Kt}$ in [10]. Hence completeness of the latter follows from our completeness results by transforming derivations bottom-up.

## 4 Completeness via Proof Search and Counter-Models

We now show how to use our calculus (without EW) for backward proof search, and how to obtain a counter-model from failed proof search, yielding completeness. For this, we separate the rules into groups, assuming an appropriate side-condition to ensure that rules are applied only when they create new formulae:

***Termination Rules:*** (id) and $\bot_L$;

***CPL Rules:*** $(\to_R)$ and $(\to_L)$. The side-conditions ensuring termination are: $A \notin \Gamma$ or $B \notin \Delta$ for $(\to_R)$, and $B \notin \Gamma$ and $A \notin \Delta$ for $(\to_L)$;

***Propagation Rules:*** $\Box^1_L$ and $\blacksquare^1_L$. These rules move subformulae to the last component. The side-condition ensuring termination is that $A \notin \Sigma$;

***Restart Rules:*** $\Box^2_L$ and $\blacksquare^2_L$. These rules make the sequent shorter. The side-condition ensuring termination is that $A \notin \Gamma$;

***Box Rules:*** $\Box^1_R$, $\Box^2_R$, $\blacksquare^1_R$ $\blacksquare^2_R$. We apply only one of these rules, even if many are applicable, and backtrack over these choices. But these rules are non-deterministic since they choose a particular formula as principal. We must also back-track over all choices of principal formula in the chosen rule.

Our *proof-search strategy* is to apply (backwards) the highest rule in the above list. Thus, assuming that the (id) rule is not applicable, our strategy first seeks to saturate the final component with the CPL-rules. Then we seek to propagate formulae from the second-final component into the final component. Then we seek to repair any incompatibilities between the final two components using the Restart rules to shorten the sequent if necessary. Only when none of these rules are applicable do we apply a Box-rule to lengthen the sequent, and backtrack over all choices of principal formula. In particular, if a node is "restarted" then we have to redo all previous Box-rule applications from this changed node.

Overall, the strategy means that the maximal modal degree, defined standardly, of a formula in a component must decrease strictly as the sequent becomes longer, and the restart rules, which shorten the sequent, do not increase this maximal modal degree. A particular component is restarted only a finite number of times because each restart adds a formula which is a strict subformula of the end-sequent, and there are only a finite number of these. Hence the proof-search terminates.

**Theorem 7 (Termination).** *Backward proof-search terminates.* ⊣

Suppose backward proof-search terminates without finding a derivation. How do we construct a counter-model that falsifies the end-sequent? Consider the search-space explored by our procedure, i.e., the space of all possible failed derivations including the various backtracking choice-points inherent in the search procedure. We visualise this search space as a single tree by conjoining the modal rules containing backtrack choices. E.g., the backtracking choices in the sequent $\epsilon \Rightarrow \Box p, \Box q, \Box r$ can be "determinised" as below where we have used "dotted" lines to indicate a meta-level conjunction which "binds" the three premisses:

$$\frac{\epsilon \Rightarrow \Box p, \Box q, \blacksquare r \nearrow \epsilon \Rightarrow p \qquad \epsilon \Rightarrow \Box p, \Box q, \blacksquare r \nearrow \epsilon \Rightarrow q \qquad \epsilon \Rightarrow \Box p, \Box q, \blacksquare r \swarrow \epsilon \Rightarrow r}{\epsilon \Rightarrow \Box p, \Box q, \blacksquare r}$$

Similarly, the sequent $\mathcal{G} \updownarrow \Gamma_1 \Rightarrow \Delta_1 \swarrow \Gamma_2 \Rightarrow \Box p, \blacksquare q$ can be determinised as:

$$\text{(a)} \quad \frac{\mathcal{G} \updownarrow \Gamma_1 \Rightarrow \Delta_1 \swarrow \Gamma_2 \Rightarrow \Box p, \blacksquare q \swarrow \epsilon \Rightarrow q}{\mathcal{G} \updownarrow \Gamma_1 \Rightarrow \Delta_1 \swarrow \Gamma_2 \Rightarrow \Box p, \blacksquare q}$$

with (a) being the pair below:

$$\Box^1_R \frac{\mathcal{G} \updownarrow \Gamma_1 \Rightarrow p, \Delta_1 \swarrow \Gamma_2 \Rightarrow \Box p, \blacksquare q \qquad \mathcal{G} \updownarrow \Gamma_1 \Rightarrow \Delta_1 \swarrow \Gamma_2 \Rightarrow \Box p, \blacksquare q \nearrow \epsilon \Rightarrow p}{(a)} \; p \notin \Delta_1$$

We dub these choice-points as *"and-nodes"* to distinguish them from the traditional *"or-nodes"* created by disjunctions [7]. We first show how we prune this search space to keep only nodes useful for building a counter-model. We then outline how the pruned search space yields a counter-model for the end-sequent.

### 4.1 Pruning Irrelevant Branches from the Search Space

Suppose the original search-space corresponds to a tree $\tau_0$, and consider some leaf to which no rule is applicable. In this search tree, delete all the rightmost components of the conclusion of a restart rule. We can do so because we know that, in the conclusion, the second-last component is incompatible with the last component precisely because its antecedent $\Gamma$ is missing $A$. So this pair of components cannot possibly be part of a counter-model.

Now consider the rule application $(\rho)$ below the restart rule. Suppose the last component of the premiss of $(\rho)$ is $\Sigma, \Box A \Rightarrow \Pi$. If deleting $\Sigma, \Box A \Rightarrow \Pi$ causes $(\rho)$ to become meaningless, then delete the last component of the conclusion of $(\rho)$. If the rule is binary or is an "and-rule" then we keep the shorter of the sequents that are returned downward by this procedure. E.g., an instance of the rule $\Box^2_L$ from Fig. 1, as below, now appears as shown below it:

$$\wedge_R \frac{\Box^2_L \dfrac{\mathcal{G} \updownarrow A, \Gamma \Rightarrow \Delta}{\mathcal{G} \updownarrow \Gamma \Rightarrow \Delta \nearrow \Box A, \Sigma \Rightarrow A} A \notin \Gamma \qquad \dfrac{\vdots}{\mathcal{G} \updownarrow \Gamma \Rightarrow \Delta \nearrow \Box A, \Sigma \Rightarrow B}}{\mathcal{G} \updownarrow \Gamma \Rightarrow \Delta \nearrow \Box A, \Sigma \Rightarrow A \wedge B}$$

$$\frac{\dfrac{\mathcal{G} \updownarrow A, \Gamma \Rightarrow \Delta}{\mathcal{G} \updownarrow \Gamma \Rightarrow \Delta} \qquad \mathcal{G}'}{\mathcal{G}''}$$

where $\mathcal{G}'$ is the pruned version of $\mathcal{G} \updownarrow \Gamma \Rightarrow \Delta \nearrow \Box A, \Sigma \Rightarrow B$ and $\mathcal{G}''$ is the shorter of $\mathcal{G} \updownarrow \Gamma \Rightarrow \Delta$ and $\mathcal{G}'$. We can do so because the shorter sequent $\mathcal{G}''$ restarts a component that is earlier in the order of expansion, hence closer to the initial sequent. Now proceed by considering the number of restarts.

**Lemma 8.** *For all $\Gamma$ and $\Delta$, if $\Gamma \Rightarrow \Delta$ is not derivable and no restart rule is ever applied then there exists a Kripke model which falsifies $\Gamma \Rightarrow \Delta$.*

*Proof.* If no restart rules are applied in backward proof-search, then every application of a Box-right-rule leads to a new component which is compatible with its parent component in that every required formula is already in the latter.

Now consider any three adjacent components of a leaf sequent, which must be of one of the following forms where the second-last component and the third-last component are separated by $\nearrow$ (we skip the similar cases when it is $\swarrow$):

$$(1)\ \mathcal{G} \updownarrow \Gamma_0, \Sigma_1, \Box\Sigma_5 \Rightarrow D_l, \Delta_0 \nearrow \Gamma_1, \blacksquare\Sigma_1, \Sigma_2, \Box\Sigma_3, \Sigma_5, \Sigma_4 \Rightarrow \Delta_1, \blacksquare A_i, \Box B_j, \Box C_k, \blacksquare D_l \swarrow \Gamma_2, \Sigma_1, \Box\Sigma_2 \Rightarrow \Delta_2, A_i \updownarrow \mathcal{H}$$

$$(2)\ \mathcal{G} \updownarrow \cdots\cdots \nearrow \cdots\cdots \nearrow \Gamma_2', \Sigma_3, \blacksquare\Sigma_4 \Rightarrow \Delta_2', C_k \updownarrow \mathcal{H}$$

$$(3)\ \mathcal{G} \updownarrow \cdots\cdots \nearrow \cdots\cdots \nearrow \Gamma_2'', \Sigma_3, \blacksquare\Sigma_4 \Rightarrow \Delta_2'', B_j \updownarrow \mathcal{H}$$

In (1), the final component is the right premiss of the rule $\blacksquare^1_R$ on $\blacksquare A_i$, so $\blacksquare A_i$ is "fulfilled". The rule $\Box^2_L$ is not applicable to the last component because $\Sigma_2$ is in the middle component. The rule $\blacksquare^1_L$ is not applicable on the middle component because $\Sigma_1$ is in the last component. The rule $\blacksquare^2_L$ is not applicable to the middle component because $\Sigma_1$ is also in the first component. The rule $\Box^1_L$ is not applicable on the first component because $\Sigma_5$ is in the middle component. The $\blacksquare D_l$ in the middle component is fulfilled because the first component contains $D_l$ via the left premiss of $\blacksquare^1_R$.

The two formulae $\Box B_j$ and $\Box C_k$ in the middle component are not fulfilled by (1). But there will be an application of $\Box^2_R$ on $\Box C_k$ shown as (2), and another similar instance on $\Box B_j$ with $C_k$ in the last component replaced by $B_j$. Rule $\blacksquare^2_L$ is not applicable on the last component because $\Sigma_4$ is in the middle one. The rule $\Box^1_L$ is not applicable on the middle component because $\Sigma_3$ is in the last one.

These arguments apply for every $\Box$-formula and for every $\blacksquare$-formula in the second-last component. Moreover, for every conjunction in the succedent of either component, at least one conjunct must be in that succedent. Similarly, for every disjunction in the succedent of either component, both disjuncts must be in that succedent. Finally, the (id) rule is not applicable to any component.

Now put the following valuation on these components: every formula in the antecedent has a value of "true" and every formula in the succedent has a value of "false". Then replace every occurrence of $\nearrow$ with $R$ and replace every occurrence of $\swarrow$ with $R^{-1}$. Thus we have the following picture:

$$\begin{array}{ccc} w_j : \Gamma_2'', \Sigma_3, \blacksquare\Sigma_4 \Rightarrow \Delta_2'', B_j & & w_k : \Gamma_2', \Sigma_3, \blacksquare\Sigma_4 \Rightarrow \Delta_2', C_k \\ R\uparrow & & \uparrow R \\ v : \Gamma_1, \blacksquare\Sigma_1, \Sigma_2, \Box\Sigma_3, \Sigma_5, \Sigma_4 & \Rightarrow & \Delta_1, \blacksquare A_i, \Box B_j, \Box C_k, \blacksquare D_l \\ \uparrow R & & R\uparrow \\ u : \Gamma_0, \Sigma_1, \Box\Sigma_5 \Rightarrow D_l, \Delta_0 & & w_i : \Gamma_2, \Sigma_1, \Box\Sigma_2 \Rightarrow \Delta_2, A_i \end{array}$$

For every world $v$, every formula $\blacksquare A_i$ and every formula $\Box C_k$ with $v \not\Vdash \blacksquare A_i$ and $v \not\Vdash \Box C_k$, there exists a predecessor world $u_i$ with $u_i R v$ and $u_i \not\Vdash A_i$, there exists a successor world $w_k$ with $v R w_k$ and $w_k \not\Vdash C_k$; for every formula $\Box B \in \Box\Sigma_2$ with $u_i \Vdash \Box B$, we have $v \Vdash B$; and for every formula $\blacksquare D \in \blacksquare\Sigma_4$ with $w_k \Vdash \blacksquare D$, we have $v \Vdash D$. Hence, the triple $u_i R v R w_k$ is mutually compatible in terms of both modalities, and each world falsifies the associated component. Similar triples exist for all the box-formulae in $v$ which are not principal in the diagram, and they all "overlap" at $v$. Hence we can "glue" them together to form the fan of R-successors and R-predecessors of $v$, maintaining global compatibility. The original sequent $\Gamma \Rightarrow \Delta$ is thus falsified at its associated world. ⊣

**Lemma 9.** *For every $\Gamma$ and $\Delta$, if the sequent $\Gamma \Rightarrow \Delta$ is not derivable, and contains restarts, then there is a Kripke model which falsifies the end-sequent.*

*Proof.* We proceed by induction on the number of restarts. If there are none, then we are done by the previous lemma. Else there are a finite number of restarts.

Consider the highest restart and suppose it is $\Box^2_L$. By our deletion strategy, it must look exactly as shown above. By the induction hypothesis, the premiss must have a counter-model. But the premiss is a strict superset of the conclusion, so the same model must falsify the conclusion. ⊣

*Example 10.* Consider the end-sequent $\epsilon \Rightarrow \Box p, \Box q, \blacksquare r$. We would need two successor worlds, falsifying $p$ and $q$ respectively, and one predecessor world falsifying $r$. One failed derivation will come from $\epsilon \Rightarrow \Box p, \Box q, \Box r \nearrow \Rightarrow p$ while another will come from $\epsilon \Rightarrow \Box p, \Box q, \blacksquare r \nearrow \Rightarrow q$, i.e., two instances of the $\Box^2_R$-rule. But there will also be a failed derivation from $\epsilon \Rightarrow \Box p, \Box q, \blacksquare r \swarrow \Rightarrow r$, i.e., an instance of the $\blacksquare^2_R$-rule. Moreover, if $r := \neg\Box r'$ then the failed derivation of this last mentioned sequent will have a backward application of $\Box^2_L$ above it, containing a failed derivation for $r' \Rightarrow \Box p, \Box q, \blacksquare\neg\Box r'$, thereby ensuring compatibility. But there will also be failed derivations for $r' \Rightarrow \Box p, \Box q, \Box r \nearrow \Rightarrow p$ and $r' \Rightarrow \Box p, \Box q, \blacksquare r \nearrow \Rightarrow q$ and the witnesses for $\Box p$ and $\Box q$ will come from these failed derivations, because the returned sequent $r' \Rightarrow \Box p, \Box q, \blacksquare\neg\Box r'$ will be shorter than the other "and-node" premisses $\epsilon \Rightarrow \Box p, \Box q, \Box r \nearrow \Rightarrow p$ and $\epsilon \Rightarrow \Box p, \Box q, \blacksquare r \nearrow \Rightarrow q$. But note that a counter-model for $r' \Rightarrow \Box p, \Box q, \blacksquare\neg\Box r'$ is also a counter-model for the end-sequent $\epsilon \Rightarrow \Box p, \Box q, \blacksquare\neg\Box r'$.

Putting Theorem 7 and Lemma 9 together we then obtain cut-free completeness:

**Theorem 11 (Cut-free Completeness).** *If backward proof-search on end-sequent $S$ fails to find a derivation then there is a counter-model for $S$.* ⊣

**Corollary 12.** *If $\varphi$ is valid then the end-sequent $\epsilon \Rightarrow \varphi$ is derivable.* ⊣

It is tempting to think that we need some sort of coherence condition as illustrated by the tree in Fig. 2.

In the lowermost application of $\wedge_R$ we choose the left premiss, and in the uppermost one the right one. Thus it seems that in the world corresponding to these last components we would need to make both $p$ and $\neg p$ true, which of course would not work. But our pruning turns this failed derivation tree into the tree in Fig. 3. Note that only the component which is not restarted survives the pruning. The previous incarnation of the component caused the restart, but the restarted node did not necessarily follow the same sequence of rule applications, once it was restarted. Indeed, the sequence may no longer be possible as it may lead to an instance of (id). Of course, if it is possible and remains open, then it will find a counter-model for a larger set, which will also suffice for the smaller set. Thus our backward proof-search procedure creates surviving successors/predecessors only when it has ensured that they will be compatible via some number of restarts. Their incarnations which are not compatible are irrelevant, and are deleted by our counter-model construction.

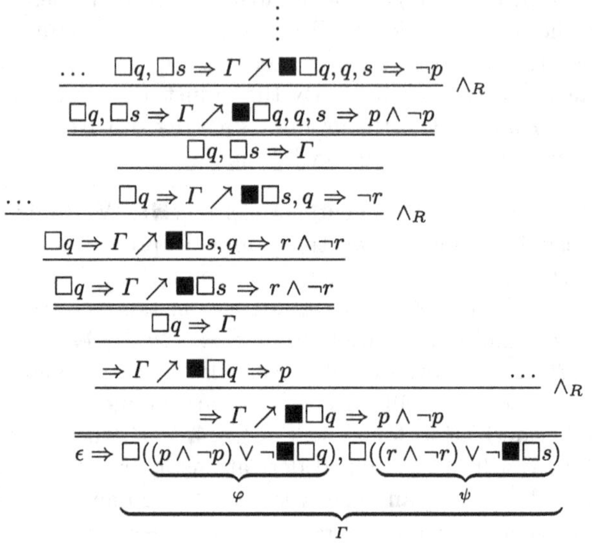

**Fig. 2.** A seemingly incoherent failed derivation tree.

## 5 Completeness via Cut Elimination

We now provide an alternative proof of cut-free completeness of our calculus via syntactic cut elimination. The proof is interesting from a technical point of view: The additional left premiss in the rules $\Box_R^1$ and $\blacksquare_R^1$ is introduced specifically as a counterpart to the restart rules $\Box_L^2$ and $\blacksquare_L^2$ to facilitate the reduction of cuts on boxed formulae to cuts of smaller complexity. However, while this enables the cut elimination proof itself, it shifts a large part of the work in the completeness proof to a perhaps unexpected place: the proof of admissibility of necessitation.

The following two lemmata are shown straightforwardly by induction on the depth of the derivation and the complexity of the formula $A$, respectively:

**Lemma 13.** *The rules below are admissible in* $\mathsf{LNS_{Kt}}$*:*

$$\frac{\mathcal{G} \updownarrow \Gamma \Rightarrow \Delta \updownarrow \mathcal{H}}{\mathcal{G} \updownarrow \Gamma, \Sigma \Rightarrow \Delta, \Pi \updownarrow \mathcal{H}}\ \mathsf{W} \qquad \frac{\mathcal{G} \updownarrow \Gamma, A, A \Rightarrow \Delta \updownarrow \mathcal{H}}{\mathcal{G} \updownarrow \Gamma, A \Rightarrow \Delta \updownarrow \mathcal{H}}\ \mathsf{C}_L \qquad \frac{\mathcal{G} \updownarrow \Gamma \Rightarrow \Delta, A, A \updownarrow \mathcal{H}}{\mathcal{G} \updownarrow \Gamma \Rightarrow \Delta, A \updownarrow \mathcal{H}}\ \mathsf{C}_R$$

⊣

**Lemma 14.** *The* generalised initial sequent rule *shown below is derivable in* $\mathsf{LNS}_{\mathsf{Kt}}$*:*

$$\overline{\mathcal{G} \updownarrow \Gamma, A \Rightarrow A, \Delta}$$

⊣

In order to introduce cuts in our framework, we need the following notion.

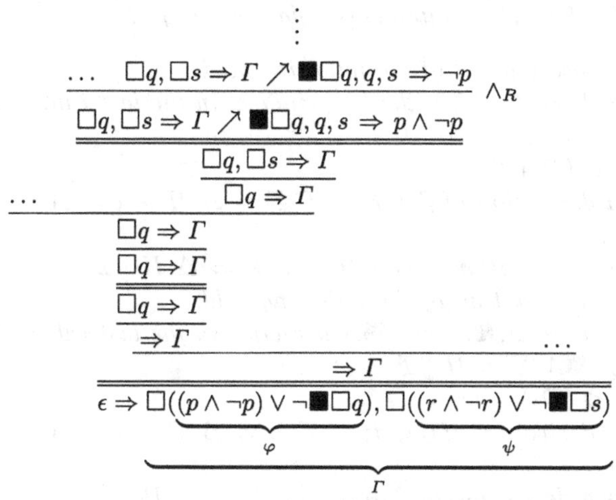

**Fig. 3.** The result of pruning the tree in Fig. 2.

**Definition 15.** *The* merge *of two linear nested sequents is defined via the following, where we assume* $\mathcal{G}, \mathcal{H}$ *to be nonempty:*

$$\begin{aligned}
(\Gamma \Rightarrow \Delta) \oplus (\Sigma \Rightarrow \Pi) &:= \Gamma, \Sigma \Rightarrow \Delta, \Pi \\
(\Gamma \Rightarrow \Delta) \oplus (\Sigma \Rightarrow \Pi \updownarrow \mathcal{H}) &:= \Gamma, \Sigma \Rightarrow \Delta, \Pi \updownarrow \mathcal{H} \\
(\Gamma \Rightarrow \Delta \updownarrow \mathcal{H}) \oplus (\Sigma \Rightarrow \Pi) &:= \Gamma, \Sigma \Rightarrow \Delta, \Pi \updownarrow \mathcal{H} \\
(\Gamma \Rightarrow \Delta \nearrow \mathcal{G}) \oplus (\Sigma \Rightarrow \Pi \nearrow \mathcal{H}) &:= \Gamma, \Sigma \Rightarrow \Delta, \Pi \nearrow (\mathcal{G} \oplus \mathcal{H}) \\
(\Gamma \Rightarrow \Delta \swarrow \mathcal{G}) \oplus (\Sigma \Rightarrow \Pi \swarrow \mathcal{H}) &:= \Gamma, \Sigma \Rightarrow \Delta, \Pi \swarrow (\mathcal{G} \oplus \mathcal{H}) .
\end{aligned}$$

Hence the merge is only defined for linear nested sequents which are *structurally equivalent*, i.e., have the same structure of the nesting operators.

Recall that we write $\mathcal{D} \vdash \mathcal{G}$ if $\mathcal{D}$ is a derivation in $\mathsf{LNS}_{\mathsf{Kt}}$ of the linear nested sequent $\mathcal{G}$, and $\vdash \mathcal{G}$ if there is a derivation $\mathcal{D}$ with $\mathcal{D} \vdash \mathcal{G}$, and that we write $\mathsf{dp}(\mathcal{D})$ for the depth of the derivation $\mathcal{D}$. The heavy lifting in the cut elimination proof is done by the following lemma, which captures the intuition that cuts are first shifted into the derivation of the left premiss of the cut until the cut formula becomes principal there. Then they are shifted into the derivation of the right premiss of the cut until they are principal here as well and can be reduced to cuts on lower complexity. The key idea is that because the calculus is end-active, the cut formula essentially always occurs in the last component of one of the premisses. As a technical subtlety, in order to shift up cuts on the principal formula of the rule $\Box^1_R$ or $\blacksquare^1_R$ we need to remember that we can eliminate the occurrence of the cut formula in the context. This is done by the additional conditions in the statements $\mathsf{SR}_\Box(n,m)$ and $\mathsf{SR}_\blacksquare(n,m)$ of the lemma, where we use $\mathsf{SL}$ and $\mathsf{SR}$ as mnemonics for "shift left" and "shift right", respectively, the latter with subscripts for the cut formula being modal or propositional:

**Lemma 16.** *The following statements hold for every $n, m$:*

($\mathsf{SR}_\Box(n,m)$) *Suppose that all of the following hold:*
- $\mathcal{D}_1 \vdash \mathcal{G} \nearrow \Gamma \Rightarrow \Delta, \Box A$ *with* $\Box A$ *principal in the last rule in* $\mathcal{D}_1$
- $\mathcal{D}_2 \vdash \mathcal{H} \nearrow \Box A, \Sigma \Rightarrow \Pi \updownarrow \mathcal{I}$
- $\mathsf{dp}(\mathcal{D}_1) + \mathsf{dp}(\mathcal{D}_2) \leq m$
- *there is a derivation of* $\mathcal{G} \oplus \mathcal{H} \nearrow \Gamma, \Sigma \Rightarrow \Delta, \Pi \nearrow \epsilon \Rightarrow A$
- $|\Box A| \leq n$.

*Then there is a derivation of* $\mathcal{G} \oplus \mathcal{H} \nearrow \Gamma, \Sigma \Rightarrow \Delta, \Pi \updownarrow \mathcal{I}$.

($\mathsf{SR}_\blacksquare(n,m)$) *Suppose that all of the following hold:*
- $\mathcal{D}_1 \vdash \mathcal{G} \swarrow \Gamma \Rightarrow \Delta, \blacksquare A$ *with* $\blacksquare A$ *principal in the last rule in* $\mathcal{D}_1$
- $\mathcal{D}_2 \vdash \mathcal{H} \swarrow \blacksquare A, \Sigma \Rightarrow \Pi \updownarrow \mathcal{I}$
- $\mathsf{dp}(\mathcal{D}_1) + \mathsf{dp}(\mathcal{D}_2) \leq m$
- *there is a derivation of* $\mathcal{G} \oplus \mathcal{H} \swarrow \Gamma, \Sigma \Rightarrow \Delta, \Pi \swarrow \epsilon \Rightarrow A$
- $|\blacksquare A| \leq n$.

*Then there is a derivation of* $\mathcal{G} \oplus \mathcal{H} \nearrow \Gamma, \Sigma \Rightarrow \Delta, \Pi \updownarrow \mathcal{I}$.

($\mathsf{SR}_p(n,m)$) *Suppose that all of the following hold where* $\mathcal{D}_1 \updownarrow \Gamma \Rightarrow \Delta, A$ *and* $\mathcal{H} \updownarrow A, \Sigma \Rightarrow \Pi$ *are structurally equivalent:*
- $\mathcal{D}_1 \vdash \mathcal{G} \updownarrow \Gamma \Rightarrow \Delta, A$ *with* $A$ *principal in the last applied rule in* $\mathcal{D}_1$
- $\mathcal{D}_2 \vdash \mathcal{H} \updownarrow A, \Sigma \Rightarrow \Pi \updownarrow \mathcal{I}$
- $\mathsf{dp}(\mathcal{D}_1) + \mathsf{dp}(\mathcal{D}_2) \leq m$
- $|A| \leq n$
- $A$ *not of the form* $\Box B$ *or* $\blacksquare B$.

*Then there is a derivation of* $\mathcal{G} \oplus \mathcal{H} \updownarrow \Gamma, \Sigma \Rightarrow \Delta, \Pi \updownarrow \mathcal{I}$.

($\mathsf{SL}(n,m)$) *If* $\mathcal{D}_1 \vdash \mathcal{G} \updownarrow \Gamma \Rightarrow \Delta, A \updownarrow \mathcal{I}$ *and* $\mathcal{D}_2 \vdash \mathcal{H} \updownarrow A, \Sigma \Rightarrow \Pi$ *with* $|A| \leq n$ *and* $\mathsf{dp}(\mathcal{D}_1) + \mathsf{dp}(\mathcal{D}_2) \leq m$, *and* $\mathcal{G} \updownarrow \Gamma \Rightarrow \Delta$ *and* $\mathcal{H} \updownarrow A, \Sigma \Rightarrow \Pi$ *are structurally equivalent, then there is a derivation of* $\mathcal{G} \oplus \mathcal{H} \updownarrow \Gamma, \Sigma \Rightarrow \Delta, \Pi \updownarrow \mathcal{I}$.

The full proof is in our arxiv paper [5]. As an immediate corollary, using the statement $\mathsf{SL}(n,m)$ from Lemma 16 for suitable $n, m$ we obtain:

**Theorem 17 (Cut elimination).** *Whenever* $\vdash \mathcal{G} \updownarrow \Gamma \Rightarrow \Delta, A$ *and* $\vdash \mathcal{H} \updownarrow A, \Sigma \Rightarrow \Pi$, *then also* $\vdash \mathcal{G} \oplus \mathcal{H} \updownarrow \Gamma, \Sigma \Rightarrow \Delta, \Pi$. ⊣

As usual, we will use cut elimination to show completeness. However, we also need to show admissibility of the *necessitation rules* $A/\Box A$ and $A/\blacksquare A$. While this is straightforward in standard calculi for modal logics, due to the additional premiss in the rules $\Box^1_R$ and $\blacksquare^1_R$, here we need to do some work:

**Theorem 18 (Admissibility of necessitation).** *If* $\epsilon \Rightarrow A$ *is derivable in* $\mathsf{LNS}_{\mathsf{Kt}}$*, then so are* $\epsilon \Rightarrow \Box A$ *and* $\epsilon \Rightarrow \blacksquare A$*.*

*Proof.* We consider the proof for $\epsilon \Rightarrow \Box A$, the other case is analogous. To refer to problematic applications of the $\blacksquare^2_R$ rule, we introduce some terminology.

**Definition 19.** *Let $\mathcal{D}$ be the derivation of the sequent $\Rightarrow A$. An application $r$ of the rule $\blacksquare^2_R$ is* critical *in $\mathcal{D}$ if its conclusion has exactly one component. The* depth *of a critical application $r$ of $\blacksquare^2_R$ is the depth of the sub-derivation of $\mathcal{D}$ ending with this rule application, written* $\mathsf{dp}(r)$*.*

Let $\mathcal{D}$ be a derivation of $\Rightarrow A$, and let $\mathsf{crit}(\mathcal{D})$ be the set of critical applications of $\blacksquare^2_R$ in $\mathcal{D}$. For every possible depth $d$ of a critical application in $\mathsf{crit}(\mathcal{D})$, fix an enumeration of all critical applications in $\mathsf{crit}(\mathcal{D})$ with this depth. We then convert the derivation $\mathcal{D}$ bottom-up into a *derivation from assumptions* of $\Rightarrow \Box A$, i.e., a derivation of $\Rightarrow \Box A$ where the leaves might be labelled with arbitrary linear nested sequents called the *assumptions.* Each of these comes from one of the critical applications of $\mathcal{D}$, i.e., we have an injection $\iota$ from $\mathsf{crit}(\mathcal{D})$ to the set of assumptions of the so far constructed derivation with assumptions. To each assumption $A$ we associate an *index*, i.e., a triple $(d, i, c)$ of natural numbers, where $d$ is the depth of the critical application $\iota^{-1}(A)$, the number $i$ is the index of $\iota^{-1}(A)$ in the enumeration of critical applications of depth $d$, and $c \leq \mathsf{dp}(\mathcal{D})$ is a number corresponding to the depth of the current position in the original derivation $\mathcal{D}$. To ensure termination of the procedure, we consider the lexicographic ordering $<_{lex}$ on the indices $(d, i, c)$, and the *multiset ordering* $\prec$ induced by $<_{lex}$ on the set of multisets of indices [3]. In particular for two such multisets $\mathcal{A}, \mathcal{B}$ we have that $\mathcal{A} \prec \mathcal{B}$ iff $\mathcal{B}$ can be obtained from $\mathcal{A}$ by replacing one or more indices $(d, i, c)$ by a finite number of indices $(d', i', c')$ with $(d', i', c') <_{lex} (d, i, c)$. It is shown in *op.cit.* that $\prec$ is well-founded.

The first ingredient in the construction of the derivation of $\Rightarrow \Box A$ is given by essentially prefixing $\epsilon \Rightarrow \Box A$ to every linear nested sequent in $\mathcal{D}$:

**Definition 20.** *Let $\mathcal{E}$ be a sub-derivation of $\mathcal{D}$ and $\Gamma \Rightarrow \Delta$ a sequent. For any natural number $n$ the derivation $(\Gamma \Rightarrow \Delta) \nearrow \mathcal{E}(n)$ is obtained by prefixing $\Gamma \Rightarrow \Delta \nearrow$ to every linear nested sequent in $\mathcal{E}$, and replacing critical applications of $\blacksquare^2_R$ with applications of $\blacksquare^1_R$ and an assumption as follows:*

$$\frac{\Sigma \Rightarrow \Pi, \blacksquare B \swarrow \epsilon \Rightarrow B}{\Sigma \Rightarrow \Pi, \blacksquare B}\ \blacksquare^2_R$$

$$\rightsquigarrow\quad \frac{\dfrac{\Gamma \Rightarrow \Delta, B}{\Gamma \Rightarrow \Delta, B \nearrow \Sigma \Rightarrow \Pi, \blacksquare B}\ \mathsf{EW} \qquad \Gamma \Rightarrow \Delta \nearrow \Sigma \Rightarrow \Pi, \blacksquare B \swarrow \epsilon \Rightarrow B}{\Gamma \Rightarrow \Delta \nearrow \Sigma \Rightarrow \Pi, \blacksquare B}\ \blacksquare^1_R$$

*The index $(d, i, n)$ of the assumption $\Gamma \Rightarrow \Delta, B$ is given by the depth $d$ of the original critical application of $\blacksquare^2_R$, its index $i$, and the number $n$.*

In the first step we obtain from $\mathcal{D}$ the derivation with assumptions $(\epsilon \Rightarrow \Box A) \nearrow \mathcal{D}(\mathsf{dp}(\mathcal{D}))$. The conclusion of this derivation is $\epsilon \Rightarrow \Box A \nearrow \epsilon \Rightarrow A$, hence applying $\Box^2_R$ we will ultimately obtain a derivation with assumptions of $\epsilon \Rightarrow \Box A$.

The next step is to construct a derivation for each assumption, starting with one of maximal index. The general idea is to copy the derivation of the premiss of the corresponding critical application of $\blacksquare^2_R$, but essentially "folding back" the second component of the original derivation into the first one of the new derivation until the linear nested sequents in the original derivation are reduced to one component again. This means that the first component of the new derivation will collect a number of second components occurring in the original derivation. To make this precise, for a sequent $\Omega \Rightarrow \Theta$, a derivation $\mathcal{E}$ with assumptions, a critical rule application $r$ and a natural number $n$, we write $(\Omega \Rightarrow \Theta) \oplus \mathcal{E}(r \leftarrow n)$ for the derivation with assumptions obtained from $\mathcal{E}$ by merging the first component of each linear nested sequent in $\mathcal{E}$ with the sequent $\Omega \Rightarrow \Theta$, and changing the indices $(d, i, c)$ of all those assumption in $\mathcal{E}$ corresponding to $r$ to $(d, i, n)$.

Take an assumption $\Gamma \Rightarrow \Delta, B$ with index $(d, i, c)$ which is maximal w.r.t. $<_{lex}$, and suppose that the corresponding critical rule application $r$ is given by:

$$\dfrac{\begin{array}{c}\vdots\ \mathcal{E}\\ \Sigma \Rightarrow \Pi, \blacksquare B \swarrow \epsilon \Rightarrow B\end{array}}{\Sigma \Rightarrow \Pi, \blacksquare B}\ \blacksquare^2_R$$

Suppose that the assumption occurs in the context

$$\begin{array}{c}\dfrac{\dfrac{\Gamma \Rightarrow \Delta, B}{\Gamma \Rightarrow \Delta, B \nearrow \Sigma \Rightarrow \Pi, \blacksquare B}\ \mathsf{EW} \qquad \begin{array}{c}\vdots\\ \Gamma \Rightarrow \Delta \nearrow \Sigma \Rightarrow \Pi, \blacksquare B \nearrow \epsilon \Rightarrow B\end{array}}{\Gamma \Rightarrow \Delta \nearrow \Sigma \Rightarrow \Pi, \blacksquare B}\ \blacksquare^1_R\\ \vdots\ \mathcal{F}\end{array}$$

where $\mathcal{F}$ is the derivation with assumptions below the conclusion of the application of $\blacksquare^1_R$. Note that all assumptions in $\mathcal{F}$ have index smaller than $(d, i, c)$. We extend this derivation upwards by applying the same rules as in the original derivation $\mathcal{E}$, until in $\mathcal{E}$ we encounter a rule $\Box^2_L$ or $\mathsf{EW}$ which shortens the sequent to only the first component again. This is straightforward unless in the original derivation we have an application of a rule in which the first component is active, i.e., an application of the rules $\blacksquare^1_L$ or $\Box^1_R$ with active first component.

The case of $\Box^1_R$ is unproblematic, replacing $\Box^1_R$ with $\Box^2_R$ and continuing upwards as in the derivation of the right premiss. Note that the first component in the original derivation stays the same.

In the case of an application of $\blacksquare^1_L$ we recreate the original first component $\Sigma \Rightarrow \Pi, \blacksquare B$ using $\mathcal{F}$. In general, this creates new copies of the assumptions in $\mathcal{F}$, in particular of other assumptions corresponding to $r$. To ensure termination we decrease the index of every assumption corresponding to $r$ to the depth of the current position in the original derivation. Hence the multiset of assumptions of the new derivation is smaller than that of the old one w.r.t. $\prec$. Suppose that we encounter an application of the rule $\blacksquare^1_L$ in the form

$$\dfrac{\begin{array}{c}\vdots\ \mathcal{G}\\ \Sigma', \blacksquare C \Rightarrow \Pi', \blacksquare B, \swarrow \Xi, C \Rightarrow \Upsilon\end{array}}{\Sigma', \blacksquare C \Rightarrow \Pi', \blacksquare B \swarrow \Xi \Rightarrow \Upsilon}\ \blacksquare^1_L$$

Since all linear nested sequents between the conclusion of this rule application and the critical rule application $r$ contain at least two components, and since when simulating applications of $\Box_R^1$ as above we never changed the first component, the first component $\Sigma', \blacksquare C \Rightarrow \Pi', \blacksquare B$ stays the same as the original first component $\Sigma \Rightarrow \Pi, \blacksquare B$. Hence we can recreate this component and continue as:

$$\dfrac{\Gamma, \Xi, C \Rightarrow \Delta, B, \Upsilon}{\Gamma, \Xi \Rightarrow \Delta, B, \Upsilon \nearrow \Sigma', \blacksquare C \Rightarrow \Pi', \blacksquare B}\ \blacksquare_L^2$$
$$\vdots\ (\Xi \Rightarrow \Upsilon) \oplus \mathcal{F}(r \leftarrow \mathsf{dp}(\mathcal{G}))$$
$$\Gamma, \Xi \Rightarrow \Delta, B, \Upsilon$$

Continuing upwards like this, in the original derivation we eventually reach initial sequents, or applications of $\Box_L^2$ or $\mathsf{EW}$ which reduce the number of components to one. In the latter case, we again recreate the original first component. E.g., suppose that in the original derivation we have an application of $\Box_L^2$ in the form

$$\vdots\ \mathcal{G}$$
$$\dfrac{\Sigma', C \Rightarrow \Pi'}{\Sigma' \Rightarrow \Pi' \nearrow \Xi, \Box C \Rightarrow \Upsilon}\ \Box_L^2$$

Then again we have that $\Sigma' \Rightarrow \Pi'$ is the same as the first component $\Sigma \Rightarrow \Pi, \blacksquare B$ of the critical rule application $r$, and hence we can recreate it and continue using

$$\vdots\ (\Gamma, \Xi, \Box C \Rightarrow \Delta, \Upsilon) \nearrow \mathcal{G}(\mathsf{dp}(\mathcal{G}))$$
$$\dfrac{\Gamma, \Xi, \Box C \Rightarrow \Delta, \Upsilon \nearrow \Sigma', C \Rightarrow \Pi'}{\Gamma, \Xi, \Box C \Rightarrow \Delta, \Upsilon \nearrow \Sigma' \Rightarrow \Pi'}\ \Box_L^1$$
$$\vdots\ (\Xi \Rightarrow \Upsilon) \oplus \mathcal{F}(r \leftarrow \mathsf{dp}(\mathcal{G}))$$
$$\Gamma, \Xi, \Box C \Rightarrow \Delta, \Upsilon$$

Note that again the multiset of indices of assumptions is decreased wrt. $\prec$. In particular, the depth of every critical rule application in $\mathcal{G}$ is smaller than the depth of the critical rule application $r$. The case for the rule $\mathsf{EW}$ is analogous.

Continuing in this way we replace every assumption by a finite multiset of smaller ones. Hence the sequence of multisets of assumptions is strictly decreasing wrt. the well-ordering $\prec$, and the procedure must terminate. When it does we obtain a derivation without assumptions, giving a derivation of $\epsilon \Rightarrow \Box A$. ⊣

**Theorem 21 (Completeness).** *The system* $\mathsf{LNS_{Kt}}$ *is cut-free complete for* $\mathsf{Kt}$.

*Proof.* It is straightforward to derive the axioms. Modus ponens is simulated as usual using cuts. The necessitation rules are simulated using Lemma 18. ⊣

## 6 Application: Linear Nested Sequents for Modal Logic KB

It is rather straightforward to adapt our system to capture modal logic KB. Semantically, KB is given as the mono-modal logic of *symmetric Kripke frames*, i.e., frames with symmetric accessibility relation. Syntactically, KB is obtained from Kt by collapsing the forwards and backwards modalities, e.g., via adding the axiom $\Box A \leftrightarrow \blacksquare A$. Correspondingly, we also collapse the structural connectives $\nearrow$ and $\swarrow$ to obtain the simpler definition of linear nested sequents for KB via the grammar $S := \Gamma \Rightarrow \Delta \mid \Gamma \Rightarrow \Delta \nearrow S$. The simplest version of the linear nested sequent calculus $\mathsf{LNS_{KB}}$ for modal logic KB then contains the propositional rules and rule EW of Fig. 1 together with the two standard rules

$$\frac{\mathcal{G} \nearrow \Gamma \Rightarrow \Delta, \Box A \nearrow \epsilon \Rightarrow A}{\mathcal{G} \nearrow \Gamma \Rightarrow \Delta, \Box A}\ \Box_R \qquad \frac{\mathcal{G} \nearrow \Gamma, \Box A \Rightarrow \Delta \nearrow \Sigma, A \Rightarrow \Pi}{\mathcal{G} \nearrow \Gamma, \Box A \Rightarrow \Delta \nearrow \Sigma \Rightarrow \Pi}\ \Box_L^1$$

found in (linear) nested sequent calculi for modal logic K and the single new rule

$$\frac{\mathcal{G} \nearrow \Gamma, A \Rightarrow \Delta}{\mathcal{G} \nearrow \Gamma \Rightarrow \Delta \nearrow \Sigma, \Box A \Rightarrow \Pi}\ \Box_L^2$$

Soundness is seen analogously to Theorem 4, and completeness follows by repeating the proofs for Kt, at each step collapsing the forwards and backwards modalities:

**Theorem 22.** *The calculus* $\mathsf{LNS_{KB}}$ *is sound and complete for modal logic* KB.⊣

In comparison with the linear nested sequent calculus for modal logic KB introduced by Parisi [17], we do not need to change the direction of the linear nested sequents, and (a variant of) our system has syntactic cut elimination. Note also that the system $\mathsf{LNS_{KB}}$ is essentially the end-active and linear version of the nested sequent calculus for KB of Brünnler and Poggiolesi [2,18] with the crucial difference that the last component is deleted in the premiss of the symmetry rule $\Box_L^2$. Since derivations of $\mathsf{LNS_{KB}}$ can be transformed straightforwardly bottom-up into derivations in the full nested sequent system considered in *op. cit.*, our completeness result implies the completeness results there.

## 7 Conclusion

We have seen that linear nested sequents are so far the minimal extension of traditional sequents needed to handle tense logics and modal logic KB. Intuitively, they provide the semantic expressive power to look both ways along the underlying Kripke reachability relation while also providing a rigorous and modular proof-theoretic framework. The main novelty to mimic traditional tableau calculi for tense logics is the addition of restart rules to maintain the compatibility between parent nodes and their children.

In future work we would like to explore the possibility of extending our calculus to capture further properties of the accessibility relation such as reflexivity, forwards or backwards directedness, or transitivity. We conjecture that suitable modifications of the rules $\Box_R^1$ and $\blacksquare_R^1$ in the spirit of the ones presented here should suffice for a cut elimination proof. It is perhaps less obvious that the proof of admissibility of necessitation goes through in these cases as well. Finally, we would like to investigate whether it is possible to use our calculi in complexity-optimal decision procedures.

## References


1. Bonnette, N., Goré, R.: A labelled sequent system for tense logic $K_t$. In: Antoniou, G., Slaney, J. (eds.) AI 1998. LNCS, vol. 1502, pp. 71–82. Springer, Heidelberg (1998). https://doi.org/10.1007/BFb0095042
2. Brünnler, K.: Deep sequent systems for modal logic. Arch. Math. Log. **48**, 551–577 (2009)
3. Dershowitz, N., Manna, Z.: Proving termination with multiset orderings. Commun. ACM **22**(8), 465–476 (1979)
4. Goré, R.: Tableau methods for modal and temporal logics. In: D'Agostino, M., Gabbay, D.M., Hähnle, R., Posegga, J. (eds.) Handbook of Tableau Methods. Kluwer, Dordrecht (1999)
5. Goré, R., Lellmann, B.: Syntactic cut-elimination and backward proof-search for tense logic via linear nested sequents (extended version). CoRR abs/1907.01270 (2019). http://arxiv.org/abs/1907.01270
6. Goré, R., Postniece, L., Tiu, A.: On the correspondence between display postulates and deep inference in nested sequent calculi for tense logics. Log. Methods Comput. Sci. **7**(2) (2011). https://doi.org/10.2168/LMCS-7(2:8)2011
7. Goré, R., Widmann, F.: Sound global state caching for $\mathcal{ALC}$ with inverse roles. In: Giese, M., Waaler, A. (eds.) TABLEAUX 2009. LNCS (LNAI), vol. 5607, pp. 205–219. Springer, Heidelberg (2009). https://doi.org/10.1007/978-3-642-02716-1_16
8. Horrocks, I., Sattler, U., Tobies, S.: Reasoning with individuals for the description logic $\mathcal{SHIQ}$. In: McAllester, D. (ed.) CADE 2000. LNCS (LNAI), vol. 1831, pp. 482–496. Springer, Heidelberg (2000). https://doi.org/10.1007/10721959_39
9. Indrzejczak, A.: Linear time in hypersequent framework. Bull. Symb. Log. **22**, 121–144 (2016)
10. Kashima, R.: Cut-free sequent calculi for some tense logics. Studia Logica **53**(1), 119–136 (1994)
11. Kuznets, R., Lellmann, B.: Interpolation for intermediate logics via hyper- and linear nested sequents. In: Bezhanishvili, G., D'Agostino, G., Metcalfe, G., Studer, T. (eds.) Advances in Modal Logic 2018, pp. 473–492. College Publications (2018)
12. Lahav, O.: From frame properties to hypersequent rules in modal logics. In: 28th Annual ACM/IEEE Symposium on Logic in Computer Science, LICS 2013, New Orleans, LA, USA, 25–28 June 2013, pp. 408–417 (2013). https://doi.org/10.1109/LICS.2013.47
13. Lellmann, B.: Linear nested sequents, 2-sequents and hypersequents. In: De Nivelle, H. (ed.) TABLEAUX 2015. LNCS (LNAI), vol. 9323, pp. 135–150. Springer, Cham (2015). https://doi.org/10.1007/978-3-319-24312-2_10
14. Lellmann, B., Pimentel, E.: Modularisation of sequent calculi for normal and non-normal modalities. ACM Trans. Comput. Logic **20**(2), 7:1–7:46 (2019)
15. Masini, A.: 2-sequent calculus: a proof theory of modalities. Ann. Pure Appl. Logic **58**, 229–246 (1992)
16. Ohnishi, M., Matsumoto, K.: Gentzen method in modal calculi I. Osaka Math. J. **9**, 113–130 (1957)
17. Parisi, A.: Second-order modal logic. Ph.D. thesis, University of Connecticut (2017)

18. Poggiolesi, F.: Gentzen Calculi for Modal Propositional Logic. Trends in Logic, vol. 32. Springer, Heidelberg (2010). https://doi.org/10.1007/978-90-481-9670-8
19. Trzesicki, K.: Gentzen-style axiomatization of tense logic. Bull. Sect. Log. **13**(2), 75–83 (1984)
20. Wansing, H.: Sequent calculi for normal modal proposisional logics. J. Log. Comput. **4**(2), 125–142 (1994)

# Combining Monotone and Normal Modal Logic in Nested Sequents – with Countermodels

Björn Lellmann(✉)

TU Wien, Vienna, Austria
lellmann@logic.at

**Abstract.** We introduce nested sequent calculi for bimodal monotone modal logic, aka. Brown's ability logic, a natural combination of non-normal monotone modal logic M and normal modal logic K. The calculus generalises in a natural way previously existing calculi for both mentioned logics, has syntactical cut elimination, and can be used to construct countermodels in the neighbourhood semantics. We then consider some extensions of interest for deontic logic. An implementation is also available.

**Keywords:** Modal logic · Non-normal modal logic · Ability logic · Nested sequents · Countermodel generation

## 1 Introduction

The *nested sequent framework* has been very successfully used to provide analytic calculi for a large number of logics. In the context of normal modal logics, it enabled modular calculi for all logics in the modal cube [3,18], for tense logics [10], and for intuitionistic and constructive modal logics [15]. One of the main advantages of this framework is that while it is a purely syntactic extension of the sequent framework with a structural connective for the modal box, the tree structure of nested sequents is also closely related to the semantics of modal logics, in particular to the underlying tree structure of Kripke models. Due to this aspect, nested sequent calculi often lend themselves to direct methods of countermodel construction: Usually, if proof search fails, it returns a saturated unprovable nested sequent from which the countermodel can be read off directly. However the full power and flexibility of this framework so far has not yet been harnessed in the context of *non-normal* modal logics. An initial attempt at doing so indeed yielded modular calculi for a number of non-normal modal logics in the framework of *linear* nested sequents [13,14]. Unfortunately, these calculi neither facilitated countermodel construction, nor was it possible to provide a formula interpretation of the linear nested sequents in the language of the logic.

Supported by WWTF project MA16-28.


S. Cerrito and A. Popescu (Eds.): TABLEAUX 2019, LNAI 11714, pp. 203–220, 2019.
https://doi.org/10.1007/978-3-030-29026-9_12

Here we propose an approach to rectify this situation by considering *bimodal* versions of the non-normal modal logics. Such logics seem to have been considered first in [1] in the form of *ability logics*. In this framework, the neighbourhood semantics of monotone modal logic is interpreted by the "can" of ability. Intuitively, the neighbourhood function maps a world to a set of neighbourhoods, which correspond to actions available to an agent. If there is an action available such that a certain proposition is true after every possible execution of this action, i.e., true in every world in the corresponding neighbourhood, then the agent can reliably bring about this proposition. This interpretation then gives rise to a second modality interpreting the "will" of ability: If a proposition is true after every available action, i.e., true in every world of every neighbourhood of a particular world, then the agent will unavoidably bring about this proposition. Crucially and very conveniently, this second modality turns out to be normal, which lets us exploit the standard connection between nesting in nested sequents and the successor relation in Kripke models. Moreover, this induced second modality does not depend on the original ability interpretation of non-normal monotone modal logic, and hence its usefulness extends far beyond that particular context.

Using this approach reformulated in terms of one of the most fundamental non-normal modal logics, *monotone modal logic* M [4,7,17], we here present a nested sequent calculus for its bimodal version biM, which combines M with normal modal logic K. Notably, the nested sequents have a formula interpretation in the bimodal language, and the calculus facilitates the construction of countermodels from failed proof search in a slightly modified version. Since biM is a reformulation of Brown's ability logic, this immediately yields a nested sequent calculus for the latter. An additional benefit is that the calculus conservatively extends both the standard nested sequent calculus for K from [3,18] and the nested sequent calculus for M from [13,14]. A prototype implementation of proof search and countermodel construction using the calculus is available under http://subsell.logic.at/bprover/nnProver/.

In terms of related work, while the presented calculus is mainly intended as a foundation for nested sequents for monotone modal logics in general, it also seems to be the first sequent-style calculus for biM resp. Brown's ability logic. There are of course a number of calculi for the monomodal logics M and K. The standard sequent calculus for M was introduced in [11], where it was also used to generate countermodels. However, due to the fact that the sequent structure is too poor to capture the information necessary to construct neighbourhood functions, the countermodel generation is rather more involved than in the nested sequent framework. Based on *op.cit.*, sequent calculi for various extensions of M were given in [8] and later converted to the prefixed tableaux framework in [9]. The latter is interesting in that successor labels in these calculi correspond to the K-modality. However, the investigated logics are still only the purely monomodal non-normal fragments. Finally, calculi for non-normal logics including M in the framework of *labelled sequents* have been introduced recently in [6,16]. They are modular, facilitate syntactic cut elimination and can be used

| | | | |
|---|---|---|---|
| (C⟨⟩) | $\langle\rangle(A \vee B) \to (\langle\rangle A \vee \langle\rangle B)$ | (RM⟨⟩) | $\vdash A \to B / \vdash \langle\rangle A \to \langle\rangle B$ |
| (V) | $\langle]( A \vee B) \to (\langle\rangle A \vee \langle] B)$ | (RM⟨]) | $\vdash A \to B / \vdash \langle] A \to \langle] B$ |
| (W) | $\langle\rangle A \to ([] B \to \langle] B)$ | (RN[]) | $\vdash A / \vdash [] A$ |

**Fig. 1.** The modal axioms and rules for biM from [1]. In addition, the full axiomatisation contains the standard axioms and rules of classical propositional logic.

| | | | |
|---|---|---|---|
| (C[]) | $[] A \wedge [] B \to []( A \wedge B)$ | (RM[]) | $\vdash A \to B / \vdash [] A \to [] B$ |
| (V′) | $\langle]( A \to B) \wedge [] A \to \langle] B$ | (RM⟨]) | $\vdash A \to B / \vdash \langle] A \to \langle] B$ |
| (W′) | $[] B \to \langle] B \vee [] C$ | (RN[]) | $\vdash A / \vdash [] A$ |

**Fig. 2.** The reformulation of the modal axioms and rules for biM.

for countermodel construction, but due to the inherent semantical character of labelled sequents and the restriction to the monomodal language they lack a formula interpretation.

The article is structured as follows. In Sect. 2 we recall bimodal monotone modal logic and introduce the base calculus. In Sect. 3 we show syntactical cut elimination in a slight variant of the calculus, and Sect. 4 contains the countermodel construction from failed proof search. Some extensions are considered in Sect. 5, followed by a short description of the implementation in Sect. 6 and the conclusion.

## 2 The Basic System

The set $\mathcal{F}$ of *formulae* of bimodal monotone modal logic is given by the following grammar, built over a set $\mathcal{V}$ of propositional variables:

$$\mathcal{F} ::= \bot \mid \mathcal{V} \mid \mathcal{F} \to \mathcal{F} \mid \langle]\mathcal{F} \mid []\mathcal{F}$$

The remaining propositional connectives are defined as usual. The semantics are given in terms of *neighbourhood semantics*, following [1,2,4,17]:

**Definition 1.** *A* neighbourhood model *is a tuple* $\mathfrak{M} = (W, \mathcal{N}, [\![ . ]\!])$ *consisting of a universe* $W$*, a* neighbourhood function $\mathcal{N} : W \to 2^{2^W}$*, and a* valuation $[\![ . ]\!] : \mathcal{V} \to 2^W$*. The* truth set of a formula $A$ *in a model, written as* $[\![A]\!]$*, extends* $[\![ . ]\!]$ *by the propositional clauses* $[\![\bot]\!] = \emptyset$ *and* $[\![A \to B]\!] = [\![A]\!]^c \cup [\![B]\!]$ *together with*

- $[\![\langle] A]\!] = \{w \in W \mid$ *exists* $\alpha \in \mathcal{N}(w)$ *s.t. for all* $v \in \alpha : v \in [\![A]\!]\}$
- $[\![[] A]\!] = \{w \in W \mid$ *for all* $\alpha \in \mathcal{N}(w)$ *and for all* $v \in \alpha : v \in [\![A]\!]\}$

*We write* $\mathfrak{M}, w \Vdash A$ *for* $w \in [\![A]\!]$ *and call* $A$ valid*, if* $[\![A]\!] = W$ *for every model.*

The dual connectives are defined via $[\rangle A \equiv \neg\langle]\neg A$ and $\langle\rangle A \equiv \neg[]\neg A$. The original axiomatisation for *bimodal monotone logic* biM from [1] (called $\mathcal{V}$ there) is given in Fig. 1, its reformulation using only $\langle]$ and $[]$ in Fig. 2. Note that for a model $\mathfrak{M} = (W, \mathcal{N}, [\![.]\!])$ and world $w \in W$ we have that $\mathfrak{M}, w \Vdash []A$ if and only if for all $v \in \bigcup \mathcal{N}(w)$ we have $\mathfrak{M}, v \Vdash A$. Hence $[]$ is a normal K-type modality. The fact that the modality $\langle]$ is a monotone modality follows immediately from the semantics, since if for $\alpha \in \mathcal{N}(w)$ we have $\alpha \subseteq [\![A]\!]$ and $[\![A]\!] \subseteq [\![B]\!]$, i.e., $A \to B$ is valid, then we also have $\alpha \subseteq [\![B]\!]$. Thus if $A \to B$ is valid, then so is $\langle]A \to \langle]B$. This can also be read off the axiomatisation in Fig. 2, since (C[]), (RM[]), (RN[]) is an axiomatisation of K, and (RM⟨]) gives monotonicity of $\langle]$.

To obtain a calculus for bimodal monotone logic, we extend the ordinary sequent structure by the two structural connectives $\langle.\rangle$ and $[.]$ in the succedent, corresponding to the logical connectives $\langle]$ and $[]$, respectively:

**Definition 2.** *A* nested sequent *is an expression*

$$\Gamma \Rightarrow \Delta, \langle \Sigma_1 \Rightarrow \Pi_1 \rangle, \ldots, \langle \Sigma_n \Rightarrow \Pi_n \rangle, [\mathcal{S}_1], \ldots, [\mathcal{S}_m]$$

*where $\Gamma, \Delta, \Sigma_i, \Pi_i$ are multisets of formulae, and the $\mathcal{S}_j$ are nested sequents. The* formula interpretation *of a nested sequent $\mathcal{S}$ is written $\iota(\mathcal{S})$ and given by*

$$\iota(\mathcal{S}) = \bigwedge \Gamma \to \bigvee \Delta \vee \bigvee\nolimits_{i=1}^{n} \langle](\bigwedge \Sigma_i \to \bigvee \Pi_i) \vee \bigvee\nolimits_{j=1}^{m} []\,\iota(\mathcal{S}_j)\,.$$

Intuitively, a nested sequent is a tree, where each node is labelled with an expression $\Gamma \Rightarrow \Delta, \langle \Sigma_1 \Rightarrow \Pi_1 \rangle, \ldots, \langle \Sigma_n \Rightarrow \Pi_n \rangle$ and is called a *component* of the nested sequent, and the successor relation corresponds to the nesting operator $[.]$. To shorten presentation we slightly abuse notation and sometimes take the succedent of a sequent to contain nested sequents as well, i.e., we might write $\Gamma \Rightarrow \Delta, [\Sigma \Rightarrow \Pi]$ instead of $\Gamma \Rightarrow \Delta', \langle \Omega \Rightarrow \Theta \rangle, [\Sigma \Rightarrow \Pi], [\Xi \Rightarrow \Upsilon]$. The rules of the nested sequent calculus intuitively then can be applied at any node of the nested sequent. Syntactically, this uses the notion of a *context* as follows.

**Definition 3 (Nested sequent context).** *A* nested sequent context *is a nested sequent with a* hole $\{.\}$, *defined by* $\mathcal{S}\{.\} ::= \{.\} \mid \Gamma \Rightarrow \Delta, [\mathcal{S}\{.\}]$.

Note that $\langle.\rangle$ never contains $\{.\}$. This ensures non-normality of its interpretation $\langle]$ by preventing application of the propositional rules inside $\langle.\rangle$.

**Definition 4.** *The rules of the nested sequent calculus $\mathcal{N}_{\mathsf{M}}$ are given in Fig. 3. A* derivation *in $\mathcal{N}_{\mathsf{M}}$ is a finite tree where each node is labelled with a nested sequent, and the label of each node results from the labels of its successors by an application of a rule from $\mathcal{N}_{\mathsf{M}}$. The* depth *of a derivation is the maximal number of nodes in its branches minus one, and the* conclusion *is the label of its root.*

Note that the fragment of $\mathcal{N}_{\mathsf{M}}$ without the rules $\langle]_R, \langle]_L, \mathsf{I}$ is the standard two-sided nested sequent calculus for K from [3,18]. Similarly, the fragment of $\mathcal{N}_{\mathsf{M}}$ without the rules $[]_L, []_R, \mathsf{I}$ is the full nested version of the linear nested sequent calculus for M from [13,14]. Hence, since the semantics are easily transferred, completeness of the full calculus implies completeness of the fragments for K and M respectively. The novel rule I corresponds to axiom (W′) and is the necessary link between those two systems. The first step is to show soundness of the system.

**Theorem 5.** *The calculus* $\mathcal{N}_{\mathsf{M}}$ *is* sound *for* biM *wrt. the formula interpretation* $\iota$*, i.e., if a nested sequent* $\mathcal{S}$ *is derivable in* $\mathcal{N}_{\mathsf{M}}$*, then* $\iota(\mathcal{S})$ *is valid in* biM.

$$\frac{}{\mathcal{S}\{\Gamma, p \Rightarrow p, \Delta\}}\ \mathsf{init} \qquad \frac{}{\mathcal{S}\{\Gamma, \bot \Rightarrow \Delta\}}\ \bot_L$$

$$\frac{\mathcal{S}\{\Gamma, A \Rightarrow \Delta, B\}}{\mathcal{S}\{\Gamma \Rightarrow \Delta, A \to B\}}\ \to_R \qquad \frac{\mathcal{S}\{\Gamma, B \Rightarrow \Delta\} \quad \mathcal{S}\{\Gamma \Rightarrow \Delta, A\}}{\mathcal{S}\{\Gamma, A \to B \Rightarrow \Delta\}}\ \to_L$$

$$\frac{\mathcal{S}\{\Gamma \Rightarrow \Delta, [\ \Rightarrow A]\}}{\mathcal{S}\{\Gamma \Rightarrow \Delta, [\,]A\}}\ [\,]_R \qquad \frac{\mathcal{S}\{\Gamma \Rightarrow \Delta, [\Sigma, A \Rightarrow \Pi]\}}{\mathcal{S}\{\Gamma, [\,]A \Rightarrow \Delta, [\Sigma \Rightarrow \Pi]\}}\ [\,]_L$$

$$\frac{\mathcal{S}\{\Gamma \Rightarrow \Delta, \langle\ \Rightarrow A\rangle\}}{\mathcal{S}\{\Gamma \Rightarrow \Delta, \langle\,]A\}}\ \langle\,]_R \qquad \frac{\mathcal{S}\{\Gamma \Rightarrow \Delta, [\Sigma, A \Rightarrow \Pi]\}}{\mathcal{S}\{\Gamma, \langle\,]A \Rightarrow \Delta, \langle\Sigma \Rightarrow \Pi\rangle\}}\ \langle\,]_L$$

$$\frac{\mathcal{S}\{\Gamma \Rightarrow \Delta, [\Sigma \Rightarrow \Pi]\}}{\mathcal{S}\{\Gamma \Rightarrow \Delta, [\,]A, \langle\Sigma \Rightarrow \Pi\rangle\}}\ \mathsf{I}$$

$$\frac{\mathcal{S}\{\Gamma, A, A \Rightarrow \Delta\}}{\mathcal{S}\{\Gamma, A \Rightarrow \Delta\}}\ \mathsf{ICL} \qquad \frac{\mathcal{S}\{\Gamma \Rightarrow \Delta, A, A\}}{\mathcal{S}\{\Gamma \Rightarrow \Delta, A\}}\ \mathsf{ICR} \qquad \frac{\mathcal{S}\{\Gamma \Rightarrow \Delta\}}{\mathcal{S}\{\Gamma, \Sigma \Rightarrow \Delta, \Pi\}}\ \mathsf{W}$$

**Fig. 3.** The nested sequent rules of the calculus $\mathcal{N}_{\mathsf{M}}$ for the bimodal system biM.

*Proof.* This follows as usual by an induction on the depth of the derivation of $\mathcal{S}$ from the fact that all rules preserve soundness wrt. the formula interpretation $\iota$. For all rules apart from $\langle\,]_L$ and I this is standard or trivial.

For $\langle\,]_L$, assume that $\mathcal{S}\{.\} = \Gamma_1 \Rightarrow \Delta_1, [\dots[\Gamma_n \Rightarrow \Delta_n, \{.\}]]\dots]$, and that $\iota(\mathcal{S}\{\Gamma, \langle\,]A \Rightarrow \Delta, \langle\Sigma \Rightarrow \Pi\rangle\})$ is falsified in $\mathfrak{M} = (W, \mathcal{N}, [\![.]\!])$ at $w$. If the contradiction comes from the context, i.e., $\iota(\mathcal{S}\{\ \Rightarrow\ \})$ is falsified at $w$, then also the interpretation of the premiss is falsified at $w$. Otherwise we have sequences $x_1, \dots, x_{n+1}$ of worlds and $\alpha_1, \dots, \alpha_{n+1}$ of neighbourhoods with

- $w = x_1$
- $x_{i+1} \in \alpha_{i+1} \in \mathcal{N}(x_i)$ for $i = 1, \dots, n$
- $\mathfrak{M}, x_i \Vdash \bigwedge \Gamma_i \wedge \neg \bigvee \iota(\Delta_i)$ for $1 \leq i \leq n$
- $\mathfrak{M}, x_{n+1} \Vdash \bigwedge \Gamma \wedge \langle\,]A \wedge \neg\iota(\Delta) \wedge [\,\rangle(\bigwedge \Sigma \wedge \neg \bigvee \Pi)$

where $\iota(\Delta)$ is the natural interpretation of $\Delta$, potentially including further nesting operators. From the last item we obtain a neighbourhood $\alpha \in \mathcal{N}(x_{n+1})$ with $\alpha \subseteq [\![A]\!]$. Due to the fact that $\mathfrak{M}, x_{n+1} \Vdash [\,\rangle(\bigwedge \Sigma \wedge \neg \bigvee \Pi)$ we then obtain a world $y \in \alpha$ such that $\mathfrak{M}, y \Vdash \bigwedge \Sigma \wedge A \wedge \neg \bigvee \Pi$. Hence we have $\mathfrak{M}, x_{n+1} \Vdash \langle\,\rangle(\bigwedge \Sigma \wedge A \wedge \neg \bigvee \Pi)$ and the formula interpretation $\iota(\mathcal{S}\{\Gamma \Rightarrow \Delta, \langle\Sigma, A \Rightarrow \Pi\rangle\})$ of the premiss of $\langle\,]_L$ is also falsified in $\mathfrak{M}, w$.

For the rule I, suppose the formula interpretation $\iota(\mathcal{S}\{\Gamma \Rightarrow \Delta, [\,]A, \langle\Sigma \Rightarrow \Pi\rangle\})$ of the conclusion is falsified in $\mathfrak{M} = (W, \mathcal{N}, [\![.]\!])$ at $w$. Then as above $\iota(\mathcal{S}\{\ \Rightarrow\ \})$ is falsified at $w$ or there is a world $v \in W$ such that $\mathfrak{M}, v$ falsifies $\bigwedge \Gamma \to \bigvee \iota(\Delta) \vee [\,]A \vee \langle\,](\bigwedge \Sigma \to \bigvee \Pi)$. Thus, in particular, we have $\mathfrak{M}, v \Vdash \langle\,\rangle\neg A$ and $\mathfrak{M}, v \Vdash [\,\rangle(\bigwedge \Sigma \wedge \neg \bigvee \Pi)$. Since $\mathfrak{M}, v \Vdash \langle\,\rangle\neg A$, there is an $\alpha \in \mathcal{N}(v)$ with $\alpha \neq \emptyset$. For this $\alpha$ then there exists a $x \in \alpha$ with $\mathfrak{M}, x \Vdash \bigwedge \Sigma \wedge \neg \bigvee \Pi$. Thus, in particular we have $\mathfrak{M}, v \Vdash \langle\,\rangle(\bigwedge \Sigma \wedge \neg \bigvee \Pi)$, and hence $\mathfrak{M}, v \nVdash [\,](\bigwedge \Sigma \to \bigvee \Pi) = \iota([\Sigma \Rightarrow \Pi])$. Hence the formula interpretation of the premiss is also falsified in $\mathfrak{M}, w$. □

We can prove completeness of the calculus in a number of different ways. Perhaps the easiest way is via a detour through the corresponding sequent calculi.

**Theorem 6.** *The calculus $\mathcal{N}_{\mathsf{M}}$ is complete for* biM, *i.e., if a formula $A$ is valid, then the nested sequent $\Rightarrow A$ is derivable in $\mathcal{N}_{\mathsf{M}}$.*

*Proof (Sketch).* First, observe that in the ordinary sequent system $\mathsf{G}_{\mathsf{biM}}$ given by the standard propositional rules of G3c of [19] together with the three rules

$$\frac{\Gamma \Rightarrow B}{\Sigma, [\,]\Gamma \Rightarrow [\,]B, \Pi} \qquad \frac{\Gamma \Rightarrow B}{\Sigma, [\,]\Gamma \Rightarrow [\,]A, \langle\,]B, \Pi} \qquad \frac{\Gamma, A \Rightarrow B}{\Sigma, [\,]\Gamma, \langle\,]A \Rightarrow \langle\,]B, \Pi}$$

and the cut rule all axioms and rules of biM are derivable. Hence $\mathsf{G}_{\mathsf{biM}}$ is complete in presence of cut. It also has cut elimination, as can be seen by straightforward adaption of the standard proof [19], or by checking that it satisfies the criteria for cut elimination from [12]. Completeness of $\mathcal{N}_{\mathsf{M}}$ then follows by simulating derivations in the sequent system in a leaf node of the nested sequents as in [13, 14]. In particular, the second and third modal rules above are simulated as follows, abbreviating multiple rule applications by a double line:

$$\dfrac{\dfrac{\dfrac{\Sigma \Rightarrow \Pi, [\Gamma \Rightarrow B]}{\Sigma, [\,]\Gamma \Rightarrow \Pi, [\ \Rightarrow B]}\ [\,]_L \text{ (double line)}}{\Sigma, [\,]\Gamma \Rightarrow [\,]A, \Pi, \langle \Rightarrow B\rangle}\ \mathsf{I}}{\Sigma, [\,]\Gamma \Rightarrow [\,]A, \langle\,]B, \Pi}\ \langle\,]_R \qquad \dfrac{\dfrac{\dfrac{\Sigma \Rightarrow \Pi, [\Gamma, A \Rightarrow B]}{\Sigma, [\,]\Gamma \Rightarrow \Pi, [A \Rightarrow B]}\ [\,]_L \text{ (double line)}}{\Sigma, [\,]\Gamma, \langle\,]A \Rightarrow \Pi, \langle \Rightarrow B\rangle}\ \langle\,]_L}{\Sigma, [\,]\Gamma, \langle\,]A \Rightarrow \langle\,]B, \Pi}\ \langle\,]_R$$

The simulation of the remaining modal rule is similar but easier. □

Note that analogously to the results in [13,14] the proof of the previous theorem further shows that completeness of the calculus is preserved if we restrict the nested sequents to be *linear*, i.e., to consist only of a single branch, and stipulate that all rules are *end-active*, i.e., only work in the last component:

**Corollary 7.** *The end-active linear version of $\mathcal{N}_{\mathsf{M}}$ is complete for* biM. □

Like the ordinary sequent calculus constructed in the proof, the end-active linear version of $\mathcal{N}_{\mathsf{M}}$ could be used to obtain an optimal PSPACE-complexity result. However, since this already follows using standard techniques and backwards proof search in the ordinary sequent system, we omit the proof.

**Theorem 8.** *The problem of deciding whether a formula is a theorem of* biM *is* PSPACE*-complete.* □

While the end-active linear version of $\mathcal{N}_{\mathsf{M}}$ is more suitable for space-efficient proof search, it is not ideal for constructing countermodels to underivable sequents. Hence in the following we consider the full nested version.

## 3 Cut Elimination

An alternative completeness proof is given by showing cut elimination. For this we move to the cumulative or *kleene'd* variant of the calculus, where all principal formulae and structures are copied into the premiss(es). The resulting *kleene'd calculus* $\mathcal{N}^k_{\mathsf{M}}$ is given in Fig. 4. Note that it contains the structural version $\mathsf{I}^s$ of the interaction rule I. To show equivalence to the base calculus, we show admissibility of the internal and external structural rules, including ICL, ICR, W. The proof for the internal rules is by standard induction on the depth of the derivation:

$$\frac{}{\mathcal{S}\{\Gamma, p \Rightarrow p, \Delta\}}\ \mathsf{init}^i \qquad \frac{}{\mathcal{S}\{\Gamma, \bot \Rightarrow \Delta\}}\ \bot^i_L$$

$$\frac{\mathcal{S}\{\Gamma, A \Rightarrow \Delta, A \to B, B\}}{\mathcal{S}\{\Gamma \Rightarrow \Delta, A \to B\}}\ \to^i_R \qquad \frac{\mathcal{S}\{\Gamma, A \to B, B \Rightarrow \Delta\} \quad \mathcal{S}\{\Gamma, A \to B \Rightarrow \Delta, A\}}{\mathcal{S}\{\Gamma, A \to B \Rightarrow \Delta\}}\ \to^i_L$$

$$\frac{\mathcal{S}\{\Gamma \Rightarrow \Delta, [\,]A, [\Rightarrow A]\}}{\mathcal{S}\{\Gamma \Rightarrow \Delta, [\,]A\}}\ [\,]^i_R \qquad \frac{\mathcal{S}\{\Gamma, [\,]A \Rightarrow \Delta, [\Sigma, A \Rightarrow \Pi]\}}{\mathcal{S}\{\Gamma, [\,]A \Rightarrow \Delta, [\Sigma \Rightarrow \Pi]\}}\ [\,]^i_L$$

$$\frac{\mathcal{S}\{\Gamma \Rightarrow \Delta, \langle\,]A, \langle \Rightarrow A\rangle\}}{\mathcal{S}\{\Gamma \Rightarrow \Delta, \langle\,]A\}}\ \langle\,]^i_R \qquad \frac{\mathcal{S}\{\Gamma, \langle\,]A \Rightarrow \Delta, \langle \Sigma \Rightarrow \Pi\rangle, [\Sigma, A \Rightarrow \Pi]\}}{\mathcal{S}\{\Gamma, \langle\,]A \Rightarrow \Delta, \langle \Sigma \Rightarrow \Pi\rangle\}}\ \langle\,]^i_L$$

$$\frac{\mathcal{S}\{\Gamma \Rightarrow \Delta, [\,]A, \langle \Sigma \Rightarrow \Pi\rangle, [\Sigma \Rightarrow \Pi]\}}{\mathcal{S}\{\Gamma \Rightarrow \Delta, [\,]A, \langle \Sigma \Rightarrow \Pi\rangle\}}\ \mathsf{I}^i \qquad \frac{\mathcal{S}\{\Gamma \Rightarrow \Delta, \langle \Sigma \Rightarrow \Pi\rangle, [\Sigma \Rightarrow \Pi], [\Omega \Rightarrow \Theta]\}}{\mathcal{S}\{\Gamma \Rightarrow \Delta, \langle \Sigma \Rightarrow \Pi\rangle, [\Omega \Rightarrow \Theta]\}}\ \mathsf{I}^s$$

**Fig. 4.** The kleene'd version $\mathcal{N}^k_{\mathsf{M}}$ of the calculus

**Lemma 9.** *The following rules are admissible in* $\mathcal{N}^k_{\mathsf{M}}$*:*

$$\frac{\mathcal{S}\{\Gamma \Rightarrow \Delta\}}{\mathcal{S}\{\Gamma, \Sigma \Rightarrow \Delta, \Pi\}} \quad \frac{\mathcal{S}\{\Gamma \Rightarrow \Delta, \langle \Sigma \Rightarrow \Pi\rangle\}}{\mathcal{S}\{\Gamma \Rightarrow \Delta, \langle \Sigma, \Omega \Rightarrow \Pi, \Theta\rangle\}} \quad \frac{\mathcal{S}\{\Gamma, A, A \Rightarrow \Delta\}}{\mathcal{S}\{\Gamma, A \Rightarrow \Delta\}} \quad \frac{\mathcal{S}\{\Gamma \Rightarrow \Delta, A, A\}}{\mathcal{S}\{\Gamma \Rightarrow \Delta, A\}}$$

**Lemma 10.** *The* merge rules *are admissible in* $\mathcal{N}^k_{\mathsf{M}}$*:*

$$\frac{\mathcal{S}\{\Gamma \Rightarrow \Delta, [\Sigma \Rightarrow \Pi], [\Omega \Rightarrow \Theta]\}}{\mathcal{S}\{\Gamma \Rightarrow \Delta, [\Sigma, \Omega \Rightarrow \Pi, \Theta]\}}\ \mathsf{mrg}_{[\,]} \qquad \frac{\mathcal{S}\{\Gamma \Rightarrow \Delta, \langle \Sigma \Rightarrow \Pi\rangle, \langle \Omega \Rightarrow \Theta\rangle\}}{\mathcal{S}\{\Gamma \Rightarrow \Delta, \langle \Sigma, \Omega \Rightarrow \Pi, \Theta\rangle\}}\ \mathsf{mrg}_{\langle\rangle}$$

*Proof.* By induction on the depth of the derivation. The only non-standard cases are for $\mathsf{mrg}_{\langle\rangle}$ with last applied rule $\langle\,]^i_L$ or $\mathsf{I}^i$. Here we apply the induction hypothesis, followed by admissibility of weakening and the original rule. □

**Lemma 11.** *The calculi* $\mathcal{N}_{\mathsf{M}} + \mathsf{mrg}_{\langle\rangle}$ *and* $\mathcal{N}^k_{\mathsf{M}}$ *are equivalent, i.e., a sequent* $\Rightarrow A$ *is derivable in* $\mathcal{N}_{\mathsf{M}}$ *plus* $\mathsf{mrg}_{\langle\rangle}$ *iff it is derivable in* $\mathcal{N}^k_{\mathsf{M}}$.

*Proof.* For one direction, using admissibility of weakening and contraction it is straightforward to transform a derivation in $\mathcal{N}_{\mathsf{M}}$ into a derivation in $\mathcal{N}^k_{\mathsf{M}}$.

For the other direction we transform derivations in $\mathcal{N}^k_{\mathsf{M}}$ into derivations in $\mathcal{N}_{\mathsf{M}}$ using contraction and merge, where $\mathsf{mrg}_{[\,]}$ is shown admissible in $\mathcal{N}_{\mathsf{M}}$ by induction on the depth of the derivation. The only tricky part is the transformation for the rule $\mathsf{l}^s$. For this we use the fact that we can permute applications of $\mathsf{l}^s$ below applications of the other rules of $\mathcal{N}^k_{\mathsf{M}}$ (the proof is rather straightforward by going through all the cases), until a formula of the shape $[\rangle A$ appears in the succedent. At this point we transform the application of $\mathsf{l}^s$ into an application of $\mathsf{l}$ creating the same formula $[\rangle A$, followed by an application of contraction. □

Note that this lemma shows equivalence only with $\mathcal{N}_{\mathsf{M}}$ extended with the merge rule $\mathsf{mrg}_{\langle\rangle}$. While it would be possible to either make this rule part of $\mathcal{N}_{\mathsf{M}}$ from the outset, or to modify $\mathcal{N}_{\mathsf{M}}$ so that it becomes admissible, the advantage of the current formulation is the direct link to the end-active linear version (Corollary 7). To state the cut rule, we use the following notion adapted from [18]:

**Definition 12.** *For two nested sequents with holes*

$$\mathcal{S}\{\ \} = \Gamma_1 \Rightarrow \Delta_1, [\dots[\Gamma_n \Rightarrow \Delta_n, \{\ \}]]\dots]$$
$$\mathcal{S}'\{\ \} = \Sigma_1 \Rightarrow \Pi_1, [\dots[\Sigma_n \Rightarrow \Pi_n, \{\ \}]]\dots]$$

*the* merge *is the nested sequent with hole*

$$(\mathcal{S} \oplus \mathcal{S}')\{\ \} := \Gamma_1, \Sigma_1 \Rightarrow \Delta_1, \Pi_1, [\dots[\Gamma_n, \Sigma_n \Rightarrow \Delta_n, \Pi_n, \{\ \}]]\dots]$$

*obtained by "zipping" together the two nested sequents along the path from the root to the hole. Note that the hole is at the same depth in both nested sequents.*

Using this notion, the *cut rule* then is the following rule:

$$\frac{\mathcal{S}\{\Gamma \Rightarrow \Delta, A\} \quad \mathcal{S}'\{A, \Sigma \Rightarrow \Pi\}}{(\mathcal{S} \oplus \mathcal{S}')\{\Gamma, \Sigma \Rightarrow \Delta, \Pi\}}\ \mathsf{cut}_1$$

In order to reduce cuts on $\langle]$-formulae, we also eliminate the *auxiliary cut rule*:

$$\frac{\mathcal{S}\{\Gamma \Rightarrow \Delta, \langle \Omega \Rightarrow \Theta, A\rangle\} \quad \mathcal{S}'\{\Sigma \Rightarrow \Pi, [A, \Xi \Rightarrow \Upsilon]\}}{(\mathcal{S} \oplus \mathcal{S}')\{\Gamma, \Sigma \Rightarrow \Delta, \Pi, \langle \Omega, \Xi \Rightarrow \Theta, \Upsilon\rangle\}}\ \mathsf{cut}_2$$

Soundness of these rules can be shown directly, but also follows from the fact that they are admissible in the cut-free calculus. Note that we only permit cut on components at the same depth of the nested sequents. While often this necessitates the addition or admissibility of certain structural rules [3,18], here the situation is simpler due to the fact that the axiomatisation of $\mathsf{biM}$ does not involve axioms of mixed modal rank such as $\mathsf{4}, \mathsf{5}$ or $\mathsf{T}$.

**Theorem 13.** *The cut rules* $\mathsf{cut}_1$ *and* $\mathsf{cut}_2$ *are admissible in the calculus* $\mathcal{N}^k_{\mathsf{M}}$.

*Proof.* The proof is for both statements simultaneously by induction on the tuples $(c, d)$ in lexicographic ordering, where $c$ is the *complexity* of the cut formula, i.e., its length, and $d$ is the *depth* of the cut, i.e., the sum of the depth of the derivations of the premisses of the cut. Call these tuples the *measure* of the corresponding application of cut. The proof of the statement for $\mathsf{cut}_1$ with measure $(c, d)$ uses the statements for $\mathsf{cut}_1$ with measure $(c, k)$ with $k < d$ and for $\mathsf{cut}_1$ and $\mathsf{cut}_2$ with measure $(c-1, n)$ for arbitrary $n$. The proof of the statement for $\mathsf{cut}_2$ with measure $(c, d)$ uses the statements for $\mathsf{cut}_2$ with measure $(c, k)$ with $k < d$ and for $\mathsf{cut}_1$ with measure $(c, n)$ with arbitrary $n$.

The general strategy is to permute applications of cut up into the left premiss until the cut formula is principal, then up into the right premiss until it is principal here as well and can be reduced. We apply *cross-cuts* to eliminate the cut formula from the context. The cases for the zero-premiss rules are standard.

**Cut Formula Contextual on the Left.** For $\mathsf{cut}_1$, the cut is permuted as usual into the premiss(es) of the last applied rule in the derivation of the left premiss of the cut and eliminated by induction hypothesis on the depth.

For $\mathsf{cut}_2$, we consider the case where the nesting $\langle . \rangle$ containing the cut formula in the left premiss of the application of $\mathsf{cut}_2$ is active in the last rule of that derivation. If the last rule is $\langle ]^i_L$, we have:

$$\dfrac{\dfrac{\mathcal{S}\{\Gamma, \langle ]B \Rightarrow \Delta, \langle \Omega \Rightarrow \Theta, A \rangle, [\Omega, B \Rightarrow \Theta, A]\}}{\mathcal{S}\{\Gamma, \langle ]B \Rightarrow \Delta, \langle \Omega \Rightarrow \Theta, A \rangle\}}\ \langle ]^i_L \qquad \mathcal{S}'\{\Sigma \Rightarrow \Pi, [A, \Xi \Rightarrow \Upsilon]\}}{(\mathcal{S} \oplus \mathcal{S}')\{\Gamma, \langle ]B, \Sigma \Rightarrow \Delta, \Pi, \langle \Omega, \Xi \Rightarrow \Theta, \Upsilon \rangle\}}\ \mathsf{cut}_2$$

We first apply $\mathsf{cut}_2$ with lower depth to the premiss of $\langle ]^i_L$ and $\mathcal{S}'\{\Sigma \Rightarrow \Pi, [A, \Xi \Rightarrow \Upsilon]\}$ to obtain $(\mathcal{S} \oplus \mathcal{S}')\{\Gamma, \langle ]B, \Sigma \Rightarrow \Delta, \Pi, \langle \Omega, \Xi \Rightarrow \Theta, \Upsilon \rangle, [\Omega, B \Rightarrow \Theta, A]\}$. Now an application of $\mathsf{cut}_1$ with possibly higher depth but the same complexity yields

$$((\mathcal{S} \oplus \mathcal{S}') \oplus \mathcal{S}')\{\Gamma, \langle ]B, \Sigma^2 \Rightarrow \Delta, \Pi^2, \langle \Omega, \Xi \Rightarrow \Theta, \Upsilon \rangle, [\Omega, B, \Xi \Rightarrow \Theta, \Upsilon]\}\,.$$

An application of $\langle ]^i_L$ followed by admissibility of contraction and merge then gives the result. The cases for the rules $\mathsf{l}^i$ and $\mathsf{l}^s$ are analogous. The cases where the nesting is not active are even simpler.

**Cut Formula Principal on the Left and Contextual on the Right.** Since the cut formula is principal on the left and no rule has a principal formula inside the nesting $\langle . \rangle$, we are dealing with the case of $\mathsf{cut}_1$ only. Thus as usual we permute the cut into the premisses of the last rule of the derivation of the right premiss of the cut and eliminate it using the induction hypothesis on the depth.

**Principal-Principal:** The cases where the cut formula is propositional are as usual. In case the cut formula is $\langle ]A$, we have:

$$\dfrac{\dfrac{\mathcal{S}\{\Gamma \Rightarrow \Delta, \langle ]A, \langle \Rightarrow A \rangle\}}{\mathcal{S}\{\Gamma \Rightarrow \Delta, \langle ]A\}}\ \langle ]^i_R \qquad \dfrac{\mathcal{S}'\{\langle ]A, \Sigma \Rightarrow \Pi, \langle \Omega \Rightarrow \Theta \rangle, [A, \Omega \Rightarrow \Theta]\}}{\mathcal{S}'\{\langle ]A, \Sigma \Rightarrow \Pi, \langle \Omega \Rightarrow \Theta \rangle\}}\ \langle ]^i_L}{(\mathcal{S} \oplus \mathcal{S}')\{\Gamma, \Sigma \Rightarrow \Delta, \Pi, \langle \Omega \Rightarrow \Theta \rangle\}}\ \mathsf{cut}_1$$

First we apply cross-cuts (i.e., the induction hypothesis on the lower depth) to the premiss of $\langle ]^i_R$ and the conclusion of $\langle ]^i_L$ and vice-versa to obtain derivations of the two nested sequents $(\mathcal{S} \oplus \mathcal{S}')\{\Gamma, \Sigma \Rightarrow \Delta, \Pi, \langle \Rightarrow A \rangle, \langle \Omega \Rightarrow \Theta \rangle\}$ and

$(\mathcal{S} \oplus \mathcal{S}')\{\Gamma, \Sigma \Rightarrow \Delta, \Pi, \langle \Omega \Rightarrow \Theta \rangle, [A, \Omega \Rightarrow \Theta]\}$. Then we apply the induction hypothesis on the smaller complexity for $\mathsf{cut}_2$ to these two to obtain

$$(\mathcal{S} \oplus \mathcal{S}') \oplus (\mathcal{S} \oplus \mathcal{S}')\{\Gamma^2, \Sigma^2 \Rightarrow \Delta^2, \Pi^2, \langle \Omega \Rightarrow \Theta \rangle, \langle \Omega \Rightarrow \Theta \rangle, \langle \Omega \Rightarrow \Theta \rangle\}$$

Now admissibility of $\mathsf{mrg}_{[]}$, $\mathsf{mrg}_{\langle\rangle}$ and contraction yields the result.

Suppose that the cut formula is $[\,]A$ with last applied rules $[\,]^i_R$ and $[\,]^i_L$:

$$\dfrac{\dfrac{\mathcal{S}\{\Gamma \Rightarrow \Delta, [\,]A, [\, \Rightarrow A]\}}{\mathcal{S}\{\Gamma \Rightarrow \Delta, [\,]A\}}\,[\,]^i_R \quad \dfrac{\mathcal{S}'\{[\,]A, \Sigma \Rightarrow \Pi, [A, \Omega \Rightarrow \Theta]\}}{\mathcal{S}'\{[\,]A, \Sigma \Rightarrow \Pi, [\Omega \Rightarrow \Theta]\}}\,[\,]^i_L}{(\mathcal{S} \oplus \mathcal{S}')\{\Gamma, \Sigma \Rightarrow \Delta, \Pi, [\Omega \Rightarrow \Theta]\}}\,\mathsf{cut}_1$$

Again, applying cross-cuts gives $(\mathcal{S} \oplus \mathcal{S}')\{\Gamma, \Sigma \Rightarrow \Delta, \Pi, [A, \Omega \Rightarrow \Theta]\}$ and $(\mathcal{S} \oplus \mathcal{S}')\{\Gamma, \Sigma \Rightarrow \Delta, \Pi, [\, \Rightarrow A], [\Omega \Rightarrow \Theta]\}$. Now an application of $\mathsf{cut}_1$ with smaller complexity gives $(\mathcal{S} \oplus \mathcal{S}') \oplus (\mathcal{S} \oplus \mathcal{S}')\{\Gamma^2, \Sigma^2 \Rightarrow \Delta^2, \Pi^2, [\Omega \Rightarrow \Theta], [\Omega \Rightarrow \Theta]\}$ and using admissibility of merge and contraction we are done.

If the cut formula is $[\,]A$ with last applied rules $\mathsf{I}^i$ and $[\,]_L$ we have

$$\dfrac{\dfrac{\mathcal{S}\{\Gamma \Rightarrow \Delta, [\,]A, \langle \Omega \Rightarrow \Theta \rangle, [\Omega \Rightarrow \Theta]\}}{\mathcal{S}\{\Gamma \Rightarrow \Delta, [\,]A, \langle \Omega \Rightarrow \Theta \rangle\}}\,\mathsf{I}^i \quad \dfrac{\mathcal{S}'\{[\,]A, \Sigma \Rightarrow \Pi, [A, \Xi \Rightarrow \Upsilon]\}}{\mathcal{S}'\{[\,]A, \Sigma \Rightarrow \Pi, [\Xi \Rightarrow \Upsilon]\}}\,[\,]^i_L}{(\mathcal{S} \oplus \mathcal{S}')\{\Gamma, \Sigma \Rightarrow \Delta, \Pi, \langle \Omega \Rightarrow \Theta \rangle, [\Xi \Rightarrow \Upsilon]\}}\,\mathsf{cut}_1$$

This is converted into

$$\dfrac{\dfrac{\mathcal{S}\{\Gamma \Rightarrow \Delta, [\,]A, \langle \Omega \Rightarrow \Theta \rangle, [\Omega \Rightarrow \Theta]\} \quad \dfrac{\mathcal{S}'\{[\,]A, \Sigma \Rightarrow \Pi, [A, \Xi \Rightarrow \Upsilon]\}}{\mathcal{S}'\{[\,]A, \Sigma \Rightarrow \Pi, [\Xi \Rightarrow \Upsilon]\}}\,[\,]^i_L}{(\mathcal{S} \oplus \mathcal{S}')\{\Gamma, \Sigma \Rightarrow \Delta, \Pi, \langle \Omega \Rightarrow \Theta \rangle, [\Omega \Rightarrow \Theta], [\Xi \Rightarrow \Upsilon]\}}\,\mathsf{cut}_1}{(\mathcal{S} \oplus \mathcal{S}')\{\Gamma, \Sigma \Rightarrow \Delta, \Pi, \langle \Omega \Rightarrow \Theta \rangle, [\Xi \Rightarrow \Upsilon]\}}\,\mathsf{I}^s$$

and we are done using the induction hypothesis on the depth. □

## 4 Completeness via Countermodel Generation

From a semantical point it is more informative to show completeness by constructing countermodels from a failed proof search. For this we slightly modify the system $\mathcal{N}^k_{\mathsf{M}}$ in two ways. First, to make the construction of a successor world more explicit, we split the nesting operator $\langle . \rangle$ into an *unfinished* version $\langle . \rangle^{\mathsf{u}}$ and a *finished* version $\langle . \rangle^{\mathsf{f}}$, adding an explicit $\mathsf{jump}$ rule which constructs a $[.]$-successor out of a finished $\langle . \rangle^{\mathsf{f}}$-successor as in [13,14]. To facilitate the construction of the neighbourhoods, we further add *annotations* to the components:

**Definition 14.** *An* annotated nested sequent *is an expression*

$$\Gamma \overset{\Xi}{\Rightarrow} \Delta, \langle \Sigma_1 \Rightarrow \Pi_1 \rangle^{\mathsf{u}}, \ldots, \langle \Sigma_n \Rightarrow \Pi_n \rangle^{\mathsf{u}},$$
$$\langle \Omega_1 \Rightarrow \Theta_1 \rangle^{\mathsf{f}}, \ldots, \langle \Omega_m \Rightarrow \Theta_m \rangle^{\mathsf{f}}, [\mathcal{S}_1], \ldots, [\mathcal{S}_k]$$

*where the* annotation $\Xi$ *is a multiset of formulae, and the* $\mathcal{S}_i$ *are annotated nested sequents. For a component* $v$ *we write* $\mathsf{an}(v)$ *for the annotation of this component.*

Again, we can identify an annotated nested sequent with a labelled tree, and we call each node labelled with an expression $\Gamma \overset{\Xi}{\Rightarrow} \Delta, \langle \Sigma_1 \Rightarrow \Pi_1 \rangle^{\mathsf{u}}, \dots, \langle \Sigma_n \Rightarrow \Pi_n \rangle^{\mathsf{u}}, \langle \Omega_1 \Rightarrow \Theta_1 \rangle^{\mathsf{f}}, \dots, \langle \Omega_m \Rightarrow \Theta_m \rangle^{\mathsf{f}}$ a *component* of the annotated nested sequent. The main intuition for the annotations is that they store information on how a component of a nested sequent was created during backwards proof search. This information will the be used in the countermodel construction to collect all successors of a component with the same annotation into one neighbourhood of the component. Finally, we drop the structural version of the interaction rule. The resulting system $\mathcal{N}^a_{\mathsf{M}}$ is given in Fig. 5. Note that the annotations only store information on how a component of a nested sequent in the proof search was created, but do not influence proof search per se. Building on this, the proof of the following Lemma shows that, modulo the structural rules, derivations in the annotated and plain systems are easily converted into each other.

$$\frac{}{\mathcal{S}\{\Gamma, p \overset{\Upsilon}{\Rightarrow} p, \Delta\}}\ \mathsf{init}^a \qquad \frac{}{\mathcal{S}\{\Gamma, \bot \overset{\Upsilon}{\Rightarrow} \Delta\}}\ \bot^a_L$$

$$\frac{\mathcal{S}\{\Gamma, A \overset{\Upsilon}{\Rightarrow} B, A \to B, \Delta\}}{\mathcal{S}\{\Gamma \overset{\Upsilon}{\Rightarrow} A \to B, \Delta\}}\ \to^a_R \qquad \frac{\mathcal{S}\{\Gamma, A \to B, B \overset{\Upsilon}{\Rightarrow} \Delta\} \quad \mathcal{S}\{\Gamma, A \to B \overset{\Upsilon}{\Rightarrow} A, \Delta\}}{\mathcal{S}\{\Gamma, A \to B \overset{\Upsilon}{\Rightarrow} \Delta\}}\ \to^a_L$$

$$\frac{\mathcal{S}\{\Gamma \overset{\Upsilon}{\Rightarrow} \Delta, [\,]A, [\ \overset{\emptyset}{\Rightarrow} A]\}}{\mathcal{S}\{\Gamma \overset{\Upsilon}{\Rightarrow} \Delta, [\,]A\}}\ [\,]^a_R \qquad \frac{\mathcal{S}\{\Gamma, [\,]A \overset{\Upsilon}{\Rightarrow} \Delta, [\Sigma, A \overset{\Xi}{\Rightarrow} \Pi]\}}{\mathcal{S}\{\Gamma, [\,]A \overset{\Upsilon}{\Rightarrow} \Delta, [\Sigma \overset{\Xi}{\Rightarrow} \Pi]\}}\ [\,]^a_L$$

$$\frac{\mathcal{S}\{\Gamma \overset{\Upsilon}{\Rightarrow} \Delta, \langle\,]A, \langle\ \Rightarrow A\rangle^{\mathsf{u}}\}}{\mathcal{S}\{\Gamma \overset{\Upsilon}{\Rightarrow} \Delta, \langle\,]A\}}\ \langle\,]^a_R \qquad \frac{\mathcal{S}\{\Gamma, \langle\,]A \overset{\Upsilon}{\Rightarrow} \Delta, \langle \Sigma \Rightarrow \Pi\rangle^{\mathsf{u}}, \langle \Sigma, A \Rightarrow \Pi\rangle^{\mathsf{f}}\}}{\mathcal{S}\{\Gamma, \langle\,]A \overset{\Upsilon}{\Rightarrow} \Delta, \langle \Sigma \Rightarrow \Pi\rangle^{\mathsf{u}}\}}\ \langle\,]^a_L$$

$$\frac{\mathcal{S}\{\Gamma \overset{\Upsilon}{\Rightarrow} \Delta, [\,]A, \langle \Sigma \Rightarrow \Pi\rangle^{\mathsf{u}}, \langle \Sigma \Rightarrow \Pi\rangle^{\mathsf{f}}\}}{\mathcal{S}\{\Gamma \overset{\Upsilon}{\Rightarrow} \Delta, [\,]A, \langle \Sigma \Rightarrow \Pi\rangle^{\mathsf{u}}\}}\ \mathsf{I}^a \qquad \frac{\mathcal{S}\{\Gamma \overset{\Upsilon}{\Rightarrow} \Delta, \langle \Sigma \Rightarrow \Pi\rangle^{\mathsf{f}}, [\Sigma \overset{\Sigma}{\Rightarrow} \Pi]\}}{\mathcal{S}\{\Gamma \overset{\Upsilon}{\Rightarrow} \Delta, \langle \Sigma \Rightarrow \Pi\rangle^{\mathsf{f}}\}}\ \mathsf{jump}^a$$

**Fig. 5.** The invertible annotated variant $\mathcal{N}^a_{\mathsf{M}}$ of the system

**Lemma 15.** *The systems $\mathcal{N}_{\mathsf{M}}$ and $\mathcal{N}^a_{\mathsf{M}}$ are equivalent, i.e.: A nested sequent $\Rightarrow A$ is derivable in $\mathcal{N}_{\mathsf{M}}$ if and only if $\overset{\emptyset}{\Rightarrow} A$ is derivable in $\mathcal{N}^a_{\mathsf{M}}$.*

*Proof.* To convert derivations in $\mathcal{N}_{\mathsf{M}}$ into derivations in $\mathcal{N}^a_{\mathsf{M}}$, we first convert them into derivations in $\mathcal{N}^k_{\mathsf{M}}$ using Lemma 11, noting that the result does not contain the rule $\mathsf{I}^s$. Hence we can convert the resulting derivation into a derivation in $\mathcal{N}^a_{\mathsf{M}}$ bottom-up, starting from the conclusion, replacing all the rules with their respective counterparts. The rules $\langle\,]^i_L$ and $\mathsf{I}^i$ are replaced with their annotated versions followed by $\mathsf{jump}^a$. In the other direction, derivations in $\mathcal{N}^a_{\mathsf{M}}$ are converted into derivations in $\mathcal{N}_{\mathsf{M}}$ by deleting all the annotations and applications of $\mathsf{jump}^a$, replacing the rules $\langle\,]^a_L$ and $\mathsf{I}^a$ by $\langle\,]_L$ and $\mathsf{I}$ respectively, and using contraction to remove additional copies of the principal formulae. □

As usual, we then construct a model from an annotated nested sequent which is saturated under the application of all the rules, defined as follows.

**Definition 16.** *An annotated nested sequent* $\mathcal{S}$ *is* saturated *if for each of its components* $\Gamma \overset{\Xi}{\Rightarrow} \Delta$ *the following hold:*

1. $\Gamma \cap \Delta \neq \emptyset$
2. $\bot \notin \Gamma$
3. $A \to B \in \Gamma$ *implies* $B \in \Gamma$ *or* $A \in \Delta$
4. $A \to B \in \Delta$ *implies* $A \in \Gamma$ *and* $B \in \Delta$
5. $\langle ] A \in \Delta$ *implies* $\langle \Rightarrow A \rangle^{\mathsf{u}} \in \Delta$
6. $\langle ] A \in \Gamma$ *and* $\langle \Sigma \Rightarrow \Pi \rangle^{\mathsf{u}} \in \Delta$ *implies* $\langle \Sigma, A \Rightarrow \Pi \rangle^{\mathsf{f}} \in \Delta$
7. $\langle \Sigma \Rightarrow \Pi \rangle^{\mathsf{f}} \in \Delta$ *implies there are* $\Omega, \Theta$ *such that* $[\Sigma, \Omega \overset{\Sigma}{\Rightarrow} \Pi, \Theta] \in \Delta$
8. $[ ] A \in \Delta$ *implies there are* $\Sigma, \Omega, \Theta$ *with* $[\Omega \overset{\Sigma}{\Rightarrow} A, \Theta] \in \Delta$
9. $[ ] A \in \Gamma$ *and* $[\Omega \overset{\Sigma}{\Rightarrow} \Theta] \in \Delta$ *implies* $A \in \Omega$.
10. $[ ] A \in \Delta$ *and* $\langle \Sigma \Rightarrow \Pi \rangle^{\mathsf{u}} \in \Delta$ *implies* $\langle \Sigma, \Omega \Rightarrow \Pi, \Theta \rangle^{\mathsf{f}} \in \Delta$ *for some* $\Omega, \Theta$.

The difficulty in building a model from a saturated nested sequent then lies in constructing the set of neighbourhoods for each world. We do this by collecting successor worlds into sets according to their annotations. A bit of care needs to be taken, depending on whether there is a formula of the form $\langle ] A$ in the succedent of the component or not. Formally:

**Definition 17.** *Let* $\mathcal{S}$ *be a saturated nested sequent. The* model generated by $\mathcal{S}$ *is the model* $\mathfrak{M}^{\mathcal{S}} = (W, \mathcal{N}, [\![ . ]\!])$ *with*

- $W$ *the set of components (nodes) of* $\mathcal{S}$
- *if* $w \in W$, *then* $w \in [\![ p ]\!]$ *iff* $w$ *is a component* $\Gamma \overset{\Sigma}{\Rightarrow} \Delta$ *with* $p \in \Gamma$
- $\mathcal{N}(w)$ *is defined as follows. Let* $\mathcal{C}_w$ *be the set of immediate successors of* $w$, *and let* $\ell[\mathcal{C}_w]$ *be the set of annotations of nodes in* $\mathcal{C}_w$. *Then let*

$$\mathcal{L}_w := \{ \{ v \in \mathcal{C}_w \mid \mathsf{an}(v) = \Sigma \} \mid \Sigma \in \ell[\mathcal{C}_w] \}$$

*Now,* $\mathcal{N}(w)$ *is defined as* $(\mathcal{L}_w \cup \{\mathcal{C}_w\}) \setminus \{\emptyset\}$ *if there is a formula* $\langle ] A \in \Delta$, *and* $\mathcal{L}_w \cup \{\mathcal{C}_w\} \cup \{\emptyset\}$ *otherwise.*

Thus, disregarding the empty set, the set of neighbourhoods of a node in a nested sequent includes the set of all its children (to make the construction work for the normal modality $[ ]$), as well as every set of children with the same annotation. Whether it contains the empty set or not depends on whether there is a formula $\langle ] A$ in its succedent. By construction we have:

**Lemma 18 (Model Lemma).** *If* $\mathcal{S}$ *is saturated, then the model generated by* $\mathcal{S}$ *is a neighbourhood model.* □

Non-derivable nested sequents then yield a saturated nested sequent via a standard proof search procedure, given as follows.

**Definition 19.** *The* proof search procedure *in $\mathcal{N}^a_{\mathsf{M}}$ is defined by application of the rules of $\mathcal{N}^a_{\mathsf{M}}$ in an arbitrary but fixed order, unless the conclusion of a potential rule application already satisfies the saturation condition corresponding to this rule. An annotated nested sequent is* minimal *if it can be obtained from an annotated nested sequent $\Gamma \overset{\emptyset}{\Rightarrow} \Delta$ by the proof search procedure.*

**Lemma 20.** *The proof search procedure terminates and either yields a derivation or a saturated annotated nested sequent.*

*Proof.* Every backwards application of a rule adds a formula or a sequent inside a nesting operator. Since the maximal modal nesting depth of formulae decreases in every nesting operator, and since by the saturation conditions no formula or sequent is created twice in the same component, the procedure terminates. □

The final ingredient for showing that the model generated from a saturated nested sequent obtained from proof search really is a model then is the following.

**Lemma 21.** *Let $\mathcal{S}$ be a minimal annotated nested sequent and $\Gamma \overset{\Xi}{\Rightarrow} \Delta$ be a component of $\mathcal{S}$. Then $\Xi \subseteq \Gamma$.*

*Proof.* Since in a minimal annotated nested sequent new components are only constructed via the $\mathsf{jump}^a$ rule which has identical label and sequent in the premiss, or via the $[]_R$ rule, which creates the empty label. □

**Lemma 22 (Truth Lemma).** *If $\mathcal{S}$ is saturated and minimal and $w$ is a component of $\mathcal{S}$ containing $\Gamma \overset{\Xi}{\Rightarrow} \Delta$, then for every formula $A$:*

1. *$A \in \Gamma$ implies $\mathfrak{M}^{\mathcal{S}}, w \Vdash A$*
2. *$A \in \Delta$ implies $\mathfrak{M}^{\mathcal{S}}, w \nVdash A$.*

*Proof.* By induction on the complexity of $A$ for both statements simultaneously.

If $A$ is atomic, an implication or $\bot$, then the statement follows as usual.

Suppose that $A = \langle]B$ and $A \in \Gamma$. We need to show that $\mathfrak{M}^{\mathcal{S}}, w \Vdash \langle]B$, i.e., that there is an $\alpha \in \mathcal{N}(w)$ with $\alpha \subseteq [\![B]\!]$.

*Case 1: There is no formula of the shape $\langle]C$ in $\Delta$.* Then by definition we have $\emptyset \in \mathcal{N}(w)$. But $\emptyset \subseteq [\![B]\!]$, and hence $\mathfrak{M}^{\mathcal{S}}, w \Vdash \langle]B$.

*Case 2: There is a formula of the shape $\langle]C$ in $\Delta$.* Then

$$\mathcal{N}(w) = \{\{v \in \mathcal{C}_w \mid \mathsf{an}(v) = \Sigma\} \mid \Sigma \in \ell[\mathcal{C}_w]\} \cup \{\mathcal{C}_w\} \cup \{\emptyset\}$$

Since $\langle]C \in \Delta$, by saturation we have $\langle \Rightarrow C\rangle^{\mathsf{u}} \in \Delta$ for that same $C$. Then again by saturation and the fact that $\langle]B \in \Gamma$ we have $\langle B \Rightarrow C\rangle^{\mathsf{f}} \in \Delta$, and hence also there are $\Omega, \Theta$ with $[B, \Omega \overset{B}{\Rightarrow} \Theta] \in \Delta$. Thus the set $\alpha := \{v \in \mathcal{C}_w \mid \mathsf{an}(v) = B\}$ is nonempty. By Lemma 21 we have for every component $[\Gamma' \overset{B}{\Rightarrow} \Delta']$ from $\alpha$ that $B \in \Gamma'$. Hence by induction hypothesis $\alpha \subseteq [\![B]\!]$, and thus $\mathfrak{M}^{\mathcal{S}}, w \Vdash \langle]B$.

Suppose that $A = \langle]B$ and $A \in \Delta$. Then by definition of $\mathcal{N}$ we have that $\emptyset \notin \mathcal{N}(w)$. We need to show that $\mathfrak{M}^{\mathcal{S}} \nVdash \langle]B$.

*Case 1: $w$ has no children.* Then $\mathcal{N}(w) = \emptyset$ and hence $\mathfrak{M}^S, w \not\Vdash \langle]B$.
*Case 2: $w$ has a child.* Then

$$\mathcal{N}(w) = (\{\{v \in \mathcal{C}_w \mid \mathsf{an}(v) = \Sigma\} \mid \Sigma \in \ell[\mathcal{C}_w]\} \cup \{\mathcal{C}_w\}) \setminus \{\emptyset\}$$

is non-empty. Let $\alpha \in \mathcal{N}(w)$. Then there is an annotation $\Sigma \in \ell[\mathcal{C}_w]$ with $\alpha = \{v \in \mathcal{C}_w \mid \mathsf{an}(v) = \Sigma\}$, or $\alpha = \mathcal{C}_w$. We need to show that there is a $v \in \alpha$ with $\mathfrak{M}^S, v \Vdash \neg B$, i.e., that $\alpha \not\subseteq [\![B]\!]$. We show this for $\alpha = \{v \in \mathcal{C}_w \mid \mathsf{an}(v) = \Sigma\}$. The statement for the second case then follows from the fact that every such set is a subset of $\mathcal{C}_w$, and that for every $v \in \mathcal{C}_w$ we have $v \in \{x \in \mathcal{C}_w \mid \mathsf{an}(x) = \mathsf{an}(v)\}$. So suppose $\alpha = \{v \in \mathcal{C}_w \mid \mathsf{an}(v) = \Sigma\}$. The only ways a successor can be created is by the rules $[]^a_R$ or $\mathsf{jump}^a$. If $\Sigma = \emptyset$, then there must be a formula $[]D \in \Delta$, since either the rule $[]^a_R$ or the rule $\mathsf{I}^a$ must have been applied. But then by saturation and the fact that both $[]D$ and $\langle]B$ are in $\Delta$, we have that $[\overset{\emptyset}{\Rightarrow} B] \in \Delta$ as well. By induction hypothesis, at this world $B$ is false, and hence we have $\alpha \not\subseteq [\![B]\!]$. If in contrast $\Sigma \neq \emptyset$, then the component must have been created by $\mathsf{jump}^a$, and hence there must be a $\langle \Sigma \Rightarrow \Pi \rangle^{\mathsf{f}} \in \Delta$. Moreover, there must be a formula $C$ with $\Sigma = C, \Sigma'$ such that $\langle]C \in \Gamma$ and $\langle \Sigma' \Rightarrow \Pi \rangle^{\mathsf{u}} \in \Delta$. Note that due to the shape of the rules we have $\Sigma' = \emptyset$. Then, since $\langle]B \in \Delta$, by saturation we also have $\langle \Rightarrow B \rangle^{\mathsf{u}} \in \Delta$, and together with the previous also $\langle \Sigma \Rightarrow B \rangle^{\mathsf{u}} \in \Delta$. Then by saturation we also have $[\Sigma, \Omega \overset{\Sigma}{\Rightarrow} B, \Theta] \in \Delta$ for some $\Omega, \Theta$. By the induction hypothesis, the formula $B$ is false at this world, and since the annotation is $\Sigma$, we have $\alpha \not\subseteq [\![B]\!]$. So in any case $\mathfrak{M}^S, w \not\Vdash \langle]B$.

Suppose that $A = []B$ and $A \in \Gamma$. We need to show that $\mathfrak{M}^S, w \Vdash []B$. If $\mathcal{N}(w) = \emptyset$ this is trivial. Assume that $\mathcal{N}(w) \neq \emptyset$ and take $\alpha \in \mathcal{N}(w)$. If $\alpha = \emptyset$, again the statement is trivial, so assume $\alpha \neq \emptyset$. By definition of $\mathcal{N}(w)$ this means that $\alpha \subseteq \mathcal{C}_w$. The children of $w$ are exactly the nested sequents $[\Omega \overset{\Sigma}{\Rightarrow} \Theta] \in \Delta$, and for these by saturation and $[]B \in \Gamma$ we have $B \in \Omega$. Thus by induction hypothesis we have $\mathcal{C}_w \subseteq [\![B]\!]$, and hence also $\alpha \subseteq [\![B]\!]$. Thus $\mathfrak{M}^S, w \Vdash []B$.

Finally, suppose that $A = []B$ and $A \in \Delta$. We need to show that $\mathfrak{M}^S, w \not\Vdash []B$. By saturation and $[]B \in \Delta$ we have that $[\Omega \overset{\Sigma}{\Rightarrow} B, \Theta] \in \Delta$ for some $\Sigma, \Omega, \Theta$. By induction hypothesis at this world $B$ is false, and since it is a member of $\mathcal{C}_w$ and $\mathcal{C}_w \in \mathcal{N}(w)$ we have that $\mathfrak{M}^S, w \not\Vdash []B$. □

Putting everything together we thus obtain:

**Theorem 23.** *Proof search on input $\overset{\emptyset}{\Rightarrow} A$ yields either a derivation or a saturated minimal nested sequent $\mathcal{S}$ with root $w$ such that $\mathfrak{M}^S, w \not\Vdash A$.* □

## 5 Extensions

A number of possible axiomatic extensions of biM have been considered in [1]. Here we highlight some of these, shown in Fig. 6 together with the corresponding semantic condition and the ordinary sequent rules beyond those of $\mathsf{G}_{\mathsf{biM}}$ obtained by converting the axioms into rules and closing the rule set under cuts as in [12].

| | | |
|---|---|---|
| $\mathsf{n}_{\langle]} : \neg\langle]\bot$ | $\emptyset \notin \mathcal{N}(w)$ | $\dfrac{\Gamma, C \Rightarrow}{[]\Gamma, \langle]C \Rightarrow}$ |
| $\mathsf{d}_{\langle]} : \neg(\langle]A \wedge []\neg A)$ | $\emptyset \notin \mathcal{N}(w)$ | $\dfrac{\Gamma, C \Rightarrow}{[]\Gamma, \langle]C \Rightarrow}$ |
| $\mathsf{d}_{[\rangle} : []A \rightarrow \langle]A$ | $\mathcal{N}(w) \neq \emptyset$ | $\dfrac{\Gamma \Rightarrow B}{[]\Gamma \Rightarrow \langle]B}$ |
| $\mathsf{d}_{[]} : \neg([]A \wedge []\neg A)$ | $\exists \alpha \in \mathcal{N}(w). \alpha \neq \emptyset$ | $\dfrac{\Gamma \Rightarrow}{[]\Gamma \Rightarrow} + \dfrac{\Gamma, C \Rightarrow}{[]\Gamma, \langle]C \Rightarrow}$ |

**Fig. 6.** Axiomatic extensions of the bimodal system from [1] with corresponding semantic conditions and direct translation into sequent rule.

$$\frac{\mathcal{S}\{\Gamma \Rightarrow \Delta, \langle \Rightarrow \rangle\}}{\mathcal{S}\{\Gamma \Rightarrow \Delta\}}\; P_{\langle]} \qquad \frac{\mathcal{S}\{\Gamma \Rightarrow \Delta, [\Sigma \Rightarrow \Pi]\}}{\mathcal{S}\{\Gamma \Rightarrow \Delta, \langle \Sigma \Rightarrow \Pi \rangle\}}\; N_{\langle]} \qquad \frac{\mathcal{S}\{\Gamma \Rightarrow \Delta, [\Rightarrow]\}}{\mathcal{S}\{\Gamma \Rightarrow \Delta\}}\; D_{[]}$$

$$\frac{\mathcal{S}\{\Gamma \Rightarrow \Delta, \langle \Rightarrow \rangle^{\mathsf{u}}\}}{\mathcal{S}\{\Gamma \Rightarrow \Delta\}}\; P^{a}_{\langle]} \quad \frac{\mathcal{S}\{\Gamma \Rightarrow \Delta, \langle \Sigma \Rightarrow \Pi \rangle^{\mathsf{u}}, \langle \Sigma \Rightarrow \Pi \rangle^{\mathsf{f}}\}}{\mathcal{S}\{\Gamma \Rightarrow \Delta, \langle \Sigma \Rightarrow \Pi \rangle^{\mathsf{u}}\}}\; N^{a}_{\langle]} \quad \frac{\mathcal{S}\{\Gamma \Rightarrow \Delta, [\overset{\emptyset}{\Rightarrow}]\}}{\mathcal{S}\{\Gamma \Rightarrow \Delta\}}\; D^{a}_{[]}$$

**Fig. 7.** The nested rules for the extensions in their plain and annotated versions.

Note that in the bimodal system the condition that $\emptyset \notin \mathcal{N}(w)$ is expressed by the two different axioms $\mathsf{n}_{\langle]}$ and $\mathsf{d}_{\langle]}$. These extensions are particularly interesting from the point of view of *deontic logic*, since they capture different readings of the "ought implies can" principle, where $\langle]A$ is read as "one ought to bring about $A$" and $[]A$ as "necessarily $A$". Note that the presence of two modalities permits a more fine-grained analysis of this principle than is possible in monomodal logics. The plain and annotated nested sequent rules are shown in Fig. 7, the kleene'd versions are as expected, copying the nesting of rule $N_{\langle]}$ into the premiss. The corresponding nested sequent calculi are given by $\mathcal{N}_{\mathsf{M}} + P_{\langle]}$ for both the axioms $\mathsf{n}_{\langle]}$ and $\mathsf{d}_{\langle]}$, by $\mathcal{N}_{\mathsf{M}} + N_{\langle]}$ for the axiom $\mathsf{d}_{[\rangle}$, and by $\mathcal{N}_{\mathsf{M}} + N_{\langle]} + D_{[]}$ for the axiom $\mathsf{d}_{[]}$. Note that we use structural versions of the rules instead of additional logical rules to enable smoother cut elimination proofs.

**Lemma 24.** *The plain rules are sound for the logics with the corresponding frame conditions under the interpretation $\iota$.*

*Proof.* For $P_{\langle]}$: Suppose that the interpretation of the conclusion is falsified in $\mathfrak{M}, w$, not due to the context. Then as in Theorem 5 there is a world $v \in W$ such that $\mathfrak{M}, v \not\Vdash \iota(\Gamma \Rightarrow \Delta)$. Since $\emptyset \notin \mathcal{N}(v)$ by assumption, we have $\mathfrak{M}, v \Vdash \neg\langle]\bot$. But since $\langle]\bot$ is equivalent to $\iota(\langle\Rightarrow\rangle)$, we have that $\mathfrak{M}, v$ falsifies $\iota(\Gamma \Rightarrow \Delta, \langle\Rightarrow\rangle)$. Hence $\mathfrak{M}, w$ falsifies the formula interpretation of the premiss.

For $N_{\langle]}$: Suppose that the interpretation of the conclusion is falsified in $\mathfrak{M}, w$, again not due to the context. Then there is a world $v$ falsifying $\iota(\Gamma \Rightarrow \Delta, \langle \Sigma \Rightarrow \Pi \rangle)$. Hence in particular we have $\mathfrak{M}, v \not\Vdash \langle](\bigwedge \Sigma \rightarrow \bigvee \Pi)$. This together with the assumption that $\mathcal{N}(v) \neq \emptyset$ yields that there is $\alpha \in \mathcal{N}(v)$ and a world $x \in \alpha$ with $\mathfrak{M}, x \Vdash \bigwedge \Sigma \wedge \neg \bigvee \Pi$. Hence $\mathfrak{M}, v \nVdash \langle](\bigwedge \Sigma \rightarrow \bigvee \Pi)$, and so the interpretation of the premiss is falsified in $\mathfrak{M}, w$ as well.

The proof for $D_{[\,]}$ is as for nested sequents for modal logic $\mathsf{KD}$. □

All the methods used to show completeness for the base calculus $\mathcal{N}_{\mathsf{M}}$ and its variants can be adapted to show completeness for the calculi for the extensions as well. First, it is straightforward to simulate the sequent rules of Fig. 6 in the plain versions of the corresponding nested calculi as in the proof of Theorem 6, giving:

**Theorem 25.** *The plain nested systems are complete.* □

Similarly, the cut elimination proof of Theorem 13 extends readily to the kleene'd versions of the calculi. The only non-trivial case is where the cut formula is contextual on the left in the conclusion of the rule $N_{\langle]}$. This is treated as the case for $\langle]^i_L$, giving:

**Theorem 26.** *The rules* $\mathsf{cut}_1, \mathsf{cut}_2$ *are admissible in the kleene'd systems.* □

Perhaps the most interesting extension is that for countermodel generation. For this we need to extend the saturation conditions of Definition 16 with the following, depending on whether the corresponding rule is in the system:

$(P_{\langle]})$ There is a $\langle \Sigma \Rightarrow \Pi \rangle^{\mathsf{u}} \in \Delta$

$(N_{\langle]})$ $\langle \Sigma \Rightarrow \Pi \rangle^{\mathsf{u}} \in \Delta$ implies there are $\Omega, \Theta$ with $\langle \Sigma, \Omega \Rightarrow \Pi, \Theta \rangle^{\mathsf{f}} \in \Delta$

$(D_{[\,]})$ $\Gamma \cup \Delta \neq \emptyset$ implies there are $\Sigma, \Omega, \Theta$ with $[\Omega \overset{\Sigma}{\Rightarrow} \Pi] \in \Delta$.

Note that the condition $(D_{[\,]})$ incorporates a loop check, preventing an infinite sequence of new components. Because of this, for the system with the rules $D_{[\,]}$ and $N_{\langle]}$ we need to slightly adapt the definition of the neighbourhood function in the generated model (Definition 17), so that $\mathcal{N}(w)$ is defined as $(\mathcal{L}_w \cup \{\mathcal{C}_w\}) \setminus \{\emptyset\}$ if there is a formula $\langle]A \in \Delta$, otherwise as $\mathcal{L}_w \cup \{\mathcal{C}_w\} \cup \{\emptyset\}$ if $\Gamma \cup \Delta = \emptyset$ and $\{\{w\}\}$ if $\Gamma \cup \Delta = \emptyset$. This ensures that when a component has no successor, the semantical condition is still met and hence the constructed model is indeed a model for the logic. Adapting the proofs for the base case accordingly, we then obtain the analogue of Theorem 23:

**Theorem 27.** *Proof search in the annotated systems produces either a derivation or a saturated minimal nested sequent yielding a countermodel.* □

## 6 Implementation

A prototype implementation of proof search and countermodel construction in the basic system $\mathcal{N}^a_{\mathsf{M}}$ is available under http://subsell.logic.at/bprover/nnProver/. The core of the program is written in SWI Prolog. It recursively performs the backwards proof search of Definition 19, at every step either returning a labelled tree representing a derivation, or a saturated nested sequent. The result is converted into a Latex file containing either the derivation or the countermodel, the latter in the form of a tikz picture. The webinterface automatically typesets this file to produce a pdf containing the derivation or countermodel.

# 7 Conclusion

In this article, we presented the calculus $\mathcal{N}_{\mathsf{M}}$, complete with a syntactic cut elimination result, countermodel construction, an implementation and some extensions. This seems to be the first sequent-style calculus for the logic biM aka Brown's ability logic. Its main interest, however, lies in the fact that it provides the key for properly treating monotone non-normal modal logics in the nested sequent framework in that the inclusion of the modality $[]$ enables a formula interpretation and facilitates direct countermodel construction. As such it should serve as a foundation both for obtaining nested sequent calculi for extensions of monotone modal logic, and for a more detailed proof-theoretic analysis of normal modal logics making use of a more fine-grained analysis of the successor states in terms of the neighbourhood function.

In line with this, it would be very interesting to extend $\mathcal{N}_{\mathsf{M}}$ to modularly capture other axioms for $\langle]$ and $[]$, in particular those of the normal modal cube [3] and the modal tesseract [14]. Further, we are planning to adapt the countermodel construction to the logics of [5] to provide certificates for the underivability statements used in the non-monotonic calculus considered there.

## References


1. Brown, M.A.: On the logic of ability. J. Philos. Log. **17**(1), 1–26 (1988)
2. Brown, M.A.: Action and ability. J. Philos. Log. **19**(1), 95–114 (1990)
3. Brünnler, K.: Deep sequent systems for modal logic. Arch. Math. Log. **48**, 551–577 (2009)
4. Chellas, B.F.: Modal Logic. Cambridge University Press, Cambridge (1980)
5. Ciabattoni, A., Gulisano, F., Lellmann, B.: Resolving conflicting obligations in Mīmāmsā: a sequent-based approach. In: Broersen, J., Condoravdi, C., Nair, S., Pigozzi, G. (eds.) DEON 2018 Proceedings, pp. 91–109. College Publications (2018)
6. Dalmonte, T., Olivetti, N., Negri, S.: Non-normal modal logics: bi-neighbourhood semantics and its labelled calculi. In: Bezanishvili, G., D'Agostino, G., Metcalfe, G., Studer, T. (eds.) Advances in Modal Logic 12, pp. 159–178. College Publications (2018)
7. Hansen, H.H.: Monotonic modal logics. University of Amsterdam, ILLC, Masters (2003)
8. Indrzejczak, A.: Sequent calculi for monotonic modal logics. Bull. Sect. Log. **34**, 151–164 (2005)
9. Indrzejczak, A.: Labelled tableau calculi for weak modal logics. Bull. Sect. Log. **36**(3/4), 159–171 (2007)
10. Kashima, R.: Cut-free sequent calculi for some tense logics. Studia Logica **53**(1), 119–135 (1994)
11. Lavendhomme, R., Lucas, T.: Sequent calculi and decision procedures for weak modal systems. Studia Logica **65**, 121–145 (2000)
12. Lellmann, B., Pattinson, D.: Constructing cut free sequent systems with context restrictions based on classical or intuitionistic logic. In: Lodaya, K. (ed.) ICLA 2013. LNCS, vol. 7750, pp. 148–160. Springer, Heidelberg (2013). https://doi.org/10.1007/978-3-642-36039-8_14

13. Lellmann, B., Pimentel, E.: Proof search in nested sequent calculi. In: Davis, M., Fehnker, A., McIver, A., Voronkov, A. (eds.) LPAR 2015. LNCS, vol. 9450, pp. 558–574. Springer, Heidelberg (2015). https://doi.org/10.1007/978-3-662-48899-7_39
14. Lellmann, B., Pimentel, E.: Modularisation of sequent calculi for normal and non-normal modalities. ACM Trans. Comput. Logic **20**(2), 7:1–7:46 (2019)
15. Marin, S., Straßburger, L.: Label-free modular systems for classical and intuitionistic modal logics. In: Goré, R., Kooi, B.P., Kurucz, A. (eds.) Advances in Modal Logic 10, pp. 387–406. College Publications (2014)
16. Negri, S.: Proof theory for non-normal modal logics. IfCoLog J. Log. Their Appl. **4**(4), 1241–1286 (2017)
17. Pacuit, E.: Neighbourhood Semantics for Modal Logic. Springer, Cham (2017). https://doi.org/10.1007/978-3-319-67149-9
18. Poggiolesi, F.: The method of tree-hypersequents for modal propositional logic. In: Makinson, D., Malinkowski, J., Wansing, H. (eds.) Towards Mathematical Philosophy. Trends In Logic, vol. 28, pp. 31–51. Springer, Dordrecht (2009). https://doi.org/10.1007/978-1-4020-9084-4_3
19. Troelstra, A.S., Schwichtenberg, H.: Basic Proof Theory. Cambridge Tracts In Theoretical Computer Science, vol. 43, 2nd edn. Cambridge University Press, Cambridge (2000)

# Semantics and Combinatorial Proofs

# On Combinatorial Proofs for Modal Logic

Matteo Acclavio[1(✉)] and Lutz Straßburger[2]

[1] Università Roma Tre, Rome, Italy
[2] Inria Saclay, Palaiseau, France
http://matteoacclavio.com/Math.html
http://www.lix.polytechnique.fr/Labo/Lutz.Strassburger/

**Abstract.** In this paper we extend Hughes' combinatorial proofs to modal logics. The crucial ingredient for modeling the modalities is the use of a self-dual non-commutative operator that has first been observed by Retoré through pomset logic. Consequently, we had to generalize the notion of skew fibration from cographs to Guglielmi's relation webs.

Our main result is a sound and complete system of combinatorial proofs for all normal and non-normal modal logics in the S4-tesseract. The proof of soundness and completeness is based on the sequent calculus with some added features from deep inference.

**Keywords:** Combinatorial proofs · Modal logic · S4-tesseract · Relation webs · Skew fibration

## 1 Introduction

During the last three decades, the proof theory of modal logics has seen enormous progress. We have now access to a systematic treatment of modal logics in display calculus [32], calculus of structures [14,26], labeled systems [22,25], hyper sequents [3,18], and nested sequents [5,20,28]. There are focused proof systems for classical and intuitionistic modal logics [7,8], and we understand the relation between display calculus and nested sequents [10] and hyper sequents [11].

The motivation for this paper is to take the natural next step in this advancement. After having developed various proof systems, using different formalisms, we are now asking the question: *When are two proofs the same?*

We are not claiming to provide a final answer to this question, but we propose an approach based on *combinatorial proofs*, introduced by Hughes [15,16] to address the question of proof identity for classical propositional logic and Hilbert's 24th problem [30,31]. Via combinatorial proofs, it is finally possible to ask the question of proof identity also for proofs in different proof formalisms; recent research has investigated this for syntactic proofs in sequent calculus [15,16], calculus of structures [29], resolution calculus, and analytic tableaux [1].

In classical propositional logic, a combinatorial proof is a *skew fibration* $f\colon \mathcal{G} \to \mathcal{F}$ from an *RB-cograph* $\mathcal{G}$, that can be seen as the "linear part" of the proof, to a *cograph* $\mathcal{F}$ that encodes the conclusion of the proof. The mapping


S. Cerrito and A. Popescu (Eds.): TABLEAUX 2019, LNAI 11714, pp. 223–240, 2019.
https://doi.org/10.1007/978-3-030-29026-9_13

$f$ precisely captures the information about what is duplicated and deleted in the proof. In terms of syntactic proof systems, this corresponds to the rules of *contraction* and *weakening*.

As an example we show below the combinatorial proof of Pierce's law. On the left we show the conclusion as formula, and on the right as cograph.

(1)

There, the regular (red) R-edges are the edges of the RB-cograph $\mathcal{G}$, and the bold (blue) B-edges represent the *linking*, corresponding to the instances of the axiom-rule in the sequent calculus. The vertical arrows (purple) represent the mapping $f$.

There is a close correspondence between cographs and formulas composed from atoms via two binary (commutative and associative) connectives, $\wedge$ (*and*) and $\vee$ (*or*): the vertices of the cograph are the atom occurrences in the formula, and there is an (undirected) edge between two atom occurrences if their first common ancestor in the formula tree is an $\wedge$, and there is no edge if it is an $\vee$.

For this reason, the cograph-approach works very well for classical propositional logic (CPL) [15,16,27,29] and for multiplicative linear logic (MLL) [24], but it is not obvious how to extend this notion to modalities, which can be seen as unary connectives.

We solve this problem by adding a third non-commutative (self-dual) operation $\lhd$ (*seq*), that has first been proposed by Retoré in *pomset logic* [23] and later been studied in the logic BV [12,13]. In the corresponding graph, we put a *directed* edge between two atoms if their first common ancestor in the formula tree is an $\lhd$. With this insight we can now represent a formula $\Box A$ (resp. $\Diamond A$) as graph by taking the graph of $A$, add a vertex labeled with $\Box$ (resp. $\Diamond$) and add a directed edge from that vertex to every vertex in $A$. This is illustrated in the example below, which is a proof in the modal logic $\mathsf{K}$, and which is a variation of the example in (1) above. As before, on the left the conclusion is written as formula, and on the right as graph.

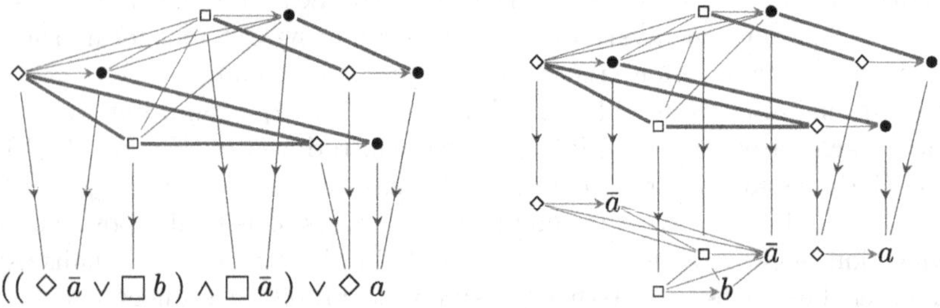


(2)

The upstairs graph is now no longer an RB-cograph but an *RGB-cograph* which additionally has directed (green) G-edges. The downstairs graph is a *relation web* which is a generalization of a cograph to more than two connectives.

The contributions of this paper can now be summarized as follows: we present a notion of combinatorial proof for the modal logics in the S4-plane (shown on the left in Fig. 1), and we show how sequent proofs are translated to combinatorial proofs, and that this translation is polynomial in the size of the proof. We then show that these results can be extended to the non-normal modal logics of the S4-tesseract [19] (shown on the right in Fig. 1).

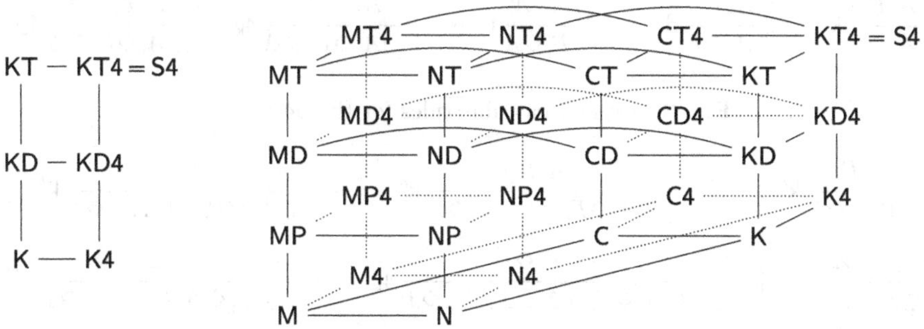


**Fig. 1. On the left:** the S4-plane and **on the right** the S4-tesseract

We begin by recalling in Sect. 2 the sequent calculus systems for the modal logics in the S4-plane. Then, in Sect. 3 we recall the notion of Guglielmi's relation webs [12]. Section 5 introduces skew fibrations on relation webs and shows that they correspond to contraction-weakening maps. In Sect. 4 we introduce the notion of RGB-cograph and show the relation to "linear" proofs in modal logics. The results of Sects. 4 and 5 are combined in Sect. 6 to define combinatorial proofs for the modal logics K and KD and show their soundness and completeness. We also show that they form a proof system in the sense of Cook and Reckhow [9]. Then, Sect. 7 shows how to treat modal logics that include the axioms T and/or 4, and finally, in Sect. 8, we show how our results can be extended to all logics in the S4-tesseract.

## 2 Sequent Calculus

We consider the class $\mathcal{K}$ of *modal formulas* (denoted by $A, B, \dots$) in negation normal form, generated by a countable set $\mathcal{A} = \{a, b, \dots\}$ of *propositional variables* and their duals $\bar{\mathcal{A}} = \{\bar{a}, \bar{b}, \dots\}$ by the following grammar:

$$A, B ::= a \mid \bar{a} \mid A \vee B \mid A \wedge B \mid \Box A \mid \Diamond A \mid \Box\bot \mid \Diamond\bot$$

where $\bot$ stands for the *empty formula*.[1] An *atom* is an element of $\mathcal{A} \cup \bar{\mathcal{A}} \cup \{\Box\bot, \Diamond\bot\}$. A sequent $\Gamma = A_1, \ldots, A_n$ is a non-empty multiset of formulas, written as list separated by comma.

$$
\frac{}{a, \bar{a}}\,\mathsf{ax} \qquad \frac{\Gamma, A, B}{\Gamma, A \vee B}\,\vee \qquad \frac{\Gamma, A \quad B, \Delta}{\Gamma, A \wedge B, \Delta}\,\wedge \qquad \frac{\Gamma}{\Gamma, A}\,\mathsf{w} \qquad \frac{\Gamma, A, A}{\Gamma, A}\,\mathsf{c}
$$

$$
\frac{A, \Gamma}{\Box A, \Diamond\Gamma}\,\mathsf{k} \qquad \frac{A, \Gamma}{\Diamond A, \Diamond\Gamma}\,\mathsf{d} \qquad \frac{\Gamma, A}{\Gamma, \Diamond A}\,\mathsf{t} \qquad \frac{A, \Gamma, \Diamond\Delta}{\Box A, \Diamond\Gamma, \Diamond\Delta}\,\mathsf{4_k} \qquad \frac{A, \Gamma, \Diamond\Delta}{\Diamond A, \Diamond\Gamma, \Diamond\Delta}\,\mathsf{4_{dk}}
$$

**Fig. 2.** Sequent calculus rules for the S4-plane

$$
\frac{\Gamma}{\Box\bot, \Diamond\Gamma}\,\mathsf{k}' \qquad \frac{A, \Gamma}{\Box A, \Diamond\Gamma, \Diamond\bot, \ldots, \Diamond\bot}\,\mathsf{k}'' \qquad \frac{\Gamma}{\Box\bot, \Diamond\Gamma, \Diamond\bot, \ldots, \Diamond\bot}\,\mathsf{k}'''
$$

$$
\frac{\Gamma, \Diamond\Delta}{\Box\bot, \Diamond\Gamma, \Diamond\Delta}\,\mathsf{4}'_\mathsf{k} \qquad \frac{A, \Gamma, \Diamond\Delta}{\Box A, \Diamond\Gamma, \Diamond\Delta, \Diamond\bot, \ldots, \Diamond\bot}\,\mathsf{4}''_\mathsf{k} \qquad \frac{\Gamma, \Diamond\Delta}{\Box\bot, \Diamond\Gamma, \Diamond\Delta, \Diamond\bot, \ldots, \Diamond\bot}\,\mathsf{4}'''_\mathsf{k}
$$

$$
\frac{\Gamma}{\Diamond\bot, \ldots, \Diamond\bot, \Diamond\Gamma}\,\mathsf{d}' \qquad \frac{\Gamma, \Diamond\Delta}{\Diamond\bot, \ldots, \Diamond\bot, \Diamond\Gamma, \Diamond\Delta}\,\mathsf{4}'_\mathsf{dk}
$$

**Fig. 3.** Extended modal rules incorporating weakening on $\bot$.

$$
\frac{\Gamma\{A\}}{\Gamma\{A \vee B\}}\,\mathsf{w}^\downarrow_{\vee 1} \qquad \frac{\Gamma\{A\}}{\Gamma\{B \vee A\}}\,\mathsf{w}^\downarrow_{\vee 2} \qquad \frac{\Gamma\{\Box\bot\}}{\Gamma\{\Box A\}}\,\mathsf{w}^\downarrow_\Box \qquad \frac{\Gamma\{\Diamond\bot\}}{\Gamma\{\Diamond A\}}\,\mathsf{w}^\downarrow_\Diamond
$$

$$
\frac{\Gamma\{A \vee A\}}{\Gamma\{A\}}\,\mathsf{c}^\downarrow \qquad \frac{\Gamma\{A\}}{\Gamma\{\Diamond A\}}\,\mathsf{t}^\downarrow \qquad \frac{\Gamma\{\Diamond\Diamond A\}}{\Gamma\{\Diamond A\}}\,\mathsf{4}^\downarrow
$$

**Fig. 4.** Deep inference rules for weakening, contraction, and the t- and 4-axioms

$$
\frac{\Gamma\{a \vee a\}}{\Gamma\{a\}}\,\mathsf{ac}^\downarrow \qquad \frac{\Gamma\{(A \wedge B) \vee (C \wedge D)\}}{\Gamma\{(A \vee C) \wedge (B \vee D)\}}\,\mathsf{m} \qquad \frac{\Gamma\{\Box A \vee \Box B\}}{\Gamma\{\Box(A \vee B)\}}\,\mathsf{m}^\downarrow_\Box \qquad \frac{\Gamma\{\Diamond A \vee \Diamond B\}}{\Gamma\{\Diamond(A \vee B)\}}\,\mathsf{m}^\downarrow_\Diamond
$$

**Fig. 5.** The atomic contraction rule (where $a$ stands for any atom) and the medial rules

$$
\frac{\Gamma\{A \wedge (B \wedge C)\}}{\Gamma\{(A \wedge B) \wedge C\}}\equiv \qquad \frac{\Gamma\{A \vee (B \vee C)\}}{\Gamma\{(A \vee B) \vee C\}}\equiv \qquad \frac{\Gamma\{B \wedge A\}}{\Gamma\{A \wedge B\}}\equiv \qquad \frac{\Gamma\{B \vee A\}}{\Gamma\{A \vee B\}}\equiv
$$

**Fig. 6.** Equality rules

[1] Note that $\bot$ is only allowed directly inside a $\Box$ or $\Diamond$. The main purpose of avoiding $\bot$ as proper formula is to avoid the empty relation web (to be introduced in the next section). However, we do need formulas $\Box\bot$ and $\Diamond\bot$ in order to allow weakenings inside a $\Box$ or $\Diamond$, which is needed to prove the decomposition theorem (stated in Theorem 2.2 below) which in turn is the basis for combinatorial proofs.

| X | K | KD | KT | K4 | KD4 | KT4 = S4 |
|---|---|---|---|---|---|---|
| $\mathsf{X}^{\mathsf{seq}}$ | $\{\mathsf{k}\}$ | $\{\mathsf{k}, \mathsf{d}\}$ | $\{\mathsf{k}, \mathsf{t}\}$ | $\{\mathsf{k}, 4_\mathsf{k}\}$ | $\{\mathsf{k}, \mathsf{d}, 4_\mathsf{k}, 4_\mathsf{dk}\}$ | $\{\mathsf{k}, \mathsf{t}, 4_\mathsf{k}\}$ |
| $\mathsf{X}^{\mathsf{LL}}$ | $\{\mathsf{k}^+\}$ | $\{\mathsf{k}^+, \mathsf{d}^+\}$ | $\{\mathsf{k}^+\}$ | $\{\mathsf{k}^+\}$ | $\{\mathsf{k}^+, \mathsf{d}^+\}$ | $\{\mathsf{k}^+\}$ |
| $\mathsf{X}^{\downarrow}$ | $\varnothing$ | $\varnothing$ | $\{\mathsf{t}^{\downarrow}\}$ | $\{4^{\downarrow}\}$ | $\{4^{\downarrow}\}$ | $\{\mathsf{t}^{\downarrow}, 4^{\downarrow}\}$ |

**Fig. 7.** Rule sets from logics

We begin our presentation with the six modal logics in the S4-plane shown on the left of Fig. 1. In Fig. 2 we show the inference rules for the sequent systems for these logics. We use $\Diamond\Gamma$ as abbreviation for $\Diamond B_1, \ldots, \Diamond B_n$ where $\Gamma = B_1, \ldots, B_n$. Then, Fig. 3 shows variations of the modal rules that are needed to obtain our decomposition theorem (Theorem 2.2 below) that will play a crucial role in the proof of soundness and completeness for combinatorial proofs. We write $\mathsf{k}^+$ (resp. $\mathsf{d}^+$, $4^+_\mathsf{k}$, $4^+_\mathsf{dk}$) for any instance in $\{\mathsf{k}, \mathsf{k}', \mathsf{k}'', \mathsf{k}'''\}$ (resp. $\{\mathsf{d}, \mathsf{d}'\}$, $\{4_\mathsf{k}, 4'_\mathsf{k}, 4''_\mathsf{k}, 4'''_\mathsf{k}\}$, $\{4_\mathsf{dk}, 4'_\mathsf{dk}\}$).

In this paper we also make use of some *deep inference* [6,12,13] rules that are shown in Fig. 4, where $\Gamma\{\ \}$ stands for a *context*, which is a sequent or a formula with a hole that takes the place of an atom. We write $\Gamma\{A\}$ when we replace the hole in $\Gamma\{\ \}$ by the formula $A$. We write $\mathsf{w}^{\downarrow}$ for the set $\{\mathsf{w}, \mathsf{w}^{\downarrow}_{\vee 1}, \mathsf{w}^{\downarrow}_{\vee 2}, \mathsf{w}^{\downarrow}_{\Box}, \mathsf{w}^{\downarrow}_{\Diamond}\}$.

For each X among the six logics K, KD, KT, K4, KD4, and KT4, we define three sets $\mathsf{X}^{\mathsf{seq}}$, $\mathsf{X}^{\mathsf{LL}}$, and $\mathsf{X}^{\downarrow}$ of inference rules as shown in Fig. 7.

We now define the following sequent systems: $\mathsf{MLL} = \{\mathsf{ax}, \vee, \wedge\}$ and $\mathsf{LK} = \mathsf{MLL} \cup \{\mathsf{w}, \mathsf{c}\}$; if X is one of the six logics in the S4-plane, then $\mathsf{MLL\text{-}X} = \mathsf{MLL} \cup \mathsf{X}^{\mathsf{LL}}$ and $\mathsf{LK\text{-}X} = \mathsf{LK} \cup \mathsf{X}^{\mathsf{seq}}$. The following theorem is well-known [33].

**Theorem 2.1.** *If* $\mathsf{X} \in \{\mathsf{K}, \mathsf{KD}, \mathsf{KT}, \mathsf{K4}, \mathsf{KD4}, \mathsf{KT4}\}$ *then* LK-X *is a sound and complete sequent system for the modal logic* X.

If $\Gamma$ is a sequent and S a sequent system, we write $\overset{\mathsf{S}}{\vdash} \Gamma$ if there is a derivation of $\Gamma$ in S. If S is s set of inference rules that all have exactly one premise, we can write $\Gamma' \overset{\mathsf{S}}{\vdash} \Gamma$ if there is a derivation from $\Gamma'$ to $\Gamma$ using only rules from S.

**Theorem 2.2.** *Let* $\mathsf{X} \in \{\mathsf{K}, \mathsf{KD}, \mathsf{KT}, \mathsf{K4}, \mathsf{KD4}, \mathsf{KT4}\}$ *and* $\Gamma$ *be a sequent. Then*

$$\overset{\mathsf{LK\text{-}X}}{\vdash} \Gamma \quad \Longleftrightarrow \quad \overset{\mathsf{MLL\text{-}X}}{\vdash} \Gamma' \overset{\mathsf{X}^{\downarrow}}{\vdash} \Gamma'' \overset{\mathsf{w}^{\downarrow}, \mathsf{c}^{\downarrow}}{\vdash} \Gamma \quad \textit{for some } \Gamma' \textit{ and } \Gamma''.$$

*Proof.* This is proved by a straightforward permutation argument. First, all instances of w (resp. c) are replaced by instances of $\mathsf{w}^{\downarrow}$ (resp. $\vee$ and $\mathsf{c}^{\downarrow}$), and then all $\mathsf{w}^{\downarrow}$- and $\mathsf{c}^{\downarrow}$-instances can be permuted down in the proof. Observe that this step introduces the rules shown in Fig. 3. Then, all t instances are also $\mathsf{t}^{\downarrow}$-instances, and all instances of $4^+_\mathsf{k}$ (resp. $4^+_\mathsf{dk}$) are replaced by instances of $\mathsf{k}^+$ (resp. $\mathsf{d}^+$) and $4^{\downarrow}$-instances. Then all $\mathsf{t}^{\downarrow}$- and $4^{\downarrow}$-instances can be permuted down. Conversely, we can first permute the instances of $\mathsf{t}^{\downarrow}$ and $4^{\downarrow}$ up and then the instances of $\mathsf{w}^{\downarrow}$ and $\mathsf{c}^{\downarrow}$ until they are not deep anymore. □

There are two reasons to use a deep contraction rule. The first is the decomposition theorem proved above, and the second is that deep contraction can be reduced to atomic form (shown on the left in Fig. 5) via the so-called (deep) *medial rules* (shown on the right in Fig. 5). We write $\mathsf{m}^{\downarrow}$ for the set $\{\mathsf{m}, \mathsf{m}^{\downarrow}_{\Box}, \mathsf{m}^{\downarrow}_{\Diamond}\}$. Additionally we make use of the *equivalence rules* shown in Fig. 6.

**Theorem 2.3.** *Let $\Gamma'$ and $\Gamma$ be sequents. Then*

$$\Gamma' \overset{\mathsf{c}^{\downarrow},\mathsf{w}^{\downarrow},\equiv}{\vdash\!\!\!-\!\!\!-} \Gamma \quad\iff\quad \Gamma' \overset{\mathsf{m}^{\downarrow},\equiv}{\vdash\!\!\!-\!\!\!-} \Delta' \overset{\mathsf{ac}^{\downarrow},\equiv}{\vdash\!\!\!-\!\!\!-} \Delta \overset{\mathsf{w}^{\downarrow},\equiv}{\vdash\!\!\!-\!\!\!-} \Gamma \ \textit{ for some sequents } \Delta, \Delta' .$$

*Proof.* For the case without modalities, this is a standard result in the calculus of structures, first proved in [6] (see also [27]). In the presence of the modalities, the proof is similar: For $\Rightarrow$ direction, we first reduce $\mathsf{c}^{\downarrow}$ to $\mathsf{ac}^{\downarrow}$ using the medial and equivalence rules, proceeding by induction on the contraction formula, as shown in Fig. 8. Note that a contraction on $\Box\bot$ (resp. $\Diamond\bot$) is already atomic.

In the next step we permute the $\mathsf{w}^{\downarrow}$ down, and finally we permute all instances of $\mathsf{ac}^{\downarrow}$ down. For the $\Leftarrow$ direction, observe that $\mathsf{ac}^{\downarrow}$ is already a special case of $\mathsf{c}^{\downarrow}$ and that all rules in $\{\mathsf{m}, \mathsf{m}^{\downarrow}_{\Box}, \mathsf{m}^{\downarrow}_{\Diamond}\}$ are derivable using $\mathsf{c}^{\downarrow}$ and $\mathsf{w}^{\downarrow}$. □

$$\frac{(A\vee B)\vee(A\vee B)}{A\vee B}\,\mathsf{c} \;\rightsquigarrow\; \cfrac{\cfrac{(A\vee B)\vee(A\vee B)}{\{A\vee A\}\vee\{B\vee B\}}\,\equiv}{A\vee B}\,\mathsf{c}^{\downarrow} \qquad \frac{\Box A\vee\Box A}{\Box A}\,\mathsf{c} \;\rightsquigarrow\; \cfrac{\cfrac{\Box A\vee\Box A}{\Box\{A\vee A\}}\,\mathsf{m}^{\downarrow}_{\Box}}{\Box A}\,\mathsf{c}^{\downarrow}$$

$$\frac{(A\wedge B)\vee(A\wedge B)}{A\wedge B}\,\mathsf{c} \;\rightsquigarrow\; \cfrac{\cfrac{(A\wedge B)\vee(A\wedge B)}{\{A\vee A\}\wedge\{B\vee B\}}\,\mathsf{m}}{A\vee B}\,\mathsf{c}^{\downarrow} \qquad \frac{\Diamond A\vee\Diamond A}{\Diamond A}\,\mathsf{c} \;\rightsquigarrow\; \cfrac{\cfrac{\Diamond A\vee\Diamond A}{\Diamond\{A\vee A\}}\,\mathsf{m}^{\downarrow}_{\Box}}{\Diamond A}\,\mathsf{c}^{\downarrow}$$

**Fig. 8.** Reducing contraction to atomic contraction via medial rules.

## 3 Relation Webs

A *directed graph* $\mathcal{G} = \langle V_{\mathcal{G}}, \overset{\mathcal{G}}{\rightharpoonup}\rangle$ is a set $V_{\mathcal{G}}$ of *vertices* equipped with a binary *edge relation* $\overset{\mathcal{G}}{\rightharpoonup} \subseteq V_{\mathcal{G}} \times V_{\mathcal{G}}$. We speak of an *undirected graph* $\mathcal{G} = \langle V_{\mathcal{G}}, \overset{\mathcal{G}}{\frown}\rangle$ if the *edge relation* $\overset{\mathcal{G}}{\frown} \subseteq V_{\mathcal{G}} \times V_{\mathcal{G}}$ is irreflexive and symmetric. A *mixed graph* is a triple $\mathcal{G} = \langle V_{\mathcal{G}}, \overset{\mathcal{G}}{\frown}, \overset{\mathcal{G}}{\rightharpoonup}\rangle$ where $\langle V_{\mathcal{G}}, \overset{\mathcal{G}}{\frown}\rangle$ is an undirected graph and $\langle V_{\mathcal{G}}, \overset{\mathcal{G}}{\rightharpoonup}\rangle$ is a directed graph, such that $\overset{\mathcal{G}}{\frown} \cap \overset{\mathcal{G}}{\rightharpoonup} = \emptyset$ and $\overset{\mathcal{G}}{\rightharpoonup}$ is irreflexive. From now on, we omit the index/superscript $\mathcal{G}$ when it is clear from the context. For two distinct vertices $v$ and $w$ in a mixed graph we use the following abbreviations:

$$\begin{aligned} v \leftharpoonup w &\iff w \rightharpoonup v \\ v \leftrightharpoons w &\iff v \rightharpoonup w \text{ or } v \leftharpoonup w \text{ or } v \frown w \\ v \smile w &\iff v \not\rightharpoonup w \text{ and } v \not\leftharpoonup w \text{ and } v \not\frown w \end{aligned} \tag{3}$$

Note that for any two vertices we have that $v \not\leftrightharpoons w$ iff $v \smile w$ or $v = w$. Furthermore, in a mixed graph, for any two vertices $v$ and $w$, exactly one of the following five statements is true:

$$v = w \quad \text{or} \quad v \smile w \quad \text{or} \quad v \frown w \quad \text{or} \quad v \rightharpoonup w \quad \text{or} \quad v \leftharpoonup w$$

When drawing a graph we use $v$—$w$ for $v \frown w$, and $v \rightarrow w$ for $v \curvearrowright w$, and for $v \smile w$ we either use $v \cdots w$ or draw no edge at all.

A *series-parallel order* is a directed graph $\mathcal{G} = \langle V_{\mathcal{G}}, \overset{\mathcal{G}}{\curvearrowright} \rangle$ where $\overset{\mathcal{G}}{\curvearrowright}$ is transitive, irreflexive, and Z-free, i.e., $\mathcal{G}$ does not contain an induced subgraph of the shape shown on the left below:

Forbidden configurations for Z-freeness:  (4)



A *cograph* is an undirected graph that is Z-free, i.e., it does not contain an induced subgraph of the shape shown on the right above.

**Definition 3.1.** A *relation web* is a mixed graph $\mathcal{G} = \langle V_{\mathcal{G}}, \overset{\mathcal{G}}{\frown}, \overset{\mathcal{G}}{\curvearrowright} \rangle$ where $\langle V_{\mathcal{G}}, \overset{\mathcal{G}}{\frown} \rangle$ is a cograph and $\langle V_{\mathcal{G}}, \overset{\mathcal{G}}{\curvearrowright} \rangle$ is a series-parallel order, and the following two configurations do not occur:

Forbidden configurations for relation webs:  (5)



**Observation 3.2.** It is easy to see that in a relation web, the undirected graph determined by the relation $\smile$ (which is symmetric and irreflexive) is also a cograph.

Let $\mathcal{G}$ and $\mathcal{H}$ be two disjoint mixed graphs. We define the following operations:

$$
\begin{aligned}
\mathcal{G} \parr \mathcal{H} &= \langle V_{\mathcal{G}} \cup V_{\mathcal{H}}\,,\ \overset{\mathcal{G}}{\frown} \cup \overset{\mathcal{H}}{\frown}\,,\ \overset{\mathcal{G}}{\curvearrowright} \cup \overset{\mathcal{H}}{\curvearrowright} \rangle \\
\mathcal{G} \lhd \mathcal{H} &= \langle V_{\mathcal{G}} \cup V_{\mathcal{H}}\,,\ \overset{\mathcal{G}}{\frown} \cup \overset{\mathcal{H}}{\frown}\,,\ \overset{\mathcal{G}}{\curvearrowright} \cup \overset{\mathcal{H}}{\curvearrowright} \cup \{(u,v) \mid u \in V_{\mathcal{G}}, v \in V_{\mathcal{H}}\} \rangle \\
\mathcal{G} \otimes \mathcal{H} &= \langle V_{\mathcal{G}} \cup V_{\mathcal{H}}\,,\ \overset{\mathcal{G}}{\frown} \cup \overset{\mathcal{H}}{\frown} \cup \{(u,v),(v,u) \mid u \in V_{\mathcal{G}}, v \in V_{\mathcal{H}}\}\,,\ \overset{\mathcal{G}}{\curvearrowright} \cup \overset{\mathcal{H}}{\curvearrowright} \rangle
\end{aligned}
\tag{6}
$$

which can be visualized as follows:

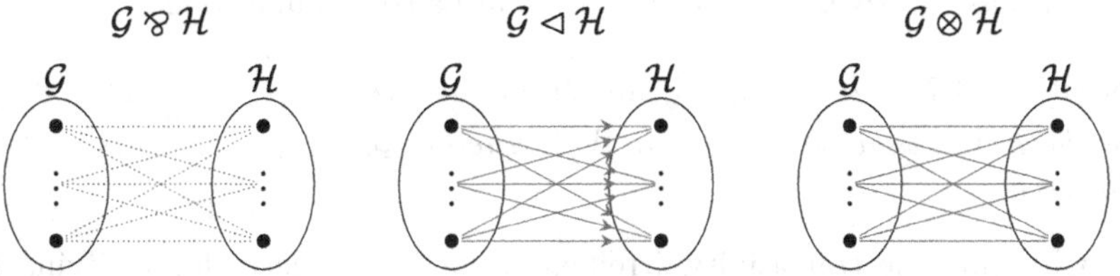



**Theorem 3.3.** *A mixed graph is a relation web if and only if it can be constructed from single vertices using the three operations defined in* (6) *above.*

*Proof.* This follows from the corresponding results on cographs and series-parallel orders, e.g. [21]. A direct proof can be found in [12]. □

A relation web is *labeled* if all its vertices carry a label selected from a label set $\mathcal{L}$. We write $l(v)$ for the label of $v$. We are now defining for a formula $F$ the labeled relation web $[\![F]\!]$ where the label set $\mathcal{L} = \mathcal{A} \cup \bar{\mathcal{A}} \cup \{\Box, \Diamond\}$. We write $\varnothing$ for the empty graph and we use the notations $\bullet_a$, $\bullet_{\bar{a}}$, $\diamond$, $\square$ for the graph consisting of a single vertex that is labeled with $a$, $\bar{a}$, $\Diamond$, $\Box$, respectively.

$$
\begin{array}{llllllll}
[\![a]\!] = \bullet_a & [\![A \wedge B]\!] = [\![A]\!] \otimes [\![B]\!] & [\![\Box A]\!] = \Box \lhd [\![A]\!] & [\![\Box\bot]\!] = \Box \\
[\![\bar a]\!] = \bullet_{\bar a} & [\![A \vee B]\!] = [\![A]\!] \parr [\![B]\!] & [\![\Diamond A]\!] = \Diamond \lhd [\![A]\!] & [\![\Diamond\bot]\!] = \Diamond
\end{array}
\tag{7}
$$

For a sequent $\Gamma = A_1, \ldots, A_n$ we define $[\![\Gamma]\!] = [\![A_1, \ldots, A_n]\!] = [\![A_1]\!] \parr \cdots \parr [\![A_n]\!]$.

**Definition 3.4.** A relation web $\mathcal{G}$ is *modalic* if for any vertices $u$, $v$, $w$ with $u \curvearrowright w$ and $v \curvearrowright w$ we have $u \curvearrowright v$ or $v \curvearrowright u$ or $u = v$, i.e., $\mathcal{G}$ does not contain the two configurations below.

Forbidden configurations for modalic relation webs: 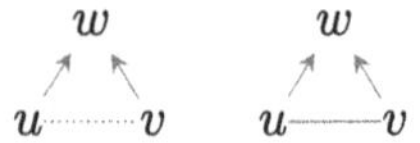 (8)

A labeled modalic relation web $\mathcal{G}$ is *properly labeled* if its label set is $\mathcal{L} = \mathcal{A} \cup \bar{\mathcal{A}} \cup \{\Box, \Diamond\}$, such that whenever there are $v, w$ with $v \curvearrowright w$ then $l(v) \in \{\Box, \Diamond\}$.

**Theorem 3.5.** *A relation web is the translation of a modal formula if and only if it is modalic and properly labeled.*

*Proof.* If $\mathcal{G} = [\![F]\!]$ for some formula $F$, then the only vertices in $\mathcal{G}$ with outgoing $\curvearrowright$-edge are the ones created in the encoding of a modal subformula and labeled with $\Box$ or $\Diamond$. If we have two distinct such vertices $u$ and $v$ with an $\curvearrowright$-edge to some vertex $w$, then one of the corresponding modal operators is in the scope of the other and we have $u \curvearrowright v$ or $v \curvearrowright u$. The converse follows from Theorem 3.3 and the fact that the operation $\lhd$ in (7) is associative. In fact, if $l(v) = \Box$ (resp. $l(v) = \Diamond$) and there is no $w$ such that $v \curvearrowright w$ then we interpret the vertex $v$ as the subformula $\Box\bot$ (resp. $\Diamond\bot$). $\Box$

**Proposition 3.6.** *For two formulas $F$ and $F'$, we have $[\![F]\!] = [\![F']\!]$ iff $F$ and $F'$ are equivalent modulo associativity and commutativity of $\wedge$ and $\vee$.*

*Proof.* By a straightforward induction, observing that the operations $\parr$, $\otimes$ and $\lhd$ in (7) are associative, and that $\parr$ and $\otimes$ are also commutative. $\Box$

**Proposition 3.7.** *Given a set $V_\mathcal{G}$ and two binary relations $\overset{\mathcal{G}}{\frown}, \overset{\mathcal{G}}{\curvearrowright} \subseteq V_\mathcal{G} \times V_\mathcal{G}$, it can be checked in time polynomial in $|V_\mathcal{G}|$, whether $\mathcal{G} = \langle V_\mathcal{G}, \overset{\mathcal{G}}{\frown}, \overset{\mathcal{G}}{\curvearrowright} \rangle$ is a modalic relation web.*

*Proof.* Checking the transitivity, irreflexivity, and symmetry for verifying that $\mathcal{G}$ is a mixed graph is trivially polynomial. Then, for checking the absence of the forbidden configurations in (4), (5), and (8) we can loop over all triples and quadruples of vertices, which is $O(|V_\mathcal{G}|^4)$. $\Box$

## 4 RGB-Cographs and Linear Proofs for K and KD

In this section we investigate when a modalic relation web does represent a proof. For this, we equip a relation web with a *linking* which is an equivalence class on its vertices. In the special case where each such equivalence class contains exactly two elements, and there are no $\curvearrowright$-edges, we have Retoré's RB-cographs [24] that with an additional correctness criterion correspond to proofs in MLL.

Here we generalize the notion of RB-cographs to the one of RGB-cographs and we give a correspondence with *linear derivations* in MLL-K and MLL-KD.[2]

$$
\frac{}{\bullet\!\!-\!\!\bullet}\ \mathsf{ax}
\qquad
\frac{\langle \mathcal{G}', \mathcal{A}, \mathcal{B} \mid \overset{\mathcal{G}}{\curlyvee} \rangle}{\langle \mathcal{G}', \mathcal{A} \parr \mathcal{B} \mid \overset{\mathcal{G}}{\curlyvee} \rangle}\ \vee
\qquad
\frac{\langle \mathcal{G}', \mathcal{A} \mid \overset{\mathcal{G}}{\curlyvee} \rangle \quad \langle \mathcal{B}, \mathcal{H}' \mid \overset{\mathcal{H}}{\curlyvee} \rangle}{\langle \mathcal{G}', \mathcal{A} \otimes \mathcal{B}, \mathcal{H}' \mid \overset{\mathcal{G}}{\curlyvee} \cup \overset{\mathcal{H}}{\curlyvee} \rangle}\ \wedge
$$

$$
\frac{\langle \mathcal{G}_{i_1}, \mathcal{G}_{i_2}, \ldots, \mathcal{G}_{i_m} \mid \overset{\mathcal{G}}{\curlyvee} \rangle}{\langle \Box \lhd \mathcal{G}_1, \Diamond \lhd \mathcal{G}_2, \ldots, \Diamond \lhd \mathcal{G}_n \mid \overset{\mathcal{G}}{\curlyvee} \cup \overset{*}{\curlyvee} \rangle}\ \mathsf{k}^+
\qquad
\frac{\langle \mathcal{G}_{i_1}, \ldots, \mathcal{G}_{i_m} \mid \overset{\mathcal{G}}{\curlyvee} \rangle}{\langle \Diamond \lhd \mathcal{G}_1, \ldots, \Diamond \lhd \mathcal{G}_n \mid \overset{\mathcal{G}}{\curlyvee} \cup \overset{*}{\curlyvee} \rangle}\ \mathsf{d}^+
$$

$$\overset{*}{\curlyvee} = \{(v, w) \mid v, w \in V^{\Box} \uplus V^{\Diamond} \text{ and } v, w \notin V_{\mathcal{G}_1} \cup \cdots \cup V_{\mathcal{G}_n}\}$$
$$i_1, \ldots, i_m \in \{1, \ldots, n\} \text{ and pairwise distinct}$$
$$\text{if } j \in \{1, \ldots, n\} \backslash \{i_1, \ldots, i_m\} \text{ then } \mathcal{G}_j = \varnothing$$

**Fig. 9.** Translating MLL-K and MLL-KD sequent proofs into RGB-cographs

**Definition 4.1.** An *RGB-cograph* is a tuple $\mathcal{G} = \langle V_{\mathcal{G}}, \overset{\mathcal{G}}{\frown}, \overset{\mathcal{G}}{\curvearrowright}, \overset{\mathcal{G}}{\curlyvee} \rangle$, where $\langle V_{\mathcal{G}}, \overset{\mathcal{G}}{\frown}, \overset{\mathcal{G}}{\curvearrowright} \rangle$ is a modalic relation web, $V_{\mathcal{G}}$ is the disjoint union of three sets $V^{\bullet}_{\mathcal{G}} \uplus V^{\Box}_{\mathcal{G}} \uplus V^{\Diamond}_{\mathcal{G}}$, and $\overset{\mathcal{G}}{\curlyvee}$ is an equivalence relation, called the *linking*, such that

- if $v \in V^{\bullet}_{\mathcal{G}}$ then for all $w \in V_{\mathcal{G}}$ we have $v \not\curvearrowright w$;
- if $v \curlyvee w$ then either $v, w \in V^{\bullet}$ or $v, w \in V^{\Box}_{\mathcal{G}} \uplus V^{\Diamond}_{\mathcal{G}}$;;
- if $v \in V^{\bullet}_{\mathcal{G}}$ then there is exactly one $w \in V^{\bullet}_{\mathcal{G}}$ with $v \curlyvee w$ and $v \neq w$.

An equivalence class of $\overset{\mathcal{G}}{\curlyvee}$ is called a *link*. The vertices in $V^{\bullet}_{\mathcal{G}}$ are called *atomic* vertices, and the vertices in $V^{\Box}_{\mathcal{G}} \uplus V^{\Diamond}_{\mathcal{G}}$ are called *modalic* vertices. An *RB-cograph* is an RGB-cograph $\mathcal{G}$ with $V^{\Box}_{\mathcal{G}} \uplus V^{\Diamond}_{\mathcal{G}} = \varnothing$.

The first condition in this definition says that if a vertex has an outgoing $\curvearrowright$-edge then it has to be in $V^{\Box}_{\mathcal{G}} \uplus V^{\Diamond}_{\mathcal{G}}$, the second condition says that vertices from $V^{\bullet}_{\mathcal{G}}$ and $V^{\Box}_{\mathcal{G}} \uplus V^{\Diamond}_{\mathcal{G}}$ cannot be linked, and the third condition says that each link on $V^{\bullet}_{\mathcal{G}}$ has exactly two elements. In an RB-cograph [24] only the last condition makes sense since $\curvearrowright$ is empty. When drawing an RGB-cograph we use bold (blue) edges $v$—$w$ when $v \neq w$ and $v \curlyvee w$.

Figure 9 shows how proofs in MLL-K and MLL-KD are translated into RGB-cographs. There, the notation $\langle \mathcal{G}_1, \mathcal{G}_2, \ldots, \mathcal{G}_n \mid \overset{\mathcal{G}}{\curlyvee} \rangle$ is used to denote the RGB-cograph whose underlying relation web is $\mathcal{G} = \mathcal{G}_1 \parr \mathcal{G}_2 \parr \cdots \parr \mathcal{G}_n$ and whose linking is $\overset{\mathcal{G}}{\curlyvee}$. The ax-rule simply produces a graph with two vertices that are linked, and $\frown$ and $\curvearrowright$ being empty. In the $\vee$-rule, premise and conclusion are the same RGB-cograph. In the $\wedge$-rule, the linking in the conclusion is the union of the linkings in the premises. These three rules behave exactly the same as in proof

[2] The logics defined by these systems can be seen as the "linear logic variants" of the standard modal logics K and KD.

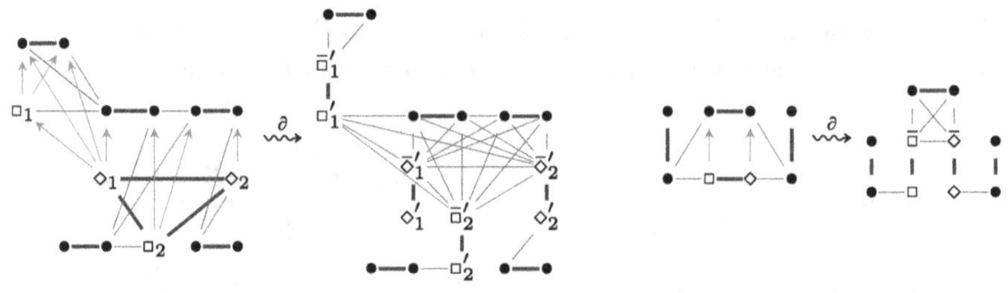

**Fig. 10.** The RGB-cographs for $F_1 = \bar{d} \vee (d \wedge \Box(\bar{b} \wedge c) \vee \bar{e} \vee (e \wedge \Diamond\bar{c}) \vee \Diamond(b \wedge \Box(a \vee \bar{a})))$ and $F_2 = b \vee (\bar{b} \wedge \Box a) \vee (\Diamond\bar{a} \wedge c) \vee \bar{c}$, and the corresponding RB-cographs $\partial(F_1)$ and $\partial(F_2)$.

nets for MLL. More interesting are the rules $\mathsf{k}^+$ and $\mathsf{d}^+$, where the linking of the conclusion is the linking of the premise together with an additional equivalence class containing all the new $\Box$- and $\Diamond$-vertices. The purpose of this section is to give a combinatorial characterization of the RGB-cographs that can be obtained via this sequent calculus translation.

**Definition 4.2.** An *alternating elementary path (æ-path)* of *length* $n$ in an RGB-cograph is a sequence of pairwise distinct vertices $x_0, x_1, \ldots, x_n$ such that we have either $x_0 \curlyvee x_1 R_1 x_2 \curlyvee x_3 R_3 x_4 \cdots x_n$ or $x_0 R_0 x_1 \curlyvee x_2 R_2 x_3 \curlyvee x_4 \cdots x_n$ with $R_i \in \{\curvearrowright, \frown\}$, i.e., an æ-path is an elementary path whose edges are alternating in $\curlyvee$ and $\curvearrowright \cup \frown$. A *chord* in an æ-path is an edge $x_i \frown x_j$ or $x_i \curvearrowright x_j$ for $i, j \in \{0, \ldots, n\}$ and $i + 2 \leq j$. A *chordless æ-path* is an æ-path without chord. An *æ-cycle* is an æ-path of even length such that $x_0 = x_n$. An RGB-cograph $\mathcal{G}$ is *æ-connected* if any two vertices are connected by a chordless æ-path, and $\mathcal{G}$ is *æ-acyclic* if it contains no chordless æ-cycle.

**Definition 4.3.** An RGB-cograph $\mathcal{G}$ is $\mathsf{K}$-*correct* (or $\{\mathsf{k}^+\}$-*correct*) if

1. $\mathcal{G}$ is æ-connected and æ-acyclic;
2. $V_{\mathcal{G}} \neq \emptyset$ and every $\curlyvee$-equivalence class in $V^{\Box} \uplus V^{\Diamond}$ contains at least one vertex $v$ such that there is a vertex $w \in V^{\bullet}$ with $v \curvearrowright w$;
3. if $w \overset{\mathcal{G}}{\curvearrowright} v$ and $v \curlyvee v'$, then there is $w' \curlyvee w$ such that $w' \overset{\mathcal{G}}{\curvearrowright} v'$; and
4. every $\curlyvee$-equivalence class in $V^{\Box} \uplus V^{\Diamond}$ contains exactly one vertex $v \in V^{\Box}$.

We say that $\mathcal{G}$ is $\mathsf{KD}$-*correct* (or $\{\mathsf{k}^+, \mathsf{d}^+\}$-*correct*) if Condition 4 is replaced by:

4. every $\curlyvee$-equivalence class in $V^{\Box} \uplus V^{\Diamond}$ contains at most one vertex $v \in V^{\Box}$.

**Theorem 4.4.** *Let* $\mathsf{X} \in \{\mathsf{K}, \mathsf{KD}\}$. *An RGB-cograph* $\mathcal{G}$ *is the translation of an* $\mathsf{MLL}$-$\mathsf{X}$ *sequent proof iff it is* $\mathsf{X}^{\mathsf{LL}}$-*correct.*

*Proof (Sketch).* For the left-to-right direction, observe that all rules in Fig. 9 preserve correctness. For the right-to-left direction, we will reuse the MLL sequentialization result for RB-cographs [24]. For this we will define for an RGB-cograph $\mathcal{G}$ an RB-cograph $\partial(\mathcal{G})$ that is æ-connected and æ-acyclic if and only if $\mathcal{G}$ is.

We define a vertex set $V^* = \{v', \bar{v}' \mid v \in V_\mathcal{G}^\square \uplus V_\mathcal{G}^\diamond\}$ and let $V_{\partial(\mathcal{G})} = V_\mathcal{G}^\bullet \uplus V^*$, i.e., we take the atomic vertices of $\mathcal{G}$, and each modalic vertex is replaced by a dual pair of atomic vertices, that are linked by $\curlyvee$ (See Fig. 10). Then we use Theorem 3.3 and Proposition 3.6 so that we can write $\mathcal{G}$ and $\partial(\mathcal{G})$ as BV-*formulas* [12]. Let $v_1, \ldots, v_n \in V_\mathcal{G}^\square \uplus V_\mathcal{G}^\diamond$ form an $\overset{\mathcal{G}}{\curlyvee}$-equivalence class. Then the formula for $\mathcal{G}$ is of shape $F\{v_1 \lhd B_1\} \cdots \{v_n \lhd B_n\}$ for some $n$-ary context $F\{\ \} \cdots \{\ \}$ (because $\mathcal{G}$ is modalic). We transform $F\{v_1 \lhd B_1\} \cdots \{v_n \lhd B_n\}$ into $(\bar{v}'_1 \otimes \cdots \otimes \bar{v}'_n \otimes \partial(B_1 \parr \cdots \parr B_n)) \parr \partial(F\{v'_1\} \cdots \{v'_n\})$ and proceed inductively for all $\overset{\mathcal{G}}{\curlyvee}$-equivalence classes. From Retoré's proof [24] we get an MLL sequentialization for $\partial(\mathcal{G})$, which we then transform back into an MLL-K or MLL-KD sequent proof for $\mathcal{G}$. □

## 5 Skew Fibrations

**Definition 5.1.** Let $\mathcal{G}$ and $\mathcal{H}$ be mixed graphs. A *skew fibration* $f \colon \mathcal{G} \to \mathcal{H}$ is a function from $V_\mathcal{G}$ to $V_\mathcal{H}$ that preserves $\frown$ and $\rightharpoonup$, i.e.,

$$v \overset{\mathcal{G}}{\frown} w \implies f(v) \overset{\mathcal{H}}{\frown} f(w) \qquad \text{and} \qquad v \overset{\mathcal{G}}{\rightharpoonup} w \implies f(v) \overset{\mathcal{H}}{\rightharpoonup} f(w) \quad , \tag{9}$$

and has the *skew lifting* property, i.e.,

$$\begin{array}{l} \text{for every } v \in V_\mathcal{G} \text{ and } w \in V_\mathcal{H} \text{ and } R \in \{\frown, \rightharpoonup\} \text{ with } \ w\, R_\mathcal{H}\, f(v) \, , \\ \text{there is a } u \in V_\mathcal{G} \text{ such that } u\, R_\mathcal{G}\, v \text{ and } w \overset{\mathcal{H}}{\not\frown} f(u) \text{ and } w \overset{\mathcal{H}}{\not\rightharpoonup} f(u). \end{array} \tag{10}$$

A skew fibration $f \colon \mathcal{G} \to \mathcal{H}$ is *modalic* if it satisfies the following condition:

$$\begin{array}{l} \text{if } u \overset{\mathcal{G}}{\smile} v \text{ and } f(u) \overset{\mathcal{H}}{\rightharpoonup} f(v) \text{, then there is a } w \in V_\mathcal{G} \text{ such that} \\ w \overset{\mathcal{G}}{\rightharpoonup} v \text{ and } f(u) = f(w) \text{, or } u \overset{\mathcal{G}}{\rightharpoonup} w \text{ and } f(v) = f(w). \end{array} \tag{11}$$

The main purpose of this definition is Theorem 5.2 which says that skew fibrations are precisely the contraction-weakening maps. This is crucial for the soundness and completeness of combinatorial proofs, to be defined in the next section.

**Theorem 5.2.** *There is a modalic skew fibration* $f \colon [\![\Gamma']\!] \to [\![\Gamma]\!]$ *iff* $\Gamma' \overset{\mathsf{c}^\downarrow, \mathsf{w}^\downarrow, \equiv}{\vdash} \Gamma$.

*Proof (Sketch).* To prove this theorem, we proceed via Theorem 2.3 and make heavy use of results from [27] and the fact that $\Gamma' \overset{\mathsf{m}^\downarrow, \equiv}{\vdash} \Gamma$ iff there is a surjective modalic skew fibration $f \colon [\![\Gamma']\!] \to [\![\Gamma]\!]$ that is bijective on atomic vertices, which is a variant of [27, Theorem 5.1] and proved in a similar way. Then we can characterize derivations $\overset{\mathsf{ac}^\downarrow, \equiv}{\vdash}$ and $\overset{\mathsf{w}^\downarrow, \equiv}{\vdash}$ as in [27, Proposition 7.6], so that we can apply Theorem 2.3 (See also [2] and [4]). □

## 6 Combinatorial Proofs for the Modal Logics K and D

**Definition 6.1.** A map $f: \mathcal{G} \to \mathcal{F}$ from an RGB-cograph $\mathcal{G}$ to a modalic and properly labeled relation web $\mathcal{F}$ is *allegiant* if the following conditions hold:

- if $v, w \in V_{\mathcal{G}}^{\bullet}$ and $v \overset{\mathcal{G}}{\curlyvee} w$ then $f(v)$ and $f(w)$ are labeled by dual atoms;
- if $v \in V_{\mathcal{G}}^{\Box}$ then $l(f(v)) = \Box$;
- if $v \in V_{\mathcal{G}}^{\Diamond}$ then $l(f(v)) = \Diamond$;

**Definition 6.2.** For $\mathsf{X} \in \{\mathsf{K}, \mathsf{KD}\}$, an $\mathsf{X}$-*combinatorial proof* of a sequent $\Gamma$ is an allegiant skew-fibration $f: \mathcal{G} \to [\![\Gamma]\!]$ from an $\mathsf{X}$-correct RGB-cograph $\mathcal{G}$ to the relation web of $\Gamma$.

The *size* $|f|$ of a combinatorial proof $f: \mathcal{G} \to [\![\Gamma]\!]$ is $|V_{\mathcal{G}}| + |\Gamma|$, where $|\Gamma|$ is the number of symbols in $\Gamma$, and the *size* $|\pi|$ of a sequent proof $\pi$ is the number of symbols in $\pi$.

**Theorem 6.3 (Completeness).** *Let* $\mathsf{X} \in \{\mathsf{K}, \mathsf{KD}\}$. *If* $\overset{\mathsf{LK\text{-}X}}{\vdash} \Gamma$ *then there is an* $\mathsf{X}$-*combinatorial proof* $f: \mathcal{G} \to [\![\Gamma]\!]$. *Furthermore, the sizes of the sequent proof and the combinatorial proof differ only by a polynomial factor.*

*Proof.* Let $\pi$ be a proof of $\Gamma$ in $\mathsf{LK\text{-}X}$. By Theorem 2.2, $\pi$ can be rewritten as $\overset{\mathsf{MLL\text{-}X}}{\vdash} \Gamma' \overset{\mathsf{w}^{\downarrow},\mathsf{c}^{\downarrow}}{\vdash} \Gamma$ for some $\Gamma'$. By Theorem 4.4, we have an $\mathsf{X}$-correct RGB-cograph $\mathcal{G}$ whose underlying relation web is $[\![\Gamma']\!]$. By Theorem 5.2, we have a skew-fibration $f: [\![\Gamma']\!] \to [\![\Gamma]\!]$, and therefore also $f: \mathcal{G} \to [\![\Gamma]\!]$, which is allegiant by construction. The size restrictions follow immediately: sequent proof and combinatorial proof are bound by the number of $\mathsf{ax}$, $\mathsf{k}^+$, and $\mathsf{d}^+$ instances. □

**Theorem 6.4 (Soundness).** *Let* $\mathsf{X} \in \{\mathsf{K}, \mathsf{KD}\}$, *and let* $f: \mathcal{G} \to [\![A]\!]$ *be an* $\mathsf{X}$-*combinatorial proof. Then* $A$ *is a theorem in the modal logic* $\mathsf{X}$.

*Proof.* We have an $\mathsf{MLL\text{-}X}$ proof of a formula $A'$ with $[\![A']\!] = \mathcal{G}$. Hence $A'$ is a theorem of $\mathsf{X}$. By Theorem 5.2 we have a derivation $A' \overset{\mathsf{c}^{\downarrow},\mathsf{w}^{\downarrow},\equiv}{\vdash} A$ in which all inferences are sound for $\mathsf{X}$, we can conclude that $A$ is also a theorem of $\mathsf{X}$. □

**Theorem 6.5.** *Let* $\Gamma$ *be a sequent,* $\mathcal{G}$ *be a mixed graph together with a linking, and let* $f$ *be a map from* $\mathcal{G}$ *to* $[\![\Gamma]\!]$. *It can be decided in polynomial time in* $|V_{\mathcal{G}}| + |\Gamma|$ *whether* $f: \mathcal{G} \to [\![\Gamma]\!]$ *is an* $\mathsf{X}$-*combinatorial proof for* $\mathsf{X} \in \{\mathsf{K}, \mathsf{KD}\}$.

*Proof.* All necessary properties (forbidden configurations (4), (5), (8) for $\mathcal{G}$ being a modalic relation web, $\mathsf{X}$-correctness conditions in Definition 4.3, preservation of $\frown$ and $\rightharpoonup$ (9) and skew lifting (10)) can be checked in polynomial time. □

These three results, together with Theorem 2.1, imply that $\mathsf{X}$-combinatorial proofs (for $\mathsf{X} = \mathsf{K}$ and $\mathsf{X} = \mathsf{KD}$) form a sound and complete proof system (in the sense of [9]) for the modal logic $\mathsf{X}$. In the remaining sections of this paper we extend this result to all logics in the $\mathsf{S4}$-tesseract.

## 7 Combinatorial Proofs for the Logics in the S4-plane

We call two vertices $v$ and $w$ in a relation web $\mathcal{G}$ *clones* if for all $u$ with $u \neq v$ and $u \neq w$ we have $uRv$ iff $uRw$ for all $R \in \{\frown, \curvearrowright, \curvearrowleft, \smile\}$. If $v = w$ then they are trivially clones.

**Definition 7.1.** Let $\mathcal{G}$ and $\mathcal{H}$ be modalic and properly labeled relation webs. A map $f \colon \mathcal{G} \to \mathcal{H}$ is a $\{4^{\downarrow}, \mathsf{t}^{\downarrow}\}$*-map* if the following conditions are fulfilled:

- if $f(v) = f(w)$ then $v$ and $w$ are clones in $\mathcal{G}$, and if also $v \neq w$ then $v \overset{\mathcal{G}}{\curvearrowright} w$ and $l(f(v)) = l(f(w)) = \Diamond$;
- if $f(v) \neq f(w)$ then $vR_{\mathcal{G}}w$ implies $f(v)R_{\mathcal{H}}f(w)$ for any $R \in \{\frown, \curvearrowright, \curvearrowleft, \smile\}$;
- if $v \in V_{\mathcal{H}}$ is not in the image of $f$ then $l(v) = \Diamond$ and $v \curvearrowright w$ for some $w \in V_{\mathcal{H}}$.

A $\{4^{\downarrow}, \mathsf{t}^{\downarrow}\}$-map is a $\{\mathsf{t}^{\downarrow}\}$*-map* if it is injective, and a $\{4^{\downarrow}\}$*-map* if it is surjective.

**Proposition 7.2.** *The composition of* $\{4^{\downarrow}, \mathsf{t}^{\downarrow}\}$*-maps is a* $\{4^{\downarrow}, \mathsf{t}^{\downarrow}\}$*-map, and every* $\{4^{\downarrow}, \mathsf{t}^{\downarrow}\}$*-map can be written as a composition of a* $\{4^{\downarrow}\}$*-map and a* $\{\mathsf{t}^{\downarrow}\}$*-map.*

**Lemma 7.3.** *For all sequents* $\Gamma$ *and* $\Gamma'$*, we have:*

- $\Gamma' \overset{4^{\downarrow},\mathsf{t}^{\downarrow},\equiv}{\longmapsto} \Gamma$ *iff there is a* $\{4^{\downarrow}, \mathsf{t}^{\downarrow}\}$*-map* $f \colon [\![\Gamma']\!] \to [\![\Gamma]\!]$*;*
- $\Gamma' \overset{4^{\downarrow},\equiv}{\longmapsto} \Gamma$ *iff there is a* $\{4^{\downarrow}\}$*-map* $f \colon [\![\Gamma']\!] \to [\![\Gamma]\!]$*;*
- $\Gamma' \overset{\mathsf{t}^{\downarrow},\equiv}{\longmapsto} \Gamma$ *iff there is a* $\{\mathsf{t}^{\downarrow}\}$*-map* $f \colon [\![\Gamma']\!] \to [\![\Gamma]\!]$*;*

*Proof.* The second and the third statement follow immediately from the definitions, and for the first statement, observe that $\Gamma' \overset{4^{\downarrow},\mathsf{t}^{\downarrow},\equiv}{\longmapsto} \Gamma$ iff $\Gamma' \overset{4^{\downarrow},\equiv}{\longmapsto} \Gamma'' \overset{\mathsf{t}^{\downarrow},\equiv}{\longmapsto} \Gamma$, and apply Proposition 7.2. □

**Definition 7.4.** Let $\mathsf{X} \in \{\mathsf{K}, \mathsf{KD}, \mathsf{KT}, \mathsf{K4}, \mathsf{KD4}, \mathsf{KT4}\}$. A map $f \colon \mathcal{G} \to \mathcal{H}$ is an $\mathsf{X}^{\downarrow}$*-skew fibration* if $f = f'' \circ f'$ for some $f' \colon \mathcal{G} \to \mathcal{G}'$ and $f'' \colon \mathcal{G}' \to \mathcal{H}$, where $f'$ is an $\mathsf{X}^{\downarrow}$-map and $f''$ is a modalic skew fibration (if $\mathsf{X}^{\downarrow} = \varnothing$ then $f'$ is the identity).

**Proposition 7.5.** *Given* $f \colon \mathcal{G} \to \mathcal{H}$ *and* $\mathsf{X}^{\downarrow} \subseteq \{\mathsf{t}^{\downarrow}, 4^{\downarrow}\}$*, it can be decided in time polynomial in* $|\mathcal{G}| + |\mathcal{H}|$ *whether* $f$ *is an* $\mathsf{X}^{\downarrow}$*-skew fibration.*

Below are three examples, a $\{\mathsf{t}^{\downarrow}\}$-skew fibration, a $\{4^{\downarrow}\}$-skew fibration, and a $\{4^{\downarrow}, \mathsf{t}^{\downarrow}\}$-skew fibration:

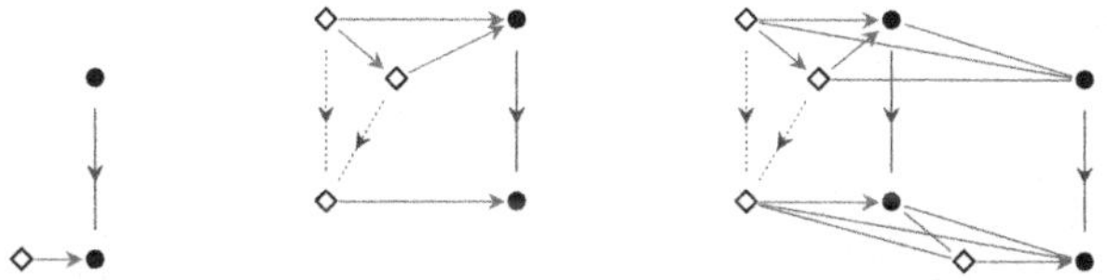

(12)

We can now easily generalize Theorem 5.2:

**Theorem 7.6.** *Let* $\mathsf{X}^{\downarrow} \subseteq \{\mathsf{t}^{\downarrow}, \mathsf{4}^{\downarrow}\}$*. There is a derivation* $\Gamma' \overset{\mathsf{X}^{\downarrow},\equiv}{\vdash} \Gamma'' \overset{\mathsf{c}^{\downarrow},\mathsf{w}^{\downarrow},\equiv}{\vdash} \Gamma$ *iff there is an* $\mathsf{X}^{\downarrow}$*-skew fibration* $f \colon [\![\Gamma']\!] \to [\![\Gamma]\!]$.

*Proof.* This follows immediately from Definition 7.4, Lemma 7.3 and Theorem 5.2. □

**Definition 7.7.** For $\mathsf{X} \in \{\mathsf{K}, \mathsf{KD}, \mathsf{KT}, \mathsf{K4}, \mathsf{KD4}, \mathsf{KT4}\}$, an $\mathsf{X}$-*combinatorial proof* of a sequent $\Gamma$ is an allegiant $\mathsf{X}^{\downarrow}$-skew-fibration $f \colon \mathcal{G} \to [\![\Gamma]\!]$ from an $\mathsf{X}^{\mathsf{LL}}$-correct RGB-cograph $\mathcal{G}$ to the relation web of $\Gamma$.

With this definition, it now follows immediately from Proposition 7.5, Theorem 7.6 and Theorem 2.2, that Theorems 6.3, 6.4 and 6.5 hold for all $\mathsf{X} \in \{\mathsf{K}, \mathsf{KD}, \mathsf{KT}, \mathsf{K4}, \mathsf{KD4}, \mathsf{KT4}\}$.

$$\frac{A, B}{\Box A, \Diamond B}\,\mathsf{m} \qquad \frac{A, B}{\Diamond A, \Diamond B}\,\mathsf{d_m} \qquad \frac{A, \Diamond B}{\Box A, \Diamond B}\,\mathsf{4_m} \qquad \frac{A, \Diamond B}{\Diamond A, \Diamond B}\,\mathsf{4_{dm}} \qquad \frac{A}{\Box A}\,\mathsf{n} \qquad \frac{A}{\Diamond A}\,\mathsf{p}$$

$$\frac{A, \Gamma}{\Box A, \Diamond \Gamma}\,\mathsf{k_c} \qquad \frac{A, \Gamma}{\Diamond A, \Diamond \Gamma}\,\mathsf{d_c} \qquad \frac{A, \Gamma, \Diamond \Delta}{\Box A, \Diamond \Gamma, \Diamond \Delta}\,\mathsf{4_c} \qquad \frac{A, \Gamma, \Diamond \Delta}{\Diamond A, \Diamond \Gamma, \Diamond \Delta}\,\mathsf{4_{dc}} \qquad |\Gamma, \Delta| > 0$$

**Fig. 11.** Sequent calculus rules for S4-tesseract logics.

$$\frac{B}{\Box\bot, \Diamond B}\,\mathsf{m}' \qquad \frac{A}{\Box A, \Diamond\bot}\,\mathsf{m}'' \qquad \frac{A}{\Diamond\bot, \Diamond A}\,\mathsf{d}'_{\mathsf{m}} \qquad \frac{\Diamond A}{\Box\bot, \Diamond A}\,\mathsf{4}'_{\mathsf{m}} \qquad \frac{\Diamond A}{\Diamond\bot, \Diamond A}\,\mathsf{4}'_{\mathsf{dm}}$$

$$\frac{\Gamma}{\Box\bot, \Diamond\Gamma}\,\mathsf{k}'_{\mathsf{c}} \qquad \frac{A, \Gamma}{\Box A, \Diamond\Gamma, \Diamond\bot, \ldots, \Diamond\bot}\,\mathsf{k}''_{\mathsf{c}} \qquad \frac{\Gamma}{\Box\bot, \Diamond\Gamma, \Diamond\bot, \ldots, \Diamond\bot}\,\mathsf{k}'''_{\mathsf{c}}$$

$$\frac{\Gamma, \Diamond\Delta}{\Box\bot, \Diamond\Gamma, \Diamond\Delta}\,\mathsf{4}'_{\mathsf{c}} \qquad \frac{A, \Gamma, \Diamond\Delta}{\Box A, \Diamond\Gamma, \Diamond\Delta, \Diamond\bot, \ldots, \Diamond\bot}\,\mathsf{4}''_{\mathsf{c}} \qquad \frac{\Gamma, \Diamond\Delta}{\Box\bot, \Diamond\Gamma, \Diamond\Delta, \Diamond\bot, \ldots, \Diamond\bot}\,\mathsf{4}'''_{\mathsf{c}}$$

$$\frac{\Gamma}{\Diamond\bot, \ldots, \Diamond\bot, \Diamond\Gamma}\,\mathsf{d}'_{\mathsf{c}} \qquad \frac{\Gamma, \Diamond\Delta}{\Diamond\bot, \ldots, \Diamond\bot, \Diamond\Gamma, \Diamond\Delta}\,\mathsf{4_{dc}}' \qquad |\Gamma, \Delta| > 0$$

$$\mathsf{m}^+ = \{\mathsf{m}, \mathsf{m}', \mathsf{m}''\} \qquad \mathsf{d}^+_{\mathsf{m}} = \{\mathsf{d_m}, \mathsf{d}'_{\mathsf{m}}\} \qquad \mathsf{4}^+_{\mathsf{m}} = \{\mathsf{4_m}, \mathsf{4}'_{\mathsf{m}}\} \qquad \mathsf{4}^+_{\mathsf{dm}} = \{\mathsf{4_{dm}}, \mathsf{4}'_{\mathsf{dm}}\}$$

$$\mathsf{k}^+_{\mathsf{c}} = \{\mathsf{k_c}, \mathsf{k}'_{\mathsf{c}}, \mathsf{k}''_{\mathsf{c}}, \mathsf{k}'''_{\mathsf{c}}\} \qquad \mathsf{d}^+_{\mathsf{c}} = \{\mathsf{d_c}, \mathsf{d}'_{\mathsf{c}}\} \qquad \mathsf{4}^+_{\mathsf{c}} = \{\mathsf{4_c}, \mathsf{4}'_{\mathsf{c}}, \mathsf{4}''_{\mathsf{c}}, \mathsf{4}'''_{\mathsf{c}}\} \qquad \mathsf{4}^+_{\mathsf{dc}} = \{\mathsf{4_{dc}}, \mathsf{4_{dc}}'\}$$

**Fig. 12.** Extended non-normal modal logic rules incorporating weakening on ⊥.

## 8 Non-normal Modal Logics

In this section, we show how to extend the results for the logics of the S4-plane to the non-normal modal logics of the S4-tesseract[3] in Fig. 1. Figure 11 shows the additional sequent rules that are needed for these logics, Fig. 12 shows the variations of these rules that are needed for obtaining the decomposition in Theorem 2.2, and Fig. 13 shows the extension of the table in Fig. 7, defining a sound and complete sequent system for each logic. We state here only the Theorem 8.1 below, and refer the reader to the exposition in [19] for more details, references, and proofs.

**Theorem 8.1.** *For all modal logics* X *of the* S4*-tesseract,* LK-X *is a sound and complete sequent system for the modal logic* X.

For our purpose, the most important observation is that the Decomposition Theorem 2.2 holds for all of these logics. This means that for defining combinatorial proofs for these logics, it suffices to define for RGB-cographs what it means to be $\mathsf{X}^{\mathsf{LL}}$-correct.

| X | M | M4 | MP | MP4 | MD | MD4 | MT | MT4 |
|---|---|---|---|---|---|---|---|---|
| $\mathsf{X}^{\mathsf{seq}}$ | $\{m\}$ | $\{m, 4_m\}$ | $\{m, p\}$ | $\{m, p, 4_m\}$ | $\{m, d_m\}$ | $\{m, d_m, 4_m\}$ | $\{m, t\}$ | $\{m, t, 4_m\}$ |
| $\mathsf{X}^{\mathsf{LL}}$ | $\{m^+\}$ | $\{m^+, 4_m^+\}$ | $\{m^+, p\}$ | $\{m^+, p\}$ | $\{m^+, d_m^+\}$ | $\{m^+, d_m^+\}$ | $\{m^+\}$ | $\{m^+\}$ |
| $\mathsf{X}^{\downarrow}$ | $\varnothing$ | $\{4^\downarrow\}$ | $\varnothing$ | $\{4^\downarrow\}$ | $\varnothing$ | $\{4^\downarrow\}$ | $\{t^\downarrow\}$ | $\{t^\downarrow, 4^\downarrow\}$ |

| X | N | N4 | NP | NP4 | ND | ND4 |
|---|---|---|---|---|---|---|
| $\mathsf{X}^{\mathsf{seq}}$ | $\{m, n\}$ | $\{m, n, 4_m\}$ | $\{m, n, p\}$ | $\{m, n, p, 4_m\}$ | $\{m, n, d_m\}$ | $\{m, n, d_m, 4_m, 4_{dm}\}$ |
| $\mathsf{X}^{\mathsf{LL}}$ | $\{m^+, n\}$ | $\{m^+, n\}$ | $\{m^+, n, p\}$ | $\{m^+, n, p\}$ | $\{m^+, n, d_m^+\}$ | $\{m^+, n, d_m^+\}$ |
| $\mathsf{X}^{\downarrow}$ | $\varnothing$ | $\{4^\downarrow\}$ | $\varnothing$ | $\{4^\downarrow\}$ | $\varnothing$ | $\{4^\downarrow\}$ |

| X | NT | NT4 | C | C4 | CD | CD4 | CT | CT4 |
|---|---|---|---|---|---|---|---|---|
| $\mathsf{X}^{\mathsf{seq}}$ | $\{m, n, t\}$ | $\{m, n, t, 4_m\}$ | $\{k_c\}$ | $\{k_c, 4_c\}$ | $\{k_c, d_c\}$ | $\{k_c, d_c, 4_c, 4_{dc}\}$ | $\{k_c, t\}$ | $\{k_c, t, 4_c\}$ |
| $\mathsf{X}^{\mathsf{LL}}$ | $\{m^+, n\}$ | $\{m^+, n\}$ | $\{k_c^+\}$ | $\{k_c^+\}$ | $\{k_c^+, d_c^+\}$ | $\{k_c^+, d_c\}$ | $\{k_c^+\}$ | $\{k_c^+\}$ |
| $\mathsf{X}^{\downarrow}$ | $\{t^\downarrow\}$ | $\{t^\downarrow, 4^\downarrow\}$ | $\varnothing$ | $\{4^\downarrow\}$ | $\varnothing$ | $\{4^\downarrow\}$ | $\{t^\downarrow\}$ | $\{t^\downarrow, 4^\downarrow\}$ |

**Fig. 13.** Rule sets for the S4-tesseract logics not given in Fig. 7

[3] Observe that all the logics in the S4-tesseract are monotone. In fact, our methods can not be applied in presence of the rule $\dfrac{A, \bar{B} \quad \bar{A}, B}{\Box A, \Diamond B}$ E. We therefore have to leave the investigation of combinatorial proofs for non-monotonic non-normal modal logics as an open problem for future research.

| X | $\mathsf{X}^{\mathsf{LL}}$ | for each $\curlyvee$-equivalence class $\rho$ in $V^{\Box} \uplus V^{\Diamond}$ |
|---|---|---|
| M, MT, M4, MT4 | $\{\mathsf{m}^+\}$ | $\rho = \{u, v\}$ with $u \in V^{\Box}$ and $v \in V^{\Diamond}$ |
| MP, MP4 | $\{\mathsf{m}^+, \mathsf{p}\}$ | $\rho = \{u, v\}$ or $\rho = \{v\}$ with $u \in V^{\Box}$ and $v \in V^{\Diamond}$ |
| MD, MD4 | $\{\mathsf{m}^+, \mathsf{d}^+_{\mathsf{m}}\}$ | $\rho = \{u, v\}$ with $v \in V^{\Diamond}$ |
| N, NT, N4, NT4 | $\{\mathsf{m}^+, \mathsf{n}\}$ | $\rho = \{u, v\}$ or $\rho = \{u\}$ with $u \in V^{\Box}$ |
| NP, NP4 | $\{\mathsf{m}^+, \mathsf{n}, \mathsf{p}\}$ | $\rho = \{u, v\}$ or $\rho = \{w\}$ with $u \in V^{\Box}$ and $v \in V^{\Diamond}$ |
| ND, ND4 | $\{\mathsf{m}^+, \mathsf{n}, \mathsf{d}^+_{\mathsf{m}}\}$ | $\rho = \{u, v\}$ or $\rho = \{w\}$ with $v \in V^{\Diamond}$ and $w \in V^{\Box}$ |
| C, CT, C4, CT4 | $\{\mathsf{k}^+_{\mathsf{c}}\}$ | $\rho = \{u, v, w_1, \ldots, w_n\}$ with $u \in V^{\Box}$ and $v, w_1, \ldots, w_n \in V^{\Diamond}$ |
| CD, CD4 | $\{\mathsf{k}^+_{\mathsf{c}}, \mathsf{d}^+_{\mathsf{c}}\}$ | $\rho = \{u, v, w_1, \ldots, w_n\}$ with $v, w_1, \ldots, w_n \in V^{\Diamond}$ |
| K, KT, K4, KT4 | $\{\mathsf{k}^+\}$ | $\rho = \{u, v_1, \ldots, v_n\}$ with $u \in V^{\Box}$ and $v_1, \ldots, v_n \in V^{\Diamond}$ |
| KD, KD4 | $\{\mathsf{k}^+, \mathsf{d}^+\}$ | $\rho = \{u, v_1, \ldots, v_n\}$ with $v_1, \ldots, v_n \in V^{\Diamond}$ |

**Fig. 14.** The fourth condition an RGB-cograph has to satisfy in order to be $\mathsf{X}^{\mathsf{LL}}$-correct.

**Definition 8.2.** Let X be a logic in the S4-tesseract. An RGB-cograph is $\mathsf{X}^{\mathsf{LL}}$-*correct*, if it obeys Conditions 1–3 of Definition 4.3, together with the corresponding version of Condition 4 shown in the table in Fig. 14.

Intuitively, the corresponding conditions in the table in Fig. 14 verify if the number of $\Box$- and $\Diamond$-vertices in an $\curlyvee$-equivalence class is compatible with the number of $\Box$- and $\Diamond$-occurrences introduced in a sequent by a single application of a sequent rule of $\mathsf{X}^{\mathsf{LL}}$.

With this we can show Theorem 4.4 for all logics in the S4-tesseract shown in Fig. 1. Therefore, X-combinatorial proofs, as defined in Definition 7.7 form a sound and complete proofs system for the modal logic X, as stated in Theorems 6.3, 6.4, and 6.5, for all logics X in the S4-tesseract shown in Fig. 1.

## 9 Conclusion and Future Work

In this paper we presented cominatorial proofs for all logics in the S4-tesseract. Since checking correctness of a combinatorial proof is polynomial in its size, they form a proof system in the sense of Cook and Reckhow [9]. Due to their combinatorial nature, they abstract away from the syntactic bureaucracy of more standard formalisms like sequent calculus or analytic tableaux. This leads naturally to the following notion of proof identity:

*Two proofs are the same iff they have the same combinatorial proof.* (13)

We conjecture that this notion of proof identity is in close correspondence to the notion of proof identity that is induced by sequent rule permutations. However, investigating the relation between the two would go beyond the scope of this paper, and we consider this to be future work. Furthermore, in order to support (13) it is necessary, not only to show how sequent proofs are related to combinatorial proofs, but also how analytic tableaux or resolution proofs or other syntactic formalisms are related to combinatorial proofs [1].

Further topics for future work include the extensions to all logics in the classical modal S5-cube, and also to intuitionistic modal logics. Another question is how our work relates to the recent development of combinatorial proofs for first-order logic [17].

Finally, from the proof theoretical perspective, the most interesting question for future research is the study of normalization of combinatorial proofs, as it has been done for propositional logic in [16,29].

## References


1. Acclavio, M., Straßburger, L.: From syntactic proofs to combinatorial proofs. In: Galmiche, D., Schulz, S., Sebastiani, R. (eds.) IJCAR 2018. LNCS (LNAI), vol. 10900, pp. 481–497. Springer, Cham (2018). https://doi.org/10.1007/978-3-319-94205-6_32
2. Acclavio, M., Straßburger, L.: On combinatorial proofs for logics of relevance and entailment. In: Iemhoff, R., Moortgat, M., de Queiroz, R. (eds.) WoLLIC 2019. LNCS, vol. 11541, pp. 1–16. Springer, Heidelberg (2019). https://doi.org/10.1007/978-3-662-59533-6_1
3. Avron, A.: The method of hypersequents in the proof theory of propositional non-classical logics. In: Logic: From Foundations to Applications, European Logic Colloquium, pp. 1–32. Oxford University Press (1994)
4. Benjamin, R., Straßburger, L.: Towards a combinatorial proof theory. In: Cerrito, S., Popescu, A., (eds.) TABLEAUX 2019. LNAI, vol. 11714, pp. 259–276 (2019)
5. Brünnler, K.: Deep sequent systems for modal logic. Arch. Math. Log. **48**(6), 551–577 (2009)
6. Brünnler, K., Tiu, A.F.: A local system for classical logic. In: Nieuwenhuis, R., Voronkov, A. (eds.) LPAR 2001. LNCS (LNAI), vol. 2250, pp. 347–361. Springer, Heidelberg (2001). https://doi.org/10.1007/3-540-45653-8_24
7. Chaudhuri, K., Marin, S., Straßburger, L.: Focused and synthetic nested sequents. In: Jacobs, B., Löding, C. (eds.) FoSSaCS 2016. LNCS, vol. 9634, pp. 390–407. Springer, Heidelberg (2016). https://doi.org/10.1007/978-3-662-49630-5_23
8. Chaudhuri, K., Marin, S., Straßburger, L.: Modular focused proof systems for intuitionistic modal logics. In: Kesner, D., Pientka, B. (eds.) FSCD 2016. LIPIcs, vol. 52, pp. 16:1–16:18. Schloss Dagstuhl-Leibniz-Zentrum fuer Informatik (2016)
9. Cook, S.A., Reckhow, R.A.: The relative efficiency of propositional proof systems. J. Symb. Logic **44**(1), 36–50 (1979)
10. Goré, R., Postniece, L., Tiu, A.: On the correspondence between display postulates and deep inference in nested sequent calculi for tense logics. Log. Methods Comput. Sci. **7**(2), 1–38 (2011)
11. Goré, R., Ramanayake, R., et al.: Labelled tree sequents, tree hypersequents and nested (deep) sequents. Adv. Modal Log. **9**, 279–299 (2012)
12. Guglielmi, A.: A system of interaction and structure. ACM Trans. Comput. Log. **8**(1), 1–64 (2007)
13. Guglielmi, A., Straßburger, L.: Non-commutativity and MELL in the calculus of structures. In: Fribourg, L. (ed.) CSL 2001. LNCS, vol. 2142, pp. 54–68. Springer, Heidelberg (2001). https://doi.org/10.1007/3-540-44802-0_5
14. Hein, R., Stewart, C.: Purity through unravelling. In: Structures and Deduction, pp. 126–143 (2005)
15. Hughes, D.: Proofs without syntax. Ann. Math. **164**(3), 1065–1076 (2006)
16. Hughes, D.: Towards Hilbert's 24$^{\text{th}}$ problem: combinatorial proof invariants: (preliminary version). Electr. Notes Theor. Comput. Sci. **165**, 37–63 (2006)

17. Hughes, D.J.D.: First-order proofs without syntax, June 2019
18. Lellmann, B.: Hypersequent rules with restricted contexts for propositional modal logics. Theor. Comput. Sci. **656**, 76–105 (2016)
19. Lellmann, B., Pimentel, E.: Modularisation of sequent calculi for normal and non-normal modalities. ACM Trans. Comput. Log. (TOCL) **20**(2), 7 (2019)
20. Marin, S., Straßburger, L.: Label-free modular systems for classical and intuitionistic modal logics. In: Advances in Modal Logic 10 (2014)
21. Möhring, R.H.: Computationally tractable classes of ordered sets. In: Rival, I. (ed.) Algorithms and Order, pp. 105–194. Kluwer, Dordrecht (1989)
22. Negri, S.: Proof analysis in modal logic. J. Philos. Log. **34**(5–6), 507 (2005)
23. Retoré, C.: Pomset logic: a non-commutative extension of classical linear logic. In: de Groote, P., Roger Hindley, J. (eds.) TLCA 1997. LNCS, vol. 1210, pp. 300–318. Springer, Heidelberg (1997). https://doi.org/10.1007/3-540-62688-3_43
24. Retoré, C.: Handsome proof-nets: perfect matchings and cographs. Theor. Comput. Sci. **294**(3), 473–488 (2003)
25. Simpson, A.K.: The proof theory and semantics of intuitionistic modal logic. Ph.D. thesis, University of Edinburgh. College of Science and Engineering (1994)
26. Stewart, C., Stouppa, P.: A systematic proof theory for several modal logics. Adv. Modal Log. **5**, 309–333 (2004)
27. Straßburger, L.: A characterization of medial as rewriting rule. In: Baader, F. (ed.) RTA 2007. LNCS, vol. 4533, pp. 344–358. Springer, Heidelberg (2007). https://doi.org/10.1007/978-3-540-73449-9_26
28. Straßburger, L.: Cut elimination in nested sequents for intuitionistic modal logics. In: Pfenning, F. (ed.) FoSSaCS 2013. LNCS, vol. 7794, pp. 209–224. Springer, Heidelberg (2013). https://doi.org/10.1007/978-3-642-37075-5_14
29. Straßburger, L.: Combinatorial flows and their normalisation. In: Miller, D. (ed.) FSCD 2017. LIPIcs, vol. 84, pp. 31:1–31:17. Schloss Dagstuhl (2017)
30. Straßburger, L.: The problem of proof identity, and why computer scientists should care about Hilbert's 24th problem. Philos. Trans. R. Soc. A **377**(2140), 20180038 (2019)
31. Thiele, R.: Hilbert's twenty-fourth problem. Am. Math. Mon. **110**, 1–24 (2003)
32. Wansing, H.: Sequent calculi for normal modal propositional logics. J. Log. Comput. **4**(2), 125–142 (1994)
33. Wansing, H.: Sequent systems for modal logics. In: Gabbay, D.M., Guenthner, F. (eds.) Handbook of Philosophical Logic. Handbook of Philosophical Logic, vol. 8, pp. 61–145. Springer, Dordrecht (2002)

# A Game Model for Proofs with Costs

Timo Lang[1], Carlos Olarte[2(✉)], Elaine Pimentel[2], and Christian G. Fermüller[1]

[1] TU-Wien, Vienna, Austria
[2] Universidade Federal do Rio Grande do Norte, Natal, Brazil
carlos.olarte@gmail.com

**Abstract.** We look at substructural calculi from a game semantic point of view, guided by certain intuitions about resource conscious and, more specifically, cost conscious reasoning. To this aim, we start with a game, where player **P** defends a claim corresponding to a (single-conclusion) sequent, while player **O** tries to refute that claim. Branching rules for additive connectives are modeled by choices of **O**, while branching for multiplicative connectives leads to splitting the game into parallel sub-games, all of which have to be won by player **P** to succeed. The game comes into full swing by adding cost labels to assumptions, and a corresponding budget. Different proofs of the same end-sequent are interpreted as more or less expensive strategies for **P** to defend the corresponding claim. This leads to a new kind of labelled calculus, which can be seen as a fragment of SELL (subexponential linear logic). Finally, we generalize the concept of costs in proofs by using a semiring structure, illustrate our interpretation by examples and investigate some proof-theoretical properties.

## 1 Introduction

Various kinds of game semantics have been introduced to characterize computational features of substructural logics, in particular fragments and variants of linear logic (LL) [11]. This line of research can be traced back to the works of Blass [5,6], Abramsky and Jagadeesan [1], Hyland and Ong [12], Lamarche [14], Japaridze [13], Melliès [17], Delande et al. [8], among several others.

Our particular view of game semantics is that it is not just a technical tool for characterizing provability in certain calculi, but rather a playground for illuminating specific semantic intuitions underlying certain proof systems. Specially, we aim at a better understanding of *resource conscious* reasoning, which is often cited as a motivation for substructural logics.

In a first step, we characterize a version of linear logic (exponential-free affine inuitionistic linear logic **aIMALL**, or, equivalently, Full Lambek Calculus with exchange and weakening FLew) by a game, where the difference between additive and multiplicative connectives is modeled as sequential versus parallel continuation in game states that directly correspond to sequents. More precisely, every

Olarte and Pimentel are funded by CNPq, CAPES and the project FWF START Y544-N23. Lang is supported by FWF project W1255-N23.


S. Cerrito and A. Popescu (Eds.): TABLEAUX 2019, LNAI 11714, pp. 241–258, 2019.
https://doi.org/10.1007/978-3-030-29026-9_14

branching rule for a multiplicative connective corresponds to a game rule that splits the current run of the game into two independent subgames. Player **P**, who seeks to establish the validity of a given sequent, has to win all the resulting subgames. In contrast, a branching rule for an additive connective is modeled by a choice of player **O** between two possible succeeding game states, corresponding to the premises of the sequent rule in question. Note that this amounts to a deviation from the paradigm "formulas as games", underlying the game semantic tradition initiated by Blass [5]. Our games are, at least structurally, closer to Lorenzen's game for intuitionistic logic [16], where a state roughly corresponds to a situation in which a proponent seeks to defend a particular statement against attacks from an opponent, who, in general, has already granted a bunch of other statements. This kind of semantics for linear logic (but without the sequential/parallel distinction) was first explored in [10].

As long as we only care about the existence of winning strategies, the distinction between sequential and parallel subgames is redundant. However, our model not only highlights the intended semantics, but it also has concrete effects once we introduce *prices* for resources (represented by formulas) into the game. This is done via unary operators $\blacktriangledown^a$ and $\triangledown^a$, $a \in \mathbb{R}^+$, which share some characteristic features with *subexponentials* in LL (SELL [7,19]). The intuition is that a formula $\triangledown^a A$ is a *single use resource with price* $a$: By paying $a$, we can "unpack" $\triangledown^a A$ to obtain the formula $A$, and $\triangledown^a A$ is destroyed in the process. On the other hand, $\blacktriangledown^a A$ denotes a *permanent resource*: From $\blacktriangledown^a A$ we can obtain $A$ as often as we want, each time paying the price $a$. We lift our game to the extended language by enriching game states with a *budget* that is decreased whenever a price is paid. Different strategies for proving the same endsequent can then be compared by the budget which they require to be run safely, i.e. without getting into debts. This form of resource consciousness not only enhances the game, but it also translates into a novel sequent system, where cost bounds for proofs are attached as labels to sequents.

We observe that, up to this point, we only considered resources in *assumptions*. This is translated to sequents by restricting *negatively* the occurrences of the modalities $\blacktriangledown^a$ and $\triangledown^a$. Thus a promotion rule is not present and the proof-theoretic properties of the proposed systems, such as cut-elimination, can be mimicked by the ones of **aIMALL**. We hence move towards two possible generalizations. First, we propose a broader notion of cost and prices (for both the game and corresponding calculi) beyond the domain of the non-negative real numbers. For this, we organize the labels/prices in a semiring structure that enables for the instantiation of several interesting concrete examples, having the same game-theoretic characterization. Second, we discuss the quest of allowing modalities also in positive contexts, showing the limitations of such approach.

*Organization and Contributions.* Section 2 defines the basic game for **aIMALL** and establishes the correspondence between winning strategies and proofs. Section 3 introduces the concept of prices and budgets into the game. The existence of cost-minimal strategies is shown in Sect. 3.1 and cut-admissibility is discussed in Sect. 3.2. In Sect. 4, the concept of prices is generalized and several

**Sequent System for $\mathcal{C}$**

$$\frac{\Gamma, A, B \longrightarrow C}{\Gamma, A \otimes B \longrightarrow C}\ \otimes_L \qquad \frac{\Delta_1 \longrightarrow A \quad \Delta_2 \longrightarrow B}{\Delta_1, \Delta_2 \longrightarrow A \otimes B}\ \otimes_R \qquad \frac{\Delta_1 \longrightarrow A \quad \Delta_2, B \longrightarrow C}{\Delta_1, \Delta_2, A \multimap B \longrightarrow C}\ \multimap_L \qquad \frac{\Gamma, A \longrightarrow B}{\Gamma \longrightarrow A \multimap B}\ \multimap_R$$

$$\frac{\Gamma, A_i \longrightarrow B}{\Gamma, A_1 \,\&\, A_2 \longrightarrow B}\ \&_{L_i} \qquad \frac{\Gamma \longrightarrow A \quad \Gamma \longrightarrow B}{\Gamma \longrightarrow A \,\&\, B}\ \&_R \qquad \frac{\Gamma, A \longrightarrow C \quad \Gamma, B \longrightarrow C}{\Gamma, A \oplus B \longrightarrow C}\ \oplus_L \qquad \frac{\Gamma \longrightarrow A_i}{\Gamma \longrightarrow A_1 \oplus A_2}\ \oplus_{R_i}$$

$$\frac{}{\Gamma, p \longrightarrow p}\ I \qquad \frac{}{\Gamma \longrightarrow \mathbf{1}}\ \mathbf{1}_R \qquad \frac{}{\Gamma, \mathbf{0} \longrightarrow A}\ \mathbf{0}_L$$

**Sequent System for $\mathcal{C}(\mathbb{R}^+)$**

$$\frac{\blacktriangledown\Gamma, \Delta_1 \longrightarrow A \quad \blacktriangledown\Gamma, \Delta_2 \longrightarrow B}{\blacktriangledown\Gamma, \Delta_1, \Delta_2 \longrightarrow A \otimes B}\ \otimes_R \qquad \frac{\blacktriangledown\Gamma, \Delta_1 \longrightarrow A \quad \blacktriangledown\Gamma, \Delta_2, B \longrightarrow C}{\blacktriangledown\Gamma, \Delta_1, \Delta_2, A \multimap B \longrightarrow C}\ \multimap_L$$

$$\frac{\Gamma, \blacktriangledown^a A, A \longrightarrow C}{\Gamma, \blacktriangledown^a A \longrightarrow C}\ \blacktriangledown_L \qquad \frac{\Gamma, A \longrightarrow C}{\Gamma, \triangledown^a A \longrightarrow C}\ \triangledown_L$$

**Fig. 1.** Sequent systems $\mathcal{C}$ and $\mathcal{C}(\mathbb{R}^+)$

examples of our interpretation of costs in proofs are presented. In Sect. 5, the challenge of extending the semantics to full subexponential linear logic is discussed. Section 6 concludes the paper. Some additional material can be found in the companion technical report [15].

## 2 A Game Model of Branching

Our starting point is a calculus for *affine intuitionistic linear logic without exponentials* (**aIMALL**) [11], whose calculus is also equivalent to FLew, the *Full Lambek calculus with exchange and weakening*. We denote this calculus simply by $\mathcal{C}$ for brevity. Formulas in $\mathcal{C}$ are built from the grammar

$$A ::= p \mid \mathbf{0} \mid \mathbf{1} \mid A_1 \multimap A_2 \mid A_1 \otimes A_2 \mid A_1 \,\&\, A_2 \mid A_1 \oplus A_2.$$

where $p$ stands for atomic propositions (variables); $\mathbf{0}/\mathbf{1}$ are the false/true units; $\multimap$ denotes linear implication; $\otimes/\&$ are the multiplicative/additive conjunctions; and $\oplus$ is the additive disjunction.

We shall use $A, B, C$ (resp. $\Gamma, \Delta$) to range over formulas (resp. multisets of formulas). The rules are in Fig. 1. Note that the cut rule is not included in our presentation of $\mathcal{C}$ and that weakening is present only implicitly, via the context $\Gamma$ in the initial sequents. Furthermore, in rule $I$, $p$ is a propositional variable. We shall write $\vdash_{\mathcal{C}} S$ if the sequent $S$ is provable in $\mathcal{C}$.

We shall characterize $\mathcal{C}$ proofs as winning strategies (w.s.) in a certain game. Usually, one can interpret bottom-up proof search in sequent systems as a game, where at any given state, player **P** first chooses a formula of a sequent and, in the next step, either **P** moves to the premise sequent of the corresponding introduction rule (if the rule has only one premise); or player **O** chooses a premise sequent in which the game continues (if the rule has more than one premise).

Alternatively, rather than letting player **O** choose the subgame, one may stipulate that the game splits into independent subgames, all of which player **P** has to win. At first glance, these two approaches might seem different. However, the difference is only of interpretation and it does not affect the (non-)existence of w.s.'s for **P**. To see this, note that, by definition of a w.s., player **P** has to be prepared to answer to every possible choice of her opponent **O**. Therefore, it does not matter whether we require **P** to actually win every subgame or whether we image **P** to play a single run where she wins irrespectively of **O**'s choices. Hence, the two interpretations are equivalent in terms of **P**'s w.s.'s but they provide different viewpoints of branching sequent rules. Going more into detail, we can see that this equivalence holds as long as the parallel games are *independent.* We will break this independence later on by introducing a budget which is shared among parallel games (see Sect. 3).

The distinguishing feature of the game $\mathcal{G}_C$ below is: branching in *additive rules* is modeled as choices of **O**, whereas in branching *multiplicative rules*, **P** splits the context into two disjoint parts, which then form the corresponding contexts of two subgames to be played in parallel. Consequently, a state of the game is represented by a *multiset of sequents*, each belonging to a separate subgame.

**Definition 1 (The game $\mathcal{G}_C$).** *$\mathcal{G}_C$ is a game of two players,* **P** *and* **O**. *Game states (denoted by $G, H$) are finite multisets of sequents. $\mathcal{G}_C$ proceeds in rounds, initiated by* **P**'*s selection of a sequent $S$ from the current game state. The successor state is determined according to rules that fit one of the following schemes:*

$$\begin{array}{lll}(1)\; G \cup \{S\} & \rightsquigarrow & G \cup \{S'\} \\ (2)\; G \cup \{S\} & \rightsquigarrow & G \cup \{S_1\} \cup \{S_2\}\end{array}$$

*In (1), the subgame $S$ changes to $S'$. In (2), the subgame $S$ splits into two subgames $S_1$ and $S_2$. Here is the complete description of a round: After* **P** *has chosen a sequent $S$ among the current game state, she chooses a principal formula in $S$ and a matching rule instance $r$ of $\mathcal{C}$ such that $S$ is the conclusion of that rule. Depending on $r$, the round proceeds as follows:*

1. *If $r$ is a unary rule with premise $S'$, then the game proceeds in the game state $G \cup \{S'\}$ (no interaction of* **O** *is required).*
2. **Parallelism:** *If $r$ is a binary rule with premises $S_1, S_2$ pertaining to a* multiplicative *connective, then the game proceeds in the game state $G \cup \{S_1\} \cup \{S_2\}$ (again, no interaction of* **O** *is required).*
3. **O-choice:** *If $r$ is a binary rule with premises $S_1, S_2$ pertaining to an* additive *connective, then* **O** *chooses $S' \in \{S_1, S_2\}$ and the game proceeds in the game state $G \cup \{S'\}$.*

*A* **winning state** *(for* **P***) is a game state consisting of initial sequents of $\mathcal{C}$ only, that is, sequents having one of the forms $(\Gamma, p \longrightarrow p)$, $(\Gamma, \mathbf{0} \longrightarrow A)$, $(\Gamma \longrightarrow \mathbf{1})$.*

*Example 2.* As an example of a round in $\mathcal{G}_C$, assume that the game starts with $\Delta \longrightarrow A \otimes B$. **P** might select $A \otimes B$ as the principal formula. For the choice of a matching instance of the rule $\otimes_R$, she also has to choose a partition $\Delta = \Delta_1 \cup \Delta_2$. The game then continues in the state $\{(\Delta_1 \longrightarrow A), (\Delta_2 \longrightarrow B)\}$.

The following definitions are standard in game theory.

**Definition 3 (Plays and strategies).** *A* **play** *of* $\mathcal{G}_\mathcal{C}$ *on a game state* $H$ *is a sequence* $H_1, H_2, \ldots, H_n$ *of game states, where* $H_1 = H$ *and each* $H_{i+1}$ *arises by playing one round on* $H_i$. *A* **strategy** *(for* **P***) on a game state* $H$ *is defined as a function telling* **P** *how to move in any given state. A strategy on* $H$ *is a* **winning strategy (w.s.)** *if all plays following it eventually reach a winning state. We shall write* $\models_{\mathcal{G}_\mathcal{C}} H$ *if* **P** *has a w.s. on the game state* $H$.

Given w.s.'s $\pi_1, \ldots, \pi_n$ for sequents $S_1, \ldots, S_n$, there is an obvious w.s. for the game state $\{S_1, \ldots, S_n\}$ which could be specified as "play according to $\pi_i$ in the subgame $S_i$". Not all w.s.'s for $\{S_1, \ldots, S_n\}$ need to arise in such a way though, since in principle it is allowed that moves in a subgame $S_i$ depend on the moves in another subgame $S_j$. Nevertheless, since in the game $\mathcal{G}_\mathcal{C}$ valid moves and the winning conditions in all subgames are independent, we can restrict to strategies of the former kind. This observation is encapsulated as follows.

**Lemma 4 (Independence).** $\models_{\mathcal{G}_\mathcal{C}} \{S_1, \ldots, S_n\}$ iff for all $i \leq n$, $\models_{\mathcal{G}_\mathcal{C}} S_i$

Strategies in a game can be pictured as trees of game states, and therefore strategies share a common form with proofs. In our case, game states are multisets of sequents. However, by virtue of the above lemma, we obtain a notation of winning strategies which uses single sequents as nodes, at least if the initial state of the game is a sequent.

**Theorem 5 (Adequacy for $\mathcal{G}_\mathcal{C}$).** *Let* $S$ *be a sequent. Then* $\models_{\mathcal{G}_\mathcal{C}} \{S\}$ *iff* $\vdash_\mathcal{C} S$.

*Proof:* ($\Leftarrow$) is a straightforward induction on the length of proofs. ($\Rightarrow$) is proved by induction on a w.s. (the maximal number of moves which can occur following it). We only present the case where Lemma 4 comes into play. Assume that the state is $\Delta_1, \Delta_2 \longrightarrow A \otimes B$ and $\pi$ tells **P** to choose the instance of $\otimes_R$ with premises $\Delta_1 \longrightarrow A$ and $\Delta_2 \longrightarrow B$. By **parallelism**, the successor state is $\{(\Delta_1 \longrightarrow A), (\Delta_2 \longrightarrow B)\}$. Since $\pi$ is a w.s., it must contain a substrategy $\pi'$ for $\{(\Delta_1 \longrightarrow A), (\Delta_2 \longrightarrow B)\}$. By Lemma 4, we may assume that $\pi'$ is of the form: "Use $\pi_1$ to play in the subgame $\Delta_1 \longrightarrow A$ and $\pi_2$ to play in $\Delta_2 \longrightarrow B$" for some w.s.'s $\pi_1, \pi_2$ for $\Delta_1 \longrightarrow A$ and $\Delta_2 \longrightarrow B$ respectively. By induction, there are $\mathcal{C}$-proofs $\Xi_1, \Xi_2$ for the sequents $\Delta_1 \longrightarrow A$ and $\Delta_2 \longrightarrow B$. Applying $\otimes_R$ below $\Xi_1$ and $\Xi_2$, we obtain a $\mathcal{C}$-proof $\Xi$ of $\Delta_1, \Delta_2 \longrightarrow A \otimes B$. □

## 3 Adding Costs

To increase the expressiveness of our framework, we now augment assumptions with costs, where assumptions are formulas occurring *negatively* on sequents. Costs will be modeled—for now—by elements of $\mathbb{R}^+$ the set of non-negative real numbers. Formally, we add the unary modal operators $\blacktriangledown^a$ and $\triangledown^a$ for each $a \in \mathbb{R}^+$ to our language and call the resulting formulas *extended formulas.* An extended formula $\triangledown^a A$ can be considered as a *single use resource with price* $a$: By paying $a$, we can "unpack" $\triangledown^a A$ to $A$ (and $\triangledown^a A$ is destroyed in the process).

On the other hand, $\blacktriangledown^a A$ is a *permanent resource*: We can obtain as many copies of $A$ from it as we want, each time paying the price $a$.

**Definition 6.** *An* extended sequent *is a sequent built from extended formulas in which subformulas* $\blacktriangledown^a A$ *and* $\triangledown^a A$ *occur only in negative polarity.*

The notion of polarity is the standard one: A subformula occurrence in the antecedent of a sequent is *negative* if it occurs in the scope of an even number (including 0) of contexts ($[\cdot] \multimap B$), and otherwise it is *positive*. For occurrences of a subformula in the consequent, one replaces "even" by "odd". For instance, $\triangledown^a p \otimes p', (\blacktriangledown^b q \multimap q') \multimap q'' \longrightarrow \blacktriangledown^c r \multimap r'$ is an extended sequent. We denote by $\blacktriangledown \Gamma$ a set of formulas prefixed with $\blacktriangledown^a$ for some (not necessarily the same) $a \in \mathbb{R}^+$. We introduce a game $\mathcal{G}_\mathcal{C}(\mathbb{R}^+)$ similarly as we did for $\mathcal{G}_\mathcal{C}$. The rules of $\mathcal{G}_\mathcal{C}(\mathbb{R}^+)$ make reference to the calculus $\mathcal{C}(\mathbb{R}^+)$ of Fig. 1. It is obtained by interpreting all sequents as extended sequents, replacing the rules $\otimes_R$ and $\multimap_L$ as indicated in Fig. 1 (for internalizing contraction) and adding the *dereliction* rules

$$\frac{\Gamma, \blacktriangledown^a A, A \longrightarrow C}{\Gamma, \blacktriangledown^a A \longrightarrow C}\ \blacktriangledown_L \qquad \frac{\Gamma, A \longrightarrow C}{\Gamma, \triangledown^a A \longrightarrow C}\ \triangledown_L$$

Note that there is no right rules for $\blacktriangledown$ and $\triangledown$ in $\mathcal{C}(\mathbb{R}^+)$ since they only appear in negative polarity.

*Remark 7.* $\mathcal{C}(\mathbb{R}^+)$ can be naturally seen as a fragment of subexponential linear logic (SELL [7]). More specifically, let $\mathbf{aSELL}(\mathbb{R}^{\mathfrak{u}}_{\mathfrak{b}})$ be a single conclusion calculus for SELL with weakening, and let $\Sigma = \langle \mathbb{R}^+ \times \{\mathfrak{b}, \mathfrak{u}\}, \preceq, \mathcal{U}\rangle$ be the subexponential signature where the set of unbounded subexponentials (that can be weakened and contracted at will) is $\mathcal{U} = \{(a, \mathfrak{u}) \mid a \in \mathbb{R}^+\}$, and $\preceq$ is any partial order on $\mathbb{R}^+ \times \{\mathfrak{b}, \mathfrak{u}\}$ in which, as standardly required in SELL, no bounded subexponential is above an unbounded one. We identify the subexponential $!^{(a,\mathfrak{b})}$ with $\triangledown^a$ and $!^{(a,\mathfrak{u})}$ with $\blacktriangledown^a$. Then $\mathcal{C}(\mathbb{R}^+)$ is precisely the subsystem of $\mathbf{aSELL}(\mathbb{R}^{\mathfrak{u}}_{\mathfrak{b}})$ given by the syntactic restriction that subexponentials occur only in negative polarity. We will exploit this relation between $\mathcal{C}(\mathbb{R}^+)$ and $\mathbf{aSELL}(\mathbb{R}^{\mathfrak{u}}_{\mathfrak{b}})$ later in Sect. 3.2. For some remarks on the system without the syntactic restriction, see Sect. 5.

Let us return to the game now. The main difference between $\mathcal{G}_\mathcal{C}$ and $\mathcal{G}_\mathcal{C}(\mathbb{R}^+)$ is that game states in the latter will involve a **budget** (modeled as a real number) which will decrease whenever rules $\blacktriangledown_L$ and $\triangledown_L$ are invoked.

**Definition 8 (The game $\mathcal{G}_\mathcal{C}(\mathbb{R}^+)$).** *$\mathcal{G}_\mathcal{C}(\mathbb{R}^+)$ is a game of two players,* **P** *and* **O**. *Game states are tuples $(H, b)$, where $H$ is a finite multiset of extended sequents and $b \in \mathbb{R}$ is a "budget". $\mathcal{G}_\mathcal{C}$ proceeds in rounds, initiated by* **P***'s selection of an extended sequent $S$ from the current game state. The successor state is determined according to rules that fit one of the two following schemes:*

$$\begin{array}{lll} (1)\ (G \cup \{S\}, b) & \rightsquigarrow & (G \cup \{S'\}, b') \\ (2)\ (G \cup \{S\}, b) & \rightsquigarrow & (G \cup \{S^1\} \cup \{S^2\}, b) \end{array}$$

*A round proceeds as follows: After* **P** *has chosen an extended sequent $S \in H$ among the current game state, she chooses a rule instance $r$ of $\mathcal{C}(\mathbb{R}^+)$ such that $S$ is the conclusion of that rule. Depending on $r$, the round proceeds as follows:*

1. *If $r$ is a unary rule different from $\blacktriangledown_L, \triangledown_L$ with premise $S'$, then the game proceeds in the game state $(G \cup \{S'\}, b)$.*
2. **Budget decrease:** *If $r \in \{\blacktriangledown_L, \triangledown_L\}$ with premise $S'$ and principal formula $\blacktriangledown^a A$ or $\triangledown^a A$, then the game proceeds in the game state $(G \cup \{S'\}, b - a)$.*
3. **Parallelism:** *If $r$ is a binary rule with premises $S_1, S_2$ pertaining to a* multiplicative *connective, then the game proceeds as $(G \cup \{S_1\} \cup \{S_2\}, b)$.*
4. **O-choice:** *If $r$ is a binary rule with premises $S_1, S_2$ pertaining to an* additive *connective, then* **O** *chooses $S' \in \{S_1, S_2\}$ and the game proceeds in the game state $(G \cup \{S'\}, b)$.*

*A* **winning state** *(for* **P***) is a game state $(H, b)$ such that all $S \in H$ are initial sequents of $\mathcal{C}(\mathbb{R}^+)$ and $b \geq 0$.*

Plays and strategies are defined as in $\mathcal{G}_\mathcal{C}$. We write $\models_{\mathcal{G}_\mathcal{C}(\mathbb{R}^+)} (H, b)$ if **P** has a w.s. in the $\mathcal{G}_\mathcal{C}(\mathbb{R}^+)$-game starting on $(H, b)$. The intuitive reading of $\models_{\mathcal{G}_\mathcal{C}(\mathbb{R}^+)} (H, b)$ is: *The budgetb suffices to win the game $H$.* From now on, we will just say "sequent" and "formula" instead of "extended sequent" and "extended formula".

*Example 9.* Consider the state $(\{\blacktriangledown^1 p, \triangledown^3 q \longrightarrow p \otimes q\}, 5)$. In a first move, **P** picks $p \otimes q$ and she finds a partition of the premises not prefixed with $\blacktriangledown$ and decides that $\triangledown^3 q$ goes to the right premise of $\otimes_R$. So by **parallelism**, the new state is $(\{(\blacktriangledown^1 p \longrightarrow p), (\blacktriangledown^1 p, \triangledown^3 q \longrightarrow q)\}, 5)$. She now chooses to pick $\blacktriangledown^1 p$ of the first component and, by **budget decrease**, her budget decreases and the next state is $(\{(\blacktriangledown^1 p, p \longrightarrow p), (\blacktriangledown^1 p, \triangledown^3 q \longrightarrow q)\}, 4)$. Now **P** picks $\triangledown^3 q$ leading to $(\{(\blacktriangledown^1 p, p \longrightarrow p), (\blacktriangledown^1 p, q \longrightarrow q)\}, 1)$. Since both components are initial sequents and *budget* $\geq 0$, this is a winning state for **P**.

Similarly to $\mathcal{G}_\mathcal{C}$, it is not necessary to consider all possible strategies in $\mathcal{G}_\mathcal{C}(\mathbb{R}^+)$: For example, **P** never needs to take the budget into account when deciding the next move. (A rule of thumb for **P** could be: always play economical, i.e. avoid the rules $\blacktriangledown_L$ and $\triangledown_L$ whenever possible.) It is easy to see that a $\mathcal{C}(\mathbb{R}^+)$-proof $\Xi$ of a sequent $S$ translates to a w.s. in $(S, b)$ for some *sufficiently large* budget $b$. Taking these observations together, one can prove the following:

**Theorem 10 (Weak adequacy for $\mathcal{G}_\mathcal{C}(\mathbb{R}^+)$).** *Let $S$ be a sequent. Then*

$$\exists b \left( \models_{\mathcal{G}_\mathcal{C}(\mathbb{R}^+)} (\{S\}, b) \right) \quad \text{iff} \quad \vdash_{\mathcal{C}(\mathbb{R}^+)} S$$

The proof is similar to the one of Theorem 5. We call this theorem *weak* adequacy since information about the budget $b$ is lost in the proof theoretic representation. In other words, the game $\mathcal{G}_\mathcal{C}(\mathbb{R}^+)$ is more expressive than the calculus $\mathcal{C}(\mathbb{R}^+)$. To overcome this discrepancy, we now introduce a labelled extension of $\mathcal{C}(\mathbb{R}^+)$ that we call $\mathcal{C}^\ell(\mathbb{R}^+)$. A $\mathcal{C}^\ell(\mathbb{R}^+)$-proof is build from labelled sequents $\Gamma \longrightarrow_b A$ where $\Gamma \longrightarrow A$ is an extended sequent and $b \in \mathbb{R}^+$. The complete system is given in Fig. 2. Our aim is to prove that $\models_{\mathcal{G}_\mathcal{C}(\mathbb{R}^+)} (\{\Gamma \longrightarrow A\}, b)$ iff $\vdash_{\mathcal{C}^\ell(\mathbb{R}^+)} \Gamma \longrightarrow_b A$.

To this end, we need an analogue of Lemma 4 (independency of subgames in $\mathcal{G}_\mathcal{C}$) for $\mathcal{G}_\mathcal{C}(\mathbb{R}^+)$. Note that crucially, the naive analogue

$$\models_{\mathcal{G}_\mathcal{C}(\mathbb{R}^+)} (\{S_1, \ldots, S_n\}, b) \qquad \text{iff} \qquad \text{for all } i \leq n,\ \models_{\mathcal{G}_\mathcal{C}(\mathbb{R}^+)} (\{S_i\}, b)$$

labelled sequent system for $\mathcal{C}^\ell(\mathbb{R}^+)$

$$\frac{\Gamma, A, B \longrightarrow_b C}{\Gamma, A \otimes B \longrightarrow_b C}\otimes_L \qquad \frac{\blacktriangledown\Gamma, \Delta_1 \longrightarrow_a A \quad \blacktriangledown\Gamma, \Delta_2 \longrightarrow_b B}{\blacktriangledown\Gamma, \Delta_1, \Delta_2 \longrightarrow_{a+b} A \otimes B}\otimes_R$$

$$\frac{\blacktriangledown\Gamma, \Delta_1 \longrightarrow_a A \quad \blacktriangledown\Gamma, \Delta_2, B \longrightarrow_b C}{\blacktriangledown\Gamma, \Delta_1, \Delta_2, A \multimap B \longrightarrow_{a+b} C}\multimap_L \qquad \frac{\Gamma, A \longrightarrow_b B}{\Gamma \longrightarrow_b A \multimap B}\multimap_R$$

$$\frac{\Gamma, A_i \longrightarrow_b B}{\Gamma, A_1 \,\&\, A_2 \longrightarrow_b B}\&_{L_i} \qquad \frac{\Gamma \longrightarrow_a A \quad \Gamma \longrightarrow_b B}{\Gamma \longrightarrow_{\max\{a,b\}} A \,\&\, B}\&_R$$

$$\frac{\Gamma, A \longrightarrow_a C \quad \Gamma, B \longrightarrow_b C}{\Gamma, A \oplus B \longrightarrow_{\max\{a,b\}} C}\oplus_L \qquad \frac{\Gamma \longrightarrow_b A_i}{\Gamma \longrightarrow_b A_1 \oplus A_2}\oplus_{R_i}$$

$$\frac{\Gamma, \blacktriangledown^a A, A \longrightarrow_c C}{\Gamma, \blacktriangledown^a A \longrightarrow_{c+a} C}\blacktriangledown_L \qquad \frac{\Gamma, A \longrightarrow_c C}{\Gamma, \triangledown^a A \longrightarrow_{c+a} C}\triangledown_L$$

$$\frac{}{\Gamma, p \longrightarrow_0 p}I \qquad \frac{}{\Gamma \longrightarrow_0 \mathbf{1}}\mathbf{1}_R \qquad \frac{}{\Gamma, \mathbf{0} \longrightarrow_0 A}\mathbf{0}_L \qquad \frac{\Gamma \longrightarrow_a A}{\Gamma \longrightarrow_b A}w_\ell(b \geq a)$$

**Fig. 2.** The labelled sequent system $\mathcal{C}^\ell(\mathbb{R}^+)$

does *not* hold: Having a w.s. in $(\{S_1, \ldots, S_n\}, b)$ is not the same as having w.s.'s in all $(\{S_i\}, b)$'s, since the budget $b$ is shared between the subgames in $\mathcal{G}_\mathcal{C}(\mathbb{R}^+)$. However, one can prove that there are strategies in $\mathcal{G}_\mathcal{C}(\mathbb{R}^+)$ which are independent up to a partition of the budget. More precisely,

**Lemma 11 (Quasi-independency of subgames in $\mathcal{G}_\mathcal{C}(\mathbb{R}^+)$).** $\models_{\mathcal{G}_\mathcal{C}(\mathbb{R}^+)}$ $(\{S_1, \ldots, S_n\}, b)$ iff
$\exists b_1, \ldots, b_n \geq 0$ s.t. $\sum_{i=1}^n b_i \leq b$ and for all $i \leq n$, $\models_{\mathcal{G}_\mathcal{C}(\mathbb{R}^+)} (\{S_i\}, b_i)$.

*Proof:* The direction from right to left is obvious. For the other direction, assume that **P** has a w.s. $\pi$ for $(\{S_1, \ldots, S_n\}, b)$. We may assume wlog that this strategy is composed of strategies $\pi_1, \ldots, \pi_n$ for the subgames $S_1, \ldots, S_n$ which are both independent from each other and from the budget. In each subgame $S_i$, let $\tau_i$ be a strategy for **O** which maximizes the cost $b_i$ (the total decrease of the budget) of playing $\pi_i$ against $\tau_i$. Then $\models_{\mathcal{G}_\mathcal{C}(\mathbb{R}^+)} (\{S_i\}, b_i)$. Furthermore, from $\tau_1, \ldots, \tau_n$ player **O** can compose a strategy $\tau$ such that when played against $\pi$ in the parallel game $\{S_1, \ldots, S_n\}$, the costs for **P** sum up to $\sum_{i=1}^n b_i$. Since $\pi$ is a w.s. for $(\{S_1, \ldots, S_n\}, b)$, it must be the case that $\sum_{i=1}^n b_i \leq b$. □

We emphasize that the game rules of $\mathcal{G}_\mathcal{C}(\mathbb{R}^+)$ do *not* force **P** to know a partition of the budget in order to play parallel subgames. Nevertheless, Lemma 11 tells us that finding such a partition is always possible *in principle* (for an omnipotent player **P**). Now we can prove the desired correspondence.

**Theorem 12: (strong adequacy for $\mathcal{G}_\mathcal{C}(\mathbb{R}^+)$).**
$\models_{\mathcal{G}_\mathcal{C}(\mathbb{R}^+)} (\{\Gamma \longrightarrow A\}, b)$ *iff* $\vdash_{\mathcal{C}^\ell(\mathbb{R}^+)} \Gamma \longrightarrow_b A$.

*Proof:* ($\Leftarrow$) By induction on the length of a proof $\Xi$ of $\Gamma \longrightarrow_b A$. We highlight two cases. Consider the following two possible ends for $\Xi$:

$$(1)\ \frac{\Gamma \longrightarrow_c C \quad \Gamma \longrightarrow_d D}{\Gamma \longrightarrow_{\max\{c,d\}} C \,\&\, D}\,\&_R \qquad (2)\ \frac{\Delta_1 \longrightarrow_c C \quad \Delta_2 \longrightarrow_d D}{\Delta_1, \Delta_2 \longrightarrow_{c+d} C \otimes D}\,\otimes_R$$

In both cases, by induction, there are w.s.'s $\pi_1$ and $\pi_2$ for: (1) the game states $(\{\Gamma \longrightarrow C\}, c)$ and $(\{\Gamma \longrightarrow D\}, d)$; and (2) the game states $(\{\Delta_1 \longrightarrow C\}, c)$ and $(\{\Delta_2 \longrightarrow D\}, d)$ respectively. The needed w.s.'s $\pi_\&$ for the game state $(\{\Gamma \longrightarrow C \,\&\, D\}, \max\{c,d\})$ and $\pi_\otimes$ for the game state $(\{\Delta_1, \Delta_2 \longrightarrow C \otimes D\}, c+d)$ are:

(1) $\pi_\&$: Choose the instance of $\&_R$ as above. By **O**-choice, the successor game state is either $(\{\Gamma \longrightarrow C\}, \max\{c,d\})$ or $(\{\Gamma \longrightarrow D\}, \max\{c,d\})$. In any case, the budget in the successor state is greater or equal than both $c$ and $d$, so **P** can continue playing according to $\pi_1$ resp. $\pi_2$.

(2) $\pi_\otimes$: Choose the instance of $\otimes_R$ as above. By **parallelism**, the successor state is $(\{\Delta_1 \longrightarrow C, \Delta_2 \longrightarrow D\}, c+d)$. Use $\pi_1$ to play the subgame $\Delta_1 \longrightarrow C$ and $\pi_2$ to play in $\Delta_2 \longrightarrow D$. By assumption on $\pi_1$ and $\pi_2$, the total costs when playing both strategies in parallel cannot exceed $c+d$.

($\Rightarrow$) By induction on the length of a strategy $\pi$. We present only the case where Lemma 11 is used. Assume that the state is $(\{\Delta_1, \Delta_2 \longrightarrow C \otimes D\}, b)$ and $\pi$ tells **P** to choose the instance of $\otimes_R$ with premises $\Delta_1 \longrightarrow C$ and $\Delta_2 \longrightarrow D$. By **parallelism**, the successor state is $(\{\Delta_1 \longrightarrow C, \Delta_2 \longrightarrow D\}, b)$. Since $\pi$ is a w.s., it must contain a substrategy $\pi'$ for this state. By Lemma 11, we may assume that $\pi'$ is composed of substrategies $\pi_1, \pi_2$ for the game states $(\{\Delta_1 \longrightarrow C\}, c)$ and $(\{\Delta_2 \longrightarrow D\}, d)$ where $c + d \leq b$. By induction, there are $\mathcal{C}$-proofs $\Xi_1, \Xi_2$ for the sequents $\Delta_1 \longrightarrow_c C$ and $\Delta_2 \longrightarrow_d D$. Applying $\otimes_R$ and $w_\ell$ below $\Xi_1$ and $\Xi_2$, we obtain a $\mathcal{C}$-proof $\Xi$ of $\Delta_1, \Delta_2 \longrightarrow_b C \otimes D$. □

Let $S_b$ denote the labelled sequent corresponding to the sequent $S$ with label $b$. Given $\Pi$ a $\mathcal{C}^\ell(\mathbb{R}^+)$-proof of $S_b$, we define the many-to-one onto *skeleton* function $\mathcal{SK}(\Pi)$ as the $\mathcal{C}(\mathbb{R}^+)$-proof $\Xi$ of $S$ obtained by removing all labels and applications of $w_\ell$ from $\Pi$. Conversely, we define the one-to-one *decoration* function $\mathcal{D}(\Xi)$ as the $\mathcal{C}^\ell(\mathbb{R}^+)$-proof $\Pi^\ell$ of $S_a$, obtained by assigning the label 0 to all initial sequents of $\Xi$ and propagating the labels downwards according to the rules of $\mathcal{C}^\ell(\mathbb{R}^+)$. We define $\mathtt{cost}(\Xi) := a$. Let $\Lambda \in \mathcal{SK}^{-1}(\Xi)$ be a proof of $S_c$. It is easy to see that $a \leq c$, that is, $\mathtt{cost}(\Xi)$ is the minimal label which can be attached to $S$ w.r.t. $\Xi$. In game theoretic terms, this means the following.

**Theorem 13.** *Given a $\mathcal{C}(\mathbb{R}^+)$-proof $\Xi$ of a sequent $S$, $\mathtt{cost}(\Xi)$ is the smallest budget which suffices to win the game $\mathcal{G}_\mathcal{C}(\mathbb{R}^+)$ on $S$ when following the strategy corresponding to $\Xi$.*

*Example 14.* Consider the following well-known riddle:

> You have white and black socks in a drawer in a completely dark room. How many socks do you have to take out blindly to be sure of having a matching pair?

We can model the matching pair by the disjunction $(w \otimes w) \oplus (b \otimes b)$, and the act of drawing a random sock by the labelled formula $\blacktriangledown^1(w \oplus b)$. The above question then becomes:

For which $n$ is the sequent $\blacktriangledown^1(w \oplus b) \longrightarrow_n (w \otimes w) \oplus (b \otimes b)$ provable?

The following proof shows that $n = 3$ suffices:

$$
\dfrac{\dfrac{\dfrac{\dfrac{\dfrac{}{G, w, w, w \oplus b \longrightarrow_0 w \otimes w}\otimes_R, I}{G, w, w, w \oplus b \longrightarrow_0 F}\oplus_R \quad \dfrac{\dfrac{\dfrac{}{G, w, b, w \longrightarrow_0 (w \otimes w)}\otimes_R, I}{G, w, b, w \longrightarrow_0 F}\oplus_R \quad \dfrac{\dfrac{}{G, w, b, b \longrightarrow_0 b \otimes b}\otimes_R, I}{G, w, b, b \longrightarrow_0 F}\oplus_R}{G, w, b, w \oplus b \longrightarrow_0 F}\oplus_L}{G, w, w \oplus b, w \oplus b \longrightarrow_0 F}\oplus_L \qquad \Xi}{G, w \oplus b, w \oplus b, w \oplus b \longrightarrow_0 F}\oplus_L}{G \longrightarrow_3 F}3 \times \blacktriangledown_L
$$

where derivation $\Xi$ is symmetric, $F = (w \otimes w) \oplus (b \otimes b)$ and $G = \blacktriangledown^1(w \oplus b)$.

## 3.1 The Spectrum of a Provable Sequent

Due to weakening on labels, many proofs in $\mathcal{C}^\ell(\mathbb{R}^+)$ of labelled sequents of the form $S_b$ correspond to one skeleton proof in $\mathcal{C}(\mathbb{R}^+)$ of the sequent $S$. On the other hand, $S$ may have, itself, many proofs in $\mathcal{C}(\mathbb{R}^+)$, each of them having a cost, uniquely determined by the decoration $\mathcal{D}$. In this section we will consider the *spectrum* of such costs and prove the existence of a minimal one.

**Definition 15.** $\mathtt{spec}(S) := \{\mathtt{cost}(\Xi) \mid \Xi$ *is a* $\mathcal{C}(\mathbb{R}^+)$*-proof of* $S\}$.

For example, $\mathtt{spec}(\blacktriangledown^1 p, \triangledown^{0.8} p, \triangledown^{0.8} p \longrightarrow p \otimes p)$ consists of the numbers $\{1.6, 1.8, 2.6\}$ and all combinations $n + k \cdot 0.8$ where $n, k$ are natural numbers and $n \geq 2, k \leq 2$.

A subset $X \subseteq \mathbb{R}$ is called *discrete* if, for every $x \in X$, there is an open interval $I \subseteq \mathbb{R}$ such that $I \cap X = \{x\}$. We can prove:

**Theorem 16.** *For any sequent* $S$*,* $\mathtt{spec}(S) \subseteq \mathbb{R}^+$ *is discrete and closed.*

*Proof:* Let $a_1, \ldots, a_n$ denote all real numbers appearing as $\blacktriangledown^a$ or $\triangledown^a$ in $S$, and let us denote by $\Omega(a_1, \ldots, a_n)$ the set of all linear combinations of $a_1, \ldots, a_n$ over $\mathbb{N}$, i.e., $\Omega(a_1, \ldots, a_n) := \{k_1 \cdot a_1 + \ldots + k_n \cdot a_n \mid k_1, \ldots, k_n \in \mathbb{N}\}$. By inspecting the rules of $\mathcal{C}^\ell(\mathbb{R}^+)$ and since $w_\ell$ is not applied in $\mathcal{D}(\Xi)$, it is easy to see that $\mathtt{cost}(\Xi) \in \Omega(a_1, \ldots, a_n)$, and hence $\mathtt{spec}(S) \subseteq \Omega(a_1, \ldots, a_n)$. It suffices to show that each bounded monotone sequence in $\Omega(a_1, \ldots, a_n)$ is eventually constant. We may assume wlog that all the $a_i$'s are nonzero. Now consider a sequence $(k_1^i \cdot a_1 + \ldots + k_n^i \cdot a_n)_{i \geq 1}$ in $\Omega(a_1, \ldots, a_n)$, and assume that $B$ is an upper bound for it (a trivial lower bound is always 0). Pick a number $K$ such that $K \cdot \min\{a_1, \ldots, a_n\} > B$. It follows that for all $i, j$ we have $k_j^i < K$. In particular, there are only finitely many different terms in the sequence, from which our claim follows. □

Since any bounded below, closed set in $\mathbb{R}$ has an minimum, we obtain:

**Corollary 17.** *If $\vdash_{\mathcal{C}(\mathbb{R}^+)} \Gamma \longrightarrow A$, then $\mathtt{spec}(\Gamma \longrightarrow A)$ has a least element. In other words, there is a smallest $b$ such that $\vdash_{\mathcal{C}^\ell(\mathbb{R}^+)} \Gamma \longrightarrow_b A$.*

Corollary 17 tells us that cost-optimal strategies for all provable sequents exist, but note that the proof is not constructive. Nevertheless, we may now define:

$$\mathtt{cost}(S) := \begin{cases} \min(\mathtt{spec}(S)) & \text{if } \vdash_{\mathcal{C}(\mathbb{R}^+)} S \\ \infty & \text{otherwise} \end{cases}$$

### 3.2 Cut Admissibility

So far, the results about our game semantics $\mathcal{G}_{\mathcal{C}}(\mathbb{R}^+)$ did not depend essentially on the chosen calculus $\mathcal{C}(\mathbb{R}^+)$. We now want to relate proof-theoretic properties of $\mathcal{C}(\mathbb{R}^+)$ and $\mathcal{C}^\ell(\mathbb{R}^+)$ to the game semantics. Recall that $\mathcal{C}(\mathbb{R}^+)$ can be seen as a fragment of **aSELL**($\mathbb{R}^{\mathfrak{u}}_{\mathfrak{b}}$), arising from the syntactic restriction that the modal operators $\nabla^a, \blacktriangledown^a$ occur only negatively in sequents (Remark 7); consequently, there is no corresponding right rule (*promotion*) in $\mathcal{C}(\mathbb{R}^+)$. This has the effect that—even though (implicit) contraction on formulas $\blacktriangledown^a A$ is present in $\mathcal{C}(\mathbb{R}^+)$—the proof theory of $\mathcal{C}(\mathbb{R}^+)$ is closer to **aIMALL** than to **aSELL**($\mathbb{R}^{\mathfrak{u}}_{\mathfrak{b}}$).

$\mathcal{C}(\mathbb{R}^+)$ inherits the admissibility of the following cut rule from **aSELL**($\mathbb{R}^{\mathfrak{u}}_{\mathfrak{b}}$)

$$\frac{\blacktriangledown\Gamma, \Delta_1 \longrightarrow A \quad \blacktriangledown\Gamma, \Delta_2, A \longrightarrow C}{\blacktriangledown\Gamma, \Delta_1, \Delta_2 \longrightarrow C}\ cut$$

Note that, appearing both in a positive and a negative context, the cut formula $A$ cannot contain any modal operator.

Now, let us extend cut admissibility to the labelled system $\mathcal{C}^\ell(\mathbb{R}^+)$. Assume that both $\blacktriangledown\Gamma, \Delta_1 \longrightarrow_a A$ and $\blacktriangledown\Gamma, \Delta_2, A \longrightarrow_b C$ are provable in $\mathcal{C}^\ell(\mathbb{R}^+)$. Forgetting labels $a$ and $b$, we can conclude, from cut-admissibility in $\mathcal{C}(\mathbb{R}^+)$, that $\vdash_{\mathcal{C}(\mathbb{R}^+)} \blacktriangledown\Gamma, \Delta_1, \Delta_2 \longrightarrow C$. But then, $\blacktriangledown\Gamma, \Delta_1, \Delta_2 \longrightarrow_c C$ is also provable in $\mathcal{C}^\ell(\mathbb{R}^+)$ with, *e.g.*, $c = \mathtt{cost}(\blacktriangledown\Gamma, \Delta_1, \Delta_2 \longrightarrow C)$ (see Corollary 17). Hence, stating cut-admissibility in $\mathcal{C}^\ell(\mathbb{R}^+)$ strongly depends on the possibility of defining a computable function $f$ relating $c$ with the labels of the premises of the cut rule. We show that $f(a, b) = a + b$ is the *minimal* such function.

**Theorem 18.** *For $f(a, b) = a+b$, the following cut rule is admissible in $\mathcal{C}^\ell(\mathbb{R}^+)$:*

$$\frac{\blacktriangledown\Gamma, \Delta_1 \longrightarrow_a A \quad \blacktriangledown\Gamma, \Delta_2, A \longrightarrow_b C}{\blacktriangledown\Gamma, \Delta_1, \Delta_2 \longrightarrow_{f(a,b)} C}\ cut_\ell$$

*Moreover, whenever $cut_\ell$ is admissible w.r.t. a given $f'$, then $a + b \leq f'(a, b)$.*

*Proof:* For cut admissibility, one can follow the standard cut reduction strategy of **aIMALL** and observe that it is compatible with the proposed labelling of the cut rule. Consider for instance the following reduction (note that $\max\{a+c, a+d\} = a + \max\{c, d\}$):

$$\dfrac{\blacktriangledown\Gamma, \Delta_1 \longrightarrow_a A \qquad \dfrac{\blacktriangledown\Gamma, \Delta_2, A \longrightarrow_c C \quad \blacktriangledown\Gamma, \Delta_2, A \longrightarrow_d D}{\blacktriangledown\Gamma, \Delta_2, A \longrightarrow_{\max\{c,d\}} C \,\&\, D}\,\&_R}{\blacktriangledown\Gamma, \Delta_1, \Delta_2 \longrightarrow_{a+\max\{c,d\}} C \,\&\, D}\,cut_\ell$$

$\rightsquigarrow$

$$\dfrac{\dfrac{\blacktriangledown\Gamma, \Delta_1, \longrightarrow_a A \quad \blacktriangledown\Gamma, \Delta_2, A \longrightarrow_c C}{\blacktriangledown\Gamma, \Delta_1, \Delta_2 \longrightarrow_{a+c} C}\,cut_\ell \qquad \dfrac{\blacktriangledown\Gamma, \Delta_1, \longrightarrow_a A \quad \blacktriangledown\Gamma, \Delta_2, A \longrightarrow_d D}{\blacktriangledown\Gamma, \Delta_1, \Delta_2 \longrightarrow_{a+d} D}\,cut_\ell}{\blacktriangledown\Gamma, \Delta_1, \Delta_2 \longrightarrow_{\max\{a+c,a+d\}} C \,\&\, D}\,\&_R$$

For the minimality, let $p, q$ be distinct propositional variables. For any $a, b \in \mathbb{R}^+$ we have proofs of $\blacktriangledown^a p \longrightarrow_a p$ and $p, \blacktriangledown^b q \longrightarrow_b p \otimes q$. Applying cut, we get $\blacktriangledown^a p, \blacktriangledown^b q \longrightarrow_c p \otimes q$. Now, $\blacktriangledown^a p, \blacktriangledown^b q \longrightarrow_c p \otimes q$ is provable (without cut) only if $a + b \leq c$. Hence if $f$ makes the cut rule admissible, $a + b \leq f(a, b)$. □

One can easily show that also weakening in the antecedent is admissible in $\mathcal{C}^\ell(\mathbb{R}^+)$ and does not lead to an increased label. Similarly, generalized axioms $\Gamma, A \longrightarrow_0 A$ are admissible: Appearing both positively and negatively, $A$ does not contain modal operators, and hence $\texttt{cost}(\Gamma, A \longrightarrow A) = \texttt{cost}(A \longrightarrow A) = 0$.

*Example 19.* Consider a labelled transition system $(T, \Longrightarrow)$ where $T$ is a set of states and $\Longrightarrow \subseteq T \times \mathbb{R}^+ \times T$ is the transition relation on states. In $(t_i, a_i, t_i') \in \Longrightarrow$, simply written as $t_i \overset{a_i}{\Longrightarrow} t_i'$, $a_i$ is interpreted as the time needed for the transition to happen. We use distinct propositional variables to represent states. Moreover, the formula $\blacktriangledown^{a_i}(t_i \multimap t_i')$ models the transition $t_i \overset{a_i}{\Longrightarrow} t_i'$. We shall use $\Delta_{\Longrightarrow}$ to denote the set of such formulas. Given two sets of states $S_{start}, S_{end} \subseteq T$, it is easy to see that the following sentences are equivalent:

1. From every state in $S_{start}$, there is a state in $S_{end}$ reachable in time $\leq a$
2. $\models_{\mathcal{G}_C(\mathbb{R}^+)} (\{\Delta_{\Longrightarrow}, \bigoplus S_{start} \longrightarrow \bigoplus S_{end}\}, a)$

Hence by Theorem 12, both are equivalent to

3. $\vdash_{\mathcal{C}^\ell(\mathbb{R}^+)} \Delta_{\Longrightarrow}, \bigoplus S_{start} \longrightarrow_a \bigoplus S_{end}$.

One common way to obtain (1) is by finding a set of intermediary states $S_i$ and a splitting of the time $a_1 + a_2 = a$ such that we can go from each state in $S_{start}$ to some state in $S_i$ in time $a_1$, and from each state in $S_i$ to some state in $S_{end}$ in time $a_2$. In terms of (3), this strategy corresponds to a cut: Assume we have proofs $\Xi_1$ and $\Xi_2$ of the sequents $\Delta_{\Longrightarrow}, \bigoplus S_{start} \longrightarrow_{a_1} \bigoplus S_i$ and $\Delta_{\Longrightarrow}, \bigoplus S_i \longrightarrow_{a_2} \bigoplus S_{end}$. By cut admissibility (Theorem 18) we obtain the desired $\Delta_{\Longrightarrow}, \bigoplus S_{start} \longrightarrow_{a_1+a_2} \bigoplus S_{end}$ as the result of the "concatenation" of the paths encoded in $\Xi_1$ with the paths encoded in $\Xi_2$.

## 4 Alternative Cost Structures

We have used non-negative real numbers for representing costs and budgets, together with basic operations for *accumulating* (+) and *comparing* ($\geq$) them.

This allowed us to give a more interesting perspective of *resource consumption* in linear logic: we know that the cost of using a formula marked with cost 3 is not the same as derelicting a formula marked with cost 7. A natural question that arises is whether it is possible to consider other systems governing the way we understand *costs* and *budgets*. In this section, we consider sequent systems $\mathcal{C}^\ell(\mathcal{K})$ in which the real numbers of $\mathcal{C}^\ell(\mathbb{R}^+)$ (see Fig. 2) are replaced by elements of a semiring $\mathcal{K}$. As we shall see, the structure of $\mathcal{K}$ determines the behavior of the system and the interpretation of costs and budgets.

A *commutative semiring* is a tuple $\mathcal{K} = \langle \mathcal{A}, +_\mathcal{A}, \times_\mathcal{A}, \bot_\mathcal{A}, \top_\mathcal{A} \rangle$ satisfying: (S1) $\mathcal{A}$ is a set and $\bot_\mathcal{A}, \top_\mathcal{A} \in \mathcal{A}$; (S2) $+_\mathcal{A}$ and $\times_\mathcal{A}$ are binary operators that make the triples $\langle A, +_\mathcal{A}, \bot_\mathcal{A} \rangle$ and $\langle A, \times_\mathcal{A}, \top_\mathcal{A} \rangle$ commutative monoids; (S3) $\times_\mathcal{A}$ distributes over $+_\mathcal{A}$ (i.e., $a \times_\mathcal{A} (b +_\mathcal{A} c) = (a \times_\mathcal{A} b) +_\mathcal{A} (a \times_\mathcal{A} c)$); and (S4) $\bot_\mathcal{A}$ is absorbing for $\times_\mathcal{A}$ (i.e., $a \times_\mathcal{A} \bot_\mathcal{A} = \bot_\mathcal{A}$); $\mathcal{K}$ is *absorptive* if it additionally satisfies (S5) $a +_\mathcal{A} (a \times_\mathcal{A} b) = a$; in absorptive semirings, $+_\mathcal{A}$ is idempotent, that is, $a +_\mathcal{A} a = a$. This allows for the definition of the following *partial order*: $a \preceq_\mathcal{A} b$ iff $a +_\mathcal{A} b = b$ (and then, $a \times_\mathcal{A} b \preceq_\mathcal{A} a$); an absorptive semiring $\mathcal{K}$ is *idempotent* whenever its $\times_\mathcal{A}$ operator is idempotent.

Absorptive semirings satisfy some additional properties [3]: $\bot_\mathcal{A}$ (resp. $\top_\mathcal{A}$) is the bottom (resp. top) of $\mathcal{A}$; $a +_\mathcal{A} \top_\mathcal{A} = \top_\mathcal{A}$; $+_\mathcal{A}$ coincides with the $lub_\mathcal{A}$ (least upper bound) operator; if $a +_\mathcal{A} b \in \{a, b\}, \forall a, b \in \mathcal{A}$ then $(\mathcal{A}, \preceq_\mathcal{A})$ is a total order; $a \times_\mathcal{A} b \preceq_\mathcal{A} glb_\mathcal{A}(a, b)$, where $glb_\mathcal{A}$ is the greatest lower bound operator; if $\mathcal{K}$ is idempotent, then $+_\mathcal{A}$ distributes over $\times_\mathcal{A}$ and $\times_\mathcal{A}$ coincides with $glb_\mathcal{A}$.

We identify costs as elements of $\mathcal{A}$. We can naturally consider $\top_\mathcal{A}$ (resp. $\bot_\mathcal{A}$) as the "best" (resp. "worst") cost. Dually, $\top_\mathcal{A}$ (resp. $\bot_\mathcal{A}$) is the "worst" (resp. "best") budget. Also, we expect the *accumulating* operator to be commutative and associative (S2). Moreover, *accumulating* costs gives rise to a "worse" cost (S5). Hence, the $\times_\mathcal{A}$ operator is used to combine costs ($+$, on $\mathbb{R}^+$, in Fig. 2). On the other hand, $+_\mathcal{A}$ is used to select which is the "best" value, in the sense that $a +_\mathcal{A} b = a$ iff $b \preceq_\mathcal{A} a$ iff $a$ is "better" than $b$ (i.e., $\preceq_\mathcal{A}$ will replace $\geq$ in Fig. 2). Finally, we generalize max (in Fig. 2) as $glb_\mathcal{A}$. As mentioned above, in the case of idempotent semirings, $\times_\mathcal{A}$ coincides with the $glb_\mathcal{A}$ while in the non-idempotent case accumulating costs often gives a "worse" result than the $glb_\mathcal{A}$.

Note that the rules $\triangledown_L^a$ and $\blacktriangledown_L^a$, in Fig. 2, the budget $c$ in the conclusion must be of the form $a + b$. In the particular case of $\mathbb{R}^+$, we know that $b = c - a$ whenever $c \geq a$. Hence, from a conclusion with budget $c$ we obtain a premise with **decreased** budget $c - a$. In the general case, we guarantee that such splitting of the budget (also present in rules $\otimes_R$ and $\multimap_L$) is possible by requiring $\mathcal{K}$ to be invertible in the following sense: $\mathcal{K}$ is *invertible* if for all $b \preceq_\mathcal{A} a$, the set $\mathcal{I}(b, a) = \{x \in \mathcal{A} \mid a \times_\mathcal{A} x = b\}$ is non-empty and admits a minimum. We then denote this minimum by $b \div_\mathcal{A} a$. Observe that, in all our examples, if $b \preceq_\mathcal{A} a$ then the set $\mathcal{I}(b, a)$ is a singleton except when $a = b = \bot_\mathcal{A}$. In that case, $\mathcal{I}(b, a) = \mathcal{A}$ and we set $\bot_\mathcal{A} \div_\mathcal{A} \bot_\mathcal{A} = \bot_\mathcal{A}$. In Remark 22 we explain and clarify this choice.

In what follows, $\mathcal{K}$ will always denote an absorptive and invertible semiring.

**Definition 20 (System $\mathcal{C}^\ell(\mathcal{K})$).** *Let $\mathcal{K} = \langle \mathcal{A}, +_\mathcal{A}, \times_\mathcal{A}, \bot_\mathcal{A}, \top_\mathcal{A} \rangle$ be an absorptive and invertible semiring. The system $\mathcal{C}^\ell(\mathcal{K})$ is obtained from $\mathcal{C}^\ell(\mathbb{R}^+)$ (Fig. 2) by* *replacing* 0 *with* $\top_{\mathcal{A}}$, + *with* $\times_{\mathcal{A}}$, max *with* $glb_{\mathcal{A}}$, *and* $\geq$ *with* $\preceq_{\mathcal{A}}$. *Similarly, we obtain* $\mathcal{C}(\mathcal{K})$ *as a generalization of* $\mathcal{C}(\mathbb{R}^{+})$ *(Fig. 1).*

Just as $\mathcal{C}(\mathbb{R}^{+})$ can be seen as a fragment of $\mathbf{aSELL}(\mathbb{R}^{\mathfrak{u}}_{\mathfrak{b}})$, the system $\mathcal{C}(\mathcal{K})$ is a fragment of $\mathbf{aSELL}(\mathcal{K}^{\mathfrak{u}}_{\mathfrak{b}})$, i.e. affine subexponential linear logic with subexponentials taken from the set $\mathcal{K} \times \{\mathfrak{u}, \mathfrak{b}\}$. We omit the (rather straightforward) formulation of the corresponding game semantics.

Next we present some instances of $\mathcal{C}^{\ell}(\mathcal{K})$ and their intended behavior.

*Example 21 (Costs).* $\mathcal{K}_c = \langle \mathbb{R}^{\infty}_{+}, \min_{\mathbb{R}}, +_{\mathbb{R}}, \infty, 0\rangle$, where $\mathbb{R}^{\infty}_{+}$ is the completion of $\mathbb{R}_{+}$ with $\infty$, reflects the meaning of costs in Sect. 3. If $a, b \neq \infty$ and $b \geq a$ (i.e., $b \preceq_{\mathcal{A}} a$), there is a unique way of splitting $b$ into $a + b'$, namely, $b' = b - a$ (i.e., $b' = b \div_{\mathcal{A}} a$). Alternatively, we may interpret the elements in $\mathcal{K}_c$ as 2D areas. Then, a label $b \neq \infty$ in a sequent can be understood as the total area available to place some objects. Each time an object of size $a$ is placed (using $\blacktriangledown^{a}{}_{L}$ or $\triangledown^{a}{}_{L}$) we observe, bottom-up, that the total area is decreased to $b - a$.

*Remark 22 (Meaning of* $\div$ *and* $\infty$*).* Consider the semiring $\mathcal{K}_c$ above (where $\bot_{\mathcal{A}} = \infty$ and $\top_{\mathcal{A}} = 0$). If the label in the sequent is $b = \bot_{\mathcal{A}}$, regardless the value $a$ in an application of $\triangledown^{a}_{L}$ or $\blacktriangledown^{a}_{L}$, the premise will be labelled with $\infty$. This is because, according to our definition, $\bot_{\mathcal{A}} \div \bot_{\mathcal{A}} = \bot_{\mathcal{A}}$. This makes sense since we select the most "generous" budget to continue the derivation. Of course, smaller suitable budgets are also allowed due to rule $w_{\ell}$. For instance, the sequent $\triangledown^{\bot}p, \triangledown^{\bot}(p \multimap q) \longrightarrow_{b} q$ is provable in $\mathcal{C}^{\ell}(\mathcal{K}_c)$ only if $b = \bot_{\mathcal{A}}$. The same sequent (removing the label $b$) is also provable in $\mathbf{aSELL}(\mathcal{K}^{\mathfrak{u}}_{\mathfrak{b}})$. Note that if we decree that $\bot_{\mathcal{A}} \div \bot_{\mathcal{A}} = \top_{\mathcal{A}}$ (as in [3]), then the sequent above would not be provable for any $b$.

*Example 23 (Protected resources).* Let $\mathcal{K}_{c/p} = \langle\{\mathtt{pub}, \mathtt{conf}\}, +, \times, \mathtt{pub}, \mathtt{conf}\rangle$ and define $a + b = \mathtt{pub}$ iff $a = b = \mathtt{pub}$ and $a \times b = \mathtt{conf}$ iff $a = b = \mathtt{conf}$. The intuition is that $\blacktriangledown^{\mathtt{pub}}F$ represents *public* information (and then not confidential) and $\blacktriangledown^{\mathtt{conf}}F$ represents *secret* information. Observe that no derivation of $\Gamma, \blacktriangledown^{\mathtt{pub}}F \longrightarrow_{\mathtt{conf}} G$ can apply $\blacktriangledown_{L}$ on $\blacktriangledown^{\mathtt{pub}}F$ (since $\mathtt{conf} \not\preceq \mathtt{pub}$). This means that only confidential (or protected) resources can be used in such a derivation. Alternatively, we can show that, if $\Gamma \longrightarrow_{\mathtt{conf}} G$ is provable then $\Gamma' \longrightarrow_{\mathtt{conf}} G'$ is also provable where $\Gamma'$ is as $\Gamma$ but replacing any formula of the form $\blacktriangledown^{\mathtt{pub}}F$ with the constant $\mathbf{1}$ (similarly for $G$ and $\triangledown^{\mathtt{pub}}F$). $\mathcal{K}_{c/p}$ is nothing less that the structure $S_c = \langle\{\mathtt{false}, \mathtt{true}\}, \vee, \wedge, \mathtt{false}, \mathtt{true}\rangle$ [4].

*Example 24 (Maximum amount of resources).* Consider now the situation where labels in sequents represents a certain amount of computational resources, e.g., RAM, available to process a series of tasks. Moreover, let us interpret $\triangledown^{c}F$ as the fact that, in order to produce $F$, $c$ resources need to be used. As expected, once $F$ is produced, the $c$ resources can be released and freed to be used in other tasks. The idea is to know what is the least amount of resources $b$ s.t. some jobs $\Gamma$ can be all of them executed, sequentially if needed.

Consider $\mathcal{K}_{\max} = \langle \mathbb{R}^{\infty}_{+}, \min, \max, \infty, 0\rangle$ where $b \div a = b$ (if $b \geq a$). Let $t_1, t_2$ be atomic propositions representing tasks and let $\Gamma = \{\triangledown^{a}t_1, \triangledown^{c}t_2\}$. Clearly, the sequents $\Gamma \longrightarrow_b t_1 \otimes t_2$ and $\Gamma \longrightarrow_b t_1 \,\&\, t_2$ are both provable if $b = \max(a, c)$. Of course, if we start with more resources, *e.g.*, $b = a + c$, the sequent is still provable (rule $w_\ell$). Interestingly, from the point of view of costs, the difference between concurrent ($\otimes$) and sequential choices ($\&$) vanishes in this particular scenario, since $\mathcal{K}_{\max}$ is idempotent (and hence $glb_{\mathcal{A}}$ and $\times_{\mathcal{A}}$ coincide).

*Example 25 (Transition systems revisited).* Consider the formulas of the shape $\blacktriangledown^{a_i}(t_i \multimap t_i')$ and the sequent $\Delta_{\Longrightarrow}, t \longrightarrow_b t'$ in Example 19. The interpretation there, of $b$ as the time needed to observe a transition from $t$ to $t'$, can be captured with the semiring $\mathcal{K}_c$ (Example 21). As expected, according to $+_{\mathcal{A}}$ (and then $\preceq_{\mathcal{A}}$), we prefer "faster" paths when there are different ways of going from $t$ to $t'$.

Another possible interpretation for $b$ is the *probability* of the different *independent* events (transitions) to happen. Hence, given a specific path from $t$ to $t'$, the possible values for $b$ must be less or equal to the product of the probabilities $a_i$ involved in that path. This behavior can be captured with the probabilistic semiring [4] $\mathcal{K}_p = \langle [0,1], \max, \times, 0, 1 \rangle$.

For yet another example of $\mathcal{K}_p$, consider the typical probabilistic choice in process calculi: the process $P +_\alpha Q$ chooses $P$ with probability $\alpha$ and $Q$ with probability $1-\alpha$. Following the process-as-formulas interpretation [9,18], relating process constructors with logical connectives and reductions with proof steps, the system $\mathcal{C}^\ell(\mathcal{K}_p)$ offers a very natural interpretation of the process $P +_\alpha Q$ as the formula $(\triangledown^\alpha P) \,\&\, (\triangledown^{1-\alpha} Q)$, that we can write as $P \,\&_\alpha\, Q$. For instance, if $\Gamma = \{t_1 \,\&_\alpha\, t_2, t_1 \multimap t_3, t_2 \multimap t_4\}$, then, the sequent $\Gamma \longrightarrow_b t_3$ (resp. $\Gamma \longrightarrow_b t_4$) is provable whenever $\alpha \geq b$ (resp. $1 - \alpha \geq b$).

## 5 Modalities in Positive Contexts

We have considered modalities appearing only in negative polarity. In this section, we show some problems and limitations that arise when trying to extend the labelled sequent approach to consider also positive occurrences of modalities as in the full system of subexponential linear logic (see e.g., [19]). Let us call $\mathcal{CP}^\ell(\mathbb{R}^+)$ the system resulting from $\mathcal{C}^\ell(\mathbb{R}^+)$ by adding the following *labelled promotion rules*

$$\frac{\Gamma^{\leq \triangledown^a} \longrightarrow_b A}{\Gamma \longrightarrow_b \triangledown^a A} \qquad \frac{\Gamma^{\leq \blacktriangledown^a} \longrightarrow_b A}{\Gamma \longrightarrow_b \blacktriangledown^a A}$$

where $\Gamma^{\leq \triangledown^a}$ denotes all formulas in $\Gamma$ which are of the form $\triangledown^c B$ or $\blacktriangledown^c B$ and $a \geq c$; and $\Gamma^{\leq \blacktriangledown^a}$ denotes all formulas in $\Gamma$ which are of the form $\blacktriangledown^c B$ where $a \geq c$. These rules follow the standard formulation of the promotion rule in SELL: the promotion of $!^a A$ requires all formulas of the context to be of the form $!^c B$ where $a \preceq c$ and $\preceq$ is the underlying preorder on the subexponential signature.

The following result shows that it is not possible to define a labelled cut rule for $\mathcal{CP}^\ell(\mathbb{R}^+)$ where the label of the conclusion depends exclusively on the labels of the premises.

**Theorem 26.** *There is no function $f : \mathbb{R}^+ \times \mathbb{R}^+ \to \mathbb{R}^+$ such that the rule*

$$\frac{\blacktriangledown\Gamma, \Delta_1 \longrightarrow_a A \quad \blacktriangledown\Gamma, \Delta_2, A \longrightarrow_b C}{\blacktriangledown\Gamma, \Delta_1, \Delta_2 \longrightarrow_{f(a,b)} C}\ cut$$

*is admissible in $\mathcal{CP}^\ell(\mathbb{R}^+)$.*

*Proof:* Let $p, q$ be different propositional variables, and let $A^{\otimes n}$ denote the $n$-fold multiplicative conjunction of a formula $A$. The sequents

$$\blacktriangledown^{1/k} p \longrightarrow_a \blacktriangledown^{1/k} p^{\otimes(k \cdot a)} \qquad \text{and} \qquad \blacktriangledown^{1/k} p^{\otimes(k \cdot a)} \longrightarrow_b p^{\otimes(k \cdot k \cdot a \cdot b)}$$

are provable in $\mathcal{CP}^\ell(\mathbb{R}^+)$ for all natural numbers $a, b, k$. The smallest label $f$ which makes their cut conclusion $\blacktriangledown^{1/k} p \longrightarrow_f p^{\otimes(k \cdot k \cdot a \cdot b)}$ provable in $\mathcal{CP}^\ell(\mathbb{R}^+)$ is $k \cdot a \cdot b$, which is not a function on the premise labels $a, b$. □

Note that Theorem 26 leaves open the possibility that cut is admissible w.r.t. a function $f$ which takes more information of the premises into account than just their labels. Please refer to the technical report [15, Appendix A.1] for a more detailed discussion.

## 6 Concluding Remarks and Future Work

We have introduced game semantics for fragments of (affine intuitionistic) linear logic with subexponentials (SELL [7,19,21]), culminating in labelled extensions of such systems so that $\Gamma \longrightarrow_b A$ is interpreted as: "Resource $A$ can be obtained from the resources $\Gamma$ with a budget $b$" or, alternatively, "The budget $b$ suffices to win the game $\Gamma \longrightarrow A$". For achieving that, we proposed a new interpretation for the *dereliction* rule, opposing to the standard controls in the *promotion rule*: derelicting on $\triangledown^a B, \blacktriangledown^a B$ means "paying $a$ to obtain (a copy of) $B$". Hence our games and systems offer a neater control of the resources appearing *negatively* on sequents.

There are several ways of extending and continuing this work. First of all, as signalized in Sect. 5, the quest of extending the cost conscious reasoning to modalities occurring *positively* in sequents is not trivial. Despite the obvious game interpretation of promotion that could be given in the style of [10], Theorem 26 shows that this would *not* be followed with a proof theoretical notion of cut-elimination, due to the impossibility of defining a functional notion of the cut-label. In [15] we discuss some possible paths to trail in this direction. On the other side, a philosophical discussion on the need of compositionally of dialogue games driven by a cut rule can also be done [20].

Finally, we expect that the study of costs of proofs and cut-elimination in labelled fragments of SELL may indicate a relationship between labels and bounds of computation [2], as well as give a different approach to study the complexity of cut-elimination process, specially in the multiplicative-(sub)exponential fragment [22,23].

## References


1. Abramsky, S., Jagadeesan, R.: Games and full completeness for multiplicative linear logic. J. Symb. Log. **59**(2), 543–574 (1994)
2. Accattoli, B., Graham-Lengrand, S., Kesner, D.: Tight typings and split bounds. PACMPL, 2(ICFP), 94:1–94:30 (2018)
3. Bistarelli, S., Gadducci, F.: Enhancing constraints manipulation in semiring-based formalisms. In: Brewka, G., Coradeschi, S., Perini, A., Traverso, P. (eds.) ECAI 2006, 17th European Conference on Artificial Intelligence, Including Prestigious Applications of Intelligent Systems (PAIS 2006), Proceedings. Frontiers in Artificial Intelligence and Applications, vol. 141, pp. 63–67. IOS Press (2006)
4. Bistarelli, S., Montanari, U., Rossi, F., Schiex, T., Verfaillie, G., Fargier, H.: Semiring-based CSPs and valued CSPs: frameworks, properties, and comparison. Constraints **4**(3), 199–240 (1999)
5. Blass, A.: A game semantics for linear logic. Ann. Pure Appl. Logic **56**(1–3), 183–220 (1992)
6. Blass, A.: Some semantical aspects of linear logic. Log. J. IGPL **5**(4), 487–503 (1997)
7. Danos, V., Joinet, J.-B., Schellinx, H.: The structure of exponentials: uncovering the dynamics of linear logic proofs. In: Gottlob, G., Leitsch, A., Mundici, D. (eds.) KGC 1993. LNCS, vol. 713, pp. 159–171. Springer, Heidelberg (1993). https://doi.org/10.1007/BFb0022564
8. Delande, O., Miller, D., Saurin, A.: Proof and refutation in MALL as a game. Ann. Pure Appl. Logic **161**(5), 654–672 (2010)
9. Deng, Y., Simmons, R.J., Cervesato, I.: Relating reasoning methodologies in linear logic and process algebra. Math. Struct. Comput. Sci. **26**(5), 868–906 (2016)
10. Fermüller, C.G., Lang, T.: Interpreting sequent calculi as client-server games. In: Schmidt, R.A., Nalon, C. (eds.) TABLEAUX 2017. LNCS (LNAI), vol. 10501, pp. 98–113. Springer, Cham (2017). https://doi.org/10.1007/978-3-319-66902-1_6
11. Girard, J.-Y.: Linear logic. Theor. Comput. Sci. **50**, 1–102 (1987)
12. Hyland, J.M.E., Ong, C.h.L.: Fair games and full completeness for multiplicative linear logic without the mix-rule (1993)
13. Japaridze, G.: A constructive game semantics for the language of linear logic. Ann. Pure Appl. Logic **85**(2), 87–156 (1997)
14. Lamarche, F.: Games semantics for full propositional linear logic. In: Proceedings, 10th Annual IEEE Symposium on Logic in Computer Science, San Diego, California, USA, 26–29 June 1995, pp. 464–473. IEEE Computer Society (1995)
15. Lang, T., Olarte, C., Pimentel, E., Fermüller, C.: A Game Model for Proofs with Costs. arXiv e-prints, arXiv:1906.11742, June 2019
16. Lorenzen, P.: Logik und agon. Atti Del XII Congresso Internazionale di Filosofia **4**, 187–194 (1960)
17. Melliès, P.-A.: Asynchronous games 4: a fully complete model of propositional linear logic. In: Proceedings of the 20th IEEE Symposium on Logic in Computer Science (LICS 2005), Chicago, IL, USA, 26–29 June 2005, pp. 386–395. IEEE Computer Society (2005)
18. Miller, D.: The π-calculus as a theory in linear logic: preliminary results. In: Lamma, E., Mello, P. (eds.) ELP 1992. LNCS, vol. 660, pp. 242–264. Springer, Heidelberg (1993). https://doi.org/10.1007/3-540-56454-3_13
19. Nigam, V., Miller, D.: Algorithmic specifications in linear logic with subexponentials. In: Porto, A., López-Fraguas, F.J. (eds.) PPDP, pp. 129–140. ACM (2009)

20. Dutilh Novaes, C., French, R.: Paradoxes and structural rules from a dialogue perspective. Philos. Issues **28**, 129–158 (2018)
21. Pimentel, E., Olarte, C., Nigam, V.: A proof theoretic study of soft concurrent constraint programming. TPLP **14**(4–5), 649–663 (2014)
22. Straßburger, L.: MELL in the calculus of structures. Theor. Comput. Sci. **309**(1–3), 213–285 (2003)
23. Straßburger, L., Guglielmi, A.: A system of interaction and structure IV: the exponentials and decomposition. ACM Trans. Comput. Log. **12**(4), 23:1–23:39 (2011)

# Towards a Combinatorial Proof Theory

Benjamin Ralph[1,2(✉)] and Lutz Straßburger[1,2]

[1] Inria Saclay, 1 rue Honoré d'Estienne d'Orves, 91120 Palaiseau, France
benjamin.ralph@inria.fr
[2] LIX, Ecole Polytechnique, Route de Saclay, 91128 Palaiseau Cedex, France

**Abstract.** The main part of a classical combinatorial proof is a skew fibration, which precisely captures the behavior of weakening and contraction. Relaxing the presence of these two rules leads to certain substructural logics and substructural proof theory. In this paper we investigate what happens if we replace the skew fibration by other kinds of graph homomorphism. This leads us to new logics and proof systems that we call combinatorial.

**Keywords:** Proof theory · Combinatorial proofs · Deep inference · Fibrations · Cographs

## 1 Introduction

Combinatorial proofs have been introduced by Hughes [17] to give a "syntax-free" presentation for proofs in classical propositional logic. In doing so, they give a possible response to Hilbert's 24th problem of identity between proofs [29]: two proofs are the same if they have the same combinatorial proof [1,18,27].

In a nutshell, a classical combinatorial proof consists of two parts: (i) a linear part, and (ii) a skew fibration. The linear part encodes a proof in multiplicative linear logic (MLL), whose conclusion is given as a cograph, together with an equivalence relation on the vertices encoding the axiom links of the proof. A combinatorial correctness criterion for this linear part can be given by Retoré's *critically corded* condition [22]. The skew fibration then maps this linear cograph to the cograph of the conclusion of the whole proof. This precisely captures the behaviour of weakening and contraction in a classical proof.

Recently, the theory of combinatorial proofs has been extended to intuitionistic propositional logic [16] and to relevant logics [2]. For intuitionistic logic, the linear part of a combinatorial proof has to be restricted to intuitionistic multiplicative linear logic (IMLL), and for relevant logic the skew fibration had to be restricted to a surjective weak fibration. This raises the question of what happens with other *substructural logics* that restrict (in their sequent calculus formulation) the use of contraction and weakening?

To answer this question, we need to address another issue first. Whereas the linear part of a combinatorial proof corresponds to well understood proof theory—there are sequent calculi and proof nets for most variants of MLL


S. Cerrito and A. Popescu (Eds.): TABLEAUX 2019, LNAI 11714, pp. 259–276, 2019.
https://doi.org/10.1007/978-3-030-29026-9_15

(intuitionistic/classical, commutative/non-commutative, with mix/without mix, etc.)—the skew fibration part is less obviously linked to established proof theory. In particular, the lack of any natural class of sequent calculus proofs equivalent to skew fibrations has lead to procrustean hacking away at either the sequent calculus side (via the *Homomorphism Sequent Calculus* in [18] or *deep inference* rules inside the sequent calculus in [1]), at the combinatorial proofs side (via *lax combinatorial proofs* in [18]), or at both (via *separated combinatorial/sequent calculus proofs* in [7]).

To address this problem, we will in this paper use a pure deep inference system [6,13–15] to deal with weakening and contraction. This leads to different degrees of freedom for the weakening and contraction rules than in the sequent calculus. Whereas in the sequent calculus we can allow or forbid the rules on the left and/or on the right of the turnstile, in deep inference we have other choices. Besides allowing or forbidding rules, we can restrict the rules to atomic formulas or not, and to shallow contexts or not. Furthermore, deep inference systems with contraction and weaking also admit the *medial* rule, which implements the classical implication

$$\mathsf{m}\,\frac{(A \wedge B) \vee (C \wedge D)}{(A \vee C) \wedge (B \vee D)}.$$

This rule is derivable if contraction and weakening are present in their general form [6]. However, as soon as one of the restrictions mentioned above is applied, medial is no longer derivable. Thus, the presence or absence of medial is another degree of freedom in the design of a logical system. This creates a rich variety of *structural proof systems*, some of which are familiar to deep inference proof theorists, and some of which are novel.

Another reason for using deep inference is the availability of *decomposition theorems* [4,5,23–25,28], which allow us to decompose a given formal derivation into several phases each using only a specific subset of the inference rules. Using techniques and results from work on *linear rewriting systems* [9–11,25] one can obtain decompositions that provide exactly the separation in the linear part and the contraction-weakening part of classical logic that is expressed by combinatorial proofs. This leads to a close correspondence between combinatorial proofs and deep inference (see also [26]).

On the other hand, we can study the effect on combinatorial proofs if we restrict the notion of skew fibration by set theoretic and graph theoretic means: we can demand injectivity, surjectivity, or bijectivity, or we can demand the skew fibration to be a proper fibration. Certain restrictions create already studied logics; some do not seem to correspond to logics studied in the literature. In fact, if the condition for the map can be checked in polynomial time (in its size) then we can speak of proof systems in the sense of Cook and Reckhow [8], and we will coin the term *combinatorial proof systems.*

As the main result of this paper we will establish a strong correspondence between these combinatorial proof systems and the structural proof systems that naturally arise from restricting contraction and weakening in a deep inference system, as described above.

We begin the paper in Sect. 2 by recalling the relation between formulas and cographs. Then, in Sect. 3, we recall standard classical combinatorial proofs, and in Sect. 4, we recall standard deep inference proof systems, using *open deduction*. Finally, in Sect. 5 we introduce all the necessary technical definitions that allow us to state our main result, Theorem 5.5, establishing the correspondence between homomophism classes on cographs and structural proof systems in open deduction. The remaining Sects. 6, 7, and 8 are dedicated to the proof of that theorem.

## 2 Formulas and Cographs

In this paper, we restrict our attention to classical and substructural propositional logics: decidable logics freely generated from atoms and the two connectives $\{\wedge, \vee\}$.

**Definition 2.1.** We define an inexhaustible set of positive and negative atoms $\mathcal{A} = \{a, \bar{a}, b, \bar{b}, \dots\}$, and a set of formulae, $\mathcal{F}$, generated from these atoms:

$$A ::= a \mid (A \wedge A) \mid (A \vee A)$$

We omit parentheses when there is no ambiguity.

A *logic* $\mathcal{L}$ is defined by a valuation function $\mathcal{V}_{\mathcal{L}} \colon \mathcal{F} \to \{0, 1\}$ selecting correct formulae. The valuation function for *Classical Propositional Logic*, $\mathcal{V}_{\mathsf{CPL}}$ is defined in the usual way.

**Definition 2.2.** A graph $G = \langle V_G, E_G \rangle$ consists of a set of vertices $V_G$ and a set of edges $E_G$ which are two-element subsets of $V_G$. We write $vw \in E_G$ for $\{v, w\} \in E_G$. Given a set $S$, a *labelled*, or $S$*-labelled*, graph adds a map $L_G \colon V_G \to S$, mapping vertices to elements of $S$. We assume graphs are irreflexive: for all $v \in V_G$, $vv \notin E_G$. A graph $H$ is an *induced subgraph* of $G$ if $V_H \subseteq V_G$ and $E_H = \{vw \mid v, w \in V_H, vw \in E_G\}$.

There is a useful correspondence between $\mathcal{F}$ and a certain class of graphs, called cographs. First, we construct the map from formulae to cographs by defining graphical equivalents for the logical connectives.

**Definition 2.3.** Let $G, H$ be disjoint graphs. We define respectively their *union* and their *join*:

$$\begin{aligned} G \vee H &= \langle V_G \cup V_H, E_G \cup E_H \rangle \\ G \wedge H &= \langle V_G \cup V_H, E_G \cup E_H \cup \{vw \mid v \in V_G, w \in V_H\} \rangle \end{aligned}$$

**Definition 2.4.** We define the map $\mathfrak{G}$ from formulae to $\mathcal{A}$-labelled graphs as follows:

- For $a \in \mathcal{A}$, $\mathfrak{G}(a) = \langle v, \emptyset \rangle$, with $L(v) = a$.
- $\mathfrak{G}(A \vee B) = \mathfrak{G}(A) \vee \mathfrak{G}(B)$, $\mathfrak{G}(A \wedge B) = \mathfrak{G}(A) \wedge \mathfrak{G}(B)$.

We call any $\mathcal{A}$-labelled graphs in the image of $\mathfrak{G}$ a *cograph*, and we denote the set of cographs as $\mathcal{G}$. For ease, we will write $V_A$ for $V_{\mathfrak{G}(A)}$ and $E_A$ for $E_{\mathfrak{G}(A)}$.

We can characterise cographs without constructing the formulae they map from.

**Definition 2.5.** A graph $G$ of four distinct vertices $V_G = \{v, w, u, z\}$ with $vw, wu, uz \in E_G$ and $vu, vz, wz \notin E_G$ is called a $P_4$ graph.

$$\begin{array}{c} v - w \\ \diagup \\ u - z \end{array}$$

**Proposition 2.6.** *Cographs are exactly those graphs that are $P_4$-free, i.e., they do not contain a $P_4$ graph as an induced subgraph.*

*Proof.* This has been proved in many places, e.g. [12]. □

**Definition 2.7.** A *context* $K\{\ \}$ is a function from $\mathcal{F} \to \mathcal{F}$, created by replacing exactly one instance of an atom in a formula with a hole $\{\ \}$.

**Definition 2.8.** We define the following two equivalence relations on formulae, with $\star \in \{\vee, \wedge\}$:

$$K\{A \star B\} \equiv_C K\{B \star A\} \qquad K\{A \star (B \star C)\} \equiv_A K\{(A \star B) \star C\}$$

The equivalence relation $\equiv$, *formula equivalence*, is the reflexive, symmetric and transitive closure of $\equiv_C \cup \equiv_A$.

This definition of formula equivalence coincides exactly with the equivalence classes induced by the cograph map $\mathfrak{G}$:

**Proposition 2.9.** *For all formulae, $A \equiv B$ iff $\mathfrak{G}(A) = \mathfrak{G}(B)$.*

*Proof.* Straightforward induction on the size of the formulas (see e.g. [12]). □

## 3 Combinatorial Proofs

Usually, the linear (or multiplicative) part of a combinatorial proof is defined graph theoretically. Here, since we are concerned chiefly with the structural part, we simply use a sequent calculus.

**Definition 3.1.** We define the logic of $\mathsf{MLL}$ by the following one-sided sequent calculus system, with multiset sequents:

$$\mathsf{ax}\,\frac{}{\vdash a, \bar{a}} \qquad \wedge r\,\frac{\vdash \Gamma, A \quad \vdash \Delta, B}{\vdash \Gamma, \Delta, A \wedge B} \qquad \vee r\,\frac{\vdash \Gamma, A, B}{\vdash \Gamma, A \vee B}$$

$\mathcal{V}_{\mathsf{MLL}}(A) = 1$ iff there is a proof of $A$ in this sequent system.

**Proposition 3.2.** *If* $A \equiv B$ *(and therefore* $\mathfrak{G}(A) = \mathfrak{G}(B)$*), then* $\mathcal{V}_{\mathsf{MLL}}(A) = \mathcal{V}_{\mathsf{MLL}}(B)$.

*Proof.* Straightforward, since sequents are multisets. □

Due to the above definition, we can define $\mathcal{V}_{\mathsf{MLL}}(G) = \mathcal{V}_{\mathsf{MLL}}(A)$ for any $A$ with $\mathfrak{G}(A) = G$.

**Definition 3.3.** A *graph homomorphism* $f\colon V_G \to V_H$ is a map such that for all $e = vw \in E_G$, there is an edge $f(e) = f(v)f(w) \in E_H$. A *skew fibration* $f\colon G \to H$ is a graph homomorphism such that for every $v \in V_G, z \in V_H$ with $f(v)z \in E_H$, there is some $vw \in E_G$ such that:

**(SF1)** $f(w)z \notin E_H$.

Figure 1 illustrates this condition and gives two examples of skew fibrations.

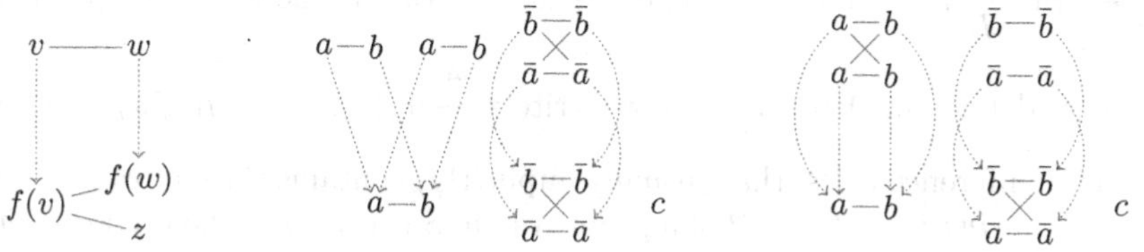

**Fig. 1.** On the left, we show a pictorial representation of the condition **SF1**. In the centre and on the right, two skew fibrations are shown, that are in fact combinatorial proofs of the formula $(a \wedge b) \vee ((\bar{a} \vee \bar{b}) \wedge (\bar{a} \vee \bar{b})) \vee c$.

With the technology that we have built up, we are able to define combinatorial proofs for classical logic.

**Definition 3.4.** Let $A \in \mathcal{F}$. $\phi = \langle G_\phi, f_\phi \rangle$ is a correct *combinatorial proof* [17] of $A$ for classical propositional logic if:

- $\mathcal{V}_{\mathsf{MLL}}(G_\phi) = 1$,
- $f_\phi : G_\phi \to \mathfrak{G}(A)$ is a skew fibration.

The two examples in Fig. 1 are combinatorial proofs.

**Remark 3.5.** A graph theoretic counterpart to the first condition is that the cograph is non-empty critically chorded R&B-cograph [22,26]. *Nicely coloured* graphs [17] correspond to theorems of MLL together with the mix rule.

## 4 Open Deduction

We introduce the deep inference proof formalism of *open deduction* [14].

**Definition 4.1.** The set of *prederivations*, $\mathcal{P} = \{\phi, \psi, \dots\}$ is generated by the following grammar:

$$\phi ::= A \mid (\phi \wedge \phi) \mid (\phi \vee \phi) \mid \frac{\phi}{\phi}$$

We define two functions *premise* and *conclusion*, $\mathsf{pr}, \mathsf{cn} \colon \mathcal{P} \to \mathcal{F}$,

- $\mathsf{pr}(A) = \mathsf{cn}(A) = A$ for $A \in \mathcal{F}$;
- $\mathsf{pr}(\phi \star \psi) = \mathsf{pr}(\phi) \star \mathsf{pr}(\psi)$, $\mathsf{cn}(\phi \star \psi) = \mathsf{cn}(\phi) \star \mathsf{cn}(\psi)$, for $\star \in \{\wedge, \vee\}$;
- $\mathsf{pr}\left(\dfrac{\phi}{\psi}\right) = \mathsf{pr}(\phi)$, $\mathsf{cn}\left(\dfrac{\phi}{\psi}\right) = \mathsf{cn}(\psi)$

**Definition 4.2.** An *inference rule* $\rho$ is a polynomial-time decidable relation on $\mathcal{F}$. We write $\dfrac{A}{B}\rho$ if $\langle A, B\rangle \in \rho$. The *equivalence rule* is the formula equivalence relation as defined in Definition 2.8, we write $\equiv \dfrac{A}{B}$ if $A \equiv B$. A *proof system* $\mathsf{S}$ is a finite set of inference rules, that (usually implicitly) contains the equivalence rule. The set of *derivations*, $\mathcal{D}_\mathsf{S} \subseteq \mathcal{P}$ of a proof system $\mathsf{S}$ is precisely the prederivations where vertical composition $\dfrac{\phi}{\psi}$ is restricted to cases where $\rho\,\dfrac{\mathsf{cn}(\phi)}{\mathsf{pr}(\psi)}$ is a correct instance for some $\rho \in \mathsf{S}$.

**Remark 4.3.** Since there is no need in this paper, we do not define proofs, i.e. derivations from no premise.

**Definition 4.4.** We define the *structural* rules of *contraction, atomic contraction, weakening* and *medial*:

$$\mathsf{c}{\downarrow}\,\frac{A \vee A}{A} \qquad \mathsf{ac}{\downarrow}\,\frac{a \vee a}{a} \qquad \mathsf{w}{\downarrow}\,\frac{B}{A \vee B} \qquad \mathsf{m}\,\frac{(A \wedge B) \vee (C \wedge D)}{(A \vee C) \wedge (B \vee D)}$$

An instance of $\mathsf{c}{\downarrow}, \mathsf{w}{\downarrow}, \mathsf{ac}{\downarrow}$ is called *shallow* if they are not in the scope of a conjunction. We denote such instances by $\mathsf{sc}{\downarrow}$, $\mathsf{sw}{\downarrow}$ and $\mathsf{sac}{\downarrow}$ respectively. In Fig. 2 we give two examples of derivations with these rules.

$$\mathsf{sw}{\downarrow}\,\frac{\mathsf{sc}{\downarrow}\,\dfrac{(a \wedge b) \vee (a \wedge b)}{a \wedge b} \vee ((\bar{a} \vee \bar{b}) \wedge (\bar{a} \vee \bar{b}))}{(a \wedge b) \vee ((\bar{a} \vee \bar{b}) \wedge (\bar{a} \vee \bar{b})) \vee c} \qquad \left(\mathsf{ac}{\downarrow}\,\frac{a \vee a}{a} \wedge \mathsf{ac}{\downarrow}\,\frac{b \vee b}{b}\right) \vee \mathsf{sw}{\downarrow}\,\frac{(a \wedge \bar{a}) \vee (b \wedge \bar{b})}{\mathsf{m}\,\dfrac{(a \wedge \bar{a}) \vee (b \wedge \bar{b})}{(a \vee b) \wedge (\bar{a} \vee \bar{b})} \vee c}$$

**Fig. 2.** Two simple derivations with the same conclusion $(a \wedge b) \vee ((\bar{a} \vee \bar{b}) \wedge (\bar{a} \vee \bar{b}))$, the first with non-atomic shallow contraction and the second with medial and deep atomic contraction.

We are now almost ready to define the class of deep inference proofs that will correspond to combinatorial proofs. First, however, we define the notion of relative strength between proof systems.

**Definition 4.5.** Let $\mathsf{S}$ be a proof system and $\rho$ an inference rule. If for every instance $\rho\frac{A}{B}$ there is a derivation $\begin{array}{c}A\\ \|\mathsf{S}\\ B\end{array}$, we say $\rho$ is *derivable* for $\mathsf{S}$. If every rule $\rho \in \mathsf{S}'$ is derivable for $\mathsf{S}$, then we write $\mathsf{S}' \preceq \mathsf{S}$, where $\preceq$ is a partial order on proof systems. We write $\mathsf{S} \simeq \mathsf{S}'$ if $\mathsf{S} \preceq \mathsf{S}'$ and $\mathsf{S}' \preceq \mathsf{S}$.

**Proposition 4.6.** $\mathsf{c}\!\downarrow$ *is derivable for* $\{\mathsf{ac}\!\downarrow, \mathsf{m}\}$, $\mathsf{m}$ *is derivable for* $\{\mathsf{w}\!\downarrow, \mathsf{c}\!\downarrow\}$. *Hence* $\{\mathsf{w}\!\downarrow, \mathsf{c}\!\downarrow\} \simeq \{\mathsf{w}\!\downarrow, \mathsf{m}, \mathsf{ac}\!\downarrow\}$ *[6].*

*Proof.* The first is proven by a simple structural induction on formulae, the key step being as follows:

$$\mathsf{c}\!\downarrow\frac{(A \wedge B) \vee (A \wedge B)}{A \vee B} \quad \longrightarrow \quad \mathsf{m}\frac{(A \wedge B) \vee (A \wedge B)}{\mathsf{c}\!\downarrow\dfrac{A \vee A}{A} \wedge \mathsf{c}\!\downarrow\dfrac{B \vee B}{B}}$$

The second needs only a simple rewrite:

$$\mathsf{m}\frac{(A \wedge B) \vee (C \wedge D)}{(A \vee C) \wedge (B \vee D)} \longrightarrow \mathsf{c}\!\downarrow\frac{\left(\mathsf{w}\!\downarrow\dfrac{A}{A \vee C} \wedge \mathsf{w}\!\downarrow\dfrac{B}{B \vee D}\right) \vee \left(\mathsf{w}\!\downarrow\dfrac{C}{A \vee C} \wedge \mathsf{w}\!\downarrow\dfrac{D}{B \vee D}\right)}{(A \vee C) \wedge (B \vee D)}$$

□

A formal correspondence between derivations in open deduction and skew fibrations has been established in other works.

**Proposition 4.7.** *Let* $\begin{array}{c}A\\ \phi\,\|\,\{\mathsf{w}\downarrow,\mathsf{m},\mathsf{ac}\downarrow\}\\ B\end{array}$ *be a derivation. Then there is a skew fibration* $f\colon \mathfrak{G}(A) \to \mathfrak{G}(B)$. *We denote this skew fibration* $\mathfrak{G}(\phi)$.

**Proposition 4.8.** *If* $A, B \in \mathcal{F}$ *and* $f\colon \mathfrak{G}(A) \to \mathfrak{G}(B)$ *is a skew fibration, then there is a derivation* $\begin{array}{c}A\\ \phi\,\|\,\{\mathsf{w}\downarrow,\mathsf{m},\mathsf{ac}\downarrow\}\\ B\end{array}$.

*Proof.* Each proposition follows from one direction of [25, Theorem 7.8] and completeness and soundness of combinatorial proofs, respectively. □

# 5 Homomorphisms Classes and Structural Proof Systems

In Sect. 2 we have shown the correspondence between formulae and cographs, and in Sect. 4 we have shown the correspondence between derivations in open deduction and skew fibrations. Now, we take this one step further, to a correspondence between deep inference proof systems and classes of homomorphisms, constructing two isomorphic lattices (partially ordered sets with meets and joins), of proof systems with respect to derivability, and homomorphism classes with respect to inclusion.

**Definition 5.1.** If $\mathsf{S} \preceq \{\mathsf{w}\downarrow, \mathsf{ac}\downarrow, \mathsf{m}\} \simeq \{\mathsf{w}\downarrow, \mathsf{c}\downarrow\}$, we say $\mathsf{S}$ is *structural.* We denote the set of structural proof systems as $\mathcal{S}_{\downarrow}$. Let $\mathsf{S}$ be a structural proof system. We define $\mathfrak{G}(\mathsf{S}) = \{\mathfrak{G}(\phi) \mid \phi \in \mathcal{D}_{\mathsf{S}}\}$, the set of cograph homomorphisms generated from derivations in $\mathsf{S}$.[1]

We now need to introduce some terminology to better characterise classes of homomorphisms.

**Definition 5.2.** A *fibration* $f\colon G \to H$ is a graph homomorphism such that for every $v \in V_G, z \in V_H$ with $f(v)z \in E_H$, there is some $vw \in E_G$ such that:

**(F1)** $f(w) = z$.
**(F2)** For all $vw'$ with $f(w') = z$ we have $w' = w$ (i.e. $w$ is unique).

A homomorphism is a *weak fibration* if it has property **(F1)**, but it may not have property **(F2)**. Due to irreflexivity, a fibration is always a weak fibration which is always a skew fibration. A graph homomorphism $f\colon G \to H$ is *full* if $f(v)f(w) \in E_H$ implies $vw \in E_G$.

The first example in Fig. 1 is a fibration, the second one is neither a fibration nor a weak fibration.

**Definition 5.3.** In Fig. 3, we define sets of graph homomorphisms, all of which are subsets of the set SkFib of skew fibrations.

| | | | |
|---|---|---|---|
| Iso | Isomorphisms | Bij | Bijections |
| Inj | Injective Skew Fibrations | Sur | Surjective Skew Fibrations |
| FInj | Full Injective Skew Fibrations | FSur | Full Surjective Skew Fibrations |
| Fib | Fibrations | SFib | Surjective Fibrations |
| FFib | Full Fibrations | FSkFib | Full Skew Fibrations |
| FIFib | Full Injective Fibrations | FSFib | Full Surjective Fibrations |
| WFib | Weak Fibrations | SWFib | Surjective Weak Fibrations |
| FWFib | Full Weak Fibrations | FSWFib | Full Surjective Weak Fibrations |

**Fig. 3.** Sets of graph homomorphisms

[1] The attentive reader might notice that $\mathfrak{G}$ can be seen as a functor between suitably defined categories. However, in order to make this paper accessible to a broader audience we decided not to use any category theoretical concepts here.

**Proposition 5.4.** $\mathfrak{G}\colon \langle \mathcal{S}_\downarrow, \preceq\rangle \to \langle 2^{\mathsf{SkFib}}, \subseteq\rangle$ *is an order-preserving injection from the set of structural proof systems into the set of subsets of cograph skew fibrations. In particular,* $\mathfrak{G}(\{\mathsf{c}\!\downarrow, \mathsf{w}\!\downarrow\}) = \mathfrak{G}(\{\mathsf{ac}\!\downarrow, \mathsf{m}, \mathsf{w}\!\downarrow\}) = \mathsf{SkFib}$.

*Proof.* We have that $\mathfrak{G}(\mathcal{S}_\downarrow) \subseteq \mathsf{SkFib}$ from Proposition 4.7, and $\mathsf{SkFib} \subseteq \mathfrak{G}(\mathcal{S}_\downarrow)$ from Proposition 4.8. Order preservation is clear from the definitions of $\preceq$ and $\mathfrak{G}$. □

In the remainder of the paper we will be study the two lattices, $\langle \mathcal{S}_\downarrow, \preceq\rangle$ and $\langle 2^{\mathsf{SkFib}}, \subseteq\rangle$; identifying which classes of homomorphisms correspond to structural proof systems and vice versa.

**Theorem 5.5.** *The diagram in Fig. 4 establishes corresponding points in the lattices* $\langle \mathcal{S}_\downarrow, \preceq\rangle$ *and* $\langle 2^{\mathsf{SkFib}}, \subseteq\rangle$*, as explained in the caption.*

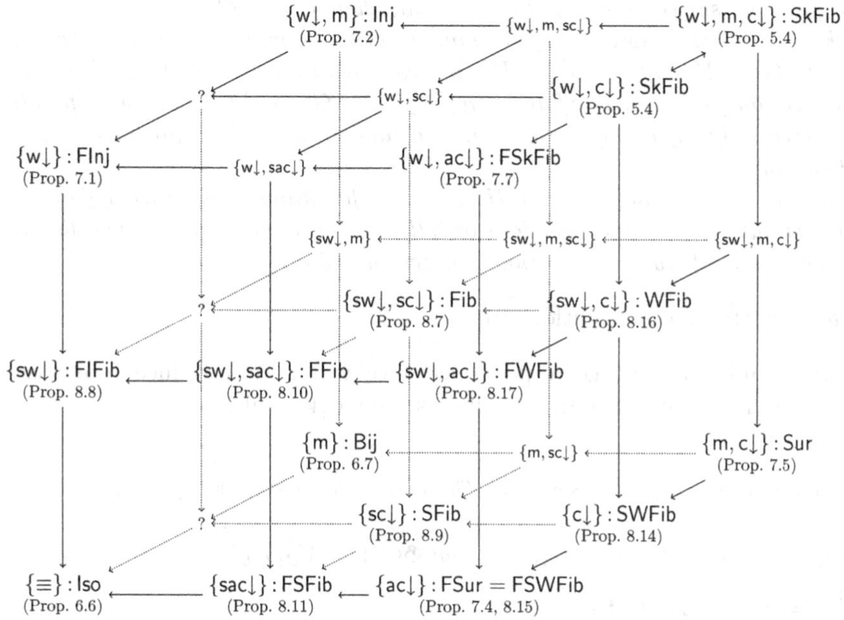


**Fig. 4.** At each point of the cube, the referenced proposition proves that $\mathfrak{G}$ maps from the proof system to the homomorphism class also at that point. Question marks refer to undefinable proof systems, and proof systems without propositions do not yet have proven homomorphism class equivalents—we do not suspect that any of these are of much interest.

## 6 Basic Correspondences

Just as we can compose derivations horizontally and vertically, we can also freely compose graph homomorphisms in corresponding ways.

**Definition 6.1.** Let $f\colon G \to H$ and $f'\colon G' \to H'$ be graph homomorphisms. We define their union $f \vee f'\colon G \vee G' \to H \vee H'$ and join $f \wedge f'\colon G \wedge G' \to H \wedge H'$ such that the restrictions of each function to $G$ or $G'$ are $f$ or $f'$ respectively. We denote the identity functions as $\mathsf{id}_G\colon G \to G$ and empty functions as $\mathsf{e}_G\colon \emptyset \to G$.

In most cases, composing homomorphisms preserves their properties.

**Proposition 6.2.** *Let $G, G', H, H'$ be cographs.*

1. *Any isomorphism $i\colon G \to H$ is a fibration and thus also a skew fibration.*
2. *The map $\mathsf{w}\!\downarrow = id_G \vee e_H\colon G \to G \vee H$, is a full injective fibration.*
3. *The map $\mathsf{c}\!\downarrow\colon G \vee G \to G$, which acts as the identity on each copy of $G$, is a full surjective fibration.*
4. *The map $\mathsf{m} : (G \wedge H) \vee (G' \wedge H') \to (G \vee G') \wedge (H \vee H')$ which acts as the identity on $G, G', H, H'$, is a bijective skew fibration (but not a fibration).*
5. *If $f\colon G \to H$ and $f'\colon G' \to H'$ are skew fibrations (respectively fibrations, injections, surjections, bijections or full) then $f \vee f'\colon G \vee G' \to H \vee H'$ is a skew fibration (respectively fibration, injection, surjection, bijection or full).*
6. *If $f\colon G \to H$ and $f'\colon G' \to H'$ are skew fibrations (respectively injections, surjections, bijections or full) then $f \wedge f'\colon G \wedge G' \to H \wedge H'$ are skew fibration (respectively injection, surjection, bijection or full). This property does not hold for fibrations.*
7. *If $f\colon G \to G'$ and $g\colon G' \to H$ are skew fibrations (respectively fibrations, injections, surjections, bijections or full), then $g \circ f\colon G \to H$ is a skew fibration (respectively fibration, injection, surjection, bijection or full).*

*Proof.* Omitted but straightforward. □

Before establishing the correspondences, we will first introduce some useful results from work investigating medial as a rewriting rule.

**Theorem 6.3.** *[25, Theorem 5.1] There is a derivation* $\begin{array}{c} A \\ \Vert \{\mathsf{m}\} \\ B \end{array}$ *iff the following properties hold of $\mathfrak{G}(A) = \langle V_A, E_A \rangle$ and $\mathfrak{G}(B) = \langle V_B, E_B \rangle$.*

1. *$V_A = V_B$ and $E_A \subseteq E_B$*
2. *For all $a, d \in V_A$, s.t. $ad \in E_B \backslash E_A$, there are $b, c \in V_A$ s.t.*

$$\begin{array}{c} a - b \\ c - d \end{array} \subseteq \mathfrak{G}(A) \qquad \begin{array}{c} a - b \\ \times \\ c - d \end{array} \subseteq \mathfrak{G}(B)$$

Using this theorem, it can be shown that skew fibrations correspond to a further stratified subclass of decomposed derivation.

**Definition 6.4.** A derivation in the following form is said to be *structurally decomposed.*

$$\begin{array}{c} A \\ \| \{\mathsf{m}\} \\ A' \\ \| \{\mathsf{ac}\!\downarrow\} \\ A'' \\ \| \{\mathsf{w}\!\downarrow\} \\ B \end{array}$$

**Theorem 6.5.** *There is a skew fibration* $\mathfrak{G}(A) \to \mathfrak{G}(B)$ *iff there is a structurally decomposed derivation* $\begin{array}{c} A \\ \| \{\mathsf{m},\mathsf{ac}\!\downarrow,\mathsf{w}\!\downarrow\} \\ B \end{array}$.

*Proof.* [25, Theorem 7.8] A refinement of Propositions 4.7 and 4.8.

Using this theorem, we can prove our first correspondences.

**Proposition 6.6.** $\mathfrak{G}(\{\equiv\}) = \mathsf{Iso}$.

*Proof.* Corollary of Proposition 2.9. □

The next proposition, that the proof system $\{\mathsf{m}\}$ corresponds to bijective skew fibrations, has been informally noted by Hughes [18, Section 9].

**Proposition 6.7.** $\mathfrak{G}(\{\mathsf{m}\}) = \mathsf{Bij}$

*Proof.* We get inclusion from Proposition 6.2 parts 4, 5, 6 and 7. For equality, we observe that if there is a bijective skew fibration from $\mathfrak{G}(A)$ to $\mathfrak{G}(B)$, then by Theorem 6.5, we must have a structurally decomposed derivation $\phi \begin{array}{c} A \\ \| \{\mathsf{m},\mathsf{ac}\!\downarrow,\mathsf{w}\!\downarrow\} \\ B \end{array}$. Since any instance of $\mathsf{ac}\!\downarrow$ in $\phi$ would break injectivity, and any instance of $\mathsf{w}\!\downarrow$ would break surjectivity, we must have that $\phi \begin{array}{c} A \\ \| \{\mathsf{m}\} \\ B \end{array}$. □

## 7 Restricting c↓ or w↓: Affine and Relevance Logic

Certain correspondences between proof systems and homomorphism classes have already been established in [18,25] and [2] and others are simple corollaries of these.

**Proposition 7.1.** *[25, Proposition 7.6]* $\mathfrak{G}(\{\mathsf{w}\!\downarrow\}) = \mathsf{FInj}$.

**Proposition 7.2.** $\mathfrak{G}(\{\mathsf{w}\!\downarrow, \mathsf{m}\}) = \mathsf{Inj}$.

*Proof.* Again, we get inclusion from Proposition 6.2. For equality, we observe that if there is an (full) injective skew fibration from $\mathfrak{G}(A)$ to $\mathfrak{G}(B)$, then by Theorem 6.5, we must have a structurally decomposed derivation $\phi \,\overset{A}{\underset{B}{\|}}$. Since any instance any instance of $\mathsf{ac}\!\downarrow$ would break injectivity, we must have that

$$\phi = \begin{array}{c} A \\ {\scriptstyle \phi_1}\,\|\,{\scriptstyle \{\mathsf{m}\}} \\ A' \\ {\scriptstyle \phi_2}\,\|\,{\scriptstyle \{\mathsf{w}\downarrow\}} \\ B \end{array}$$

This gives us that $\mathfrak{G}(\{\mathsf{w}\!\downarrow, \mathsf{m}\}) = \mathsf{Inj}$. For $\mathfrak{G}(\{\mathsf{w}\!\downarrow\}) = \mathsf{FInj}$, we observe that if $\phi$ contains an instance of medial, then, by Theorem 6.3, there is some $(\mathfrak{G}(\phi_1)(v))(\mathfrak{G}(\phi_1)(w)) \in E_{A'}$ such that $vw \notin E_A$. Since weakenings do not alter edges between existing vertices in the cographs, $vw \in E_B$, and therefore $\mathfrak{G}(\phi)$ is not full. □

**Remark 7.3.** Allowing weakening but not contraction gives us *affine logic.* Therefore, insisting that the skew fibrations of combinatorial proofs are fully injective leads to combinatorial proofs for affine logic [18].

**Proposition 7.4.** *[25, Proposition 7.6]* $\mathfrak{G}(\{\mathsf{ac}\!\downarrow\}) = \mathsf{FSur} = \mathsf{FSWFib}$.

**Proposition 7.5.** $\mathfrak{G}(\{\mathsf{ac}\!\downarrow, \mathsf{m}\}) = \mathsf{Sur}$.

*Proof.* Once more, we get inclusion from Proposition 6.2. For equality, we observe that if there is a bijective skew fibration from $\mathfrak{G}(A)$ to $\mathfrak{G}(B)$, then by Theorem 6.5, we must have a structurally decomposed derivation $\phi \,\overset{A}{\underset{B}{\|}}$. Since any instance of $\mathsf{w}\!\downarrow$ would break surjectivity, we must have if $\mathfrak{G}(\phi)$ is a surjection, then

$$\phi = \begin{array}{c} A \\ {\scriptstyle \phi_1}\,\|\,{\scriptstyle \{\mathsf{m}\}} \\ A' \\ {\scriptstyle \phi_2}\,\|\,{\scriptstyle \{\mathsf{ac}\downarrow\}} \\ B \end{array}$$

This gives us $\mathfrak{G}(\{\mathsf{ac}\!\downarrow, \mathsf{m}\}) = \mathsf{Sur}$. For $\mathfrak{G}(\{\mathsf{w}\!\downarrow\}) = \mathsf{FSur}$, we observe that if $\phi$ contains an instance of medial, then, by Theorem 6.3, there is some $(\mathfrak{G}(\phi_1)(v))(\mathfrak{G}(\phi_1)(w)) \in E_{A'}$ such that $vw \notin E_A$. Since atomic contractions only contract vertices with no edge between them, the images of $v$ and $w$ under $\mathfrak{G}(\phi)$ are distinct and $(\mathfrak{G}(\phi)(v))(\mathfrak{G}(\phi)(w)) \in E_B$, and so $\mathfrak{G}(\phi)$ is not full. □

**Remark 7.6.** Adding contraction but not weakening to MLL gives us *relevance logic.* Therefore, insisting that the skew fibrations of combinatorial proofs are surjective leads to combinatorial proofs for relevance logic [2,18].

Just leaving out medial, we get full skew fibrations.

**Proposition 7.7.** $\mathfrak{G}(\{\mathsf{ac}{\downarrow}, \mathsf{w}{\downarrow}\}) = \mathsf{FSkFib}$.

*Proof.* Inclusion follows from Proposition 6.2. For equality, let $f\colon \mathfrak{G}(A) \to \mathfrak{G}(B)$ be a full skew fibration, and consider its corresponding structurally decomposed derivation:

$$\phi = \begin{array}{c} A \\ \phi_1 \,\|\, \{\mathsf{m}\} \\ A' \\ \phi_2 \,\|\, \{\mathsf{ac}{\downarrow}, \mathsf{w}{\downarrow}\} \\ B \end{array}$$

Assume there is some $\mathfrak{G}(\phi)(vw) \in E_B$ with $vw \notin E_{A'}$. From the inclusion result, we have that $\mathfrak{G}(\phi_2)$ is a full skew fibration. In particular, since $\mathfrak{G}(\phi)(vw) \in E_B$, we have $\mathfrak{G}(\phi_1)(vw) \in E'_A$. Therefore $\phi_1$ must contain at least one medial rule. □

## 8 Restricting to Shallow Inference: A Logic of Fibrations

We now come on to logics not yet studied: what happens if we do not insist on either injectivity or surjectivity, but that the skew fibration is a graph fibration? It is instructive to turn to the simplest possible examples that are skew fibrations but not fibrations, in Fig. 5. The left hand derivation fails condition **F1** and the right hand derivation fails **F2**. In both cases, it is precisely the *deepness* of the rules that breaks each condition: if the contraction or weakening was in a disjunction with $b$, both would still be fibrations.

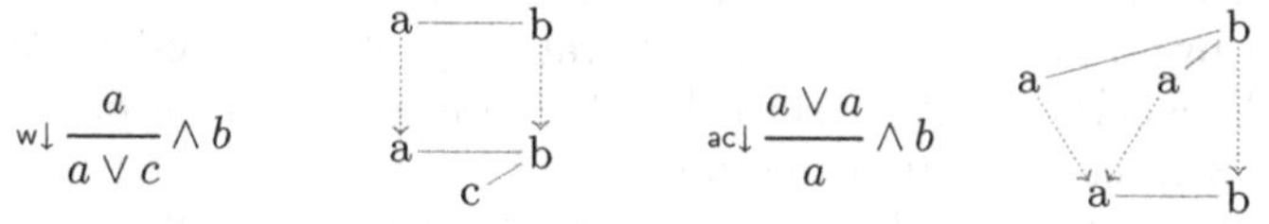


**Fig. 5.** Simple examples of skewed fibrations

**Definition 8.1.** A path $v_0 \ldots v_n$ in a graph $G$ is a sequence of vertices such that $v_i v_{i+1} \in E_G$ for $0 \leq i < n$. Two vertices of a graph $v, w \in V_G$ are *connected* if there is a path from $v$ to $w$. A graph (or subgraph) $G$ is *connected* if any two vertices in $G$ are connected. A subset $V' \subseteq V_G$ is connected if there is a path between any two vertices in $V'$ (the path does not need to stay in $V'$). A maximal connected subset of vertices is called a *component*.

**Proposition 8.2.** *Let $G$ be a cograph. If $v$ and $w$ are connected, then either $vw \in E_G$, or there is some $z \in V_G$ with $vz, zw \in E_G$.*

*Proof.* If the shortest path between $v$ and $w$ has three edges or more, then the first four vertices in that path will form a $P_4$ subgraph. □

**Proposition 8.3.** *Let $f\colon G \to H$ be a fibration between cographs. If $v, w \in G$ and $v \neq w$ and $f(v) = f(w)$, then $v$ and $w$ are in different components.*

*Proof.* Assume $v$ and $w$ are connected, so, since $f$ is a homomorphism, and $f(v)f(w) \notin E_G$ due to irreflexivity, there must be some $z \in G$ with $vz, zw \in G$, breaking the uniqueness property **(F2)** at $z$. □

**Proposition 8.4.** *Let $f\colon G \to H$ be a fibration between cographs. If $v, w \in H$ are connected, then either they are both in the image of $f$ or both not.*

*Proof.* We assume (WLOG) that $vw \in H$, $v$ is in the image of $f$, but $w$ is not, the fibration property **(F1)** breaks at $v$. □

**Proposition 8.5.** *If $f\colon G \to H$ is a fibration, and $H_1$ is a component of $H$, then $f^{-1}(H_1)$ is the union of zero or more copies of $H_1$.*

*Proof.* If some vertex in $H_1$ has a non-empty pre-image in $f$ then, by Proposition 8.4, every vertex does. Let $v' \in H_1$. By Proposition 8.3, each vertex $v \in f^{-1}(v')$ is in a different component, and for each edge $v'w'_i$ and vertex $v$ there is a unique pre-image $vw_i$. Thus we can progressively recreate the whole component from a single vertex. □

**Example 8.6.** In Fig. 2, the $\mathsf{c}\!\downarrow$ in the left hand proof is shallow, but the two instances of $\mathsf{ac}\!\downarrow$ in the right hand proof are not.

**Proposition 8.7.** $\mathfrak{G}(\{\mathsf{sc}\!\downarrow, \mathsf{sw}\!\downarrow\}) = \mathsf{Fib}$.

*Proof.* Inclusion from Proposition 6.2, as usual, noting that we forbid horizontal composition by $\wedge$. For equality, take a fibration $f\colon \mathfrak{G}(A) \to \mathfrak{G}(B)$. Write $B$ as $B_1 \vee \ldots \vee B_n$, where each $B_i$ is such that $\mathfrak{G}(B_i)$ is a component of $B$. Following Proposition 8.5, let $m_i \geq 0$ be the number of pre-images $\mathfrak{G}(B_i)$ has in $B$. We can rearrange the $B_i$ such that there is some $k$ with $n_i = 0$ iff $i \leq k$. Then, we can construct the following derivation:

$$\phi = \ \mathsf{sw}\!\downarrow^{k-1}\frac{\mathsf{sc}\!\downarrow^{n}\dfrac{B_{k+1} \vee \ldots \vee B_{k+1}}{B_{k+1}} \vee \ldots \vee \mathsf{sc}\!\downarrow^{n}\dfrac{B_n \vee \ldots \vee B_n}{B_n}}{B_1 \vee \ldots \vee B_k \vee \ldots \vee B_{k+1} \vee \ldots \vee B_n}$$

where $\mathfrak{G}(\phi) = f$. □

**Proposition 8.8.** $\mathfrak{G}(\{\mathsf{sw}\!\downarrow\}) = \mathsf{FIFib}$.

**Proposition 8.9.** $\mathfrak{G}(\{\mathsf{sc}\!\downarrow\}) = \mathsf{SFib}$.

**Proposition 8.10.** $\mathfrak{G}(\{\mathsf{sw}\!\downarrow, \mathsf{sac}\!\downarrow\}) = \mathsf{FFib}$.

**Proposition 8.11.** $\mathfrak{G}(\{\mathsf{sac}\!\downarrow\}) = \mathsf{FSFib}$.

*Proof.* All four are straightforward corollaries of Proposition 8.7. □

We can adapt the above approach slightly for the case with weak fibrations.

**Definition 8.12.** Let $f: \mathfrak{G}(A) \to \mathfrak{G}(B)$ be a weak fibration, if $C$ is a subformula of $B$ with $\mathfrak{G}(C)$ a connected subgraph but $f^{-1}(\mathfrak{G}(C))$ not a connected subgraph, we say that $C$ is a *contracted* subformula. If there is no larger subfomula of $B$ with this property, we say that $C$ is a *maximal* contracted subformula.

**Proposition 8.13.** *Let $f: \mathfrak{G}(A) \to \mathfrak{G}(B)$ be a surjective weak fibration, with $B = K_B\{C\}$ and $C$ a maximal contracted subformula of $B$. Define $B' = K_B\{C \vee C\}$. Then we can find a surjective weak fibration $f': \mathfrak{G}(A) \to \mathfrak{G}(B')$.*

*Proof.* If $\mathfrak{G}(C)$ is a component of $\mathfrak{G}(B)$, then it is straightforward. If not, then since $C$ is a contracted subformula, $f^{-1}(\mathfrak{G}(C))$ is a disconnected subgraph of $\mathfrak{G}(A)$. Denote the components of $f^{-1}(\mathfrak{G}(C))$ as $C_1, \ldots, C_n$, and define $C_l = \{v_l \mid v \in C\}$ and $C_r = \{v_r \mid v \in C\}$. We have:

$$V_A = V_{K_A} \cup \bigcup_1^n (V_{C_i}) \quad V_B = V_{K_B} \cup V_C \quad V'_B = V_{K_B} \cup V_{C_l} \cup V_{C_r}$$

We define the homomorphisms $f': \mathfrak{G}(A) \to \mathfrak{G}(B)$ and $c: \mathfrak{G}(B') \to \mathfrak{G}(B)$ (Fig. 6):

$$f'(v) = \left\{ \begin{array}{ll} f(v) : v \in V_{K_A} \\ v_l \quad : v \in V_{C_1} \\ v_r \quad : v \in V_{C_i}, i > 1 \end{array} \right\} \qquad c(v) = \left\{ \begin{array}{l} v \ : v \in V_{K_B} \\ w : v \in V_{C_l}, v = w_l \\ w : v \in V_{C_r}, v = w_r \end{array} \right\}$$

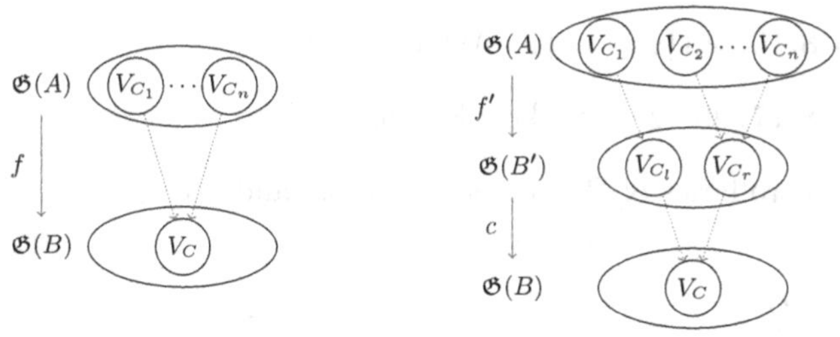


**Fig. 6.** Constructing $f'$ in Proposition 8.13

Since $f$ is a homomorphism and a surjective weak fibration, $f(C_i) = C$ for every $1 \leq i \leq n$, so $f = cf'$ and $f'$ is surjective. We now need to show that $f'$ is a weak fibration. Let $f'(w)z \in E_{B'}$, we need to show that there is some $\hat{z}$ with $w\hat{z} \in E_A$ and $f'(\hat{z}) = z$. The case where both $f'(w), z \in V_{K_B}$ is trivial, as are the cases where $f'(w), z \in V_{C_l}$ or $f'(w), z \in V_{C_r}$.

If $f'(w) \in V_{C_l}$ (WLOG) and $z \in V_{K_{B'}}$, then we have $f(w)z \in E_B$, and since $f$ is a weak fibration, we have $\hat{z}$ with $f(\hat{z}) = z$ and $w\hat{z} \in E_A$. As $z \in V_{K_B}$, we also have that $f'(\hat{z}) = z$.

Finally, if $f'(w) \in V_{K_{B'}}$ and $z \in V_{C_l}$ (WLOG), then we have $f(w)c(z) \in E_B$. As $f$ is a weak fibration, then we have some $\hat{z}_i \in C_i$ with $f(\hat{z}_i) = z$, $z_i \in C_i$ and $\hat{z}_i w \in E_A$. If $i = 1$ we are done. If not, we need to show that $\hat{z}_1 w \in E_A$. Since $C$ is maximal, we must have that $f^{-1}(V_C \cup \{f(w)\} = \bigcup_1^n (V_{C_i}) \cup f^{-1}(f(w))$ is connected. Therefore, by Proposition 8.2, since $\hat{z}_1 \hat{z}_i \notin E_A$ we must have $\hat{z}_1 w', w' \hat{z}_i$, for some $w'$ where $f(w') = f(w)$. But then, for $\hat{z}_1, \hat{z}_i, w, w'$ not to be a $P_4$ subgraph, we need $\hat{z}_1 \hat{z}_i$, $ww'$ or $\hat{z}_1 w$ to be in $V_A$. Since $f(w) = f(w')$ and $f(\hat{z}_1) = f(\hat{z}_i)$, we must have $\hat{z}_1 w \in E_A$. Therefore $f'$ is a weak fibration. □

**Proposition 8.14.** $\mathfrak{G}(\{\mathsf{c}\downarrow\}) = \mathsf{SWFib}$.

*Proof.* Inclusion from Proposition 6.2, noting that horizontal composition of derivations only violates condition **(F1)** if the derivation contains weakenings. For equality, consider a surjectve weak fibration $f\colon \mathfrak{G}(A) \to \mathfrak{G}(B)$. We build the derivation $\begin{array}{c} A \\ \| \{\mathsf{c}\downarrow\} \\ B \end{array}$ working up by contracting on maximal contracted subformulae of $B$ using Proposition 8.13. □

We can now prove a simple but purely graph theoretic result using the correspondence with structural proof systems.

**Proposition 8.15.** *Every full surjective skew fibration is a weak fibration, i.e.* $\mathsf{FSur} = \mathsf{FSWFib}$.

*Proof.* By definition $\mathsf{FSWFib} \subseteq \mathsf{FSur}$. Since $\mathfrak{G}(\mathsf{ac}\downarrow) = \mathsf{FSur}$, $\{\mathsf{ac}\downarrow\} \preceq \{\mathsf{c}\downarrow\}$ and $\mathfrak{G}(\mathsf{c}\downarrow) = \mathsf{SWFib}$, we must have that $\mathsf{FSur} \subseteq \mathsf{SWFib}$. □

**Proposition 8.16.** $\mathfrak{G}(\{\mathsf{sw}\downarrow, \mathsf{c}\downarrow\}) = \mathsf{WFib}$.

**Proposition 8.17.** $\mathfrak{G}(\{\mathsf{sw}\downarrow, \mathsf{ac}\downarrow\}) = \mathsf{FWFib}$.

*Proof.* Simple corollaries of Propositions 7.4, 8.7 and 8.14. □

## 9 Conclusion

Cographs can describe formulas without using a syntax tree. Even though this concept has been known for more than 50 years, these *formulas without syntax* have been used for proof theoretical considerations first by Retoré [20–22] and Guglielmi [13]. Hughes [17] provided the next natural step by studying combinatorial proofs as *proofs without syntax*, as they describe proofs without the syntax of a proof tree. In this paper we have generalized this further to *proof systems without syntax*, using graph homomorphism classes instead of inference rules.

Summarizing the main theorem leads to the following slogans relating homomorphism classes and proof systems:

| | | |
|---:|---|---|
| No Weakening | = | Surjectivity |
| No Contraction | = | Injectivity |
| Atomic Contraction | = | Fullness |
| Shallow Inference | = | Fibrations |
| Deep Inference | = | Skew Fibrations |

An important line of future research is the extension of these results to modal logics [3] and first-order logic [19].

**Acknowledgements.** This research was supported by the ANR-FWF International Project ANR-15-CE25-0014-01 *FISP*. We would also like to thank the anonymous referees for their helpful comments.

## References

1. Acclavio, M., Straßburger, L.: From syntactic proofs to combinatorial proofs. In: Galmiche, D., Schulz, S., Sebastiani, R. (eds.) IJCAR 2018. LNCS (LNAI), vol. 10900, pp. 481–497. Springer, Cham (2018). https://doi.org/10.1007/978-3-319-94205-6_32
2. Acclavio, M., Straßburger, L.: On combinatorial proofs for logics of relevance and entailment. In: Iemhoff, R., Moortgat, M., de Queiroz, R. (eds.) WoLLIC 2019. LNCS, vol. 11541, pp. 1–16. Springer, Berlin Heidelberg (2019). https://doi.org/10.1007/978-3-662-59533-6_1
3. Acclavio, M., Straßburger, L.: On Combinatorial proofs for modal logic. In: Cerrito, S., Popescu, A. (eds.) TABLEAUX 2019. LNAI 11714, pp. 223–240 (2019)
4. Aler Tubella, A., Guglielmi, A., Ralph, B.: Removing cycles from proofs. In: Goranko, V., Dam, M. (eds.) 26th EACSL Annual Conference on Computer Science Logic (CSL 2017). Leibniz International Proceedings in Informatics (LIPIcs), vol. 82, pp. 9:1–9:17. Schloss Dagstuhl-Leibniz-Zentrum fuer Informatik (2017)
5. Brünnler, K.: Locality for classical logic. Notre Dame J. Form. Log. **47**(4), 557–580 (2006)
6. Brünnler, K., Tiu, A.F.: A local system for classical logic. In: Nieuwenhuis, R., Voronkov, A. (eds.) LPAR 2001. LNCS (LNAI), vol. 2250, pp. 347–361. Springer, Heidelberg (2001). https://doi.org/10.1007/3-540-45653-8_24
7. Carbone, A.: A new mapping between combinatorial proofs and sequent calculus proofs read out from logical flow graphs. Inf. Comput. **208**(5), 500–509 (2010)
8. Cook, S.A., Reckhow, R.A.: The relative efficiency of propositional proof systems. J. Symb. Log. **44**(01), 36–50 (1979)
9. Das, A.: Rewriting with linear inferences in propositional logic. In: RTA, pp. 158–173 (2013)
10. Das, A., Straßburger, L.: No complete linear term rewriting system for propositional logic. In: 26th International Conference on Rewriting Techniques and Applications (RTA), Leibniz International Proceedings in Informatics (LIPIcs). Schloss Dagstuhl–Leibniz-Zentrum für Informatik (2015)
11. Das, A., Straßburger, L.: On linear rewriting systems for Boolean logic and some applications to proof theory. Log. Methods Comput. Sci. **12**(4:9), 1–27 (2016)

12. Duffin, R.J.: Topology of series-parallel networks. J. Math. Anal. Appl. **10**(2), 303–318 (1965)
13. Guglielmi, A.: A system of interaction and structure. ACM Trans. Comput. Log. **8**(1), 1–64 (2007)
14. Guglielmi, A., Gundersen, T., Parigot, M.: A proof calculus which reduces syntactic bureaucracy. In: Proceedings of the 21st International Conference on Rewriting Techniques and Applications (RTA 2010), pp. 6:135–150 (2010)
15. Guglielmi, A., Straßburger, L.: Non-commutativity and MELL in the calculus of structures. In: Fribourg, L. (ed.) CSL 2001. LNCS, vol. 2142, pp. 54–68. Springer, Heidelberg (2001). https://doi.org/10.1007/3-540-44802-0_5
16. Heijltjes, W., Hughes, D.J.D., Straßburger, L.: Intuitionistic combinatorial proofs. In: LICS (2019)
17. Hughes, D.J.D.: Proofs without syntax. Ann. Math. **164**(3), 1065–1076 (2006)
18. Hughes, D.J.D.: Towards Hilbert's 24th problem: combinatorial proof invariants: (preliminary version). Electron. Notes Theor. Comput. Sci. **165**, 37–63 (2006)
19. Hughes, D.J.D.: First-order proofs without syntax. arXiv:1906.11236 [math], June 2019
20. Retoré, C.: Pomset logic: a non-commutative extension of classical linear logic. In: de Groote, P., Roger Hindley, J. (eds.) TLCA 1997. LNCS, vol. 1210, pp. 300–318. Springer, Heidelberg (1997). https://doi.org/10.1007/3-540-62688-3_43
21. Retoré, C.: Handsome proof-nets: R&B-graphs, perfect matchings and series-parallel graphs. Ph.D. thesis, INRIA (1999)
22. Retoré, C.: Handsome proof-nets: perfect matchings and cographs. Theor. Comput. Sci. **294**(3), 473–488 (2003)
23. Straßburger, L.: Linear logic and noncommutativity in the calculus of structures. Ph.D. thesis, Technischen Universität Dresden (2003)
24. Straßburger, L.: MELL in the calculus of structures. Theor. Comput. Sci. **309**(1), 213–285 (2003)
25. Straßburger, L.: A characterization of medial as rewriting rule. In: Baader, F. (ed.) RTA 2007. LNCS, vol. 4533, pp. 344–358. Springer, Heidelberg (2007). https://doi.org/10.1007/978-3-540-73449-9_26
26. Straßburger, L.: Combinatorial flows and their normalisation. In: LIPIcs-Leibniz International Proceedings in Informatics. Schloss Dagstuhl-Leibniz-Zentrum fuer Informatik, vol. 84 (2017)
27. Straßburger, L.: The problem of proof identity, and why computer scientists should care about Hilbert's 24th problem. Philos. Trans. R. Soc. A **377**(2140), 20180038 (2019)
28. Straßburger, L., Guglielmi, A.: A system of interaction and structure IV: the exponentials and decomposition. ACM Trans. Comput. Log. **12**(4), 1–39 (2011)
29. Thiele, R.: Hilbert's twenty-fourth problem. Am. Math. Mon. **110**(1), 1–24 (2003)

# Birkhoff Completeness for Hybrid-Dynamic First-Order Logic

Daniel Găină[1,2] and Ionuţ Ţuţu[3,4(✉)]

[1] Institute of Mathematics for Industry, Kyushu University, Fukuoka, Japan
daniel@imi.kyushu-u.ac.jp
[2] Department of Mathematics and Statistics, La Trobe University, Melbourne, Australia
[3] Simion Stoilow Institute of Mathematics of the Romanian Academy, Bucharest, Romania
ittutu@gmail.com
[4] Department of Computer Science, Royal Holloway University of London, Egham, UK

**Abstract.** Hybrid-dynamic first-order logic is a kind of modal logic obtained by enriching many-sorted first-order logic with features that are common to hybrid and to dynamic logics. This provides us with a logical system with an increased expressive power thanks to a number of distinctive attributes: first, the possible worlds of Kripke structures, as well as the nominals used to identify them, are endowed with an algebraic structure; second, we distinguish between rigid symbols, which have the same interpretation across possible worlds – and thus provide support for the standard rigid quantification in modal logic – and flexible symbols, whose interpretation may vary; third, we use modal operators over dynamic-logic actions, which are defined as regular expressions over binary nominal relations. In this context, we propose a general notion of hybrid-dynamic Horn clause and develop a proof calculus for the Horn-clause fragment of hybrid-dynamic first-order logic. We investigate soundness and compactness properties for the syntactic entailment system that corresponds to this proof calculus, and prove a Birkhoff-completeness result for hybrid-dynamic first-order logic.

## 1 Introduction

The dynamic-reconfiguration paradigm is a most promising approach in the development of highly complex and integrated systems of interacting 'components', which now often evolve dynamically, at run time, in response to internal or external stimuli. More than ever, we are witnessing a continuous increase in the number of applications with reconfigurable features, many of which have aspects that are safety- or security-critical. This calls for suitable formal-specification and verification technologies, and there is already a significant body of research on this topic; hybrid(ized) logics [2,5,17], first-order dynamic logic [15], and modal $\mu$-calculus [14] are three prominent examples, among many others.


S. Cerrito and A. Popescu (Eds.): TABLEAUX 2019, LNAI 11714, pp. 277–293, 2019.
https://doi.org/10.1007/978-3-030-29026-9_16

The application domain of the work reported in this contribution refers to a broad range of reconfigurable systems whose states or configurations can be presented explicitly, based on some kind of context-independent data types, and for which we distinguish the computations performed at the local/configuration level from the dynamic evolution of the configurations. This suggests a two-layered approach to the design and analysis of reconfigurable systems, involving *a local view*, which amounts to describing the structural properties of configurations, and *a global view*, which corresponds to a specialized language for specifying and reasoning about the way system configurations evolve.

In this paper, we develop sound and complete proof calculi for a new modal logic (recently proposed in [11]) that provides support for the reconfiguration paradigm. The logic, named hybrid-dynamic first-order logic, is obtained by enriching first-order logic (FOL) – regarded as a parameter for the whole construction – with both hybrid and dynamic features. This means that we model reconfigurable systems as Kripke structures (or transition systems), where:

- from a local perspective, we consider a dedicated FOL-signature for configurations, and hence capture configurations as first-order structures; and
- from a global perspective, we consider a second FOL-signature for the possible worlds of the Kripke structure; the terms over that signature are *nominals* used to identify configurations, and the binary nominal relations are regarded as *modalities*, which capture the transitions between configurations.

Sentences are build from equations and relational atoms over the two first-order signatures mentioned above (one pertaining to data, and the other to possible worlds) by using Boolean connectives, quantifiers, standard hybrid-logic operators such as *retrieve* and *store*, and dynamic-logic operators such as *necessity* over structured actions, which are defined as regular expressions over modalities. In practice, actions are used to capture specific patterns of reconfigurability.

The construction is reminiscent of the hybridization of institutions from [7,17] and of the hybrid-dynamic logics presented in [1,16], but it departs fundamentally from any of those studies due to the fact that the possible worlds of the Kripke structures that we consider here have an algebraic structure. This special feature of the logic that we put forward is extremely important for dealing with reconfigurable systems whose states are obtained from initial configurations by applying constructor operations; see, e.g. [12]. In this context, we advance a general notion of Horn clause, which allows the use of implications, universal quantifiers, as well as the hybrid- and dynamic-logic operators listed above.

Besides the fact that it relies on an algebraic structure for possible worlds, the notion of Horn clause that we use in this paper also allows structured actions for (*a*) the conditions of logical implication, and (*b*) the arguments of the *necessity* operator. This feature distinguishes the present work from [8], where the first author reported a Birkhoff completeness result for hybrid logics. That is, the Horn clauses that we study in this paper are strictly, and significantly, more expressive than those considered before; this poses a series of new challenges in developing a completeness result. We show that any set of Horn clauses has an initial model despite the fact that the structured actions alone do not have this property. In addition, we provide proof rules to reason formally about the properties of those Kripke structures that are specified using Horn clauses. To conclude, the main result of the paper is a completeness theorem for the Horn-clause fragment of hybrid-dynamic first-order logic.

A brief comparison with the work recently reported in [11] is also in order: both papers deal with properties of hybrid-dynamic first-order logic (with [11] being the contribution in which we introduced the logic); and in both papers we examine Horn clauses; but the results that we develop are complementary: in [11], we focused on an initiality result and on Herbrand's theorem, whereas here we advance proof calculi for the logic. This latter endeavour is much more complex, because it deals with syntactic entailment instead of semantic entailment.

The paper is structured as follows: Sect. 2 is devoted to the definition of hybrid-dynamic first-order logic. Then, in Sect. 3, we discuss entailment systems and present the problem we aim to solve. Once the preliminaries are set, we proceed in a layered fashion, in the sense that we consider progressively more complex entailment relations, which are adequate for different fragments of hybrid-dynamic first-order logic. In Sect. 4 we study completeness for the atomic fragment of the logic. Building on that result, in Sect. 5 we develop a quasi-completeness result for entailments where the left-hand side is an arbitrary set of Horn clauses, but the right-hand side is only an atomic sentence or an action relation. Finally, in Sect. 6, we generalize completeness to the full Horn-clause fragment of hybrid-dynamic first-order logic. Proofs of the lemmas and propositions that support the main results can be found in [10].

## 2 Hybrid-Dynamic First-Order Logic

The hybrid-dynamic first-order logic with user-defined sharing[1] (HDFOLS) that we examine in this work is based on ideas that are similar to those used to define hybrid first-order logic [2] and hybrid first-order logic with rigid symbols [5,7].

We present HDFOLS from an institutional perspective [13], meaning that we focus on signatures and signature morphisms (though, for the purpose of this paper, inclusions would suffice), Kripke structures and homomorphisms, sentences, and the (local) satisfaction relation and condition that relate the syntax and the semantics of the logic. However, other than the notations used, the text requires no prior knowledge of institution theory, and should be accessible to readers with a general background in modal logic and first-order model theory. In order to establish some of the notations used in the rest of the paper, we briefly recall the notion of (many-sorted) first-order signature: a FOL-*signature* is a triple $(S, F, P)$, where $S$ is a set of *sorts*, $F$ is a family $\{F_{ar\to s}\}_{ar\in S^*,s\in S}$ of sets of *operation symbols*, indexed by *arities* $ar \in S^*$ and *sorts* $s \in S$, and $P$ is family $\{P_{ar}\}_{ar\in S^*}$ of sets of *relation symbols*, indexed by *arities* $ar \in S^*$.

[1] This last attribute is meant to indicate the fact that users have control over the symbols that should be interpreted the same across the worlds of a Kripke structure.

*Signatures.* The *signatures* of HDFOLS are tuples $\Delta = (\Sigma^n, \Sigma^r \subseteq \Sigma)$, where:

1. $\Sigma^n = (S^n, F^n, P^n)$ is a FOL-signature of *nominals* such that $S^n = \{\star\}$,
2. $\Sigma^r = (S^r, F^r, P^r)$ is a FOL-signature of so-called *rigid symbols*, and
3. $\Sigma = (S, F, P)$ is a FOL-signature of both *rigid* and *flexible* symbols.

We let $S^f = S \setminus S^r$, and $F^f$ and $P^f$ be the sub-families of $F$ and $P$ that consist of *flexible symbols* (obtained by removing rigid symbols). In general, we denote by $\Delta$ or $\Delta'$ signatures of the form $(\Sigma^n, \Sigma^r \subseteq \Sigma)$ or $(\Sigma'^n, \Sigma'^r \subseteq \Sigma')$, respectively.

*Signature Morphisms.* A *signature morphism* $\varphi\colon \Delta \to \Delta'$ consists of a pair of FOL-signature morphisms $\varphi^n\colon \Sigma^n \to \Sigma'^n$ and $\varphi\colon \Sigma \to \Sigma'$ such that $\varphi(\Sigma^r) \subseteq \Sigma'^r$.

*Kripke Structures.* The *models* of a signature $\Delta$ are pairs $(W, M)$, where:

1. $W$ is a $\Sigma^n$-model, for which we denote by $|W|$ the carrier set of the sort $\star$;
2. $M = \{M_w\}_{w \in |W|}$ is a family of $\Sigma$-models, indexed by *worlds* $w \in |W|$, such that the rigid symbols[2] have the same interpretation across possible worlds; i.e., $M_{w_1,\varsigma} = M_{w_2,\varsigma}$ for all worlds $w_1, w_2 \in |W|$ and all symbols $\varsigma$ in $\Sigma^r$.

*Kripke Homomorphisms.* A *morphism* $h\colon (W, M) \to (W', M')$ is also a pair $(W \xrightarrow{h} W', \{M_w \xrightarrow{h_w} M'_{h(w)}\}_{w \in |W|})$ consisting of first-order homomorphisms such that $h_{w_1,s} = h_{w_2,s}$ for all possible worlds $w_1, w_2 \in |W|$ and all rigid sorts $s \in S^r$.

*Actions.* As in dynamic logic, HDFOLS supports structured actions obtained from atoms using sequential composition, union, and iteration. The set $A^n$ of *actions* over $\Sigma^n$ is defined in an inductive fashion, according to the grammar: $\mathfrak{a} ::= \lambda \mid \mathfrak{a} \mathbin{;} \mathfrak{a} \mid \mathfrak{a} \cup \mathfrak{a} \mid \mathfrak{a}^*$, where $\lambda \in P^n_{\star\star}$ is a binary nominal relation. Given a natural number $n > 0$, we denote by $\mathfrak{a}^n$ the composition $\mathfrak{a} \mathbin{;} \cdots \mathbin{;} \mathfrak{a}$ (where the action $\mathfrak{a}$ occurs $n$ times); and we let $\mathfrak{a}^0(k_1, k_2)$ denote the equation $k_1 = k_2$.

Actions are interpreted in Kripke structures as *accessibility relations* between possible worlds. This is done by extending the interpretation of binary modalities (from $P^n_{\star\star}$): $W_{\mathfrak{a}_1 ; \mathfrak{a}_2} = W_{\mathfrak{a}_1} \mathbin{;} W_{\mathfrak{a}_2}$ (diagrammatic composition of relations), $W_{\mathfrak{a}_1 \cup \mathfrak{a}_2} = W_{\mathfrak{a}_1} \cup W_{\mathfrak{a}_2}$ (union), and $W_{\mathfrak{a}^*} = (W_{\mathfrak{a}})^*$ (reflexive *&* transitive closure).

*Hybrid Terms.* For every $\Sigma^n$-model $W$, the family $T^W = \{T^W_w\}_{w \in |W|}$ of sets of hybrid terms over $W$ is defined inductively according to the following rules:

$$(1)\ \frac{w_0 \in |W| \quad \overline{\tau} \in T^W_{w_0,ar}}{\sigma(\overline{\tau}) \in T^W_{w,s}}\ [\,\sigma \in F^r_{ar\to s}\,] \qquad (2)\ \frac{w_0 \in |W| \quad \overline{\tau} \in T^W_{w_0,ar}}{\sigma(w_0; \overline{\tau}) \in T^W_{w,s}}\ [\,\sigma \in F^f_{ar\to s},\ s \in S^r\,] \qquad (3)\ \frac{w \in |W| \quad \overline{\tau} \in T^W_{w,ar}}{\sigma(w; \overline{\tau}) \in T^W_{w,s}}\ [\,\sigma \in F^f_{ar\to s},\ s \in S^f\,]$$

Notice that flexible operation symbols receive a possible world $w \in |W|$ as a parameter, while rigid operation symbols keep their initial arity. It is easy to check that the hybrid terms of rigid sorts are shared across the worlds.

**Fact 1.** $T^W_{w_1,s} = T^W_{w_2,s}$ for all possible worlds $w_1, w_2 \in |W|$ and all sorts $s \in S^r$.

[2] By *symbol* we usually refer to sorts as well, not only to operation/relation symbols.

Given a world $w \in |W|$, the $S$-sorted set $T^W_w$ can be regarded as a $\Sigma$-model by interpreting every rigid operation symbol $\sigma\colon ar \to s$ as the function that maps (tuples of) hybrid terms $\bar{\tau} \in T^W_{w,ar}$ to $\sigma(\bar{\tau}) \in T^W_{w,s}$, every flexible operation symbol $\sigma\colon ar \to s$ as the function that maps hybrid terms $\bar{\tau} \in T^W_{w,ar}$ to $\sigma(w;\bar{\tau}) \in T^W_{w,s}$, and every relation symbol (either rigid or flexible) as the empty set.

**Lemma 2 (Hybrid-term model and its freeness).** *For every $\Sigma^n$-model $W$, $(W, T^W)$ is a $\Delta$-model. Moreover, for any $\Delta$-model $(W', M')$ and first-order $\Sigma^n$-homomorphism $f\colon W \to W'$, there exists a unique $\Delta$-homomorphism $h\colon (W, T^W) \to (W', M')$ that agrees with $f$ on $W$.* □

*Standard Term Model.* When $W$ is the first-order term model $T_{\Sigma^n}$, by Lemma 2 we obtain the standard hybrid-term model over $\Delta$, denoted $(T_{\Sigma^n}, \{T^\Delta_k\}_{k\in T_{\Sigma^n}})$.

The initiality of the standard term model provides a straightforward interpretation of hybrid terms in $\Delta$-models $(W, M)$: for every hybrid term $t \in T^\Delta_k$, we denote by $(W, M)_t$ or $M_{h(k),t}$ the image of $t$ under the function $h_k$, where $h$ is the unique homomorphism $(T_{\Sigma^n}, T^\Delta) \to (W, M)$.

*Reachable Hybrid-Term Models.* We say that a first-order $\Sigma^n$-model $W$ is *reachable* if the unique homomorphism $T_{\Sigma^n} \to W$ is surjective. In a similar manner, for HDFOLS, we say that a $\Delta$-model $(W, M)$ is *reachable* if the unique homomorphism $h\colon (T_{\Sigma^n}, T^\Delta) \to (W, M)$ is (componentwise) surjective. In order to avoid naming the homomorphism, we make the following notation.

**Notation 3.** If a $\Delta$-model $(W, M)$ is reachable, then we may denote by $[\_]$ the unique homomorphism $(T_{\Sigma^n}, T^\Delta) \to (W, M)$ given by the initiality of $(T_{\Sigma^n}, T^\Delta)$.

**Proposition 4 (Reachability of hybrid term models).** *If $W$ is a* reachable *first-order model of $\Sigma^n$, then $(W, T^W)$ is* reachable *for the signature $\Delta$.* □

*Sentences.* The *atomic sentences* $\rho$ defined over a signature $\Delta$ are given by:

$$\rho ::= k_1 = k_2 \mid \lambda(\overline{k'}) \mid t_1 =_{k,s} t_2 \mid \varpi(\bar{t}) \mid \pi(k;\bar{t})$$

where $k, k_i \in T_{\Sigma^n}$ are nominal terms, $\overline{k'}$ is a tuple of terms corresponding to the arity of $\lambda \in P^n$, $t_i \in T^\Delta_{k,s}$ and $\bar{t} \in T^\Delta_{k,ar}$ are (tuples of) hybrid terms,[3] $\varpi \in P^r_{ar}$, and $\pi \in P^f_{ar}$. We refer to these sentences, in order, as *nominal equations*, *nominal relations*, *hybrid equations*, *rigid hybrid relations*, and *non-rigid/flexible hybrid relations*, respectively. When there is no danger of confusion, we may drop one or both subscripts $k, s$ from the notation $t_1 =_{k,s} t_2$. *Full sentences* over $\Delta$ are built from atomic sentences according to the following grammar:

$$\gamma ::= \rho \mid \mathfrak{a}(k_1, k_2) \mid @_k\,\gamma \mid \neg\gamma \mid \bigwedge\Gamma \mid {\downarrow} z \cdot \gamma' \mid \forall X \cdot \gamma'' \mid [\mathfrak{a}]\gamma \mid \langle o \rangle\,\gamma$$

where $k, k_i \in T_{\Sigma^n}$ are nominal terms, $\mathfrak{a} \in A^n$ is an action, $\Gamma$ is a finite set of sentences, $z$ is a nominal variable, $\gamma'$ is a sentence over the signature $\Delta[z]$

[3] Note that, by Fact 1, if the arity $ar$ is rigid, then the sets $\{T^\Delta_{k,ar}\}_{k\in T_{\Sigma^n}}$ coincide.

obtained by adding $z$ as a new constant to $F^n$, $X$ is a set of nominal and/or rigid variables, $\gamma''$ is a a sentence over the signature $\Delta[X]$ obtained by adding the elements of $X$ as new constants to $F^n$ and $F^r$, and $o \in F^n_{\star\to\star}$. Other than the first two kinds of sentences (*atoms* and *action relations*), we refer to the sentence-building operators, in order, as *retrieve*, *negation*, *conjunction*, *store*, *universal quantification*, *necessity*, and *next*, respectively. Notice that *necessity* and *next* are parameterized by actions and by unary nominal operations, respectively.

We denote by $\mathsf{Sen}^{\mathsf{HDFOLS}}(\Delta)$ the set of all HDFOLS-sentences over $\Delta$.

*The Local Satisfaction Relation.* Given a $\Delta$-model $(W, M)$ and a world $w \in |W|$, we define the *satisfaction of $\Delta$-sentences at $w$* by structural induction as follows:

1. *For atomic sentences:*
   - $(W, M) \models^w k_1 = k_2$ iff $W_{k_1} = W_{k_2}$ for all nominal equations $k_1 = k_2$;
   - $(W, M) \models^w \lambda(\bar{k})$ iff $W_{\bar{k}} \in W_\lambda$ for all nominal relations $\lambda(\bar{k})$;
   - $(W, M) \models^w t_1 =_k t_2$ iff $M_{W_k, t_1} = M_{W_k, t_2}$ for all hybrid equations $t_1 =_k t_2$;
   - $(W, M) \models^w \varpi(\bar{t})$ iff $(W, M)_{\bar{t}} \in M_{w,\varpi}$ for all rigid relations $\varpi(\bar{t})$;
   - $(W, M) \models^w \pi(k; \bar{t})$ iff $(W, M)_{\bar{t}} \in M_{W_k,\pi}$ for flexible relations $\pi(k; \bar{t})$.
2. *For full sentences:*
   - $(W, M) \models^w \mathfrak{a}(k_1, k_2)$ iff $(W_{k_1}, W_{k_2}) \in W_\mathfrak{a}$ for all action relations $\mathfrak{a}(k_1, k_2)$;
   - $(W, M) \models^w @_k \gamma$ iff $(W, M) \models^{w'} \gamma$, where $w' = W_k$;
   - $(W, M) \models^w \neg\gamma$ iff $(W, M) \not\models^w \gamma$;
   - $(W, M) \models^w \bigwedge \Gamma$ iff $(W, M) \models^w \gamma$ for all $\gamma \in \Gamma$;
   - $(W, M) \models^w \downarrow z \cdot \gamma$ iff $(W, M)^{z \leftarrow w} \models^w \gamma$, where $(W, M)^{z \leftarrow w}$ is the unique $\Delta[z]$-expansion[4] of $(W, M)$ that interprets the variable $z$ as $w$;
   - $(W, M) \models^w \forall X \cdot \gamma$ iff $(W', M') \models^w \gamma$ for all $\Delta[X]$-expansion[6] $(W', M')$ of $(W, M)$;
   - $(W, M) \models^w [\mathfrak{a}]\gamma$ iff $(W, M) \models^{w'} \gamma$ for all $w' \in |W|$ such that $(w, w') \in W_\mathfrak{a}$;
   - $(W, M) \models^w (o)\,\gamma$ iff $(W, M) \models^{w'} \gamma$, where $w' = W_o(w)$.

**Fact 5.** The following two properties can be checked with ease:

1. The satisfaction of atoms and of action relations $\rho$ does not depend on the possible worlds: $(W, M) \models^w \rho$ iff $(W, M) \models^{w'} \rho$ for all $w, w' \in |W|$.
2. The satisfaction of atoms and of action relations $\rho$ is preserved by homomorphisms: if $(W, M) \models \rho$ and $h\colon (W, M) \to (W', M')$ then $(W', M') \models \rho$.

To state the *satisfaction condition* – and thus finalize the presentation of HDFOLS – let us first notice that every signature morphism $\varphi\colon \Delta \to \Delta'$ induces appropriate *translations of sentences* and *reductions of models*, as follows: every $\Delta$-sentence $\gamma$ is translated to a $\Delta'$-sentence $\varphi(\gamma)$ by replacing (usually in an inductive manner) the symbols in $\Delta$ with symbols from $\Delta'$ according to $\varphi$; and every $\Delta'$-model $(W', M')$ is reduced to a $\Delta$-model $(W', M')\restriction_\varphi$ that interprets every symbol $x$ in $\Delta$ as $(W', M')_{\varphi(x)}$. When $\varphi$ is an inclusion, we usually denote $(W', M')\restriction_\varphi$ by $(W', M')\restriction_\Delta$ – in this case, the model reduct simply forgets the interpretation of those symbols in $\Delta'$ that do not belong to $\Delta$.

[4] In general, by a $\Delta[X]$-expansion of $(W, M)$ we understand a $\Delta[X]$-model $(W', M')$ that interprets all symbols in $\Delta$ in the same way as $(W, M)$.

The following satisfaction condition can be proved by induction on the structure of $\Delta$-sentences. Its argument is essentially identical to those developed for several other variants of hybrid logic presented in the literature (see, e.g. [5]).

**Proposition 6 (Local satisfaction condition for signature morphisms).** *For every signature morphism $\varphi\colon \Delta \to \Delta'$, $\Delta'$-model $(W', M')$, world $w' \in |W'|$, and $\Delta$-sentence $\gamma$, we have: $(W', M') \models^{w} \varphi(\gamma)$ iff $(W', M')\upharpoonright_{\varphi} \models^{w} \gamma$.*[5] □

*Substitutions.* Consider two signature extensions $\Delta[X]$ and $\Delta[Y]$ with sets of variables, and let $X = X^n \cup X^r$ and $Y = Y^n \cup Y^r$ be the partitions of $X$ and $Y$ into sets of nominal variables and rigid variables. A *$\Delta$-substitution* $\theta\colon X \to Y$ consists of a pair of functions $\theta^n\colon X^n \to T_{\Sigma^n[Y^n]}$ and $\theta^r\colon X^r \to T_k^{\Delta[Y]}$, where $k$ is a nominal term – note that, since the sorts of the hybrid variables are rigid, by Fact 1, it does not matter which nominal term $k$ we choose.

Similarly to signature morphisms, $\Delta$-substitutions $\theta\colon X \to Y$ determine translations of $\Delta[X]$-sentences into $\Delta[Y]$-sentences, and reductions of $\Delta[Y]$-models to $\Delta[X]$-models. The proofs of the next two propositions are similar to the ones given in [9] for hybrid substitutions.

**Proposition 7 (Local satisfaction condition for substitutions).** *For every $\Delta$-substitution $\theta\colon X \to Y$, every $\Delta[Y]$-model $(W, M)$, world $w \in |W|$, and $\Delta[X]$-sentence $\gamma$, we have: $(W, M) \models^{w} \theta(\gamma)$ iff $(W, M)\upharpoonright_{\theta} \models^{w} \gamma$.* □

**Fact 8.** Let $\theta_{z\leftarrow k}\colon \{z\} \to \varnothing$ be the substitution that maps the nominal variable $z$ to the term $k$. Then $(W, M)\upharpoonright_{\theta_{z\leftarrow k}} = (W, M)^{z\leftarrow k}$ for every model $(W, M)$.

Propositions 7 and 9 (below) have an important technical role in the Birkhoff completeness proofs presented in the later sections of the paper.

**Proposition 9 (Subst. generated by expansions of reachable models).** *If $(W, M)$ is reachable, then for every $\Delta[X]$-expansion $(W', M')$ of $(W, M)$ there exists a $\Delta$-substitution $\theta\colon X \to \varnothing$ such that $(W, M)\upharpoonright_{\theta} = (W', M')$.* □

*Expressive Power.* Fact 5 highlights one of the main distinguishing features of HDFOLS: the satisfaction of atomic sentences, whether they involve flexible symbols or not, does not depend on the possible world where the sentences are evaluated. This contrasts the standard approach in hybrid logic, where each nominal is regarded as an atomic sentence satisfied precisely at the world that corresponds to the interpretation of that nominal. In HDFOLS, the dependence of the satisfaction of sentences on possible worlds is explicit rather than implicit, and is achieved through the *store* operator. Following the lines of [9, Section 4.3], one can show that even without considering action relations, HDFOLS is strictly more expressive than other standard hybrid logics constructed from the same base logic such as the hybrid first-order logic with rigid symbols [5,7].

[5] By the definition of reducts, $(W', M')$ and $(W', M')\upharpoonright_{\varphi}$ have the same possible worlds.

## 3 Entailment

Let $\Gamma$ and $\Gamma'$ be two sets of sentences over a signature $\Delta$. We say that $\Gamma$ *semantically entails* $\Gamma'$, or that $\Gamma'$ is a *semantic consequence* of $\Gamma$, and we write $\Gamma \models_\Delta \Gamma'$, when every $\Delta$-model that satisfies $\Gamma$ satisfies $\Gamma'$ too. When the set $\Gamma'$ is a singleton $\{\gamma\}$, we simplify the notation to $\Gamma \models_\Delta \gamma$. Moreover, we usually drop the subscript $\Delta$ when the signature can be easily inferred from the context.

*Horn Clauses.* The problem we propose to address in this paper is that of finding a suitable syntactic characterisation of *entailments* of the form $\Gamma \models \gamma$, where both $\Gamma$ and $\gamma$ correspond to the Horn-clause fragment of HDFOLS.

By *Horn clause*, we mean a sentence obtained from atomic sentences by repeated applications of the following sentence-building operators, in any order: $(a)$ *retrieve* $(b)$ *implication* such that the condition is a conjunction of atomic sentences or action relations, $(c)$ *store*, $(d)$ *universal quantification*, $(e)$ *necessity*, and $(f)$ *next*. We denote by HDCLS the Horn-clause fragment of HDFOLS, and by $\mathtt{Sen}^{\mathsf{HDCLS}}(\Delta)$ the set of all Horn clauses over the signature $\Delta$.

In the next sections, we develop a series of *syntactic entailment relations*, whose corresponding entailments are denoted by $\Gamma \vdash \gamma$. All of them are *sound*, in the sense that $\Gamma \vdash \gamma$ implies $\Gamma \models \gamma$; and some are also *compact*, which means that, whenever $\Gamma \vdash \gamma$, there exists a finite subset $\Gamma_f \subseteq \Gamma$ such that $\Gamma_f \vdash \gamma$.

As in previous studies on Birkhoff completeness [4,8], we follow a layered approach. This means that we distinguish the atomic layer of HDCLS from the layer of general Horn clauses. The former is intrinsically dependent on the details of HDCLS, whereas the latter is in essence logic-independent, and can easily be adapted to other hybrid-dynamic logics, not necessarily based on first-order logic. The same ideas apply, for example, to hybrid-dynamic propositional logic.

*Nominal Replacement.* In order to capture syntactically relations between hybrid terms that correspond to different nominals, we introduce a way to replace nominals with nominals within hybrid terms. Given two nominals $k_1$ and $k_2$, let $f\colon T_{\Sigma^n} \to T_{\Sigma^n}$ be the function that maps $k_1$ to $k_2$ and leaves the other nominals unchanged. We define the family $\{\delta_{k_1/k_2,k}\colon T_k^\Delta \to T_{f(k)}^\Delta\}_{k\in T_{\Sigma^n}}$ by induction:

1. $\delta_{k_1/k_2,k}(\sigma(\bar{t})) = \sigma(\delta_{k_1/k_2,k_0}(\bar{t}))$ when $\sigma \in F^r_{ar\to s}$ and $\bar{t} \in T^\Delta_{k_0,ar}$;
2. $\delta_{k_1/k_2,k}(\sigma(k_0;\bar{t})) = \sigma(f(k_0);\delta_{k_1/k_2,k_0}(\bar{t}))$ when $\sigma \in F^f_{ar\to s}$, $s \in S^r$, $\bar{t} \in T^\Delta_{k_0,ar}$;
3. $\delta_{k_1/k_2,k}(\sigma(k;\bar{t})) = \sigma(f(k);\delta_{k_1/k_2,k}(\bar{t}))$ when $\sigma \in F^f_{ar\to s}$, $s \in S^f$, and $\bar{t} \in T^\Delta_{k,ar}$.

We usually drop the subscript $k$, and denote the map $\delta_{k_1/k_2,k}$ simply by $\delta_{k_1/k_2}$.

## 4 Atomic Completeness

In this section, we focus on a completeness result for the atomic fragment of HDCLS. There are two major advancements that distinguish the work presented herein from previous contributions (see, e.g. [8]): $(a)$ the state space of every Kripke model is equipped with a full algebraic structure, and $(b)$ the signatures can have flexible sorts – instead of being restricted to rigid sorts only.

To start, let $\vdash$ be the syntactic entailment relation generated by the rules listed in Fig. 1. The following soundness and compactness result can be proved in essentially the same way as in [8]. In particular, the compactness property follows from the fact that all rules have a finite number of premises.

**Proposition 10 (Atomic soundness & compactness).** *The atomic syntactic entailment relation* $\vdash$ *is both sound and compact.* □

As it is often the case, completeness is much more difficult to prove, and relies on a number of preliminary results. For the developments presented in this section, we make use of a specific notion of congruence on a Kripke structure.

$$(\mathsf{R}^n)\ \frac{}{\Gamma \vdash k = k} \qquad (\mathsf{S}^n)\ \frac{\Gamma \vdash k_1 = k_2}{\Gamma \vdash k_2 = k_1} \qquad (\mathsf{T}^n)\ \frac{\Gamma \vdash k_1 = k_2 \quad \Gamma \vdash k_2 = k_3}{\Gamma \vdash k_1 = k_3}$$

$$(\mathsf{F}^n)^a\ \frac{\Gamma \vdash \overline{k_1} = \overline{k_2}}{\Gamma \vdash o(\overline{k_1}) = o(\overline{k_2})} \qquad (\mathsf{P}^n)\ \frac{\Gamma \vdash \lambda(\overline{k_1}) \quad \Gamma \vdash \overline{k_1} = \overline{k_2}}{\Gamma \vdash \lambda(\overline{k_2})} \qquad (\mathsf{W}^h)\ \frac{\Gamma \vdash k = k'}{\Gamma \vdash t =_{k,s} \delta_{k'/k}(t)}\ [\, s \in S^r \,]$$

$$(\mathsf{W}^r)\ \frac{\Gamma \vdash t_1 =_{k_1,s} t_2}{\Gamma \vdash t_1 =_{k_2,s} t_2}\ [\, s \in S^r \,] \qquad (\mathsf{W}^f)\ \frac{\Gamma \vdash k = k' \quad \Gamma \vdash t_1 =_k t_2}{\Gamma \vdash \delta_{k/k'}(t_1) =_{k'} \delta_{k/k'}(t_2)}$$

$$(\mathsf{R}^h)\ \frac{}{\Gamma \vdash t = t} \qquad (\mathsf{S}^h)\ \frac{\Gamma \vdash t_1 = t_2}{\Gamma \vdash t_2 = t_1} \qquad (\mathsf{T}^h)\ \frac{\Gamma \vdash t_1 =_{k,s} t_2 \quad \Gamma \vdash t_2 =_{k,s} t_3}{\Gamma \vdash t_1 =_{k,s} t_3}$$

$$(\mathsf{F}^r)\ \frac{\Gamma \vdash \overline{t_1} =_{k,ar} \overline{t_2}}{\Gamma \vdash \sigma(\overline{t_1}) =_{k,s} \sigma(\overline{t_2})}\ [\, \sigma \in F^r_{ar \to s} \,] \qquad (\mathsf{F}^f)\ \frac{\Gamma \vdash \overline{t_1} =_{k,ar} \overline{t_2}}{\Gamma \vdash \sigma(k; \overline{t_1}) =_{k,s} \sigma(k; \overline{t_2})}\ [\, \sigma \in F^f_{ar \to s} \,]$$

$$(\mathsf{P}^r)\ \frac{\Gamma \vdash \overline{t_1} =_k \overline{t_2} \quad \Gamma \vdash \pi(\overline{t_1})}{\Gamma \vdash \pi(\overline{t_2})}\ [\, \pi \in P^r \,] \qquad (\mathsf{P}^f)\ \frac{\Gamma \vdash \overline{t_1} =_k \overline{t_2} \quad \Gamma \vdash \pi(k; \overline{t_1})}{\Gamma \vdash \pi(k; \overline{t_2})}\ [\, \pi \in P^f \,]$$

$$(\mathsf{P}^h)\ \frac{\Gamma \vdash k_1 = k_2 \quad \Gamma \vdash \pi(k_1; \overline{t_1})}{\Gamma \vdash \pi(k_2; \delta_{k_1/k_2}(\overline{t_1}))}\ [\, \pi \in P^f \,] \qquad (\mathsf{Ret_0})\ \frac{\Gamma \vdash @_k\, \rho}{\Gamma \vdash \rho}\ [\, \rho \text{ is atomic} \,]$$

[a] For brevity, $\Gamma \vdash \overline{k_1} = \overline{k_2}$ stands for $\Gamma \vdash k_{1,i} = k_{2,i}$ for all indexes $i$ of the two tuples.

**Fig. 1.** Proof rules for atomic sentences

**Definition 11 (Congruence).** *Let* $\Delta = (\Sigma^n, \Sigma^r \subseteq \Sigma)$ *be a* HDCLS*-signature, and* $(W, M)$ *a Kripke structure for it. A* $\Delta$-congruence *on* $(W, M)$ *is a family* $\equiv\, = \{\equiv_w\}_{w \in |W|}$ *of* $\Sigma$*-congruences* $\equiv_w$ *on* $M_w$*, for each possible world* $w \in |W|$*, such that* $(\equiv_{w_1,s}) = (\equiv_{w_2,s})$ *for all worlds* $w_1, w_2 \in |W|$ *and rigid sorts* $s \in S^r$.

The next construction is a straightforward generalization of its first-order counterpart, and has been studied in several other papers in the literature (see, e.g. [8]). For that reason, we include it for further reference without a proof.

**Proposition 12 (Quotient model).** *Every* $\Delta$*-congruence* $\equiv$ *on* $(W, M)$ *determines a* quotient-model homomorphism $(\_/\!\equiv)\colon (W, M) \to (W, M/\!\equiv)$ *that acts as an identity on* $W$*, and for which* $(M/\!\equiv)_w$ *is the quotient* $\Sigma$*-model* $M_w/\!\equiv_w$.

*Moreover,* $(\_/\equiv)$ *has the following* universal property*: for any Kripke homomorphism* $h\colon (W, M) \to (W', M')$ *such that* $\equiv\ \subseteq \ker(h)$,[6] *there exists a unique homomorphism* $h'\colon (W, M/\equiv) \to (W', M')$ *such that* $(\_/\equiv) \mathbin{;} h' = h$.[7] □

We prove the atomic completeness of HDCLS in two steps: first, for nominal equations only; then, for arbitrary atomic sentences (both nominal and hybrid). According to the lemma below, every set of nominal equations $\Gamma^n$ admits a 'least' Kripke structure $(W^n, M^n)$ that encapsulates the formal deduction of equations.

**Lemma 13 (Least Kripke structure of a set of nominal equations).** *For every set* $\Gamma^n$ *of* nominal equations *over a signature* $\Delta$*, there exists a reachable initial model* $(W^n, M^n)$ *such that* $\Gamma^n \vdash \rho$ *if and only if* $(W^n, M^n) \models \rho$*, for all nominal or hybrid equations* $\rho$ *over the signature* $\Delta$. □

The following proposition shows that a set $\Gamma$ of (nominal or hybrid) equations generates a congruence on a reachable Kripke structure $(W, M)$ when $\Gamma$ entails all the equations satisfied by $(W, M)$. In particular, the result holds when $\Gamma$ includes the set of all equations that are satisfied by $(W, M)$.

**Proposition 14 (Congruence generated by a set of equations).** *Consider a set* $\Gamma$ *of equations over a signature* $\Delta$*, and a reachable* $\Delta$*-model* $(W, M)$ *such that* $\Gamma \vdash \rho$ *for all equations* $\rho$ *satisfied by* $(W, M)$*. For all* $w \in |W|$*, let* $\equiv_w$ *be the relation on* $M_w$ *defined by* $\tau_1 \equiv_w \tau_2$ *whenever* $\Gamma \vdash t_1 =_k t_2$ *for some* $k \in T_{\Sigma^n}$ *and* $t_1, t_2 \in T_k^{\Delta}$ *such that* $w = W_k$*, and* $\tau_i = M_{w,t_i}$*. Then:*

*P1.* $[t_1] \equiv_{[k]} [t_2]$ *iff* $\Gamma \vdash t_1 =_k t_2$*, for all* $k \in T_{\Sigma^n}$ *and* $t_1, t_2 \in T_k^{\Delta}$*;*
*P2.* $\equiv$ *is a* $\Delta$*-congruence on* $(W, M)$. □

Now we can finally prove the completeness result for atomic sentences.

**Theorem 15 (Atomic completeness).** *Every set* $\Gamma$ *of atomic sentences over a signature* $\Delta$ *has a reachable initial model* $(W^{\Gamma}, M^{\Gamma})$ *such that* $\Gamma \vdash \rho$ *if and only if* $(W^{\Gamma}, M^{\Gamma}) \models \rho$*, for all atomic sentences* $\rho$ *over* $\Delta$.

*Proof.* Let $\Gamma^n$ be the subset of nominal equations in $\Gamma$. By Lemma 13, there exists a initial model $(W^n, M^n)$ of $\Gamma^n$ such that $\Gamma^n \vdash \rho$ iff $(W^n, M^n) \models \rho$ for all equations $\rho$ over $\Delta$. Then $(W^n, M^n)$ satisfies the hypotheses of Proposition 14 with respect to the set of all (nominal or hybrid) equations in $\Gamma$. It follows that the relation $\equiv$ defined by $[t_1] \equiv_{[k]} [t_2]$ whenever $\Gamma \vdash t_1 =_k t_2$, for all nominals $k$ and all terms $t_1, t_2 \in T_{k,s}^{\Delta}$, is a congruence on $(W^n, M^n)$. We define $(W^{\Gamma}, M^{\Gamma})$ as the model obtained from $(W^n, M^n/\equiv)$ by interpreting:

- each nominal relation symbol $\lambda \in P^n$ as $W_{\lambda}^{\Gamma} = \{[\bar{k}] \in |W^n| \mid \Gamma \vdash \lambda(\bar{k})\}$;
- each relation symbol $\varpi \in P^r$ as $M_{[k],\varpi}^{\Gamma} = \{[\bar{t}]/{\equiv_{[k]}} \in M_{[k]}^{\Gamma} \mid \Gamma \vdash \varpi(\bar{t})\}$;
- each relation symbol $\pi \in P^f$ as $M_{[k],\pi}^{\Gamma} = \{[\bar{t}]/{\equiv_{[k]}} \in M_{[k]}^{\Gamma} \mid \Gamma \vdash \pi(k; \bar{t})\}$.

[6] This means that $h_{w,s}(a_1) = h_{w,s}(a_2)$ for all $a_1, a_2 \in M_{w,s}$ such that $a_1 \equiv_{w,s} a_2$.
[7] Note that we use the diagrammatic notation for function composition.

Note that the interpretations of $\varpi \in P^r$ and $\pi \in P^f$ are independent of the choice of the nominal $k$. For example, for flexible relation symbols, if $[k] = [k']$ then $\Gamma \vdash k = k'$; therefore, if $\Gamma \vdash \pi(k; \bar{t})$, we also have $\Gamma \vdash \pi(k'; \bar{t'})$ by $(\mathsf{P}^h)$, where $\bar{t'} = \delta_{k/k'}(\bar{t})$ is a tuple of hybrid terms that satisfies $[\bar{t}] \equiv_{[k]} [\bar{t'}]$.

The fact that $(W^\Gamma, M^\Gamma)$ is a reachable model of $\Gamma$ follows in a straightforward manner by construction. Therefore, we focus on the initiality property. Let $(W, M)$ be a $\Delta$-model that satisfies $\Gamma$. In particular, $(W, M)$ satisfies all nominal equations in $\Gamma$. By Lemma 13, we deduce that there exists a unique homomorphism $h\colon (W^n, M^n) \to (W, M)$. We also know that $(W, M)$ satisfies all hybrid equations in $\Gamma$, which implies that $\equiv\, \subseteq \ker(h)$. By Proposition 12, this means that there exists a unique Kripke homomorphism $h'\colon (W^n, M^n/{\equiv}) \to (W, M)$ such that $(\_/{\equiv}) \mathbin{;} h' = h$. To finalize this part of the proof, we need to ensure that $h'$ preserves the interpretation of all relation symbols (nominal or hybrid) satisfied by $(W^\Gamma, M^\Gamma)$. We only consider the case of flexible relation symbols. Nominal relations and rigid relations can be treated in a similar manner. Suppose $\pi \in P^f_{ar}$ and $\bar{\tau} \in M^\Gamma_{[k],\pi}$, for an arbitrary but fixed nominal $k \in T_{\Sigma^n}$. Then:

| | | |
|---|---|---|
| 1 | $\Gamma \vdash \pi(k; \bar{t})$ for some tuple of terms $\bar{t} \in T^\Delta_{k,ar}$ such that $\bar{\tau} = [\bar{t}]/{\equiv_{[k]}}$ | by the definition of $M^\Gamma_{[k],\pi}$ |
| 2 | $\Gamma \models \pi(k; \bar{t})$ | by Proposition 10 |
| 3 | $(W, M) \models \pi(k; \bar{t})$ | since $(W, M) \models \Gamma$ |
| 4 | $M_{w,\bar{t}} \in M_{w,\pi}$ for $w = W_k$ | by the definition of $\models$ |
| 5 | $h'(\bar{\tau}) \in M_{w,\pi}$ | since $h'(\bar{\tau}) = h'([\bar{t}]/{\equiv_{[k]}}) = M_{w,\bar{t}}$ |

Lastly, we show that $\Gamma \vdash \rho$ iff $(W^\Gamma, M^\Gamma) \models \rho$, for all atomic sentences $\rho$. The 'only if' part is straightforward since $(W^\Gamma, M^\Gamma)$ is a model of $\Gamma$. For the 'if' part, we proceed by case analysis on the structure of $\rho$. The more interesting cases are those of relational atoms. Suppose, for instance, that $(W^\Gamma, M^\Gamma) \models \pi(k; \bar{t})$, where $\pi \in P^f_{ar}$, $k \in T_{\Sigma^n}$, and $\bar{t} \in T^\Delta_{k,ar}$. If follows that:

| | | |
|---|---|---|
| 1 | $[\bar{t}]/{\equiv_{[k]}} \in M^\Gamma_{[k],\pi}$ | by the definition of $\models$ |
| 2 | $\Gamma \vdash \pi(k; \bar{t'})$ for some tuple of terms $\bar{t'} \in T^\Delta_{k,ar}$ such that $[\bar{t'}] \equiv_{[k]} [\bar{t}]$ | by the definition of $M^\Gamma_{[k],\pi}$ |
| 3 | $\Gamma \vdash \bar{t} =_{k,ar} \bar{t'}$ | by Proposition 14 |
| 4 | $\Gamma \vdash \pi(k; \bar{t})$ | by the proof rule $(\mathsf{P}^f)$ |

□

## 5 Quasi-completeness

The main contribution in this section is the construction, for any set of Horn clauses, of an initial model that encapsulates the syntactic deduction of atomic sentences and action relations. An initiality result is obtained in [11] as well, but in that paper it is based on the semantic entailment. In contrast, the present result is based on syntactic deduction, which requires a higher level of complexity, and it is developed in the context of a modular approach to completeness.

$$(\mathsf{Comp})\ \frac{\Gamma \vdash \mathfrak{a}_1(k_1,k_2) \quad \Gamma \vdash \mathfrak{a}_2(k_2,k_3)}{\Gamma \vdash (\mathfrak{a}_1 \mathbin{;} \mathfrak{a}_2)(k_1,k_3)} \qquad (\mathsf{Union})\ \frac{\Gamma \vdash \mathfrak{a}_i(k_1,k_2)}{\Gamma \vdash (\mathfrak{a}_1 \cup \mathfrak{a}_2)(k_1,k_2)}\ [\, i \in \{1,2\}\,]$$

$$(\mathsf{Refl})\ \frac{\Gamma \vdash k_1 = k_2}{\Gamma \vdash \mathfrak{a}^*(k_1,k_2)} \qquad (\mathsf{Star})\ \frac{\Gamma \vdash \mathfrak{a}(k_i,k_{i+1}) \text{ for } 0 \leqslant i < n}{\Gamma \vdash \mathfrak{a}^*(k_0,k_n)} \qquad (\mathsf{Ret}_{\mathsf{a}})\ \frac{\Gamma \vdash @_k\, \mathfrak{a}(k_1,k_2)}{\Gamma \vdash \mathfrak{a}(k_1,k_2)}$$

**Fig. 2.** Proof rules for action relations

$$(\mathsf{Ret}_{@})\ \frac{\Gamma \vdash @_{k_1} @_{k_2} \gamma}{\Gamma \vdash @_{k_2} \gamma} \qquad (\mathsf{Ret}_{\mathsf{I}})\ \frac{\Gamma \vdash \gamma}{\Gamma \vdash @_k \gamma} \qquad (\mathsf{Imp}_{\mathsf{E}})\ \frac{\Gamma \vdash @_k (\bigwedge H \Rightarrow \gamma)}{\Gamma \cup H \vdash @_k \gamma}$$

$$(\mathsf{Store}_{\mathsf{E}})^a\ \frac{\Gamma \vdash @_k \downarrow z \cdot \gamma}{\Gamma \vdash @_k\, \theta_{z \leftarrow k}(\gamma)} \qquad (\mathsf{Subst}_{\mathsf{q}})^b\ \frac{\Gamma \vdash @_k \forall X \cdot \gamma}{\Gamma \vdash @_k\, \theta(\gamma)}$$

$$(\mathsf{Nec}_{\mathsf{E}})\ \frac{\Gamma \vdash @_{k_1} [\mathfrak{a}]\gamma \quad \Gamma \vdash \mathfrak{a}(k_1,k_2)}{\Gamma \vdash @_{k_2} \gamma} \qquad (\mathsf{Next}_{\mathsf{E}})\ \frac{\Gamma \vdash @_k \langle o \rangle \gamma}{\Gamma \vdash @_{o(k)} \gamma}$$

[a] Recall that $\theta_{z \leftarrow k} \colon \{z\} \to \varnothing$ is the substitution that maps $z$ to the nominal $k$.
[b] $\theta \colon X \to \varnothing$ is a ground substitution.

**Fig. 3.** Proof rules for Horn clauses

This means that the present results are applicable to other modal logics, where some of the sentence-building operators considered here may be disregarded.

We focus on entailments of the form $\Gamma \vDash \rho$, where $\Gamma$ is an arbitrary set of Horn clauses, and $\rho$ is either an atomic sentence, or an action relation. To that end, let $\vdash$ be the syntactic entailment relation generated by the rules listed in Figs. 1, 2 and 3. The soundness and compactness result presented in Sect. 4 can be generalized with ease for the entailment relation $\vdash$ that we consider here.

**Proposition 16.** *The entailment relation* $\vdash$ *is sound and compact.* □

**Fact 17 (Retrieve redundancies).** For all nominals $k_1, k_2 \in T_{\Sigma^n}$ and all sentences $\gamma$ over a signature $\Delta$, the sentences $@_{k_1} @_{k_2} \gamma$ and $@_{k_2} \gamma$ are both syntactically and semantically equivalent. Moreover, if $\rho$ is atomic or an action relation, then $@_{k_n} \rho$ is syntactically and semantically equivalent to $\rho$.

To prove that $\vdash$ is also complete, we first extend Theorem 15 to entailments $\Gamma \vdash \rho$ for which $\Gamma$ is a set of atoms and $\rho$ is either atomic or an action relation.

**Proposition 18 (Extending atomic completeness).** *Let* $\Gamma$ *be a set of atomic sentences over a signature* $\Delta$, *and* $(W^\Gamma, M^\Gamma)$ *a reachable initial model of* $\Gamma$ *as in Theorem 15. Then* $\Gamma \vdash \rho$ *if and only if* $(W^\Gamma, M^\Gamma) \vDash \rho$, *for all atomic sentences or action relations* $\rho$ *over the signature* $\Delta$. □

The result below shows that, in order to obtain an initial model of a set $\Gamma$ of clauses, it suffices to consider the initial model $(W^{\Gamma_0}, M^{\Gamma_0})$ of the set $\Gamma_0$ of atoms entailed by $\Gamma$. Moreover, $(W^{\Gamma_0}, M^{\Gamma_0})$ satisfies all clauses entailed by $\Gamma$.

**Theorem 19 (Initiality preserves formal deductions).** *Let $\Gamma$ be a set of clauses over a signature $\Delta$, $\Gamma_0 = \{\rho \in \mathsf{Sen}^{\mathsf{HDCLS}}(\Delta) \mid \Gamma \vdash \rho \text{ \& } \rho \text{ is atomic}\}$, and $(W^{\Gamma_0}, M^{\Gamma_0})$ a reachable initial model of $\Gamma_0$ as in Theorem 15. Then $\Gamma \vdash \gamma$ implies $(W^{\Gamma_0}, M^{\Gamma_0}) \models \gamma$ for all Horn clauses $\gamma$ over $\Delta$.*

*Proof.* Since the model $(W^{\Gamma_0}, M^{\Gamma_0})$ is reachable, it suffices to prove that $\Gamma \vdash @_k \gamma$ implies $(W^{\Gamma_0}, M^{\Gamma_0}) \models @_k \gamma$ for all nominals $k \in T_{\Sigma^n}$ and Horn clauses $\gamma \in \mathsf{Sen}^{\mathsf{HDCLS}}(\Delta)$. We proceed by structural induction on $\gamma$.

*For the base case*, assume $\Gamma \vdash @_k \gamma$, where $\gamma$ is atomic. It follows that:

| | | |
|---|---|---|
| 1 | $\Gamma \vdash \gamma$ | by $(\mathsf{Ret}_0)$ in Figure 1 |
| 2 | $\gamma \in \Gamma_0$ | by the definition of $\Gamma_0$ |
| 3 | $\Gamma_0 \vdash \gamma$ | by the monotonicity of $\vdash$ |
| 4 | $(W^{\Gamma_0}, M^{\Gamma_0}) \models \gamma$ | by Theorem 15 |
| 5 | $(W^{\Gamma_0}, M^{\Gamma_0}) \models @_k \gamma$ | by Fact 17 |

*For the induction step*, we proceed by case analysis on the topmost sentence-building operator of $\gamma$. We only present the case corresponding to the *necessity* operator. Proofs for the remaining cases can be found in [10].

[ $\Gamma \vdash @_k [\mathfrak{a}]\gamma$ ] Let $w = W_k^{\Gamma_0}$. We want to show that $(W^{\Gamma_0}, M^{\Gamma_0}) \models^{w'} \gamma$ for all possible worlds $w'$ such that $(w, w') \in W_{\mathfrak{a}}^{\Gamma_0}$. Given such a possible world, since the model $(W^{\Gamma_0}, M^{\Gamma_0})$ is reachable, we know that there exists a nominal $k'$ such that $w' = W_{k'}^{\Gamma_0}$. It follows that:

| | | |
|---|---|---|
| 1 | $(W^{\Gamma_0}, M^{\Gamma_0}) \models \mathfrak{a}(k, k')$ | since $(w, w') \in W_{\mathfrak{a}}^{\Gamma_0}$ |
| 2 | $\Gamma_0 \vdash \mathfrak{a}(k, k')$ | by Proposition 18 |
| 3 | $\Gamma_f \vdash \mathfrak{a}(k_1, k_2)$ for some finite $\Gamma_f \subseteq \Gamma_0$ | since $\vdash$ is compact |
| 4 | $\Gamma \vdash \mathfrak{a}(k, k')$ | since $\Gamma \vdash \Gamma_f$ and $\Gamma_f \vdash \mathfrak{a}(k_1, k_2)$ |
| 5 | $\Gamma \vdash @_{k'} \gamma$ | by $(\mathsf{Nec}_E)$ |
| 6 | $(W^{\Gamma_0}, M^{\Gamma_0}) \models @_{k'} \gamma$ | by the induction hypothesis |
| 7 | $(W^{\Gamma_0}, M^{\Gamma_0}) \models^{w'} \gamma$ | since $w' = W_{k'}^{\Gamma_0}$. |

□

We are now finally ready to tackle the quasi-completeness of HDCLS: the initial model of a set of Horn clauses encapsulates the formal deduction of both atomic sentences and action relations. Note that, in general, action relations are not Horn clauses; nonetheless, we discuss their case too because it provides an important technical tool for the final completeness result.

**Corollary 20 (Quasi-completeness).** *Under the notations and hypotheses of Theorem 19, $(W^{\Gamma_0}, M^{\Gamma_0})$ is also an initial model of $\Gamma$. Moreover, for all atomic sentences or action relations $\rho$, the following statements are equivalent:*

*1.* $\Gamma \models \rho$ *2.* $(W^{\Gamma_0}, M^{\Gamma_0}) \models \rho$ *3.* $\Gamma \vdash \rho$ □

## 6 Horn-Clause Completeness

This final technical section deals with Birkhoff completeness, which corresponds to the existence of a syntactic characterization for the semantic entailment relation of HDCLS. This is practically very useful, because Horn clauses facilitate the development of an operational semantics of formal specifications based on rewriting. For example, action relations can provide logical support for the *rewriting rules* used in Maude [3], or for the *transitions* from CafeOBJ [6].

In order to generalize completeness to arbitrary Horn clauses, we need to consider additional rules, which are particular to different kinds of clauses. We say that a sentence is *action-free* if it contains no occurrences of any of the action-building operators (*composition*, *union*, or *transitive closure*), and that it is *star-free* if it contains no occurrences of the *transitive-closure* operator.

**Notation 21.** Consider the following fragments of HDFOLS. Each of them is obtained through a specific restriction on sentences:

$\mathsf{HDFOLS}^{(1)}$ – corresponding to action-free Horn clauses;
$\mathsf{HDFOLS}^{(2)}$ – corresponding to star-free Horn clauses and action relations;
$\mathsf{HDFOLS}^{(3)}$ – corresponding to Horn clauses and action relations.

$$(\mathsf{Ret_E})\ \frac{\Gamma \vdash_{\Delta[z]} @_z\,\gamma}{\Gamma \vdash_{\Delta} \gamma} \qquad (\mathsf{Imp_I})\ \frac{\Gamma \cup H \vdash @_k\,\gamma}{\Gamma \vdash @_k\,(\bigwedge H \Rightarrow \gamma)} \qquad (\mathsf{Store_I})\ \frac{\Gamma \vdash @_k\,\theta_{z\leftarrow k}(\gamma)}{@_k\downarrow z\cdot\gamma}$$

$$(\mathsf{Quant_I})\ \frac{\Gamma \vdash_{\Delta[X]} @_k\,\gamma}{\Gamma \vdash_{\Delta} @_k\,\forall X\cdot\gamma} \qquad (\mathsf{Nec_I})\ \frac{\Gamma \cup \{\mathfrak{a}(k,z)\} \vdash_{\Delta[z]} @_z\,\gamma}{\Gamma \vdash_{\Delta} @_k\,[\mathfrak{a}]\gamma} \qquad (\mathsf{Next_I})\ \frac{\Gamma \vdash @_{o(k)}\,\gamma}{\Gamma \vdash @_k\,\langle o\rangle\,\gamma}$$

**Fig. 4.** Additional proof rules for Horn clauses

$$(\mathsf{Comp_I})\ \frac{E \cup \{\mathfrak{a}_1(k_1,z), \mathfrak{a}_2(z,k_2)\} \vdash^{(2)}_{\Delta[z]} e}{E \cup \{(\mathfrak{a}_1 \mathbin{;} \mathfrak{a}_2)(k_1,k_2)\} \vdash^{(2)}_{\Delta} e}\ [\, E \cup \{e\} \subseteq \mathsf{Sen}^{\mathsf{HDFOLS}}(\Delta)\,]$$

$$(\mathsf{Union_I})\ \frac{E \cup \{\mathfrak{a}_i(k_1,k_2)\} \vdash^{(2)} e \text{ for } i \in \{1,2\}}{E \cup \{(\mathfrak{a}_1 \cup \mathfrak{a}_2)(k_1,k_2)\} \vdash^{(2)} e}\ [\, E \cup \{e\} \subseteq \mathsf{Sen}^{\mathsf{HDFOLS}}(\Delta)\,]$$

$$(\mathsf{Star_I})^a\ \frac{E \cup \{\mathfrak{a}^n(k_1,k_2)\} \vdash^{(3)} e \text{ for all } n \in \mathbb{N}}{E \cup \{\mathfrak{a}^*(k_1,k_2)\} \vdash^{(3)} e}\ [\, E \cup \{e\} \subseteq \mathsf{Sen}^{\mathsf{HDFOLS}}(\Delta)\,]$$

[a] Note that this rule is infinitary; we only use it in the final result in Section 6.

**Fig. 5.** Additional proof rules for action relations

Notice that $\mathsf{HDFOLS}^{(3)}$ is the richest fragment, and that $\gamma$ is a clause in $\mathsf{HDFOLS}$ iff it is a Horn clause in $\mathsf{HDFOLS}^{(3)}$. We also define three entailment relations:

1. $\vdash^{(1)}$ is generated by the proof rules in Figs. 1, 2, 3 and 4, but restricts the applications of ($\mathsf{Nec_I}$) to situations where $\mathfrak{a}$ is a modality (i.e., an atomic action);
2. $\vdash^{(2)}$ is generated by the proof rules in Figs. 1, 2, 3, 4 and 5, except ($\mathsf{Star_I}$), and restricted to applications of ($\mathsf{Comp_I}$) and ($\mathsf{Union_I}$) to star-free sentences;
3. $\vdash^{(3)}$ is generated by all proof rules in Figs. 1, 2, 3, 4 and 5.

Notice also that $\vdash^{(3)}$ is the most general one. Given a set of Horn clauses, $\vdash^{(3)}$ can be used to derive arbitrary Horn clauses from it, whereas $\vdash^{(2)}$ can only be used to derive star-free Horn clauses, and $\vdash^{(1)}$ only action-free Horn clauses.

It is easy to check that all these entailment relations are sound – similarly to Propositions 10 and 22, along the lines of [8]. Compactness, however, holds only for the first two. That is because the rule ($\mathsf{Star_I}$) in Fig. 5 is infinitary.

**Proposition 22 (Soundness & compactness).** *The entailment relation* $\vdash^{(x)}$ *is sound, for all* $x \in \{1, 2, 3\}$. *Moreover,* $\vdash^{(1)}$ *and* $\vdash^{(2)}$ *are also compact.* □

Our approach to completeness relies on the introduction rules in Figs. 4 and 5. These allow us to simplify, for example, the action relations that may appear in the left-hand side of the turnstile symbol during the proof process.

**Theorem 23 (Birkhoff completeness).** *Let* $x \in \{1, 2, 3\}$. *For every set* $\Gamma$ *of Horn clauses in* $\mathsf{HDFOLS}$, *and for every clause* $\gamma$ *in* $\mathsf{HDFOLS}^{(x)}$,

$$\Gamma \models \gamma \qquad \textit{implies} \qquad \Gamma \vdash^{(x)} \gamma.$$

*Proof.* Notice that $\Gamma \models \gamma$ implies $\Gamma \models @_k\, \gamma$, for any nominal $k$. Therefore, given the proof rule ($\mathsf{Ret_E}$), it suffices to prove that $\Gamma \models @_k\, \gamma$ implies $\Gamma \vdash^{(x)} @_k\, \gamma$. We proceed by induction on the structure of the sentence $\gamma$.

*For the base case*, where $\gamma$ is an atomic sentence, the conclusion follows by Fact 17, Corollary 20, and the fact that $\Gamma \vdash \gamma$ implies $\Gamma \vdash^{(x)} \gamma$.

*For the induction step*, we consider only the case where $\gamma$ is universally quantified. The remaining cases can be proved in a similar fashion; see [10].

[ $\Gamma \models @_k \forall X \cdot \gamma$ ] Then:

| | | |
|---|---|---|
| 1 | $\Gamma \models_{\Delta[X]} @_k\, \gamma$ | by the general properties of $\models$ |
| 2 | $\Gamma \vdash^{(x)}_{\Delta[X]} @_k\, \gamma$ | by the induction hypothesis |
| 3 | $\Gamma \vdash^{(x)}_{\Delta} @_k \forall X \cdot \gamma$ | by ($\mathsf{Quant_I}$) |

□

To come to an end, notice that the entailment relation $\vdash^{(3)}$ is sound (by Proposition 22) and complete (by Theorem 23), but it is not compact, since the rule ($\mathsf{Star_I}$) is infinitary. The next proposition shows this is the best result we can obtain, because the semantic entailment relation in $\mathsf{HDCLS}$ is not compact.

**Proposition 24 (Lack of compactness).** HDCLS *is not compact.*

*Proof (sketch).* It suffices to consider a signature $\Delta$ with two nominals, $k$ and $k'$, and two modalities, $\lambda$ and $\alpha$, and the set $\Gamma = \{\lambda^n(k, k') \Rightarrow \alpha(k, k') \mid n \in \mathbb{N}\}$ of Horn clauses over $\Delta$. Then the following properties hold:

1. $\Gamma \models \lambda^*(k, k') \Rightarrow \alpha(k, k')$;
2. There is no finite subset $\Gamma_f \subseteq \Gamma$ such that $\Gamma_f \models \lambda^*(k, k') \Rightarrow \alpha(k, k')$. □

## 7 Conclusions

The hybrid-dynamic first-order logic that we have studied in this paper is obtained by enriching first-order logic with a unique combination of features that are specific to hybrid and to dynamic logics. This provides a language that is particularly well suited for specifying and reasoning about reconfigurable systems. More precisely, it allows us to capture reconfigurable systems as Kripke structures whose possible worlds (a) have an algebraic structure, which supports operations on configurations, and (b) are labelled with constrained first-order models that capture the local structure of configurations. From a syntactic perspective, we define nominals and hybrid terms to refer to possible worlds and to the elements of the first-order structures associated to those worlds. Terms are then used to form nominal and hybrid equations, as well as relational atoms, from which we build complex sentences using Boolean connectives, quantifiers, hybrid-logic operators such as *retrieve* and *store*, and dynamic-logic operators such as *necessity* over actions, i.e., regular expressions over modalities.

In this context, we have developed a layered approach towards a Birkhoff completeness result for hybrid-dynamic first-order logic. There are three major layers to consider: first, the atomic layer, which deals with entailments where both the premises and the conclusion are atomic sentences; second, a mixed layer, which deals with entailments where the premises are Horn clauses, but the conclusion is only an atomic sentence or an action relation; and third, the general, Horn-clause layer, which deals with entailments where both the premises and the conclusion are Horn clauses. For each of these layers, we have developed sound and complete proof systems. Moreover, for the first two layers, the proof systems considered have also been shown to be compact.

The third layer deserves more attention. In that case, we distinguish between two main proof systems: (a) one that is compact, but complete only for entailments whose conclusion is a star-free clause; and (b) one that is not compact, but it is complete for all entailments. To conclude this line of developments, we have shown that this is the best result one can obtain for hybrid-dynamic logic.

As mentioned already, thanks to its features and expressive power, hybrid-dynamic first-order logic is a promising formalism for reasoning about reconfigurable systems. The work reported in this paper provides a rigorous foundation for that purpose. Therefore, an important task to pursue further is the development of a language, specification methodology, and appropriate tool support (that implements the proof systems presented here) for this new logic.

## References


1. Bohrer, B., Platzer, A.: A hybrid, dynamic logic for hybrid-dynamic information flow. In: Proceedings of the 33rd Annual ACM/IEEE Symposium on Logic in Computer Science, LICS 2018, Oxford, UK, 09–12 July 2018, pp. 115–124. ACM (2018)
2. Braüner, T.: Hybrid Logic and its Proof-Theory. Applied Logic Series, vol. 37. Springer, Dordrecht (2011). https://doi.org/10.1007/978-94-007-0002-4
3. Clavel, M., et al.: All About Maude - A High-Performance Logical Framework: How to Specify, Program, and Verify Systems in Rewriting Logic. LNCS, vol. 4350. Springer, Heidelberg (2007). https://doi.org/10.1007/978-3-540-71999-1
4. Codescu, M., Găină, D.: Birkhoff completeness in institutions. Logica Universalis **2**(2), 277–309 (2008)
5. Diaconescu, R.: Quasi-varieties and initial semantics for hybridized institutions. J. Log. Comput. **26**(3), 855–891 (2016)
6. Diaconescu, R., Futatsugi, K.: CafeOBJ Report - The Language, Proof Techniques, and Methodologies for Object-Oriented Algebraic Specification. AMAST Series in Computing, vol. 6. World Scientific, Singapore (1998)
7. Diaconescu, R., Madeira, A.: Encoding hybridized institutions into first-order logic. Math. Struct. Comput. Sci. **26**(5), 745–788 (2016)
8. Găină, D.: Birkhoff style calculi for hybrid logics. Form. Asp. Comput. **29**(5), 805–832 (2017)
9. Găină, D.: Foundations of logic programming in hybrid logics with user-defined sharing. Theor. Comput. Sci. **686**, 1–24 (2017)
10. Găină, D., Ţuţu, I.: Birkhoff completeness for hybrid-dynamic first-order logic (extended version). Technical report, Kyushu University, Japan, & Royal Holloway, University of London, UK (2019). https://pure.royalholloway.ac.uk/portal/files/34167903/BCHDL.pdf
11. Găină, D., Ţuţu, I.: Horn clauses in hybrid-dynamic first-order logic. CoRR abs/1905.04146 (2019). http://arxiv.org/abs/1905.04146
12. Găină, D., Ţuţu, I., Riesco, A.: Specification and verification of invariant properties of transition systems. In: 25th Asia-Pacific Software Engineering Conference, APSEC 2018, Nara, Japan, 4–7 December 2018. IEEE (2018)
13. Goguen, J.A., Burstall, R.M.: Institutions: abstract model theory for specification and programming. J. ACM **39**(1), 95–146 (1992)
14. Groote, J.F., Mousavi, M.R.: Modeling and Analysis of Communicating Systems. MIT Press, Cambridge (2014)
15. Harel, D., Kozen, D., Tiuryn, J.: Dynamic logic. SIGACT News **32**(1), 66–69 (2001)
16. Hennicker, R., Madeira, A., Knapp, A.: A hybrid dynamic logic for event/data-based systems. In: Hähnle, R., van der Aalst, W. (eds.) FASE 2019. LNCS, vol. 11424, pp. 79–97. Springer, Cham (2019). https://doi.org/10.1007/978-3-030-16722-6_5
17. Martins, M.A., Madeira, A., Diaconescu, R., Barbosa, L.S.: Hybridization of institutions. In: Corradini, A., Klin, B., Cîrstea, C. (eds.) CALCO 2011. LNCS, vol. 6859, pp. 283–297. Springer, Heidelberg (2011). https://doi.org/10.1007/978-3-642-22944-2_20

# Non-Wellfounded Proof Systems

# Infinets: The Parallel Syntax for Non-wellfounded Proof-Theory

Abhishek De[1] and Alexis Saurin[2(✉)]

[1] IRIF, Université de Paris, Paris, France
[2] IRIF, CNRS, Université de Paris, Paris, France
{ade,alexis.saurin}@irif.fr

**Abstract.** Logics based on the $\mu$-calculus are used to model inductive and coinductive reasoning and to verify reactive systems. A well-structured proof-theory is needed in order to apply such logics to the study of programming languages with (co)inductive data types and automated (co)inductive theorem proving. The traditional proof system suffers some defects, non-wellfounded (or infinitary) and circular proofs have been recognized as a valuable alternative, and significant progress have been made in this direction in recent years. Such proofs are non-wellfounded sequent derivations together with a global validity condition expressed in terms of *progressing threads*.

The present paper investigates a discrepancy found in such proof systems, between the sequential nature of sequent proofs and the parallel structure of threads: various proof attempts may have the exact threading structure while differing in the order of inference rules applications. The paper introduces infinets, that are proof-nets for non-wellfounded proofs in the setting of multiplicative linear logic with least and greatest fixed-points ($\mu\mathsf{MLL}^{\infty}$) and study their correctness and sequentialization.

**Keywords:** Circular proofs · Non-wellfounded proofs · Fixed points · $\mu$-calculus · Linear logic · Proof-nets · Induction and coinduction

## 1 Introduction

*Inductive and coinductive reasoning* is pervasive in computer science to specify and reason about infinite data as well as reactive properties. Developing appropriate proof systems amenable to automated reasoning over (co)inductive statements is therefore important for designing programs as well as for analyzing computational systems. Various logical settings have been introduced to reason about such inductive and coinductive statements, both at the level of the logical languages modelling (co)induction (such as Martin Löf's inductive predicates or fixed-point logics, also known as $\mu$-calculi) and at the level of the

This project has received funding from the European Union's Horizon 2020 research and innovation programme under the Marie Skłodowska-Curie grant agreement No 754362. Partially funded by ANR Project RAPIDO, ANR-14-CE25-0007.


S. Cerrito and A. Popescu (Eds.): TABLEAUX 2019, LNAI 11714, pp. 297–316, 2019.
https://doi.org/10.1007/978-3-030-29026-9_17

proof-theoretical framework considered (finite proofs with explicit (co)induction rules *à la* Park [27] or infinite, non-wellfounded proofs with fixed-point unfoldings) [1,2,5,8–10]. Moreover, such proof systems have been considered over classical logic [8,10], intuitionistic logic [11], linear-time or branching-time temporal logic [15,17,18,23,24,29,30] or linear logic [4,5,17,19,28].

*Logics based on the $\mu$-calculus* have been particularly successful in modelling inductive and coinductive reasoning and for the verification of reactive systems. While the model-theory of the $\mu$-calculus has been well-studied, its proof-theory still deserves further investigations. Indeed, while explicit induction rules are simple to formulate (For instance, Fig. 1 shows the introduction rule *à la* Park for a coinductive property) the treatment of (co)inductive reasoning brings some highly complex proof objects.

$$\frac{\vdash \Gamma, S \qquad \vdash S^{\perp}, F[S/X]}{\vdash \Gamma, \nu X.F}\ (\nu_{\text{inv}})$$

**Fig. 1.** Coinduction rule

At least two fundamental technical shortcomings prevent the application of traditional $\mu$-calculus-based proof-systems for the study of programming languages with (co)inductive data types and automated (co)inductive theorem proving and call for alternative proposals of proof systems supporting (co)induction. Firstly, the fixed point introduction rules break the subformula property which is highly problematic for automated proof construction: at each coinduction rule, one shall guess an invariant (in the same way as one has to guess an appropriate induction hypothesis in usual mathematical reasoning). Secondly, $(\nu_{\text{inv}})$ actually hides a cut rule that *cannot* be eliminated, which is problematic for extending the Curry-Howard correspondence to fixed-point logics.

$$\dfrac{\dfrac{\dfrac{\vdots}{\vdash \mu X.X}}{\vdash \mu X.X}\,(\mu) \qquad \dfrac{\dfrac{\vdots}{\vdash \nu X.X, \Gamma}}{\vdash \nu X.X, \Gamma}\,(\nu)}{\vdash \Gamma}\ (\mathsf{cut})$$

**Fig. 2.** An unsound proof

*Non-wellfounded proof systems* have been proposed as an alternative [8–10] to explicit (co)induction. By having the coinduction rule with simple fixed-point unfoldings and allowing for non-wellfounded branches, those proof systems address the problem of the subformula property for the cut-free systems: the set of subformula is then known as Fischer-Ladner subformulas, incorporating fixed-point unfolding but preserving finiteness of the subformula space. Moreover, the cut-elimination dynamics for inductive-coinductive rules becomes much simpler. A particularly interesting subclass of non-wellfounded proofs, is that of circular, or cyclic proofs, that have infinite but *regular* derivations trees: they have attracted a lot of attention for retaining the simplicity of the inferences of non-wellfounded proof systems but finitely representable making it possible to have an algorithmic treatment of such proof objects. However, in those proof systems when considering all possible infinite, non-wellfounded derivations (*a.k.a.* pre-proofs), it is straightforward to derive any sequent $\Gamma$ (see Fig. 2). Such pre-proofs are therefore unsound: one needs to impose a validity criterion to sieve the logically valid proofs from the unsound ones. This condition will actually reflect the inductive and coinductive nature of our two fixed-point connectives: a standard approach [4,8–10,28] is to consider a pre-proof to be valid if every infinite branch is supported by an infinitely progressing thread. As a result, the logical correctness of circular proofs becomes non-local, much in the spirit of correctness criteria for proof-nets [14,20].

However the structure of non-wellfounded proofs has to be further investigated: the present work stems from the observation of a discrepancy between the sequential nature of sequent proofs and the parallel structure of threads. An immediate consequence is that various proof attempts may have the exact same threading structure but differ in the order of inference rule applications; moreover, cut-elimination is known to fail with more expressive thread conditions [3]. This paper proposes a theory of proof-nets for $\mu\mathsf{MLL}^\infty$ non-wellfounded proofs.

*Organization of the Paper.* In Sect. 2, we recall the necessary background from [4] on linear logic with least and greatest fixed points and its non-wellfounded proofs, we only present the unit-free multiplicative setting which is the framework in which we will define our proof-nets. In Sect. 3 we adapt Curien's proof-nets [12] to a very simple extension of MLL, $\mu\mathsf{MLL}^*$, in which fixed-points inferences are unfoldings and only wellfounded proofs are allowed; this allows us to set the first definitions of proof-nets and extend correctness criterion, sequentialization and cut-elimination to this setting but most importantly it sets the proof-net formalism that will be used for the extension to non-wellfounded derivations. Infinets are introduced in Sect. 4 as an extension of the $\mu\mathsf{MLL}^*$ proof-nets of the previous section. A correctness criterion is defined in Sect. 5 which is shown to be sound (every proof-nets obtained from a sequent (pre-)proof is correct). The completeness of the criterion (*i.e.* sequentialization theorem) is addressed in Sect. 6. We quotient proofs differing in the order of rule application in Sect. 7 and give a partial cut elimination result in Sect. 8. We conclude in Sect. 9 and comment on related works and future directions.

*Notation.* For any sequence $S$, let $\mathsf{Inf}(S)$ be the terms of $S$ that appears infinitely often in $S$. Given a finite alphabet $\Sigma$, $\Sigma^*$ and $\Sigma^\omega$ are the set of finite and infinite words over $\Sigma$ *resp.* Let $\Sigma^\infty = \Sigma^* \cup \Sigma^\omega$. We denote the empty word by $\epsilon$. Given two words $u, u'$ (finite or infinite) we denote by $u \sqcap u'$ the greatest common prefix of $u$ and $u'$ and $u \sqsubseteq u'$ if $u$ is a prefix of $u'$. Given a language, $\mathcal{L} \subseteq \Sigma^\infty$, $\overline{\mathcal{L}} \subseteq \Sigma^\infty$ is the set of all prefixes of the words in $\mathcal{L}$.

## 2 Background

We denote the multiplicative additive fragment of linear logic by MALL and the multiplicative fragment by MLL. The non-wellfounded extension of MALL with least and greatest fixed points operators, $\mu\mathsf{MALL}^\infty$, was introduced in [4,17]. Proof-nets for additives and units are quite cumbersome [7,21], so, in the current presentation, we will only consider the unit-free multiplicative fragment which we denote by $\mu\mathsf{MLL}^\infty$.

**Definition 1.** *Given an infinite set of atoms, $\mathcal{A} = \{A, B, \dots\}$, and an infinite set of propositional variables, $\mathcal{V} = \{X, Y, \dots\}$, s.t. $\mathcal{A} \cap \mathcal{V} = \emptyset$, $\mu$MLL* **pre-formulas** *are given by the following grammar:*

$$\phi, \psi ::= A \mid A^\perp \mid X \mid \phi \parr \psi \mid \phi \otimes \psi \mid \sigma X.\phi$$

*where $A \in \mathcal{A}$ and $X \in \mathcal{V}$, and $\sigma \in \{\mu, \nu\}$; $\sigma$ binds the variable $X$ in $\phi$. When a pre-formula is closed (i.e. no free variables), we simply call it a* **formula**.

Note that negation is not a part of the syntax, so that we do not need any positivity condition on the fixed-point expressions. We define negation, $(\bullet)^\perp$, as a meta-operation on the pre-formulas and will use it only on formulas.

**Definition 2. Negation** *of a pre-formula $\phi$, $\phi^\perp$, is the involution satisfying:*

$$(\phi \otimes \psi)^\perp = \psi^\perp \parr \phi^\perp,\ X^\perp = X,\ (\mu X.\phi)^\perp = \nu X.\phi^\perp.$$

*Example 1.* As a running example, we will consider the formulas $\phi = A \parr A^\perp \in$ MLL and $\psi = \nu X.X \otimes \phi \in \mu\mathsf{MLL}^\infty$. Observe that $\phi^\perp = A^\perp \otimes A$ as usual in MLL and by Definition 2, $\psi^\perp = \mu X.X \parr \phi^\perp$.

The reader may find it surprising to define $X^\perp = X$, but it is harmless since our proof system only deals with formulas. Note that $(F[G/X])^\perp = F^\perp[G^\perp/X]$.

**Definition 3.** *An* **(infinite) address** *is a finite (resp. infinite) word in $\{l, r, i\}^\infty$. Negation extends over addresses as the morphism satisfying $l^\perp = r$, $r^\perp = l$, and $i^\perp = i$. We say that $\alpha'$ is a* **sub-address** *of $\alpha$ if $\alpha' \sqsubseteq \alpha$. We say that $\alpha$ and $\beta$ are disjoint if $\alpha \sqcap \beta$ is not equal to $\alpha$ or $\beta$.*

**Definition 4.** *A* **formula occurrence** *(denoted by $F, G, \dots$) is given by a formula, $\phi$, and a finite address, $\alpha$, and written $\phi_\alpha$. Let $\mathsf{addr}(\phi_\alpha) = \alpha$. We say that occurrences are disjoint when their addresses are. Operations on formulas are extended to occurrences as follows: ${\phi_\alpha}^\perp = \phi^\perp_{\alpha^\perp}$, for any $\star \in \{\parr, \otimes\}$, $F \star G = (\phi \star \psi)_\alpha$ if $F = \phi_{\alpha l}$ and $G = \psi_{\alpha r}$, and for $\sigma \in \{\mu, \nu\}$, $\sigma X.F = (\sigma X.\phi)_\alpha$ if $F = \phi_{\alpha i}$. Substitution of occurrences forgets addresses i.e. $(\phi_\alpha)[\psi_\beta/X] = (\phi[\psi/X])_\alpha$. Finally, we use $\lceil\bullet\rceil$ to denote the address erasure operation on occurrences.*

Fixed-points logics come with a notion of subformulas (and suboccurrences) slightly different from usual:

**Definition 5.** *The* **Fischer-Ladner closure** *of a formula occurrence $F$, $\mathsf{FL}(F)$, is the least set of formula occurrences s.t. $F \in \mathsf{FL}(F)$, $G_1 \star G_2 \in \mathsf{FL}(F) \implies G_1, G_2 \in \mathsf{FL}(F)$ for $\star \in \{\parr, \otimes\}$, and $\sigma X.G \in \mathsf{FL}(F) \implies G[\sigma X.G/X] \in \mathsf{FL}(F)$ for $\sigma \in \{\mu, \nu\}$. We say that $G$ is a* **FL-suboccurrence** *of $F$ (denoted $G \leq F$) if $G \in \mathsf{FL}(F)$ and $G$ is an* **immediate FL-suboccurrence** *of $F$ (denoted $G \lessdot F$) if $G \leq F$ and for every $H$ s.t. $G \leq H \leq F$ either $H = G$ or $H = F$. The* **FL-subformulas** *of $F$ are elements of $\{\phi \mid \phi = \lceil G \in \mathsf{FL}(F)\rceil\}$.*

Clearly, we could have defined Fischer-Ladner closure on the level of formulas. By abuse of notation, we will sometimes use $\mathsf{FL}(\bullet), \leq, \lessdot$ on formulas.

$$\frac{\lceil F\rceil = \lceil G\rceil^{\perp}}{\vdash F, G}\ (\mathsf{ax}) \qquad \frac{\vdash F, \Delta_1 \quad \vdash F^{\perp}, \Delta_2}{\vdash \Delta_1, \Delta_2}\ (\mathsf{cut}) \qquad \frac{\vdash F, G, \Delta}{\vdash F \parr G, \Delta}\ (\parr)$$

$$\frac{\vdash F, \Delta_1 \quad \vdash G, \Delta_2}{\vdash F \otimes G, \Delta_1, \Delta_2}\ (\otimes) \qquad \frac{\vdash G[\mu X.G/X], \Delta}{\vdash \mu X.G, \Delta}\ (\mu) \qquad \frac{\vdash G[\nu X.G/X], \Delta}{\vdash \nu X.G, \Delta}\ (\nu)$$

**Fig. 3.** Inference rules for $\mu\mathsf{MLL}^{\infty}$

*Remark 1.* Observe that for any $F$, the number of FL-subformulas of $F$ is finite.

The usual notion of subformula (say in MLL) is obtained by traversing the syntax tree of a formula. In the same way, the notion of FL-subformula can be obtained by traversing the graph of the formula (*resp.* occurrence).

**Definition 6.** *The* **FL-graph** *of a formula $\phi$, denoted $\mathfrak{G}(\phi)$, is the graph obtained from $\mathsf{FL}(\phi)$ by identifying the nodes of bound variable occurrences with their binders (i.e. $\phi \rightarrow \psi$ if $\phi < \psi$).*

*Example 2.* The graphs of the formulas $\phi$ and $\psi$ of Example 1 are the following:

Observe that the graph of a MLL formula is acyclic corresponding to the usual syntax tree but the graph of a $\mu\mathsf{MLL}^{\infty}$ formula could potentially contain a cycle.

As usual with classical linear logic $\Gamma, \phi \vdash \Delta$ is provable iff the sequent $\Gamma \vdash \phi^{\perp}, \Delta$ is provable. Hence, it is enough to consider the one-sided proof system. A one-sided $\mu\mathsf{MLL}^{\infty}$ sequent is an expression $\vdash \Delta$ where $\Delta$ is a finite set of pairwise disjoint formula occurrences.

**Definition 7.** *A* **pre-proof** *of $\mu\mathsf{MLL}^{\infty}$ is a possibly infinite tree generated from the inference rules given in Fig. 3.*

**Definition 8.** *A* **thread** *of a formula occurrence, $F$, is a sequence, $t = \{F_i\}_{i \in I}$, where $I \in \omega + 1$, $F_0 = F$, and for every $i \in I$ s.t. $i + 1 \in I$ either $F_i$ is suboccurrence of $F_{i+1}$ or $F_i = F_{i+1}$. We denote by $\lceil t\rceil$ the sequence $\{\lceil F_i\rceil\}_{i \in I}$ where $t = \{F_i\}_{i \in I}$. A thread, $t$, is said to be* **valid** *if $\min(\mathsf{Inf}(\lceil t\rceil))$ is a $\nu$-formula where minimum is taken in the $\leq$ ordering.*

*Remark 2.* Observe that for any infinite thread $t$ of a formula occurrence $F$, $\mathsf{Inf}(\lceil t\rceil)$ is non-empty since $F$ has finitely many FL-subformulas.

**Definition 9.** *A $\mu\mathsf{MLL}^{\infty}$* **proof** *is a pre-proof in which every infinite branch contains a valid thread. A* **circular** *pre-proof is a regular $\mu\mathsf{MLL}^{\infty}$ pre-proof i.e. one which has a finite number of distinct subtrees.*

*Example 3.* The following non-wellfounded pre-proof of the sequent $\vdash \psi_\alpha$ ($\alpha$ is an arbitrary address) is circular and is a proof because the only infinite thread $\{\psi_{\alpha(il)^n}\}_{n=0}^{\infty}$ is valid.

$$\dfrac{\dfrac{\dfrac{\star}{\vdash \psi_{\alpha il}} \qquad \dfrac{\dfrac{}{\vdash A_{\alpha irl}, A^{\perp}_{\alpha irr}}\,(\mathsf{ax})}{\vdash A \parr A^{\perp}_{\alpha ir}}\,(\parr)}{\vdash \psi \otimes (A \parr A^{\perp})_{\alpha i}}\,(\otimes)}{\star \vdash \psi_\alpha}\,(\nu)$$

## 3 A First Taste of Proof-Nets in Logics with Fixed Points

Proof-nets are a geometrical method of representing proofs, introduced by Girard that eliminates two forms of bureaucracy which differentiate sequent proofs: irrelevant syntactical features and the order of rules. As a stepping stone, we first consider proof nets in $\mu\mathsf{MLL}^*$ which is the proof system with the same inference rules as $\mu\mathsf{MLL}^\infty$ (Fig. 3) but with finite proofs. $\mu\mathsf{MLL}^*$ is strictly weaker than $\mu\mathsf{MLL}^\infty$.

Proof-nets are usually defined as vertex labelled, edge labelled directed multi-graphs. In this presentation a proof structure is "almost" a forest (*i.e.* a collection of trees) with the leaves joined by axioms or cuts. We use a different presentation due to Curien [12] to separate the forest of syntax trees and the space of axiom links for reasons that will become clearer later.

**Definition 10.** *A* **syntax tree** *of a formula occurrence, $F$, is the (possibly infinite) unfolding tree of $\mathfrak{G}(F)$. The syntax tree induces a prefix closed language, $\mathcal{L}_F \subset \{l, r, i\}^\infty$ s.t. there is a natural bijection between the finite (resp. infinite) words in $\mathcal{L}_F$ and the finite (resp. infinite) paths of the tree. A* **partial syntax tree,** *$F^U$, is a subtree of the syntax tree of the formula occurrence, $F$, such that the set of words, $U \subseteq \mathcal{L}_F$, represents a "frontier" of the syntax tree of $F$ i.e. any $u, u' \in U$ are pairwise disjoint and for every $uav \in U$, there is a $v'$ s.t. $ua^{\perp}v' \in U$. For a finite $u \in \overline{U}$, we denote by $(F, u)$ the unique suboccurrence of $F$ with the address $\mathsf{addr}(F).u$.*

*Example 4.* The syntax tree of $\psi$ is the unfolding of $\mathfrak{G}(\psi)$ and induces the language $\overline{i(li)^*r(l+r)} + (il)^\omega$. Further, given an arbitrary address $\alpha$, $\psi_\alpha^{\{ili,irl,irr\}}$ is a partial syntax tree whereas $\psi_\alpha^{\{ilil,irl,irr\}}$ is not. If $u = ililir$ then $(\psi_\alpha, u) = A \parr A^{\perp}_{\alpha ililir}$.

**Definition 11.** *A* **proof structure** *is given by $[\Theta']\{B_i^{U_i}\}_{i\in I}[\Theta]$, where,*

- *$I$ is a finite index set;*
- *for every $i \in I$, $B_i$ is a formula occurrence, $B_i^{U_i}$ is a partial syntax tree with $U_i \subset \{l, r, i\}^*$;*
- *$\Theta'$ is a (possibly empty) collection of disjoint subsets of $\{B_i\}_{i\in I}$ of the form $\{C, C^{\perp}\}$;*

$$\dfrac{\dfrac{\pi_1}{\vdash \Gamma, F} \quad \dfrac{\pi_2}{\vdash \Delta, F^\perp}}{\vdash \Gamma, \Delta}\,(\mathsf{cut}) \qquad \dfrac{\dfrac{\pi_1}{\vdash \Gamma, F} \quad \dfrac{\pi_2}{\vdash \Delta, G}}{\vdash \Gamma, \Delta, F \otimes G}\,(\otimes) \qquad \dfrac{\dfrac{\pi_0}{\vdash \Gamma, F, G}}{\vdash \Gamma, F \parr G}\,(\parr) \qquad \dfrac{\dfrac{\pi_0}{\vdash \Gamma, F[\mu X.F/X]}}{\vdash \Gamma, \mu X.F}\,(\mu)$$

(a) (b) (c) (d)

**Fig. 4.** Induction cases for definition 12.

- $\Theta$ *is a partition of* $\bigcup_{i \in I}\{\alpha_i u_i \mid \mathsf{addr}(B_i) = \alpha_i, u_i \in U_i\}$ *s.t. the partitions are of the form* $\{\alpha_i u_i, \alpha_j u_j\}$ *with* $\lceil (B_i, u_i) \rceil = \lceil (B_j, u_j) \rceil^\perp$.

Each class of $\Theta$ represents an axiom, each of class of $\Theta'$ represents a cut, and $\{B_i\}_{i\in I} \setminus \bigcup_{\theta \in \Theta'} \theta$ are the conclusions of the proof structure.

**Definition 12.** *Let* $\pi$ *be a* $\mu\mathsf{MLL}^*$ *proof.* **Desequentialization** *of* $\pi$*, denoted* $\mathsf{Deseq}(\pi)$*, is defined by induction on the structure of the proof:*

- *The base case is a proof with only an* ax *rule, say* $\dfrac{}{F, G^\perp}\,(\mathsf{ax})$. *Then*

$$\mathsf{Deseq}(\pi) = [\emptyset]\{F^{\{\epsilon\}}, (G^\perp)^{\{\epsilon\}}\}[\{\{\mathsf{addr}(F), \mathsf{addr}(G^\perp)\}\}]$$

- *If* $\mathsf{Deseq}(\pi_1) = [\Theta_1']\Gamma_1 \cup \{F^U\}[\Theta_1]$ *and* $\mathsf{Deseq}(\pi_2) = [\Theta_2']\Gamma_1 \cup \{F^{\perp^{U'}}\}[\Theta_2]$, *then* $\mathsf{Deseq}(\pi) = [\Theta_1' \cup \Theta_2' \cup \{F, F^\perp\}]\Gamma_1 \cup \Gamma_2[\Theta_1 \cup \Theta_2]$ *where* $\pi$ *is Fig. 4(a).*
- *If* $\mathsf{Deseq}(\pi_1) = [\Theta_1']\Gamma_1 \cup \{F^U\}[\Theta_1]$ *with* $\mathsf{addr}(F) = \alpha l$ *and* $\mathsf{Deseq}(\pi_2) = [\Theta_2']\Gamma_1 \cup \{G^{U'}\}[\Theta_2]$ *with* $\mathsf{addr}(G) = \alpha r$, *then* $\mathsf{Deseq}(\pi) = [\Theta_1' \cup \Theta_2']\Gamma_1 \cup \Gamma_2 \cup \{F \otimes G^{l\cdot U + r\cdot U'}\}[\Theta_1 \cup \Theta_2]$ *with* $\mathsf{addr}(F \otimes G) = \alpha$ *where* $\pi$ *is Fig. 4(b).*
- *If* $\mathsf{Deseq}(\pi_0) = [\Theta_0']\Gamma_0 \cup \{F^U, G^{U'}\}[\Theta_0]$ *with* $\mathsf{addr}(F) = \alpha l, \mathsf{addr}(G) = \alpha r$ *then* $\mathsf{Deseq}(\pi) = [\Theta_0]\Gamma_0 \cup \{F \parr G^{l\cdot U + r\cdot U'}\}[\Theta_0]$ *with* $\mathsf{addr}(F \parr G) = \alpha$ *where* $\pi$ *is Fig. 4(c).*
- *If* $\mathsf{Deseq}(\pi_0) = [\Theta_0']\Gamma_0 \cup \{F[\mu X.F/X]^U\}[\Theta_0]$ *with* $\mathsf{addr}(F[\mu X.F/X]) = \alpha i$ *then* $\mathsf{Deseq}(\pi) = [\Theta_0]\Gamma_0 \cup \{\mu X.F^{i\cdot U}\}[\Theta_0]$ *with* $\mathsf{addr}(\mu X.F) = \alpha$ *where* $\pi$ *is Fig. 4(d).*
- *The case for* $\nu$ *follows exactly as* $\mu$.

*Example 5.* Consider the following proof $\pi$ of the sequent $\vdash \nu X.X \parr \mu X.X$.

$$\dfrac{\dfrac{\dfrac{}{\vdash \nu X.X_{\alpha l}, \mu Y.Y_{\beta i}}\,(\mathsf{ax})}{\vdash \nu X.X_{\alpha l}, \mu Y.Y_\beta}\,(\mu) \quad \dfrac{\dfrac{}{\vdash \nu Y.Y_{\beta^\perp i}, \mu X.X_{\alpha r}}\,(\mathsf{ax})}{\vdash \nu Y.Y_{\beta^\perp}, \mu X.X_{\alpha r}}\,(\nu)}{\dfrac{\vdash \nu X.X_{\alpha l}, \mu X.X_{\alpha r}}{\vdash \nu X.X \parr \mu X.X_\alpha}\,(\parr)}\,(\mathsf{cut})$$

We choose $\alpha, \beta$ *s.t.* they are disjoint. We have that $\mathsf{Deseq}(\pi) = [\Theta']\Gamma[\Theta]$ *s.t.*

$$\Theta' = \{\{\mu Y.Y_\beta, \nu Y.Y_{\beta^\perp}\}\} \qquad \Theta = \{\{\alpha l, \beta i\}, \{\alpha r, \beta^\perp i\}\}$$

$$\Gamma = \left\{\nu X.X \parr \mu X.X_\alpha^{\{l,r\}}, \mu Y.Y_\beta^{\{i\}}, \nu Y.Y_{\beta^\perp}^{\{i\}}\right\}$$

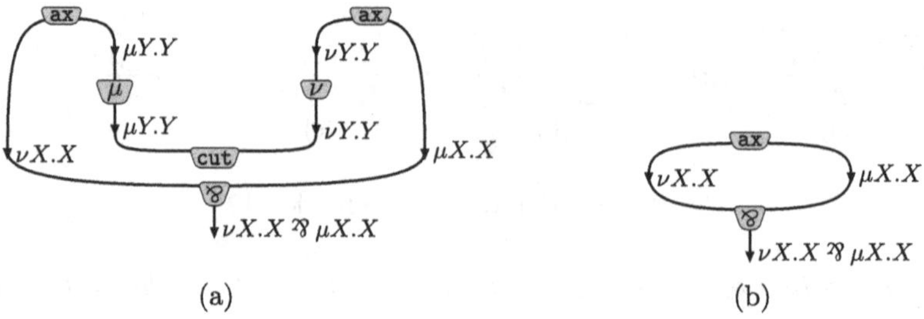


**Fig. 5.** Graph of $\mu\mathsf{MLL}^\infty$ proof structures

**Definition 13 (Graph of proof structure).** *Let $\mathcal{S} = [\Theta']\{B_i^{U_i}\}_{i\in I}[\Theta]$ be a proof structure. The graph of $\mathcal{S}$, denoted $\mathsf{Gr}(\mathcal{S})$, is the graph formed by:*

- *taking the transpose (i.e. reversal of every edge) of the partial syntax tree $\{B_i^{U_i}\}_{i\in I}$;*
- *for each $\{B_i, B_j\} \in \Theta'$, adding a node labelled* cut *with two incoming edges from $(B_i, \epsilon)$ and $(B_j, \epsilon)$;*
- *for each $\{\alpha_i u_i, \alpha_j u_j\} \in \Theta$, adding a node labelled* ax *with two outgoing edges to $(B_i, u_i)$ and $(B_j, u_j)$ where $\mathsf{addr}(B_i)$ and $\mathsf{addr}(B_j)$ is $\alpha_i$ and $\alpha_j$ resp.*

*Example 6.* The graph of the proof structure in Example 5 is Fig. 5a.

$\mathsf{Gr}(\mathcal{S})$ are exactly the proof structures that we obtain from directly lifting the formalism of MLL proof nets à la Girard to $\mu\mathsf{MLL}^*$.

As usual in the theory of proof nets, we need a correctness criterion on the $\mu\mathsf{MLL}^*$ proof structures to exactly characterize the class of proof nets. The following correctness criterion lifts to $\mu\mathsf{MLL}^*$ a criterion first investigated by Danos and Regnier [14]. We present it in a slightly different syntax using the notion of orthogonal partitions [13,14].

**Definition 14.** *Let $P_1$ and $P_2$ be partitions of a set $S$. The graph induced by $P_1$ and $P_2$ is defined as the undirected bipartite multigraph, $(P_1, P_2, E)$, s.t. for every $p \in P_1$ and $p' \in P_2$, $(p, p') \in E$ if $p \cap p' \neq \emptyset$. Finally, $P_1$ and $P_2$ are said to be* **orthogonal** *to each other if the graph induced by them is acyclic and connected.*

**Definition 15.** *Given a proof structure, $\mathcal{S} = [\Theta']\{B_i^{U_i}\}_{i\in I}[\Theta]$, define a set of* **switchings** *of $\mathcal{S}$, $sw = \{sw_i\}_{i\in I}$ s.t. for every $i \in I$, $sw_i : P_i \to \{l, r\}$ is a function over $P_i$, the ⅋ nodes of $B_i^{U_i}$. The* **switching graph** *$\mathcal{S}^{sw}$ associated with $sw$ is formed by:*

- *taking the partial syntax tree $\{B_i^{U_i}\}_{i\in I}$ as an undirected graph;*
- *for each $\{B_i, B_j\} \in \Theta'$, adding a node labelled cut with two edges to $(B_i, \epsilon)$ and $(B_j, \epsilon)$;*
- *for each node $(B_i, u) \in P_i$, removing the edge between $(B_i, u)$ and $(B_i, u \cdot sw((B_i, u)))$.*

*Let* $\Theta_S^{sw}$ *be the partition over* $\bigcup_{i\in I}\{\alpha_i u_i \mid \mathsf{addr}(B_i) = \alpha_i, u_i \in U_i\}$ *induced by the connected component of* $\mathcal{S}^{sw}$.

**Definition 16.** *A proof structure,* $\mathcal{S}$, *is said to be* **OR-correct** *if for any switching* $sw$, $\Theta_S^{sw}$ *and* $\Theta$ *is orthogonal. The graph induced by* $\Theta_S^{sw}$ *and* $\Theta$ *is called a* **correction graph** *of* $\mathcal{S}$.

**Proposition 1.** *Let* $\pi$ *be a* $\mu\mathsf{MLL}^*$ *proof. Then* $\mathsf{Deseq}(\pi)$ *is an OR-correct proof structure. Conversely, given an OR-correct* $\mu\mathsf{MLL}^*$ *proof structure, it can be sequentialized into a* $\mu\mathsf{MLL}^*$ *sequent proof.*

**Definition 17.** $\mu\mathsf{MLL}^*$ **cut-reduction rules** *is obtained by adding the following rule to the usual cut-reduction rules for* MLL *proof nets:*

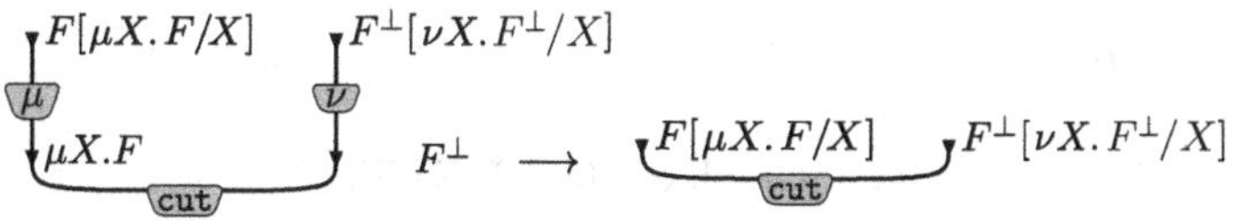


**Proposition 2.** *Cut elimination on* $\mu\mathsf{MLL}^*$ *proof-nets preserves correctness and is strongly normalizing and confluent.*

The proofs of Propositions 1 and 2 are straightforward extensions from MLL.

*Example 7.* The proof structure in Example 5 after cut-elimination produces the proof structure in Fig. 5b.

*Remark 3.* Now the question is how this translates to non-wellfounded proofs. Consider the proof in Example 3. Firstly observe that there is no finite proof of this sequent *i.e.* it is not provable in $\mu\mathsf{MLL}^*$. Now, if we naively translate it into a proof structure using the same recipe as Definition 12 (except allowing for infinite partial syntax trees), we have

$$[\emptyset]\left\{\psi_\alpha^{\{i(li)^* r(l+r)+(il)^\omega\}}\right\}[\{\alpha i(li)^n rl, i(li)^n rr\}_{n\geq 0}].$$

Observe that $(il)^\omega$ is not in any partition. In fact, it represents a thread in an infinite branch and must be accounted for. Hence the partition should account for the threads invariant by an infinite branch in a proof (in particular, in the example above there should be a singleton partition, $\{(il)^\omega\}$). This is also the reason we will not use the graphical presentation for non-wellfounded proof-nets since we would potentially need to join two infinite paths by a node which is unclear graph-theoretically. However we will sometimes draw the "graph" of non-wellfounded proof-nets for ease of presentation by using ellipsis points (for example Fig. 6b represents the proof-net we discussed above).

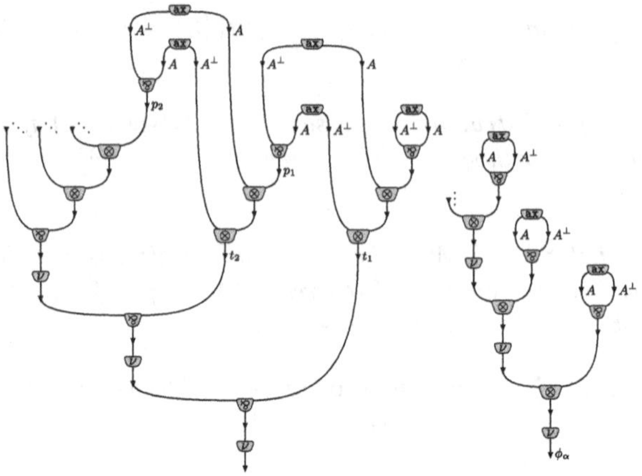


(a) (b)

**Fig. 6.** Graph of $\mu$MLL$^\infty$ NWFPS

## 4 Infinets

We will now lift our formalism for defining proof nets for $\mu$MLL$^*$ to $\mu$MLL$^\infty$.

**Definition 18.** *A* **non-wellfounded proof structure** *(NWFPS) is given by* $[\Theta']\{B_i^{U_i}\}_{i\in I}[\Theta]$*, where*

- *$I$ is a possibly infinite index set;*
- *for every $i \in I$, $B_i$ is a formula occurrence, $B_i^{U_i}$ is a partial syntax tree;*
- *$\Theta'$ is a (possibly empty) collection of disjoint subsets of $\{B_i\}_{i\in I}$ of the form $\{C, C^\perp\}$;*
- *$\Theta$ is a partition of $\bigcup_{i\in I}\{\alpha_i u_i \mid \mathsf{addr}(B_i) = \alpha_i, u_i \in U_i\}$ s.t. the partitions are one of the following forms:*
  - *$\{\alpha_i u_i, \alpha_j u_j\}$ s.t. $u_i, u_j$ are finite and $\lceil(B_i, u_i)\rceil = \lceil(B_j, u_j)\rceil^\perp$.*
  - *It contains an elements of the form $\alpha_i u_i$ s.t. $u$ is an infinite address;*
- *$\{B_i\}_{i\in I} \setminus \bigcup_{\theta\in\Theta'} \theta$ is necessarily finite.*

Intuitively, each class of $\Theta$ represents either an axiom or an infinite branch in a sequentialization. In fact, the infinite addresses in a partition correspond exactly to the infinite threads in a proof. Hence it is also straightforward to define a valid NWFPS.

**Definition 19.** *Let $\pi$ be a pre-proof of the $\mu$MLL$^\infty$ sequent $\vdash \Gamma$ and $\mathsf{addr}(\pi) \subseteq \{l, r, i\}^\infty$ be the set of all addresses occurring in $\pi$ and all infinite addresses such that all their strict prefixes are addresses occurring in $\pi$. Desequentialization of $\pi$, denoted $\mathsf{Deseq}(\pi)$, is the NWFPS, $[\Theta']\Gamma'[\Theta]$, s.t. $\Theta'$ are the cut formulas in $\pi$, $B_i^{U_i} \in \Gamma'$ where $B_i \in \Gamma$, $U_i = \mathsf{addr}(B_i)^{-1}\mathsf{addr}(\pi)$, to any finite maximal branch* *of $\pi$, associate a partition in $\Theta$ containing the addresses of the occurrences that are the conclusion of the corresponding axiom rule in $\pi$ and to any infinite branch, $\beta$, of $\pi$, associate a partition in $\Theta$ such that a finite address is in the partition if it is belongs to infinitely many sequents of $\beta$ and an infinite address is in the partition if all its strict prefixes belong to $\beta$. A NWFPS that is the desequentialization of a $\mu\mathsf{MLL}^{\infty}$ (pre-)proof is called an (valid)* **infinet.**

*Example 8.* As expected from the discussion in Remark 3, desequentialization of the proof in Example 3 is

$$[\emptyset]\left\{\psi_{\alpha}^{\{i(li)^{*}r(l+r)+(il)^{\omega}\}}\right\}[\{\alpha i(li)^{n}rl, i(li)^{n}rr\}_{n\geq 0}, \{(il)^{\omega}\}].$$

*Remark 4.* The reader might think that there is discrepancy in the way desequentialization of wellfounded and non-wellfounded proofs are defined in Definitions 12 and 19 *resp.* Note that Definition 12 can be reformulated à la Definition 19 but not vice versa. However, we choose to inductively define wellfounded desequentialization since it is closer to the standard definition in proof-net theory.

## 5 Correctness Criteria

The OR-correctness of a NWFPS is defined as in Definitions 15 and 16 (up to the fact that the switching can be an infinite set of switching functions). However this straightforward translation is not enough to ensure soundness.

*Example 9.* Consider the following sequent proof with infinitely many cuts.

$$\dfrac{\dfrac{}{\mu X.X, \nu Z.Z}(\mathsf{ax}) \qquad \dfrac{\star \vdash \mu Z.Z, \nu Y.Y}{\vdash \mu Z.Z, \nu Y.Y}(\nu)}{\star \vdash \mu X.X, \nu Y.Y}(\mathsf{cut})$$

Observe that this structure is not OR-correct:

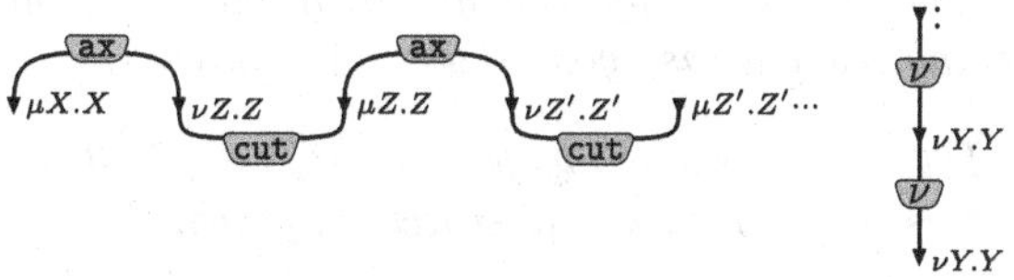


Consequently, we restrict ourselves to NWFPS with at most finitely many cuts. The proof structures discussed in the rest of the paper have finitely many cuts unless otherwise mentioned.

*Example 10.* Consider the graph of proof structure of the sequent $\vdash \nu X.X \parr (A^{\perp}\otimes(A\otimes\ (A^{\perp} \parr A)))$ in Fig. 6a. Note that for the sake of readability, edge labels have been concealed. This proof structure is OR-correct but it is not sequentializable. Consider the $\otimes$ node labelled $t_1$. In any sequentialization it should be above $p_1$, which should be above $t_2$, which in turn should be above $p_2$ and so on. This is absurd since even in a non-wellfounded proof every rule is executed at a finite depth.

Hence we impose a "lock-free" condition (borrowing the terminology from concurrent programming) on NWFPS.

**Definition 20.** *Let $[\Theta']\{B_i^{U_i}\}_{i\in I}[\Theta]$ be a NWFPS. For any $u_i \in U_i, u_j \in U_j$, we say that $(u_i, u_j)$ is a* **coherent pair** *if there exists $\theta \in \Theta, B_i, B_j$ s.t. $\{\alpha_i u_i, \alpha_j u_j\} \subseteq \theta$ $\mathsf{addr}(B_i) = \alpha_i$ and $\mathsf{addr}(B_j) = \alpha_j$.*

**Definition 21.** *A* **switching path** *is an undirected path in a partial syntax tree s.t. it does not go consecutively through two premises of a ⅋ formula. A* **strong switching path** *is a switching path whose first edge is not the premise of a ⅋ node. We denote by $\mathsf{src}(\bullet), \mathsf{tgt}(\bullet)$ the source and target of a switching path resp. Two switching paths $\gamma, \gamma'$ are said to be* **compatible** *if $\gamma'$ is strong and $\mathsf{tgt}(\gamma) = \mathsf{src}(\gamma')$.*

**Proposition 3.** *If $\gamma, \gamma'$ are compatible switching paths, then their concatenation $\gamma \cdot \gamma'$ is a switching path. Furthermore, if $\gamma$ is strong, then $\gamma \cdot \gamma'$ is also strong.*

The underlying undirected path of any path in a partial syntax tree is a switching path. We call such paths ***straight switching paths***. In particular, the path from any vertex, $v$, to the root is a straight switching path. We denote it by $\delta(v)$. By abuse of notation, we will also sometimes write $\delta((B_i, u))$ where $u$ is infinite to mean the infinite path from the root of $B_i^{U_i}$ following $u$, although technically $(B_i, u)$ is not a node per se. Observe that any straight switching path in a partial syntax tree, $F^U$, can be represented by a pair of words $(u, u') \in \overline{U}^2$ *s.t.* $u \sqsubset u'$. Intuitively, it means that the path is from $(F, u)$ to $(F, u')$.

**Definition 22.** *A* **switching sequence** *is a sequence $\sigma = \{\gamma_i\}_{i=1}^n$ s.t. $\gamma_i$s are disjoint switching paths and for every $i \in \{1, 2, \ldots, n-1\}$, either $\gamma_i, \gamma_{i+1}$ are compatible or they are straight and the word pairs corresponding to them, $(u_i, u_i')$ and $(u_{i+1}, u_{i+1}')$, are s.t. $(u_i', u_{i+1}')$ is a coherent pair. Two vertices, $v$ and $v'$, are said to be connected by the switching sequence, $\sigma$, if $\mathsf{src}(\gamma_1) = v$ and $\mathsf{tgt}(\gamma_n) = v'$. We say the switching sequence is cyclic if $\mathsf{src}(\gamma_1) = \mathsf{tgt}(\gamma_n)$.*

**Proposition 4.** *Let $\gamma$ be a switching path in $B_j^{U_j} \in \Gamma$. Then there exists a switching sw s.t. $\gamma$ is also a path in the switching graph, $\mathcal{S}^{sw}$.*

**Proposition 5.** *If $\mathcal{S}$ is a NWFPS containing a cyclic switching sequence, then there is switching of $\mathcal{S}$, s.t. the corresponding correction graph is contains a cycle.*

**Definition 23.** *Let $\mathcal{S} = [\Theta']\{B_i^{U_i}\}[\Theta]$ be a proof structure. Let $T = \{(B_i, u_i) \mid u_i \in \overline{U_i}; (B_i, u_i)$ is a $\otimes$ formula$\}$ and let $P = \{(B_i, u_i) \mid u_i \in \overline{U_i}; (B_i, u_i)$ is a ⅋ formula$\}$. The* **dependency graph** *of $\mathcal{S}$, $D(\mathcal{S})$, is the directed graph $(V, E)$ s.t. $V = T \uplus P$, for every $v \in V$ and $p \in P$, $(p, v) \in E$ if the premises of $p$ are connected by a switching sequence containing $v$, and, for every $v, v' \in V$, $(v, v') \in E$ if $v' \in \mathsf{FL}(v)$.*

**Proposition 6 (Bagnol *et al.* [6]).** *If $\mathcal{S}$ is OR-correct then $D(\mathcal{S})$ is acyclic.*

From Proposition 6, we can impose an order on the nodes of an OR-correct proof structure, $\mathcal{S}$, namely, $n_1 <_{D(\mathcal{S})} n_2$ if $n_1 \rightarrow n_2$ in $D(\mathcal{S})$.

**Definition 24.** *A NWFPS, $\mathcal{S}$, is said to be* **deeply lock-free** *if $<_{D(\mathcal{S})}$ has no infinite descending chains.*

*Example 11.* Consider the proof structure, $\mathcal{S} = [\emptyset]\{\nu X.X \parr X_\alpha^L, A \otimes B_\beta^{\{l,r\}}\}[\Theta]$ where, $L = (i(l+r))^\omega$ , $\Theta = \{\{\alpha(il)^\omega, \beta l\}, \alpha \cdot (L \setminus (il)^\omega) \cup \{\beta r\}\}$.

Observe that $\mathcal{S}$ is OR-correct and deeply lock-free. But $\mathcal{S}$ cannot be sequentialized into a sequent proof, because a potential sequentialization has a $\otimes$ rule at a finite depth, then either there are no subsoccurences of $\nu X.X \parr X_\alpha$ in the left premise in which case $A$ cannot reside with only the left-branch in $\Theta$, or, there are no subsoccurences of $\nu X.X \parr X_\alpha$ in the left premise in which case $A$ cannot reside with any infinite branch in $\Theta$.

**Definition 25.** *A NWFPS, $\mathcal{S} = [\Theta']\{B_i^{U_i}\}_{i \in I}[\Theta]$, is said to be* **widely lock-free** *if there is a function $f : \mathbb{N} \rightarrow \mathbb{N}$ s.t. for every $(B_i, u) \in P$ and $(B_j, v) \in T$ if $((B_i, u), (B_j, v)) \in E$, $f(|v|) \geq |u|$ where $D(\mathcal{S}) = (T \uplus P, E)$. We call such a function a* **wait function** *of $\mathcal{S}$. A proof structure is simply called* **lock-free** *if it is both deeply and widely lock-free.*

*Remark 5.* The wait function of a NWFPS need not be unique (if one exists).

**Proposition 7.** *An infinet is an OR-correct lock-free NWFPS.*

## 6 Sequentialization

In this section we show that any NWFPS satisfying the correctness criterion introduced in Sect. 5 is indeed sequentializable. Since we deal with finitely many cuts, without loss of generality, we can assume that we have cut-free proof structures due to the standard trick shown in Fig. 7.

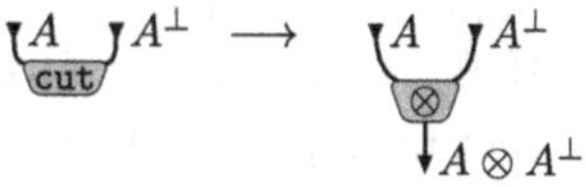


**Fig. 7.** Translating cuts to tensors

So, in this section, we will write NWFPS without the left component. We try to adapt the standard proof for MLL but the straightforward adaptation is not *fair* since we may never explore one branch by forever prioritizing the sequentialization of another infinite branch. We restore fairness by a time-stamping algorithm.

**Definition 26.** *Let $\mathcal{S} = \Gamma[\Theta]$ be an OR-correct NWFPS. The root, $B_i$, of a tree in $\Gamma$ is said to be* **splitting** *if:*

- *$\Gamma = \{B_i^\epsilon, B_j^\epsilon\}$,*
- *$B_i$ is a $\parr, \mu$ or $\nu$ formula, or,*
- *$B_i$ is a $\otimes$ formula and there exists $\Theta_1, \Theta_2$ s.t. $\Theta = \Theta_1 \uplus \Theta_2$ and $\mathcal{S}_1 = \Gamma_1[\Theta_1]$, $\mathcal{S}_2 = \Gamma_2[\Theta_2]$ are OR-correct NWFPS where $\Gamma_1 = \Gamma \setminus \{B_i^{U_i}\} \cup \{(B_i, l)^{U_l}\}$, $\Gamma_2 = \Gamma \setminus \{B_i^{U_i}\} \cup \{(B_i, r)^{U_r}\}$ and $U_i = lU_l + rU_r$.*

**Proposition 8.** *Let $\mathcal{S} = \Gamma[\Theta]$ be an OR-correct NWFPS and $B_i$ be a splitting $\otimes$ formula in $\mathcal{S}$. If $\mathcal{S}$ is lock-free then so is $\mathcal{S}_1$ and $\mathcal{S}_2$ as defined in Definition 26.*

*Dated Sequentialization Process.* We time-stamp each node of $\Gamma$ to indicate the time when it will be sequentialized. Formally, we have $(\mathcal{S}, \tau)$ where $\tau$ is a function *s.t.* $\tau : \{(B_i, u) | u \in U_i\}_{i \in I} \to \mathbb{N} \cup \{\infty\}$ where $\Gamma = \{B_i^{U_i}\}_{i \in I}$ and $\infty > n$ for all $n \in \mathbb{N}$. Define the minimal finite image, min, as

$$\min(\tau) := \min\{n \in \mathbb{N} \mid \exists i \in I, u \in U_i \ s.t.\ \tau((B_i, u)) = n\}.$$

We will describe the sequentialization process. Suppose we are given $\mathcal{S}(= \Gamma[\Theta], \tau)$. We maintain the following invariant:

$$\begin{array}{c} \mathcal{S} \text{ is cut-free, OR-correct and lock-free;} \\ \tau((B_i, u)) \neq \infty \text{ iff } (B_i, u) \text{ is splitting in } \mathcal{S}. \end{array} \qquad (\star)$$

Assume that $\Gamma$ contains a splitting root, $B_j$, st. $\tau(B_j) = \min(\tau)$.

- If $\Gamma = \{B_i^\epsilon, B_j^\epsilon\}$ then we stop successfully with the proof reduced to an ax.
- If $B_j$ is a $⅋$, (co)recursively apply the sequentialization process to $\mathcal{S}_0(= \Gamma_0[\Theta], \tau_0)$ where $\Gamma_0 = \Gamma \setminus \{B_j^{U_j}\} \cup \{(B_j, l)^{U_l}, (B_j, r)^{U_r}\}$, $U_j = lU_l + rU_r$, and
$$\tau_0((B_i, u)) = \begin{cases} t & \text{if } (B_i, u) \text{ is splitting in } \mathcal{S}_0; \\ \tau((B_i, u)) & \text{otherwise.} \end{cases}$$
where for each splitting $(B_i, u)$, $t$ is arbitrarily chosen to be any natural number greater than $\tau(B_j)$. We apply a $⅋$ rule on the obtained proof.
- If $B_j$ is a $\mu$(*resp.* $\nu$) formula, (co)recursively apply the sequentialization process to $(\mathcal{S}_0 = \Gamma_0[\Theta], \tau_0)$ where $\Gamma_0 = \Gamma \setminus \{B_j^{U_j}\} \cup \{(B_j, i)^{U_i}\}$, $U_j = iU_i$, and $\tau_0$ is defined as above. We apply a $\mu$ (*resp.* $\nu$) rule on the obtained proof.
- If $B_j$ is a $\otimes$ formula we (co)recursively apply the sequentialization process to $(\mathcal{S}_1, \tau_1)$ and $(\mathcal{S}_2, \tau_2)$ where $\mathcal{S}_1, \mathcal{S}_2$ are as defined in Definition 26 and $\tau_1, \tau_2$ are defined as above. We apply a $\otimes$ rule on the two obtained proofs.

Observe that the invariant $(\star)$ is maintained in this (co)recursive process. To start the sequentialization, we initialize $\tau$ by assigning arbitrary natural numbers to splitting nodes and $\infty$ to the other nodes.

**Proposition 9.** *Let $T$ be a non-splitting conclusion in an OR-correct NWFPS. Then there exists a $⅋$ formula, $P$, s.t. there exists disjoint switching sequences, $\sigma, \sigma'$, from $T$ to $P$ which both start with a premise of $T$ and end with a premise of $P$. We call $(P, \sigma, \sigma')$ the* **witness** *for $T$.*

**Lemma 1.** *Let $\mathcal{S}$ be a cut-free OR-correct NWFPS. $\mathcal{S}$ contains a splitting root.*

**Lemma 2.** *The sequentialization assigns a finite natural number to every formula i.e. $\tau((B_i, u)) \neq \infty$ after some iterations of the process described above.*

Lemma 1 crucially uses OR-correctness and Lemma 2 crucially uses lock-freeness. Lemma 1 ensures productivity of the aforementioned sequentialization process while Lemma 2 ensures that every inference in a NWFPS is ultimately executed. From that, we conclude the following theorem.

**Theorem 1.** *Let $\mathcal{S} = [\Theta']\Gamma[\Theta]$ be an OR-correct lock-free NWFPS s.t. $\Theta' = \emptyset$. Then $\mathcal{S}$ is an infinet.*

*Remark 6.* Observe that the choice of the time-stamping function at each step of our sequentialization is non-deterministic. By considering appropriate time-stamping functions we can generate all sequentializations. The detailed study is beyond the scope of the present paper.

## 7 Canonicity

We started investigating proof nets for non-wellfounded proofs since we expected that the proof net formalism would quotient sequent proofs that are equivalent up to a permutation of inferences. At this point, we carry out that sanity check.

Consider the following proofs $\pi_1$ and $\pi_1'$.

$$\dfrac{\dfrac{\dfrac{\pi_2}{\vdash \Gamma, F[\mu X.F/X], A}}{\vdash \Gamma, \mu X.F, A}(\mu) \quad \dfrac{\pi_3}{\vdash B, \Delta}}{\pi_1 \vdash \Gamma, \mu X.F, A \otimes B, \Delta}(\otimes) \qquad \dfrac{\dfrac{\dfrac{\pi_2}{\vdash \Gamma, F[\mu X.F/X], A} \quad \dfrac{\pi_3}{\vdash B, \Delta}}{\vdash \Gamma, F[\mu X.F/X], A \otimes B, \Delta}(\otimes)}{\pi_1' \vdash \Gamma, \mu X.F, A \otimes B, \Delta}(\mu)$$

We say that $\pi \rightsquigarrow_{(\mu,\otimes_L)} \pi'$ if $\pi$ is a proof with $\pi_1$ as a subproof at a finite depth and $\pi'$ is $\pi$ where $\pi_1'$ has been replaced by $\pi_1'$. Observe that we can define $\rightsquigarrow_\square$ for every $\square \in P \times P$ where $P = \{\mu, \nu, \parr, \otimes_\star, \mathsf{cut}_\star \mid \star \in \{L, R\}\}$. Let $\rightsquigarrow_\blacksquare = \bigcup_{\square \in S} \rightsquigarrow_\square$.

Observe that the usual notion of equivalence by permutation, *viz.* $\sim = (\rightsquigarrow_\blacksquare)^*$ does not characterize equivalence by infinets. Consider the following two proofs, $\pi_1$ and $\pi_2$, *s.t.* $\pi_1 \not\sim \pi_2$ which have the same infinet,

$$[\emptyset]\{\mu X.X_\alpha^{\{i^\omega\}}, \nu X.X_\beta^{\{i^\omega\}}\}[\{\{\alpha i^\omega, \beta i^\omega\}\}].$$

$$\dfrac{\dfrac{\pi_1 \vdash \mu X.X, \nu X.X}{\vdash \mu X.X, \nu X.X}(\mu)}{\pi_1 \vdash \mu X.X, \nu X.X}(\nu) \qquad \dfrac{\dfrac{\pi_2 \vdash \mu X.X, \nu X.X}{\vdash \mu X.X, \nu X.X}(\nu)}{\pi_2 \vdash \mu X.X, \nu X.X}(\mu)$$

Suppose we allow infinite permutations. We say that $\pi(\rightsquigarrow_\blacksquare)^\omega\pi'$ if there exists a proof $\pi''$ (not necessarily different from $\pi, \pi'$) and two sequence of proofs, $\{\pi_i\}_{i=0}^\infty$ and $\{\pi_i'\}_{i=0}^\infty$, *s.t.* $\pi_0 = \pi$, $\pi_0' = \pi'$, for every $i$, $\pi_i \rightsquigarrow_\blacksquare \pi_{i+1}$ and $\pi_i' \rightsquigarrow_\blacksquare \pi_{i+1}'$, and $d(\pi_i, \pi'') \to 0$, $d(\pi_i', \pi'') \to 0$ as $i \to \infty$. Consider the following proofs.

$$\dfrac{\dfrac{\pi}{\vdash A} \quad \dfrac{\vdots}{\vdash B, \nu Y.Y}(\nu)}{\vdash A \otimes B, \nu Y.Y}(\otimes) \quad (\sim_{\blacksquare})^{\omega} \quad \dfrac{\vdots}{\vdash A \otimes B, \nu Y.Y}(\nu) \quad (\sim_{\blacksquare})^{\omega} \quad \dfrac{\dfrac{\pi'}{\vdash A} \quad \dfrac{\vdots}{\vdash B, \nu Y.Y}(\nu)}{\vdash A \otimes B, \nu Y.Y}(\otimes)$$

Note that equating these proofs is absurd since $\pi$ and $\pi'$ can have different computation behaviour (for example, $A = (X^{\perp} \parr X^{\perp}) \parr (X \otimes X)$ and $\pi$ corresponds to true while $\pi$' corresponds to false). To exactly capture equivalence by infinets we need to refine this equivalence. To do that we introduce the notion of an active occurrence. We say that for a permutation step $\leadsto_{(r_i, r_i')}$, the formula occurrence $F_i$ introduced by the rule $r_i'$ is the **active occurrence** in that step.

Given two node-labelled trees $T_1$ and $T_2$, we define $d(T_1, T_2) = \frac{1}{2^{\delta}}$ where $\delta$ is the minimal depth of the nodes at which they differ. We say that $\pi(\sim_{\blacksquare})^{\omega}_{\mathsf{fair}}\pi'$ if there exists a sequence of proofs $\{\pi_i\}_{i=0}^{\infty}$ *s.t.* $\pi_0 = \pi$, for every $i$, $\pi_i \leadsto_{(r_i, r_i')} \pi_{i+1}$, the sequence of addresses of the active occurrences occurring infinitely often is empty, *i.e.* $\mathsf{Inf}(\{\mathsf{addr}(F_i)\}_{i=0}^{\infty}) = \emptyset$, and $d(\pi_i, \pi') \to 0$ as $i \to \infty$. Let $\sim^{\infty} = (\sim_{\blacksquare})^{*} \cup (\sim_{\blacksquare})^{\omega}_{\mathsf{fair}}$.

**Proposition 10.** $\pi_1 \sim^{\infty} \pi_2$ *iff* $\mathsf{Deseq}(\pi_1) = \mathsf{Deseq}(\pi_2)$.

## 8 Cut Elimination

In this section we provide cut elimination results albeit with two crucial restrictions: firstly, we consider only finitely many cuts as in the rest of the paper and secondly, we consider proofs with no axioms and no atoms. An infinet $\mathcal{S} = [\Theta']\Gamma[\Theta]$ is said to be ***$\eta^{\infty}$-expanded*** if it does not contain any axioms or atoms *i.e.* every $\theta \in \Theta$ contains only infinite addresses. Any infinet can be made $\eta^{\infty}$-expanded in a way akin to $\eta$-expansion of axioms in MLL. There are two issues to be resolved to obtain the result: first, to specify the notion of a normal form and second, formulate how to reach that.

**Proposition 11.** *Let $\mathcal{S} = [\Theta']\Gamma[\Theta]$ be an $\eta^{\infty}$-expanded infinet. Let $\{C, C^{\perp}\} \in \Theta$ and $B_i^{U_i}, B_j^{U_j} \in \Gamma$ s.t. $B_i = C = B_j{}^{\perp}$. Then, $U_i = U_j{}^{\perp}$ i.e. $u \in U_i$ iff $u^{\perp} \in U_j$.*

*Proof (Sketch).* Since $B_i = B_j{}^{\perp}$, their syntax trees are orthogonal. Since $\mathcal{S}$ is $\eta^{\infty}$-expanded, $U_i$(resp. $U_j$) is actually the full syntax tree. Hence $U_i = U_j{}^{\perp}$.

**Definition 27.** *Let $\mathcal{S}_0 = [\Theta_0']\Gamma_0[\Theta_0]$ be a $\eta^{\infty}$-expanded infinet. Let $\{C, C^{\perp}\} \in \Theta_0'$ and $B_i^{U_i}, B_j^{U_j} \in \Gamma$ s.t. $B_i = C = B_j{}^{\perp}$. A* **big-step** $\{C, C^{\perp}\}$ **elimination** *on $\mathcal{S}_0$ produces non-wellfounded proof-structure $\mathcal{S}_1 = [\Theta_1']\Gamma_1[\Theta_1]$ where,*

- $\Theta_1' = \Theta_0' \setminus \{\{C, C^{\perp}\}\}$
- $\Gamma_1 = \Gamma_0 \setminus \{B_i^{U_i}, B_j^{U_j}\}$
- *If $\theta \in \Theta_0$ s.t. $\theta \cap U_i = \emptyset$ and $\theta \cap U_j = \emptyset$, then $\theta \in \Theta_1$. If $u \in \theta \cap U_i$ then $\theta \cup \theta' \setminus \{u, u^{\perp}\} \in \Theta_1$ where $\theta' \in \Theta_0$ and $u^{\perp} \in \theta' \cap U_j$.*

*Remark 7.* Definition 27 is well-defined because of Proposition 11.

**Proposition 12.** *A big-step operation on a valid infinet produces a valid infinet.*

Given $\mathcal{S} = [\Theta']\Gamma[\Theta]$, an $\eta^\infty$-expanded infinet, we can extend the definition of a big-step $\{C, C^\perp\}$ elimination on $\mathcal{S}$, for any $\{C, C^\perp\} \in \Theta'$, to a big-step $\mathcal{C}$ elimination on $\mathcal{S}$, for $\mathcal{C} \subseteq \Theta'$. We call the big-step $\Theta'$ elimination on $\mathcal{S}$ the **normal form** of $\mathcal{S}$ and denote it by $[\![\mathcal{S}]\!]$.

The idea now is to show that local cut-elimination indeed produces in the limit the normal form defined above. For this we need to define a metric, $d$, over infinets with the same normal form so that we can formalize the limit of infinite reduction sequences. See [16] for details.

**Lemma 3.** *The set of all valid infinets with the same normal form together with the distance, d, forms a metric space.*

We can now define the limit of an infinite sequence of valid infinets with the same normal form in the standard way: we say that $\{\mathcal{S}_i\}_{i=0}^{\infty}$ converges to $\mathcal{S}$ if $d(\mathcal{S}_i, \mathcal{S}) \to 0$ as $i \to \infty$.

**Definition 28.** *A sequence of infinets, $\{\mathcal{S}_i\}_{i=0}^{\infty}$, is called a* **reduction sequence** *if for every $i > 0$, $\mathcal{S}_i \to \mathcal{S}_{i+1}$ by the cut reduction rules in Definition 17. A reduction sequence is said to be* **fair** *if for every $i$, for every cut $\{C, C^\perp\}$ in $\mathcal{S}_i$, there is a $j > i$ such that $C'$ is a suboccurrence of $C$ where $\{C', C'^\perp\}$ is the cut being reduced in the step $\mathcal{S}_j \to \mathcal{S}_{j+1}$.*

**Theorem 2.** *Let $\{\mathcal{S}_i\}_{i=0}^{\infty}$ be a fair reduction sequence s.t. $\mathcal{S}_0$ is valid. Then, it converges to $[\![\mathcal{S}_0]\!]$.*

**Corollary 1.** *If two reduction sequences starting from a valid $\eta^\infty$-expanded infinet, $\mathcal{S}$, converges to $\mathcal{S}_1$ and $\mathcal{S}_2$, then all fair reduction sequences starting from $\mathcal{S}_1$ and $\mathcal{S}_2$ resp. converge to $[\![\mathcal{S}]\!]$.*

# 9 Conclusion

In this paper, we introduced infinitary proof-nets for $\mu\mathsf{MLL}^\infty$. We defined a correctness criterion and showed its soundness and completeness in characterizing those proof structures which come from non-wellfounded sequent (pre)proofs. We also gave a partial cut elimination result. Currently, our results are subject to the restriction that non-wellfounded shall only contain finitely many cut inferences.

For the non-wellfounded correctness criterion, we extended the Danos-Regnier criterion from the finitary case. Other more efficient criteria (like the parsing criterion [22]) are impossible to adapt since any reasonable operation over non-wellfounded structures should necessarily be of a bottom-up nature (unlike the parsing criterion).

*Related and Future Works.* The closest works we know of are Montelatici's polarized proof nets with cycles [26] and Mellies' work on higher-order parity automata [25] which considers a $\lambda$Y-calculus and an infinitary $\lambda$-calculus endowed with parity conditions, therefore quotienting some of the non-determinism of sequent-calculus albeit in the case of intuitionistic logic.

Our work is a first step in developing a general theory of non-wellfounded and circular proof-nets:

- We plan to extend our framework and strengthen the correctness criterion in several directions to capture more proofs: we shall strengthen the criterion to capture proofs with infinitely many cuts as well as to extend our formalism to the additives, capturing $\mu\mathsf{MALL}^\infty$.
- Once our framework can handle proofs with infinitely many cuts, we plan to investigate how the so-called bouncing thread criterion [3] can be captured in proof nets. This is indeed one of the motivation of our work to solve the discrepancy between the sequential nature of proofs and the parallel nature of threads which is especially problematic when relaxing validity conditions as in [3].
- Finally we plan to carry an investigation of the notion of circularity in proof-nets: while one can capture circular proofs as finitely representable proof nets, there are non-wellfounded proofs which are not circular but which have finitely representable desequentialization. The simplest example is the proof of $\vdash \nu X.X \parr X$ which contains sequents of unbounded size and is therefore not circular. Not only is the study of circular infinets interesting from a programming perspective but also it would be possible to do a complexity analysis on such finitely representable proof-nets to better understand the cost of checking correctness, sequentialization and cut-elimination.

**Acknowledgement.** We are indebted to anonymous reviewers for providing insightful comments which has immensely enhanced the presentation of the paper.

## References

1. Baelde, D.: On the proof theory of regular fixed points. In: Giese, M., Waaler, A. (eds.) TABLEAUX 2009. LNCS (LNAI), vol. 5607, pp. 93–107. Springer, Heidelberg (2009). https://doi.org/10.1007/978-3-642-02716-1_8
2. Baelde, D.: Least and greatest fixed points in linear logic. ACM Trans. Computat. Logic (TOCL) **13**(1), 2 (2012)
3. Baelde, D., Doumane, A., Kuperberg, D., Saurin, A.: Bouncing threads for infinitary and circular proofs. Manuscript, June 2019
4. Baelde, D., Doumane, A., Saurin, A.: Infinitary proof theory: the multiplicative additive case. In: 25th EACSL Annual Conference on Computer Science Logic, CSL 2016, Marseille, France, 29 August–1 September 2016. LIPIcs, vol. 62, pp. 42:1–42:17. Schloss Dagstuhl - Leibniz-Zentrum fuer Informatik (2016)
5. Baelde, D., Miller, D.: Least and greatest fixed points in linear logic. In: Dershowitz, N., Voronkov, A. (eds.) LPAR 2007. LNCS (LNAI), vol. 4790, pp. 92–106. Springer, Heidelberg (2007). https://doi.org/10.1007/978-3-540-75560-9_9

6. Bagnol, M., Doumane, A., Saurin, A.: On the dependencies of logical rules. In: Pitts, A. (ed.) FoSSaCS 2015. LNCS, vol. 9034, pp. 436–450. Springer, Heidelberg (2015). https://doi.org/10.1007/978-3-662-46678-0_28
7. Blute, R.F., Cockett, J.R.B., Seely, R.A.G., Trimble, T.H.: Natural deduction and coherence for weakly distributive categories. J. Pure Appl. Algebr. **113**(3), 229–296 (1996)
8. Brotherston, J.: Sequent calculus proof systems for inductive definitions. Ph.D. thesis, University of Edinburgh, November 2006
9. Brotherston, J., Simpson, A.: Complete sequent calculi for induction and infinite descent. In: Proceedings of the 22nd IEEE Symposium on Logic in Computer Science (LICS 2007), Wroclaw, Poland, 10–12 July 2007, pp. 51–62. IEEE Computer Society (2007)
10. Brotherston, J., Simpson, A.: Sequent calculi for induction and infinite descent. J. Log. Comput. **21**(6), 1177–1216 (2011)
11. Clairambault, P.: Least and greatest fixpoints in game semantics. In: de Alfaro, L. (ed.) FoSSaCS 2009. LNCS, vol. 5504, pp. 16–31. Springer, Heidelberg (2009). https://doi.org/10.1007/978-3-642-00596-1_3
12. Curien, P.-L.: Introduction to linear logic and ludics, Part II (2006)
13. Danos, V.: Une application de la logique linéaire á l'étude des processus de normalisation (principalement du $\lambda$-calcul). Thèse de doctorat, Université Denis Diderot, Paris 7 (1990)
14. Danos, V., Regnier, L.: The structure of multiplicatives. Arch. Math. Log. **28**, 181–203 (1989)
15. Dax, C., Hofmann, M., Lange, M.: A proof system for the linear time $\mu$-calculus. In: Arun-Kumar, S., Garg, N. (eds.) FSTTCS 2006. LNCS, vol. 4337, pp. 273–284. Springer, Heidelberg (2006). https://doi.org/10.1007/11944836_26
16. De, A., Saurin, A.: Infinets: the parallel syntax for non-wellfounded proof-theory. Long version, June 2019
17. Doumane, A.: On the infinitary proof theory of logics with fixed points. (Théorie de la démonstration infinitaire pour les logiques à points fixes). Ph.D. thesis, Paris Diderot University, France (2017)
18. Doumane, A., Baelde, D., Hirschi, L., Saurin, A.: Towards completeness via proof search in the linear time $\mu$-calculus, January 2016. Accepted for publication at LICS
19. Fortier, J., Santocanale, L.: Cuts for circular proofs: semantics and cut-elimination. In: Della Rocca, S.R., (ed.) Computer Science Logic 2013 (CSL 2013), CSL, Torino, Italy, 2–5 September 2013. LIPIcs, vol. 23, pp. 248–262. Schloss Dagstuhl - Leibniz-Zentrum fuer Informatik (2013)
20. Girard, J.-Y.: Linear logic. Theor. Comput. Sci. **50**, 1–102 (1987)
21. Girard, J.-Y.: Proof-nets: the parallel syntax for proof theory (1996)
22. Guerrini, S.: A linear algorithm for mll proof net correctness and sequentialization. Theor. Comput. Sci. **412**(20), 1958–1978 (2011)
23. Kaivola, R.: A simple decision method for the linear time mu-calculus. In: Desel, J. (ed.) Structures in Concurrency Theory. Workshops in Computing, pp. 190–204. Springer, London (1995). https://doi.org/10.1007/978-1-4471-3078-9_13
24. Kozen, D.: Results on the propositional mu-calculus. Theor. Comput. Sci. **27**, 333–354 (1983)
25. Melliès, P.-A.: Higher-order parity automata. In: 32nd Annual ACM/IEEE Symposium on Logic in Computer Science, LICS 2017, Reykjavik, Iceland, 20–23 June 2017, pp. 1–12. IEEE Computer Society (2017)

26. Montelatici, R.: Polarized proof nets with cycles and fixpoints semantics. In: Hofmann, M. (ed.) TLCA 2003. LNCS, vol. 2701, pp. 256–270. Springer, Heidelberg (2003). https://doi.org/10.1007/3-540-44904-3_18
27. Park, D.: Fixpoint induction and proofs of program properties. Mach. Intell. **5**, 59–78 (1969)
28. Santocanale, L.: A calculus of circular proofs and its categorical semantics. In: Nielsen, M., Engberg, U. (eds.) FoSSaCS 2002. LNCS, vol. 2303, pp. 357–371. Springer, Heidelberg (2002). https://doi.org/10.1007/3-540-45931-6_25
29. Walukiewicz, I.: On completeness of the mu-calculus. In: Proceedings of the Eighth Annual Symposium on Logic in Computer Science (LICS 1993), Montreal, Canada, 19–23 June 1993, pp. 136–146. IEEE Computer Society (1993)
30. Walukiewicz, I.: Completeness of Kozen's axiomatisation of the propositional mu-calculus. In: Proceedings of the 10th Annual IEEE Symposium on Logic in Computer Science, San Diego, California, USA, 26–29 June 1995, pp. 14–24. IEEE Computer Society (1995)

# PSPACE-Completeness of a Thread Criterion for Circular Proofs in Linear Logic with Least and Greatest Fixed Points

Rémi Nollet[1,2,3(✉)], Alexis Saurin[1,2,3,4], and Christine Tasson[1,2,3]

[1] Université de Paris, Paris, France
[2] IRIF, Paris, France
{nollet,saurin,tasson}@irif.fr
[3] CNRS, Paris, France
[4] INRIA $\pi r^2$, Paris, France
https://www.irif.fr/~nollet, https://www.irif.fr/~saurin,
https://www.irif.fr/~tasson

**Abstract.** In the context of logics with least and greatest fixed points, circular (i.e. non-wellfounded but regular) proofs have been proposed as an alternative to induction and coinduction with explicit invariants. However, those proofs are not wellfounded and to recover logical consistency, it is necessary to consider a validity criterion which differentiates valid proofs among all preproofs (i.e. infinite derivation trees).

The paper focuses on circular proofs for MALL with fixed points. It is known that given a finite circular representation of a non-wellfounded preproof, one can decide in PSPACE whether this preproof is valid with respect to the thread criterion. We prove that the problem of deciding thread-validity for $\mu$MALL is in fact PSPACE-complete.

Our proof is based on a deeper exploration of the connection between thread-validity and the size-change termination principle, which is usually used to ensure program termination.

**Keywords:** Sequent calculus · Non-wellfounded proofs · Circular proofs · Induction · Coinduction · Fixed points · Linear logic · mu-MALL · Size-change · PSPACE-complete · Complexity

## 1 Introduction

The search for proofs of formulas or theorems is one of the fundamental and difficult tasks in proof theory. In the usual setting, those proofs should be easy to check and thus finite. Induction and coinduction principles have been used in order to provide such a finite proof theory for reasoning on formulas with least

Supported by ANR RAPIDO. An extended version of this article is available at https://hal.archives-ouvertes.fr/hal-02173207.


S. Cerrito and A. Popescu (Eds.): TABLEAUX 2019, LNAI 11714, pp. 317–334, 2019.
https://doi.org/10.1007/978-3-030-29026-9_18

or greatest fixed points (see Kozen [10,11] and Baelde [2]). However in those finite systems, the inference rule for greatest fixed points does not preserve the subformula property. As a consequence, the proof search cannot be driven by the formula that we aim to prove. This is one reason why infinite proofs have been considered for logic with fixed points. The price to pay is that the consistency of the logical system is broken and that a validity criterion has to be added in order to ensure consistency. However, checking the validity criterion might be complex and the purpose of this paper is to show that it is PSPACE-complete. Let us get into more details.

*Circular proofs,* which are infinite proofs satisfying the validity criteria, have thus been proposed as an alternative to induction and coinduction with explicit invariants. Circular proofs present the advantage over explicit induction or coinduction to offer a framework in which it is possible to recover the good structural properties of sequent calculus, such as cut-elimination, subformula property and focusing, making them a more suitable tool to automated proof search. Indeed, cut-elimination and focusing have recently been extended to non well-founded proofs for $\mu$MALL by Baelde, Doumane and Saurin [3,6].

Circular proofs have already proved useful in implementing efficient automatic provers, *e.g.* the Cyclist prover [1]. However, the complexity avoided in the search, thanks to the subformula property and the fact that we need not guess invariants, is counterbalanced by the complexity of the validity criterion at the time of proof checking.

There are already polynomial-space and exponential-time methods to decide thread validity criterions in several settings, but there was no lower bound on its complexity and the exact complexity of checking the thread criterion was still unknown.

The contribution of this work is to show that, in the setting of linear logic with least and greatest fixed point, the decidability of thread criterion is PSPACE-complete.

*Thread Validity and Size-Change Termination.* Our proof takes a lot of inspiration from the proof of PSPACE-completeness of size-change termination by Lee, Jones and Ben Amram [12]: in order to prove that deciding size-change termination is PSPACE-complete, they define a notion of boolean program and use the fact that the following set is complete in PSPACE:

$$\mathcal{B} = \{b \mid b \text{ is a boolean program and } b \text{ terminates.}\}$$

then they reduce $\mathcal{B}$ to the problem of size-change termination. We adapt their method by reducing $\mathcal{B}$ to the problem of thread-validity in circular $\mu\text{MALL}^\omega$ preproof.

It would be very interesting to get a more precise understanding of the relation between threads in circular proofs and size-change termination.

*Organization of the Paper.* In Sect. 2 we recall the formulas and rules of linear logic with least and greatest fixed points, as well as the notions of preproofs and the thread validity criterion, and we recall that the thread criterion is effectively decidable in PSPACE. The main section of the article is Sect. 3, in which we show the PSPACE-completeness of the thread criterion for $\mu\text{MALL}^\omega$, in Theorem 1. Section 4 is devoted to a discussion of our approach and a comparison with related works. We conclude in Sect. 5.

## 2 Background on Circular Proofs and Thread Validity

In this section, we recall the definition of the logic $\mu\text{MALL}^\omega$.

### 2.1 Formulas

Formulas of $\mu\text{MALL}^\omega$ are selected among a set of preformulas. Preformulas of $\mu\text{MALL}^\omega$ are obtained by taking the usual formulas of MALL and adding two monadic second order binders, $\mu$ and $\nu$:

**Definition 1 ($\mu\text{MALL}^\omega$ preformulas).**

$$A, B ::= X \mid A \otimes B \mid A \parr B \mid \mathbf{1} \mid \bot \mid A \oplus B \mid A \,\&\, B \mid \mathbf{0} \mid \top \mid \mu X A \mid \nu X A$$

*where $X$ ranges over an infinite set of propositional variables.*

As usual, preformulas are considered modulo renaming of bound variables. For instance, $\nu X(X \otimes X)$ and $\nu Y(Y \otimes Y)$ denote the same preformula.

**Definition 2 ($\mu\text{MALL}^\omega$ formulas).** *A* formula *is a closed preformula. We denote by $\mathcal{F}$ the set of all formulas.*

**Definition 3 ($\mu\text{MALL}^\omega$ negation).** *An involutive negation $\cdot^\perp$ is defined on every $\mu\text{MALL}^\omega$ preformula, inductively specified by:*

$$(A \otimes B)^\perp = A^\perp \parr B^\perp \qquad \mathbf{1}^\perp = \bot \qquad X^\perp = X$$
$$(A \oplus B)^\perp = A^\perp \,\&\, B^\perp \qquad \mathbf{0}^\perp = \top \qquad (\mu X A)^\perp = \nu X A^\perp$$

*Example 1.* If $A$ is any formula and $F = \nu X(\mu Y((A \otimes X) \parr Y))$ then $F^\perp = \mu X(\nu Y((A^\perp \parr X) \otimes Y))$.

*Remark 1.* It may be counterintuitive that $X^\perp = X$. Yet, in practice negation will only be applied to formulas, which are *closed* preformulas. This simple hack allows us to avoid the use of negative atoms $\overline{X}, \overline{Y}, \ldots$ The fact that we have only positive atoms guarantees in turn that bound variables can only appear in covariant position, thus avoiding the need for a positivity condition when forming a fixed point formula.

### 2.2 Sequents and Preproofs

Proofs of $\mu$MALL$^\omega$ are selected among a set of preproofs. Preproofs of $\mu$MALL$^\omega$ are circular objects, defined by adding back-edges to ordinary proof-trees.

In this article, a sequent is a list of formulas. The inference rules of $\mu$MALL$^\omega$ are given below

$$\frac{}{\vdash A, A^\perp}\,(\text{id}) \qquad \frac{\vdash A, \Gamma \quad \vdash A^\perp, \Delta}{\vdash \Gamma, \Delta}\,(\text{cut}) \qquad \frac{\vdash A_{\sigma(0)}, \ldots, A_{\sigma(n-1)}}{\vdash A_0, \ldots, A_{n-1}}\,(\text{exc})$$

$$\frac{\vdash \Gamma, A \quad \vdash \Delta, B}{\vdash \Gamma, \Delta, A \otimes B}\,(\otimes) \qquad \frac{\vdash \Gamma, A, B}{\vdash \Gamma, A \parr B}\,(\parr) \qquad \frac{}{\vdash \mathbf{1}}\,(1) \qquad \frac{\vdash \Gamma}{\vdash \Gamma, \bot}\,(\bot)$$

$$\frac{\vdash \Gamma, A}{\vdash \Gamma, A \oplus B}\,(\oplus^1) \qquad \frac{\vdash \Gamma, B}{\vdash \Gamma, A \oplus B}\,(\oplus^2) \qquad \frac{\vdash \Gamma, A \quad \vdash \Gamma, B}{\vdash \Gamma, A \,\&\, B}\,(\&) \qquad \frac{}{\vdash \Gamma, \top}\,(\top)$$

$$\frac{\vdash \Gamma, A[\mu X A[X]]}{\vdash \Gamma, \mu X A[X]}\,(\mu) \qquad \frac{\vdash \Gamma, A[\nu X A[X]]}{\vdash \Gamma, \nu X A[X]}\,(\nu)$$

Note that, in the exchange rule (exc), $\sigma$ must be a permutation of $\{0, 1, \ldots, n-1\}$. The (exc) rules are generally left implicit in descriptions of proof trees.

**Definition 4** (**$\Pi_0(\mu\text{MALL}^\omega)$: preproofs**). *A $\mu$MALL$^\omega$ preproof consists of a finite proof tree $\pi$, composed using the rules given above, and which may have open sequents*[1]*, together with a function* **back**, *which associate to each occurrence $s$ of an open sequent in $\pi$, an occurrence* **back**$(s)$ *of the same sequent in $\pi$, such that* **back**$(s)$ *is strictly below $s$ in $\pi$ (i.e. closer to the root).*

*We denote by $\Pi_0(\mu\text{MALL}^\omega)$ the set of all $\mu$MALL$^\omega$ preproofs.*

*Example 2.* Let $\pi$ be the following proof tree, with three open sequents, and let us denote by $s_0, \ldots, s_8$ its occurrences of sequents, as indicated:

$$\dfrac{\dfrac{\dfrac{\dfrac{s_4 = \vdash \nu X(X \parr X), \mu XX \quad \dfrac{s_6 = \vdash \nu XX, \nu X(X \parr X), \mu XX}{s_5 = \vdash \nu XX, \nu X(X \parr X), \mu XX}\,(\nu)}{s_3 = \vdash \nu X(X \parr X), \nu X(X \parr X), \mu XX}\,(\text{cut})}{s_2 = \vdash \nu X(X \parr X) \parr \nu X(X \parr X), \mu XX}\,(\parr)}{s_1 = \vdash \nu X(X \parr X), \mu XX}\,(\nu) \quad \dfrac{s_8 = \vdash \nu XX}{s_7 = \vdash \nu XX}\,(\nu)}{s_0 = \vdash \nu X(X \parr X)}\,(\text{cut})$$

[1] We call an occurrence of open sequent any occurrence of sequent which is not the conclusion of an inference. In Example 2, $s_4$, $s_6$ and $s_8$ are occurrences of open sequents.

then $(\pi, \{s_4 \mapsto s_1, s_6 \mapsto s_5, s_8 \mapsto s_7\})$ is a preproof of $\mu\mathrm{MALL}^\omega$, that we will more simply denote by:

$$
\dfrac{\dfrac{\dfrac{\dfrac{\overset{(0)}{\vdash \nu X(X \parr X), \mu XX} \qquad \dfrac{\overset{(1)}{\vdash \nu XX, \nu X(X \parr X), \mu XX}}{(1) \vdash \nu XX, \nu X(X \parr X), \mu XX}\,(\nu)}{\vdash \nu X(X \parr X), \nu X(X \parr X), \mu XX}\,(\mathrm{cut})}{\vdash \nu X(X \parr X) \parr \nu X(X \parr X), \mu XX}\,(\parr)}{(0) \vdash \nu X(X \parr X), \mu XX}\,(\nu) \qquad \dfrac{\overset{(2)}{\vdash \nu XX}}{(2) \vdash \nu XX}\,(\nu)}{\vdash \nu X(X \parr X)}\,(\mathrm{cut})
$$

## 2.3 Proofs

The validity criterion used to distinguish proofs among preproofs is given in Definition 11 and can be stated as: "every infinite branch must contain a valid thread". To make this formal, we will first define how a preproof induces two graphs and then define the "branches" and "threads" of a preproof as infinite paths in these graphs. Note that:

- In the following definitions, a "graph" always means a directed pseudograph, *i.e.* a directed graph which may have loops and in which there may be several edges between any pair of vertices.
- If $(\pi, \mathbf{back})$ is a preproof, we say that an occurrence of a sequent in $\pi$ is "closed" when it is not an open sequent *i.e.* it is the conclusion of some inference in $\pi$.

**Definition 5 ($G_{\mathrm{branch}}$, branch graph of a preproof).** *Let $(\pi, \mathbf{back})$ be a $\mu\mathrm{MALL}^\omega$ preproof. Its branch graph is the graph $G_{\mathrm{branch}}$ defined as follows. The vertices of $G_{\mathrm{branch}}$ are the occurrences of closed sequents in $\pi$. For each inference $I$ with conclusion $s$ in $\pi$ and for each premise $s'$ of $I$, there is an edge in $G_{\mathrm{branch}}$, from $s$ to $s'$ if $s'$ is a closed occurrence of sequent in $\pi$, and from $s$ to $\mathbf{back}(s')$ if $s'$ is an open occurrence of sequent in $\pi$.*

To clarify the following definition, remember that in every proof tree $\pi$, for every inference $I$ in $\pi$, every occurrence of formula $\alpha$ in a premise of $I$ has a unique immediate descendant in the conclusion of $I$, except if $I$ is a cut and $\alpha$ is a cut formula, in which case $\alpha$ has no immediate descendant.

**Definition 6 ($G_{\mathrm{thread}}$, thread graph of a preproof).** *Let $(\pi, \mathbf{back})$ be a $\mu\mathrm{MALL}^\omega$ preproof. Its thread graph is the graph $G_{\mathrm{thread}}$ defined as follows. The vertices of $G_{\mathrm{thread}}$ are the occurrences of formulas in the closed sequents of $\pi$. For each inference $I$ with conclusion $s$ in $\pi$, for each premise $s'$ of $I$ and for each occurrence of formula $\beta$ in $s'$ which has an immediate descendant $\alpha$ in $s$, there is an edge in $G_{\mathrm{thread}}$, from $\alpha$ to $\beta$ if $s'$ is a closed occurrence of sequent in $\pi$, and from $\alpha$ to the occurrence of the formula corresponding to $\beta$ in $\mathbf{back}(s')$ if $s'$ is an open occurrence of sequent in $\pi$.*

**Definition 7 (Infinite branch).** *If* $(\pi, \mathbf{back})$ *is a preproof and* $G_{\text{branch}}$ *is its branch graph, we call an infinite branch of this preproof any infinite path in* $G_{\text{branch}}$ *starting from the root of* $\pi$.

*Example 3.* The infinite branches of the preproof of Example 2 are $s_0(s_7)^\omega$, $s_0(s_1s_2s_3)^\omega$ and all elements of $\{s_0(s_1s_2s_3)^k(s_5)^\omega \mid k \in \mathbf{N}\}$.

Note that, in order to be totally rigorous, we should

1. not only give the vertices of the paths but also the edges, *i.e.* when an inference has several premises, indicate explicitly which one was chosen;
2. include the implicit (exc) rules.

These details are omitted here for concision; they will cause no ambiguity on the validity of this preproof.

**Definition 8 (Thread).** *A thread in a preproof is simply a path (finite or infinite) in* $G_{\text{thread}}$.

*Example 4.* Let us denote by $\{\alpha, \beta, \gamma, \ldots, \mu\}$ the vertices of $G_{\text{thread}}$ for the preproof shown on Example 2, as indicated here:

$$\dfrac{\dfrac{\dfrac{\dfrac{\begin{matrix}(0)\\ \vdash \nu X(X \parr X), \mu XX\end{matrix} \quad \dfrac{\begin{matrix}(1)\\ \vdash \nu XX, \nu X(X \parr X), \mu XX\end{matrix}}{(1) \vdash \nu XX_\iota, \nu X(X \parr X)_\kappa, \mu XX_\lambda}\,(\nu)}{\vdash \nu X(X \parr X)_\zeta, \nu X(X \parr X)_\eta, \mu XX_\theta}\,(\text{cut})}{\vdash \nu X(X \parr X) \parr \nu X(X \parr X)_\delta, \mu XX_\epsilon}\,(\parr)}{(0) \vdash \nu X(X \parr X)_\beta, \mu XX_\gamma}\,(\nu) \qquad \dfrac{\begin{matrix}(2)\\ \vdash \nu XX\end{matrix}}{(2) \vdash \nu XX_\mu}\,(\nu)}{\vdash \nu X(X \parr X)_\alpha}\,(\text{cut})$$

The maximal threads of this preproof are $(\mu)^\omega$, $\gamma\epsilon\theta(\lambda)^\omega$, $(\iota)^\omega$, $\alpha(\beta\delta\zeta)^\omega$ and the elements of $\{\alpha(\beta\delta\zeta)^k\beta\delta\eta(\kappa)^\omega \mid k \in \mathbf{N}\}$.

Once again, in order to be totally rigorous, we should explicitly include the occurrences of formulas in the sequents that are hidden by the elision of the (exc) rules.

**Definition 9 ($\mathbf{U}$: $G_{\text{thread}} \to G_{\text{branch}}$).** *For any preproof, there is an obvious graph morphism from* $G_{\text{thread}}$ *to* $G_{\text{branch}}$, *associating to every occurrence of a formula the sequent occurrence it belongs to. We denote this graph morphism by* $\mathbf{U}$. *If* $t$ *is a path in* $G_{\text{thread}}$ *(i.e. a thread), we will also denote by* $\mathbf{U}(t)$ *the corresponding path in* $G_{\text{branch}}$.

*Remark 2.* Even when $t$ is an infinite thread, $\mathbf{U}(t)$ may not be an infinite branch because it may not start at the root of the preproof. However, if $t$ is an infinite thread, then $\mathbf{U}(t)$ is a suffix of an infinite branch.

*Example 5.* The images, by this morphism, of the threads of Example 4 are

$$\mathbf{U}((\mu)^\omega) = (s_7)^\omega \qquad \mathbf{U}(\gamma\epsilon\theta(\lambda)^\omega) = s_1 s_2 s_3 (s_5)^\omega \qquad \mathbf{U}((\iota)^\omega) = (s_5)^\omega$$

$$\mathbf{U}(\alpha(\beta\delta\zeta)^\omega) = s_0(s_1 s_2 s_3)^\omega$$

$$\forall k \in \mathbf{N}, \mathbf{U}(\alpha(\beta\delta\zeta)^k \beta\delta\eta(\kappa)^\omega) = s_0(s_1 s_2 s_3)^{k+1}(s_5)^\omega$$

The following lemma is the key to the notion of a valid thread, which is defined right after it. If $s$ is an occurrence of formula in a proof tree, we denote by $\mathbf{fml}(s) \in \mathcal{F}$ the associated formula.

**Lemma 1.** *Let* $t = (s_n)_{n \in \mathbf{N}}$ *be an infinite thread in a preproof. Let* $\inf(t) = \{A \in \mathcal{F} \mid \forall n_0 \in \mathbf{N}, \exists n \geqslant n_0, s_n \text{ is principal and } \mathbf{fml}(s_n) = A\}$ *i.e. the set of formulas that are infinitely often principal in* $t$*.*

*If* $\inf(t) \neq \emptyset$*, i.e. if* $t$ *encounters infinitely often principal formulas, then it contains a smallest infinitely principal formula, and this formula is a fixed point formula:* $\exists \sigma \in \{\mu, \nu\}, \exists C, \sigma X C \in \inf(t)$ *and* $\forall A \in \inf(t), \sigma X C$ *is a subformula of* $A$*. As a minimum, this formula is unique.*

**Definition 10 (Valid thread).** *An infinite thread* $t$ *is valid if* $\inf(t)$ *is non-empty and the smallest formula in* $\inf(t)$ *is a* $\nu$*-formula (cf. Lemma 1 just above).*

*Example 6.* Among the threads of Example 4:

- $(\mu)^\omega$ and $(\iota)^\omega$ are valid: their smallest infinitely principal formula is $\nu X X$;
- $\alpha(\beta\delta\zeta)^\omega$ is valid: its smallest infinitely principal formula is $\nu X(X \parr X)$;
- $\gamma\epsilon\theta(\lambda)^\omega$ is not valid: it has no principal formula;
- $\forall k \in \mathbf{N}, \alpha(\beta\delta\zeta)^k\beta\delta\eta(\kappa)^\omega$ is not valid: it has no principal formula after the last occurrence of $\beta$.

**Definition 11 ($\Pi(\mu\text{MALL}^\omega)$: proofs).** *We say that an infinite branch* $b$ *of a preproof* $\varpi$ *is valid if there is a valid infinite thread* $t$ *of* $\varpi$ *such that* $\mathbf{U}(t)$ *is a suffix of* $b$*.*

*A* $\mu\text{MALL}^\omega$ *preproof* $\varpi$ *is a proof if all its infinite branches are valid.*

*We denote by* $\Pi(\mu\text{MALL}^\omega)$ *the set of all* $\mu\text{MALL}^\omega$ *proofs and we denote by* $\overline{\Pi(\mu\text{MALL}^\omega)}$ *its complement in* $\Pi_0(\mu\text{MALL}^\omega)$*, i.e. the set of all invalid preproofs.*

*Example 7.* The preproof of Example 2 is a proof:

- the branch $s_0(s_7)^\omega$ contains the valid thread $(\mu)^\omega$;
- the branch $s_0(s_1 s_2 s_3)^\omega$ contains the valid thread $(\beta\gamma\zeta)^\omega$;
- $\forall k \in \mathbf{N}$, the branch $s_0(s_1 s_2 s_3)^k(s_5)^\omega$ contains the valid thread $(\iota)^\omega$.

## 2.4 Deciding Thread Validity in PSPACE

In this section, we recall the fact that the problem $\Pi(\mu\text{MALL}^\omega)$ is in PSPACE. Several algorithms can be used for that. Here we reduce this problem to the problem of deciding equality of languages for parity $\omega$-automata, which is known to be in PSPACE. More precisely, given a preproof $\varpi$, we define two parity automata: the language of the first one is the set of infinite branches of $\varpi$ and the language of the second one is the set of *valid* infinite branches of $\varpi$.

Let $\varpi = (\pi, \mathbf{back})$ be a preproof. Let $A = E_{\text{branch}}$, the set of edges of $G_{\text{branch}}$; this will be the input alphabet of our automata.

The first $\omega$-automaton is $\mathcal{A}_{\text{branch}} = \langle Q_{\text{branch}}, i_{\text{branch}}, T_{\text{branch}} \rangle$, where:

- the set of states is $Q_{\text{branch}} = V_{\text{branch}}$, the set of vertices of $G_{\text{branch}}$
- the initial state $i_{\text{branch}}$ is the root of $\pi$
- the set of transitions is

$$T_{\text{branch}} = \{s \stackrel{e}{\longrightarrow} s' \mid e \text{ is an edge from } s \text{ to } s' \text{ in } G_{\text{branch}}\}$$

and the acceptance condition is trivial: every infinite run is accepted. With that definition, the following lemma is immediate:

**Lemma 2.** *The language $\mathcal{L}(\mathcal{A}_{\text{branch}})$ is the set of infinite branches of $\varpi$.*

For our second automaton, we need a priority assignment $\Omega\colon \mathcal{F} \to \mathbf{N}$ with two properties:

1. if $A$ is a subformula of $B$ then $\Omega(A) \leqslant \Omega(B)$;
2. $\forall A, \Omega(\mu X A)$ is even and $\Omega(\nu X A)$ is odd.

Such a function is not difficult to construct. From now on we assume that one has been chosen.

Our second automaton is a parity $\omega$-automaton, with priorities in $\mathbf{N} \cup \{\infty\}$, defined as $\mathcal{A}_{\text{thread}} = \langle Q_{\text{thread}}, i_{\text{thread}}, T_{\text{thread}} \rangle$, where:

- the set of states is $Q_{\text{thread}} = V_{\text{thread}} + \{\bot_s \mid s \in V_{\text{branch}}\}$, *i.e.* the vertices of $G_{\text{thread}}$ plus one extra vertex for each vertex of $G_{\text{branch}}$
- the initial state is $i_{\text{thread}} = \bot_r$ where $r$ is the root of $\pi$
- the set of transitions is

$$\begin{aligned} T_{\text{thread}} = \quad & \{\bot_s \stackrel{e\,:\,\infty}{\longrightarrow} \bot_{s'} \mid e \text{ is an edge from } s \text{ to } s' \text{ in } G_{\text{branch}}\} \\ \cup \, \{ & \alpha \xrightarrow{\mathbf{U}(e)\,:\,\Omega(\alpha)} \beta \mid \\ & e \text{ is an edge from } \alpha \text{ to } \beta \text{ in } G_{\text{thread}} \text{ and } \alpha \text{ is principal}\} \\ \cup \, \{ & \alpha \xrightarrow{\mathbf{U}(e)\,:\,\infty} \beta \mid \\ & e \text{ is an edge from } \alpha \text{ to } \beta \text{ in } G_{\text{thread}} \text{ and } \alpha \text{ is not principal}\} \\ \cup \, & \{\bot_s \stackrel{e\,:\,\infty}{\longrightarrow} \alpha \mid s = \mathbf{U}(\alpha)\} \end{aligned}$$

where $q \stackrel{e\,:\,i}{\longrightarrow} q'$ denote a transition from state $q \in Q_{\text{thread}}$ to state $q' \in Q_{\text{thread}}$ with label $e \in A$ and priority $i \in \mathbf{N} \cup \{\infty\}$.

The acceptance condition is: a run is accepted if the smallest priority appearing infinitely often is *odd* ($\infty$ being even).

Once again, it should be clear from Definitions 10 and 11 that:

**Lemma 3.** *The language $\mathcal{L}(\mathcal{A}_{\text{thread}})$ is the set of* valid *infinite branches of $\varpi$.*

From these two lemmas it is immediate that

**Proposition 1.** *We have the inclusion $\mathcal{L}(\mathcal{A}_{\text{thread}}) \subseteq \mathcal{L}(\mathcal{A}_{\text{branch}})$ and the pre-proof $\varpi$ is valid iff. this inclusion is an equality.*

Deciding this equality can be done in PSPACE, and the constructions of these automata are obviously PSPACE, so:

**Proposition 2.** *The problem $\Pi(\mu\text{MALL}^{\omega})$ is in* PSPACE.

# 3 PSPACE-Completeness

## 3.1 Outline of the PSPACE-Completeness Proof

We now aim at proving that $\Pi(\mu\text{MALL}^{\omega})$ is PSPACE-complete for LOGSPACE reductions. As it is already known that $\Pi(\mu\text{MALL}^{\omega}) \in$ PSPACE, it remains to prove that we have PSPACE $\leqslant_{\text{L}} \Pi(\mu\text{MALL}^{\omega})$.

We follow the same methodology as Lee, Jones and Ben Amram [12]: in order to prove that deciding size-change termination is PSPACE-complete, they define a notion of boolean program (see Definition 12) and use the fact that the following problem is complete in PSPACE:

$$\mathcal{B} = \{b \mid b \text{ is a boolean program and } b \text{ terminates.}\}$$

then they reduce $\mathcal{B}$ to the decidability of size-change termination.

We try to adapt their method by reducing $\mathcal{B}$ to $\Pi(\mu\text{MALL}^{\omega})$.

## 3.2 Defining the Reduction

Let us first introduce boolean programs.

**Definition 12** ($\texttt{BOOLE}_{false}$ **and** $\mathcal{B}_{false}$). *A boolean program in* `BOOLE` *is a sequence of instructions $b = \texttt{1:I}_1\ \texttt{2:I}_2\ \dots\ \texttt{m:I}_m$ where an instruction can have one of the two following forms:*

$$\texttt{I} ::= \texttt{X} := \neg\texttt{X} \mid \text{if } \texttt{X} \text{ then goto } \ell' \text{ else goto } \ell''$$

*where $X$ ranges over a finite set of variable names and labels $\ell', \ell''$ range overs $\{0, \dots, m\}$.*

*The semantics is as expected: a program is executed together with a store assigning values to variables which shall initially assign all variables to* `false` *at the beginning of the execution (this is the* initial store*). More precisely, an* *execution is a sequence of pairs $(\ell, s)$ of a label and a store subject to the expected transitions $(\ell_1, s_1) \to (\ell_2, s_2)$ if $\ell_1 \neq 0$ and:*

- *if $I_{\ell_1} = X := \neg X$, then $\ell_2 = \ell_1 + 1 \pmod{m+1}$ and $s_2(Y) = s_1(Y)$ for all variable $Y \neq X$ and $S_2(X) = \neg(s_1(X))$;*
- *if $I_{\ell_1} = \text{if}\ \mathtt{X}\ \text{then goto}\ \ell'\ \text{else goto}\ \ell''$ then $s_2 = s_1$ and $\ell_2 = \ell'$ if $s_1(X) = \mathtt{true}$ and $\ell_2 = \ell''$ otherwise.*

*The program terminates when the label reaches* 0*, the current store at termination is the* final store.

*A program in* $\mathtt{BOOLE}_{false}$ *is a program in* $\mathtt{BOOLE}$ *such that, if it terminates, its final store is such that all variables have value false. We also denote the set of terminating* $\mathtt{BOOLE}_{false}$ *programs as:*

$$\mathcal{B}_{false} = \{b \in \mathtt{BOOLE}_{false} \mid b \textit{ terminates}\}.$$

*Remark 3.* The constraint on the values of the variables at the end of the program will be useful when reducing it to $\Pi(\mu\text{MALL}^\omega)$. This circular preproof will encode the fact that the program $b$ is terminating by connecting the final state to the initial one, hence the necessity that its initial and terminal states are the same.

**Lemma 4.** $\mathcal{B}_{false}$ *is* PSPACE*-hard under* LOGSPACE*-reductions:*

$$\text{PSPACE} \leqslant_{\text{L}} \mathcal{B}_{false}$$

*Proof.* We reduce from the problem of termination for a more expressive language, which has been defined and proved PSPACE-complete by Jones in [9], under the name of BOOLE.

The following definition will be used in the proof of Proposition 3:

**Definition 13 (Call graph of a program).** *Assume a boolean program $b$ with variables $\mathtt{X}_1, \ldots, \mathtt{X}_k$ and instructions $\mathtt{1} : \mathtt{I}_1, \ldots, \mathtt{m} : \mathtt{I}_m$. Define the call graph of $b$ to be $G = (V, E)$ with*

- $V = \{0, 1, \ldots, m\}$
- $E = \{0 \xrightarrow{0} 1\}$

$$\cup \{\ell \xrightarrow{\ell} ((\ell+1) \bmod (m+1)) \mid \mathtt{I}_\ell = \text{“}\mathtt{X} := \mathtt{not\ X}\text{”}\}$$

$$\cup \{\ell \xrightarrow{\ell^+} \ell', \ell \xrightarrow{\ell^-} \ell'' \mid \mathtt{I}_\ell = \text{“}\mathtt{if\ X\ goto}\ \ell'\ \mathtt{else}\ \ell''\text{”}\}$$

**Definition 14** ($[\![\cdot]\!]$: $\mathtt{BOOLE}_{false} \to \Pi_0(\mu\text{MALL}^\omega)$)**.** *For every boolean program $b \in \mathtt{BOOLE}_{false}$, we define a preproof $[\![b]\!] \in \Pi_0(\mu\text{MALL}^\omega)$. Let $\mathtt{X}_1, \ldots, \mathtt{X}_k$ be the variables of $b$ and $\mathtt{1} : \mathtt{I}_1, \ldots, \mathtt{m} : \mathtt{I}_m$ its instructions. We first give names to the formulas that will appear in $[\![b]\!]$: we define a unary operation ¿, three* *formulas* $A, B, C$*, a family of unary operations* $(¿_n)$ *and two families of formulas* $(D_n), (E_n)$*:*

$$A = ¿(\nu X ¿X) \qquad B = \nu X(\bot \oplus X) \qquad C = \mu X(B \parr X) \qquad E_n = ¿_n(\nu X ¿_n X)$$

$$¿F = \mu X(F \oplus (\bot \oplus (X \parr X))) \qquad ¿_n F = \mu X(\bot \oplus (X \parr (\underbrace{F \parr \cdots \parr F}_{n-1})))$$

$$D_n = \mu X(\underbrace{X \& \cdots \& X}_{n})$$

*We now define* $[\![b]\!]$ *to be the preproof*

$$\dfrac{\dfrac{[\![0:]\!]}{\vdash A^{2k}, B, C, D_2, D_m, E_m^m} \quad \dfrac{[\![1 : \mathtt{I}_1]\!]}{\vdash A^{2k}, B, C, D_2, D_m, E_m^m} \quad \cdots \quad \dfrac{[\![\mathtt{m} : \mathtt{I}_\mathtt{m}]\!]}{\vdash A^{2k}, B, C, D_2, D_m, E_m^m}}{(\textsc{Root}) \vdash A^{2k}, B, C, D_2, \underline{D_m}, E_m^m}\ (\mu), (\&)^{m-1} \tag{1}$$

*where* $\Gamma^n$ *is an abbreviation for* $\underbrace{\Gamma, \ldots, \Gamma}_{n}$*.*

*The root of the preproof* $[\![b]\!]$ *is constructed by translating each pair* $\ell : I_\ell$ *of a label and an instruction into a finite segment of branch of preproof, as defined in Eq.* (1)*, with each subderivation* $[\![\ell : I_\ell]\!]$ *defined in Fig. 2 and each subderivation* $[\![\ell : \mathtt{goto}\ \ell']\!]$ *in Fig. 1.*

$$[\![\ell : \mathtt{goto}\ \ell']\!] \quad = \quad \dfrac{\dfrac{\dfrac{\text{Back-edge to } (\textsc{Root})}{\vdash A^{2k}, B, C, D_2, D_m, E_m, \ldots, E_m}\ (\nu)^{m-1}}{\vdash A^{2k}, B, C, D_2, D_m, \underline{(\nu X ¿_m X)}^{\ell'-1}, E_m, \underline{(\nu X ¿_m X)}^{m-\ell'}}\ (\mu), (\oplus^2), (\parr)^{m-1}}{\dfrac{\vdash A^{2k}, B, C, D_2, D_m, \underline{E_m}}{\vdash A^{2k}, B, C, D_2, D_m, \underline{E_m}^{\ell-1}, E_m, \underline{E_m}^{m-\ell}}\ ((\mu), (\oplus^1), (\bot))^{m-1}}$$

**Fig. 1.** Back-edges of the preproof

*Remark 4 (Implicit vs. explicit exchange rules).* Notice that in the translation of the previous definition, our derivations make an implicit use of the exchange rule. In order to make explicit the exchange, it is enough to add an exchange rule at the conclusion of every inference in the proof, simply doubling the size of the proof. This will therefore have no impact on the forthcoming reductions and completeness proofs that will be studied in the remaining of the paper.

*Remark 5 (Infinite branches of* $[\![b]\!] \simeq E^\omega$*).* The preproof $[\![b]\!]$ constructed from $b$ by the reduction $[\![\cdot]\!]$ of Definition 14 is a finite tree with back-edges in which every finite branch ends with a back-edge to the root. This finite tree has exactly as many branches, and, consequently, as many back-edges to the root as the number

**Card** $E$ of edges in the call-graph of $b$ (Definition 13). This in turn entails that the set of infinite branches of the preproof $[\![b]\!]$ is in one-to-one correspondence with the set $E^\omega$ of infinite words on $E$. Note however that an infinite word $\overline{u} \in E^\omega$ has no reason *a priori* to be a path in $G$.

From now on, we will refer directly to infinite branches of the preproof by words $\overline{u} \in E^\omega$.

### 3.3 Main Theorem

We now prove that $\Pi(\mu\mathrm{MALL}^\omega)$ is PSPACE-complete.

*Remark 6 (Thread groups).* We need to be more precise about the occurrences of formulas in the conclusion sequent of preproof $[\![b]\!]$:

$$\underbrace{A, \dots, A}_{2k}, B, C, D_2, D_m, \underbrace{E_m, \dots, E_m}_{m}$$

Let us label the occurrences of $A$ in this sequent as follows:

$$A_1^+, A_1^-, \dots, A_k^+, A_k^-, B, C, D_2, D_m, \underbrace{E_m, \dots, E_m}_{m}$$

so that we can talk precisely about them. It can be seen by examining the definition of $[\![\cdot]\!]$ (Definition 14) that a valid thread in the preproof cannot pass

$$[\![\texttt{0:}]\!] \quad = \quad \dfrac{\dfrac{\dfrac{\dfrac{\dfrac{\dfrac{\begin{array}{c}[\![\texttt{0:goto 1}]\!]\\ \vdash (A, A)^k, B, C, D_2, D_m, E_m^m\end{array}}{\vdash (\underline{\nu X \text{¿} X}, A)^k, B, C, D_2, D_m, E_m^m}\,(\nu)^k}{\vdash (\underline{A}, A)^k, B, C, D_2, D_m, E_m^m}\,((\mu), (\oplus^1))^k}{\vdash \underline{A}^k, B, C, D_2, D_m, E_m^m}\,((\mu), (\oplus^2), (\oplus^2), (⅋))^k}{\vdash (\underline{A}, A)^k, B, C, D_2, D_m, E_m^m}\,((\mu), (\oplus^2), (\oplus^1), (\bot))^k}{\vdash A^{2k}, \underline{C}, D_2, D_m, E_m^m}\,(\mu), (⅋)}{\vdash A^{2k}, \underline{B}, C, D_2, D_m, E_m^m}\,(\nu), (\oplus^1), (\bot)$$

$$[\![\ell\texttt{:X}_i\ \texttt{:= not X}_i]\!] \quad = \quad \dfrac{\dfrac{\begin{array}{c}[\![\ell\texttt{:goto }(\ell + 1 \bmod m + 1)]\!]\\ \vdash A^{2k}, B, C, D_2, D_m, E_m^m\end{array}}{\vdash A^{2(i-1)}, \underline{A}, \underline{A}, A^{2(k-i)}, B, C, D_2, D_m, E_m^m}\,(\text{exc})}{\vdash A^{2k}, \underline{B}, C, D_2, D_m, E_m^m}\,(\nu), (\oplus^2)$$

$$[\![\ell\texttt{:if X}_i\ \texttt{then goto }\ell'\ \texttt{else }\ell'']\!] \quad =$$

$$\dfrac{\dfrac{\dfrac{\dfrac{\begin{array}{c}[\![\ell\texttt{:goto }\ell']\!]\\ \vdash A^{2(i-1)}, A, A, A^{2(k-i)}, B, C, D_2, D_m, E_m^m\end{array}}{\vdash A^{2(i-1)}, A, \underline{\nu X(\text{¿}X)}, A^{2(k-i)}, B, C, D_2, D_m, E_m^m}\,(\nu)}{\vdash A^{2(i-1)}, A, \underline{A}, A^{2(k-i)}, B, C, D_2, D_m, E_m^m}\,(\mu), (\oplus^1) \qquad \dfrac{\dfrac{\begin{array}{c}[\![\ell\texttt{:goto }\ell'']\!]\\ \vdash A^{2(i-1)}, A, A, A^{2(k-i)}, B, C, D_2, D_m, E_m^m\end{array}}{\vdash A^{2(i-1)}, \underline{\nu X(\text{¿}X)}, A, A^{2(k-i)}, B, C, D_2, D_m, E_m^m}\,(\nu)}{\vdash A^{2(i-1)}, \underline{A}, A, A^{2(k-i)}, B, C, D_2, D_m, E_m^m}\,(\mu), (\oplus^1)}{\vdash A^{2k}, B, C, \underline{D_2}, D_m, E_m^m}\,(\mu), (\&)}{\vdash A^{2k}, \underline{B}, C, D_2, D_m, E_m^m}\,(\nu), (\oplus^2)$$

**Fig. 2.** Premises $p_\ell$ of the preproof

through $D_2$ or $D_m$, which contain no $\nu$, and that the remaining formulas are divided into $k + 2$ groups

$$\underbrace{A_1^+, A_1^-}, \ldots, \underbrace{A_k^+, A_k^-}, \underbrace{B, C}, \underbrace{E_m, \ldots, E_m}$$

which cannot thread-interact with each other, in the sense that, for instance, no thread can contain a $B$ and a $E_m$, or a $A_\ell^\epsilon$ and a $A_{\ell'}^{\epsilon'}$ if $\ell \neq \ell'$.

**Lemma 5.** *An infinite branch $\overline{u} \in E^\omega$ in the preproof contains a validating thread*

- *in the $E_m$ group iff. no suffix of $\overline{u}$ is a valid path in $G$.*
- *in the $B, C$ group iff. 0 occurs only finitely in $\overline{u}$.*

*Proof (Proof sketch).* By case on the instructions involved.

In order to prove the first part of the statement, that is that an infinite branch $\overline{u} \in E^\omega$ in the preproof contains a validating thread in the $E_m$ group iff. no suffix of $\overline{u}$ is a valid path in $G$, we reason by case on the instructions involved and remark that the $E_m$ formulas are touched only in the $[\![\ell\!:\!\texttt{goto}\ \ell']\!]$ parts of the preproof.

In order to prove the second part of the statement, that is that an infinite branch $\overline{u} \in E^\omega$ in the preproof contains a validating thread in the $B, C$ group iff. 0 occurs only finitely in $\overline{u}$, we reason by case on the instructions involved.

*Remark 7.* Because of Lemma 5, the only infinite branches of $[\![b]\!]$ whose validity is not known in advance are the $\overline{u} \in E^\omega$ which are valid paths in $G$ going infinitely many times through edge 0, and we know that these infinite branches may have validating threads only in one of the $k$ groups $\{A_i^+, A_i^-\}_{1 \leqslant i \leqslant k}$. Such an infinite branch can always be factorized into $u_0 0 u_1 0 u_2 0 \cdots$ where the $u_n$ do not contain 0. As the edge $0 \in E$ has source and target $0 \xrightarrow{0} 1$, and because of the hypothesis that $\overline{u}$ is a path in $G$, for $n \geqslant 1$ every $u_n$ has source and target $1 \xrightarrow{u_n}{}^+ 0$.

**Lemma 6.** *Assume $1 \xrightarrow{u}{}^+ \ell$, which does not contain the edge 0. If $u$ is a prefix of the execution of $b$ then the threads of $\{A_i^+, A_i^-\}$ in $0 \xrightarrow{0u}{}^+ \ell$ are*

$$
\begin{array}{ccc}
A_i^+ & & A_i^- \\
& \searrow & \mid \\
& \nu X_i X & A \\
& & \searrow \mid \\
A_i^+ & & A_i^-
\end{array}
$$

*if $\mathbf{X}_i = false$ at the end of $u$ and*

$$
\begin{array}{ccc}
A_i^+ & & A_i^- \\
& \searrow & \mid \\
& A & \nu X_i X \\
& \searrow & \mid \\
A_i^+ & & A_i^-
\end{array}
$$

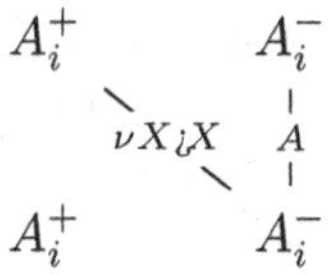

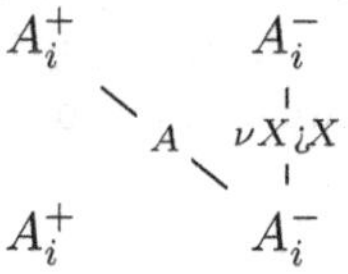

*if* $\mathtt{X}_i = $ *true at the end of* $u$*; and if* $u$ *is not a prefix of the execution of* $b$ *then there is an* $i \in [\![1, m]\!]$ *such that the threads of* $\{A_i^+, A_i^-\}$ *in* $0 \xrightarrow{0u}{}^{+} \ell$ *are*

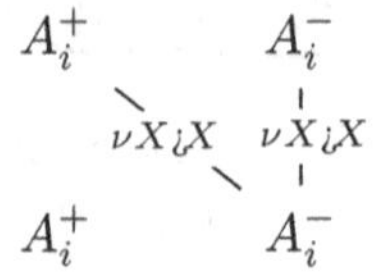


*Proof (Proof sketch).* The proof goes by induction on the length of $u$.

The diagrams we use here are sketches of the thread structure of a segment of branch. For instance the first of these diagrams should be read as: the occurrence $A_i^-$ in the conclusion sequent is a descendant of both occurrences $A_i^+$ and $A_i^-$ in the sequent at the top of the segment of branch we consider. The smallest principal formula along the segment of thread from the lower $A_i^-$ to the upper $A_i^+$ is $\nu X ¿ X$ and the smallest principal formula along the segment of thread from the lower $A_i^-$ to the upper $A_i^-$ is $A$. The occurrence $A_i^+$ in the lower sequent is not a descendant of any of the occurrences $A_i^+$ nor $A_i^-$ in the upper sequent.

**Proposition 3.** $[\![\cdot]\!]$ *is a* LOGSPACE *reduction from* $\overline{\Pi(\mu\mathrm{MALL}^\omega)}$ *to* $\mathcal{B}_{false}$.

*Proof.* For the LOGSPACE character: the only data that need to be remembered while constructing the preproof are integers like $k$, $m$, $\ell$, $\ell'$. As $\ell, \ell' \leqslant m$ and the entry has size $\Omega(k+m)$, this takes a space at most logarithmic in the size of the entry.

As for the fact that it is indeed a reduction: let us assume a $b \in \mathtt{BOOLE}_{false}$ and prove that $[\![b]\!] \notin \Pi(\mu\mathrm{MALL}^\omega) \Leftrightarrow b \in \mathcal{B}_{false}$. Let $G = (V, E)$ be the call-graph of $b$, as defined in Definition 13. Following Remark 5, we denote by elements of $E^\omega$ the infinite branches of $[\![b]\!]$. There are two cases: either $b \in \mathcal{B}_{false}$ and we have to prove that $p \notin \Pi(\mu\mathrm{MALL}^\omega)$, or $b \notin \mathcal{B}_{false}$ and we have to prove that $p \in \Pi(\mu\mathrm{MALL}^\omega)$. First case: if $b \in \mathcal{B}_{false}$: the execution of $b$ induces a finite path $u = 1 \rightarrow^* 0$ in $G$. This finite path can be completed into $v = 0 \xrightarrow{0} 1 \xrightarrow{u}{}^{*} 0$. Then $v^\omega$ is an invalid branch of $[\![p]\!]_\omega$. Here we use the fact that when $b$ terminates, every variable has value *false*. Second case: if $b \notin \mathcal{B}_{false}$: let $\mathcal{P}_1 = \{vw_\infty \mid v \in E^*, w_\infty \in E^\omega$ and $w_\infty$ is a path in $G\}$ and $\mathcal{P}_2 = \{v_\infty \in \mathcal{P}_1 \mid 0$ occurs infinitely in $v_\infty\}$. By construction, $\mathcal{P}_2 \subseteq \mathcal{P}_1 \subseteq E^\omega$. We will prove three facts: that every branch $v_\infty \in E^\omega \setminus \mathcal{P}_1$ is thread-valid, that every branch $v_\infty \in \mathcal{P}_1 \setminus \mathcal{P}_2$ is thread-valid and that every branch $v_\infty \in \mathcal{P}_2$ is thread-valid. These three facts, together with the fact that $(E^\omega \setminus \mathcal{P}_1) \cup (\mathcal{P}_1 \setminus \mathcal{P}_2) \cup \mathcal{P}_2 = E^\omega$, are enough to conclude that every branch $v_\infty \in E^\omega$ is thread-valid. The first fact, that every branch $v_\infty \in E^\omega \setminus \mathcal{P}_1$ is thread-valid, is due to the thread going through the $E_m$. The second fact, that every branch $v_\infty \in \mathcal{P}_1 \setminus \mathcal{P}_2$ is thread-valid, is due to the thread going through $B$. The third fact, that every branch $v_\infty \in \mathcal{P}_2$ is thread-valid, is due to the fact that $b$ is non-terminating and that, because of that, one of the $2k$ threads going through the $A$ is valid.

**Theorem 1.** *The problem $\Pi(\mu\text{MALL}^\omega)$ is* PSPACE*-hard under* LOGSPACE *reductions:*

$$\text{PSPACE} \leqslant_{\text{L}} \Pi(\mu\text{MALL}^\omega)$$

*Proof.* We reduce from $\mathcal{B}_{false}$, which is PSPACE-complete by Lemma 4. More precisely, we reduce $\mathcal{B}_{false}$ to $\overline{\Pi(\mu\text{MALL}^\omega)}$, the complement of $\Pi(\mu\text{MALL}^\omega)$. This is enough because PSPACE is closed under complements, in the same way as all deterministic classes. The reduction $[\![\cdot]\!] : \texttt{BOOLE}_{false} \to \Pi_0(\mu\text{MALL}^\omega)$ is defined in Definition 14. It is indeed a LOGSPACE reduction by Proposition 3.

*Remark 8.* In fact, since our construction do not use the (cut) rule, the cut-free fragment of $\Pi(\mu\text{MALL}^\omega)$ is already PSPACE-hard.

*Remark 9.* Our result extends to $\mu$LJ, $\mu$LK, $\mu$LK$\bigcirc$ and $\mu$LK$\Box\Diamond$ and we conjecture that the method we illustrate here on $\mu$MALL can apply as well to the guarded cases of $\mu$-calculi with modalities.

## 4 Comments on Our Approach and Discussion of Related Works

Our proof for the PSPACE-completeness of the thread criterion is an encoding and an adaptation to our setting of the proof used by Lee, Jones and Ben Amram to prove that size-change termination is PSPACE-complete [12]. We reduce, as they do, from the problem of termination of boolean programs and the thread diagrams that we have used to describe the preproof generated by the reduction are very similar to the size-change graphs generated by their reduction; this is in fact what has guided the design of this preproof: formula $A$ mimicks the $\texttt{X}_i, \overline{\texttt{X}}_i$ part of their graphs and formulas $B$ and $C$ adapt the $\texttt{Z}$ part of their graphs. We had to add the formulas $D_2$ and $D_m$ in order to have branching rules in the preproof. One of the main novelties of our reduction, compared to the reduction of Lee, Jones and Ben Amram for size-change termination, lies in the $E_m$ and $[\![\ell\texttt{:goto }\ell']\!]$ part of the constructed preproof, which has no equivalent in the size-change graphs obtained by their reduction. This part of our construction allows us to construct a preproof which is a tree with back-edges, hence proving that the thread criterion is PSPACE-complete even when preproofs are represented by trees with back-edges. We could in fact drop the $E_m$ and $[\![\ell\texttt{:goto }\ell']\!]$ part of the construction by constructing $[\![b]\!]$ as a rooted graph instead of a tree with back-edges. The constructions proofs are still correct—and shorter. The caveat is that it only proves the thread-criterion to be PSPACE-hard in graph-shaped preproofs and not in tree-with-back-edges-shaped preproofs. Furthermore, we could not have filled this gap by simply unfolding the graph into a tree with back-edges, for it could lead, as shown in the following example, to an exponential blow-up in size, which would prevent the reduction to be LOGSPACE, or even PTIME. The following boolean program:

```
1:if X then goto 2 else goto 2
2:if X then goto 3 else goto 3
        ⋮
n:if X then goto n+1 else goto n+1
```

will be translated to a graph-shaped preproof of size $\Theta(n)$ but the unfolding of this preproof into a tree-with-back-edges-shaped preproof will have size $\Theta(2^n)$. Therefore we had to be clever in order to target trees with back-edges by *simulating* several vertices with a single one; this is accomplished by the $E_m$ and $[\![\ell\texttt{:goto } \ell']\!]$.

This improvement of the reduction of Lee, Jones and Ben Amram could in fact be adapted in the other direction, to show that size-change termination is already PSPACE-complete even when restricted to programs with only one function (in the terminology of [12]), that is when the corresponding call graph/control flow graph has only one vertex.

If, as it is commonly believed, NP $\neq$ PSPACE, our result implies that there is no way to add a polynomial quantity of information to a preproof so that its thread-validity can be checked in polynomial time. This can be seen as a problem, both for the complexity of proof search and proof verification. It suggests trying to find restrictions of the thread criterion which will be either decidable or certifiable in polynomial time, while keeping enough expressivity to validate interesting proofs. A first step in this direction has already been done in [14].

We recalled in Sect. 2.4 that thread validity is decidable in PSPACE, and we did so by reducing to the problem of language inclusion for $\omega$-parity-automata. The original size-change article [12] gives two different methods to check size-change termination, the first one is based on reducing to inclusions of $\omega$-languages defined by finite automata while the second one is a direct, graph-based approach. It is in fact possible to use this more direct method to decide the thread criterion, and this has already been done in [5] by Dax, Hofmann and Lange, who remark furthermore that this method leads to a more efficient implementation than the automata-based one.

## 5 Conclusion

In the present paper, we analyzed the complexity of deciding the validity of circular proofs in $\mu$MALL logic: while the problem was already known to be in PSPACE, we established here its PSPACE-completeness. In doing so, we drew inspiration from the PSPACE-completeness proof of size-change termination even though we depart at some crucial points in order to build our reduction to take into account the specific form of circular proofs.

We conjecture that our proof adapts straightforwardly to a number of other circular proof systems based on sequent calculus such as intutionistic or classical proof systems in addition to the linear case on which we have focused here.

While our result can be seen as negative for circular proofs, it does not prevent actual implementations from being tractable and usable in many situations as exemplified by the Cyclist prover for instance. In such systems, validity checking does not seem to be the bottleneck in circular proof construction compared with the complexity that is inherent in exploring and backtracking in the search tree [4,15,16].

Our work suggests deep connections between thread-validity and size-change termination, which we only touched upon in the previous section. This confirms connections previously hinted by other authors [5,7,8,13] that we plan to investigate further in the future.

**Acknowledgements.** A special thanks must go to Anupam Das and Reuben Rowe, and to the anonymous reviewer, for their very complete and most relevant comments.

## References

1. The cyclist theorem prover. http://www.cyclist-prover.org/
2. Baelde, D.: Least and greatest fixed points in linear logic. ACM Trans. Comput. Log. **13**(1), 2:1–2:44 (2012). https://doi.org/10.1145/2071368.2071370
3. Baelde, D., Doumane, A., Saurin, A.: Infinitary proof theory: the multiplicative additive case. In: Talbot, J., Regnier, L. (eds.) 25th EACSL Annual Conference on Computer Science Logic, CSL 2016. LIPIcs, Marseille, France, August 29–September 1 2016, vol. 62, pp. 42:1–42:17. Schloss Dagstuhl - Leibniz-Zentrum fuer Informatik (2016). https://doi.org/10.4230/LIPIcs.CSL.2016.42
4. Brotherston, J., Gorogiannis, N., Petersen, R.L.: A generic cyclic theorem prover. In: Jhala, R., Igarashi, A. (eds.) APLAS 2012. LNCS, vol. 7705, pp. 350–367. Springer, Heidelberg (2012). https://doi.org/10.1007/978-3-642-35182-2_25
5. Dax, C., Hofmann, M., Lange, M.: A proof system for the linear time $\mu$-Calculus. In: Arun-Kumar, S., Garg, N. (eds.) FSTTCS 2006. LNCS, vol. 4337, pp. 273–284. Springer, Heidelberg (2006). https://doi.org/10.1007/11944836_26
6. Doumane, A.: On the infinitary proof theory of logics with fixed points. (Théorie de la démonstration infinitaire pour les logiques à points fixes). Ph.D. thesis, Paris Diderot University, France (2017). https://tel.archives-ouvertes.fr/tel-01676953
7. Hyvernat, P.: The size-change termination principle for constructor based languages. Log. Methods Comput. Sci. **10**(1) (2014). https://doi.org/10.2168/LMCS-10(1:11)2014
8. Hyvernat, P.: The size-change principle for mixed inductive and coinductive types. CoRR abs/1901.07820 (2019). http://arxiv.org/abs/1901.07820
9. Jones, N.D.: Computability and Complexity - from a Programming Perspective. Foundations of Computing Series. MIT Press, Cambridge (1997)
10. Kozen, D.: On induction vs. *-continuity. In: Kozen, D. (ed.) Logic of Programs 1981. LNCS, vol. 131, pp. 167–176. Springer, Heidelberg (1982). https://doi.org/10.1007/BFb0025782
11. Kozen, D.: Results on the propositional mu-calculus. Theor. Comput. Sci. **27**, 333–354 (1983). https://doi.org/10.1016/0304-3975(82)90125-6

12. Lee, C.S., Jones, N.D., Ben-Amram, A.M.: The size-change principle for program termination. In: Hankin, C., Schmidt, D. (eds.) Conference Record of POPL 2001: The 28th ACM SIGPLAN-SIGACT Symposium on Principles of Programming Languages, London, UK, 17–19 January 2001, pp. 81–92. ACM (2001). https://doi.org/10.1145/360204.360210. http://doi.acm.org/10.1145/360204.360210
13. Lepigre, R., Raffalli, C.: Practical subtyping for curry-style languages. ACM Trans. Program. Lang. Syst. **41**(1), 5:1–5:58 (2019). https://doi.org/10.1145/3285955
14. Nollet, R., Saurin, A., Tasson, C.: Local validity for circular proofs in linear logic with fixed points. In: Ghica, D.R., Jung, A. (eds.) 27th EACSL Annual Conference on Computer Science Logic, CSL 2018. LIPIcs, Birmingham, UK, 4–7 September 2018, vol. 119, pp. 35:1–35:23. Schloss Dagstuhl - Leibniz-Zentrum fuer Informatik (2018). https://doi.org/10.4230/LIPIcs.CSL.2018.35
15. Rowe, R.N.S., Brotherston, J.: Automatic cyclic termination proofs for recursive procedures in separation logic. In: Bertot, Y., Vafeiadis, V. (eds.) Proceedings of the 6th ACM SIGPLAN Conference on Certified Programs and Proofs, CPP 2017, Paris, France, 16–17 January 2017, pp. 53–65. ACM (2017). https://doi.org/10.1145/3018610.3018623
16. Tellez, G., Brotherston, J.: Automatically verifying temporal properties of pointer programs with cyclic proof. In: de Moura, L. (ed.) CADE 2017. LNCS (LNAI), vol. 10395, pp. 491–508. Springer, Cham (2017). https://doi.org/10.1007/978-3-319-63046-5_30

# A Non-wellfounded, Labelled Proof System for Propositional Dynamic Logic

Simon Docherty[1(✉)] and Reuben N. S. Rowe[2]

[1] Department of Computer Science, University College London, London, UK
simon.docherty@ucl.ac.uk
[2] School of Computing, University of Kent, Canterbury, UK
r.n.s.rowe@kent.ac.uk

**Abstract.** We define an infinitary labelled sequent calculus for PDL, $\mathbf{G3PDL}^{\infty}$. A finitarily representable cyclic system, $\mathbf{G3PDL}^{\omega}$, is then given. We show that both are sound and complete with respect to standard models of PDL and, further, that $\mathbf{G3PDL}^{\infty}$ is cut-free complete. We additionally investigate proof-search strategies in the cyclic system for the fragment of PDL without tests.

## 1 Introduction

Fischer and Ladner's Propositional Dynamic Logic (PDL) [14], which is the propositional variant of Pratt's Dynamic Logic [34], is perhaps *the* quintessential modal logic of action. While (P)DL arose initially as a modal logic for reasoning about program execution its impact as a formalism for extending 'static' logical systems with 'dynamics' via composite actions [22, p. 498] has been felt broadly across logic. This is witnessed in extensions and variants designed for reasoning about games [31], natural language [21], cyber-physical systems [33], epistemic agents [19], XML [1], and knowledge representation [11], among others.

Much of the proof theoretic work on PDL, and logics extending it, focuses on Hilbert-style axiomatisations, which are not amenable to automation. Outside of this, proof systems for PDL itself can broadly be characterised as one of two sorts. Falling into the first category are a multitude of infinitary systems [16, 24, 35] employing either infinitely-wide $\omega$-proof rules, or (equivalently) allowing countably infinite contexts. In the other category are tableau-based algorithms for deciding PDL-satisfiability [18, 20]. While these are (neccessarily) finitary, they employ a great deal of auxillary structure tailored to the decision procedure itself.

In the proof theory of modal logic, a high degree of uniformity and modularity has been achieved through labelled systems. The idea of using labels as syntactic representatives of Kripke models in modal logic proof systems can be traced back to Kanger [25], but perhaps has been most famously deployed by Fitting [15].

S. Docherty—supported by EPSRC grant no. EP/S013008/1.
R. N. S. Rowe—supported by EPSRC grant no. EP/N028759/1.


S. Cerrito and A. Popescu (Eds.): TABLEAUX 2019, LNAI 11714, pp. 335–352, 2019.
https://doi.org/10.1007/978-3-030-29026-9_19

A succinct history of the use of labelled systems is provided by Negri [29]. Negri's work [28] is the high point of the technique, giving a procedure to transform frame conditions for Kripke models into labelled sequent calculi rules preserving structural properties of the proof system, given they are defined as *coherent axioms*.

The power of this rule generation technique is of particular interest because it enables the specification of sound and complete systems for classes of Kripke frames that are first-order, but not modally, definable. In the context of PDL-type logics, this is of interest because of common additional program constructs like intersection which have a non-modally definable intended interpretation [32]. However, even with this expressive power, such a framework on its own cannot account for program modalities involving iteration. In short, formulae involving these modalities are interpreted via the reflexive-transitive closure of accessibility relations, and this closure is not first-order (and therefore, not coherently) definable. Something more must be done to capture the PDL family of logics.

In this paper we provide the first step towards a uniform proof theory of the sort that is currently missing for this family of logics by giving two new proof systems for PDL. We combine two ingredients from modern proof theory that have hitherto remained separate: labelled deduction à la Negri and non-wellfounded (in particular, *cyclic*) sequent calculi.

We first construct a labelled sequent calculus $\mathbf{G3PDL}^{\infty}$, extending that of Negri [28], in which proofs are permitted to be infinitely tall. For this system soundness (via descending counter-models) and cut-free completeness (via counter-model construction) are proved in a similar manner to Brotherston and Simpson's infinitary proof theory for first-order logic with inductive definitions [6]. Next we restrict attention to *regular* proofs, meaning only those infinite proof trees that are finitely representable (i.e. only have a finite number of distinct subtrees), obtaining the cyclic system $\mathbf{G3PDL}^{\omega}$. This can be done by permitting the forming of backlinks (or, cycles) in the proof tree, granted a (decidable) trace condition guaranteeing soundness can be established. We then show that the axiomatisation of PDL [23] can be derived in $\mathbf{G3PDL}^{\omega}$, obtaining completeness. We finish the paper with an investigation of proof-search in the cyclic system for a sub-class of sequents, and conjecture cut-free completeness for the test-free fragment of PDL.

There are a number of advantages to setting up PDL's proof theory in this manner. Most crucially, through the cyclic system we obtain a finitary sequent calculus with natural, declarative proof rules, in which requirements on Kripke models and traces are elegantly handled with the labels. Such a system is amenable to automation through, for example, the Cyclist [5] theorem prover. We also conjecture (see Sect. 5) that labels can be used to compute bounds to determine termination of proof-search.

We believe this work can be built upon in two complementary directions. First, towards a uniform proof theory of PDL-type logics. We conjecture the presence of labels should facilitate the extension of the system with rules for additional program constructs. Second, this work gives a case study for the extension of the expressivity of Negri-style modal proof theory. Our system thus indicates the viability of constructing an analogous general framework that naturally captures modal logics interpreted on classes of Kripke frames defined by logics more expressive than first-order logic (for example, epistemic logics with a common knowledge modality). We discuss this, and other ideas for future work, in the conclusion.

For space reasons we elide proofs, but these can be found in an extended version of this paper available online [13].

*Related Work.* Beyond the proof systems outlined above, the most significant related work can be found in Das and Pous' [9,10] cyclic proof systems for deciding Kleene algebra (in)equalities. Das and Pous' insight that iteration can be handled in a cyclic sequent calculus is essential to our work here, although there are additional complications involved in formulating a system for PDL because of the interaction between programs (which form a Kleene algebra with tests) and formulae. We also note that Goré and Widmann's tableau procedure also utilises the formation of cycles in proof trees. Our proof of cut-free completeness of the infinitary system also follows that of Brotherston and Simpson [6] for first-order logic with inductive definitions.

Recent work by Cohen and Rowe [8] gives a cyclic proof system for the extension of first-order logic with a transitive closure operator and we conjecture that our labelled cyclic system (and labelled cyclic systems for modal logics more generally) can be formalised within it. This idea echoes van Benthem's suggestion that the most natural frame language for many modal logics is not first-order logic, but in fact first-order logic with a least fixed point operator [4].

Cyclic proof systems have also been defined for some modal logics with similar model properties to PDL, including the logic of common knowledge [40] and Gödel-Löb logic [37]. The idea of cyclic proof can be traced to modal $\mu$-calculus [30]. Indeed, it can be shown that the logic of common knowledge [2], Gödel-Löb logic [4,39] and PDL [4,7] can be faithfully interpreted in the modal $\mu$-calculus, indicating that perhaps cyclic proof was the right approach for PDL all along.

## 2 PDL: Syntax and Semantics

The syntax of PDL formulas is defined as follows. We assume countably many atomic *propositions* (ranged over by $p$, $q$, $r$), and countably many atomic *programs* (ranged over by $a$, $b$, $c$).

**Definition 1 (Syntax of PDL).** *The set of* formulas *($\varphi$, $\psi$, ...) and the set of* programs *($\alpha$, $\beta$, ...) are defined mutually by the following grammar:*

$$\varphi, \psi ::= \bot \mid p \mid \varphi \wedge \psi \mid \varphi \vee \psi \mid \varphi \rightarrow \psi \mid [\alpha]\varphi$$
$$\alpha, \beta, \gamma ::= a \mid \alpha \,;\, \beta \mid \alpha \cup \beta \mid \varphi? \mid \alpha^*$$

We briefly reprise the semantics of PDL (see [23, §5.2]). A PDL model $\mathfrak{m} = (\mathcal{S}, \mathcal{I})$ is a Kripke model consisting of a set $\mathcal{S}$ of *states* and an *interpretation* function $\mathcal{I}$ that assigns: a subset of $\mathcal{S}$ to each atomic proposition; and a binary relation on $\mathcal{S}$ to each atomic program. We inductively construct an extension of the interpretation function, denoted $\mathcal{I}_{\mathfrak{m}}$, that operates on the full set of propositions and programs.

**Definition 2 (Semantics of PDL).** *Let $\mathfrak{m} = (\mathcal{S}, \mathcal{I})$ be a PDL model. We define the extended interpretation function $\mathcal{I}_{\mathfrak{m}}$ inductively as follows:*

$$\begin{aligned} \mathcal{I}_{\mathfrak{m}}(\bot) &= \emptyset & \mathcal{I}_{\mathfrak{m}}(a) &= \mathcal{I}(a) \\ \mathcal{I}_{\mathfrak{m}}(p) &= \mathcal{I}(p) & \mathcal{I}_{\mathfrak{m}}(\alpha\,;\beta) &= \mathcal{I}_{\mathfrak{m}}(\alpha) \circ \mathcal{I}_{\mathfrak{m}}(\beta) \\ \mathcal{I}_{\mathfrak{m}}(\varphi \wedge \psi) &= \mathcal{I}_{\mathfrak{m}}(\varphi) \cap \mathcal{I}_{\mathfrak{m}}(\psi) & \mathcal{I}_{\mathfrak{m}}(\alpha \cup \beta) &= \mathcal{I}_{\mathfrak{m}}(\alpha) \cup \mathcal{I}_{\mathfrak{m}}(\beta) \\ \mathcal{I}_{\mathfrak{m}}(\varphi \vee \psi) &= \mathcal{I}_{\mathfrak{m}}(\varphi) \cup \mathcal{I}_{\mathfrak{m}}(\psi) & \mathcal{I}_{\mathfrak{m}}(\varphi?) &= \mathsf{Id}(\mathcal{I}_{\mathfrak{m}}(\varphi)) \\ \mathcal{I}_{\mathfrak{m}}(\varphi \rightarrow \psi) &= (\mathcal{S} \setminus \mathcal{I}_{\mathfrak{m}}(\varphi)) \cup \mathcal{I}_{\mathfrak{m}}(\psi) & \mathcal{I}_{\mathfrak{m}}(\alpha^*) &= \bigcup_{k \geq 0} \mathcal{I}_{\mathfrak{m}}(\alpha)^k \end{aligned}$$

$$\mathcal{I}_{\mathfrak{m}}([\alpha]\varphi) = \mathcal{S} \setminus \Pi_1(\mathcal{I}_{\mathfrak{m}}(\alpha) \circ \mathsf{Id}(\mathcal{S} \setminus \mathcal{I}_{\mathfrak{m}}(\varphi)))$$

*where $\circ$ denotes relational composition, $R^n$ denotes the composition of $R$ with itself $n$ times, $\Pi_1$ returns a set by projecting the first component of each tuple in a relation, and $\mathsf{Id}(X)$ denotes the identity relation over the set $X$.*

We write $\mathfrak{m}, s \models \varphi$ to mean $s \in \mathcal{I}_{\mathfrak{m}}(\varphi)$, and $\mathfrak{m} \models \varphi$ to mean that $\mathfrak{m}, s \models \varphi$ for all states $s \in \mathcal{S}$. A PDL formula $\varphi$ is *valid* when $\mathfrak{m} \models \varphi$ for all models $\mathfrak{m}$.

## 3 An Infinitary, Labelled Sequent Calculus

We now define a sequent calculus for deriving theorems (i.e. valid formulas) of PDL. This proof system has two important features. The first is that it is a *labelled* proof system. Thus sequents contain assertions about the structure of the underlying Kripke models and formulas are labelled with atoms denoting specific states in which they should be interpreted. Secondly, we allow proofs of infinite height.

We assume a countable set $\mathcal{L}$ of *labels* (ranged over by $x$, $y$, $z$) that we will use to denote particular states. A *relational atom* is an expression of the form $x\ R_a\ y$, where $x$ and $y$ are labels and $a$ is an atomic program. A *labelled formula* is an expression of the form $x : \varphi$, where $x$ is a label and $\varphi$ is a formula. We define a label substitution operation by $z\{x/y\} = y$ when $x = z$, and $z\{x/y\} = z$ otherwise. We lift this to relational atoms and labelled formulas by: $(z\ R_a\ z')\{x/y\} = z\{x/y\}\ R_a\ z'\{x/y\}$ and $(z : \varphi)\{x/y\} = z\{x/y\} : \varphi$.

Sequents are expressions of the form $\Gamma \Rightarrow \Delta$, where $\Gamma$ and $\Delta$ are finite sets of relational atoms and labelled formulas. We denote an arbitrary member of such a set using $A$, $B$, etc. As usual, $\Gamma, A$ and $A, \Gamma$ both denote the set $\{A\} \cup \Gamma$, and $\Gamma\{z/y\}$ denotes the application of the (label) substitution $\{x/y\}$ to all the elements in $\Gamma$. We denote by $[\alpha]\Gamma$ the set of formulas obtained from $\Gamma$ by prepending the modality $[\alpha]$ to every labelled formula. That is, we define $[\alpha]\Gamma = \{x\ R_a\ y \mid x\ R_a\ y \in \Gamma\} \cup \{x : [\alpha]\varphi \mid x : \varphi \in \Gamma\}$. $\mathsf{labs}(\Gamma)$ denotes the set of all labels occurring in the relational atoms and labelled formulas in $\Gamma$.

We interpret sequents with respect to PDL models using label *valuations* $v$, which are functions from labels to states. We write $\mathfrak{m}, v \models x \; R_a \; y$ to mean that $(v(x), v(y)) \in \mathcal{I}_{\mathfrak{m}}(a)$. We write $\mathfrak{m}, v \models x : \varphi$ to mean $\mathfrak{m}, v(x) \models \varphi$. For a sequent $\Gamma \Rightarrow \Delta$, denoted by $S$, we write $\mathfrak{m}, v \models S$ to mean that $\mathfrak{m}, v \models B$ for *some* $B \in \Delta$ whenever $\mathfrak{m}, v \models A$ for *all* $A \in \Gamma$. We write $\mathfrak{m}, v \not\models S$ whenever this is not the case, i.e. when $\mathfrak{m}, v \models A$ for all $A \in \Gamma$ and $\mathfrak{m}, v \not\models B$ for all $B \in \Delta$. We say $S$ is *valid*, and write $\models S$, when $\mathfrak{m}, v \models S$ for all models $\mathfrak{m}$ and valuations $v$ that map each label to some state of $\mathfrak{m}$.

$$(\text{Ax}): \frac{}{A \Rightarrow A} \qquad (\bot): \frac{}{x : \bot \Rightarrow} \qquad (\text{WL}): \frac{\Gamma \Rightarrow \Delta}{A, \Gamma \Rightarrow \Delta} \qquad (\text{WR}): \frac{\Gamma \Rightarrow \Delta}{\Gamma \Rightarrow \Delta, A}$$

$$(\wedge\text{L}): \frac{x : \varphi, x : \psi, \Gamma \Rightarrow \Delta}{x : \varphi \wedge \psi, \Gamma \Rightarrow \Delta} \qquad (\wedge\text{R}): \frac{\Gamma \Rightarrow \Delta, x : \varphi \quad \Gamma \Rightarrow \Delta, x : \psi}{\Gamma \Rightarrow \Delta, x : \varphi \wedge \psi}$$

$$(\vee\text{L}): \frac{x : \varphi, \Gamma \Rightarrow \Delta \quad x : \psi, \Gamma \Rightarrow \Delta}{x : \varphi \vee \psi, \Gamma \Rightarrow \Delta} \qquad (\vee\text{R}): \frac{\Gamma \Rightarrow \Delta, x : \varphi, x : \psi}{\Gamma \Rightarrow \Delta, x : \varphi \vee \psi}$$

$$(\to\text{L}): \frac{\Gamma \Rightarrow \Delta, x : \varphi \quad x : \psi, \Gamma \Rightarrow \Delta}{x : \varphi \to \psi, \Gamma \Rightarrow \Delta} \qquad (\to\text{R}): \frac{x : \varphi, \Gamma \Rightarrow \Delta, x : \psi}{\Gamma \Rightarrow \Delta, x : \varphi \to \psi}$$

$$(\Box\text{L}): \frac{y : \varphi, \Gamma \Rightarrow \Delta}{x : [a]\varphi, x \; R_a \; y, \Gamma \Rightarrow \Delta} \qquad (\Box\text{R}): \frac{x \; R_a \; y, \Gamma \Rightarrow \Delta, y : \varphi}{\Gamma \Rightarrow \Delta, x : [a]\varphi} \; (y \text{ fresh})$$

$$(;\text{L}): \frac{x : [\alpha][\beta]\varphi, \Gamma \Rightarrow \Delta}{x : [\alpha \, ; \beta]\varphi, \Gamma \Rightarrow \Delta} \qquad (;\text{R}): \frac{\Gamma \Rightarrow \Delta, x : [\alpha][\beta]\varphi}{\Gamma \Rightarrow \Delta, x : [\alpha \, ; \beta]\varphi}$$

$$(\cup\text{L}): \frac{x : [\alpha]\varphi, x : [\beta]\varphi, \Gamma \Rightarrow \Delta}{x : [\alpha \cup \beta]\varphi, \Gamma \Rightarrow \Delta} \qquad (\cup\text{R}): \frac{\Gamma \Rightarrow \Delta, x : [\alpha]\varphi \quad \Gamma \Rightarrow \Delta, x : [\beta]\varphi}{\Gamma \Rightarrow \Delta, x : [\alpha \cup \beta]\varphi}$$

$$(?\text{L}): \frac{\Gamma \Rightarrow \Delta, x : \varphi \quad x : \psi, \Gamma \Rightarrow \Delta}{x : [\varphi?]\psi, \Gamma \Rightarrow \Delta} \qquad (?\text{R}): \frac{x : \varphi, \Gamma \Rightarrow \Delta, x : \psi}{\Gamma \Rightarrow \Delta, x : [\varphi?]\psi}$$

$$(*\text{L}): \frac{x : \varphi, x : [\alpha][\alpha^*]\varphi, \Gamma \Rightarrow \Delta}{x : [\alpha^*]\varphi, \Gamma \Rightarrow \Delta} \qquad (*\text{R}): \frac{\Gamma \Rightarrow \Delta, x : \varphi \quad \Gamma \Rightarrow \Delta, x : [\alpha][\alpha^*]\varphi}{\Gamma \Rightarrow \Delta, x : [\alpha^*]\varphi}$$

$$(\text{Subst}): \frac{\Gamma \Rightarrow \Delta}{\Gamma\{x/y\} \Rightarrow \Delta\{x/y\}} \qquad (\text{Cut}): \frac{\Gamma \Rightarrow \Delta, A \quad A, \Sigma \Rightarrow \Pi}{\Gamma, \Sigma \Rightarrow \Delta, \Pi}$$

**Fig. 1.** Inference rules of $\mathbf{G3PDL}^{\infty}$

The sequent calculus $\mathbf{G3PDL}^{\infty}$ is defined by the inference rules in Fig. 1. A *pre-proof* is a possibly infinite derivation tree built from these inference rules.

**Definition 3 (Pre-proof).** *A pre-proof is a possibly infinite (i.e. non-well-founded) derivation tree formed from inference rules. A path in a pre-proof is a possibly infinite sequence of sequents $s_0, s_1, \ldots (, s_n)$ such that $s_0$ is the root sequent of the proof, and $s_{i+1}$ is a premise of $s_i$ for each $i < n$.*

Not all pre-proofs derive sound judgements.

*Example 1.* The following pre-proof derives an invalid sequent.

$$\dfrac{\dfrac{\begin{matrix}\vdots\\ \Rightarrow x : [a^*]p\end{matrix}}{\Rightarrow x : [a^*]p, x : p}\,(\text{WR}) \qquad \dfrac{\begin{matrix}\vdots\\ \Rightarrow x : [a^*]p\end{matrix}}{\Rightarrow x : [a^*]p, x : [a][a^*]p}\,(\text{WR})}{\Rightarrow x : [a^*]p}\,(*\text{R})$$

Note that, since our sequents consist of *sets* of formulas, each instance of the $(*\text{R})$ rule incorporates a contraction

To distinguish pre-proofs deriving valid sequents, we define the notion of a trace through a pre-proof. Traces consist of trace values, which (uniquely) identify particular modalities within labelled formulas. $\boldsymbol{\alpha}_n$ denotes a sequence $\alpha_1, \ldots, \alpha_n$, and $\varepsilon$ denotes the empty sequence. We sometimes omit the subscript indicating length, writing $\boldsymbol{\alpha}$, when irrelevant or evident from the context.

**Definition 4 (Trace Value).** *A* trace value *$\tau$ is a tuple $(x, \boldsymbol{\alpha}_n, \beta, \varphi)$ consisting of a label $x$, a (possibly empty) sequence $\boldsymbol{\alpha}_n$ of $n$ programs, a program $\beta$, and a formula $\varphi$. We call $\boldsymbol{\alpha}$ the* spine *of $\tau$, and $\beta$ the* focus *of $\tau$. We write $[\gamma]\tau$ for the trace value $(x, \gamma \cdot \boldsymbol{\alpha}_n, \beta, \varphi)$, and $y : \tau$ for the trace value $(y, \boldsymbol{\alpha}_n, \beta, \varphi)$. In an abuse of notation we also use $\tau$ to denote the corresponding labelled formula $x : [\alpha_1] \ldots [\alpha_n][\beta^*]\varphi$.*

Trace values in the conclusion of an inference rule are related to trace values in its premises as follows.

**Definition 5 (Trace Pairs).** *Let $\tau$ and $\tau'$ be trace values, with sequents $\Gamma \Rightarrow \Delta$ and $\Gamma' \Rightarrow \Delta'$ (respectively denoted by $s$ and $s'$) the conclusion and a premise, respectively, of an inference rule $r$; we say that $(\tau, \tau')$ is a trace pair for $(s, s')$ when $\tau \in \Delta$ and $\tau' \in \Delta'$ and the following conditions hold.*

*(1) If $\tau$ is the principal formula of the rule instance, then $\tau'$ is its immediate ancestor and moreover if the rule is an instance of:*
- ($\Box$R) *then $\tau = x : [a]\tau'$, where $x$ is the label of the principal formula;*
- (?R) *then $\tau = [\varphi?]\tau'$;*
- (;R) *then $\tau = [\alpha\, ; \beta]\tau''$ and $\tau' = [\alpha][\beta]\tau''$ for some trace value $\tau''$;*
- ($\cup$R) *then there is some $\tau''$ such that: $\tau = [\alpha \cup \beta]\tau''$; $\tau' = [\alpha]\tau''$ if $s'$ is the left-hand premise; and $\tau' = [\beta]\tau''$ if $s'$ is the right-hand premise;*
- ($*$R) *then $\tau = [\alpha^*]\tau'$ if $s'$ is the left-hand premise, and $\tau' = [\alpha]\tau$ if $s'$ is the right-hand premise.*

(2) *If $\tau$ is not the principal formula of the rule then $\tau = x : \tau'$ if the rule is an instance of (Subst) and $x$ is the label substituted, and $\tau = \tau'$ otherwise.*

*If $\tau$ is the principal formula of the rule instance and the spine of $\tau$ is empty, then we say that the trace pair is* progressing.

Notice that when a trace pair is progressing for $(s, s')$, it is necessarily the case that the corresponding rule is an instance of $(*\mathrm{R})$ and that $s'$ is the right-hand premise (although, not necessarily vice versa).

Traces along paths in a pre-proof consist of consecutive pairs of trace values for each corresponding step of the path.

**Definition 6 (Trace).** *A* trace *is a (possibly infinite) sequence of trace values. We say that a trace $\tau_1, \tau_2, \dots (, \tau_n)$* follows *a path $s_1, s_2, \dots (, s_m)$ in a pre-proof when there exists some $k \geq 0$ such that each consecutive pair of trace values $(\tau_i, \tau_{i+1})$ is a trace pair for $(s_{i+k}, s_{i+k+1})$; when $k = 0$, we say that the trace* covers *the path. We say that the trace* progresses *at $i$ if $(\tau_i, \tau_{i+1})$ is progressing, and say the trace is infinitely progressing if it progresses at infinitely many points.*

Proofs are pre-proofs that satisfy a well-formedness condition, called the global trace condition.

**Definition 7 (Infinite Proof).** *A* **G3PDL$^\infty$** proof *is a pre-proof in which every infinite path is followed by some infinitely progressing trace.*

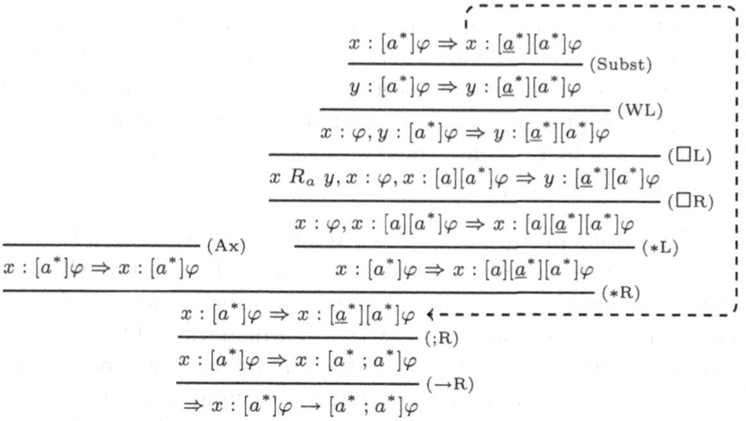


**Fig. 2.** Representation of a **G3PDL$^\infty$** proof of $[a^*]\varphi \to [a^* \,;\, a^*]\varphi$. (Color figure online)

*Example 2.* Figure 2 shows a finite representation of a **G3PDL$^\infty$** proof of the formula $[a^*]\varphi \to [a^* \,;\, a^*]\varphi$. The full infinite proof can be obtain by unfolding the cycle an infinite number of times. An infinitely progressing trace following

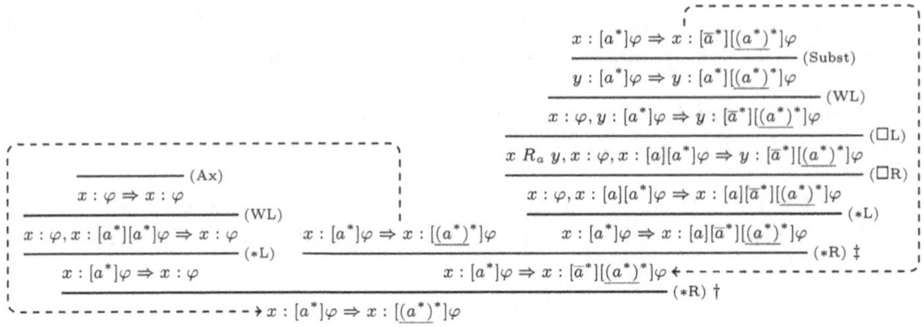


**Fig. 3.** Representation of a **G3PDL**$^\infty$ proof of $x : [a^*]\varphi \Rightarrow x : [(a^*)^*]\varphi$. (Color figure online)

the (unique) infinite path in this proof is indicated by the underlined programs highlighted in blue, which denote the focus of the trace value in each sequent. The progression point is the (only) instance of the ($*$R) rule.

Figure 3 shows a finite representation of a **G3PDL**$^\infty$ proof of the sequent $x : [a^*]\varphi \Rightarrow x : [(a^*)^*]\varphi$. This proof is more complex than that of Fig. 2, and involves two overlapping cycles. This proof contains more than one infinite path (in fact, it contains an infinite number of infinite paths). However, they fall into three categories: (1) those that eventually traverse only the upper cycle; (2) those that eventually traverse only the lower cycle; and (3) those that traverse both cycles infinitely often. Infinite paths of the first variety have an infinitely progressing trace indicated by the overlined programs highlighted in red. The progression point is the upper instance of ($*$R) rule, marked by (‡). The remaining infinite paths have a trace indicated by the underlined programs highlighted in blue. This trace does not progress around the upper cycle (for those paths that traverse it), but does progress once around each lower cycle at the instance of the ($*$R) rule marked by (†). Since these paths traverse this lower cycle infinitely often, the trace is infinitely progressing.

*Remark 1.* The notion of trace in the system for Kleene Algebra of Das and Pous [9,10] appears simpler than ours: a sequence of formulas (on the left) connected by ancestry, with such a trace being valid if it is principal for a (left) unfolding rule infinitely often. In fact, we can show that our definition of trace is equivalent to an analogous formulation of this notion for our system. However, our definition allows for a direct, semantic proof of soundness via infinite descent. In contrast, the soundness proof in [10] relies on cut-admissibility and an inductive proof-theoretic argument for the soundness of the cut-free fragment. It is unclear that a similar technique can be used to show soundness of the cut-free fragment of our system. Furthermore, the cut-free fragment of the system of Das and Pous is notable in that it admits a simpler trace condition than the full system: namely, that every infinite path is fair for the (left) unfolding rule [10, prop. 8]. Our system does not satisfy this property, due to the ability to perform contraction and weakening, as demonstrated in Example 1.

The proof system is sound since, for invalid sequents, we can map traces to decreasing sets of counter-examples in (finitely branching) models.

A path in a model $\mathfrak{m}$ is a sequence of states $s_1, \dots, s_n$ in $\mathfrak{m}$ such that each successive pair of states satisfies $(s_i, s_{i+1}) \in \mathcal{I}_\mathfrak{m}(a)$ for some $a$. A path in $\mathfrak{m}$ is called *loop-free* if it does not contain any repeated states. If $\boldsymbol{s}$ and $\boldsymbol{s}'$ are paths in $\mathfrak{m}$, we write $\boldsymbol{s} \sqsubseteq \boldsymbol{s}'$ to denote that $\boldsymbol{s}$ is a prefix of $\boldsymbol{s}'$.

An $m$-partition of a path $\boldsymbol{s}_n$ is a sequence of $m$ increasing indices $k_1 \leq \dots \leq k_m \leq n$. A path in $\mathfrak{m}$ for a trace value $\tau = (x, \boldsymbol{\alpha}_n, \beta, \varphi)$ with respect to a valuation $v$ is a path $\boldsymbol{s}_m$ in $\mathfrak{m}$ with $s_1 = v(x)$ having an $n$-partition $k_1, \dots, k_n$ satisfying $(s_{k_i}, s_{k+1}) \in \mathcal{I}_\mathfrak{m}(\alpha_{i+1})$ for each $0 \leq i < n$ and $(s_{k_n}, s_m) \in \mathcal{I}_\mathfrak{m}(\beta^*)$, where we take $k_0 = 1$ (i.e. $s_{k_0} = s_1$). The $n$-partition $k_1, \dots, k_n$ is called a partition of $\boldsymbol{s}_m$ for $\tau$. A counter-example in $\mathfrak{m}$ for a trace value $\tau$ at $v$ is simply a path $\boldsymbol{s}_m$ in $\mathfrak{m}$ for $\tau$ w.r.t. $v$ such that $\mathfrak{m}, s_m \not\models \varphi$.

A given path in $\mathfrak{m}$ for $\tau$ at $v$ can, in general, have many different partitions. A partition $\boldsymbol{k}_n$ of a path $\boldsymbol{s}_m$ for $\tau$ at $v$ is called *maximal* if the length of its final segment $s_{k_n}, \dots, s_m$ is maximal among all such partitions. We define the *weight* of a path $\boldsymbol{s}$ in $\mathfrak{m}$ for $\tau$ at $v$ to be the length of the final segment(s) of its maximal partition(s). We denote this by $\mu_{(\mathfrak{m},v)}(\boldsymbol{s}, \tau)$. If $\Pi$ is a set of paths in $\mathfrak{m}$ for $\tau$ at $v$, we define the *measure* of $\Pi$, denoted $\mu_{(\mathfrak{m},v)}(\Pi, \tau)$, to be the multiset of weights of the paths it contains; that is $\mu_{(\mathfrak{m},v)}(\Pi, \tau) = \{\mu_{(\mathfrak{m},v)}(\boldsymbol{s}, \tau) \mid \boldsymbol{s} \in \Pi\}$.

The measure for trace values in a model $\mathfrak{m}$ at a valuation $v$, then, is simply the measure of the set of all of its 'nearest' counter-examples.

**Definition 8 (Trace Value Measure).** *Let $\mathcal{C}_{(\mathfrak{m},v)}(\tau)$ denote the set of all loop-free counter-examples $\boldsymbol{s}$ in $\mathfrak{m}$ for $\tau$ at $v$ such that there is no counter-example $\boldsymbol{s}'$ in $\mathfrak{m}$ for $\tau$ at $v$ with $\boldsymbol{s}' \sqsubseteq \boldsymbol{s}$. The measure of $\tau$ in $\mathfrak{m}$ at $v$ is defined as $\mu_{(\mathfrak{m},v)}(\tau) = \mu_{(\mathfrak{m},v)}(\mathcal{C}_{(\mathfrak{m},v)}(\tau), \tau)$.*

For finitely branching models $\mathfrak{m}$, it is clear that trace value measures are always finite. Note that finite multisets $M$ of elements of a well-ordering can be well-ordered using, e.g., the Dershowitz-Manna ordering $<_{\mathrm{DM}}$ [12]. This means that we have the following property.

**Lemma 1 (Descending Counter-models).** *Let $\Gamma \Rightarrow \Delta$, denoted $S$, be the conclusion of an instance of an inference rule, and suppose there is a finitely branching model $\mathfrak{m}$ and valuation $v$ such that $\mathfrak{m}, v \not\models S$, then there is a premise $\Gamma' \Rightarrow \Delta'$ of the rule instance, denoted $S'$, and a valuation $v'$ such that $\mathfrak{m}, v' \not\models S'$ and for each trace pair $(\tau, \tau')$ for $(S, S')$, $\mu_{(\mathfrak{m},v')}(\tau') \leq_{DM} \mu_{(\mathfrak{m},v)}(\tau)$ and also $\mu_{(\mathfrak{m},v')}(\tau') <_{DM} \mu_{(\mathfrak{m},v)}(\tau)$ if $(\tau, \tau')$ is progressing.*

This entails the soundness of our proof system, since PDL has the finite model property [14, Thm. 3.2]. This property states that, if a PDL formula is satisfiable, then it is satisfiable in a finite (and thus finitely branching) model. Thus, if a sequent is not valid then there is a finitely branching model that falsifies it. If a $\mathbf{G3PDL}^\infty$ proof $\mathcal{P}$ were to derive an invalid sequent, then by Lemma 1 it would contain an infinite path $\Gamma_1 \Rightarrow \Delta_1, \Gamma_2 \Rightarrow \Delta_2, \dots$ for which there exists a finite model $\mathfrak{m}$ and a matching sequence of valuations $v_1, v_2, \dots$

that invalidate each sequent in the path. Moreover, these invalidating valuations ensure that the measures of the trace values in any trace pair along the path is decreasing, and strictly so for progressing trace pairs. However, since $\mathcal{P}$ is a proof, it satisfies the global trace condition. This means that there would be an infinitely progressing trace following the path $\Gamma_1 \Rightarrow \Delta_1, \Gamma_2 \Rightarrow \Delta_2, \ldots$ and thus we would be able to construct an infinitely descending chain of (finite) trace value measures. Because the set of finite trace value measures is well-founded, this is impossible and so the derived sequent must in fact be valid.

**Theorem 1 (Soundness).** **G3PDL**$^\infty$ *derives only valid sequents.*

The *cyclic* system **G3PDL**$^\omega$ is obtained by restricting consideration to only those proofs of **G3PDL**$^\infty$ that are *regular*, i.e. have only a finite number of distinct subtrees.

**Definition 9 (Cyclic Pre-proof).** *A* cyclic *pre-proof is a pair* $(P, f)$ *consisting of a finite derivation tree* $P$ *possibly containing open leaves called* buds, *and a function* $f$ *assigning to each bud an internal node of the tree, called its* companion, *with a syntactically identical sequent.*

We usually represent a cyclic pre-proof as the graph induced by identifying each bud with its companion (as in Figs. 2 and 3). The infinite unfolding of a cyclic pre-proof is the **G3PDL**$^\infty$ pre-proof obtained as the limit of the operation that replaces each bud with a copy of the subderivation concluding with its companion an infinite number of times. A cyclic proof is a cyclic pre-proof whose infinite unfolding satisfies the global trace condition. As in other cyclic systems (e.g. [6,8,36,38]) it is decidable whether or not this is the case via a construction involving complementation of Büchi automata. This means that decidability of the global trace condition for **G3PDL**$^\omega$ pre-proofs is PSPACE-complete.

Since every **G3PDL**$^\omega$ is also a **G3PDL**$^\infty$ proof, soundness of the cyclic system is an immediate corollary of Theorem 1.

**Corollary 1.** *If* $\Gamma \Rightarrow \Delta$ *is derivable in* **G3PDL**$^\omega$, *then* $\Gamma \Rightarrow \Delta$ *is valid.*

## 4 Completeness

In this section, we give completeness results for our systems. We show that the full system, **G3PDL**$^\infty$, is cut-free complete. On the other hand, if we allow instances of the (Cut) rule, then every valid theorem of PDL has a proof in the cyclic subsystem **G3PDL**$^\omega$.

### 4.1 Cut-Free Completeness of G3PDL$^\infty$

We use a standard technique of defining a pre-proof that encodes an exhaustive search for a cut-free proof (as used in, e.g., [6,8]). For invalid sequents, this results in a pre-proof from which we can construct a counter-model, using the formulas that occur along a particular path.

A *schedule* $\sigma$ is an enumeration of labelled non-atomic formulas in which each labelled formula occurs infinitely often. The $i^{\text{th}}$ element of $\sigma$ is written $\sigma_i$.

**Definition 10 (Search Tree).** *Given a sequent $\Gamma \Rightarrow \Delta$ and a schedule $\sigma$, we can define an infinite sequence $\mathcal{D}$ of open derivations inductively. Taking $\mathcal{D}_0 = \Gamma \Rightarrow \Delta$, we construct each $\mathcal{D}_{i+1}$ from its predecessor $\mathcal{D}_i$ by:*

1. *firstly closing any open leaves $\Gamma' \Rightarrow \Delta'$ for which $x : \bot \in \Gamma$ for some $x$ or $\Gamma \cap \Delta \neq \emptyset$ by applying weakening rules leading to an instance of $(\bot)$ or an axiom $A \Rightarrow A$ for some $A \in \Gamma \cap \Delta$ (thus the antecedent of each remaining open node is disjoint from its consequent);*
2. *then replacing each remaining open node $\Gamma' \Rightarrow \Delta'$ in which $\sigma_i$ occurs with applications of the rule for which $\sigma_i$ is principal in the following way.*
   - *If $\sigma_i = x : [a]\varphi \in \Delta'$, then we pick a label $y$ not occurring in $\Gamma' \Rightarrow \Delta'$, and replace the open node with the following derivation.*

$$\frac{x \; R_a \; y, \Gamma' \Rightarrow \Delta', y : \varphi}{\Gamma' \Rightarrow \Delta', x : [a]\varphi} \; (\Box\text{R})$$

   - *If $\sigma_i = x : [a]\varphi \in \Gamma'$ then, letting $\{y_1, \ldots, y_n\}$ be the set of all $y_i$ such that $x \; R_a \; y_i \in \Gamma'$, we replace the open node with the following derivation.*

$$\cfrac{\cfrac{\begin{array}{c} x : [a]\varphi, \{y_1 : \varphi, \ldots, y_n : \varphi\}, \Gamma' \Rightarrow \Delta' \\ \vdots \\ x : [a]\varphi, \{y_1 : \varphi, y_2 : \varphi\}, \Gamma' \Rightarrow \Delta' \end{array}}{x : [a]\varphi, \{y_1 : \varphi\}, \Gamma' \Rightarrow \Delta'} \; (\Box\text{L})}{x : [a]\varphi, \Gamma' \Rightarrow \Delta'} \; (\Box\text{L})$$

   - *In all other cases, we replace the open node with an application of the appropriate rule $(r)$ as follows, where $\Gamma_i'$ and $\Delta_i'$, $i \in \{1, 2\}$, are the sets of left and right immediate ancestors of $\sigma_i$, respectively, for the appropriate premise.*

$$\frac{\Gamma_1', \Gamma' \Rightarrow \Delta', \Delta_1' \quad (\Gamma_2', \Gamma' \Rightarrow \Delta', \Delta_2')}{\Gamma' \Rightarrow \Delta'} \; (r)$$

*Since each $\mathcal{D}_i$ is a prefix of $\mathcal{D}_{i+1}$, there is a smallest derivation containing each $\mathcal{D}_i$ as a prefix. We call this derivation a* search tree *for $\Gamma \Rightarrow \Delta$ (w.r.t. $\sigma$).*

Notice that search trees do not contain instances of the (Cut) or (Subst) rules. Moreover, when a given search tree $\mathcal{D}$ is not a valid proof, we may extract from it two sets of labelled formulas and relational atoms that we can use to construct a countermodel. If $\mathcal{D}$ is not a valid proof, then either it contains an open node to which no schedule element applies or it contains an infinite path that does not satisfy the global trace condition (an *untraceable* branch). For a search tree $\mathcal{D}$, we say that a pair $(\Gamma, \Delta)$ is a *template induced by $\mathcal{D}$* when either: (i) $\Gamma \Rightarrow \Delta$ is an open node of $\mathcal{D}$; or (ii) $\Gamma = \bigcup_{i>0} \Gamma_i$ and $\Delta = \bigcup_{i>0} \Delta_i$, where

$\Gamma_1 \Rightarrow \Delta_1, \Gamma_2 \Rightarrow \Delta_2, \ldots$ is an untraceable branch in $\mathcal{D}$. Notice that, due to the construction of search trees, the component sets of a template are necessarily disjoint. Given a template, we construct a PDL model as follows.

**Definition 11 (Countermodel Construction).** *Let $P = (\Gamma, \Delta)$ be a template induced by a search tree. The PDL model determined by the template $P$ is given by $\mathfrak{m}_P = (\mathcal{L}, \mathcal{I}_P)$, where $\mathcal{I}_P$ is the following interpretation function:*

1. *$\mathcal{I}_P(p) = \{x \mid x : p \in \Gamma\}$ for each atomic proposition $p$; and*
2. *$\mathcal{I}_P(a) = \{(x, y) \mid x\ R_a\ y \in \Gamma\}$ for each atomic program $a$.*

*We write $\mathfrak{v}$ for the valuation defined by $\mathfrak{v}(x) = x$ for each label $x$.*

PDL models determined by templates have the following property.

**Lemma 2.** *Let $P = (\Gamma, \Delta)$ be a template induced by a search tree. Then we have $\mathfrak{m}_P, \mathfrak{v} \models A$ for all $A \in \Gamma$ and $\mathfrak{m}_P, \mathfrak{v} \not\models B$ for all $B \in \Delta$.*

Lemma 2 entails the cut-free completeness of $\mathbf{G3PDL}^{\infty}$.

**Theorem 2 (Completeness of $\mathbf{G3PDL}^{\infty}$).** *If $\Gamma \Rightarrow \Delta$ is valid, then it has a cut-free $\mathbf{G3PDL}^{\infty}$ proof.*

### 4.2 Completeness of $\mathbf{G3PDL}^{\omega}$ for PDL

We show that the cyclic system $\mathbf{G3PDL}^{\omega}$ can derive all theorems of PDL by demonstrating that it can derive each of the axiom schemas and inference rules in Fig. 4, which (along with the axiom schemas of classical propositional logic) constitute a complete axiomatisation of PDL [23, §7.1].

$$[\alpha](\varphi \to \psi) \to ([\alpha]\varphi \to [\alpha]\psi) \quad (1) \qquad [\alpha](\varphi \wedge \psi) \to ([\alpha]\varphi \wedge [\alpha]\psi) \quad (2)$$

$$[\alpha \cup \beta]\varphi \leftrightarrow [\alpha]\varphi \wedge [\beta]\varphi \quad (3) \qquad [\alpha\,;\beta]\varphi \leftrightarrow [\alpha][\beta]\varphi \quad (4)$$

$$[\psi?]\varphi \leftrightarrow (\psi \to \varphi) \quad (5) \qquad \varphi \wedge [\alpha][\alpha^*]\varphi \leftrightarrow [\alpha^*]\varphi \quad (6)$$

$$\varphi \wedge [\alpha^*](\varphi \to [\alpha]\varphi) \to [\alpha^*]\varphi \quad (7)$$

$$\frac{\varphi \quad \varphi \to \psi}{\psi}\ (\mathsf{MP}) \qquad \frac{\varphi}{[\alpha]\varphi}\ (\mathsf{Nec})$$

**Fig. 4.** Axiomatisation of PDL.

The derivation of the axioms of classical propositional logic is standard, and axioms (3) to (6) are immediately derivable via the left and right proof rules for their corresponding syntactic constructors. Each such derivation is finite, and thus trivially a $\mathbf{G3PDL}^{\omega}$ proof. Axioms (1), (2), (7) and (NEC) require the following lemma showing that a general form of necessitation is derivable.

**Lemma 3 (Necessitation).** *For any labelled formula $x : \varphi$, program $\alpha$, and finite set $\Gamma$ of labelled formulas such that* $\mathsf{labs}(\Gamma) = \{x\}$, *there exists a* **G3PDL**$^\omega$ *derivation concluding with the sequent $[\alpha]\Gamma \Rightarrow x : [\alpha]\varphi$ and containing open leaves of the form $\Gamma \Rightarrow x : \varphi$ such that:*

*(i) for each trace value $\tau = x : \varphi$, every path from the conclusion to an open leaf is covered by a trace $[\alpha]\tau, \ldots, \tau$; and*
*(ii) every infinite path is followed by an infinitely progressing trace.*

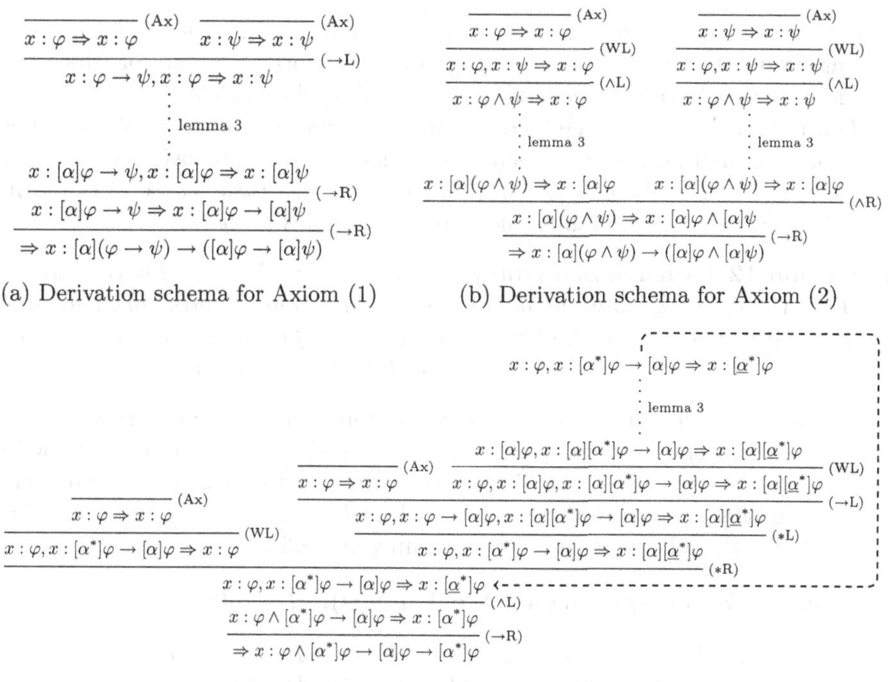


**Fig. 5.** **G3PDL**$^\omega$ derivation schemata for the distribution and induction axioms.

Schemas for deriving Axioms (1), (2) and (7) are shown in Fig. 5. Any infinite paths which exist in the schemas for deriving axioms (1) and (2) are followed by infinitely progressing traces by Lemma 3. Thus, they are **G3PDL**$^\omega$ proofs. In the schema for axiom (7), the open leaves of the subderivation constructed via Lemma 3 are converted into buds, the companion of each of which is the conclusion of the instance of the (∗R) rule. Condition (i) of Lemma 3 guarantees that each infinite path along these cycles has an infinitely progressing trace. We thus have the following completeness result.

**Theorem 3.** *If $\varphi$ is valid then $\Rightarrow x : \varphi$ is derivable in* **G3PDL**$^\omega$.

It should be noted that Theorem 3 is not a deductive completeness result, i.e. it does not say that any sequent $\Gamma \Rightarrow \Delta$ is only valid if there is a **G3PDL**$^\omega$ proof for it. This is no major restriction, as a finitary syntactic consequence relation cannot capture semantic consequence in PDL: due to the presence of iteration, PDL is not compact. This can only be rectified by allowing infinite sequents in the proof system, which yields a system that is not amenable to automation.

## 5 Proof Search for Test-Free, Acyclic Sequents

In this section, we describe a cut-free proof-search procedure for sequents containing formulas without tests (i.e. programs of the form $\varphi?$), and for which the relational atoms in the antecedents do not entail cyclic models.

Our approach relies on the following notion of normal form for sequents. For a set of relational atoms and labelled formulas, we write $*\text{-}\mathsf{labs}(\Gamma)$ for the set $\{x \mid x : [\alpha^*]\varphi \in \Gamma\}$. We call formulas of the form $[a]\varphi$ *basic*, those of the form $[\alpha^*]\varphi$ *iterated*, and the remaining non-atomic formulas *composite*.

**Definition 12 (Normal Sequents).** *A sequent $\Gamma \Rightarrow \Delta$ is called* normal *when: (1) $\Gamma \cap \Delta = \emptyset$; (2) $\Delta$ contains only labelled atomic and iterated formulas; and (3) $\Gamma$ contains only relational atoms, labelled atomic formulas, and labelled basic formulas $x : [a]\varphi$ for which there is no $y$ such that also $x\ R_a\ y \in \Gamma$.*

We say that $x$ *reaches* $y$ (or $y$ is *reachable* from $x$) in $\Gamma$ when there are labels $z_1, \ldots, z_n$ and atomic programs $a_1, \ldots, a_{n-1}$ such that $x = z_1$ and $y = z_n$ with $z_i\ R_{a_i}\ z_{n+1} \in \Gamma$ for each $i < n$. We say that a sequent $\Gamma \Rightarrow \Delta$ is *cyclic* if there is some $x \in \mathsf{labs}(\Gamma)$ such that $x$ reaches itself in $\Gamma$; otherwise it is called *acyclic*.

Crucially, the following forms of weakening are validity-preserving.

**Lemma 4 (Validity-preserving Weakenings).** *The following hold.*

*(1) If $\Gamma \Rightarrow \Delta, x\ R_a\ z$ is valid and $x\ R_a\ z \notin \Gamma$, then $\Gamma \Rightarrow \Delta$ is valid.*
*(2) If normal $\Gamma \Rightarrow \Delta, x : p$ is valid with $x \notin *\text{-}\mathsf{labs}(\Delta)$, then $\Gamma \Rightarrow \Delta$ is valid.*
*(3) If normal $\Gamma, x : \varphi \Rightarrow \Delta$ is valid with $x \notin \mathsf{labs}(\Delta)$, then $\Gamma \Rightarrow \Delta$ is valid.*
*(4) If normal $\Gamma, x\ R_a\ y \Rightarrow \Delta$ is valid, $z \in \mathsf{labs}(\Delta)$ for all $z : \varphi \in \Gamma$, $x \notin \mathsf{labs}(\Delta)$ and $x$ not reachable in $\Gamma$ from any $z \in \mathsf{labs}(\Delta)$, then $\Gamma \Rightarrow \Delta$ is valid.*

An *unwinding* of a sequent $\Gamma \Rightarrow \Delta$ is a possibly open derivation of $\Gamma \Rightarrow \Delta$ obtained by applying left and right logical rules as much as possible, and satisfying the properties that: no trace progresses more than once; and all rule instances consume the active labelled formula of their conclusion, but preserve in the premise any relational atoms. A *capped* unwinding is an unwinding for which: (a) weakening rules and (Ax) and ($\bot$) have been applied to all open leaves $\Gamma \Rightarrow \Delta$ with $\bot \in \Gamma$ or $\Gamma \cap \Delta \neq \emptyset$; and (b) the sequence of weakenings in Lemma 4 have been exhaustively applied to all other open leaves.

**Lemma 5.** *Let $\mathcal{D}$ be a capped unwinding for $\Gamma \Rightarrow \Delta$ (denoted $S$) and $\Gamma' \Rightarrow \Delta'$ an open leaf (denoted $S'$) of $\mathcal{D}$. The following hold: (1) $\Gamma' \Rightarrow \Delta'$ is normal; (2) if $\Gamma \Rightarrow \Delta$ is valid, then so are all the open leaves of $\mathcal{D}$; and (3) For every trace $\tau_n$ covering the path from $S$ to $S'$, if $\tau_1 = (x, \varepsilon, \beta, \varphi)$ is a sub-formula of $\tau_n$, then the trace is progressing.*

We call a sequent *test-free* if it does not contain any programs of the form $\varphi?$. A crucial property for termination of the proof-search is the following.

**Lemma 6.** *Let $\mathcal{D}$ be a capped unwinding for a test-free, acyclic sequent; then $\mathcal{D}$ is finite, and $\mathsf{labs}(\Gamma') \subseteq \mathsf{labs}(\Delta') \subseteq *\text{-}\mathsf{labs}(\Delta')$ for all open leaves $\Gamma' \Rightarrow \Delta'$ of $\mathcal{D}$.*

Both cyclicity and the presence of tests can cause Lemma 6 to fail, since then it is possible for there to be a path of ancestry between two occurrences of an antecedent formula $x : [\alpha^*]\varphi$ that traverses an instance of the $(*\mathrm{L})$ rule. That is, antecedent formulas may be infinitely unfolded. Moreover, in the presence of tests or cyclicity, the weakenings of Lemma 4(4) do not result in $\mathsf{labs}(\Gamma') \subseteq \mathsf{labs}(\Delta')$ for open leaves $\Gamma' \Rightarrow \Delta'$.

We define a function $*\text{-}\mathsf{max}$ on test-free sequents (details are given in the appendix), whose purpose is to provide a bound ensuring termination of proof-search. We have conjectured that it satisfies the following property.

*Conjecture 1.* Let $\mathcal{D}$ be a capped unwinding of test-free, acyclic $\Gamma \Rightarrow \Delta$. Then:

1. $|\{x : \varphi \in \Delta' \mid \varphi \text{ non-atomic}\}| \leq *\text{-}\mathsf{max}(\Gamma \Rightarrow \Delta)$.
2. $*\text{-}\mathsf{max}(\Gamma' \Rightarrow \Delta') \leq *\text{-}\mathsf{max}(\Gamma \Rightarrow \Delta)$ for all open leaves $\Gamma' \Rightarrow \Delta'$ of $\mathcal{D}$.

Proof-search proceeds by iteratively building capped unwindings for open leaves. All formulas encountered in the search are in the (finite) Fischer-Ladner closure of the initial sequent, and validity and acyclicity are preserved throughout the procedure. Lemma 6 and Conjecture 1 will ensure that the number of distinct open leaves (modulo relabelling) encountered during proof-search is bounded, so we may apply substitutions to form back-links during proof-search. Lemma 5(3) ensures that the resulting pre-proof satisfies the global trace condition. For invalid sequents, proof-search produces atomic sequents that are not axioms. We thus conjecture cut-free regular completeness for test-free PDL.

## 6 Conclusion

In this paper we have given two new non-wellfounded proof systems for PDL. $\mathbf{G3PDL}^{\infty}$ allows proof trees to be infinitely tall, and $\mathbf{G3PDL}^{\omega}$ restricts to the proofs of $\mathbf{G3PDL}^{\infty}$ that are finitely representable as cyclic graphs satisfying a trace condition. Soundness and completeness of both systems was shown, in particular, cut-free completeness of $\mathbf{G3PDL}^{\infty}$ and a strategy for cut-free completeness of $\mathbf{G3PDL}^{\omega}$ for test-free PDL.

There is much further work to be done. Of immediate interest is the verification of cut-free regular completeness for test-free PDL, and the extension of the argument to the full logic. We would also like to consider *additional* program constructs. Some, like converse, can already be treated through De Giacomo's [17] efficient translation of Converse PDL into PDL. It may be more desirable, however, to represent the program construct directly, to aid in the modular combination of different constructs. One construct that is particularly notorious is Intersection. Despite the modal definability of its dual, Choice, the intended interpretation of Intersection is *not* modally definable, and the completeness (and existence) of an axiomatisation for it remained open until Balbiani and Vakarelov [3]. An earlier, and significantly simpler, solution to this problem was the augmentation of PDL with nominals, denoted Combinatory DL [32]. We conjecture that the presence of labels in our system enables us to perform a similar trick, without contaminating the syntax of the logic itself. However we should note that a key prerequisite of our soundness proof, namely that we can restrict attention to finitely branching models (guaranteed by the finite model property of PDL), is an assumption that may no longer hold for particular combinations of program constructs. Weakening this assumption will aid in the goal of giving a truly uniform proof theory for PDL-type logics.

Our work should be seen as a part of a wider program of research to give a uniform and modular proof theory for a larger group of modal logics, including what we have denoted PDL-type logics. One source of modularity and uniformity is the existing Negri labelled system our calculi extend. This allows us to freely add proof rules corresponding to first-order frame axioms defining Kripke models. A wider class of modal logics than those directly covered by Negri's framework are those with accessibility relations that are defined to be wellfounded or arise as transitive closures of other accessibility relations (we note Negri is able to treat the specific case of Gödel-Löb logic due to its special interpretation of $\Box$, but not the general class we describe). We believe an appropriate framework to uniformly capture *these* logics as well is cyclic labelled deduction. We are encouraged in this pursuit by recent work of Cohen and Rowe [8] in which first-order logic with a transitive closure operator is given a cyclic proof theory. We may think of labelled deduction as a way of giving a proof theoretic analysis of the first-order theory of Kripke models and their modal satisfaction relations. Labelled cyclic deduction, we conjecture, can capture the first-order-with-least-fixpoint theory of Kripke models and modal satisfaction relations.

Finally, and somewhat more speculatively, with the cyclic system in hand we intend to investigate the hitherto open problem of interpolation for PDL. This has seen no satisfactory resolution in the years since PDL was first formulated, with the only attempted proofs strongly disputed [27] or withdrawn [26]. It would be interesting to see if the existence of a straightforward proof system for the logic opens up any new lines of attack on the problem. For example, Lyndon interpolation has been proved for Gödel-Löb logic using a cyclic system [37].

## References


1. Afanasiev, L., et al.: PDL for ordered trees. J. Appl. Non-Classical Logics **15**(2), 115–135 (2005)
2. Alberucci, L.: The Modal $\mu$-calculus and the Logic of Common Knowledge. Ph.D. thesis, Universität Bern (2002)
3. Balbiani, P., Vakarelov, D.: PDL with intersection of programs: a complete axiomatization. J. Appl. Non-Classical Logics **13**(3–4), 231–276 (2003)
4. Van Benthem, J.: Modal frame correspondences and fixed-points. Studia Logica **83**(1), 133–155 (2006)
5. Brotherston, J., Gorogiannis, N., Petersen, R.L.: A generic cyclic theorem prover. In: Jhala, R., Igarashi, A. (eds.) APLAS 2012. LNCS, vol. 7705, pp. 350–367. Springer, Heidelberg (2012). https://doi.org/10.1007/978-3-642-35182-2_25
6. Brotherston, J., Simpson, A.: Sequent calculi for induction and infinite descent. J. Log. Comput. **21**(6), 1177–1216 (2011)
7. Carreiro, F., de Venema, Y.: PDL inside the $\mu$-calculus: a syntactic and an automata-theoretic characterization. In: Goré, R., Kooi, B., Kurucz, A. (eds.) Advances in Modal Logic, vol. 10, pp. 74–93. CSLI Publications, San Diego (2014)
8. Cohen, L., Rowe, R.N.S.: Uniform inductive reasoning in transitive closure logic via infinite descent. In: 27th EACSL Annual Conference on Computer Science Logic, CSL 2018, Birmingham, UK, 4–7 September 2018, pp. 17:1–17:16 (2018)
9. Das, A., Pous, D.: A cut-free cyclic proof system for Kleene algebra. In: Automated Reasoning with Analytic Tableaux and Related Methods - 26th International Conference, Proceedings, TABLEAUX 2017, Brasília, Brazil, 25–28 September 2017, pp. 261–277 (2017)
10. Das, A., Pous, D.: Non-wellfounded proof theory for (Kleene+action)(algebras+lattices). In: 27th EACSL Annual Conference on Computer Science Logic, CSL 2018, Birmingham, UK, 4–7 September 2018, pp. 19:1–19:18 (2018)
11. De Giacomo, G., Lenzerini, M.: Boosting the correspondence between description logics and propositional dynamic logics. In: Proceedings of the Twelfth National Conference on Artificial Intelligence, AAAI 1994, vol. 1, pp. 205–212. American Association for Artificial Intelligence, Menlo Park (1994)
12. Dershowitz, N., Manna, Z.: Proving termination with multiset orderings. Commun. ACM **22**(8), 465–476 (1979)
13. Docherty, S., Rowe, R.N.S.: A non-wellfounded, labelled proof system for propositional dynamic logic. CoRR, abs/1905.06143 (2019)
14. Fischer, M.J., Ladner, R.E.: Propositional dynamic logic of regular programs. J. Comput. Syst. Sci. **18**(2), 194–211 (1979)
15. Fitting, M.: Proof Methods for Modal and Intuitionistic Logics. Synthese Library. Springer, Dordrecht (1983). https://doi.org/10.1007/978-94-017-2794-5
16. Frittella, S., Greco, G., Kurz, A., Palmigiano, A.: Multi-type display calculus for propositional dynamic logic. J. Log. Comput. **26**(6), 2067–2104 (2014)
17. De Giacomo, G.: Eliminating "converse" from converse PDL. J. Log. Lang. Inf. **5**(2), 193–208 (1996)
18. De Giacomo, G., Massacci, F.: Combining deduction and model checking into tableaux and algorithms for converse-PDL. Inf. Comput. **162**(1), 117–137 (2000)
19. Girard, P., Seligman, J., Liu, F.: General dynamic dynamic logic. In: Bolander, T., Braüner, T., Ghilardi, S., Moss, L. (eds.) Advances in Modal Logic, vol. 9, pp. 239–260. CSLI Publications, Stanford (2012)

20. Goré, R., Widmann, F.: An optimal on-the-fly tableau-based decision procedure for PDL-satisfiability. In: Schmidt, R.A. (ed.) CADE 2009. LNCS (LNAI), vol. 5663, pp. 437–452. Springer, Heidelberg (2009). https://doi.org/10.1007/978-3-642-02959-2_32
21. Groenendijk, J., Stokhof, M.: Dynamic predicate logic. Linguist. Philos. **14**(1), 39–100 (1991)
22. Harel, D.: Dynamic logic. In: Gabbay, D., Guenthner, F. (eds.) Handbook of Philosophical Logic: Volume II: Extensions of Classical Logic, vol. 165, pp. 497–604. Springer, Dordrecht (1984). https://doi.org/10.1007/978-94-009-6259-0_10
23. Harel, D., Tiuryn, J., Kozen, D.: Dynamic Logic. MIT Press, Cambridge (2000)
24. Hill, B., Poggiolesi, F.: A contraction-free and cut-free sequent calculus for propositional dynamic logic. Stud. Log. Int. J. Symb. Log. **94**(1), 47–72 (2010)
25. Kanger, S.: Provability in Logic. Almqvist & Wiksell, Stockholm (1957)
26. Kowalski, T.: Retraction note for "PDL has interpolation". J. Symb. Log. **69**(3), 935–936 (2004)
27. Kracht, M.: Dynamic logic. In: Tools and Techniques in Modal Logic, volume 142 of Studies in Logic and the Foundations of Mathematics, pp. 497–533. Elsevier (1999)
28. Negri, S.: Proof analysis in modal logic. J. Philos. Log. **34**(5), 507 (2005)
29. Negri, S.: Proof theory for modal logic. Philos. Compass **6**(8), 523–538 (2011)
30. Niwiński, D., Walukiewicz, I.: Games for the $\mu$-calculus. Theor. Comput. Sci. **163**(1), 99–116 (1996)
31. Parikh, R.: Propositional game logic. In: Proceedings of the 24th Annual Symposium on Foundations of Computer Science, SFCS 1983, pp. 195–200. IEEE Computer Society, Washington, DC (1983)
32. Passy, S., Tinchev, T.: An essay in combinatory dynamic logic. Inf. Comput. **93**(2), 263–332 (1991)
33. Platzer, A.: Differential dynamic logic for hybrid systems. J. Autom. Reason. **41**(2), 143–189 (2008)
34. Pratt, V.R.: Semantical consideration on Floyd-Hoare logic. In: 17th Annual Symposium on Foundations of Computer Science (SFCS 1976), pp. 109–121, October 1976
35. de Lavalette, G.R., Kooi, B., Verbrugge, R.: Strong completeness and limited canonicity for PDL. J. Log. Lang. Inf. **17**(1), 69–87 (2008)
36. Rowe, R.N.S., Brotherston, J.: Automatic cyclic termination proofs for recursive procedures in separation logic. In: Proceedings of the 6th ACM SIGPLAN Conference on Certified Programs and Proofs, CPP 2017, Paris, France, 16–17 January 2017, pp. 53–65 (2017)
37. Shamkanov, D.S.: Circular proofs for the Gödel-Löb provability logic. Math. Notes **96**(3), 575–585 (2014)
38. Tellez, G., Brotherston, J.: Automatically verifying temporal properties of pointer programs with cyclic proof. In: Automated Deduction - CADE 26–26th International Conference on Automated Deduction, Proceedings, Gothenburg, Sweden, 6–11 August 2017, pp. 491–508 (2017)
39. Visser, A.: Löb's Logic Meets the $\mu$-Calculus, pp. 14–25. Springer, Heidelberg (2005)
40. Wehbe, R.: Annotated Systems for Common Knowledge. Ph.D. thesis, Universität Bern (2010)

# Automated Theorem Provers

# Herbrand Constructivization for Automated Intuitionistic Theorem Proving

Gabriel Ebner(✉)

TU Wien, Vienna, Austria
gebner@gebner.org

**Abstract.** We describe a new method to constructivize proofs based on Herbrand disjunctions by giving a practically effective algorithm that converts (some) classical first-order proofs into intuitionistic proofs. Together with an automated classical first-order theorem prover such a method yields an (incomplete) automated theorem prover for intuitionistic logic. Our implementation of this prover approach, Slakje, performs competitively on the ILTP benchmark suite for intuitionistic provers: it solves 1674 out of 2670 problems (1290 proofs and 384 claims of non-provability) with Vampire as a backend, including 800 previously unsolved problems.

## 1 Introduction

Intuitionistic logic is a logic of particular practical importance. Many interactive theorem provers use intuitionistic logic as a foundation, like Coq [3], Agda [6], or Lean [26]. In some foundational frameworks the law of excluded middle is even provably false, such as in homotopy type theory[1] [36]. Automating first-order intuitionistic logic thus has immediate practical applications in these systems.

That intuitionistic proofs are often similar to classical proofs of the same formula is a folklore observation, stated e.g. by Otten [29]. Hence it is reasonable to approach automated theorem proving in intuitionistic logic by adapting proofs from classical theorem provers. This general idea of proof constructivization has recently been described and evaluated by Cauderlier [8] and Gilbert [16]; both transform detailed proofs (natural deduction resp. sequent calculus) using essentially local rewriting operations. However their constructivization procedures are hard to apply to state-of-the-art automated theorem provers as these provers typically do not produce sequent calculus or natural deduction proofs.

Integrations of (classical) first-order theorem provers in higher-order theorem provers—so-called "hammers"—typically use a similar general approach: passing a (sometimes even unsound) translation of the input problem to a classical prover, and then reconstructing the proof in the higher-order system [5,10,18].

[1] Considering of course the law of excluded middle for arbitrary types, not just mere propositions.


S. Cerrito and A. Popescu (Eds.): TABLEAUX 2019, LNAI 11714, pp. 355–373, 2019.
https://doi.org/10.1007/978-3-030-29026-9_20

In this framework, proof constructivization also uses an unsound translation: one that maps each formula to itself (but reinterprets it in a different logic).

We present a new and different proof constructivization method based on Herbrand disjunctions. Herbrand's theorem [7,17] captures the insight that the classical validity of a quantified formula is characterized by the existence of a tautological finite set of quantifier-free instances. In its simplest case, the validity of a purely existential formula $\exists x\, \varphi(x)$ is characterized by the existence of a tautological disjunction of instances $\varphi(t_1) \vee \cdots \vee \varphi(t_n)$, a Herbrand disjunction. We say that $t_i$ is a quantifier instance term for $\exists x\, \varphi(x)$. To store such Herbrand disjunctions for general non-prenex formulas, we use an elegant data structure called expansion trees, which also generalize this result to simply-typed higher-order logic in the form of elementary type theory [23].

- We describe a new and effective procedure to constructivize classical proofs into intuitionistic proofs based on Herbrand disjunctions.
- We have implemented the intuitionistic first-order theorem prover Slakje based on this procedure using the GAPT [15] system for proof theory, and show that it performs competitively on the ILTP benchmark suite.
- We show that the prover is complete on a practically relevant class of formulas.

We start out in Sect. 2 by giving an overview of the Slakje prover. In the following sections we explain the technical details. Expansion trees, the central data structure to represent classical proofs, are introduced in Sect. 3. In Sect. 4 we describe the SAT-based procedure that constructivizes expansion proofs and produces intuitionistic proofs. Key optimizations are discussed in Sect. 5, and completeness for a large class of problems including Horn problems and purely equational problems is shown in Sect. 6. Finally, we evaluate the prover on the ILTP benchmark suite in Sect. 7.

## 2 Overview of the Prover

We consider intuitionistic first-order logic with the connectives $\rightarrow, \wedge, \vee, \bot, \top$ and the quantifiers $\exists, \forall$. The abbreviation $\neg\varphi$ stands for $\varphi \rightarrow \bot$. Let us first give a short overview of the resulting intuitionistic first-order prover. Given an input formula $\varphi$, it proceeds in three big phases:

1. Call classical prover (e.g. Vampire [19]) with $\varphi$[2]
(if the result is "satisfiable", immediately return "non-theorem")
2. Convert proof output into (classical) expansion proof
3. Produce intuitionistic proof from expansion proof

The only phase that is specific to this prover is the third one; in our implementation, phases 1 and 2 are part of the general-purpose external prover interface that produces expansion proofs available in the GAPT [13,15] framework. We use expansion proofs as a compact intermediate format for classical proofs, which

[2] Internally, Vampire, and in general most classical provers then refute $\neg\varphi$.

only contain the quantifier instance terms but not the propositional reasoning in the proof. Many automated theorem proving paradigms generate proofs that directly contain the same essential data as expansion proofs, e.g. the terms used for heuristic instantiation in SMT solvers, or the global substitution used in tableaux or connection proofs. Resolution and superposition proofs also contain this information after grounding. This direct correspondence applies for formulas in clause normal form; in the general case we also need to treat strong quantifiers, which are Skolemized in classical provers.

While there are normal forms similar to CNF in intuitionistic logic which avoid Skolemization [24,25], it makes little sense to use them here: the main difference is that they produce a different kind of "clauses", such as $(\forall x\, P(x,y)) \to Q(y)$, which we cannot pass to classical theorem provers. But the use of Skolemization as a preprocessing step is (in general) not sound in intuitionistic logic: for example $(\neg\forall x\; P(x)) \to \exists x\; \neg P(x)$ is an intuitionistic non-theorem, while its Skolemization $(\neg P(c)) \to \exists x\; \neg P(x)$ is a theorem. Hence we use *deskolemization* [1] to eliminate Skolemization from classical proofs (which is a natural operation on expansion proofs).

The way we construct an intuitionistic proof from the expansion proof is by a bottom-up proof construction in an intuitionistic multi-succedent sequent calculus. We make use of a SAT solver to organize this proof construction. While SAT solvers—as the name implies—can decide satisfiability of a propositional formula $\varphi$ in classical logic, only the part where we need to prove $\varphi \to \bigwedge C$ for the CNF $\bigwedge C$ requires classical logic. If the CNF $\bigwedge C$ is unsatisfiable, then $\bigwedge C$ is already provable in intuitionistic logic: proofs produced by SAT solvers can be translated to resolution proofs; and resolution is just the cut inference, which is sound for intuitionistic logic. (See also Theorem 4 for a different explanation.) In our setting, a SAT solver hence decides the following question: "is the sequent $\Gamma \vdash \Delta$ derivable from a set of sequents $\mathcal{T}$ using only cut and weakening?"

If the SAT solver cannot derive this sequent, we obtain an assignment which corresponds to a leaf in this bottom-up proof construction and we apply the inferences that cannot be encoded as clauses (e.g. the right-rule for implication). This technique of using SAT solvers to support intuitionistic reasoning has already been successfully used in the Intuit [9] prover, albeit only for propositional logic, and their implementation does not produce proofs.

## 3 Expansion Proofs

The proof formalism of expansion trees was introduced in [23] to describe Herbrand disjunctions in classical higher-order logic. In first-order logic, they provide an elegant data structure to describe Herbrand disjunctions for non-prenex formulas, storing the quantifier instance terms. The central idea is that each expansion tree $E$ comes with a *shallow formula* $\mathrm{sh}(E)$ and a quantifier-free *deep formula* $\mathrm{dp}(E)$. The deep formula corresponds to the quantifier-free Herbrand disjunction, and the shallow formula is the quantified formula that we want to prove. If the deep formula is a quasi-tautology (a tautology modulo equality), then the shallow formula is valid in classical logic.

Expansion trees have two polarities, $-$ and $+$. We write $-p$ for the inverse polarity of $p$, i.e. $-- = +$ and $-+ = -$. Polarity only changes on the left side of the connective $\rightarrow$. This distinction is important since we must instantiate positive occurrences of $\forall$ (resp. negative occurrences of $\exists$, called "strong quantifiers") with an eigenvariable, while we can instantiate the negative ones with whatever terms we want ("weak quantifiers"). An atom is a predicate such as $P(x, y)$ or an equality; the formulas $\top, \bot$ are not atoms.

**Definition 1.** *The set* $\mathrm{ET}^p(\varphi)$ *of* expansion trees *with polarity* $p \in \{+, -\}$ *and* shallow formula $\varphi$ *is inductively defined as the smallest set containing:*

$$\frac{A \textit{ atom}/\top/\bot}{A^p \in \mathrm{ET}^p(A)} \qquad \frac{E_1 \in \mathrm{ET}^p(\varphi) \qquad E_2 \in \mathrm{ET}^p(\psi)}{E_1 \wedge E_2 \in \mathrm{ET}^p(\varphi \wedge \psi)}$$

$$\frac{E_1 \in \mathrm{ET}^p(\varphi) \qquad E_2 \in \mathrm{ET}^p(\psi)}{E_1 \vee E_2 \in \mathrm{ET}^p(\varphi \vee \psi)} \qquad \frac{E_1 \in \mathrm{ET}^{-p}(\varphi) \qquad E_2 \in \mathrm{ET}^p(\psi)}{E_1 \rightarrow E_2 \in \mathrm{ET}^p(\varphi \rightarrow \psi)}$$

$$\frac{E \in \mathrm{ET}^+(\varphi[x\backslash\alpha])}{\forall x\, \varphi +^{\alpha}_{\mathrm{ev}} E \in \mathrm{ET}^+(\forall x\, \varphi)} \qquad \frac{E_1 \in \mathrm{ET}^-(\varphi[x\backslash t_1]) \quad \cdots \quad E_n \in \mathrm{ET}^-(\varphi[x\backslash t_n])}{\forall x\, \varphi +^{t_1} E_1 \cdots +^{t_n} E_n \ \in \ \mathrm{ET}^-(\forall x\, \varphi)}$$

$$\frac{E \in \mathrm{ET}^-(\varphi[x\backslash\alpha])}{\exists x\, \varphi +^{\alpha}_{\mathrm{ev}} E \in \mathrm{ET}^-(\exists x\, \varphi)} \qquad \frac{E_1 \in \mathrm{ET}^+(\varphi[x\backslash t_1]) \quad \cdots \quad E_n \in \mathrm{ET}^+(\varphi[x\backslash t_n])}{\exists x\, \varphi +^{t_1} E_1 \cdots +^{t_n} E_n \ \in \ \mathrm{ET}^+(\exists x\, \varphi)}$$

Each expansion tree $E$ has a uniquely determined shallow formula and polarity, we write $\mathrm{sh}(E)$ for its shallow formula, and $\mathrm{pol}(E)$ for its polarity. Given an expansion tree $E = Qx\, \varphi +^{\alpha}_{\mathrm{ev}} E'$ where $Q \in \{\forall, \exists\}$, we say that $\alpha$ is the eigenvariable of $E$.

*Example 1.* Consider the formula[3] $\varphi := \forall x\, P(x) \rightarrow (\forall x\, P(f(x)) \rightarrow Q) \rightarrow Q$. The expansion tree $\mathrm{ET}^+(\varphi) \ni E := (\forall x\, P(x) +^{f(\alpha)} P(f(\alpha))) \rightarrow (\forall x\, P(f(x)) +^{\alpha}_{\mathrm{ev}} P(f(\alpha)) \rightarrow Q) \rightarrow Q$ has the shallow formula $\mathrm{sh}(E) = \varphi$, and its deep formula $\mathrm{dp}(E) = (P(f(\alpha)) \rightarrow (P(f(\alpha)) \rightarrow Q) \rightarrow Q)$ is tautological. The quantifier instance terms here are $f(\alpha)$ and $\alpha$ (written in superscript after the $+$). An instructive way to think about expansion proofs is that they are a compressed form of cut-free sequent calculus proofs where we only store the quantifier inferences. The following proof uses exactly the same terms, $f(\alpha)$ and $\alpha$, in the quantifier inferences $\forall_l$ and $\forall_r$, resp.

$$\dfrac{\dfrac{\dfrac{\dfrac{\dfrac{P(f(\alpha)) \vdash P(f(\alpha))}{\forall x\, P(x) \vdash P(f(\alpha))}\,\forall_l}{\forall x\, P(x) \vdash \forall x\, P(f(x))}\,\forall_r \qquad Q \vdash Q}{\forall x\, P(x), \forall x\, P(f(x)) \rightarrow Q \vdash Q}\,\rightarrow_l}{\forall x\, P(x) \vdash (\forall x\, P(f(x)) \rightarrow Q) \rightarrow Q}\,\rightarrow_r}{\vdash \forall x\, P(x) \rightarrow (\forall x\, P(f(x)) \rightarrow Q) \rightarrow Q}\,\rightarrow_r$$

[3] We use the convention that the quantifiers $\forall, \exists$ bind stronger than $\rightarrow, \wedge, \vee$. That is, $\forall x\, P(x) \rightarrow Q$ is the same formula as $(\forall x\, P(x)) \rightarrow Q$. Furthermore, $\rightarrow$ is right-associative, that is, $P \rightarrow Q \rightarrow R$ is the same formula as $P \rightarrow (Q \rightarrow R)$.

**Definition 2.** *Let $E$ be an expansion tree. We define the* deep formula $\mathrm{dp}(E)$ *recursively as follows:*

$$\mathrm{dp}(A^p) = A, \quad \mathrm{dp}(\top^p) = \top, \quad \mathrm{dp}(\bot^p) = \bot, \quad \mathrm{dp}(E_1 \wedge E_2) = \mathrm{dp}(E_1) \wedge \mathrm{dp}(E_2)$$

$$\mathrm{dp}(E_1 \vee E_2) = \mathrm{dp}(E_1) \vee \mathrm{dp}(E_2), \quad \mathrm{dp}(E_1 \rightarrow E_2) = \mathrm{dp}(E_1) \rightarrow \mathrm{dp}(E_2)$$

$$\mathrm{dp}(\forall x\, \varphi +^y_{\mathrm{ev}} E) = \mathrm{dp}(E), \quad \mathrm{dp}(\forall x\, \varphi +^{t_1} E_1 \cdots +^{t_n} E_n) = \mathrm{dp}(E_1) \wedge \cdots \wedge \mathrm{dp}(E_n)$$

$$\mathrm{dp}(\exists x\, \varphi +^y_{\mathrm{ev}} E) = \mathrm{dp}(E), \quad \mathrm{dp}(\exists x\, \varphi +^{t_1} E_1 \cdots +^{t_n} E_n) = \mathrm{dp}(E_1) \vee \cdots \vee \mathrm{dp}(E_n)$$

The deep formula corresponds to the Herbrand disjunction. In an expansion proof, the eigenvariables need to be acyclic. This restriction is similar to the eigenvariable condition in sequent calculi and the acceptability condition for substitutions in matrices [4]. Let $\mathrm{FV}(\varphi)$ be the set of free variables of a formula $\varphi$.

**Definition 3.** *Let $E$ be an expansion tree. The dependency relation $<_E$ is a binary relation on eigenvariables where $\alpha <_E \beta$ iff $E$ contains a subtree $E'$ such that $\alpha \in FV(\mathrm{sh}(E'))$ and $\beta$ is an eigenvariable of a subtree of $E'$.*

**Definition 4.** *An* expansion proof[4] *$E$ of $\varphi$ is an $E \in \mathrm{ET}^+(\varphi)$ such that:*

1. *$<_E$ is acyclic (i.e., can be extended to a linear order) and there are no duplicate eigenvariables, and*
2. *$\mathrm{dp}(E)$ is a quasi-tautology*

**Theorem 1 ([23, Theorems 4.1 and 4.2]).** *A formula $\varphi$ is a theorem of classical first-order logic if and only if there exists an expansion proof $E \in \mathrm{ET}^+(\varphi)$.*

*Example 2.* The formula $\exists x\, (p(c) \vee p(d) \rightarrow p(c))$ has $E_1 = \exists x\, (p(c) \vee p(d) \rightarrow p(c)) +^c (p(c)^- \vee p(d)^- \rightarrow p(c)^+)$ as an expansion proof. The deep formula $\mathrm{dp}(E_1) = p(c) \vee p(d) \rightarrow p(c)$ is a tautology. This is not the only possible expansion proof of this formula: we could also use the instance $d$ instead of $c$.

Expansion proofs are closely related to the matrix characterization for classical first-order logic [4] used by connection-based theorem provers [30]. Both separate the proof into two layers: the quantifier inference terms, and the propositional proof. In connection proofs, the quantifier instance terms are stored implicitly as the result of the unifier induced by the connections, while expansion proofs contain these terms explicitly. In the classical setting, the multiplicity in a connection proof corresponds essentially to the number of children in the weak quantifier nodes of an expansion tree. (In the intuitionistic setting, the multiplicity also constrains the amount of contraction in a corresponding sequent calculus proof, i.e., how often subformulas can be used. There is no such constraint in our expansion-proof based method.) Integrating equality into connection proofs is hard as it requires simultaneous rigid E-unification [38], and

[4] Proof systems (for propositional logic) are typically required to be polynomial-time checkable, as certificates to the coNP-complete validity problem. Expansion proofs are coNP-checkable certificates for the undecidable first-order validity problem.

connection provers such as leanCoP hence perform equational reasoning during proof search by adding axioms for reflexivity, transitivity, and congruence of equality during preprocessing.

By contrast, expansion proofs work modulo equality. We do not need to add explicit axioms for equality. Instead, the handling of equality is part of verifying that the deep formula is a quasi-tautology, and can be done using off-the-shelf SMT solvers (which are typically the fastest tools to decide validity of quantifier-free formulas). In principle, this could also be extended to other decidable (and for our purposes, intuitionistically valid) theories used in SMT solvers such as Presburger arithmetic.

## 4 Proof Constructivization

Our proof constructivization method operates on the level of expansion proofs. That is, it takes an expansion proof and (if successful) produces an intuitionistic proof using (at most) the quantifier inferences indicated by the expansion proof. The expansion proof only restricts the quantifier instance *terms* in the proof; not how often a subformula is used (i.e. contraction). We want to find a proof in the multi-succedent intuitionistic sequent calculus mLJ as shown in Fig. 1, where all eigenvariables and quantifier instances occur in the expansion proof and there are no duplicate eigenvariables along any branch of the proof.

**Definition 5.** *An mLJ-proof $\pi$* realizes *an expansion proof $E$ iff every quantifier instance term in $\pi$ is contained in $E$, i.e.: (and analogously for $\exists$)*

- *If* $\dfrac{\Gamma \vdash \varphi(\alpha)}{\Gamma \vdash \forall x\, \varphi(x)}$ *is a subproof of $\pi$, then $\forall x\, \varphi(x) +^{\alpha}_{\mathrm{ev}} E'$ is a subtree of $E$ (for some $E'$)*
- *If* $\dfrac{\varphi(t), \Gamma \vdash \Delta}{\forall x\, \varphi(x), \Gamma \vdash \Delta}$ *is a subproof of $\pi$, then $\forall x\, \varphi(x) +^{t} E' \cdots$ is a subtree of $E$ (for some $E'$)*

Note that Definition 5 ignores the ancestor relationship of formulas in a proof: if two subtrees $E_1, E_2$ of $E$ have the same shallow formula, then their instances can be be used interchangeably in $\pi$.

**Definition 6.** *An mLJ-proof $\pi$ is called* weakly regular *iff for all subproofs of the following form, $\alpha$ does not occur as the eigenvariable of an inference in $\pi'$:*

$$\dfrac{\begin{matrix}(\pi')\\ \Gamma \vdash \varphi(\alpha)\end{matrix}}{\Gamma \vdash \forall x\, \varphi(x)}\,\forall_r \qquad \text{or} \qquad \dfrac{\begin{matrix}(\pi')\\ \varphi(\alpha), \Gamma \vdash \Delta\end{matrix}}{\exists x\, \varphi(x), \Gamma \vdash \Delta}\,\exists_l$$

Our algorithm will have the property that whenever a cut-free weakly regular mLJ-proof of $\mathrm{sh}(E)$ exists which realizes $E$, the algorithm will succeed and return an intuitionistic proof. The restriction of cut-free weak regularity is due to the intuitionistic logic; in classical logic, we can always find a cut-free weakly regular proof realizing $E$, provided that $\mathrm{dp}(E)$ is quasi-tautological.

$$\frac{}{\varphi \vdash \varphi}\,\mathrm{ax} \qquad \frac{\Gamma \vdash \Delta}{\varphi, \Gamma \vdash \Delta}\,w_l \qquad \frac{\Gamma \vdash \Delta}{\Gamma \vdash \Delta, \varphi}\,w_r \qquad \frac{\Gamma \vdash \Delta, \varphi \qquad \varphi, \Pi \vdash \Lambda}{\Gamma, \Pi \vdash \Delta, \Lambda}\,\mathrm{cut}$$

$$\frac{}{\vdash t = t}\,\mathrm{rfl} \qquad \frac{\Gamma \vdash \Delta, \varphi(t)}{\Gamma, t = s \vdash \Delta, \varphi(s)}\,\mathrm{eq}_r^{\rightarrow} \qquad \frac{\Gamma \vdash \Delta, \varphi(s)}{\Gamma, t = s \vdash \Delta, \varphi(t)}\,\mathrm{eq}_r^{\leftarrow}$$

$$\frac{\varphi(t), \Gamma \vdash \Delta}{\varphi(s), \Gamma, t = s \vdash \Delta}\,\mathrm{eq}_l^{\rightarrow} \qquad \frac{\varphi(s), \Gamma \vdash \Delta}{\varphi(t), \Gamma, t = s \vdash \Delta}\,\mathrm{eq}_l^{\leftarrow}$$

$$\frac{}{\vdash \top}\,\top_r \qquad \frac{}{\bot \vdash}\,\bot_l \qquad \frac{\Gamma \vdash \Delta, \varphi, \psi}{\Gamma \vdash \Delta, \varphi \vee \psi}\,\vee_r \qquad \frac{\varphi, \Gamma \vdash \Delta \qquad \psi, \Gamma \vdash \Delta}{\varphi \vee \psi, \Gamma \vdash \Delta}\,\vee_l$$

$$\frac{\varphi, \psi, \Gamma \vdash \Delta}{\varphi \wedge \psi, \Gamma \vdash \Delta}\,\wedge_l \qquad \frac{\Gamma \vdash \Delta, \varphi \qquad \Gamma \vdash \Delta, \psi}{\Gamma \vdash \Delta, \varphi \wedge \psi}\,\wedge_r$$

$$\frac{\Gamma, \varphi \vdash \psi}{\Gamma \vdash \varphi \rightarrow \psi}\,\rightarrow_r \qquad \frac{\Gamma \vdash \Delta, \varphi \qquad \psi, \Gamma \vdash \Delta}{\varphi \rightarrow \psi, \Gamma \vdash \Delta}\,\rightarrow_l$$

$$\frac{\Gamma \vdash \Delta, \varphi(t)}{\Gamma \vdash \Delta, \exists x\, \varphi(x)}\,\exists_r \qquad \frac{\varphi(\alpha), \Gamma \vdash \Delta}{\exists x\, \varphi(x), \Gamma \vdash \Delta}\,\exists_l \qquad \frac{\varphi(t), \Gamma \vdash \Delta}{\forall x\, \varphi(x), \Gamma \vdash \Delta}\,\forall_l \qquad \frac{\Gamma \vdash \varphi(\alpha)}{\Gamma \vdash \forall x\, \varphi(x)}\,\forall_r$$

**Fig. 1.** The multi-succedent calculus mLJ for intuitionistic first-order logic (variant of L'J first introduced by Maehara [20] but using sets instead of sequences, see also mG1i in Troelstra and Schwichtenberg's classification [37]). A sequent $\Gamma \vdash \Delta$ consists of two sets of formulas $\Gamma$ and $\Delta$ and is interpreted as the formula $\bigwedge \Gamma \rightarrow \bigvee \Delta$. The variable $\alpha$ in the $\exists_l$ and $\forall_r$ inferences is called an eigenvariable, and must not occur in $\Gamma, \Delta$ as a free variable. The proof system is cut-free complete for intuitionistic first-order logic with equality. Note that the rules $\rightarrow_r, \forall_r$ do not have any extra formulas $\Delta$ in the succedent: this is the only difference to the classical calculus.

*Example 3.* Consider the expansion proof $\vdash \neg\neg(\forall x\,(p \vee \neg p) +^{\alpha}_{\mathrm{ev}} (p^+ \vee \neg p^-))$. This expansion proof cannot be realized by a weakly regular cut-free mLJ-proof of $\vdash \neg\neg\forall x\,(p \vee \neg p)$, since we would need to use two $\forall_r$ inferences but the expansion proof only contains one eigenvariable. The natural proof would use the eigenvariable $\alpha$ twice. (Note that this example requires that $\alpha$ does not occur in $p$: the formula $\neg\neg\forall x\,(q(x) \vee \neg q(x))$ is not an intuitionistic theorem.)

The SAT-based bottom-up proof construction is done in the Construct and Solve procedures shown in Algorithm 1. The main function Solve applies the inference rules $\exists_l, \forall_r$, and $\rightarrow_r$ and calls itself recursively with the premise of these inferences. However it does this in a loop where it first extends the given sequent to a maximal sequent by obtaining a model from the SAT solver. Such a sequent corresponds to a leaf in a restricted bottom-up search, which only uses inferences (all except for $\rightarrow_r, \forall_r, \exists_l$—these inferences have an eigenvariable or single-conclusion restriction) that we have encoded as clauses in the SAT solver (in the Construct function). There may be multiple such leaves (e.g. corresponding to two $\forall_r$ inferences in different branches of the proof), hence Solve iterates over the models in a loop.

We use an incremental interface to the SAT solver: the solver internally maintains a set of clauses $\mathcal{C}$. A clause is a set of literals. If $l$ is a literal, then $-l$

---

**Algorithm 1.** SAT-based proof constructivization

---

1: **procedure** CONSTRUCT($E$)     ▷ returns true if intuitionistic proof of sh($E$) found
2:     ASSERT($\|\top\|$);   ASSERT($-\|\bot\|$)
3:     ASSERT($-\|\text{sh}(E_1)\|, \|\text{sh}(E_1 \vee E_2)\|$) **for each** subtree $E_1 \vee E_2$ of $E$
4:     ASSERT($-\|\text{sh}(E_2)\|, \|\text{sh}(E_1 \vee E_2)\|$) **for each** subtree $E_1 \vee E_2$ of $E$
5:     ASSERT($-\|\text{sh}(E_1 \vee E_2)\|, \|\text{sh}(E_1)\|, \|\text{sh}(E_2)\|$) **for each** subtree $E_1 \vee E_2$ of $E$
6:     ASSERT($-\|\text{sh}(E_1 \wedge E_2)\|, \|\text{sh}(E_1)\|$) **for each** subtree $E_1 \wedge E_2$ of $E$
7:     ASSERT($-\|\text{sh}(E_1 \wedge E_2)\|, \|\text{sh}(E_2)\|$) **for each** subtree $E_1 \wedge E_2$ of $E$
8:     ASSERT($-\|\text{sh}(E_1)\|, -\|\text{sh}(E_2)\|, \|\text{sh}(E_1 \wedge E_2)\|$) **for each** subtree $E_1 \wedge E_2$ of $E$
9:     ASSERT($-\|\text{sh}(E_2)\|, \|\text{sh}(E_1 \rightarrow E_2)\|$) **for each** subtree $E_1 \rightarrow E_2$ of $E$
10:     ASSERT($-\|\text{sh}(E_1)\|, -\|\text{sh}(E_1 \rightarrow E_2)\|, \|\text{sh}(E_2)\|$) **for each** subtree
11:                             $E_1 \rightarrow E_2$ of $E$
12:     ASSERT($-\|\text{sh}(E')|, \|\text{sh}(E_i)\|$) **for each** subtree
13:                 $E' = \forall x\, \varphi(x) +^{t_1} E_1 \cdots +^{t_n} E_n$ of $E$ and $1 \le i \le n$
14:     ASSERT($-\|\text{sh}(E_i)\|, \|\text{sh}(E')\|$) **for each** subtree
15:                 $E' = \exists x\, \varphi(x) +^{t_1} E_1 \cdots +^{t_n} E_n$ of $E$ and $1 \le i \le n$
16:     **return** SOLVE($E; \emptyset; \vdash \text{sh}(E)$)
17: **procedure** SOLVE($E; \Sigma; \Gamma \vdash \Delta$)     ▷ $\Sigma$ is the set of already used eigenvariables
18:                         ▷ returns true if we have found an intuitionistic proof of $\Gamma \vdash \Delta$
19:     **while** ISESATISFIABLE($\|\Gamma\|, -\|\Delta\|$) **do**
20:         $I = $ GETMODEL($\|\Gamma\|, -\|\Delta\|$)
21:         $(\Gamma', \Delta') = (\{\varphi \mid I(\|\varphi\|) = 1\}, \{\varphi \mid I(\|\varphi\|) = 0\})$     ▷ $\Gamma' \supseteq \Gamma$ and $\Delta' \supseteq \Delta$
22:         **for each** $\varphi \rightarrow \psi$ in $\Delta'$ such that $\varphi \notin \Gamma'$ **do**
23:             **if** SOLVE($E; \Sigma; \Gamma', \varphi \vdash \psi$) **then**
24:                 $C = $ UNSATCORE($\|\Gamma'\|, \|\varphi\|, -\|\psi\|$)
25:                 ASSERT($-C \setminus \{\|\psi\|, -\|\varphi\|\}, \|\varphi \rightarrow \psi\|$)
26:                 **continue outer loop**
27:         **for each** $\forall x\, \varphi$ in $\Delta'$ and subtree $\forall x\, \varphi +^{\alpha}_{\text{ev}} \ldots$ in $E$ with $\alpha \notin \Sigma$ **do**
28:             $\Gamma'_{\alpha} = \{\psi \in \Gamma' \mid \alpha \notin \text{FV}(\psi)\}$
29:             **if** SOLVE($E; \Sigma \cup \{\alpha\}; \Gamma'_{\alpha} \vdash \varphi(\alpha)$) **then**
30:                 $C = $ UNSATCORE($\|\Gamma'_{\alpha}\|, -\|\varphi(\alpha)\|$)
31:                 ASSERT($-C \setminus \{\|\varphi(\alpha)\|\}, \|\forall x\, \varphi(x)\|$)
32:                 **continue outer loop**
33:         **for each** $\exists x\, \varphi$ in $\Gamma'$ and subtree $\exists x\, \varphi +^{\alpha}_{\text{ev}} \ldots$ in $E$ with $\alpha \notin \Sigma$ **do**
34:             $(\Gamma'_{\alpha}, \Delta'_{\alpha}) = (\{\psi \in \Gamma' \mid \alpha \notin \text{FV}(\psi)\}, \{\psi \in \Delta' \mid \alpha \notin \text{FV}(\psi)\})$
35:             **if** SOLVE($E; \Sigma \cup \{\alpha\}; \Gamma'_{\alpha}, \varphi(\alpha) \vdash \Delta'_{\alpha}$) **then**
36:                 $C = $ UNSATCORE($\|\Gamma'_{\alpha}\|, \|\varphi(\alpha)\|, -\|\Delta'_{\alpha}\|$)
37:                 ASSERT($-C \setminus \{-\|\varphi(\alpha)\|\}, -\|\exists x\, \varphi(x)\|$)
38:                 **continue outer loop**
39:         **return** false
40:     **return** true
41: **procedure** ISESATISFIABLE($A$)     ▷ returns true iff satisfiable modulo equality
42:     **while** ISSATISFIABLE($A$) **do**     ▷ implemented using congrence closure
43:         $I = $ GETMODEL($A$)
44:         **if** $\{\varphi \mid I(\|\varphi\|) = 1\} \vdash \{\varphi \mid I(\|\varphi\|) = 0\}$ is provable with cut, w, rfl, eq **then**
45:             ASSERT(end-sequent of equality proof)
46:         **else return** true
47:     **return** false

---

is the negated literal. The function ASSERT adds a clause to this set. Given a set of literals $A$ (short for assumptions), the function ISSATISFIABLE($A$) returns true iff $\bigwedge C \wedge \bigwedge A$ is satisfiable. If it is satisfiable, the function GETMODEL($A$) returns a model. If it is unsatisfiable, then UNSATCORE($A$) returns a minimal subset $A' \subseteq A$ such that $\bigwedge C \wedge \bigwedge A'$ is still unsatisfiable.

Concretely we associate to the shallow formula $\varphi$ of every subtree of $E$ a variable $\|\varphi\|$ in the SAT solver. Given a set of formulas $\Gamma$, we also define $\|\Gamma\| = \{\|\varphi\| \mid \varphi \in \Gamma\}$ (in particular $\|\emptyset\| = \emptyset$). We only call ASSERT($-\|\Gamma\|, \|\Delta\|$) if we have an intuitionistic mLJ-proof of $\Gamma \vdash \Delta$ (that is, $\bigwedge \Gamma \to \bigvee \Delta$ is an intuitionistic theorem). A model $I$ returned by the SAT solver corresponds to the sequent $\{\varphi \mid I(\|\varphi\|) = 1\} \vdash \{\varphi \mid I(\|\varphi\|) = 0\}$.

By asserting specific clauses, we can ensure that these sequents obtained from models are closed under inferences: for example if we call ASSERT($-\|\varphi \wedge \psi\|, \|\psi\|$) then any model $I$ satisfies $I(\|\psi\|) = 1$ if $I(\|\varphi \wedge \psi\|) = 1$. Hence, the sequent $\Gamma \vdash \Delta$ corresponding to the model has $\psi \in \Gamma$ if $\varphi \wedge \psi \in \Gamma$ and is closed under (part of) the $\wedge_l$ rule (read bottom-up). We add these clauses in the CONSTRUCT function.

The other inferences, $\to_r, \forall_r, \exists_l$, are handled in the SOLVE function. For example, lines 23–26 handle an $\to_r$ inference inferring $\Gamma' \vdash \varphi \to \psi$ from $\Gamma', \varphi \vdash \psi$ in the following way: the recursive SOLVE-call first tries to prove $\Gamma', \varphi \vdash \psi$. The set of literals $C$ returned by UNSATCORE then corresponds to a minimal subset $\Gamma'' \subseteq \Gamma'$ such that $\Gamma'', \varphi \vdash \psi$ (the minimization is purely an optimization, albeit an important one). The correspondence is that $\|\Gamma''\| = C \setminus \{\|\varphi\|, -\|\psi\|\}$. We then assert $-\|\Gamma''\|, \|\varphi \to \psi\|$ (computed using $C$ in line 25) corresponding to the provable sequent $\Gamma'' \vdash \varphi \to \psi$. Note that the polarities of the SAT solver variables are inverse in the *assumptions* passed to UNSATCORE and the *clause* passed to ASSERT: UNSATCORE($A$) returns a minimal subset of $A' \subseteq A$ such that $-A'$ (regarded as a clause) is implied by the current clauses.

Algorithm 1 terminates, since each recursive call of SOLVE either increases the size of $\Sigma$ or increases the size of the antecedent $\Gamma$ while keeping $\Sigma$ the same; the while-loop in SOLVE never iterates over the same model twice since the ASSERT-calls in SOLVE add clauses that are not true in the current model.

*Example 4.* Consider again the expansion proof from Example 1 and abbreviate $\psi = \forall x\, P(f(x)) \to Q$. Then CONSTRUCT asserts the following clauses:

$$
\begin{array}{ll}
\|\top\| & -\|\varphi\|, -\|\forall x\, P(x)\|, \|\psi \to Q\| \\
-\|\bot\| & -\|\psi \to Q\|, -\|\psi\|, \|Q\| \\
-\|\psi \to Q\|, \|\varphi\| & -\|\psi\|, -\|\forall x\, P(f(x))\|, \|Q\| \\
-\|Q\|, \|\psi \to Q\| & -\|\forall x\, P(x)\|, \|P(f(\alpha))\|
\end{array}
$$

And SOLVE proceeds in the following recursive call tree:

- SOLVE($E$; $\emptyset$; $\vdash \varphi$)
  - Obtained model: $\|\top\|$ (list of all $p$ with $I(p) = 1$)
  - SOLVE($E$; $\emptyset$; $\top, \forall x\, P(x) \vdash \psi \to Q$) (line 23 for $\varphi$)
    - Obtained model: $\|\top\|, \|\forall x\, P(x)\|, \|P(f(\alpha))\|$
    - SOLVE($E$; $\emptyset$; $\top, \forall x\, P(x), P(f(\alpha)), \psi \vdash Q$) (line 23 for $\psi \to Q$)

· Obtained model: $\|\top\|, \|\forall x\, P(x)\|, \|P(f(\alpha))\|, \|\psi\|$
· $\textsc{Solve}(E; \{\alpha\}; \top, \forall x\, P(x), \psi \vdash P(f(\alpha)))$ (line 29 for $\forall x\, P(f(x))$)
· $\textsc{Assert}(-\|\forall x\, P(x)\|, \|\forall x\, P(f(x))\|)$
* $\textsc{Assert}(-\|\forall x\, P(x)\|, \|\psi \to Q\|)$

- $\textsc{Assert}(\|\varphi\|)$

**Theorem 2.** *If* Construct*(E) returns true, then there is an mLJ-proof of* sh$(E)$ *realizing $E$ (and* sh$(E)$ *is an intuitionistic theorem).*

*Proof.* We store an mLJ-proof $\Gamma \vdash \Delta$ for every clause $-\|\Gamma\|, \|\Delta\|$ that is passed to Assert: these all have straightforward proofs in mLJ. Whenever IsESatisfiable$(\|\Gamma\|, -\|\Delta\|)$ returns false for a sequent $\Gamma \vdash \Delta$, we have an mLJ-proof of $\Gamma \vdash \Delta$ by combining the previously stored proofs using cuts as in the resolution refutation returned by the SAT solver. □

Let us now prove completeness, i.e., Construct returns true if the expansion proof $E$ is realized by a weakly regular proof $\pi$ of sh$(E)$ in mLJ. Intuitively, the procedure succeeds because it can just pick the same inferences as in $\pi$. In a sense, the function Solve proceeds upwards through the proof $\pi$, the SAT solver jumps over all inferences except $\exists_l, \forall_r, \to_r$, and the model $\Gamma' \vdash \Delta'$ that we consider in Solve corresponds to an $\exists_l, \forall_r$ or $\to_r$ inference in $\pi$.

Formally, we capture the required properties for the model obtained from the SAT solver as "maximal" sequents. Whenever a proof ends in a sub-sequent of a maximal sequent, we can trace the proof upwards to find a $\exists_l, \forall_r$ or $\to_r$ inference also ending in a sub-sequent of the maximal sequent.

**Definition 7.** *A sequent $\Gamma \vdash \Delta$ is called maximal (for an expansion proof $E$) iff all of the following are true:*

- *$\Gamma \cup \Delta$ is the set of all shallow formulas of subtrees of $E$*
- *$\Gamma \cap \Delta = \emptyset$*
- *$\bot \in \Delta$*
- *$\top \in \Gamma$*
- *If $\varphi \wedge \psi \in \Gamma$, then $\varphi, \psi \in \Gamma$*
- *If $\varphi \wedge \psi \in \Delta$, then $\varphi \in \Gamma$ or $\psi \in \Gamma$*
- *If $\varphi \vee \psi \in \Delta$, then $\varphi, \psi \in \Delta$*
- *If $\varphi \vee \psi \in \Gamma$, then $\varphi \in \Gamma$ or $\psi \in \Gamma$*
- *If $\varphi \to \psi \in \Gamma$, then $\varphi \in \Delta$ or $\psi \in \Gamma$*
- *If $\forall x\, \varphi(x) \in \Gamma$ and $\forall x\, \varphi(x) +_{t_1} \cdots +_{t_n}$ is a subtree of $E$, then $\varphi(t_1), \ldots, \varphi(t_n) \in \Gamma$.*
- *If $\exists x\, \varphi(x) \in \Delta$ and $\exists x\, \varphi(x) +_{t_1} \cdots +_{t_n}$ is a subtree of $E$, then $\varphi(t_1), \ldots, \varphi(t_n) \in \Delta$*

**Lemma 1.** *The sequent $\Gamma' \vdash \Delta'$ obtained in line 21 from the model returned by* GetModel *in* Solve*$(E; \Sigma; \Gamma \vdash \Delta)$ is maximal for $E$.*

*Proof.* The set $\Gamma' \cup \Delta'$ contains all shallow formulas, and we have $\Gamma' \cap \Delta' = \emptyset$ because $\Gamma' \vdash \Delta'$ corresponds to a model. Each of the other conditions in Definition 7 is then ensured by a clause that is asserted in the Construct function: e.g. $\bot \in \Delta'$ is ensured by $\textsc{Assert}(-\|\bot\|)$. □

**Lemma 2.** *Let $\mathcal{S}$ be maximal for $E$, and $\pi$ be an mLJ-proof of $\mathcal{S}'$ realizing $E$, such that $\mathcal{S}'$ is a subsequent of $\mathcal{S}$. Then there is a subproof $\pi'$ of $\pi$ such that the end-sequent of $\pi'$ is also a sub-sequent of $\mathcal{S}$, and $\pi'$ ends in a $\exists_l, \forall_r$ or $\to_r$ inference.*

*Proof.* By straightforward induction on $\pi$. For illustration, let us prove the case where $\pi$ ends in an $\wedge_r$-inference:

$$\dfrac{\overset{(\pi_1)}{\Gamma \vdash \Delta, \varphi} \qquad \overset{(\pi_2)}{\Gamma \vdash \Delta, \psi}}{\Gamma \vdash \Delta, \varphi \wedge \psi} \wedge_r$$

Let $\mathcal{S} = (\Gamma' \vdash \Delta')$. Note that $\Gamma \vdash \Delta, \varphi \wedge \psi$ is a subsequent of $\Gamma' \vdash \Delta'$ by assumption and hence $\varphi \wedge \psi \in \Delta'$. The sequent $\mathcal{S}$ is maximal, so $\varphi \in \Delta'$ or $\psi \in \Delta'$. First consider $\varphi \in \Delta'$; then $\Gamma \vdash \Delta, \varphi$ is a subsequent of $\mathcal{S}$ and we can apply the induction hypothesis. The case $\psi \in \Delta'$ is symmetric. □

**Theorem 3.** *If $E$ can be realized by a weakly regular cut-free mLJ-proof of* sh($E$)*, then* Construct*($E$) returns true.*

*Proof.* We use the following invariant for Solve($E; \Sigma; \Gamma \vdash \Delta$): if there is a weakly regular cut-free mLJ-proof $\pi$ of a subsequent of $\Gamma \vdash \Delta$ realizing $E$ such that the eigenvariables in $\pi$ are disjoint from $\Sigma$, then Solve returns true.

In Solve, let $\pi$ be the subproof described above, and let $\Gamma' \vdash \Delta'$ be the sequent constructed from the model in line 21. Then $\Gamma' \vdash \Delta'$ is a maximal sequent by Lemma 1. There is a subproof $\pi'$ of $\pi$ whose end-sequent is a subsequent of $\Gamma' \vdash \Delta'$ and that ends in a $\exists_l, \forall_r$ or $\to_r$ inference by Lemma 2. At least one of the recursive calls then corresponds to this inference, and invokes Solve with the premise of the inference, which hence returns true. Weak regularity of $\pi$ ensures that the precondition for $\Sigma$ is satisfied. The clause passed to Assert corresponds to the conclusion of the $\exists_l, \forall_r, \to_r$-inference, and we continue to the next iteration of the loop. □

For quantifier-free formulas, the constructivization method is a decision procedure since mLJ is cut-free complete and proofs of quantifier-free formulas are trivially weakly regular and realize the expansion proof. The author believes that it should be possible to give a similar proof-theoretic completeness proof for the Intuit prover [9] (their paper gives a proof based on Kripke models).

**Corollary 1.** *If* sh($E$) *is a quantifier-free formula, then* Construct*($E$) returns true if and only if* sh($E$) *is an intuitionistic theorem.*

## 5 Optimizations

For performance reasons, we implement several optimizations in the Solve procedure. The first one was already described in [9].

*Caching Unsolvable Cases.* By using a SAT solver, we already have a cache for the solvable cases: whenever Solve($E; \Sigma; \Gamma \vdash \Delta$) returns true, the SAT solver remembers the conflict clause and all subsequent calls to Solve will terminate after just one call to IsESatisfiable. However if we cannot find a proof, then we would need to repeat the costly recursive backtracking procedure. Hence we store all pairs $(\Sigma; \Gamma' \vdash \Delta')$ where the result is false ($\Gamma' \vdash \Delta'$ is the model obtained in line 21). At the beginning of Solve we check if we have already stored a pair $(\Sigma''; \Gamma'' \vdash \Delta'')$ such that $\Sigma \subseteq \Sigma''$ and $\Gamma \supseteq \Gamma''$ and $\Delta \supseteq \Delta''$, and return false if there is such a pair. (Pairs are stored in a trie-like data structure.)

*Classical Quasi-Tautology Check.* If a sequent $\Gamma \vdash \Delta$ is not even classically provable, then it cannot be intuitionistically provable either. This easy observation allows us to prune large branches of the backtracking search in Solve. The function IsESatisfiable$(\|\Gamma\|, -\|\Delta\|)$ returns false if there is a weakly regular cut-free mLJ-proof of $\Gamma \vdash \Delta$ realizing $E$ *without* the inferences $\rightarrow_r, \forall_r, \exists_l$. We add a fresh variable $\mathsf{c}$ in such a way that IsESatisfiable$(\mathsf{c}, \|\Gamma\|, -\|\Delta\|)$ returns false iff there is a classical proof of $\Gamma \vdash \Delta$ realizing $E$. The classical calculus differs from mLJ only in the rules $\rightarrow_r, \forall_r$. (Concretely, we assert the clauses $\mathsf{c} \vdash \|\varphi \rightarrow \psi\|, \|\varphi\|$ and $\mathsf{c}, \|\varphi(\alpha)\| \vdash \|\forall x\, \varphi(x)\|$ and $\mathsf{c}, \|\exists x\, \varphi(x)\| \vdash \|\varphi(\alpha)\|$ for the corresponding shallow formulas of subtrees of $E$.)

*Invertible Occurrences of $\exists_l$.* In some cases we can avoid backtracking with existential quantifiers in the antecedent. This is the case if we have a subtree $\exists x\, \varphi(x) +^{\alpha}_{\mathrm{ev}} E_1$, where all free variables in $\exists x\, \varphi(x)$ are already in $\Sigma$. In this case we immediately apply the corresponding $\exists_l$ inference, and skip all the loops in the Solve procedure. This is correct because we can permute the $\exists_l$ inference downward in the realizing proof. Consider e.g. the expansion proof $\forall x\, \exists y\, R(x,y) +^{\alpha} (\cdots +^{\beta}_{\mathrm{ev}} R(\alpha,\beta)^-) +^{f(\alpha)} (\cdots +^{\gamma}_{\mathrm{ev}} R(f(\alpha),\gamma)^-) \vdash \forall x\, \exists y\, \exists z\, (R(x,y) \wedge R(f(x),z)) +^{\alpha}_{\mathrm{ev}} \cdots +^{\beta} \cdots +^{\gamma} (R(\alpha,\beta)^+ \wedge R(f(\alpha),\gamma)^+)$. In line 29, we first introduce the $\alpha$ eigenvariable, and then (in recursive calls) the $\beta$ and $\gamma$ eigenvariables in line 35. Without the optimization, we would then try *every permutation* of $\beta, \gamma$ if Solve returned false (which will be expensive if there are more eigenvariables). With the optimization, we only need to consider a single permutation.

## 6 Completeness on Subclasses

In general, our proof constructivization-based approach to intuitionistic theorem proving is incomplete. For example, $\forall x\, (p(x) \vee \neg p(x)) \vdash \neg\neg p(c) \rightarrow p(c)$ is an intuitionistic theorem where our approach will fail—we clearly need the assumption $\forall x\, (p(x) \vee \neg p(x))$, but virtually all classical theorem provers will discard it immediately and never use it. For decidability assumptions such as $\forall x\, (p(x) \vee \neg p(x))$ we can use heuristic instantiation as a pre-processing step, adding all instances of the formula for subterms occurring in the expansion proof.

However there are some classes of formulas where our approach is complete. These are classes of first-order formulas where intuitionistic provability is equivalent to classical provability, such classes are called Glivenko classes and were e.g. studied by Orevkov. See also [27] for a more modern presentation.

**Definition 8** ([28]). Class 1 *is the set of sequents which do not have positive occurrences of* $\to$ *or* $\forall$.

*Example 5.* The sequent $\forall x\,(p(x) \to q(x) \vee \neg r(x)), p(c) \wedge r(c) \vdash q(c)$ is in Class 1; $(p \to q) \to p \vdash$ and $(\forall x\, p(x)) \to q \vdash$ are not in Class 1 since they have a positive occurrence of $\to$ and $\forall$, resp.

Many practically relevant problems are in Class 1; all Horn problems, all rewriting problems, and all problems in CNF are in Class 1. It is instructive to look at the proof that intuitionistic provability is equivalent to classical provability for all problems in Class 1:

**Theorem 4** ([28]). *Let* $\Gamma \vdash \Delta$ *be a sequent in Class 1. If* $\Gamma \vdash \Delta$ *is provable in classical logic, then it is provable in intuitionistic logic as well.*

*Proof.* Let $\pi$ be a cut-free proof of $\Gamma \vdash \Delta$ in the sequent calculus LK (which is cut-free complete for classical logic). Then $\pi$ does not contain any of the inferences $\to_r$ or $\forall_r$ by the subformula property (these are the only inferences that are different between LK and mLJ), and is hence a proof in mLJ. □

**Corollary 2.** *Let* $E$ *be an expansion proof such that* $\mathrm{sh}(E)$ *is in Class 1, and* $\mathrm{dp}(E)$ *is a quasi-tautology. Then* CONSTRUCT*(E) returns true.*

*Proof.* There is a weakly regular proof in LK of $\mathrm{sh}(E)$ realizing $E$ since $\mathrm{dp}(E)$ is a quasi-tautology, with the observation in Theorem 4 this proof is in mLJ. Hence CONSTRUCT succeeds by Theorem 3. □

Corollary 2 shows that our prover as a whole is complete for sequents in Class 1. A similar result also holds for Orevkov's Class 2 (no positive occurrences of $\to$ and no negative occurrences of $\vee$). The constructivization procedure of Gilbert [16] was also shown to be complete for Class 2 (called F in their paper).

## 7 Experimental Evaluation

We have implemented and evaluated this constructivization approach in the open source GAPT[5] system for proof theory [15], version 2.14. Many of its features are centered around a computational implementation of Herbrand's theorem and expansion trees, such as lemma generation [14], inductive theorem proving [12], deskolemization, and proof import [33].

The intuitionistic first-order prover based on this constructivization procedure is called Slakje, and provides a command-line program reading input problems in TPTP format [35]. Since GAPT is written in Scala and distributed as a platform-neutral tarball, we want to avoid external dependencies and prefer to use libraries available on the JVM: as a SAT solver we use Sat4j [2], for equality reasoning we wrote a simple congruence closure implementation.

[5] Open source, and freely available at https://logic.at/gapt.

By default, Slakje prints the generated mLJ proof in the TPTP derivation format, writing each sequent in the proof on a separate line. Other output options are supported as well: the `--prooftool` option displays the mLJ proof tree in the graphical ProofTool user interface [11]. The `--lj` option transforms the mLJ proof to a cut-free proof in the single-conclusion intuitionistic LJ calculus.

GAPT already contains a reliable interface to external theorem provers that produces expansion proofs and supports many first-order provers, including Vampire [19], E [34], SPASS [40], leanCoP [30], Prover9 [21], as well as others. GAPT also includes a simple built-in superposition prover called Escargot, which is mainly used for proof replay and small-scale automation in tactic proofs. For the experimental evaluation, we used Vampire 4.2.2, E 2.2, and Escargot as backends for Slakje. We do not call the external provers for quantifier-free problems: there are no quantifier instances that we could import, and furthermore the constructivization procedure is already a decision problem in the quantifier-free case by Corollary 1. The prover interface in GAPT supports most external provers (including Vampire and E) using proof replay, which reconstructs the proofs line-by-line by reproving each inference as a first-order problem using Escargot. There is special support for the Avatar [39] splitting inferences produced by Vampire. The interface operates on the level of clauses, the clausification and Skolemization is performed inside GAPT. Parsing and importing the Skolemization and clausification steps of all supported provers would be a tremendous amount of work, since every prover (and sometimes different versions of the same prover) use different proof output for these steps.

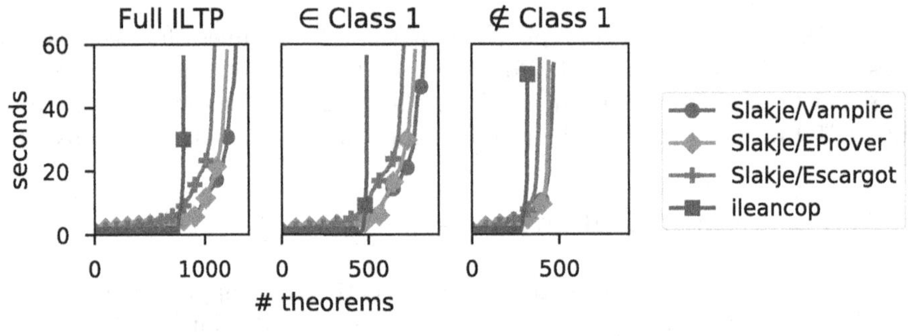


**Fig. 2.** Cactus plot of the prover runtime on proved theorems.

We evaluated the Slakje prover on the problems in the first-order section of the Intuitionistic Logic Theorem Proving library [32], version 1.1.2. The ILTP library contains a mixture of problems from the TPTP, as well as problems designed for intuitionistic provers in the `GEJ` (constructive geometry), `GPJ` (group theory), and `SYJ` (intuitionistic syntactic) categories. The ILTP also contains benchmarking results for a number of intuitionistic theorem provers from 2006. In those results, ileanCoP [30] solves the largest number of problems by a significant margin.

The provers imogen [22] and WhaleProver [31] were also benchmarked on the ILTP, performing competitively with ileanCoP. According to [22], imogen solves 857 problems, improving on the 690 problems solved by ileanCoP 1.0. (On our hardware, the newest ileanCoP 1.2 version now solves 891 problems.) Unfortunately we were not able to build a working version of imogen, so we could not benchmark it on our hardware. WhaleProver is not publicly available at all, according to [31] it solves 811 problems.

In our evaluation we compared Slakje against the current ileanCoP 1.2 version, running both Slakje and ileanCoP on a Debian Linux system with an Intel i5-4570 CPU and 8 GiB RAM.

The ILTP contains 2670 problems in the first-order section. (There are only 2550 problems according to the ILTP website, however the archive file[6] contains 2670 problem files.). Slakje solves 1674 of these problems (1290 theorems and 384 non-theorems) with Vampire 4.2.2 as a backend (total time limit of 60 s, no command-line options). ileanCoP 1.2 solves 891 (813 theorems and 78 non-theorems). 905 of the problems solved by Slakje were not solved by ileanCoP (546 of which are theorems, and 359 non-theorems). (Including the ILTP benchmark results as well, Slakje solves 800 problems not solved by ileanCoP in our benchmarks, or any prover in the 2006 ILTP benchmarks.) Slakje could not solve 122 problems that were solved by ileanCoP; 69 of these problems are intuitionistic non-theorems. For the other 53 intuitionistic theorems, in one case Slakje fails due to a timeout, and 24 could be solved with a different backend (Escargot or E).

The runtime of Slakje with the three backends (Escargot, E, and Vampire), and ileanCoP is shown as a cactus plot in Fig. 2. Slakje is leading in the number of proven theorems with any of the three backends; the most theorems are obtained using Vampire (1290 thms. and 384 nonthms.), followed by E (1210 thms. and 370 nonthms.), and Escargot (1096 thms. and 363 nonthms.).

While Slakje can prove many difficult problems, it has a high overhead: the median runtime for solved problems is 3199 ms, compared to 46 ms for ileanCoP. Within a time limit of one second, Slakje can only prove a single theorem, while ileanCoP proves 734. This overhead is likely due to multiple factors: since Slakje runs on the JVM, it takes some time for the just-in-time compiler to compile the code. Furthermore, the interface to the external provers such as Vampire was designed to be generic and is not highly optimized, e.g. we use the first-order Escargot prover to reconstruct every inference that Vampire produces.

We might assume that the success of Slakje is due to benchmark set: that the ILTP contains many Horn problems or purely equational problems. However this is not the case. Only 650 of the problems in the ILTP are in Class 1 (recall Definition 8). If we remove formulas that were not used in the classical proofs, then 980 of the problems are in Class 1. Looking at the runtime plots for Class 1 problems vs. non-Class 1 problems, there does not seem to be a large difference and Slakje is also leading even for the non-Class 1 problems, which should be harder for Slakje as it is not complete there.

[6] available at http://iltp.de/download/ILTP-v1.1.2-firstorder.tar.gz.

We have also run Vampire directly on the problems in the ILTP (in CASC mode with a time limit of 60 s) to obtain a realistic upper limit on how many problems we can expect to solve intuitionistically. In this configuration Vampire solves 2468 problems (2079 proofs and 389 satisfiable). When used via GAPT's prover interface, Vampire solves 1938 out of the 2421 non-quantifier-free problems, returning 1585 proofs in the textual TPTP derivation format and 353 satisfiable results (for which Slakje can immediately return non-theorem). Proof replay then produces 1541 resolution proofs, which are converted to 1526 expansion proofs, ultimately yielding 1098 intuitionistic proofs in mLJ. (The remaining 192 theorems are quantifier-free formulas, which we directly passed to the constructivization procedure.) In each step we lose a few proofs due to the time limit. The largest difference is in the initial step of running the external theorem prover. We believe that this is mainly due to two reasons: first, the TPTP parser in GAPT is less efficient and takes a long time to parse larger problems. Second, we run Vampire in the default mode instead of the CASC mode, since the CASC mode produces proofs that GAPT cannot parse reliably, making it less effective in our interface.

## 8 Conclusion

First-order theorem proving seems to be fundamentally easier in classical logic than in intuitionistic logic. We can use Skolemization, and have CNFs as a simple normal form. The practical proof constructivization procedure that we have presented allows to reuse some of these advantages of classical logic. In a sense, we are learning from classical proofs to produce intuitionistic ones. In our setting, we are learning the quantifier instances. On an empirical level, we have shown that these instances as captured by expansion trees provide enough information to produce intuitionistic proofs.

This proof constructivization technique is so effective that we obtain a highly competitive automated intuitionistic first-order theorem prover by combining it with a classical theorem prover. This prover, Slakje, performs very well on the ILTP benchmark library for intuitionistic theorem provers: it proves 1290 out of 2670 problems with Vampire as a backend. This is significantly more than other state-of-the-art provers such as ileanCoP (proving 813 problems).

However, this intuitionistic prover is incomplete since the classical theorem prover may not produce enough quantifier instances for an intuitionistic proof. One idea to fix this incompleteness that was already suggested in [9] is to add a complete instantiation strategy akin to the support for first-order reasoning in SMT solvers. Another approach would be to investigate variants of sound (semantic) translations of intuitionistic logic into classical logic which are optimized for automated theorem provers, and constructivize proofs of these translations.

As future work, we intend to integrate this prover approach in interactive theorem provers and evaluate its use for proof automation and as a strong reconstruction tactic for hammers in proof assistants based on intuitionistic logic.

**Acknowledgements.** The author would like to thank the anonymous reviewers for their suggestions which have led to a considerable improvement of this paper. This work has been supported by the Vienna Science and Technology Fund (WWTF) project VRG12-004.

## References


1. Baaz, M., Hetzl, S., Weller, D.: On the complexity of proof deskolemization. J. Symb. Log. **77**(2), 669–686 (2012)
2. Berre, D.L., Parrain, A.: The Sat4j library, release 2.2. JSAT **7**(2–3), 59–64 (2010)
3. Bertot, Y., Castéran, P.: Interactive Theorem Proving and Program Development – Coq'Art: The Calculus of Inductive Constructions. Texts in Theoretical Computer Science. Springer, Heidelberg (2004). https://doi.org/10.1007/978-3-662-07964-5
4. Bibel, W.: Matings in matrices. Commun. ACM **26**(11), 844–852 (1983)
5. Blanchette, J.C., Bulwahn, L., Nipkow, T.: Automatic proof and disproof in isabelle/HOL. In: Tinelli, C., Sofronie-Stokkermans, V. (eds.) FroCoS 2011. LNCS (LNAI), vol. 6989, pp. 12–27. Springer, Heidelberg (2011). https://doi.org/10.1007/978-3-642-24364-6_2
6. Bove, A., Dybjer, P., Norell, U.: A brief overview of agda – a functional language with dependent types. In: Berghofer, S., Nipkow, T., Urban, C., Wenzel, M. (eds.) TPHOLs 2009. LNCS, vol. 5674, pp. 73–78. Springer, Heidelberg (2009). https://doi.org/10.1007/978-3-642-03359-9_6
7. Buss, S.R.: On Herbrand's theorem. In: Leivant, D. (ed.) LCC 1994. LNCS, vol. 960, pp. 195–209. Springer, Heidelberg (1995). https://doi.org/10.1007/3-540-60178-3_85
8. Cauderlier, R.: A rewrite system for proof constructivization. In: Dowek, G., Licata, D.R., Alves, S. (eds.) 11th Workshop on Logical Frameworks and Meta-Languages: Theory and Practice. LFMTP, pp. 2:1–2:7. ACM (2016)
9. Claessen, K., Rosén, D.: SAT modulo intuitionistic implications. In: Davis, M., Fehnker, A., McIver, A., Voronkov, A. (eds.) LPAR 2015. LNCS, vol. 9450, pp. 622–637. Springer, Heidelberg (2015). https://doi.org/10.1007/978-3-662-48899-7_43
10. Czajka, L., Kaliszyk, C.: Hammer for Coq: automation for dependent type theory. J. Autom. Reason. **61**(1–4), 423–453 (2018)
11. Dunchev, C., et al.: PROOFTOOL: a GUI for the GAPT framework. In: Kaliszyk, C., Lüth, C. (eds.) Proceedings 10th International Workshop On User Interfaces for Theorem Provers (UITP) 2012. EPTCS, vol. 118, pp. 1–14 (2012)
12. Eberhard, S., Hetzl, S.: Inductive theorem proving based on tree grammars. Ann. Pure Appl. Log. **166**(6), 665–700 (2015)
13. Ebner, G.: Extracting expansion trees from resolution proofs with splitting and definitions (2018). Preprint https://gebner.org/pdfs/2018-01-29_etimport.pdf
14. Ebner, G., Hetzl, S., Leitsch, A., Reis, G., Weller, D.: On the generation of quantified lemmas. J. Autom. Reason. **63**(1), 95–126 (2018)
15. Ebner, G., Hetzl, S., Reis, G., Riener, M., Wolfsteiner, S., Zivota, S.: System description: GAPT 2.0. In: Olivetti, N., Tiwari, A. (eds.) IJCAR 2016. LNCS (LNAI), vol. 9706, pp. 293–301. Springer, Cham (2016). https://doi.org/10.1007/978-3-319-40229-1_20
16. Gilbert, F.: Automated constructivization of proofs. In: Esparza, J., Murawski, A.S. (eds.) FoSSaCS 2017. LNCS, vol. 10203, pp. 480–495. Springer, Heidelberg (2017). https://doi.org/10.1007/978-3-662-54458-7_28

17. Herbrand, J.: Recherches sur la théorie de la démonstration. Ph.D. thesis, Université de Paris (1930)
18. Kaliszyk, C., Urban, J.: Learning-assisted automated reasoning with Flyspeck. J. Autom. Reason. **53**(2), 173–213 (2014)
19. Kovács, L., Voronkov, A.: First-order theorem proving and VAMPIRE. In: Sharygina, N., Veith, H. (eds.) CAV 2013. LNCS, vol. 8044, pp. 1–35. Springer, Heidelberg (2013). https://doi.org/10.1007/978-3-642-39799-8_1
20. Maehara, S.: Eine Darstellung der intuitionistischen Logik in der Klassischen. Nagoya Math. J. **7**, 45–64 (1954)
21. McCune, W.: Prover9 and Mace4 (2005–2010). http://www.cs.unm.edu/~mccune/prover9/
22. McLaughlin, S., Pfenning, F.: Efficient intuitionistic theorem proving with the polarized inverse method. In: Schmidt, R.A. (ed.) CADE 2009. LNCS (LNAI), vol. 5663, pp. 230–244. Springer, Heidelberg (2009). https://doi.org/10.1007/978-3-642-02959-2_19
23. Miller, D.A.: A compact representation of proofs. Studia Logica **46**(4), 347–370 (1987)
24. Mints, G.: Gentzen-type systems and resolution rules part I propositional logic. In: Martin-Löf, P., Mints, G. (eds.) COLOG 1988. LNCS, vol. 417, pp. 198–231. Springer, Heidelberg (1990). https://doi.org/10.1007/3-540-52335-9_55
25. Mints, G.: Gentzen-type system and resolution rules. Part II: Propositional logic. In: Logic Colloquium 1990 (1993)
26. de Moura, L., Kong, S., Avigad, J., van Doorn, F., von Raumer, J.: The lean theorem prover (system description). In: Felty, A.P., Middeldorp, A. (eds.) CADE 2015. LNCS (LNAI), vol. 9195, pp. 378–388. Springer, Cham (2015). https://doi.org/10.1007/978-3-319-21401-6_26
27. Negri, S.: Glivenko sequent classes in the light of structural proof theory. Arch. Math. Log. **55**(3–4), 461–473 (2016)
28. Orevkov, V.P.: On Glivenko sequent classes. Trudy Matematicheskogo Instituta imeni V. A. Steklova **98**, 131–154 (1968)
29. Otten, J.: Clausal connection-based theorem proving in intuitionistic first-order logic. In: Beckert, B. (ed.) TABLEAUX 2005. LNCS (LNAI), vol. 3702, pp. 245–261. Springer, Heidelberg (2005). https://doi.org/10.1007/11554554_19
30. Otten, J.: leanCoP 2.0 and ileanCoP 1.2: high performance lean theorem proving in classical and intuitionistic logic (system descriptions). In: Armando, A., Baumgartner, P., Dowek, G. (eds.) IJCAR 2008. LNCS (LNAI), vol. 5195, pp. 283–291. Springer, Heidelberg (2008). https://doi.org/10.1007/978-3-540-71070-7_23
31. Pavlov, V., Pak, V.: WhaleProver: first-order intuitionistic theorem prover based on the inverse method. In: Petrenko, A.K., Voronkov, A. (eds.) PSI 2017. LNCS, vol. 10742, pp. 322–336. Springer, Cham (2018). https://doi.org/10.1007/978-3-319-74313-4_23
32. Raths, T., Otten, J., Kreitz, C.: The ILTP problem library for intuitionistic logic. J. Autom. Reason. **38**(1–3), 261–271 (2007)
33. Reis, G.: Importing SMT and connection proofs as expansion trees. In: Fourth Workshop on Proof eXchange for Theorem Proving, PxTP, pp. 3–10 (2015)
34. Schulz, S.: System description: E 1.8. In: McMillan, K., Middeldorp, A., Voronkov, A. (eds.) LPAR 2013. LNCS, vol. 8312, pp. 735–743. Springer, Heidelberg (2013). https://doi.org/10.1007/978-3-642-45221-5_49
35. Sutcliffe, G.: The TPTP problem library and associated infrastructure: the FOF and CNF parts, v3.5.0. J. Autom. Reason. **43**(4), 337–362 (2009)

36. The Univalent Foundations Program: Homotopy Type Theory: Univalent Foundations of Mathematics. Institute for Advanced Study (2013). https://homotopytypetheory.org/book
37. Troelstra, A.S., Schwichtenberg, H.: Basic Proof Theory. Cambridge Tracts in Theoretical Computer Science. Cambridge University Press, Cambridge (2000)
38. Voronkov, A.: Proof-search in intuitionistic logic with equality, or back to simultaneous rigid E-unification. J. Autom. Reason. **30**(2), 121–151 (2003)
39. Voronkov, A.: AVATAR: the architecture for first-order theorem provers. In: Biere, A., Bloem, R. (eds.) CAV 2014. LNCS, vol. 8559, pp. 696–710. Springer, Cham (2014). https://doi.org/10.1007/978-3-319-08867-9_46
40. Weidenbach, C., Dimova, D., Fietzke, A., Kumar, R., Suda, M., Wischnewski, P.: SPASS version 3.5. In: Schmidt, R.A. (ed.) CADE 2009. LNCS (LNAI), vol. 5663, pp. 140–145. Springer, Heidelberg (2009). https://doi.org/10.1007/978-3-642-02959-2_10

# ENIGMAWatch: ProofWatch Meets ENIGMA

Zarathustra Goertzel, Jan Jakubův, and Josef Urban(✉)

Czech Technical University in Prague, Prague, Czech Republic
josef.urban@gmail.com

**Abstract.** In this work we describe a new learning-based proof guidance – ENIGMAWatch – for saturation-style first-order theorem provers. ENIGMAWatch combines two guiding approaches for the given-clause selection implemented for the E ATP system: ProofWatch and ENIGMA. ProofWatch is motivated by the watchlist (hints) method and based on symbolic matching of multiple related proofs, while ENIGMA is based on statistical machine learning. The two methods are combined by using the evolving information about symbolic proof matching as additional characterization of the saturation-style proof search for the statistical learning methods. The new system is evaluated on a large set of problems from the Mizar library. We show that the added proof-matching information is considered important by the statistical machine learners, and that it leads to improved performance over ProofWatch and ENIGMA.

## 1 Introduction

This work describes a new learning-based proof guidance – *ENIGMAWatch* – for saturation-style first-order theorem provers. ENIGMAWatch[1] is the combination of two previous guidance methods implemented for the E theorem prover [35]: ProofWatch [11] and ENIGMA [16,17]. Both ProofWatch and ENIGMA learn to guide E's proof search for a new conjecture based on related proofs.

ProofWatch uses the hints (watchlist) mechanism, which is a form of precise symbolic memory that can allow inference chains done in a former proof to be replayed in the current proof search. It uses standard symbolic subsumption to check which clauses subsume clauses in related proofs. In addition to boosting the priority of these clauses, the completion ratios of the related proofs are computed, and the proof search is biased towards the most completed ones.

ENIGMA uses fast statistical machine learning to learn from related proof-searches to identify good and bad (positive and negative) clauses for the current

[1] The E version used in this paper can be found at https://github.com/ai4reason/eprover/tree/devel, and the library for running ENIGMA with E can be found at https://github.com/ai4reason/enigma.

J. Urban—Supported by the *AI4REASON* ERC Consolidator grant number 649043, and by the Czech project AI&Reasoning CZ.02.1.01/0.0/0.0/15_003/0000466 and the European Regional Development Fund.


S. Cerrito and A. Popescu (Eds.): TABLEAUX 2019, LNAI 11714, pp. 374–388, 2019.
https://doi.org/10.1007/978-3-030-29026-9_21

conjecture. ENIGMA chooses the given clauses based only on features of the problem's conjecture, which is static throughout the whole proof search. This seems suboptimal: as the proof search evolves, information about the work done so far should influence the selection of the next given clauses.

ENIGMAWatch combines the two approaches by giving the ENIGMA's learner the ProofWatch completion ratios of the related proofs as an evolving vectorial characterization of the current proof search state. This allows E's machine learning guidance to have more information about how the proof search is unfolding.

An early version of ENIGMAWatch was tested on the MPTP Challenge[2] [36,39] benchmark. It contains 252 first-order problems extracted from the Mizar Mathematical Library (MML) [14], used in Mizar to prove the Bolzano-Weierstrass theorem. Initially, ENIGMAWatch could not be run on a larger dataset, such as the 57897 Mizar40 [21] benchmark, in a reasonable time. Since then, ENIGMA implemented dimensionality reduction using feature hashing [6], extending its applicability to large corpora. We have additionally improved watchlist mechanism in E through enhanced indexing, first time presented in this work in Sect. 4. This allows also ENIGMAWatch to be applied to larger corpora.

The rest of the paper is organized as follows. Section 2 provides an introduction to saturation-based theorem proving and briefly describes ENIGMA and ProofWatch. Section 3 explains how ENIGMA and ProofWatch are combined into ENIGMAWatch, and how watchlists can be selected. Section 4 describes our improved watchlist indexing in E. Both ENIGMAWatch and the improved watchlist indexing are evaluated in Sect. 5.

## 2 Guiding the Given Clause Selection in ATPs

### 2.1 Automated Theorem Proving and Machine Learning

State-of-the-art saturation-based automated theorem provers (ATPs) for first-order logic (FOL), such as E [33] and Vampire [25] employ the *given clause algorithm*, translating the input FOL problem $T \cup \{\neg C\}$ into a refutationally equivalent set of clauses. The search for a contradiction is performed maintaining sets of *processed* ($P$) and *unprocessed* ($U$) clauses (the *proof state* $\Pi$). The algorithm repeatedly selects a *given clause* $g$ from $U$, moves $g$ to $P$, and extends $U$ with all clauses inferred with $g$ and $P$. This process continues until a contradiction is found, $U$ becomes empty, or a resource limit is reached.

The search space of this loop grows quickly and it is a well-known fact that the selection of the right given clause is crucial for success. Machine learning from a large number of proofs and proof searches [1–4,7–10,15,16,19,20,22,26,29,31,32,38,40,41] may help guide the selection of the given clauses.

[2] http://tptp.cs.miami.edu/~tptp/MPTPChallenge/.

### 2.2 ENIGMA: Learning from Successful Proof Searches

ENIGMA [6,16–18] (*Efficient learNing-based Internal Guidance MAchine*) is our method for guiding given clause selection in saturation-based ATPs. The method needs to be efficient because it is internally applied to every generated clause. ENIGMA uses E's capability to analyze successful proof searches, and to output lists of given clauses annotated as either *positive* or *negative* training examples. Each processed clause which is present in the final proof is classified as positive. On the other hand, processing of clauses not present in the final proof was redundant, hence they are classified as negative. ENIGMA's goal is to learn such classification (possibly conditioned on the problem and its features) in a way that generalizes and allows solving new related problems.

**ENIGMA Learning and Models.** Given a set of problems $\mathcal{P}$, we can run E with a strategy $\mathcal{S}$ and obtain positive and negative training data $\mathcal{T}$ from each of the successful proof searches. Various machine learning methods can be used to learn the clause classification given by $\mathcal{T}$, each method yielding a *classifier* or a (classification) *model* $\mathcal{M}$. In order to use the model $\mathcal{M}$ in E, $\mathcal{M}$ is used as a function that computes clause weights. This weight function is then used to guide future E runs.

First-order clauses need to be represented in a format recognized by the selected learning method. While neural networks have been very recently practically used for internal guidance with ENIGMA [6], the strongest setting currently uses manually engineered *clause features* and fast non-neural state-of-the-art gradient boosted trees libraries such as XGBoost [5]. The model $\mathcal{M}$ produced by XGBoost consists of a set (*ensemble* [30]) of decision trees. Given a clause $C$, the model $\mathcal{M}$ yields the probability that $C$ represents a positive clause. When using $\mathcal{M}$ as a weight function in E, the probabilities are turned into binary classification, assigning weight 1.0 for probabilities $\geq 0.5$ and weight 10.0 otherwise.

**Clause Features.** Clause features represent a finite set of various syntactic properties of clauses, and are used to encode clauses by a fixed-length numeric vector. Various machine learning methods can handle numeric vectors and their success heavily depends on the selection of correct clause features. Various possible choices of efficient clause features for theorem prover guidance have been experimented with [16,17,22,23]. The original ENIGMA [16] uses term-tree walks of length 3 as features, while the second version [17] reaches better results by employing various additional features.

Since there are only finitely many features in any training data, the features can be serially numbered. This numbering is fixed for each experiment. Let $n$ be the number of different features appearing in the training data. A clause $C$ is translated to a feature vector $\varphi_C$ whose $i$-th member counts the number of occurrences of the $i$-th feature in $C$. Hence every clause is represented by a sparse numeric vector of length $n$. Additionally, we embed information about the conjecture currently being proved in the feature vector, yielding vectors of length $2n$. See [6,17] for more details.

**Feature Hashing.** Experiments revealed that XGBoost is capable of dealing with vectors up to the length of $10^5$ with a reasonable performance. In experiments with the whole translated Mizar Mathematical Library, the feature vector length can easily grow over $10^6$. This significantly increases both the training and the clause evaluation times. To handle such larger data sets, a simple *hashing* method has previously been implemented to decrease the dimension of the vectors.

Instead of serially numbering all features, we represent each feature $f$ by a unique string and apply a general-purpose string hashing function to obtain a number $n_f$ within a required range (between 0 and an adjustable *hash base*). The value of $f$ is then stored in the feature vector at the position $n_f$. If different features get mapped to the same vector index, the corresponding values are summed up. See [6] for more details.

### 2.3 ProofWatch: Proof Guidance by Clause Subsumption

In this section we explain the ProofWatch guiding mechanisms. Unlike the statistical approach in ENIGMA, ProofWatch implements a form of *symbolic* memory and guidance. It produces a notion of *proof-state vector* that is dynamically created and updated.

**Standard Watchlist Guidance.** The watchlist (hint list) mechanism itself does not perform any statistical machine learning. It steers given clause selection via symbolic matching between generated clauses and a set of clauses called a *watchlist*. This technique has been originally developed by Veroff [42] and implemented in Otter [27] and Prover9 [28]. Since then, it has been extensively used in the AIM project [24] for obtaining long and advanced proofs of open algebraic conjectures. The watchlist mechanism is nowadays implemented also in E. All the above implementations use only a single watchlist, as opposed to ProofWatch discussed below.

Recall that a clause $C$ *subsumes* a clause $D$, written $C \sqsubseteq D$, when there exists a substitution $\sigma$ such that $C\sigma \subseteq D$ (where clauses are considered to be sets of literals). The watchlist guidance then works as follows. Every generated clause $C$ is checked for subsumption with every watchlist clause $D \in W$. When $C$ subsumes at least one of the watchlist clauses, then $C$ is considered important for the proof search and is processed with high priority. The idea behind this is that the watchlist $W$ contains clauses which were processed during a previous successful proof search of a related conjecture. Hence processing of similar clauses may lead to success again.

In E, the watchlist mechanism is implemented using a priority function[3] which takes precedence over the weight function used to select the next given clause. Priority functions assign the priority to each clause, and clauses with higher priority are selected as given before clauses with lower priority[4].

[3] See the priority function `PreferWatchlist` in the E manual.

[4] Numerically the lower the priority, the better. Hence 0 is the best priority.

When clauses from previous proofs are put on a watchlist, E thus prefers to follow steps from the previous proofs whenever it can.

**ProofWatch.** Our approach [11, Sect. 5] extends standard watchlist guidance by allowing for multiple watchlists $W_1,\ldots,W_n$, for example, one corresponding to each related proof found before. We say that a generated clause $C$ *matches* the watchlist $W_i$, written $C \sqsubseteq W_i$, iff $C$ subsumes some clause $D \in W_i$ ($C \sqsubseteq D$). Similarly, the above watchlist clause $D$ is said to be *matched* by $C$.

The reason to include multiple watchlists is that during a proof search, clauses from some watchlists might get matched more often than clauses from others. The more clauses are matched from some watchlist $W_i$, the more the current proof search resembles $W_i$, and hence $W_i$ might be more relevant for this proof search. Thus the idea of ProofWatch is to prioritize clauses that match more relevant watchlists (proofs).

Watchlist *relevance* is dynamically computed as follows. We define $progress(W_i)$ to be the count of clauses from $W_i$ that have been matched in the proof search thus far. The *completion ratio*, $c_i = \frac{progress(W_i)}{|W_i|}$, measures how much of the watchlist $W_i$ has been matched. The *dynamic relevance* of each generated clause $C$ is defined as the maximum completion ratio over all the watchlists $W_i$ that $C$ matches:

$$relevance(C) = \max_{W \in \{W_i : C \sqsubseteq W_i\}} \left( \frac{progress(W)}{|W|} \right)$$

The higher the dynamic relevance $relevance(C)$, the higher the priority of $C$. The dynamic watchlist mechanism is implemented using the E priority function.[5] The results of experiments in [11, Sect. 6.3] on the same dataset as this work (Mizar40 [21]) indicate that dynamic relevance improves performance over an ensemble of strategies, whereas the single watchlist approach is stronger on each individual strategy.

When using a large problem library such as Mizar40, it is practically useful to choose only some proofs for watchlists. First, E's speed decreases with each additional proof on the watchlist, so if working on a large dataset, loading all available proofs as watchlists will lead to a large slowdown (cf. Sect. 4). Second, it's not guaranteed that all proofs will help E with proving the problem at hand.

## 3 ENIGMAWatch: ProofWatch Meets ENIGMA

### 3.1 Completion Ratios as Semantic Embeddings of the Proof Search

The watchlist completion ratios $(c_0, \ldots, c_N)$ ($N$ ranges over the watchlist proofs) at each step in E's proof search can be taken as a vectorial representation of the current proof state $\Pi$. The general motivation for this approach is to come up with an *evolving* characterization of the saturation-style proof state $\Pi$, preferably in a vectorial form $\varphi_\Pi$ suitable for machine learning tools, such as ENIGMA.

[5] See `PreferWatchlistRelevant` in [11].

Recall that the proof state $\Pi$ is a set of processed clauses $P$ and unprocessed clauses $U$. The vector of watchlist completion ratios thus maintains a running tally of where clauses in $P \cup U$ match the different related proofs. In general, this could be replaced, e.g., by a vector of more abstract similarities of the current proof state to other proofs measured in various (possibly approximate) ways. In ENIGMAWatch we use the ProofWatch based *proof-state vector* for a proof state $\Pi$ defined by the completion ratios, i.e., $\varphi_\Pi = (c_0, \ldots, c_N)$. This is the first practical implementation of the general idea: using *semantic embeddings* (i.e., representations in $R^n$) of the proof state $\Pi$ for guiding statistical learning methods. ENIGMAWatch uses the proof-state vectors $\varphi_\Pi$ as follows. The positive $\mathcal{C}^+$ and negative $\mathcal{C}^-$ given clauses are output along with $\varphi_\Pi$, the proof-state vector at the time of their selection, and used as added features of the proof state when training ENIGMA-style classifiers.

**Table 1.** Example of the proof-state vector for 8 (of 32) (serially numbered) proofs loaded to guide the proof of `YELLOW_5:36`. The three columns are the watchlist $i$, the completion ratio of $i$, and $progress(W_i)/|W_i|$.

| | | | | | | | | | | | |
|---|---|---|---|---|---|---|---|---|---|---|---|
| 0 | 0.438 | 42/96 | 1 | 0.727 | 56/77 | 2 | 0.865 | 45/52 | 3 | 0.360 | 9/25 |
| 4 | 0.750 | 51/68 | 5 | 0.259 | 7/27 | 6 | 0.805 | 62/77 | 7 | 0.302 | 73/242 |

Table 1 shows a sample proof-state vector based on 32 related proofs[6] for the Mizar theorem **YELLOW 5:36**[7] (De Morgan's law[8]) at the end of the proof search. Note that some related proofs, such as #2, were almost fully matched, while others, such as #7 were mostly not matched in the proof search.

### 3.2 Proof Vector Construction

**Data Construction.** In the ProofWatch [11] experiments, the best method for selecting related proofs (watchlists) was to use k-nearest neighbor (k-NN) to recommend 32 proofs per problem. The watchlists there are thus problem specific. In ENIGMAWatch, we want the watchlists to be globally fixed across the whole library, so that the proof completion ratios have the same meaning in all proofs. To construct the proof vectors, we first use a strong E strategy to produce a set of initial proofs (14882 over the 57897 Mizar40 problems). Then we run E with ProofWatch and the same strategy over the full 57897 problems with the 14882 proofs loaded into the watchlist. The time limit for both runs was *T60-G10000*, which means that E stops after 60 s or 10000 generated clauses. This data provides information on how often each watchlist was encountered in each successful proof search. The training data then

[6] The proofs were chosen via k-NN. See [11, Sec. 6.1] for details.
[7] http://grid01.ciirc.cvut.cz/~mptp/7.13.01_4.181.1147/html/yellow_5#T36.
[8] $\neg(P \vee Q) \iff (\neg P) \wedge (\neg Q)$.

consists of a proof vector for each given clause (for each conjecture/problem): $(conjecture, given\text{-}clause, proof\text{-}state\ vector)$.

**Dimensionality Reduction.** Next, we experiment with various pre-processing methods to reduce the *proof-state vector* dimension and thus decrease the number of watchlists loaded in E. For each problem we compute the mean of proof-state vectors over all given clauses $g$: $\frac{1}{\#g}\sum_g \varphi_{\Pi_g}$. This vector consists of the averaged completion ratios for each watchlist, which will be higher if the watchlist was matched earlier in the proof. This results in the mean proof-state matrix $M$ consisting of row vectors (*mean-proof-vector*) (one for each conjecture/problem).

The following are methods experimented with in this paper for constructing the globally fixed vector of 512 watchlists from matrix $M$:

- *Mean*: compute the mean of $M$ across the rows to obtain a mean proof-state vector that contains for each watchlist its average use across all problems. Then we take the top 512 watchlists.
- *Corr*: compute the Pearson correlation matrix[9] based on (the transpose of) $M$, and find a relatively uncorrelated set of 512 watchlists.
- *Var*: compute the variance (across the rows) of each column in $M$, and take the 512 watchlists with the highest variance. The intuition is that watchlists whose completion ratio vary more over the problem corpus may be more useful for learning.
- *Rand*: randomly select 512 watchlists.

## 4 Multi-indices Subsumption Indexing

In order to determine whether a generated clause matches a watchlist, the generated clause must be checked for subsumption with every watchlist clause. A major limitation of previous work [11,12] was the slowdown of E as the watchlist size increased beyond 4000 clauses. Including more than 128 proofs was impractical. This section describes a method we have developed to speed up watchlist matching.

E already implements feature vector indexing [34] used also for the purpose of watchlist matching. The watchlist clauses are inserted into an indexing data structure and various properties of clauses are used to prune possible subsumption candidates. In this way, the number of possibly expensive subsumption calls is reduced. We build upon this, and further limit the number of required subsumption checks by using multiple indices instead of a single index.[10]

We take advantage of the fact that a clause $C$ cannot subsume a clause $D$ if the top-level predicate symbols do not match. In particular, $C \sqsubseteq D$ can only hold if all the predicate symbols from $C$ also appear in $D$, because substitution can neither introduce nor remove predicate symbols from a clause.

[9] https://docs.scipy.org/doc/numpy/reference/generated/numpy.corrcoef.html.

[10] Even with multiple watchlists, all the watchlist clauses are inserted into a single index, and only the name of the original watchlist is additionally stored.

We define the *code* of a clause $C$, denoted code($C$), as the set of predicate symbols with their logical signs (either + for positive predicates, or − for negated ones). For example, the code of the clause "$P(a) \vee \neg P(b) \vee P(f(x))$" is the set $\{+P, -P\}$. The following holds because codes are preserved under substitution.

**Lemma 1.** *Given clauses $C$ and $D$, $C \sqsubseteq D$ implies* $\text{code}(C) \subseteq \text{code}(D)$.

We create a separate index for every different clause code. Each watchlist clause $D$ is inserted only to the index corresponding to code($D$). In order to check whether some clause $C$ matches a watchlist, we only need to search in the indices whose codes are supersets of (or equal to) code($C$). Each index is implemented using E's native feature vector indexing structure. Evaluation of this simple indexing method is provided in Sect. 5.1.

**Table 2.** Evaluation of multi-indices subsumption indexing.

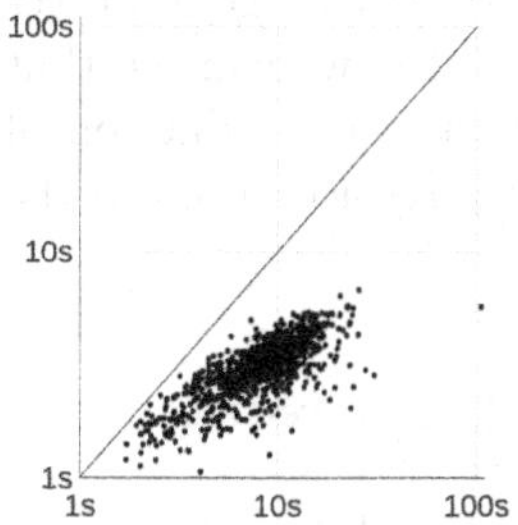


| | runtime (left graph ←) | | |
|---|---|---|---|
| | single | multi | speedup |
| avg | 9.23s | 3.16s | 2.9× |
| best | 105.3s | 5.7s | 18.5× |
| worst | 2.26s | 2.09s | 1.08× |
| | subsumptions (right →) | | |
| | single | multi | reduction |
| avg | 2328k | 52k | 44.1× |
| best | 3059 | 1 | 3059× |
| worst | 709k | 367k | 1.9× |

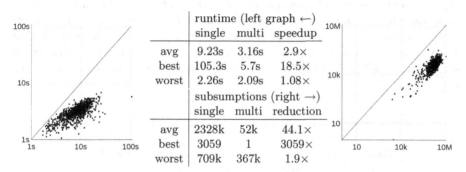

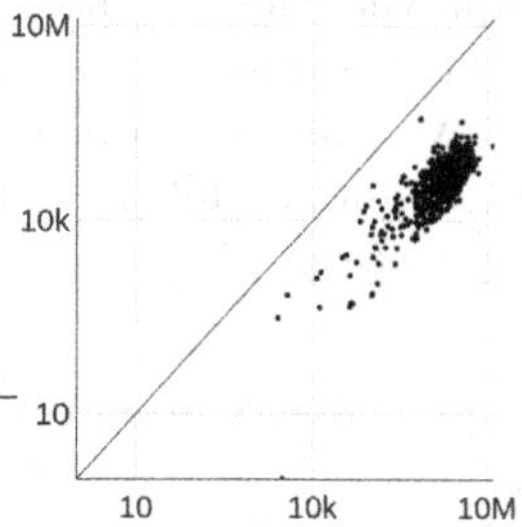


## 5 Experiments

This section describes the experimental evaluation[11] of

1. the improved watchlist mechanism from Sect. 4
2. the watchlist selection for ENIGMAWatch from Sect. 3

### 5.1 Multi-indices Subsumption Indexing Evaluation

We propose a simple experiment to evaluate our implementation of multi-indices subsumption indexing from Sect. 4. We take a random sample of 1000 problems from the Mizar40 [21] data set and create a watchlist with around 60 k clauses coming from proofs of problems similar to the sample problems. We then run E

[11] Experiments code and data are available at https://github.com/ai4reason/eprover-data/tree/master/TABLEAUX-19
All experiments are run on the same hardware: Intel(R) Xeon(R) Gold 6140 CPU @ 2.30 GHz with 188 GB RAM.

on the sample problems with a fixed limit of 1000 generated clauses. This gives us a measure of how fast the single-index and multi-indices versions are, that is, how fast they can generate the first 1000 clauses. As the watchlist indexing does not influence the proof search, both versions process the same clauses and output the same result. Each generated clause has to be checked for watchlist subsumption and hence the limit on generated clauses is also the limit on different watchlist checks. We expect the number of clause-to-clause subsumption checks to decrease with multi-indices, as the method prunes possible subsumption candidates.

The results of the experiments are presented in Table 2. For each problem, we measure the runtime (left graph) and the number of different clause subsumption calls (right graph). The suffix "s" stands for seconds, "k" stands for thousands, and "M" stands for millions. Although subsumption is also used for purposes other than watchlist matching, we should be able to observe a decrease in the number of calls. Each point in the graphs corresponds to one sample problem, and is drawn at the position $(x, y)$ corresponding to the results of single-index $(x)$ and multi-indices $(y)$ versions. Hence points below the diagonal signify an improvement. Also note logarithmic axes. The table shows the average improvement, and also the best and the worst cases. From the results, we can see that an average speed-up is almost 3 times. Furthermore, the average reduction of subsumption calls is more than 44 times and the number is reduced even in the worst case.

**Table 3.** ProofWatch evaluation: Problems solved by different versions.

| Baseline | Mean | Var | Corr | Rand | Baseline ∪ Mean | Total |
|---|---|---|---|---|---|---|
| 1140 | 1357 | 1345 | 1337 | 1352 | 1416 | 1483 |

**Table 4.** ENIGMAWatch evaluation: Problems solved and the effect of looping.

| loop | ENIGMA | Mean | Var | Corr | Rand | ENIGMA ∪ Mean | Total |
|---|---|---|---|---|---|---|---|
| 0 | 1557 | 1694 | 1674 | 1665 | 1690 | 1830 | 1974 |
| 1 | 1776 | 1815 | 1812 | 1812 | 1847 | 1983 | 2131 |
| 2 | 1871 | 1902 | 1912 | 1882 | 1915 | 2058 | 2200 |
| 3 | 1931 | 1954 | 1946 | 1920 | 1926 | 2110 | 2227 |

The number of watchlist clauses in the experiments was 61501, and the multi-indices version used 11442 different indices. This means that there were less than 6 clauses per index in average, although the count of clauses in different indices varied from 1 to 3837. The most crowded index was for the code $\{+ =\}$, that is, for positive equality clauses. Finally, 6955 indices contained only a single clause.

### 5.2 Experimental Evaluation of ENIGMAWatch

The experiments are done on a random subset of 5000 Mizar40 [21] problems. The time limit of 60 s and 30000 generated clauses is used to allow a comparison to be done without regard for the differences in clause processing speed. The 30000 is approximately the average number of clauses that the baseline strategy generates in 10 s. Table 3 provides the evaluation of different watchlist selection mechanims using ProofWatch (without ENIGMA) and making use of the improved watchlist indexing. The last two columns show the number of problems solved by (1) the Baseline together with Mean, and by (2) all the five methods. This shows the relative complementarity of the methods. We can see that the Mean method yields the best results, reaching more than 15% improvement over the baseline strategy. The Rand method is however quite competitive.

Table 4 provides the evaluation of ENIGMAWatch and its comparison to ENIGMA. The experiments are done in multiple loops, where in each loop all the proof-runs in prior loops can be used as training data. This way ENIGMA can learn increasingly effective models.

We can see that ENIGMAWatch can attain superior performance to ENIGMA. The relation of looping and results is interesting. The largest absolute improvement over ENIGMA is in loop 0 – 8.8% by the Mean method. This however drops to 1.2% in loop 4. In loops 1 and 2, Rand is the strongest, but Mean ends up being the best in loop 3. In total, all the ENIGMA and ENIGMAWatch methods solve together nearly twice as many problems as the baseline strategy. Figure 1 shows the results of running ENIGMA and Mean for 13 loops. The rate of improvement slows down, both methods eventually converge to a similar level of performance, and the union of the two is ca. 150 problems better.

**Table 5.** ENIGMA and ENIGMAWatch: Model and training statistics.

| Model | Pos. acc | Neg. acc | Features | Watchlist F | Train size | Train time |
|---|---|---|---|---|---|---|
| ENIGMA0 | 99.12% | 92.16% | 5061 | 0 | 0.4 GB | 14 min |
| ENIGMA1 | 97.39% | 86.82% | 7071 | 0 | 0.8 GB | 31 min |
| ENIGMA2 | 96.13% | 83.92% | 8089 | 0 | 1.4 GB | 55 min |
| ENIGMA3 | 95.39% | 82.5% | 8662 | 0 | 2.0 GB | 85 min |
| Mean0 | 99.05% | 92.59% | 5424 | 308 | 2.9 GB | 19 min |
| Mean1 | 96.92% | 88.16% | 6950 | 316 | 6.2 GB | 29 min |
| Mean2 | 95.75% | 86.46% | 7809 | 331 | 9.6 GB | 38 min |
| Mean3 | 95.04% | 85.24% | 8313 | 330 | 13.0 GB | 39 min |

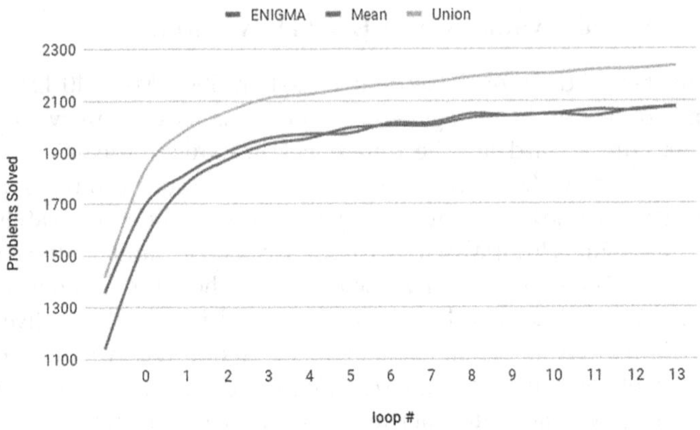


**Fig. 1.** Convergence: The improvement of ENIGMA and Mean decreases over 13 loops, and their performance converges. The Union is consistently ca. 150 problems better.

### 5.3 Training, Model Statistics and Analysis

The XGBoost models used in our experiments are trained with a maximum tree depth of 9 and 200 rounds (which means 200 trees are learned). There are 300000 features in the 5000 problem dataset hashed into $2^{15}$ buckets. Combining clause and conjecture features with the watchlist completion ratios, XGBoost makes its predictions based on 66048 features ($2 \cdot 2^{15}$ plus the count of completion ratios).

Table 5 provides various training and model statistics of the ENIGMA and ENIGMAWatch models and their loops. The columns "Pos. Acc." and "Neg. Acc." describe the training accuracy of the models on positive and negative training examples. The column "Features" presents the number of features referenced in the decision trees. We see that the models use a small fraction of all the 66048 available features. The column "Watchlist F." provides the number of watchlist features out of all the used features. Finally, "Train Size" and "Train Time" specify the size of the input training file (in GB) and training times (in minutes). The XGBoost models after the training are smaller than 4 MB.

We can see that the accuracy decreases with the increase of the training data size, but the number of theorems proved increases. About 62% of the watchlists are judged as useful by XGBoost and used in the decision trees. Figure 2 shows the root of the first decision tree of the Mean model in loop 3. Green means "yes" (the condition holds), red means "no", and blue means that the feature is not present. The multi line box is a (shortened) bucket of features, and single line boxes correspond to watchlists (#194, etc.). We can see that ENIGMAWatch uses a watchlist feature for the very first decision when judging newly generated clauses. This shows that the features that characterize the evolving proof state are indeed considered very significant by the methods that automatically learn given clause guidance.

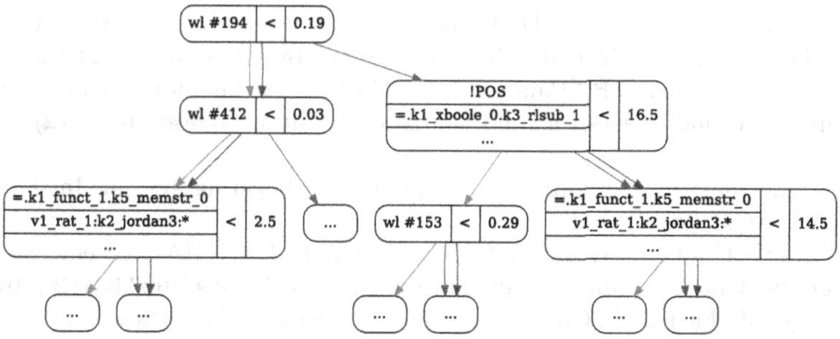


**Fig. 2.** Example of an XGBoost decision tree.

## 6 Conclusion and Future Work

We have produced and evaluated the first practically usable version of the ENIGMAWatch system which can now be efficiently used over large mathematical datasets. The previous experiments with the first prototype on the small MPTP Challenge [12] demonstrated that ENIGMAWatch can find proofs faster (in terms of how many processed clauses are needed). The work presented here shows that with improved subsumption indexing, feature hashing, and suitable global watchlist selection, ENIGMAWatch outperforms ENIGMA on the large Mizar40 dataset. In particular, ENIGMAWatch significantly outperforms both ProofWatch and ENIGMA when used without looping. With several MaLARea-style [37,40] iterations of proving and learning, the difference to ENIGMA gets smaller, however the two methods are still quite complementary, providing solutions to a large number of different problems. In total, all the ENIGMA and ENIGMAWatch methods (Table 4) together solve almost twice as many problems as the baseline strategy after four iterations of learning and proving.

The system is ready to be used on hard problems and to expand the set of Mizar problems for which an ATP proof has been found. Future work includes refining the watchlist selection, defining more sophisticated methods of computing the proof completion ratios, analyzing the learned decision tree models to see which watchlists are the most useful, and also defining further and more abstract meaningful representations and embeddings of saturation-style proof search.

## References

1. Alama, J., Heskes, T., Kühlwein, D., Tsivtsivadze, E., Urban, J.: Premise selection for mathematics by corpus analysis and kernel methods. J. Autom. Reasoning **52**(2), 191–213 (2014)
2. Alemi, A.A., Chollet, F., Eén, N., Irving, G., Szegedy, C., Urban, J.: DeepMath - deep sequence models for premise selection. In: Lee, D.D., Sugiyama, M., Luxburg, U.V., Guyon, I., Garnett, R. (eds.) Advances in Neural Information Processing Systems 29: Annual Conference on Neural Information Processing Systems 2016, 5–10 December 2016, Barcelona, Spain, pp. 2235–2243 (2016)

3. Blanchette, J.C., Greenaway, D., Kaliszyk, C., Kühlwein, D., Urban, J.: A learning-based fact selector for Isabelle/HOL. J. Autom. Reasoning **57**(3), 219–244 (2016)
4. Bridge, J.P., Holden, S.B., Paulson, L.C.: Machine learning for first-order theorem proving - learning to select a good heuristic. J. Autom. Reasoning **53**(2), 141–172 (2014)
5. Chen, T., Guestrin, C.: Xgboost: a scalable tree boosting system. In: KDD, pp. 785–794. ACM (2016)
6. Chvalovský, K., Jakubuv, J., Suda, M., Urban, J.: ENIGMA-NG: efficient neural and gradient-boosted inference guidance for E. CoRR, abs/1903.03182 (2019)
7. Denzinger, J., Fuchs, M., Goller, C., Schulz, S.: Learning from Previous Proof Experience. Technical Report AR99-4, Institut für Informatik, Technische Universität München (1999)
8. Ertel, W., Schumann, J., Suttner, C.B.: Learning heuristics for a theorem prover using back propagation. In: Retti, J., Leidlmair, K. (eds.) Österreichische Artificial Intelligence-Tagung, Igls, Tirol, vol. 208, pp. 87–95. Springer, Heidelberg (1989). https://doi.org/10.1007/978-3-642-74688-8_10
9. Färber, M., Brown, C.: Internal Guidance for Satallax. In: Olivetti, N., Tiwari, A. (eds.) IJCAR 2016. LNCS (LNAI), vol. 9706, pp. 349–361. Springer, Cham (2016). https://doi.org/10.1007/978-3-319-40229-1_24
10. Gauthier, T., Kaliszyk, C.: Premise selection and external provers for HOL4. In: Certified Programs and Proofs (CPP 2015). LNCS. Springer, Berlin (2015). https://doi.org/10.1145/2676724.2693173
11. Goertzel, Z., Jakubův, J., Schulz, S., Urban, J.: ProofWatch: watchlist guidance for large theories in E. In: Avigad, J., Mahboubi, A. (eds.) ITP 2018. LNCS, vol. 10895, pp. 270–288. Springer, Cham (2018). https://doi.org/10.1007/978-3-319-94821-8_16
12. Goertzel, Z., Jakubuv, J., Urban, J.: ProofWatch meets ENIGMA: first experiments. In: Barthe, G., Korovin, K., Schulz, S., Suda, M., Sutcliffe, G., Veanes, M. (eds.) LPAR-22 Workshop and Short Paper Proceedings, vol. 9, pp. 15–22, Kalpa Publications in Computing. EasyChair (2018)
13. Gottlob, G., Sutcliffe, G., Voronkov, A. (eds.): Global Conference on Artificial Intelligence, GCAI 2015, Tbilisi, Georgia, October 16–19, 2015, vol. 36, EPiC Series in Computing. EasyChair (2015)
14. Grabowski, A., Korniłowicz, A., Naumowicz, A.: Mizar in a nutshell. J. Formalized Reasoning **3**(2), 153–245 (2010)
15. Jakubův, J., Urban, J.: Hierarchical invention of theorem proving strategies. AI Commun. **31**(3), 237–250 (2018)
16. Jakubův, J., Urban, J.: ENIGMA: efficient learning-based inference guiding machine. In: Geuvers, H., England, M., Hasan, O., Rabe, F., Teschke, O. (eds.) CICM 2017. LNCS (LNAI), vol. 10383, pp. 292–302. Springer, Cham (2017). https://doi.org/10.1007/978-3-319-62075-6_20
17. Jakubův, J., Urban, J.: Enhancing ENIGMA given clause guidance. In: Rabe, F., Farmer, W.M., Passmore, G.O., Youssef, A. (eds.) CICM 2018. LNCS (LNAI), vol. 11006, pp. 118–124. Springer, Cham (2018). https://doi.org/10.1007/978-3-319-96812-4_11
18. Jakubuv, J., Urban, J.: Hammering Mizar by learning clause guidance. CoRR, abs/1904.01677 (2019)
19. Kaliszyk, C., Urban, J.: Learning-assisted automated reasoning with Flyspeck. J. Autom. Reasoning **53**(2), 173–213 (2014)

20. Kaliszyk, C., Urban, J.: FEMaLeCoP: fairly efficient machine learning connection prover. In: Davis, M., Fehnker, A., McIver, A., Voronkov, A. (eds.) LPAR 2015. LNCS, vol. 9450, pp. 88–96. Springer, Heidelberg (2015). https://doi.org/10.1007/978-3-662-48899-7_7
21. Kaliszyk, C., Urban, J.: MizAR 40 for Mizar 40. J. Autom. Reasoning **55**(3), 245–256 (2015)
22. Kaliszyk, C., Urban, J., Michalewski, H., Olšák, M.: Reinforcement learning of theorem proving. In: Advances in Neural Information Processing Systems 31: Annual Conference on Neural Information Processing Systems 2018, NeurIPS 2018, Montréal, Canada, 3–8 December 2018, pp. 8836–8847 (2018)
23. Kaliszyk, C., Urban, J., Vyskocil, J.: Efficient semantic features for automated reasoning over large theories. In: IJCAI, pp. 3084–3090. AAAI Press (2015)
24. Kinyon, M., Veroff, R., Vojtěchovský, P.: Loops with abelian inner mapping groups: an application of automated deduction. In: Bonacina, M.P., Stickel, M.E. (eds.) Automated Reasoning and Mathematics. LNCS (LNAI), vol. 7788, pp. 151–164. Springer, Heidelberg (2013). https://doi.org/10.1007/978-3-642-36675-8_8
25. Kovács, L., Voronkov, A.: First-order theorem proving and VAMPIRE. In: Sharygina, N., Veith, H. (eds.) CAV 2013. LNCS, vol. 8044, pp. 1–35. Springer, Heidelberg (2013). https://doi.org/10.1007/978-3-642-39799-8_1
26. Loos, S.M., Irving, G., Szegedy, C., Kaliszyk, C.: Deep network guided proof search. In: Eiter, T., Sands, D. (eds.) LPAR-21, 21st International Conference on Logic for Programming, Artificial Intelligence and Reasoning, Maun, Botswana, 7–12 May 2017. EPiC Series in Computing, vol. 46, pp. 85–105. EasyChair (2017)
27. McCune, W., Wos, L.: Otter: the CADE-13 competition incarnations. J. Autom. Reasoning **18**(2), 211–220 (1997). Special Issue on the CADE 13 ATP System Competition
28. McCune, W.W.: Prover9 and Mace4. http://www.cs.unm.edu/~mccune/prover9/, 2005–2010. Acccessed 29 Mar 2016
29. Piotrowski, B., Urban, J.: ATPBOOST: learning premise selection in binary setting with ATP feedback. In: Galmiche, D., Schulz, S., Sebastiani, R. (eds.) IJCAR 2018. LNCS (LNAI), vol. 10900, pp. 566–574. Springer, Cham (2018). https://doi.org/10.1007/978-3-319-94205-6_37
30. Polikar, R.: Ensemble based systems in decision making. IEEE Circuits Syst. Mag. **6**(3), 21–45 (2006)
31. Schäfer, S., Schulz, S.: Breeding theorem proving heuristics with genetic algorithms. In: Gottlob et al. [13], pp. 263–274 (2015)
32. Schulz, S.: Learning search control knowledge for equational deduction. In: DISKI, vol. 230, Infix Akademische Verlagsgesellschaft (2000)
33. Schulz, S.: E - a brainiac theorem prover. AI Commun. **15**(2–3), 111–126 (2002)
34. Schulz, S.: Simple and efficient clause subsumption with feature vector indexing. In: Bonacina, M.P., Stickel, M.E. (eds.) Automated Reasoning and Mathematics. LNCS (LNAI), vol. 7788, pp. 45–67. Springer, Heidelberg (2013). https://doi.org/10.1007/978-3-642-36675-8_3
35. Schulz, S.: System description: E 1.8. In: McMillan, K., Middeldorp, A., Voronkov, A. (eds.) LPAR 2013. LNCS, vol. 8312, pp. 735–743. Springer, Heidelberg (2013). https://doi.org/10.1007/978-3-642-45221-5_49
36. Urban, J.: MPTP 0.2: Design, implementation, and initial experiments. J. Autom. Reasoning **37**(1–2), 21–43 (2006)
37. Urban, J.: MaLARea: a metasystem for automated reasoning in large theories. In: Sutcliffe, G., Urban, J., Schulz, S. (eds.) ESARLT, vol. 257, CEUR Workshop Proceedings. CEUR-WS.org (2007)

38. Urban, J.: BliStr: the blind strategymaker. In: Gottlob et al. [13], pp. 312–319 (2013)
39. Urban, J., Sutcliffe, G.: ATP cross-verification of the mizar MPTP challenge problems. In: Dershowitz, N., Voronkov, A. (eds.) LPAR 2007. LNCS (LNAI), vol. 4790, pp. 546–560. Springer, Heidelberg (2007). https://doi.org/10.1007/978-3-540-75560-9_39
40. Urban, J., Sutcliffe, G., Pudlák, P., Vyskočil, J.: MaLARea SG1 - machine learner for automated reasoning with semantic guidance. In: Armando, A., Baumgartner, P., Dowek, G. (eds.) IJCAR 2008. LNCS (LNAI), vol. 5195, pp. 441–456. Springer, Heidelberg (2008). https://doi.org/10.1007/978-3-540-71070-7_37
41. Urban, J., Vyskočil, J., Štěpánek, P.: MaLeCoP machine learning connection prover. In: Brünnler, K., Metcalfe, G. (eds.) TABLEAUX 2011. LNCS (LNAI), vol. 6793, pp. 263–277. Springer, Heidelberg (2011). https://doi.org/10.1007/978-3-642-22119-4_21
42. Veroff, R.: Using hints to increase the effectiveness of an automated reasoning program: case studies. J. Autom. Reasoning **16**(3), 223–239 (1996)

# Logics for Program or System Verification

# Behavioral Program Logic

Eduard Kamburjan(✉)

Department of Computer Science, Technische Universität Darmstadt, Darmstadt, Germany
kamburjan@cs.tu-darmstadt.de

**Abstract.** We present Behavioral Program Logic (BPL), a dynamic logic for trace properties that incorporates concepts from behavioral types and allows reasoning about non-functional properties within a sequent calculus. BPL uses *behavioral modalities* $[\mathtt{s} \Vdash \tau]$, to verify statements s against *behavioral specifications* $\tau$. Behavioral specifications generalize postconditions and behavioral types. They can be used to specify other static analyses, e.g., data flow analyses. This enables deductive reasoning about the results of multiple analyses on the same program, potentially implemented in different formalisms. Our calculus for BPL verifies the behavioral specification gradually, as common for behavioral types. This vastly simplifies specification, calculus and composition of local results. We present a sequent calculus for object-oriented actors with futures that integrates a pointer analysis and bridges the gap between behavioral types and deductive verification.

## 1 Introduction

When reasoning about concurrent programs, the intermediate states of an execution are of more relevance than when reasoning about sequential programs. In an object-oriented setting, it does not suffice to specify pre- and postcondition of some method m. Instead, the *traces* generated by m must be specified.

Recently, dynamic logics for trace properties have been developed [3,7,12] to leverage well-established verification techniques from dynamic logic [1] to a concurrent setting. The application of these approaches to real world models of distributed systems [13,25] revealed two shortcomings: (1) the composition of method-local verification results to a guarantee for the whole system is not automatic and (2) the specification of trace properties is too complex. Thus, the current approaches are deemed as not practical for serious verification efforts.

Another group of verification techniques, *behavioral types*, aim "*to describe properties associated with the behavior of programs and in this way also describe how a computation proceeds.*" [20]. For object-oriented languages, behavioral types can also be seen as specifications of traces of methods. Behavioral types, especially session types [19], are restricted in their expressive power to easily compose their local results to global guarantees, and are natural specifications

This work is supported by the FormbaR project, part of AG Signalling/DB RailLab.


S. Cerrito and A. Popescu (Eds.): TABLEAUX 2019, LNAI 11714, pp. 391–408, 2019.
https://doi.org/10.1007/978-3-030-29026-9_22

for protocols. However, they lack precision when handling state [5] or require additional static analyses [23]. For Active Objects [9] (object-oriented actors with futures), a translation from session types to a trace logic has been given [23].

We introduce *Behavioral Program Logic* (BPL) to combine precise state reasoning from program logics with the relative simplicity of behavioral types and enable the integration of static analyses into deductive reasoning. The main difference to previous approaches in dynamic logic for trace properties is the *behavioral modality* $[\mathtt{s} \Vdash \tau]$, which expresses that all traces of statement s satisfy specification $\tau$. The specification $\tau$ is not a formula, as the postcondition of modalities in classical dynamic logic, but is a specification translated into a monadic second order formula over traces. Similarly to behavioral types, $\tau$ may contain syntactic elements and allows to syntactically match with s. Sequent calculi for BPL may reduce s *and* $\tau$ in one rule. Contrary to previous dynamic logics for traces, behavioral specifications are more succinct and easier to compose and decompose by, e.g., using the projection mechanism of session types.

We distinguish between behavioral types, that have a sequent calculus of the above kind, and *behavioral specifications*, which do not. Behavioral specifications interface with external properties, such as a data-flow points-to analysis. Beyond integrating external analyses into the sequent calculus, this modularizes the sequent calculus by expressing different properties with different behavioral specifications. Behavioral specifications are clear interfaces that allow to close proofs once more context is known and generalize proof repositories [6].

Our main contributions are (1) BPL, a trace program logic that integrates deductive reasoning with static analyses (2) *method types*, a behavioral type in BPL that generalizes method contracts, object invariants and local types for Active Objects. Due to space constraints, we do not give (de-)compositions and full semantics and refer to [23] and our technical report [22] for full details. We introduce our programming language in Sect. 2 and BPL in Sect. 3. In Sect. 4 we introduce method types. Sect. 5 summarizes previous approaches and concludes.

## 2 Preliminaries: An Actor Language with Futures

We introduce Behavioral Program Logic using a Core Active Object language [9] (*CAO*) with futures, *CAO* uses strong encapsulation (i.e., all fields are object-private) and cooperative scheduling. *CAO* is based on ABS [21] and we use a locally abstract, globally concrete (LAGC) semantics [11]. An LAGC semantics consists of two layers: A locally abstract (LA) layer for statements and methods, and a globally concrete (GC) layer for objects and systems. The LA layer is a denotational semantics that abstractly describes the behavior of a method in every possible context, while the GC layer is an operational semantics that concretizes the LA semantics of processes in a concrete context. LA semantics enable one to analyze a method in isolation and Active Objects allow us to demonstrate that BPL is suited for complex concurrency models.

**Definition 1 (Syntax).** *Let* $\sim$ *range over* `&&`, `||`, `+`, `-`, `*`, `/`, `>=`, `>`, `<`, `<=`, `v` *over variables,* `f` *over fields,* `C` *over class names,* `m` *over method names,* $i$ *and* `n` *over* $\mathbb{N}$. *The* syntax *of CAO is defined in Fig. 1, where* $\overrightarrow{\cdot}$ *denotes (possibly empty) lists.*

```
Prgm ::= →Class Main   Main ::= main{si}   Class ::= class C (→(C f)){→Field →Meth}   Field ::= D f = e;
Meth ::= D m(→(D v)){s;return e;}     D ::= Int | Bool | Fut<D>     si ::= C v = C(→v); si | v!m(→e)
  s ::= [D] l = e | [D] v = e.get_i | [D] v = f!m(→e) | skip | while(e){s} | if(e){s}else{s} | s;s
  e ::= l | n | True | False | e ~ e | !e | -e       l ::= this.f | v
```

**Fig. 1.** Syntax of *CAO*.

A program consists of a set of classes and a main block. The main block contains object instantiations and a method call to initialize the communication. All objects are created at once, not in the order of their instantiations. A class contains (1) parameter fields which reference other objects, (2) fields for data, initialized upon creation and (3) methods. Multiple instances can share their parameters. Parameters cannot be reassigned. As data types we use integers, booleans and parametric futures. Each class has a `run` method that is started upon creation. We omit `run` if it is empty.

```
class T(Comp S, Log L){
 Int test(Int i){
  Fut<Int> f = S!cmp(i);
  Int r = f.get_0;
  if(r < 0){
   r = -r;
   f = L!log(i);
  }
  return r;
 }
}
```

**Fig. 2.** An example method

Statement `v = f!m(`$\overrightarrow{\texttt{e}}$`)` calls method `m` asynchronously on the object `f` with parameters $\overrightarrow{\texttt{e}}$. A fresh future is generated and stored in `v`. This future identifies the called process. We say that the called process will *resolve* the future by executing **return** `e` and storing the value of `e` in the future. The synchronizing statement `v = e.`**get**$_i$ reads from the future in `e` into `v`. Until the future is resolved, the reading process blocks its object. The identifier $i$ is used to distinguish multiple synchronization points. The other statements, expressions and methods are standard.

*Example 1.* Figure 2 shows a simple method that passes its input to `Comp.cmp` and reads the result. If the result is negative, its sign is inverted and the original input data is logged by `Log.log`. The possibly inverted result is returned.

The semantics of a method is a set of symbolic traces, to describe the behavior of the method in every possible context, i.e., for every possible heap, call parameters and accessed futures. Additionally to semantic values (semantic values are, e.g., object identifiers, rationals, futures etc.), symbolic traces contain symbolic expressions. Structurally, symbolic expressions mirror syntactic expressions, but do not contain variables or fields. Instead they contain symbolic values and symbolic fields. Symbolic values have no operations defined on them and act as placeholders. They are replaced by semantic values once the method is running and the context is known. Symbolic fields are special symbolic values that contain the name of the field they are abstracting.

**Definition 2 (Symbolic Expressions).** *Let $v$ range over semantic values, $\underline{v}$ over symbolic values, $i$ over $\mathbb{N}$ and $\underline{\textbf{this}.\mathtt{f}}_i$ over symbolic fields.* Symbolic expressions $\underline{\mathtt{e}}$ *are defined below. We highlight symbolic elements by underlining.*

$$\underline{\mathtt{e}} ::= \underline{\mathtt{e}} \sim \underline{\mathtt{e}} \mid \mathtt{!}\underline{\mathtt{e}} \mid \mathtt{-}\underline{\mathtt{e}} \mid \underline{v} \mid v \mid \underline{\textbf{this}.\mathtt{f}}_i$$

To model the points where processes, objects and futures interact, traces contain events as markers for visible communication.

**Definition 3 (Events).** Events *are defined by the following grammar.*

$$\mathsf{ev} ::= \mathsf{invEv}(\underline{\mathtt{X}}, \underline{\mathtt{X}}', \underline{f}, \mathtt{m}, \overrightarrow{\underline{\mathtt{e}}}) \mid \mathsf{invREv}(\underline{\mathtt{X}}, \underline{f}, \mathtt{m}, \overrightarrow{\underline{\mathtt{e}}}) \mid \mathsf{futEv}(\underline{\mathtt{X}}, \underline{f}, \mathtt{m}, \underline{\mathtt{e}}) \mid \mathsf{futREv}(\underline{\mathtt{X}}, \underline{f}, \underline{\mathtt{m}}, \underline{\mathtt{e}}, i) \mid \mathsf{noEv}$$

Event $\mathsf{invEv}(\underline{\mathtt{X}}, \underline{\mathtt{X}}', \underline{f}, \mathtt{m}, \overrightarrow{\underline{\mathtt{e}}})$ models a call from $\underline{\mathtt{X}}$ to $\underline{\mathtt{X}}'$ on method $\mathtt{m}$ with future $\underline{f}$ and call parameters $\overrightarrow{\underline{\mathtt{e}}}$. The future and the callee may be symbolic: locally it is not possible to know the used future and the called object. Event $\mathsf{invREv}(\underline{\mathtt{X}}, \underline{f}, \mathtt{m}, \overrightarrow{\underline{\mathtt{e}}})$ is the callee view on a call. The object $\underline{\mathtt{X}}$ here is the callee, the caller is not visible to the callee. Event $\mathsf{futEv}(\underline{\mathtt{X}}, \underline{f}, \mathtt{m}, \underline{\mathtt{e}})$ models the termination of a process for future $\underline{f}$, computing method $\mathtt{m}$ in object $\underline{\mathtt{X}}$ and returning $\underline{\mathtt{e}}$. Event $\mathsf{futREv}(\underline{\mathtt{X}}, \underline{f}, \underline{\mathtt{m}}, \underline{\mathtt{e}}, i)$ models a $\mathbf{get}_i$ statement in object $\underline{\mathtt{X}}$ on the future $\underline{f}$, which was computed by $\mathtt{m}$ and returned $\underline{\mathtt{e}}$. Finally, $\mathsf{noEv}$ models an internal step.

Local traces consist of a *selection condition*, a set of symbolic expressions that express when a trace executes and a *history*, a sequence of events and states.

**Definition 4 (Local Semantics and Traces).** *A* heap $\rho$ *maps from fields to symbolic expressions and a* local state $\sigma$ *maps variables to symbolic expressions. Pairs of local states and heaps are* object states *and we write* $\binom{\sigma}{\rho}$. *The evaluation function* $[\![\mathtt{e}]\!]_{\binom{\sigma}{\rho}}$ *maps a syntactic expression to a symbolic expression.*

*A local trace $\theta$ has the form* $\mathsf{sc} \triangleright \mathsf{hs}$*, where* $\mathsf{sc}$ *is a set of symbolic expressions, called* selection condition*, and* $\mathsf{hs}$ *is a non-empty sequence, called* history*, such that every odd-indexed element is an object state and every even-indexed element an event. The semantics of methods and statements is defined by a function* $[\![\cdot]\!]_{\mathtt{X},f,\mathtt{m},\binom{\sigma}{\rho}}$ *where* $\mathtt{X}$ *is the object name,* $f$ *the future the method is resolving,* $\mathtt{m}$ *the method name and* $\binom{\sigma}{\rho}$ *the current object state. Future, object name and state may be symbolic. The semantics of a method* $\mathtt{m}$ *with body* $\mathtt{s}$ *is, for a symbolic* $\binom{\sigma}{\rho}$*:*

$$[\![\mathtt{m}]\!]_{\underline{\mathtt{X}},\underline{f},\mathtt{m},\binom{\sigma}{\rho}} = \left\{\emptyset \triangleright \left\langle \begin{pmatrix}\sigma\\\rho\end{pmatrix}, \mathsf{invREv}(\underline{\mathtt{X}}, \underline{f}, \mathtt{m}, \overrightarrow{\mathtt{e}}) \right\rangle \circ \theta \mid \theta \in [\![\mathtt{s}]\!]_{\underline{\mathtt{X}},\underline{f},\mathtt{m},\binom{\sigma}{\rho}} \right\}$$

*where* $\overrightarrow{\mathtt{e}}$ *is extracted from the parameter names in* $\sigma$*. E.g., for a method* `Int m(Int a, Rat b)` *we set* $\overrightarrow{\mathtt{e}} = \langle\sigma(\mathtt{a}), \sigma(\mathtt{b})\rangle$*. Figure 3 shows selected rules. All variables are initialized, futures with* **no***, a special future that is never resolved.*

The rules for assignment both update the local state or the heap, and add a noEv event. The rule for branching evaluates both branches and adds the corresponding guard evaluation to the selection condition. The rule for **get** is similar to the variable assignment, but receives a fresh symbolic value and stores it in the local state. As the event, a resolving reaction event is added, which stores the accessed future and the fresh symbolic value. The rule for method calls is analogous, but uses a fresh future for the call instead of a fresh read value. The added event is an invocation event with the evaluated parameters. Figure 4 shows the two traces in the semantics of Example 1.

$$\llbracket \mathtt{v\ =\ e} \rrbracket_{\mathtt{X},f,\mathtt{m},\left(\begin{smallmatrix}\sigma\\\rho\end{smallmatrix}\right)} = \left\{ \emptyset \triangleright \left\langle \begin{pmatrix}\sigma\\\rho\end{pmatrix}, \mathsf{noEv}, \begin{pmatrix}\sigma[\mathtt{v} \mapsto \llbracket \mathtt{e} \rrbracket_{\left(\begin{smallmatrix}\sigma\\\rho\end{smallmatrix}\right)}]\\\rho\end{pmatrix} \right\rangle \right\} \qquad \llbracket \mathbf{this}.\mathtt{f} \rrbracket_{\left(\begin{smallmatrix}\sigma\\\rho\end{smallmatrix}\right)} = \rho(\mathtt{f})$$

$$\llbracket \mathtt{v} \rrbracket_{\left(\begin{smallmatrix}\sigma\\\rho\end{smallmatrix}\right)} = \sigma(\mathtt{v})$$

$$\llbracket \mathbf{this}.\mathtt{f\ =\ e} \rrbracket_{\mathtt{X},f,\mathtt{m},\left(\begin{smallmatrix}\sigma\\\rho\end{smallmatrix}\right)} = \left\{ \emptyset \triangleright \left\langle \begin{pmatrix}\sigma\\\rho\end{pmatrix}, \mathsf{noEv}, \begin{pmatrix}\sigma\\\rho[\mathtt{f} \mapsto \llbracket \mathtt{e} \rrbracket_{\left(\begin{smallmatrix}\sigma\\\rho\end{smallmatrix}\right)}]\end{pmatrix} \right\rangle \right\} \qquad \llbracket \mathtt{-e} \rrbracket_{\left(\begin{smallmatrix}\sigma\\\rho\end{smallmatrix}\right)} = \begin{cases} -\llbracket \mathtt{e} \rrbracket_{\left(\begin{smallmatrix}\sigma\\\rho\end{smallmatrix}\right)} & \text{if } \llbracket \mathtt{e} \rrbracket_{\left(\begin{smallmatrix}\sigma\\\rho\end{smallmatrix}\right)} \text{ symbolic} \\ \left\llbracket -\llbracket \mathtt{e} \rrbracket_{\left(\begin{smallmatrix}\sigma\\\rho\end{smallmatrix}\right)} \right\rrbracket & \text{otherwise} \end{cases}$$

$$\llbracket \mathbf{return}\ \mathtt{e} \rrbracket_{\mathtt{X},f,\mathtt{m},\left(\begin{smallmatrix}\sigma\\\rho\end{smallmatrix}\right)} = \left\{ \emptyset \triangleright \left\langle \begin{pmatrix}\sigma\\\rho\end{pmatrix}, \mathsf{futEv}\left(\mathtt{X}, f, \mathtt{m}, \llbracket \mathtt{e} \rrbracket_{\left(\begin{smallmatrix}\sigma\\\rho\end{smallmatrix}\right)}\right), \begin{pmatrix}\sigma\\\rho\end{pmatrix} \right\rangle \right\} \qquad \llbracket \mathbf{skip} \rrbracket_{\mathtt{X},f,\mathtt{m},\left(\begin{smallmatrix}\sigma\\\rho\end{smallmatrix}\right)} = \left\{ \emptyset \triangleright \left\langle \begin{pmatrix}\sigma\\\rho\end{pmatrix} \right\rangle \right\}$$

$$\begin{aligned} \llbracket \mathbf{if}\mathtt{(e)\{s\}}\mathbf{else}\mathtt{\{s'\}} \rrbracket_{\mathtt{X},f,\mathtt{m},\left(\begin{smallmatrix}\sigma\\\rho\end{smallmatrix}\right)} = & \left\{ \mathsf{sc} \cup \{\llbracket \mathtt{e} \rrbracket_{\left(\begin{smallmatrix}\sigma\\\rho\end{smallmatrix}\right)}\} \triangleright \mathsf{hs} \,\middle|\, \mathsf{sc} \triangleright \mathsf{hs} \in \llbracket \mathtt{s} \rrbracket_{\mathtt{X},f,\mathtt{m},\left(\begin{smallmatrix}\sigma\\\rho\end{smallmatrix}\right)} \right\} \\ & \cup \left\{ \mathsf{sc} \cup \{\llbracket \mathtt{!e} \rrbracket_{\left(\begin{smallmatrix}\sigma\\\rho\end{smallmatrix}\right)}\} \triangleright \mathsf{hs} \,\middle|\, \mathsf{sc} \triangleright \mathsf{hs} \in \llbracket \mathtt{s'} \rrbracket_{\mathtt{X},f,\mathtt{m},\left(\begin{smallmatrix}\sigma\\\rho\end{smallmatrix}\right)} \right\} \end{aligned}$$

$$\llbracket \mathtt{v\ =\ e.}\mathbf{get}_i \rrbracket_{\mathtt{X},f,\mathtt{m},\left(\begin{smallmatrix}\sigma\\\rho\end{smallmatrix}\right)} = \left\{ \emptyset \triangleright \left\langle \begin{pmatrix}\sigma\\\rho\end{pmatrix}, \mathsf{futREv}\left(\mathtt{X}, \llbracket \mathtt{e} \rrbracket_{\left(\begin{smallmatrix}\sigma\\\rho\end{smallmatrix}\right)}, \underline{\mathtt{m}}, \underline{v}, i\right), \begin{pmatrix}\sigma[\mathtt{v} \mapsto \underline{v}]\\\rho\end{pmatrix} \right\rangle \right\} \text{ where } \underline{v} \text{ is fresh}$$

$$\llbracket \mathtt{v\ =\ f!n(}\overrightarrow{\mathtt{e}}\mathtt{)} \rrbracket_{\mathtt{X},f,\mathtt{m},\left(\begin{smallmatrix}\sigma\\\rho\end{smallmatrix}\right)} = \left\{ \emptyset \triangleright \left\langle \begin{pmatrix}\sigma\\\rho\end{pmatrix}, \mathsf{invEv}\left(\mathtt{X}, \llbracket \mathtt{f} \rrbracket_{\left(\begin{smallmatrix}\sigma\\\rho\end{smallmatrix}\right)}, \underline{v}, \mathtt{n}, \overrightarrow{\llbracket \mathtt{e} \rrbracket_{\left(\begin{smallmatrix}\sigma\\\rho\end{smallmatrix}\right)}}\right), \begin{pmatrix}\sigma[\mathtt{v} \mapsto \underline{v}]\\\rho\end{pmatrix} \right\rangle \right\} \text{ where } \underline{v} \text{ is fresh}$$

$$\llbracket \mathbf{while}\mathtt{(e)\{s\}} \rrbracket_{\mathtt{X},f,\mathtt{m},\left(\begin{smallmatrix}\sigma\\\rho\end{smallmatrix}\right)} = \llbracket \mathbf{if}\mathtt{(e)\{s;}\,\mathbf{while}\mathtt{(e)\{s\};}\mathbf{skip}\mathtt{\}}\mathbf{else}\mathtt{\{}\mathbf{skip}\mathtt{\}} \rrbracket_{\mathtt{X},f,\mathtt{m},\left(\begin{smallmatrix}\sigma\\\rho\end{smallmatrix}\right)}$$

$$\llbracket \mathtt{s;s'} \rrbracket_{\mathtt{X},f,\mathtt{m},\left(\begin{smallmatrix}\sigma\\\rho\end{smallmatrix}\right)} = \left\{ \mathsf{sc} \cup \mathsf{sc'} \triangleright \mathsf{hs} \circ \mathsf{hs'} \,\middle|\, \mathsf{sc} \triangleright \mathsf{hs} \circ \left\langle \begin{pmatrix}\sigma'\\\rho'\end{pmatrix} \right\rangle \in \llbracket \mathtt{s} \rrbracket_{\mathtt{X},f,\mathtt{m},\left(\begin{smallmatrix}\sigma\\\rho\end{smallmatrix}\right)}, \mathsf{sc'} \triangleright \mathsf{hs'} \in \llbracket \mathtt{s'} \rrbracket_{\mathtt{X},f,\mathtt{m},\left(\begin{smallmatrix}\sigma'\\\rho'\end{smallmatrix}\right)} \right\}$$

**Fig. 3.** Selected rules of the LA semantics of statements and expression. Evaluation $\llbracket \mathtt{e} \rrbracket$ of semantic values has its natural definition.

$$\begin{aligned} &\{\underline{v}\mathtt{<0}\}\triangleright \\ &\quad \langle(\sigma \cup \{\mathtt{f} \mapsto \mathbf{no}, \mathtt{r} \mapsto 0\}, \rho), \mathsf{invREv}(\underline{X}, \underline{f}, \mathtt{T.test}, \langle\underline{i}\rangle)\rangle \circ \\ &\quad \langle(\sigma \cup \{\mathtt{f} \mapsto \mathbf{no}, \mathtt{r} \mapsto 0\}, \rho), \mathsf{invEv}(\underline{X}, \underline{S}, \underline{f'}, \mathtt{Comp.cmp}, \langle\underline{i}\rangle)\rangle \circ \\ &\quad \langle(\sigma \cup \{\mathtt{f} \mapsto \underline{f'}, \mathtt{r} \mapsto 0\}, \rho), \mathsf{futREv}(\underline{X}, \underline{f'}, \mathtt{Comp.cmp}, \underline{v}, 0)\rangle \circ \\ &\quad \langle(\sigma \cup \{\mathtt{f} \mapsto \underline{f'}, \mathtt{r} \mapsto \underline{v}\}, \rho), \mathsf{noEv}\rangle \circ \\ &\quad \langle(\sigma \cup \{\mathtt{f} \mapsto \underline{f'}, \mathtt{r} \mapsto -\underline{v}\}, \rho), \mathsf{invEv}(\underline{X}, \underline{L}, \underline{f''}, \mathtt{Log.log}, \langle\underline{i}\rangle)\rangle \circ \\ &\quad \langle(\sigma \cup \{\mathtt{f} \mapsto \underline{f''}, \mathtt{r} \mapsto -\underline{v}\}, \rho), \mathsf{futEv}(\underline{X}, \underline{f}, \mathtt{T.test}, -\underline{v})\rangle \circ \\ &\quad \langle(\sigma \cup \{\mathtt{f} \mapsto \underline{f''}, \mathtt{r} \mapsto -\underline{v}\}, \rho)\rangle \end{aligned}$$

$$\begin{aligned} &\{\underline{v}\mathtt{>=0}\}\triangleright \\ &\quad \langle(\sigma \cup \{\mathtt{f} \mapsto \mathbf{no}, \mathtt{r} \mapsto 0\}, \rho), \mathsf{invREv}(\underline{X}, \underline{f}, \mathtt{T.test}, \langle\underline{i}\rangle)\rangle \circ \\ &\quad \langle(\sigma \cup \{\mathtt{f} \mapsto \mathbf{no}, \mathtt{r} \mapsto 0\}, \rho), \mathsf{invEv}(\underline{X}, \underline{S}, \underline{f'}, \mathtt{Comp.cmp}, \langle\underline{i}\rangle)\rangle \circ \\ &\quad \langle(\sigma \cup \{\mathtt{f} \mapsto \underline{f'}, \mathtt{r} \mapsto 0\}, \rho), \mathsf{futREv}(\underline{X}, \underline{f'}, \mathtt{Comp.cmp}, \underline{v}, 0)\rangle \circ \\ &\quad \langle(\sigma \cup \{\mathtt{f} \mapsto \underline{f'}, \mathtt{r} \mapsto \underline{v}\}, \rho), \mathsf{futEv}(\underline{X}, \underline{f}, \mathtt{T.test}, \underline{v})\rangle \circ \\ &\quad \langle(\sigma \cup \{\mathtt{f} \mapsto \underline{f'}, \mathtt{r} \mapsto \underline{v}\}, \rho)\rangle \end{aligned}$$

**Fig. 4.** LA semantics of `T.test`, with $\sigma = \{\mathtt{i} \mapsto \underline{i}\}, \rho = \{\mathtt{S} \mapsto \underline{S}, \mathtt{L} \mapsto \underline{L}\}$.

Symbolic traces represent a set of concrete traces, which contain only semantic values and correspond to possible behaviors of the statement. The set of concrete traces represented by a symbolic trace is a vast overapproximation and we only consider *selected* traces: concrete traces used in some terminating run of a given program. For a formal definition and the GC semantics, we refer to [22].

**Definition 5 (Selected Traces).** *A trace $\theta$ continues trace $\theta'$, written $\theta \preceq \theta'$, if its history is a suffix of the history of $\theta'$, with all symbolic elements replaced by concrete values, such that this substitution evaluates all expressions in the selection condition to true. A trace $\theta$ is* selected *in a program* Prgm*, if it is used during some run of* Prgm*. Let* $\mathtt{m}$ *be the method containing* $\mathtt{s}$.

$$[\![\mathtt{s}]\!]^{\mathsf{Prgm}}_{\mathtt{X},f,\mathtt{m},\binom{\sigma}{\rho}} = \{\theta \in [\![\mathtt{s}]\!]_{\mathtt{X},f,\mathtt{m},\binom{\underline{\sigma}}{\underline{\rho}}} \mid \exists \theta' \in [\![\mathtt{m}]\!]_{\underline{\mathtt{X}},\underline{f},\mathtt{m},\binom{\sigma}{\rho}}.\ \theta \preceq \theta' \wedge \theta \textit{ used in a terminating run of } \mathsf{Prgm}\}$$

We use a first-order state (FOS) logic to express properties of states and a monadic second-order (MSO) logic to express properties of traces. The MSO logic embeds the FOS by using FOS formulas similar to predicates on states. Similarly, it uses terms that allow to specify events.

**Definition 6 (FOS Syntax).** *Let $p$ range over predicate symbols, $f$ over function symbols, $x$ over logical variable names and $S$ over sorts. As sorts we take all data types* $\mathtt{D}$*, all class names and additionally $\mathbb{N}$ and* **Heap**. *The logical heaps are functions from field names to semantic values. Formulas $\varphi$ and terms* $\mathsf{t}$ *are defined by the following grammar, where* $\mathtt{v}$ *are program variables, consisting of local variables and the special variables* `heap` *and* `result`*, and* $\mathtt{f}$ *are all field names.*

$$\varphi ::= p(\overrightarrow{\mathsf{t}}) \mid \mathsf{t} \doteq \mathsf{t} \mid \varphi \vee \varphi \mid \neg\varphi \mid \exists x \in S.\ \varphi \qquad \mathsf{t} ::= x \mid \mathtt{v} \mid \mathtt{f} \mid f(\overrightarrow{\mathsf{t}})$$

We demand the usual constants, (e.g., 0, `True`) and that each operator defined in syntactic expressions $\mathtt{e}$ is a function symbol, so one can directly translate a syntactic expression into a FOS term. We additionally assume the following function symbols to handle heaps: $\mathsf{select}(\mathsf{t},\mathsf{t}) \mid \mathsf{store}(\mathsf{t},\mathsf{t},\mathsf{t})$, where $\mathsf{select}(h,\mathtt{f})$ reads field $\mathtt{f}$ from heap $h$ and $\mathsf{store}(h,\mathtt{f},\mathsf{t})$ stores the value of $\mathsf{t}$ in field $\mathtt{f}$ of heap $h$. As only one object is considered, we do not require an object parameter.

**Definition 7 (FOS Semantics).** *Interpretation $I$ maps function names to functions and predicate names to predicates. Assignment $\beta$ maps logical variable to semantic values of the resp. sort. Evaluation of terms in state $\binom{\sigma}{\rho}$ is defined as a function $[\![\mathsf{t}]\!]_{\binom{\sigma}{\rho},I,\beta}$ and satisfiability of formulas by a relation $\binom{\sigma}{\rho},\beta,I \models \varphi$.*

For the special variable `heap` we set $[\![\mathtt{heap}]\!]_{\mathtt{X},f,\mathtt{m},\binom{\sigma}{\rho}} = \rho$ and for the heap functions we follow JavaDL [1] and demand, e.g., the following connection axiom for all heaps $h$, all fields $\mathtt{f}$ and terms $\mathsf{t}$: $I(\mathsf{select})(I(\mathsf{store})(h,\mathtt{f},\mathsf{t}),\mathtt{f}) = \mathsf{t}$.

The models for the MSO logic are local traces and the whole semantic domain. This allows to quantify over method names etc. – it is not a logic over finite sequences. In additional to standard MSO constructs, we use $[\mathsf{t_{tr}}] \doteq \mathsf{t}$ to say that the event at position $\mathsf{t_{tr}}$ of the trace is equal to the term $\mathsf{t}$. Similarly, $[\mathsf{t_{tr}}] \vdash \varphi$ expresses that the state at position $\mathsf{t_{tr}}$ is a model for the FOS formula $\varphi$.

**Definition 8 (MSO Syntax).** *Let p,f,x range over the same sets as before, S over sorts. As sorts we take all data types* D *and additionally* **I**, *the set of trace indices,* O, *the set of all object names,* **Fut**, *the set of all futures of all types, the supertype* Any, *the set of all well-typed expressions and* M, *the set of all method names. Formulas* $\psi$ *are defined as follows. Terms* $\mathsf{t_{tr}}$ *are standard.*

$$\psi ::= p(\overrightarrow{\mathsf{t_{tr}}}) \mid \psi \vee \psi \mid \neg\psi \mid \mathsf{t_{tr}} \subseteq \mathsf{t_{tr}} \mid \exists x \in S.\ \psi \mid \exists X \subseteq S.\ \psi \mid [\mathsf{t_{tr}}] \doteq \mathsf{t_{tr}} \mid [\mathsf{t_{tr}}] \vdash \varphi$$

The predicate $\mathsf{isEvent}(i)$ that holds iff $\theta[i]$ is an event. For each type of event, there is a function symbol that maps its parameters to an event of its type and a predicate that holds iff the given position is an event of that kind, e.g.,

$$\mathsf{isfutEv}(i) \iff \exists f \in \mathbf{Fut}.\ \exists o \in \mathsf{O}.\ \exists m \in \mathsf{M}.\ \exists v \in \mathtt{Any}.\ [i] \doteq \mathsf{futEv}(o,f,m,v)$$

**Definition 9 (MSO Semantics).** *The semantics of terms and event terms is defined by a function* $[\![\cdot]\!]_{I,\beta}$. *The satisfiability of MSO-formulas is defined by a relation* $\theta, I, \beta \models \psi$. *The semantics of our extensions follows.*

$$\theta, I, \beta \models [\mathsf{t_{tr1}}] \doteq \mathsf{t_{tr2}} \iff 1 \leq [\![\mathsf{t_{tr1}}]\!]_{I,\beta} \leq |\theta| \wedge \theta[[\![\mathsf{t_{tr1}}]\!]_{I,\beta}] = [\![\mathsf{t_{tr2}}]\!]_{I,\beta}$$
$$\theta, I, \beta \models [\mathsf{t_{tr}}] \vdash \varphi \iff 1 \leq [\![\mathsf{t_{tr}}]\!]_{I,\beta} \leq |\theta| \wedge \theta[[\![\mathsf{t_{tr}}]\!]_{I,\beta}] \textit{ is a state } \wedge \theta[[\![\mathsf{t_{tr}}]\!]_{I,\beta}], I, \beta \models \varphi$$

*Example 2.* Let `r = f.get`$_0$ be the statement from Example 1. The following MSO-formula expresses that if all values read from futures of `cmp` is positive, and every future read at point 0 is from `cmp`, then after the read the value of `r` is positive.

$$\begin{aligned}&(\forall i \in \mathbf{I}.\ (\forall v \in \mathtt{Int}.\ [i] \doteq \mathsf{futREv}(\_,\_,\mathtt{cmp},v,\_) \rightarrow v > 0)\\ &\wedge \forall i \in \mathbf{I}.\ (\forall m \in \mathsf{M}.\ [i] \doteq \mathsf{futREv}(\_,\_,m,\_,0) \rightarrow m \doteq \mathtt{cmp}))\\ &\rightarrow \forall i \in \mathbf{I}.\ ([i] \doteq \mathsf{futREv}(\_,\_,\_,\_,0) \rightarrow [i+1] \vdash \mathtt{r > 0})\end{aligned}$$

Relativization [17], an established technique in abstract model theory [15], syntactically restricts a formula $\psi$ on a substructure defined by another formula $\psi'$. It is denoted $\psi[x \in S \setminus \psi']$, where $x$ is a free variable in $\psi'$ of $S$ sort. Each quantifier of $S$ sort is restricted to elements that fulfill $\psi'$.

*Example 3.* Formula $\varphi$ expresses that every trace-element is either an event, or a state with `r > 0`. The relativization with $\psi$ expresses that $\varphi$ holds for every index above 9. Both traces of Fig. 2 satisfy $\varphi[j \in \mathbf{I} \setminus \psi]$, neither satisfies $\varphi$.

$$\varphi = \forall i \in \mathbf{I}.\ \mathsf{isEvent}(i) \vee [i] \vdash \mathtt{r > 0} \qquad \psi = j \geq 9$$
$$\varphi[j \in \mathbf{I} \setminus \psi] = \forall i \in \mathbf{I}.\ i \geq 9 \rightarrow (\mathsf{isEvent}(i) \vee [i] \vdash \mathtt{r > 0})$$

We use common abbreviations, e.g., $\forall x \in S.\ \varphi$ for $\neg\exists x \in S.\ \neg\varphi$ and $\mathsf{true}$ and shorten comparisons of `Bool` terms by writing, e.g., $i > j$ instead of $i > j \doteq \mathbf{True}$.

## 3 Behavioral Program Logic

Behavioral Program Logic (BPL) is an extension of FOS with *behavioral modalities* $[\mathtt{s} \Vdash^{\alpha} \tau]$ that contain a statement s and a behavioral specification $(\tau, \alpha)$. A behavioral specification consists of (1) a syntactic component (the type $\tau$) and (2) a translation $\alpha$ of the type into an MSO formula that has to hold for all traces generated by the statement. Behavioral specifications can be seen as representations of a certain class of MSO formulas, which are deemed useful for verification of distributed systems. For the rest of this section, we assume fixed parameters Prgm, X, $f$, m for evaluation.

**Definition 10 (Behavioral Program Logic).** *A behavioral specification* $\mathbb{T}$ *is a pair* $(\tau_{\mathbb{T}}, \alpha_{\mathbb{T}})$*, where* $\alpha_{\mathbb{T}}$ *maps elements of* $\tau_{\mathbb{T}}$ *to MSO formulas.*

BPL-*formulas* $\varphi$*, terms* t *and updates* $U$ *are defined by the following grammar, which extends Definition 6. The meta variables range as in Definition 6. Additionally let* s *range over statements and* $(\tau_{\mathbb{T}}, \alpha_{\mathbb{T}})$ *over behavioral specifications.*

$$\varphi ::= \ldots \mid [\mathtt{s} \Vdash^{\alpha_{\mathbb{T}}} \tau_{\mathbb{T}}] \mid \{U\}\varphi \quad \mathtt{t} ::= \ldots \mid \{U\}\mathtt{t} \quad U ::= \epsilon \mid U||U \mid \{U\}U \mid \mathtt{v} := \mathtt{t}$$

$$[\![\{U\}t]\!]_{\binom{\sigma}{\rho},I,\beta} = [\![t]\!]_{[\![U]\!]_{\binom{\sigma}{\rho},I,\beta},I,\beta} \quad [\![\epsilon]\!]_{\binom{\sigma}{\rho},I,\beta}(x) = x \quad \binom{\sigma}{\rho}, I, \beta \models \{U\}\varphi \Leftrightarrow [\![U]\!]_{\binom{\sigma}{\rho},I,\beta}, I, \beta \models \varphi$$

$$[\![\mathtt{v} := \mathtt{t}]\!]_{\binom{\sigma}{\rho},I,\beta}\left(\binom{\sigma'}{\rho'}\right) = \begin{cases} \binom{\sigma'}{\rho''} & \text{if } \mathtt{v} = \mathbf{heap},\ \rho'' = [\![\mathtt{t}]\!]_{\binom{\sigma}{\rho},I,\beta} \\ \binom{\sigma''}{\rho'} & \text{otherwise, } \sigma'' = \sigma'[\mathtt{v} \mapsto [\![\mathtt{t}]\!]_{\binom{\sigma}{\rho},I,\beta}] \end{cases}$$

$$[\![U||U']\!]_{\binom{\sigma}{\rho},I,\beta}(x) = [\![U']\!]_{\binom{\sigma}{\rho},I,\beta}\left([\![U]\!]_{\binom{\sigma}{\rho},I,\beta}(x)\right) \qquad [\![\{U\}U']\!]_{\binom{\sigma}{\rho},I,\beta} = [\![U']\!]_{[\![U]\!]_{\binom{\sigma}{\rho},I,\beta},I,\beta}$$

$$\binom{\sigma}{\rho}, I, \beta \models [\mathtt{s} \Vdash^{\alpha_{\mathbb{T}}} \tau_{\mathbb{T}}] \Leftrightarrow \forall \theta \in [\![\mathtt{s}]\!]^{\mathsf{Prgm}}_{\mathtt{X},f,\mathtt{m},\binom{\sigma}{\rho}}.\ \theta, I, \beta \models \alpha_{\mathbb{T}}(\tau_{\mathbb{T}})$$

**Fig. 5.** Semantics of BPL. The satisfiability relation on the right of the semantics of behavioral modalities is the one of MSO.

The semantics of a behavioral modality $[\mathtt{s} \Vdash^{\alpha_{\mathbb{T}}} \tau_{\mathbb{T}}]$ is that all traces generated by s *selected within* Prgm are models for $\alpha_{\mathbb{T}}(\tau_{\mathbb{T}})$. We use updates [1,2] to keep track of state changes, their semantics is a state transition. Update $\mathtt{v} := \mathtt{t}$ changes the state by updating v to t. The parallel update $U||U'$ applies $U$ and $U'$ in parallel, with $U'$ winning in case of clashes. $\epsilon$ is the empty update and application $\{U\}$ evaluates the term (resp. formula) in the state after applying $U$.

**Definition 11 (Semantics of BPL).** *The semantical extension of FOS to* BPL *is given in Fig. 5. The interpretation I has the properties described above. A formula* $\varphi$ *is* valid *if every* $\binom{\sigma}{\rho}$ *and every* $\beta$ *make it true.*

Object, program, method name, resolved future and type of `result` are implicitly known, but we omit them for readability's sake. We use a sequent calculus to reason about BPL (resp. FOS).

**Definition 12 (Sequents and Rules).** *Let $\Delta, \Gamma$ be sets of* BPL*-formulas. A* sequent $\Gamma \Rightarrow \Delta$ *has the semantics of* $\bigwedge \Gamma \rightarrow \bigvee \Delta$. *$\Gamma$ is called the antecedent and $\Delta$ the succedent. Let $C, P_i$ be sequents. A rule has the form*

$$\textsf{(name)}\ \frac{P_1 \quad \dots \quad P_n}{C}\ cond$$

*Where $C$ is called the conclusion and $P_i$ the premise, while cond is a side-condition. Side-conditions are always decidable. For readability's sake, we apply side conditions containing equalities directly in the premises.*

Rules may contain, in addition to expressions, schematic variables. Their handling is standard [1]. We assume the usual FO rules for the FOS part of BPL handling all FO operators such as quantifiers.

**Definition 13 (Soundness).** *A rule is* sound *if validity of all premisses implies validity of the conclusion.*

Soundness implicitly refers to a program $\mathsf{Prgm}$, as behavioral modalities are defined over $\mathsf{Prgm}$-selectable traces. Rewrite rules $\tau_1 \leftrightsquigarrow \tau_2$ syntactically replace one type $\tau_1$ by another, $\tau_2$ (and vice versa) and are sound if $\alpha(\tau_1) \equiv \alpha(\tau_2)$.

*Discussion.* Before we introduce method types, a particular behavorial specification, we illustrate BPL with further examples. To reason about postconditions, as standard modal logics, we define a behavioral specification that only uses the last state of a trace (denoted by the function symbol *last*) for its semantics.

*Example 4.* The specification for postconditions is the pair of the set of all FOS sentences and the function $\mathsf{pst}$, defined below. $\mathsf{T}$ is the type of `result`. The first case accesses the return value stored in the $\mathsf{futEv}$ when `result` is used.

$$\mathsf{pst}(\varphi) = \begin{cases} \exists v \in \mathsf{T}.\ [last-1] \doteq \mathsf{futEv}(\_,\_,\_,v) \wedge [last] \vdash \varphi[\mathtt{result} \backslash v] & \text{if } \varphi \text{ contains } \mathtt{result} \\ [last] \vdash \varphi & \text{otherwise} \end{cases}$$

A Hoare triple $\{\varphi\}\mathtt{s}\{\psi\}$ has the same semantics as the formula $\varphi \rightarrow [\mathtt{s} \Vdash^{\mathsf{pst}} \psi]$. A standard dynamic logic modality $[\mathtt{s}]\psi$ has the same semantics as the behavioral modality $[\mathtt{s} \Vdash^{\mathsf{pst}} \psi]$[1]. Behavioral modalities generalize these systems and can be used to express any (MSO) trace property, independent of the form of its verification system. The following defines a points-to analysis for futures [14] (for the next statement), normally implemented in a data-flow framework.

*Example 5 (Points-To).* The behavioral specification of a *points-to analysis* specifies that the next statement reads a future resolved by a method from set $M$.

$$\mathbb{T}_{\mathsf{p2}} = (\mathcal{P}(\mathsf{M}), \mathsf{p2}) \text{ with}$$

$$\mathsf{p2}(M) = \exists \mathtt{X} \in \mathsf{O}.\exists f \in \mathbf{Fut}.\exists \mathtt{m} \in \mathsf{M}.\exists v \in \mathsf{Any}.\exists i \in \mathbb{N}.\ [1] \doteq \mathsf{futREv}(\mathtt{X}, f, \mathtt{m}, v, i) \wedge \bigvee_{\mathtt{m}' \in M} \mathtt{m} \doteq \mathtt{m}'$$

[1] This justifies our use of the term "modality". Contrary to standard modalities, behavioral modalities are not formulas that express modal statements about *formulas*, but formulas that express a modal statement about *more general specifications*.

The following formula expresses that the **get** statement reads a positive number, if the future is resolved by `Comp.cmp`. This is the case if `Comp.cmp` always returns positive values. The identifier connects the two modalities semantically.

$$\varphi_p = [\texttt{r = f.get}_0 \overset{\mathsf{p2}}{\Vdash} \{\texttt{Comp.cmp}\}] \rightarrow [\texttt{r = f.get}_0 \overset{\mathsf{pst}}{\Vdash} \texttt{r} > 0]$$

It is not necessary to include postcondition reasoning. Rule (ex1) in Fig. 6 expresses that if the next read from `s` is from some set $E'$ and it is required to show that the next read is from $E$, it suffices to check whether $E$ is a subset of $E'$. Rule (ex-$\Vdash$) connects two analyses and generalizes Example 2: one may assume some formula $\psi$ for a read value, if this synchronization always reads from method `Comp.cmp` and that the method body of `Comp.cmp` establishes $\psi$.

The above example illustrates the difference between modalites and typing judgments. Modalites are formulas and can be used for deductive reasoning about a type judgment (which, in our case, is encoded into $\Vdash$). While a calculus for pst is easily carried over from other sequent calculi, this is not possible for all behavioral specifications. The proof can still be closed in two ways.

- There may be some rules, such as (ex1) above, that enable to reason about the analysis without reducing the statement at all.
- If the proof contains only open branches containing behavioral specification, one may run a static analysis to evaluate them to true or false directly. E.g., if for the formula $\varphi_p$ above the pointer analysis returns that the synchronization point 0 reads from `L.log`, the first behavioral modality evaluates to false and the whole formula to true.

$$\textsf{(ex1)}\ \frac{}{\Gamma, [\texttt{s} \overset{\mathsf{p2}}{\Vdash} E'] \Rightarrow [\texttt{s} \overset{\mathsf{p2}}{\Vdash} E], \Delta}\ E \subseteq E' \qquad \textsf{(ex-}\Vdash\textsf{)}\ \frac{\begin{array}{c}\Gamma, \psi(v) \Rightarrow \{\texttt{v} := v\}[\texttt{s} \overset{\mathsf{pst}}{\Vdash} \varphi], \Delta \\ \Rightarrow [\texttt{v = f.get}_0 \overset{\mathsf{p2}}{\Vdash} \{\texttt{Comp.cmp}\}] \wedge [\texttt{s}_{\texttt{Comp.cmp}} \overset{\mathsf{pst}}{\Vdash} \psi]\end{array}}{\Gamma \Rightarrow [\texttt{v = f.get}_0\texttt{;s} \overset{\mathsf{pst}}{\Vdash} \varphi], \Delta}\ v \text{ fresh}$$

**Fig. 6.** Two example rules for behavioral specifications. $\psi(v)$ replaces **`result`** by $v$ and we assume that $\psi$ contains no fields.

Using external analyses increases modularity: (1) the BPL-calculus is simpler because it does not need to encode the implementation and (2) one may verify functional correctness of a method *up to its context*. Open branches are then a description of the context which the method requires. This may be verified once more context is known, thus extending proof repositories [6] to external analyses.

## 4 A Sequent Calculus for BPL: Behavioral Types

In this section we characterize behavioral types as behavioral specifications with a set of sequent calculus rules and a constraint on the proof obligations of the methods within a program. Before we formalize this in general, we introduce method types [23,24], a behavioral type for Active Objects that suffices to generalize method contracts and object invariants by integrating the behavioral specifications for postcondition reasoning and points-to analysis. The method type of a method describes the local view of a method on a protocol.

**Definition 14.** *The local protocol* $\mathbf{L}$ *and method type* $\mathsf{L}$ *of a method are defined by the grammar below. The behavioral specification for method types is* $\mathbb{T}_{\mathsf{met}} = (\mathsf{L}, \alpha_{\mathsf{met}})$. *Let* $\mathtt{X}_0, \ldots, \mathtt{X}_n$ *be roles, and* $\mathtt{f}_{\mathtt{X}_0}, \ldots, \mathtt{f}_{\mathtt{X}_n}$ *fields of fitting type.* $\alpha_{\mathsf{met}}(\mathsf{L})$ *is defined as* $\exists \mathtt{X}_0, \ldots, \mathtt{X}_n \in \mathsf{O}.\ \bigwedge_{i \leq n} \mathtt{X}_i \doteq \mathtt{f}_{\mathtt{X}_i} \wedge \alpha'_{\mathsf{met}}(\mathsf{L})$. *The first part models the (generated [24]) assignment of roles (as function symbols) to fields.*

$$\mathbf{L} ::= ?\mathtt{m}(\varphi).\mathsf{L} \qquad \mathsf{L} ::= \mathtt{X}!\mathtt{m}(\varphi) \mid \downarrow(\varphi) \mid \mathsf{skip} \mid \mathsf{L}.\mathsf{L} \mid \mathsf{L}^* \mid \oplus\{\mathsf{L}_i\}_{i \in I} \mid \&(\overrightarrow{\mathtt{m}}, \varphi)\{\mathsf{L}, \mathsf{L}\}$$

The local protocol of a method contains the receiving action $?\mathtt{m}(\varphi)$, which models that the parameters satisfy the predicate $\varphi$. The method body is checked against the method type – there is no statement corresponding to receiving. Roles keep track of an object through the protocol. We stress that statements and method types share syntactic elements – it is possible to pattern match on statements/expressions on one side and a method type on the other side in rules.

Calls are specified with the call action $\mathtt{X}!\mathtt{m}(\varphi)$, where $\mathtt{X.m}$ is the receiver and the predicate $\varphi$ has to hold. Here, $\varphi$ does not only specify the sent data but also local variables and fields. It can express properties such as "the sent data is larger then some field". The termination action $\downarrow(\varphi)$ models termination in a state satisfying $\varphi$ (which again may include `result`). The empty action $\mathsf{skip}$ models no visible actions and $\mathsf{L}_1.\mathsf{L}_2$ to sequential composition: all interactions in $\mathsf{L}_1$ must happen before $\mathsf{L}_2$. Repetition $\mathsf{L}^*$ corresponds to the Kleene star (and loops) and models zero or more repetitions of the interactions in $\mathsf{L}$.

There are two choice operators: $\oplus\{\mathsf{L}_i\}_{i \in I}$ is the active choice, the method must select one branch $\mathsf{L}_i$. It is not necessary to implement all branches, the method may choose to never select some branches. The index set $I$ must not be empty. $\&(\overrightarrow{\mathtt{m}}, \varphi)\{\mathsf{L}_1, \mathsf{L}_2\}$ is the passive choice: some other method made a choice and this method has to follow the protocol according to this choice. The choice is communicated via a future which has to be resolved by one of the methods in $\overrightarrow{\mathtt{m}}$. If the choice condition $\varphi$, which may only include the program variable `result`, is fulfilled by the read data, $\mathsf{L}_1$ has to be followed, otherwise $\mathsf{L}_2$ has to be followed. Both branches have to be implemented.

The semantics of the call and termination actions specify a trace with at least three elements with the correct event on second position and a state fulfilling the given predicate on the third position. Every other event is $\mathsf{noEv}$. The semantics of the empty action and active choice are straightforward. Sequential composition uses relativization: some position $i$ is chosen, such that the left translation holds before $i$ and the right translation afterwards. Note that $i$ is included in both relativization, to uphold the invariant that a trace always starts and ends with a state. The semantics of repetition are the only point where we require second order quantifiers: set $I$ is a set of indices, such that the first and last position are included and for every consecutive pair $k, l$ of elements of $I$, the translation of the repeated type holds in the relativization between $k$ and $l$. Passive choice specifies that the first event is a read on a correct future (i.e., resolved by the correct method) and the suffix afterwards follows the communicated choice correctly.

*Example 6.* The following formalizes the behavior described informally in Example 2:

$$?\mathtt{T.test(true).S!Comp.cmp(data} \doteq \mathtt{i).\&(\{Comp.cmp\}, result} < 0)\left\{\begin{matrix}\mathtt{L!Log.log(data} \doteq \mathtt{i)}, \\ \mathsf{skip}\end{matrix}\right\}.\downarrow(\mathtt{result} \geq 0)$$

The **result** variable in the guard of the passive choice is referring to the result of the read value, not the specified method.

We define behavioral types from a program logic perspective[2] by a type system, which is a set of sequent calculus rules that match on behavioral modalities and a obligation scheme, that maps every method to a proof obligation

**Definition 15 (Behavioral Types).** *A* behavioral type $\mathbb{T}$ *is a behavioral specification* $(\tau_\mathbb{T}, \alpha_\mathbb{T})$ *extended with* $(\gamma_\mathbb{T}, \iota_\mathbb{T})$.

*The* obligation scheme $\iota_\mathbb{T}$ *maps method names* $\mathtt{m}$ *to* proof obligations, *sequents of the form* $\varphi_\mathtt{m} \Rightarrow [\mathtt{s_m} \Vdash^{\alpha_\mathbb{T}} \tau_\mathtt{m}]$, *which have to be proven.* $\mathtt{s_m}$ *is the method body of* $\mathtt{m}$. *The* type system $\gamma_\mathbb{T}$ *is a set of rewrite rules for* $\tau_\mathbb{T}$ *and sequent calculus rules with conclusions matching the sequent* $\Gamma \Rightarrow \{U\}[\mathtt{s} \Vdash^{\alpha_\mathbb{T}} \tau_\mathbb{T}], \Delta$.

We demand that obligation schemes are consistent, i.e., proof obligations do not contradict each other. This would be the case if, for example a method is

$$\alpha'_{\mathsf{met}}(\mathtt{X!m}(\varphi)) = \forall i \in \mathbf{I}.\ \mathsf{isEvent}(i) \wedge [i] \not\doteq \mathsf{noEv} \rightarrow [i] \doteq \mathsf{invEv}(x, \mathtt{X}, f, \mathtt{m}, \overrightarrow{\mathsf{e}}) \wedge [i-1] \vdash \varphi(\overrightarrow{\mathsf{e}})$$
$$\wedge\ \exists i \in \mathbf{I}.\ [i] \not\doteq \mathsf{noEv} \wedge \mathsf{isEvent}(i)$$
$$\alpha'_{\mathsf{met}}(\downarrow(\varphi)) = \forall i \in \mathbf{I}.\ \mathsf{isEvent}(i) \wedge [i] \not\doteq \mathsf{noEv} \rightarrow [i] \doteq \mathsf{futEv}(x, f, m, \mathsf{e}) \wedge [i-1] \vdash \varphi[\mathtt{result} \setminus \mathsf{e}]$$
$$\wedge\ \exists i \in \mathbf{I}.\ [i] \not\doteq \mathsf{noEv} \wedge \mathsf{isEvent}(i)$$

where $\varphi(\overrightarrow{\mathsf{e}})$ replaces its free variables by $\overrightarrow{\mathsf{e}}$. $\varphi[\mathtt{result} \setminus \mathsf{e}]$ replaces **result** by $\mathsf{e}$.

$$\alpha'_{\mathsf{met}}(\mathsf{skip}) = \forall l \in \mathbf{I}.\ [l] \doteq \mathsf{noEv} \vee [l] \vdash \mathtt{true} \qquad \alpha'_{\mathsf{met}}(\oplus\{\mathsf{L}_i\}_{i\in I}) = \bigvee_{i\in I} \alpha'_{\mathsf{met}}(\mathsf{L}_i)$$
$$\alpha'_{\mathsf{met}}(\mathsf{L}_1.\mathsf{L}_2) = \exists i \in \mathbf{I}.\ \alpha'_{\mathsf{met}}(\mathsf{L}_1)[n \in \mathbf{I} \setminus n \leq i] \wedge \alpha'_{\mathsf{met}}(\mathsf{L}_2)[n \in \mathbf{I} \setminus n \geq i]$$
$$\alpha'_{\mathsf{met}}(\mathsf{L}^*) = \exists I \subseteq \mathbf{I}.\ \exists a, b \in I.\ a < b \wedge$$
$$\forall k \in \mathbf{I}.\ ((k < a \wedge \mathsf{isEvent}(i) \rightarrow [i] \not\doteq \mathsf{noEv}) \vee (a \leq k \wedge k \leq b)) \wedge$$
$$\forall i_1, i_2 \in I.\ \big((\forall l \in I.\ l \leq i_1 \wedge i_2 \leq l) \rightarrow \alpha'_{\mathsf{met}}(\mathsf{L})[n \in \mathbf{I} \setminus i_1 \leq n \wedge n \leq i_2)]\big)$$

$$\alpha'_{\mathsf{met}}(\&(\{\mathtt{m}_l\}_{l\in I}, \varphi)\{\mathsf{L}_1, \mathsf{L}_2\}) = \exists i, j, k \in \mathbf{I}.\ i < j \wedge j < k \wedge$$
$$(\forall l \in \mathbf{I}.\ l \doteq j \vee l \geq k \vee (l \leq i \wedge ([l] \doteq \mathsf{noEv} \vee [l] \vdash \mathtt{true})) \wedge [j] \doteq \mathsf{futREv}(x, m, f, \mathsf{e}, n) \wedge$$
$$\bigvee_{l\in I} m \doteq \mathtt{m}_l \wedge ([k] \vdash \varphi \rightarrow \alpha'_{\mathsf{met}}(\mathsf{L}_1)[n \in \mathbf{I} \setminus n \geq k]) \wedge ([k] \not\vdash \varphi \rightarrow \alpha'_{\mathsf{met}}(\mathsf{L}_2)[n \in \mathbf{I} \setminus n \geq k])$$

**Fig. 7.** Semantics for $\mathbb{T}_{\mathsf{met}}$. Unbound variables are implicitly existentially quantified.

[2] Behavioral types are sometimes (informally) distinguished from data types by having a subject reduction theorem where the typing relation is preserved, but not the type itself [10]. In BPL this would correspond to the property that one of the rules has a premise where the type in the behavioral modality is different than in the conclusion.

called and its precondition $\varphi$ is checked caller-side, then $\varphi$ must truly be used as a precondition by the proof obligation for the called method.

**Definition 16.** *Let* $\mathbf{L}_{\mathtt{m}} = ?\mathtt{m}(\varphi_{\mathtt{m}}).\mathsf{L}_{\mathtt{m}}$ *be the local protocols in* $\mathsf{Prgm}$*. We require that all* $\mathbf{L}_{\mathtt{m}}$ *are consistent: If* $\mathtt{m}$ *is called in the method type of any other method* $\mathtt{m}'$*, then the call condition implies* $\varphi_{\mathtt{m}}$*. Furthermore,* $\varphi_{\mathtt{X.run}} = \mathtt{true}$*.*

*The extension of the behavioral specification* $\mathbb{T}_{\mathsf{met}}$ *of method types to a behavioral type is given by the calculus in Fig. 8 and* $\iota_{\mathsf{met}}(\mathtt{m}) = \varphi_{\mathtt{m}} \wedge \Phi \Rightarrow [\mathtt{s}_{\mathtt{m}} \Vdash^{\alpha_{\mathsf{met}}} \mathsf{L}_{\mathtt{m}}]$*. Formula* $\Phi = \bigwedge_{\mathtt{X}} \mathtt{X} \doteq \mathit{select}(\mathtt{heap}, \mathtt{f}_{\mathtt{X}})$ *encodes the assignment of roles to fields.*

The call condition may contain fields of the other objects, but this is not an issue when checking consistency, as the precondition only contains fields of the own object and the fields are simply uninterpreted function symbols. The method in Fig. 2 can be typed with the type in Example 6.

Rule (met-V) translates a variable-assignment into an update and (met-F) is analogous for fields. Rule (met-get) has three premises: one premise checks via $\mathbb{T}_{\mathsf{p2}}$ that the correct methods are synchronized with. The two others use a fresh constant $v$ for the read value and assign it to the target variable. The two

$$\text{(met-V)}\ \frac{\Gamma \Rightarrow \{U\}\{\mathtt{v} := \mathtt{e}\}[\mathtt{s} \overset{\alpha_{\mathsf{met}}}{\Vdash} \mathsf{L}], \Delta}{\Gamma \Rightarrow \{U\}[\mathtt{v = e;\ s} \overset{\alpha_{\mathsf{met}}}{\Vdash} \mathsf{L}], \Delta} \qquad \text{(met-F)}\ \frac{\Gamma \Rightarrow \{U\}\{\mathtt{heap} := \mathsf{store}(\mathtt{heap}, \mathtt{f}, \mathtt{e})\}[\mathtt{s} \overset{\alpha_{\mathsf{met}}}{\Vdash} \mathsf{L}], \Delta}{\Gamma \Rightarrow \{U\}[\mathbf{this}.\mathtt{f = e;\ s} \overset{\alpha_{\mathsf{met}}}{\Vdash} \mathsf{L}], \Delta}$$

$$\text{(met-get)}\ \frac{\begin{array}{c}\Gamma \Rightarrow \{U\}\{\mathtt{v} := v\}(\varphi(v) \rightarrow [\mathtt{s} \overset{\alpha_{\mathsf{met}}}{\Vdash} \mathsf{L}_1]), \Delta \\ \Gamma \Rightarrow \{U\}\{\mathtt{v} := v\}(\neg\varphi(v) \rightarrow [\mathtt{s} \overset{\alpha_{\mathsf{met}}}{\Vdash} \mathsf{L}_2]), \Delta \qquad \Rightarrow [\mathtt{v = e.}\mathbf{get}_i\mathtt{;s} \overset{\mathsf{p2}}{\Vdash} \{\overrightarrow{\mathtt{m}}\}]\end{array}}{\Gamma \Rightarrow \{U\}[\mathtt{v = e.}\mathbf{get}_i\mathtt{;s} \overset{\alpha_{\mathsf{met}}}{\Vdash} \&(\overrightarrow{\mathtt{m}}, \varphi)\{\mathsf{L}_1, \mathsf{L}_2\}], \Delta}\ v \text{ fresh}$$

$$\text{(met-while)}\ \frac{\Gamma \Rightarrow \{U\}I, \Delta \qquad I, \mathtt{e} \Rightarrow [\mathtt{s} \overset{\mathsf{pst}}{\Vdash} I] \qquad I, \mathtt{e} \Rightarrow [\mathtt{s} \overset{\alpha_{\mathsf{met}}}{\Vdash} \mathsf{L}] \qquad I, \neg\mathtt{e} \Rightarrow [\mathtt{s'} \overset{\alpha_{\mathsf{met}}}{\Vdash} \mathsf{L}'], \Delta}{\Gamma \Rightarrow \{U\}[\mathbf{while}\ \mathtt{e}\ \mathbf{do}\ \mathtt{s}\ \mathbf{od}\ \mathtt{s'} \overset{\alpha_{\mathsf{met}}}{\Vdash} \mathsf{L}^*.\mathsf{L}'], \Delta}$$

$$\text{(met-if)}\ \frac{\begin{array}{c}\Gamma \Rightarrow \{U\}(\mathtt{e} \rightarrow [\mathtt{s;s''} \overset{\alpha_{\mathsf{met}}}{\Vdash} \oplus\{\mathsf{L}_i\}_{i \in I_1}]), \Delta \\ \Gamma \Rightarrow \{U\}(\neg\mathtt{e} \rightarrow [\mathtt{s';s''} \overset{\alpha_{\mathsf{met}}}{\Vdash} \oplus\{\mathsf{L}_i\}_{i \in I_2}]), \Delta\end{array}}{\Gamma \Rightarrow \{U\}[\mathbf{if}\ \mathtt{e}\ \mathbf{then}\ \mathtt{s}\ \mathbf{else}\ \mathtt{s'}\ \mathbf{fi}\ \mathtt{s''} \overset{\alpha_{\mathsf{met}}}{\Vdash} \oplus\{\mathsf{L}_i\}_{i \in I}], \Delta}\ I_1 \cup I_2 \subseteq I$$

$$\text{(met-call)}\ \frac{\Gamma \Rightarrow \{U\}\left(\varphi(\mathtt{e}) \wedge \mathit{select}(\mathtt{heap}, \mathtt{f}) \doteq \mathtt{X}\right), \Delta \qquad \Gamma \Rightarrow \{U\}\{\mathtt{v} := f\}[\mathtt{s} \overset{\alpha_{\mathsf{met}}}{\Vdash} \mathsf{L}], \Delta}{\Gamma \Rightarrow \{U\}[\mathtt{v = f!m(e);\ s} \overset{\alpha_{\mathsf{met}}}{\Vdash} \mathtt{X!m}(\varphi).\mathsf{L}], \Delta}\ f \text{ fresh}$$

$$\text{(met-return)}\ \frac{\Gamma \Rightarrow \{U\}\{\mathtt{result} := \mathtt{e}\}\varphi, \Delta}{\Gamma \Rightarrow \{U\}[\mathbf{return}\ \mathtt{e} \overset{\alpha_{\mathsf{met}}}{\Vdash} \downarrow(\varphi)], \Delta} \qquad \text{(met-skip)}\ \frac{}{\Gamma \Rightarrow \{U\}[\mathbf{skip} \overset{\alpha_{\mathsf{met}}}{\Vdash} \mathsf{skip}], \Delta}$$

$$\mathsf{L} \leftrightsquigarrow \oplus\{\mathsf{L}\} \qquad \mathsf{skip.L} \leftrightsquigarrow \mathsf{L} \qquad \mathsf{L.skip} \leftrightsquigarrow \mathsf{L}$$

**Fig. 8.** Rules for $\mathbb{T}_{\mathsf{met}}$. We remind that the sets $I_1, I_2$ are defined as non-empty. For simplicity, we assume that every branch and every loop body implicitly ends in **skip**.

premises differ in the branch that is checked afterwards, depending on whether or not the choice condition holds. Rule (**met-while**) is a standard loop invariant rule. An invariant $I$ holds before the first iteration and is preserved by the loop to remove all other information afterwards. The loop body is checked against the repeated type and the continuation against the continuation of the type. Method types have no special action for the end of a statement, so $\mathbb{T}_{\mathsf{pst}}$ is used for checking that the loop preserves its invariant. Rule (**met-if**) splits the set of possible choices into two and checks each branch against one of these sets. These sets may overlap and do not need to cover all original choices, but may not be empty. Rule (**met-call**) checks the annotated condition of the called method and the correct target explicitly and that the correct method is called by matching call type and call statement. We remind that references are not reassigned, so call targets can be verified locally. The other rules are straightforward.

*Contracts and Invariants.* Method types generalize method contracts and object invariants as follows. An object invariant is encoded by adding it to the formula in the receiving and terminating actions of all method in an object – except the constructor `run`, where it is only added to the terminating action. A method contract (consisting of a precondition on the parameters and a postcondition) is encoded analogously by adding the precondition to the receiving and the post-condition to the terminating actions. However, one additional step is required: Method types are generated by projection of global types [23], so to use them for object invariants or method contracts requires to infer a method type first. This is done by mapping every call to a call action, every branching to an active choice, every loop to a repetition, termination to a terminating action and using **true** at every position where a formula is required, before adding precondition, post-condition or object invariant. The most complex construct is synchronization. Each such read is mapped to a passive choice with all methods as the method set and **true** as the choice condition. The following code is added in the first branch.

```
class T(Comp S, Log L){
 Int nr = 0;
 Int test(Int i){
  Fut<Int> f = S!cmp(i);
  this.nr = this.nr + 1;
  Int r = f.get₀;
  if(r < 0 && i > 0){
   r = -r; f = L!log(i);
  }
  return r;
 }
}
```

Precondition: $\mathtt{i} \geq 0$
Postcondition: $\mathtt{result} \geq 0$
Invariant: $\mathbf{this}.\mathtt{nr} \geq 0$

$$\mathsf{L}_1 = \ ?\mathtt{T.test}(\mathtt{i} \geq 0 \wedge \mathbf{this}.\mathtt{nr} \geq 0).\mathtt{S!Comp.cmp}(\varphi_{\mathtt{cmp}}) \\ .\ \&(\mathsf{M}, \mathsf{true})\left\{ \begin{array}{l} \oplus \left\{ \begin{array}{l} \mathtt{L!Log.log}(\varphi_{\mathtt{log}}), \\ \mathsf{skip} \end{array} \right\}, \\ \mathsf{skip} \end{array} \right\} \\ .\ \downarrow(\mathtt{result} \geq 0 \wedge \mathbf{this}.\mathtt{nr} \geq 0)$$

$$\mathsf{L}_2 = \ ?\mathtt{T.test}(\mathtt{i} \geq 0 \wedge \mathbf{this}.\mathtt{nr} \geq 0).\mathtt{S!Comp.cmp}(\mathtt{data} \doteq \mathtt{i} \wedge \varphi_{\mathtt{cmp}}) \\ .\ \&(\{\mathtt{Comp.cmp}\}, \mathtt{result} < 0)\left\{ \begin{array}{l} \mathtt{L!Log.log}(\mathtt{data} \doteq \mathtt{i} \wedge \varphi_{\mathtt{log}}), \\ \mathsf{skip} \end{array} \right\} \\ .\ \downarrow(\mathtt{result} \geq 0) \wedge \mathbf{this}.\mathtt{nr} \geq 0$$

**Fig. 9.** An example method and two method types for method contracts and invariants.

The second branch is skip. Invariants require fields in the precondition and a fitting notion of consistency, which was developed in [23].

*Example 7.* Consider the code in Fig. 9, a variation of our running example. It tracks the number of calls to `T.test` and inverts the result if the input is positive. It adheres to the contract with precondition $\texttt{i} \geq 0$ and postcondition $\texttt{result} \geq 0$ and the invariant $\textbf{this}.\texttt{nr} \geq 0$. The algorithm above derives the following type:

$$?\texttt{T.test}(\mathsf{true})\ .\ \texttt{S!Comp.cmp}(\mathsf{true})\ .\ \&(\mathsf{M},\mathsf{true})\left\{\begin{array}{l}\oplus\left\{\begin{array}{l}\texttt{L!Log.log}(\mathsf{true}),\\ \mathsf{skip}\end{array}\right\},\\ \mathsf{skip}\end{array}\right\}.\downarrow(\mathsf{true})$$

Let $\varphi_{\texttt{cmp}}$ and $\varphi_{\texttt{log}}$ be the preconditions of the called methods. The final specification, after adding the contract and the invariant, is shown on the right in Fig. 9 as $\mathsf{L}_1$. The inferred type is not the one we gave in Example 6: For one, it differs in its shape (two choice operators). For another, it neither keeps track of the passed data, nor specifies the relation between the return value of `Comp.cmp` and the taken branch. These properties are typical for protocol specifications and require a global view, contrary to the local view of method contracts and object invariants. However, one can add the pre- and postcondition and the object invariant also to the type given in Example 6 and combine local and global specification. The result is shown as $\mathsf{L}_2$ in Fig. 9. $\mathsf{L}_2$ expresses that the method follows the protocol and adheres to contract and object invariant.

**Theorem 1.** *$\mathbb{T}_{\mathsf{met}}$ is sound for every program.*

The proof is standard [22]. Consistency of the obligation scheme is required to establish that all selected traces are models for the type of their method. The first two elements are not described by the method type and, thus, removed.

**Corollary 1.** *If (1) for every method* m *with type $?\texttt{m}(\varphi).\mathsf{L}_{\texttt{m}}$ the formula $\iota_{\mathsf{met}}(\texttt{m})$ is valid and (2) the obligation scheme is consistent, then for every selected trace $\theta$ of any method* m*, the trace after the invocation reaction event follows its type:*

$$\theta[2..|\theta|], I, \emptyset \models \alpha_{\mathsf{met}}(\mathsf{L}_{\texttt{m}})$$

## 5 Conclusion and Related Work

This work presents BPL, a program logic for object-oriented distributed programs that enables deductive reasoning about the results of static analyses and integrates concepts from behavioral types by pattern-matching statement and specification. The *method type* behavioral type generalizes method contracts, session types and object invariants. In the following, we discuss related work.

*Dynamic Logics.* Beckert and Bruns [3] use LTL formulas in dynamic logic modalities in their Dynamic Trace Logic (DTL) for Java. Given an LTL formula $\varphi$, the DTL-formula $[\texttt{s}]\varphi$ expresses that $\varphi$ describes all traces of s. DTL uses a restricted form of pattern matching: its three loop invariant rules depend on the outermost operator of $\varphi$ and other rules may consume a "next" operator. DTL does not use events and specifies patterns of state changes, not of interactions.

The Abstract Behavior Specification Dynamic Logic (ABSDL) of Din and Owe [12] is for the ABS language [21]. In ABSDL, a formula $[\mathtt{s}]\varphi$, where $\varphi$ is a first-order formula over the program state, has the standard meaning that $\varphi$ holds after $\mathtt{s}$ is executed. ABSDL uses a special program variable to keep track of the visible events. Its rules are tightly coupled with object-invariant reasoning. This makes it impossible to specify the state at arbitrary interactions.

Bubel et al. [7] define dynamic logic with coinductive traces (DLCT). In DLCT, a formula $[\mathtt{s}]\varphi$, where $\varphi$ is a trace modality formula, containing symbolic trace formulas, has the meaning that every trace of $\mathtt{s}$ is a model for $\varphi$. Contrary to ABSDL, DLCT keeps track of the whole trace, not just the events. DLCT is not able to specify the property that between two states, some form of event does *not* occur, as symbolic trace formulas are not closed under negation.

*Behavioral Types.* A number of behavioral types deals with assertions [4,5] or Actors [16,18,26]. Stateful Behavioral Types for Active Objects (STAO) [23] uses both and defines the judgment $\varphi, \mathtt{s}' \vdash \mathtt{s} : \tau$, that expresses that all traces of $\mathtt{s}$ are models for the translation of $\tau$. $\varphi$ and $\mathtt{s}'$ keep track of the chosen path so far. STAO is not able to reason about multiple judgments, but relies on external analyses for precision. Reasoning about these results happens on a meta-level.

Finally, Propositions-as-Types theorems (PaT) have been established [8,27] between session types for the $\pi$-calculus and intuitionistic linear logic. They are specific to this setting and do not characterize general behavioral types. To our best knowledge, Definition 15 is the first formal characterization of behavioral types.

*Future Work.* An implementation of BPL for full ABS is ongoing and as future work, we plan to investigate further types and concurrency models, in particular systems with shared memory and effect type systems.

## References

1. Ahrendt, W., Beckert, B., Bubel, R., Hähnle, R., Schmitt, P.H., Ulbrich, M. (eds.): Deductive Software Verification - The KeY Book - From Theory to Practice. LNCS, vol. 10001. Springer, Cham (2016). https://doi.org/10.1007/978-3-319-49812-6
2. Beckert, B.: A dynamic logic for the formal verification of Java Card programs. In: Attali, I., Jensen, T. (eds.) JavaCard 2000. LNCS, vol. 2041, pp. 6–24. Springer, Heidelberg (2001). https://doi.org/10.1007/3-540-45165-X_2
3. Beckert, B., Bruns, D.: Dynamic logic with trace semantics. In: Bonacina, M.P. (ed.) CADE 2013. LNCS (LNAI), vol. 7898, pp. 315–329. Springer, Heidelberg (2013). https://doi.org/10.1007/978-3-642-38574-2_22
4. Berger, M., Honda, K., Yoshida, N.: Completeness and logical full abstraction in modal logics for typed mobile processes. In: Aceto, L., Damgård, I., Goldberg, L.A., Halldórsson, M.M., Ingólfsdóttir, A., Walukiewicz, I. (eds.) ICALP 2008. LNCS, vol. 5126, pp. 99–111. Springer, Heidelberg (2008). https://doi.org/10.1007/978-3-540-70583-3_9

5. Bocchi, L., Lange, J., Tuosto, E.: Three algorithms and a methodology for amending contracts for choreographies. Sci. Ann. Comput. Sci. **22**(1), 61–104 (2012)
6. Bubel, R., et al.: Proof repositories for compositional verification of evolving software systems - managing change when proving software correct. In: Steffen, B. (ed.) Transactions on Foundations for Mastering Change I. LNCS, vol. 9960, pp. 130–156. Springer, Cham (2016). https://doi.org/10.1007/978-3-319-46508-1_8
7. Bubel, R., Din, C.C., Hähnle, R., Nakata, K.: A dynamic logic with traces and coinduction. In: De Nivelle, H. (ed.) TABLEAUX 2015. LNCS (LNAI), vol. 9323, pp. 307–322. Springer, Cham (2015). https://doi.org/10.1007/978-3-319-24312-2_21
8. Caires, L., Pfenning, F.: Session types as intuitionistic linear propositions. In: Gastin, P., Laroussinie, F. (eds.) CONCUR 2010. LNCS, vol. 6269, pp. 222–236. Springer, Heidelberg (2010). https://doi.org/10.1007/978-3-642-15375-4_16
9. de Boer, F.S., et al.: A survey of active object languages. ACM Comput. Surv. **50**(5), 76:1–76:39 (2017)
10. Dezani-Ciancaglini, M.: Personal Communication, 19 October 2018
11. Din, C.C., Hähnle, R., Johnsen, E.B., Pun, K.I., Tapia Tarifa, S.L.: Locally abstract, globally concrete semantics of concurrent programming languages. In: Schmidt, R.A., Nalon, C. (eds.) TABLEAUX 2017. LNCS (LNAI), vol. 10501, pp. 22–43. Springer, Cham (2017). https://doi.org/10.1007/978-3-319-66902-1_2
12. Din, C.C., Owe, O.: A sound and complete reasoning system for asynchronous communication with shared futures. J. Log. Algebraic Methods Program. **83**(5–6), 360–383 (2014)
13. Din, C.C., Tapia Tarifa, S.L., Hähnle, R., Johnsen, E.B.: History-based specification and verification of scalable concurrent and distributed systems. In: Butler, M., Conchon, S., Zaïdi, F. (eds.) ICFEM 2015. LNCS, vol. 9407, pp. 217–233. Springer, Cham (2015). https://doi.org/10.1007/978-3-319-25423-4_14
14. Flores-Montoya, A.E., Albert, E., Genaim, S.: May-happen-in-parallel based deadlock analysis for concurrent objects. In: Beyer, D., Boreale, M. (eds.) FMOODS/-FORTE -2013. LNCS, vol. 7892, pp. 273–288. Springer, Heidelberg (2013). https://doi.org/10.1007/978-3-642-38592-6_19
15. García-Matos, M., Väänänen, J.: Abstract model theory as a framework for universal logic. In: Beziau, J.-Y. (ed.) Logica Universalis, pp. 19–33. Basel, Birkhäuser Basel (2005)
16. Giachino, E., Johnsen, E.B., Laneve, C., Pun, K.I.: Time complexity of concurrent programs. In: Braga, C., Ölveczky, P.C. (eds.) FACS 2015. LNCS, vol. 9539, pp. 199–216. Springer, Cham (2016). https://doi.org/10.1007/978-3-319-28934-2_11
17. Henkin, L.: Relativization with respect to formulas and its use in proofs of independence. Compositio Mathematica **20**, 88–106 (1968)
18. Henrio, L., Laneve, C., Mastandrea, V.: Analysis of synchronisations in stateful active objects. In: Polikarpova, N., Schneider, S. (eds.) IFM 2017. LNCS, vol. 10510, pp. 195–210. Springer, Cham (2017). https://doi.org/10.1007/978-3-319-66845-1_13
19. Honda, K., Yoshida, N., Carbone, M.: Multiparty asynchronous session types. JACM **63**, 9:1–9:67 (2016)
20. Hüttel, H., et al.: Foundations of session types and behavioural contracts. ACM Comput. Surv. **49**(1), 3:1–3:36 (2016)
21. Johnsen, E.B., Hähnle, R., Schäfer, J., Schlatte, R., Steffen, M.: ABS: a core language for abstract behavioral specification. In: Aichernig, B.K., de Boer, F.S., Bonsangue, M.M. (eds.) FMCO 2010. LNCS, vol. 6957, pp. 142–164. Springer, Heidelberg (2011). https://doi.org/10.1007/978-3-642-25271-6_8

22. Kamburjan, E. Behavioral program logic and LAGC semantics without continuations (technical report). CoRR abs/1904.13338 (2019)
23. Kamburjan, E., Chen, T.-C.: Stateful behavioral types for active objects. In: Furia, C.A., Winter, K. (eds.) IFM 2018. LNCS, vol. 11023, pp. 214–235. Springer, Cham (2018). https://doi.org/10.1007/978-3-319-98938-9_13
24. Kamburjan, E., Din, C.C., Chen, T.-C.: Session-based compositional analysis for actor-based languages using futures. In: Ogata, K., Lawford, M., Liu, S. (eds.) ICFEM 2016. LNCS, vol. 10009, pp. 296–312. Springer, Cham (2016). https://doi.org/10.1007/978-3-319-47846-3_19
25. Kamburjan, E., Hähnle, R.: Deductive verification of railway operations. In: Fantechi, A., Lecomte, T., Romanovsky, A.B. (eds.) RSSRail 2017. LNCS, vol. 10598, pp. 131–147. Springer, Uk (2017). https://doi.org/10.1007/978-3-319-68499-4_9
26. Neykova, R., Yoshida, N.: Multiparty session actors. Log. Methods Comput. Sci. **13**, 1 (2017)
27. Wadler, P.: Propositions as types. Commun. ACM **58**(12), 75–84 (2015)

# Prenex Separation Logic with One Selector Field

Mnacho Echenim[1], Radu Iosif[2], and Nicolas Peltier[1](✉)

[1] Univ. Grenoble Alpes, CNRS, LIG, 38000 Grenoble, France
Nicolas.Peltier@imag.fr
[2] Univ. Grenoble Alpes, CNRS, VERIMAG, 38000 Grenoble, France

**Abstract.** We show that infinite satisfiability can be reduced to finite satisfiability for all prenex formulas of Separation Logic with $k \geq 1$ selector fields ($\mathsf{SL}^k$). This fact entails the decidability of the finite and infinite satisfiability problems for the class of prenex formulas of $\mathsf{SL}^1$, by reduction to the first-order theory of a single unary function symbol and an arbitrary number of unary predicate symbols. We also prove that the complexity of this fragment is not elementary recursive, by reduction from the first-order theory of one unary function symbol. Finally, we prove that the Bernays-Schönfinkel-Ramsey fragment of prenex $\mathsf{SL}^1$ formulas with quantifier prefix in the language $\exists^*\forall^*$ is PSPACE-complete.

## 1 Introduction

Separation Logic [8,11] (SL) is a logical framework used to describe properties of the heap memory, such as the placement of pointer variables within the topology of complex data structures (lists, trees, etc.). The features that make SL attractive for program verification are the ability of defining (i) weakest pre- and post-condition calculi that capture the semantics of programs with pointers, and (ii) compositional verification methods, based on inferring local specifications of methods and threads independently of the context in which they evolve. The search for automated push-button program verification methods motivates the understanding of the decidability, complexity and expressive power of various dialects of SL, used as assertion languages in Hoare-style proofs [8], or logic-based abstract domains in static analysis [3].

Formal definitions are provided later, but essentially, SL can be viewed as the first order theory of one partial finite function from $\mathfrak{U} \to \mathfrak{U}^k$, called a *heap*, where $\mathfrak{U}$ denotes the universe of memory locations (i.e., addresses), to which two non-classical connectives are added: (i) the *separating conjunction* $\phi_1 * \phi_2$, that asserts a split of the heap into disjoint heaps satisfying $\phi_1$ and $\phi_2$ respectively, and (ii) the *separating implication* or *magic wand* $\phi_1 \mathrel{-\!\!*} \phi_2$, stating that each extension of the heap by a disjoint heap satisfying $\phi_1$ must satisfy $\phi_2$. The number $k$ denotes the number of selector fields and we use the notation $\mathsf{SL}^k$ to make this number explicit. Quantification over elements of $\mathfrak{U}$ is allowed. A fragment of separation


S. Cerrito and A. Popescu (Eds.): TABLEAUX 2019, LNAI 11714, pp. 409–427, 2019.
https://doi.org/10.1007/978-3-030-29026-9_23

logic that is practically relevant in verification is when $k = 1$, i.e., every allocated cell points to a unique cell. This fragment allows, e.g., to describe simply linked lists.

As a simple example of application, let us consider the following Hoare triple with left-hand side that is the weakest precondition of an arbitrary formula $\phi$ with respect to a selector update in a program handling lists:

$$\{\exists x \,.\, \mathsf{i} \mapsto x * (\mathsf{i} \mapsto \mathsf{j} \mathrel{-\!\!*} \phi)\} \quad \mathsf{i.next} = \mathsf{j} \quad \{\phi\}$$

Informally, the formula $\exists x \,.\, \mathsf{i} \mapsto x * (\mathsf{i} \mapsto \mathsf{j} \mathrel{-\!\!*} \phi)$ holds when the heap can be separated into disjoint parts, one in which cell $i$ is allocated (the formula $\mathsf{i} \mapsto x$ states that the heap maps i to $x$), and one that, when extended by allocating cell $i$ to $j$, satisfies $\phi$. In other words, the formula states that cell $i$ is allocated and that $\phi$ holds after $i$ is redirected to $j$. A typical verification condition checks whether this formula is entailed by another precondition $\psi$, generated by a program verifier or supplied by the user. The entailment $\psi \models \exists x \,.\, \mathsf{i} \mapsto x * (\mathsf{i} \mapsto \mathsf{j} \mathrel{-\!\!*} \phi)$ is valid if and only if the formula $\theta \stackrel{\mathsf{def}}{=} \psi \wedge \forall x \,.\, \neg(\mathsf{i} \mapsto x * (\mathsf{i} \mapsto \mathsf{j} \mathrel{-\!\!*} \phi))$ is unsatisfiable. In addition, if $\phi$ and $\psi$ are formulas in prenex form[1] then, because the assertions $\mathsf{i} \mapsto x$ and $\mathsf{i} \mapsto \mathsf{j}$ unambiguously define a specific part of the heap (the cell corresponding to i), the quantifiers of $\phi$ can be hoisted outside of the separating conjunction and implication, and the formula $\theta$ can be written in prenex form.

Deciding the satisfiability of (prenex) SL formulas is thus an important ingredient for push-button program verification. Unlike first order logic, some SL formulas do not have a prenex form (see Example 2 on Page 7). Moreover, satisfiability is decidable (and PSPACE-complete) for quantifier-free SL-formulas, but it is undecidable for first-order SL-formulas, even when $k = 1$. In fact $\mathsf{SL}^1$ is as expressive as second-order logic in the presence of $*$ and $\mathrel{-\!\!*}$ whereas the fragment of $\mathsf{SL}^1$ without $\mathrel{-\!\!*}$ is decidable but not elementary recursive [2]. In [6], we investigated the Bernays-Schönfinkel-Ramsey fragment of $\mathsf{SL}^k$, i.e., the fragment containing formulas of the form $\exists x_1, \ldots, x_n \forall y_1, \ldots, y_m \,.\, \phi$ where $\phi$ is a quantifier-free formula of $\mathsf{SL}^k$. We proved that for $k > 1$, satisfiability is undecidable in general and decidable if $\mathrel{-\!\!*}$ only occurs in the scope of an odd number of negations. However, nothing is known concerning the prenex fragment of $\mathsf{SL}^1$. In this paper we fill in this gap and show that:

1. the prenex fragment of $\mathsf{SL}^1$ is decidable but not elementary recursive, and
2. the Bernays-Schönfinkel-Ramsey fragment of $\mathsf{SL}^1$ is PSPACE-complete.

The results are established using reductions to and from the fragment of first order logic with one monadic function symbol [1]. The decidability of this fragment is a consequence of the celebrated Rabin Tree Theorem [10], which established the decidability of monadic second order logic of infinite binary tree (S2S). As in our previous work [6] and unlike most existing approaches, we consider both the finite and infinite satisfiability problems (other approaches

[1] $Q_1 x_1 \ldots Q_n x_n \,.\, \varphi$, where $Q_1, \ldots, Q_n$ are the first order quantifiers $\exists$ or $\forall$ and $\varphi$ is quantifier-free.

usually assume that the universe is infinite). Essential to our reductions to and from this fragment is a result (proven in [6]) stating that each quantifier-free $\mathsf{SL}^k$ formula, for $k \geq 1$, is equivalent to a boolean combination of formulas of some specific forms, called *test formulas*. Similar translations exist for quantifier-free $\mathsf{SL}^1$ [2,9] and for $\mathsf{SL}^1$ with one quantified variable [5]. In addition we show in the present paper that the infinite satisfiability reduces to the finite satisfiability for quantified boolean quantifications of test formulas.

## 2 Preliminaries

In this section, we briefly review some usual definitions and notations (missing definitions can be found in, e.g., [7] or [1]). We denote by $\mathbb{Z}$ the set of integers and by $\mathbb{N}$ the set of positive integers including zero. We define $\mathbb{Z}_\infty = \mathbb{Z} \cup \{\infty\}$ and $\mathbb{N}_\infty = \mathbb{N} \cup \{\infty\}$, where for each $n \in \mathbb{Z}$ we have $n + \infty = \infty$ and $n < \infty$. For two positive integers $m \leq n$, we denote by $[\![m \mathinner{..} n]\!]$ the set $\{m, m+1, \ldots, n\}$. For a countable set $S$ we denote by $||S|| \in \mathbb{N}_\infty$ the cardinality of $S$. A decision problem is in $(\mathsf{N})\mathsf{SPACE}(n)$ if it can be decided by a (nondeterministic) Turing machine in space $\mathcal{O}(n)$ and in $\mathsf{PSPACE}$ if it is in $\mathsf{SPACE}(n^c)$ for some input-independent integer $c \geq 1$.

### 2.1 First Order Logic

**Syntax.** Let $\mathsf{Var}$ be a countable set of *variables*, denoted by $x, y, z$ and $B$ and $U$ be distinct *sorts*, where $B$ denotes booleans and $U$ denotes memory locations. A *function symbol* $f$ has $\#(f) \geq 0$ arguments of sort $U$ and a sort $\sigma(f)$, which is either $B$ or $U$. If $\#(f) = 0$, we call $f$ a *constant*. We use $\bot$ and $\top$ for the boolean constants false and true, respectively. First-order ($\mathsf{FO}$) *terms* $t$ and *formulas* $\varphi$ are defined by the following grammar:

$$t := x \mid f(\underbrace{t, \ldots, t}_{\#(f)}) \qquad \varphi := \bot \mid \top \mid \varphi \wedge \varphi \mid \neg\varphi \mid \exists x \,.\, \varphi \mid t \approx t \mid p(\underbrace{t, \ldots, t}_{\#(p)})$$

where $x \in \mathsf{Var}$, $f$ and $p$ are function symbols, $\sigma(f) = U$ and $\sigma(p) = B$. We write $\varphi_1 \vee \varphi_2$ for $\neg(\neg\varphi_1 \wedge \neg\varphi_2)$, $\varphi_1 \rightarrow \varphi_2$ for $\neg\varphi_1 \vee \varphi_2$, $\varphi_1 \leftrightarrow \varphi_2$ for $\varphi_1 \rightarrow \varphi_2 \wedge \varphi_2 \rightarrow \varphi_1$ and $\forall x \,.\, \varphi$ for $\neg\exists x \,.\, \neg\varphi$. The *size* of a formula $\varphi$, denoted by $\mathsf{size}(\varphi)$, is the number of occurrences of symbols in $\varphi$. A variable is *free* in $\varphi$ if it occurs in $\varphi$ but not in the scope of a quantifier. We denote by $\mathrm{fv}(\varphi)$ the set of variables that are free in $\varphi$. A *sentence* is a formula $\varphi$ such that $\mathrm{fv}(\varphi) = \emptyset$. The *Bernays-Schönfinkel-Ramsey fragment* of $\mathsf{FO}$ [$\mathsf{BSR}(\mathsf{FO})$] is the set of sentences of the form $\exists x_1 \ldots \exists x_n \forall y_1 \ldots \forall y_m \,.\, \varphi$, where $\varphi$ is a quantifier-free formula in which all function symbols $f$ of arity $\#(f) > 0$ have sort $\sigma(f) = B$. We denote by $\mathsf{FO}^1$ the set of formulas built on a signature containing only one function symbol of arity 1, the equality predicate and an arbitrary number of unary predicate symbols[2].

[2] The fragment $\mathsf{FO}^1$ is denoted by $[all, (\omega), (1)]_=$ in [1].

**Semantics.** First-order formulas are interpreted over FO-*structures* (called structures, when no confusion arises) $\mathcal{S} = (\mathfrak{U}, \mathfrak{s}, \mathfrak{i})$, where $\mathfrak{U}$ is a nonempty countable set, called the *universe*, the elements of which are called *locations*, $\mathfrak{s} : \mathsf{Var} \to \mathfrak{U}$ is a function mapping variables to locations called a *store*, and $\mathfrak{i}$ interprets each function symbol $f$ by a function $f^{\mathfrak{i}} : \mathfrak{U}^{\#(f)} \to \mathfrak{U}$, if $\sigma(f) = U$ and $f^{\mathfrak{i}} : \mathfrak{U}^{\#(f)} \to \{\bot, \top\}$ if $\sigma(f) = B$. A structure $(\mathfrak{U}, \mathfrak{s}, \mathfrak{i})$ is *finite* when $||\mathfrak{U}|| \in \mathbb{N}$ and *infinite* otherwise. We write $\mathcal{S} \models \varphi$ iff $\varphi$ is true when interpreted in $\mathcal{S}$. This relation is defined recursively on the structure of $\varphi$, as usual. When $\mathcal{S} \models \varphi$, we say that $\mathcal{S}$ is a *model* of $\varphi$. A formula is *satisfiable* when it has a model. We write $\varphi_1 \models \varphi_2$ when every model of $\varphi_1$ is also a model of $\varphi_2$ and by $\varphi_1 \equiv \varphi_2$ we mean $\varphi_1 \models \varphi_2$ and $\varphi_1 \models \varphi_2$. The *(in)finite satisfiability problem* asks, given a formula $\varphi$, whether a (in)finite model exists for this formula.

We now recall and refine an essential known result concerning the satisfiability problem for formulas in $\mathsf{FO}^1$:

**Theorem 1.** *The finite satisfiability problem is decidable for first-order formulas in* $\mathsf{FO}^1$. *Furthermore, the problem is nonelementary even if the formula contains no unary predicate symbols.*

*Proof.* The decidability result is proven in [1, Corollary 7.2.12, page 341]. The complexity lower bound is established in [1, Theorem 7.2.15, page 342] for arbitrary domains, however a careful analysis of the proof reveals that it also holds for finite domains. Indeed, the proof goes by showing that a domino problem of nonelementary complexity can be polynomially reduced to the satisfiability problem for a first-order formula $\varphi$ satisfying the conditions of the lemma. The initial domino problem is not important here and its definition is omitted. To establish the desired result, we only have to prove that satisfiability is actually equivalent to finite satisfiability for the obtained formula $\varphi$. The formula $\varphi$ output of the reduction is of the following form (see [1, Page 345]): $\varphi = \alpha \wedge \gamma \wedge \eta'[D(x)/\delta(x), P_i(x,y)/\pi_i(x,y)$, where:

- $\alpha = \exists x \forall y \,.\, f(x) \approx x \wedge f^{n+1}(y) \approx x$. This formula states that the domain can be viewed as a tree of height at most $n+1$, where the (necessarily unique) element corresponding to the variable $x$ is the root of the tree, and where $f$ maps every other node to its parent.
- The formula $\delta$ is based on an equivalence relation $E_{n-1}$ on nodes in a (possibly infinite) tree, which is inductively defined as follows:
  - All nodes are $E_0$-equivalent.
  - For $m > 1$, two nodes are $E_m$-equivalent if for every $E_{m-1}$-equivalence class $K$, either both nodes have no child in $K$ or both nodes have a child in $K$.

  The formula $\delta(x)$ states that $x$ is a child of the root with at most one child in each $E_{n-1}$-equivalence class. We also denote by $E$ the intersection $\bigcap_{i=1}^{n} E_i$.
- $\gamma = \forall x, y \,.\, \delta(x) \wedge \delta(y) \wedge \beta_n(x,y) \to x \approx y$, where $\beta_n(x,y)$ is a formula stating that $x$ and $y$ have height at most $n$ and are $E_n$-equivalent.

- For $i = 0, \ldots, r$, $\pi_i(x, y)$ is a formula stating that there exists a $z$ satisfying the following property denoted by $\mathcal{P}(i, a, b)$: $z$ is a child of the root and for every $E_{n-1}$-equivalence class $K$ and for all $j, k \in \{0, 1\}$, if $x$ and $y$ have exactly $j$ and $k$ children in $K$ respectively, then $z$ has exactly $2 + 4i + 2j + k$ children in $K$.
- $\eta'$ is equivalent to a closed formula defined over a signature containing a unary predicate symbol $D$ and $r + 1$ binary predicate symbols $P_0, \ldots, P_r$, in which every quantification ranges over elements $x$ satisfying $D(x)$. It is thus of the form $\exists x \,.\, D(x) \wedge \psi$ or $\forall x \,.\, D(x) \to \psi$.
- $\eta'[D(x)/\delta(x), P_i(x, y)/\pi_i(x, y)]$ denotes the formula $\eta'$ in which every occurrence of a formula $D(x)$ (resp. $P_i(x, y)$) is replaced by $\delta(x)$ (resp. $\pi_i(x, y)$). Thus it is equivalent to a formula in which every quantification ranges over elements $x$ satisfying $\delta(x)$.

The formal definitions of $\eta'$, $\delta(x)$ and $\pi_i(x, y)$ are unimportant and omitted.

Let $\mathcal{I} = (\mathfrak{U}, \mathfrak{s}, \mathfrak{i})$ be a model of $\varphi$, with $\mathfrak{f} = f^{\mathfrak{i}}$. We denote by $\mathfrak{r}$ the root of the tree, i.e., the unique element of $\mathfrak{U}$ with $(\mathfrak{U}, \mathfrak{s}[x \mapsto \mathfrak{r}], \mathfrak{i}) \models \forall y \,.\, f(x) \approx x \wedge f^{n+1}(y) \approx x$. Given $i \in [0, r]$ and $a, b \in \mathfrak{U}$, if $(\mathfrak{U}, \mathfrak{s}[x \mapsto a, y \mapsto b], \mathfrak{i}) \models \pi_i(x, y)$, then we denote by $\mu(i, a, b)$ a set containing an arbitrarily chosen element $z$ satisfying $\mathcal{P}(i, a, b)$ in the definition of $\pi_i(x, y)$ along with all the children of $z$, otherwise $\mu(i, a, b)$ is empty. Observe that $\mu(i, a, b)$ is always finite because the number of children of $z$ in each equivalence class is bounded by $2+4\times i+2+1 \leq 2+4\times r+2+1$, moreover the number of $E$-equivalence classes is finite [1, bottom of Page 343].

We show that $\varphi$ admits a finite model $\mathcal{I}'$. The set $B$ of elements $b$ such that $(\mathfrak{U}, \mathfrak{s}[x \mapsto b], \mathfrak{i}) \models \delta(x)$ is finite [1, Page 344, Lines 21–22]. Let $\Pi$ be the set: $\Pi = \bigcup\{\mu(i, a, b) \mid a, b \in B, i \in [0, r]\}$. Since $B$ is finite and every set $\mu(i, a, b)$ is finite, $\Pi$ is also finite. With each element $a \in \mathfrak{U}$ and each $E$-equivalence class $K$, we associate a set $\nu(a, K)$ containing exactly one child of $a$ in $K$ if such a child exists, otherwise $\nu(a, K)$ is empty. We now consider the subset $\mathfrak{U}'$ of $\mathfrak{U}$ defined as the set of elements $a$ such that for every $m \in \mathbb{N}$, $\mathfrak{f}^m(a)$ occurs either in $\{\mathfrak{r}\} \cup B \cup \Pi$ or in a set $\nu(b, K)$, where $b \in \mathfrak{U}$ and $K$ is an $E$-equivalence class. Note that $\mathfrak{r} \in \mathfrak{U}'$ and that if $a \in \mathfrak{U}'$ then necessarily $\mathfrak{f}(a) \in \mathfrak{U}'$. Furthermore, if $\mathfrak{f}(b) \in \mathfrak{U}'$ and $b \in \nu(\mathfrak{f}(b), K)$ then $b \in \mathfrak{U}'$.

It is easy to check that $\mathfrak{U}'$ is finite. Indeed, since $(\mathfrak{U}, \mathfrak{s}, \mathfrak{i}) \models \alpha$ and no new node or edge is added, all nodes are of height less or equal to $n+1$. Furthermore, all nodes have at most $||B|| + ||\Pi|| + \#K$ children in $\mathfrak{U}'$, where $\#K$ denotes the number of $E$-equivalence classes.

We denote by $\mathcal{I}' = (\mathfrak{U}', \mathfrak{s}, \mathfrak{i}')$ the restriction of $\mathcal{I}$ to the elements of $\mathfrak{U}'$ (we may assume that $\mathfrak{s}$ is a store on $\mathfrak{U}'$ since $\varphi$ is closed). We prove that $\mathcal{I}' \models \varphi$.

- Since $\mathfrak{U}'$ contains the root, and $\mathcal{I} \models \alpha$, we must have $\mathcal{I}' \models \alpha$.
- Observe that $\mathfrak{U}'$ necessarily contains $\nu(b, K)$, for every $b \in \mathfrak{U}'$, since by definition the parent of the (unique) element of $\nu(b, K)$ is $b$. Thus at least one child of $b$ is kept in each equivalence class. Thus the relations $E_m$ on elements of $\mathfrak{U}'$ are preserved in the transformation: for every $a, b \in \mathfrak{U}'$, $a, b$ are

$E_m$-equivalent in the structure $\mathcal{I}$ iff they are equivalent in the structure $\mathcal{I}'$. Further, the height of the nodes cannot change. Therefore, for every $a, a' \in U'$:

$$(\mathfrak{U}', \mathfrak{s}[x \mapsto a, y \mapsto a'], \mathfrak{i}') \models \beta_n(x, y) \text{ iff } (\mathfrak{U}, \mathfrak{s}[x \mapsto a, y \mapsto a'], \mathfrak{i}) \models \beta_n(x, y)$$

By definition, for every $a \in B$ and $m \in \mathbb{N}$, $\mathfrak{f}^m(a) \in \{a, \mathfrak{r}\}$, thus $B \subseteq \mathfrak{U}'$. Because no new edges are added, we deduce:

$$(\mathfrak{U}', \mathfrak{s}[x \mapsto a], \mathfrak{i}') \models \delta(x) \Leftrightarrow (\mathfrak{U}, \mathfrak{s}[x \mapsto a], \mathfrak{i}) \models \delta(x) \Leftrightarrow a \in B$$

Consequently, since $\mathcal{I} \models \gamma$, we have $\mathcal{I}' \models \gamma$.

- All elements in $\mu(i, a, a')$ with $a, a' \in B$ occur in $\mathfrak{U}'$ (because if $b \in \mu(i, a, a')$ and $m \in \mathbb{N}$ then $\mathfrak{f}^m(b) \in \{\mathfrak{r}\} \cup B \cup \mu(i, a, a')$), thus, for all $a, a' \in B$:

$$(\mathfrak{U}', \mathfrak{s}[x \mapsto a, y \mapsto a'], \mathfrak{i}') \models \pi_i(x, y) \Leftrightarrow (\mathfrak{U}, \mathfrak{s}[x \mapsto a, y \mapsto a'], \mathfrak{i}) \models \pi_i(x, y)$$

Since all quantifications in $\eta'$ range over elements in $B$, we deduce, by a straightforward induction on the formula, that $\mathcal{I}$ and $\mathcal{I}'$ necessarily agree on the formula $\eta'[D(x)/\delta(x), P_i(x, y)/\pi_i(x, y)]$. Consequently, $\mathcal{I}' \models \eta'[D(x)/\delta(x), P_i(x, y)/\pi_i(x, y)]$. □

### 2.2 Separation Logic

**Syntax.** Let $k \in \mathbb{N}$ be a strictly positive integer. The logic $\mathsf{SL}^k$ is the set of formulas generated by the grammar:

$$\varphi := \bot \mid \top \mid \mathsf{emp} \mid x \approx y \mid x \mapsto (y_1, \ldots, y_k) \mid \varphi \wedge \varphi \mid \neg\varphi \mid \varphi * \varphi \mid \varphi \mathrel{-\!\!*} \varphi \mid \exists x \,.\, \varphi$$

where $x, y, y_1, \ldots, y_k \in \mathsf{Var}$. The connectives $*$ and $\mathrel{-\!\!*}$ are respectively called the *separating conjunction* and *separating implication* (*magic wand*). The symbols $\vee$, $\rightarrow$, $\leftrightarrow$ and $\forall$ are defined as in first-order logic, and in addition, we write $\varphi_1 \multimap \varphi_2$ for $\neg(\varphi_1 \mathrel{-\!\!*} \neg\varphi_2)$ ($\multimap$ is called *septraction*).

A tuple $(y_1, \ldots, y_k) \in \mathsf{Var}^k$ is sometimes denoted by $\mathbf{y}$. The *size* and *free variables* of an $\mathsf{SL}^k$ formula $\varphi$ are defined as for first-order formulas. The *prenex fragment* of $\mathsf{SL}^k$ (denoted by $\mathsf{PRE}(\mathsf{SL}^k)$) is the set of sentences $Q_1 x_1 \ldots Q_n x_n \,.\, \phi$, where $Q_1, \ldots, Q_n \in \{\exists, \forall\}$ and $\phi$ is a quantifier-free $\mathsf{SL}^k$ formula. The *Bernays-Schönfinkel-Ramsey fragment* of $\mathsf{SL}^k$ [$\mathsf{BSR}(\mathsf{SL}^k)$] is the set of sentences $\exists x_1 \ldots \exists x_n \forall y_1 \ldots \forall y_m \,.\, \phi$, where $\phi$ is a quantifier-free $\mathsf{SL}^k$ formula. Since there are no function symbols of arity greater than zero in $\mathsf{SL}^k$, there are no restrictions, other than the form of the quantifier prefix, defining $\mathsf{BSR}(\mathsf{SL}^k)$.

**Semantics.** $\mathsf{SL}^k$ formulas are interpreted over $\mathsf{SL}$-*structures* (called structures when no confusion arises) $\mathcal{I} = (\mathfrak{U}, \mathfrak{s}, \mathfrak{h})$, where $\mathfrak{U}$ and $\mathfrak{s}$ are defined as for first-order formulas[3] and $\mathfrak{h} : \mathfrak{U} \rightharpoonup_{fin} \mathfrak{U}^k$ is a finite partial mapping of locations to

[3] In contrast to most existing work in Separation Logic, we do not assume that $\mathfrak{U}$ is infinite.

$k$-tuples of locations, called a *heap*. A structure $(\mathfrak{U}, \mathfrak{s}, \mathfrak{h})$ is finite when $||\mathfrak{U}|| \in \mathbb{N}$ and infinite otherwise (note that the heap is always finite, but that the universe may be finite or infinite).

Given a heap $\mathfrak{h}$, we denote by $\mathsf{dom}(\mathfrak{h})$ the domain of the heap, by $\mathsf{img}(\mathfrak{h}) \stackrel{\text{def}}{=} \{\ell_i \mid \exists \ell \in \mathsf{dom}(\mathfrak{h}), \mathfrak{h}(\ell) = (\ell_1, \ldots, \ell_k), i \in [\![1 \mathinner{..} k]\!]\}$ its range and we let $\mathrm{elems}(\mathfrak{h}) \stackrel{\text{def}}{=} \mathsf{dom}(\mathfrak{h}) \cup \mathsf{img}(\mathfrak{h})$. A element $x$ is *allocated* in $(\mathfrak{U}, \mathfrak{s}, \mathfrak{h})$ if it belongs to $\mathsf{dom}(\mathfrak{h})$. For a store $\mathfrak{s}$, we define its range $\mathsf{img}(\mathfrak{s}) \stackrel{\text{def}}{=} \{\ell \mid x \in \mathsf{Var},\ \mathfrak{s}(x) = \ell\}$. If $\mathbf{x} = (x_1, \ldots, x_n)$ is a vector of pairwise distinct variables and $\mathbf{e} = (e_1, \ldots, e_n)$ is a vector of elements of $\mathfrak{U}$ of the same length as $\mathbf{x}$, then $\mathfrak{s}[\mathbf{x} \mapsto \mathbf{e}]$ denotes the store that maps $x_i$ to $e_i$ (for all $i \in [\![1 \mathinner{..} n]\!]$) and coincides with $\mathfrak{s}$ on every variable distinct from $x_1, \ldots, x_n$. Two heaps $\mathfrak{h}_1$ and $\mathfrak{h}_2$ are *disjoint* if and only if $\mathsf{dom}(\mathfrak{h}_1) \cap \mathsf{dom}(\mathfrak{h}_2) = \emptyset$, in which case $\mathfrak{h}_1 \uplus \mathfrak{h}_2$ denotes their union ($\mathfrak{h}_1 \uplus \mathfrak{h}_2$ is undefined if $\mathfrak{h}_1$ and $\mathfrak{h}_2$ are not disjoint). The relation $(\mathfrak{U}, \mathfrak{s}, \mathfrak{h}) \models \varphi$ is defined inductively, as follows:

$$
\begin{array}{ll}
(\mathfrak{U}, \mathfrak{s}, \mathfrak{h}) \models \mathsf{emp} & \Leftrightarrow \mathfrak{h} = \emptyset \\
(\mathfrak{U}, \mathfrak{s}, \mathfrak{h}) \models x \approx y & \Leftrightarrow \mathfrak{s}(x) = \mathfrak{s}(y) \\
(\mathfrak{U}, \mathfrak{s}, \mathfrak{h}) \models x \mapsto (y_1, \ldots, y_k) & \Leftrightarrow \mathfrak{h}(\mathfrak{s}(x)) = (\mathfrak{s}(y_1), \ldots, \mathfrak{s}(y_k)) \wedge \mathsf{dom}(\mathfrak{h}) = \{\mathfrak{s}(x)\} \\
(\mathfrak{U}, \mathfrak{s}, \mathfrak{h}) \models \varphi_1 \wedge \varphi_2 & \Leftrightarrow (\mathfrak{U}, \mathfrak{s}, \mathfrak{h}) \models \varphi_1 \text{ and } (\mathfrak{U}, \mathfrak{s}, \mathfrak{h}) \models \varphi_2 \\
(\mathfrak{U}, \mathfrak{s}, \mathfrak{h}) \models \neg\varphi & \Leftrightarrow (\mathfrak{U}, \mathfrak{s}, \mathfrak{h}) \not\models \varphi \\
(\mathfrak{U}, \mathfrak{s}, \mathfrak{h}) \models \exists x \,.\, \varphi & \Leftrightarrow \text{there exists } e \in \mathfrak{U} \text{ s.t. } (\mathfrak{U}, \mathfrak{s}[x \mapsto e], \mathfrak{h}) \models \varphi \\
(\mathfrak{U}, \mathfrak{s}, \mathfrak{h}) \models \varphi_1 * \varphi_2 & \Leftrightarrow \text{there exist disjoint heaps } \mathfrak{h}_1, \mathfrak{h}_2 \text{ such that } \mathfrak{h} = \mathfrak{h}_1 \uplus \mathfrak{h}_2 \\
& \quad \text{and } (\mathfrak{U}, \mathfrak{s}, \mathfrak{h}_i) \models \varphi_i, \text{ for } i = 1, 2 \\
(\mathfrak{U}, \mathfrak{s}, \mathfrak{h}) \models \varphi_1 -\!\!* \varphi_2 & \Leftrightarrow \text{for all heaps } \mathfrak{h}' \text{ disjoint from } \mathfrak{h} \text{ such that } (\mathfrak{U}, \mathfrak{s}, \mathfrak{h}') \models \varphi_1, \\
& \quad \text{we have } (\mathfrak{U}, \mathfrak{s}, \mathfrak{h}' \uplus \mathfrak{h}) \models \varphi_2
\end{array}
$$

Satisfiability, entailment and equivalence are defined for $\mathsf{SL}^k$ as for $\mathsf{FO}$ formulas. The finite [resp. infinite] satisfiability problem for $\mathsf{SL}^k$ asks whether a finite [resp. an infinite] model exists for a given formula. We write $\phi \equiv^{fin} \psi$ [$\phi \equiv^{inf} \psi$] whenever $(\mathfrak{U}, \mathfrak{s}, \mathfrak{h}) \models \phi \Leftrightarrow (\mathfrak{U}, \mathfrak{s}, \mathfrak{h}) \models \psi$ for every finite [infinite] structure $(\mathfrak{U}, \mathfrak{s}, \mathfrak{h})$.

As stated in the introduction, $\mathsf{SL}$ formulas do not admit prenex forms in general, because the quantifiers cannot be shifted outside of separating connectives. This is an essential difference with $\mathsf{FO}$, where each formula is equivalent to a linear-size formula in prenex form. In particular, the equivalences $\phi * \forall x \,.\, \psi(x) \Leftrightarrow \forall x \,.\, \phi * \psi(x)$ and $\phi -\!\!* \exists x \,.\, \psi(x) \Leftrightarrow \exists x \,.\, \phi -\!\!* \psi(x)$ do not always hold.

*Example 2.* For instance, the formula $(\forall x \,.\, x \mapsto x) * \top$ is satisfiable only on universes of cardinality 1 (because $\forall x \,.\, x \mapsto x$ entails that the domain of the heap is of size 1 and contains all locations), but the formula $\forall x \,.\, (x \mapsto x * \top)$ is satisfiable if and only if the universe is finite and each location points to itself. ∎

## 2.3 Test Formulas for $\mathsf{SL}^k$

This section presents the definitions and results from [6], needed for self-containment.

**Definition 3.** *The following patterns are called* test formulas *of* $\mathsf{SL}^k$, *for any* $k \geq 1$:

$$x \hookrightarrow \mathbf{y} \stackrel{\mathsf{def}}{=} x \mapsto \mathbf{y} * \top \qquad |U| \geq n \stackrel{\mathsf{def}}{=} \top \multimap |h| \geq n,\ n \in \mathbb{N}$$

$$\mathsf{alloc}(x) \stackrel{\mathsf{def}}{=} x \mapsto \underbrace{(x, \ldots, x)}_{k \text{ times}} \multimap\!\!* \bot \qquad |h| \geq |U| - n \stackrel{\mathsf{def}}{=} |h| \geq n+1 \multimap\!\!* \bot, n \in \mathbb{N}$$

$$x \approx y \qquad |h| \geq n \stackrel{\mathsf{def}}{=} \begin{cases} |h| \geq n-1 * \neg\mathsf{emp}, & \text{if } n > 0 \\ \top, & \text{if } n = 0 \end{cases}$$

*where* $x, y \in \mathsf{Var}$, $\mathbf{y} \in \mathsf{Var}^k$ *is a* $k$*-tuple of variables and* $n \in \mathbb{N}$ *is a positive integer. A* literal *is a test formula or its negation and a* minterm *is any conjunction of literals.*

The semantics of test formulas is intuitive: $x \hookrightarrow \mathbf{y}$ holds when $x$ denotes a location and $\mathbf{y}$ is the image of that location in the heap, $\mathsf{alloc}(x)$ holds when $x$ denotes a location in the domain of the heap (allocated), $|h| \geq n$, $|U| \geq n$ and $|h| \geq |U| - n$ are cardinality constraints involving the size of the heap, denoted by $|h|$ and that of the universe, denoted by $|U|$. We recall that $|h|$ ranges over $\mathbb{N}$, whereas $|U|$ is always interpreted as a number larger than $|h|$ and possibly infinite. The truth value of the test formulas of the form $|U| \geq n$ and $|h| \geq |U| - n$ depend on the universe $\mathfrak{U}$, hence such test formulas are called *universe-dependent*. The truth value of the other test formulas depend only on the store and heap, thus they are called *universe-independent*. Clearly, all universe-dependent test formulas are trivially equivalent to true (for $|U| \geq n$) or false (for $|h| \geq |U| - n$) when interpreted over an infinite universe. Observe that not all atoms of $\mathsf{SL}^k$ are test formulas, for instance $x \mapsto \mathbf{y}$ and $\mathsf{emp}$ are not test formulas. However, it is easy to check that any atom may be written as a boolean combination of test formulas, for instance $x \mapsto \mathbf{y}$ is equivalent to $x \hookrightarrow \mathbf{y} \wedge \neg|h| \geq 2$ and $\mathsf{emp}$ is equivalent to $\neg|h| \geq 1$.

The following result establishes a translation of quantifier-free $\mathsf{SL}^k$ formulas into boolean combinations of test formulas. A *literal* is a test formula or its negation and a *minterm* is any conjunction of literals.

**Lemma 4.** *Given a quantifier-free* $\mathsf{SL}^k$ *formula* $\phi$, *there exist finite sets of minterms* $\mu^{fin}(\phi)$ *and* $\mu^{inf}(\phi)$ *such that* $\phi \equiv^{fin} \bigvee_{M \in \mu^{fin}(\phi)} M$ *and* $\phi \equiv^{inf} \bigvee_{M \in \mu^{inf}(\phi)} M$. *Furthermore, the size of every* $M \in \mu^{fin}(\phi) \cup \mu^{inf}(\phi)$ *is polynomial w.r.t.* $\mathsf{size}(\phi)$, *and given a minterm* $M$, *the problem of checking whether* $M \in \mu^{fin}(\phi)$ *[resp.* $M \in \mu^{inf}(\phi)$*] is in* PSPACE.

*Proof.* See [6]. □

Given a quantifier-free $\mathsf{SL}^k$ formula $\phi$, the number of minterms in $\mu^{fin}(\phi)$ [resp. in $\mu^{inf}(\phi)$] is exponential in the size of $\phi$, in the worst case. An optimal decision procedure does not generate and store these sets explicitly, but rather enumerate minterms lazily.

*Example 5.* The formula $x \mapsto y * y \mapsto x * \neg\mathsf{emp}$ is equivalent to the minterm: $x \hookrightarrow y \wedge y \hookrightarrow x \wedge x \not\approx y \wedge |h| \geq 3$. Indeed, because the atoms $x \mapsto y$, $y \mapsto x$ and $\neg\mathsf{emp}$ must be satisfied on disjoint heaps, the initial formula entails that $x, y$ are distinct and that the heap contains at least 3 allocated elements ($x$, $y$ and an additional element distinct from $x$ and $y$). The formula $x \mapsto y \mathrel{-\!\!*} x \mapsto z$ is equivalent to the disjunction of minterms $\mathsf{alloc}(x) \vee (\neg|h| \geq 1 \wedge y \approx z)$. Indeed, if $x$ is allocated then the heap cannot be extended by a disjoint heap satisfying $x \mapsto y$ hence the separating implication trivially holds, otherwise the implication holds iff the heap is empty and $y \approx z$. ■

## 3 From Infinite to Finite Satisfiability

We begin by showing that for prenex SL-formulas, the infinite satisfiability problem can be reduced to the finite satisfiability problem. The intuition is that two SL-structures defined on the same heap and store can be considered as equivalent if both have enough locations outside of the heap.

**Definition 6.** *Let $X$ be a set of variables and let $n \in \mathbb{N}$. Two SL-structures $\mathcal{I} = (\mathfrak{U}, \mathfrak{s}, \mathfrak{h})$ and $\mathcal{I}' = (\mathfrak{U}', \mathfrak{s}', \mathfrak{h}')$ are $(X, n)$-similar (written $\mathcal{I} \sim_X^n \mathcal{I}'$) iff the following conditions hold:*

1. *$\mathfrak{h} = \mathfrak{h}'$.*
2. *For every $x \in X$, if $\mathfrak{s}(x) \in \mathrm{elems}(\mathfrak{h})$ or $\mathfrak{s}'(x) \in \mathrm{elems}(\mathfrak{h}')$ then $\mathfrak{s}(x) = \mathfrak{s}'(x)$.*
3. *$||\mathfrak{U} \setminus \mathrm{elems}(\mathfrak{h})|| \geq n + ||X||$ and $||\mathfrak{U}' \setminus \mathrm{elems}(\mathfrak{h})|| \geq n + ||X||$.*
4. *For all $x, y \in X$, $\mathcal{I} \models x \approx y$ iff $\mathcal{I}' \models x \approx y$.*

Condition 1 entails that $\mathrm{elems}(\mathfrak{h}) \subseteq \mathfrak{U} \cap \mathfrak{U}'$. We prove that any two SL-structures that are $(\mathrm{fv}(\phi), m)$-similar are indistinguishable by any formula $\phi$ prefixed by $m$ quantifiers.

**Proposition 7.** *Let $\phi = \mathsf{Q}_1 x_1 \dots \mathsf{Q}_m x_m \,.\, \psi$ be a prenex $\mathsf{SL}^k$ formula, with $\mathsf{Q}_i \in \{\forall, \exists\}$ for all $i = 1, \dots, m$, where $\psi$ is a quantifier-free boolean combination of universe-independent test formulas. If $\mathcal{I} \sim_{\mathrm{fv}(\phi)}^m \mathcal{I}'$ and $\mathcal{I} \models \phi$ then $\mathcal{I}' \models \phi$.*

*Proof.* Let $\mathcal{I} = (\mathfrak{U}, \mathfrak{s}, \mathfrak{h})$ and $\mathcal{I}' = (\mathfrak{U}', \mathfrak{s}', \mathfrak{h}')$. Assume that $\mathcal{I} \sim_{\mathrm{fv}(\phi)}^m \mathcal{I}'$ and $\mathcal{I} \models \phi$. By Condition 1 in Definition 6 we have $\mathfrak{h} = \mathfrak{h}'$. We prove that $\mathcal{I}' \models \phi$ by induction on $m$.

- If $m = 0$, then we have $\phi = \psi$, we show that $\mathcal{I}$ and $\mathcal{I}'$ agree on every atomic formula in $\phi$, which entails by an immediate induction that they agree on $\phi$. By Condition 4 in Definition 6, we already have that $\mathcal{I}$ and $\mathcal{I}'$ agree on every atom $x \approx x'$ with $x, x' \in \mathrm{fv}(\phi)$. By Condition 1, $\mathcal{I}$ and $\mathcal{I}'$ agree on all atoms $|h| \geq n$. Consider an atom $\ell \in \{y_0 \hookrightarrow (y_1, \dots, y_k), \mathsf{alloc}(y_0)\}$, with $y_0, \dots, y_k \in \mathrm{fv}(\phi)$. If for every $i \in [\![0 \,..\, k]\!]$ we have $\mathfrak{s}(y_i) \in \mathrm{elems}(\mathfrak{h})$ then by Condition 2 we deduce that $\mathfrak{s}'$ and $\mathfrak{s}$ coincide on $y_0, \dots, y_k$ hence $\mathcal{I}$ and $\mathcal{I}'$ agree on $\ell$ because they share the same heap. The same holds if $\mathfrak{s}'(y_i) \in \mathrm{elems}(\mathfrak{h})$, $\forall i \in [\![0 \,..\, k]\!]$. If both conditions are false, then we must have $\mathcal{I} \not\models \ell$ and $\mathcal{I}' \not\models \ell$, by definition of $\mathrm{elems}(\mathfrak{h})$, thus $\mathcal{I}$ and $\mathcal{I}'$ also agree on $\ell$ in this case.

- Assume that $m \geq 1$ and $Q_1 = \exists$, i.e., $\phi = \exists x_1 . \phi'$. Then there exists $e \in \mathfrak{U}$ such that $(\mathfrak{U}, \mathfrak{s}[x_1 \mapsto e], \mathfrak{h}) \models \phi'$. We construct an element $e' \in \mathfrak{U}'$ as follows. If $e = \mathfrak{s}(y)$, for some $y \in \text{fv}(\phi)$, then we let $e' = \mathfrak{s}'(y)$. If $\forall y \in \text{fv}(\phi), e \neq \mathfrak{s}(y)$ and if $e \in \text{elems}(\mathfrak{h})$ then we let $e' = e$. Otherwise, $e'$ is an arbitrarily chosen element in $\mathfrak{U}' \setminus (\mathfrak{s}'(\text{fv}(\phi)) \cup \text{elems}(\mathfrak{h}))$. Such an element necessarily exists, because by Condition 3 in Definition 6, $\mathfrak{U}'$ contains at least $m + ||\text{fv}(\phi)|| \geq 1 + ||\mathfrak{s}(\text{fv}(\phi))||$ elements distinct from those in elems($\mathfrak{h}$). Let $\mathcal{J} = (\mathfrak{U}, \mathfrak{s}[x_1 \mapsto e], \mathfrak{h})$ and $\mathcal{J}' = (\mathfrak{U}, \mathfrak{s}[x_1 \mapsto e], \mathfrak{h})$, we prove that $\mathcal{J} \sim^{m-1}_{\text{fv}(\phi)\cup\{x_1\}} \mathcal{J}'$. This entails the required results since by the induction hypothesis we deduce $\mathcal{J}' \models \phi'$, so that $\mathcal{I}' \models \phi$.
  - Condition 1 trivially holds.
  - For Condition 2, assume that there exists a variable $x \in \text{fv}(\phi) \cup \{x_1\}$ such that either $\mathfrak{s}[x_1 \mapsto e](x) \in \text{elems}(\mathfrak{h})$ or $\mathfrak{s}'[x_1 \mapsto e'](x) \in \text{elems}(\mathfrak{h})$, and $\mathfrak{s}[x_1 \mapsto e](x) \neq \mathfrak{s}'[x_1 \mapsto e'](x)$. Since $\mathcal{I} \sim^m_{\text{fv}(\phi)} \mathcal{I}'$, if $x \in \text{fv}(\phi)$ then $[\mathfrak{s}(x) \in \text{elems}(\mathfrak{h}) \vee \mathfrak{s}'(x) \in \text{elems}(\mathfrak{h})] \Rightarrow \mathfrak{s}(x) = \mathfrak{s}'(x)$, thus necessarily $x = x_1$. In this case, $\mathfrak{s}[x_1 \mapsto e](x) = e$ and $\mathfrak{s}'[x_1 \mapsto e'](x) = e'$. Since $e \neq e'$ by hypothesis, there can be no $y \in \text{fv}(\phi)$ such that $\mathfrak{s}(y) = e$ because otherwise by construction we would have $e = \mathfrak{s}(y) = \mathfrak{s}'(y) = e'$. By definition of $e'$ we cannot have $e \in \text{elems}(\mathfrak{h})$ either, so $e'$ is necessarily in $\mathfrak{U}' \setminus (\mathfrak{s}'(\text{fv}(\phi)) \cup \text{elems}(\mathfrak{h}))$ and the disjunction $e \in \text{elems}(\mathfrak{h}) \vee e' \in \text{elems}(\mathfrak{h})$ cannot hold.
  - Condition 3 follows from the fact that $\mathcal{I} \sim^m_{\text{fv}(\phi)} \mathcal{I}'$ because we have $m - 1 + ||\text{fv}(\phi) \cup \{x_1\}|| = m + ||\text{fv}(\phi)||$.
  - We now establish Condition 4. Let $x, x' \in \text{fv}(\phi) \cup \{x_1\}$. If $x, x' \in \text{fv}(\phi)$ then $\mathfrak{s}[x_1 \mapsto e]$ and $\mathfrak{s}'[x_1 \mapsto e']$ coincide with $\mathfrak{s}$ and $\mathfrak{s}'$ respectively on $x$ and $x'$, hence $\mathcal{J}$ and $\mathcal{J}'$ must agree on $x \approx x'$ since $\mathcal{I} \sim^m_{\text{fv}(\phi)} \mathcal{I}'$. The result also trivially holds when $x = x' = x_1$. Now assume that $x = x_1$ and $x' \neq x_1$. If $e = \mathfrak{s}(y)$ for some $y \in \text{fv}(\phi)$, then $\mathcal{J} \models x \approx x'$ iff $\mathcal{I} \models y \approx x'$. By definition of $e'$, we also have $e' = \mathfrak{s}'(y)$, hence $\mathcal{J}' \models x \approx x'$ iff $\mathcal{I}' \models y \approx x'$. Since both $y$ and $x'$ are in $\text{fv}(\phi)$, we have $\mathcal{J} \models x \approx x' \Leftrightarrow \mathcal{I} \models y \approx x' \Leftrightarrow \mathcal{I}' \models y \approx x' \Leftrightarrow \mathcal{J}' \models x \approx x'$. If the previous condition does not hold then necessarily $e \neq \mathfrak{s}(x')$, and $\mathcal{J} \not\models x_1 \approx x'$. If $e \in \text{elems}(\mathfrak{h})$, then by definition of $e'$, we have $e' = e$. If $\mathcal{J}' \models x_1 \approx x'$ then we must have $\mathfrak{s}'(x') = \mathfrak{s}'(x_1) = e' = e \in \text{elems}(\mathfrak{h})$, which by Condition 2 entails that $\mathfrak{s}'(x') = \mathfrak{s}(x') = e$, hence $\mathcal{J} \models x_1 \approx x'$, a contradiction. Finally, if $e \notin \text{elems}(\mathfrak{h})$, then by definition of $e'$, $e'$ cannot occur in $\mathfrak{s}'(\text{fv}(\phi))$, thus $\mathcal{J}' \not\models x_1 \approx x'$.
- Finally, assume that $m \geq 1$ and $Q_1 = \forall$. Then $\phi = \forall x_1 . \phi'$. Let $\phi_2 = \exists x_1 . \phi'_1$, where $\phi'_1$ denotes the nnf of $\neg\phi'$. Assume that $\mathcal{I}' \not\models \phi$, then $\mathcal{I}' \models \phi_2$, because $\neg\phi \equiv \exists x_1 . \neg\phi' \equiv \exists x_1 . \phi'_1 = \phi_2$. By the previous case, using the symmetry of $\sim^m_{\text{fv}(\phi)}$ and the fact that $\phi$ and $\phi_2$ have exactly the same free variables and number of quantifiers, we have $\mathcal{I} \models \phi_2$, i.e. $\mathcal{I} \not\models \phi$, a contradiction. □

We define the following shorthands:

$$
\begin{array}{rcl}
x \in h & \stackrel{\mathsf{def}}{=} & \exists y_0, y_1, \dots y_k \,.\, y_0 \hookrightarrow (y_1, \dots, y_k) \wedge \bigvee_{i=0}^{k} x \approx y_i \\
\mathsf{dist}(x_1, \dots, x_n) & \stackrel{\mathsf{def}}{=} & \bigwedge_{i=1}^{n} \bigwedge_{j=1}^{i-1} \neg(x_i \approx x_j) \\
\lambda_p & \stackrel{\mathsf{def}}{=} & \exists x_1, \dots, x_p \,.\, (\mathsf{dist}(x_1, \dots, x_p) \wedge \bigwedge_{i=1}^{p} \neg x_i \in h)
\end{array}
$$

It is clear that $(\mathfrak{U}, \mathfrak{s}, \mathfrak{h}) \models \lambda_p$ iff $||\mathfrak{U} \setminus \mathrm{elems}(\mathfrak{h})|| \geq p$. In particular, $\lambda_p$ is always true on an infinite universe. Observe, moreover, that $\lambda_p$ belongs to the $\mathsf{PRE}(\mathsf{SL}^k)$ fragment, for any $p \geq 2$ and any $k \geq 1$.

The following lemma reduces the infinite satisfiability problem to the finite version of this problem. This is done by adding an axiom ensuring that there are enough locations outside of the heap. Note that there is no need to consider test formulas of the form $|U| \geq n$ [resp. $|h| \geq |U| - n$] because they always evaluate to true [resp. false] on infinite SL-structures.

**Theorem 8.** *Let $\phi = \mathrm{Q}_1 x_1 \dots \mathrm{Q}_m x_m \,.\, \psi$ be a prenex $\mathsf{SL}^k$ formula, where $\mathrm{Q}_i \in \{\forall, \exists\}$ for $i = 1, \dots, m$ and $\mathrm{fv}(\phi) = \emptyset$. Assume that $\psi$ is a boolean combination of universe-independent test formulas. The two following assertions are equivalent.*

1. *$\phi$ admits an infinite model.*
2. *$\phi \wedge \lambda_m$ admits a finite model.*

*Proof.* (1) $\Rightarrow$ (2): Assume that $\phi$ admits an infinite model $(\mathfrak{U}, \mathfrak{s}, \mathfrak{h})$. Let $\mathfrak{U}'$ be a finite subset of $\mathfrak{U}$ containing $\mathrm{elems}(\mathfrak{h})$ and $m$ additional elements. It is clear that $(\mathfrak{U}, \mathfrak{s}, \mathfrak{h}) \sim_{\emptyset}^{m} (\mathfrak{U}', \mathfrak{s}, \mathfrak{h})$. Indeed, Condition 1 holds since the two structures share the same heap, Conditions 4 and 2 trivially hold since the considered set of variables is empty, and Condition 3 holds since $\mathfrak{U}$ is infinite and the additional elements in $\mathfrak{U}'$ do not occur in $\mathrm{elems}(\mathfrak{h})$. Thus $(\mathfrak{U}', \mathfrak{s}, \mathfrak{h}) \models \phi$ by Proposition 7, and $(\mathfrak{U}', \mathfrak{s}, \mathfrak{h}) \models \lambda_m$, by definition of $\mathfrak{U}'$.

(2) $\Rightarrow$ (1): Assume that $\phi \wedge \lambda_m$ has a finite model $(\mathfrak{U}, \mathfrak{s}, \mathfrak{h})$. Let $\mathfrak{U}'$ be any infinite set containing $\mathfrak{U}$. Again, we have $(\mathfrak{U}, \mathfrak{s}, \mathfrak{h}) \sim_{\emptyset}^{m} (\mathfrak{U}', \mathfrak{s}, \mathfrak{h})$. As in the previous case, Conditions 1, 2 and 4 trivially hold, and Condition 3 holds since $\mathfrak{U}'$ is infinite and $(\mathfrak{U}, \mathfrak{s}, \mathfrak{h}) \models \lambda_m$. By Proposition 7, we deduce that $(\mathfrak{U}', \mathfrak{s}, \mathfrak{h}) \models \phi$. □

## 4 $\mathsf{PRE}(\mathsf{SL}^1)$ is Decidable but Not Elementary Recursive

Using Lemma 4 and Theorem 8 we shall prove that the satisfiability problem is decidable for the prenex fragment of $\mathsf{SL}^1$. This shows that $\mathsf{PRE}(\mathsf{SL}^1)$ is strictly less expressive than $\mathsf{SL}^1$, because $\mathsf{SL}^1$ has an undecidable satisfiability problem [2]. For this purpose, we first define a translation of quantified boolean combination of test formulas into FO that is sat-preserving on finite structures. Let $\mathfrak{d}$ be a unary predicate symbol and for $i = 1, \dots, k$, let $\mathfrak{f}_i$ be a unary function symbol.

We define the following transformation from quantified boolean combinations of test formulas into first order formulas:

$$\begin{aligned}
\Theta(x \approx y) &\stackrel{\text{def}}{=} x \approx y \\
\Theta(x \hookrightarrow (y_1, \dots, y_k)) &\stackrel{\text{def}}{=} \mathfrak{d}(x) \wedge \bigwedge_{i=1}^{k} y_i \approx \mathfrak{f}_i(x) \\
\Theta(\mathsf{alloc}(x)) &\stackrel{\text{def}}{=} \mathfrak{d}(x) \\
\Theta(|U| \geq n) &\stackrel{\text{def}}{=} \exists x_1, \dots, x_n \,.\, \mathsf{dist}(x_1, \dots, x_n) \\
\Theta(|h| \geq n) &\stackrel{\text{def}}{=} \exists x_1, \dots, x_n \,.\, \mathsf{dist}(x_1, \dots, x_n) \wedge \bigwedge_{i=1}^{n} \mathfrak{d}(x_i) \\
\Theta(|h| \geq |U| - n) &\stackrel{\text{def}}{=} \exists x_1, \dots, x_n \forall y \,.\, \bigwedge_{i=1}^{n} y \not\approx x_i \rightarrow \mathfrak{d}(y) \\
\Theta(\neg \phi) &\stackrel{\text{def}}{=} \neg \Theta(\phi) \\
\Theta(\phi_1 \wedge \phi_2) &\stackrel{\text{def}}{=} \Theta(\phi_1) \wedge \Theta(\phi_2) \\
\Theta(\exists x \,.\, \phi) &\stackrel{\text{def}}{=} \exists x \,.\, \Theta(\phi)
\end{aligned}$$

**Proposition 9.** *Let $\phi$ be a quantified boolean combination of test formulas. The formula $\phi$ has a finite* SL *model if and only if $\Theta(\phi)$ has a finite* FO *model.*

*Proof.* An FO-structure $\mathcal{I} = (\mathfrak{U}, \mathfrak{s}, \mathfrak{i})$ on the signature $\mathfrak{d}$, $\mathfrak{f}_1, \dots, \mathfrak{f}_k$ *corresponds* to an SL-structure $\mathcal{I}' = (\mathfrak{U}', \mathfrak{s}', \mathfrak{h})$ iff $\mathfrak{U} = \mathfrak{U}'$, $\mathfrak{s} = \mathfrak{s}'$, $\mathfrak{d}^{\mathfrak{i}} = \mathsf{dom}(\mathfrak{h})$ and for every $j \in [\![1 \mathinner{..} k]\!]$, $\mathfrak{f}_j^{\mathfrak{i}}(x) = y_j$ if $\mathfrak{h}(x) = (y_1, \dots, y_k)$. It is clear that for every finite first-order structure $\mathcal{I}$ there exists a finite SL-structure $\mathcal{I}'$ such that $\mathcal{I}$ corresponds to $\mathcal{I}'$ and vice-versa. Furthermore, if $\mathcal{I}$ corresponds to $\mathcal{I}'$ then it is straightforward to check that $\mathcal{I}' \models \phi \Leftrightarrow \mathcal{I} \models \Theta(\phi)$. □

If $\phi$ is an $\mathsf{SL}^1$ formula, then clearly $\Theta(\phi)$ is in $\mathsf{FO}^1$, with one monadic boolean function symbol $\mathfrak{d}$ and one function symbol $\mathfrak{f}_1$ of sort $\sigma(f) = U$. This yields the following result:

**Theorem 10.** *The finite and infinite satisfiability problems are decidable for* $\mathsf{PRE}(\mathsf{SL}^1)$.

*Proof.* Given a formula $\psi = Q_1 x_1 \dots Q_n x_n \,.\, \phi$ of $\mathsf{PRE}(\mathsf{SL}^1)$, where $\phi$ is quantifier-free, let $\mu \stackrel{\text{def}}{=} \bigvee_{M \in \mu^{inf}(\phi)} M$ be the infinite-domain equivalent expansion of $\phi$ as a disjunction of minterms. We have $\psi \equiv^{inf} Q_1 x_1 \dots Q_n x_n \,.\, \mu$ (Lemma 4) and $Q_1 x_1 \dots Q_n x_n \,.\, \mu$ admits an infinite model if and only if $Q_1 x_1 \dots Q_n x_n \,.\, \mu \wedge \lambda_n$ admits a finite model (Theorem 8; note that $\mu$ contains no occurrence of universe-dependent formulas, as such formulas are always true or false in infinite universes). But $Q_1 x_1 \dots Q_n x_n \,.\, \mu \wedge \lambda_n$ has a finite SL model if and only if $\Theta(Q_1 x_1 \dots Q_n x_n \,.\, \mu \wedge \lambda_n)$ has a finite FO model (Proposition 9). Since the latter formula belongs to $\mathsf{FO}^1$, its finite satisfiability problem is decidable (Theorem 1). The finite case is similar. □

The complexity lower bound is established thanks to the following proposition.

**Proposition 11.** *There is a polynomial reduction of the finite satisfiability problem for first-order formulas with one monadic function symbol $f$ and no predicate symbols other than $\approx$ to the finite [resp. infinite] satisfiability problem for quantified boolean quantifications of test formulas in* $\mathsf{SL}^1$.

*Proof.* By flattening we may assume that all the equations occurring in the considered first-order formula are of the form $f(x) \approx y$ or $x \approx y$, where $x, y$ are variables. For finite domains, the reduction is immediate: it suffices to add the axiom $\forall x \,.\, \mathsf{alloc}(x)$, stating that the heap is a total function, and to replace all equations of the form $f(x) \approx y$ by $x \hookrightarrow y$. It is straightforward to check that satisfiability is preserved ($f$ is encoded in the heap). For infinite domains, it is not possible to add the axiom $\forall x \,.\, \mathsf{alloc}(x)$ as the resulting formula is unsatisfiable[4], so the first-order formula is translated on one that holds on the (finite) domain of the heap. We thus add the axiom $\neg\mathsf{emp} \wedge \forall x, y \,.\, x \hookrightarrow y \rightarrow \mathsf{alloc}(y)$, and we replace every quantification $\forall x \,.\, \phi$ (resp. $\exists x \,.\, \phi$) by a quantification over the domain of the heap: $\forall x \,.\, \mathsf{alloc}(x) \rightarrow \phi$ (resp. $\exists x \,.\, \mathsf{alloc}(x) \wedge \phi$). It is straightforward to check that satisfiability is preserved. Note that infinite satisfiability is equivalent to finite satisfiability, since the quantifications range over elements occurring in the heap. □

Note there is no obvious reduction from the usual first-order satisfiability problem (i.e., on arbitrary models), because the heap is always finite in SL-structures. This explains why we had to refine in Theorem 1 the complexity lower bound from [1] to cope with finite satisfiability.

**Theorem 12.** *The finite and infinite satisfiability problems are not elementary recursive for* $\mathsf{PRE}(\mathsf{SL}^1)$.

*Proof.* The proof follows immediately from the lower bound complexity result of Theorem 1 and from the reductions in Proposition 11. □

## 5 The $\mathsf{BSR}(\mathsf{SL}^1)$ Fragment is PSPACE-complete

The last result concerns the tight complexity of the $\mathsf{BSR}(\mathsf{SL}^1)$ fragment. For $k \geq 2$, we showed that $\mathsf{BSR}(\mathsf{SL}^k)$ is undecidable, in general, and PSPACE-complete if the positive occurrences of the magic wand are forbidden[5] [6]. Here we show that $\mathsf{BSR}(\mathsf{SL}^1)$ is PSPACE-complete. The result does not directly follow from the $\Sigma_2^p$-complexity of the satisfiability problem for $\exists^*\forall^*$ first-order formulas with one unary function symbol[6] because only partial finite functions are considered in our context. The proof is based on the following definitions and results.

**Definition 13.** *A model* $(\mathfrak{U}, \mathfrak{s}, \mathfrak{h})$ *of a formula* $\varphi$ *is* minimal *if* $\varphi$ *admits no model of the form* $(\mathfrak{U}', \mathfrak{s}', \mathfrak{h}')$ *with* $\mathfrak{U}' \subsetneq \mathfrak{U}$.

**Proposition 14.** *Let* $\varphi = \forall y_1, \ldots, y_m \,.\, \phi$ *be a prenex formula with free variables* $x_1, \ldots, x_n$ *(with* $n > 0$*) where* $\phi$ *is a boolean combination of universe-independent test formulas, and let* $\mathcal{I} = (\mathfrak{U}, \mathfrak{s}, \mathfrak{h})$ *be a minimal model of* $\varphi$*. Then* $\mathfrak{U} = \{\mathfrak{h}^j(\mathfrak{s}(x_i)) \mid i \in [\![1 \,..\, n]\!], j \in \mathbb{N}\}$.

4 Since the domain of the heap is finite.

5 For infinite satisfiability, it is enough to forbid positive occurrences of the magic wand containing universally quantified variables only.

6 See [1, Theorem 6.4.19].

*Proof.* Let $\mathfrak{U}' = \{\mathfrak{h}^j(\mathfrak{s}(x_i)) \mid i \in [\![1 \mathrel{..} n]\!], j \in \mathbb{N}\}$ and assume that $\mathfrak{U}' \neq \mathfrak{U}$; note that $\mathfrak{U}' \neq \emptyset$ since $n > 0$. Let $\mathfrak{s}'$ be a store on $\mathfrak{U}'$ coinciding with $\mathfrak{s}$ on $x_1, \ldots, x_n$ and let $\mathfrak{h}'$ be the restriction of $\mathfrak{h}$ to $\mathfrak{U}'$. Both $\mathfrak{s}'$ and $\mathfrak{h}'$ are well-defined by construction of $\mathfrak{U}'$, and $\mathfrak{h}'$ is a heap on $\mathfrak{U}'$. Since $\mathfrak{U}$ is minimal, $(\mathfrak{U}', \mathfrak{s}', \mathfrak{h}') \not\models \varphi$, thus there exist $b_1, \ldots, b_m \in \mathfrak{U}'$ such that by letting $\mathfrak{s}_1' \stackrel{\text{def}}{=} \mathfrak{s}'[y_i \mapsto b_i \mid i \in [\![1 \mathrel{..} m]\!]]$, we have $(\mathfrak{U}', \mathfrak{s}_1', \mathfrak{h}') \models \neg\phi$. Since the atomic formulas in $\phi$ are universe-independent, we deduce that $(\mathfrak{U}, \mathfrak{s}_1', \mathfrak{h}') \models \neg\phi$. Further, $\mathfrak{s}_1'$ and $\mathfrak{s}[y_i \mapsto b_i \mid i \in [\![1 \mathrel{..} m]\!]]$ coincide on all the variables $x_1, \ldots, x_n, y_1, \ldots, y_m$ that are free in $\phi$, thus $(\mathfrak{U}, \mathfrak{s}[y_i \mapsto b_i \mid i \in [\![1 \mathrel{..} m]\!]], \mathfrak{h}') \models \neg\phi$. Finally, $\mathfrak{h}$ and $\mathfrak{h}'$ coincide on every element of $\mathfrak{U}'$ and by definition we have $\mathfrak{s}(x_i), b_j \in \mathfrak{U}'$ for $i \in [\![1 \mathrel{..} n]\!]$ and $j \in [\![1 \mathrel{..} m]\!]$, hence $(\mathfrak{U}, \mathfrak{s}[y_i \mapsto b_i \mid i \in [\![1 \mathrel{..} m]\!]], \mathfrak{h}) \models \neg\phi$, and $(\mathfrak{U}, \mathfrak{s}, \mathfrak{h}) \not\models \varphi$, which contradicts our assumption.

**Definition 15.** *Let $\varphi$ be a formula with free variables $x_1, \ldots, x_n$ and let $\mathcal{I} = (\mathfrak{U}, \mathfrak{s}, \mathfrak{h})$ be a structure. A* line *for $(\mathcal{I}, \varphi)$ is a sequence of pairwise distinct elements $a_1, \ldots, a_\ell$ in $\mathfrak{U}$ such that:*

1. $\forall i \in [\![1 \mathrel{..} \ell - 1]\!]$, $a_{i+1} = \mathfrak{h}(a_i)$.
2. $\forall i \in [\![1 \mathrel{..} \ell - 1]\!]$, $\forall e \in \mathfrak{U}$, *if* $\mathfrak{h}(e) = a_{i+1}$ *then* $e = a_i$.
3. $\forall i \in [\![1 \mathrel{..} \ell]\!]$, $\forall j \in [\![1 \mathrel{..} n]\!]$, $a_i \neq \mathfrak{s}(x_j)$.

The next proposition shows that there is a bound on the length of any line in a minimal model.

**Proposition 16.** *Let $\varphi = \forall y_1, \ldots, y_m \,.\, \phi$ be a prenex formula with free variables $x_1, \ldots, x_n$ where $\phi$ is a boolean combination of domain-independent test formulas, and let $\mathcal{I} = (\mathfrak{U}, \mathfrak{s}, \mathfrak{h})$ be a model of $\varphi$. If $(\mathcal{I}, \varphi)$ admits a line of length strictly greater than $m + 2$ then $\mathcal{I}$ is not minimal.*

*Proof.* Let $a_1, \ldots, a_l$ be a sequence of elements satisfying the conditions of Definition 15 with $l > m + 2$. Let $\mathcal{I}' = (\mathfrak{U}', \mathfrak{s}', \mathfrak{h}')$, where $\mathfrak{U}' \stackrel{\text{def}}{=} \mathfrak{U} \setminus \{a_2\}$, $\mathfrak{s}'$ is a store on $\mathfrak{U}'$ coinciding with $\mathfrak{s}$ on all variables $x$ such that $\mathfrak{s}(x) \in \mathfrak{U}'$, $\mathsf{dom}(\mathfrak{h}') \stackrel{\text{def}}{=} \mathsf{dom}(\mathfrak{h}) \setminus \{a_2\}$, $\mathfrak{h}'(a_1) \stackrel{\text{def}}{=} a_3$ and $\mathfrak{h}'(x) \stackrel{\text{def}}{=} \mathfrak{h}(x)$ if $x \in \mathsf{dom}(\mathfrak{h}') \setminus \{a_1\}$. Note that $\mathfrak{s}$ and $\mathfrak{s}'$ coincide on all variables $x_1, \ldots, x_n$ free in $\varphi$ since $\forall i \in [\![1 \mathrel{..} n]\!]$, $a_2 \neq \mathfrak{s}(x_i)$, by Definition 15 (3). Since $\mathcal{I}$ is minimal, necessarily $\mathcal{I}' \not\models \varphi$, thus there exist $b_1', \ldots, b_m' \in \mathfrak{U}'$ such that by letting $\mathfrak{s}_1' \stackrel{\text{def}}{=} \mathfrak{s}'[y_j \mapsto b_j' \mid j \in [\![1 \mathrel{..} m]\!]]$, we have $(\mathfrak{U}', \mathfrak{s}_1', \mathfrak{h}') \models \neg\phi$. Since $l > m + 2$ and $a_1, \ldots, a_l$ are distinct by Definition 15, there exists $i \in [\![2 \mathrel{..} l - 1]\!]$ such that $a_{i+1} \notin \{b_1', \ldots, b_m'\}$. We define a sequence $b_1, \ldots, b_m \in \mathfrak{U}$ as follows. For every $j \in [\![1 \mathrel{..} m]\!]$, if there exists $o \in [\![3 \mathrel{..} i]\!]$ such that $b_j' = a_o$, then we let $b_j \stackrel{\text{def}}{=} a_{o-1}$; otherwise, $b_j \stackrel{\text{def}}{=} b_j'$. Note that $b_j$ is well-defined, because $a_1, \ldots, a_l$ are distinct, hence there exists at most one $o$ satisfying the above condition.

We emphasize some useful consequences of the above definitions before proving that $(\mathfrak{U}', \mathfrak{s}_1', \mathfrak{h}') \models \phi$. Let $V = \{x_1, \ldots, x_n\} \cup \{y_j \mid j \in [\![1 \mathrel{..} m]\!], \mathfrak{s}_1'(y_j) \notin \{a_3, \ldots, a_i\}\}$. By definition $\mathfrak{s}$ and $\mathfrak{s}'$ coincide on $x_1, \ldots, x_n$, and $\mathfrak{s}_1(y_j) = b_j = b_j' = \mathfrak{s}_1'(y_j)$ if $b_j' \notin \{a_3, \ldots, a_i\}$, hence $\mathfrak{s}_1'$ and $\mathfrak{s}[y_j \mapsto b_j \mid j \in [\![1 \mathrel{..} m]\!]]$ coincide on every variable in $V$. Furthermore, for every variable $x \in V$,

$\mathfrak{s}_1(x) \in \mathfrak{U} \setminus \{a_2, \ldots, a_i\}$. Indeed, either $x \in \{x_1, \ldots, x_n\}$ and in this case $\mathfrak{s}(x) \notin \{a_1, \ldots, a_n\}$ by Definition 15 (3); or $x = y_j$ for some $j \in [\![1 \mathinner{..} m]\!]$, and then $\mathfrak{s}_1(x) = b_j = b'_j \notin \{a_3, \ldots, a_i\}$, so that $\mathfrak{s}_1(x) \in \mathfrak{U}' \setminus \{a_3, \ldots, a_i\} = \mathfrak{U} \setminus \{a_2, \ldots, a_i\}$. Finally, if $x$ occurs in $\phi$ and $x \notin V$ then $x = y_j$ for some $j \in [\![1 \mathinner{..} m]\!]$ such that $b'_j = a_o$, with $o \in [\![3 \mathinner{..} i]\!]$, thus $\mathfrak{s}'_1(x) = a_o$ and $\mathfrak{s}_1(x) = b_j = a_{o-1}$, and therefore $\mathfrak{s}'_1(x) \in \{a_3, \ldots, a_i\}$ and $\mathfrak{s}_1(x) \in \{a_2, \ldots, a_{i-1}\}$. Let $\mathfrak{s}_1 \stackrel{\text{def}}{=} \mathfrak{s}[y_j \mapsto b_j \mid j \in [\![1 \mathinner{..} m]\!]]$; we show that $(\mathfrak{U}', \mathfrak{s}'_1, \mathfrak{h})$ and $(\mathfrak{U}, \mathfrak{s}_1, \mathfrak{h})$ coincide on every test formula $\ell$ in $\phi$.

$\ell = x \approx y$. If $x, y \in V$ then the proof is immediate since $\mathfrak{s}_1$ and $\mathfrak{s}'_1$ coincide on $x$ and $y$. If $x \in V$ and $y \notin V$ then $\mathfrak{s}_1(x) = \mathfrak{s}'_1(x) \in \mathfrak{U} \setminus \{a_2, \ldots, a_i\}$ and $\mathfrak{s}_1(y), \mathfrak{s}'_1(y) \in \{a_2, \ldots, a_i\}$ hence $x \approx y$ is false in both structures. The proof is symmetric if $x \notin V$ and $y \in V$. If $x, y \notin V$ then $\mathfrak{s}'_1(x) = a_o$, $\mathfrak{s}'_1(y) = a_{o'}$, with $\mathfrak{s}_1(x) = a_{o-1}$ and $\mathfrak{s}_1(y) = a_{o'-1}$. Since the $a_1, \ldots, a_l$ are pairwise distinct we have $\mathfrak{s}'_1(x) = \mathfrak{s}'_1(y) \Leftrightarrow o = o' \Leftrightarrow o - 1 = o' - 1 \Leftrightarrow \mathfrak{s}_1(x) = \mathfrak{s}_1(y)$.
$\ell = \mathsf{alloc}(x)$. If $x \in V$ then $\mathfrak{s}_1(x) = \mathfrak{s}'_1(x) \neq a_2$ Thus $\mathfrak{s}_1(x) \in \mathsf{dom}(\mathfrak{h}) \Leftrightarrow \mathfrak{s}'_1(x) \in \mathsf{dom}(\mathfrak{h}) \Leftrightarrow \mathfrak{s}'_1(x) \in \mathsf{dom}(\mathfrak{h}')$. If $x \notin V$ then $\mathfrak{s}'_1(x) \in \{a_3, \ldots, a_i\}$ and $\mathfrak{s}_1(x) \in \{a_2, \ldots, a_{i-1}\}$ (with $i < l$) thus $\mathsf{alloc}(x)$ is true in both structures.
$\ell = x \hookrightarrow y$. We distinguish several cases.

- If $x, y \in V$ then $\mathfrak{s}_1(x) = \mathfrak{s}'_1(x)$ and $\mathfrak{s}_1(y) = \mathfrak{s}'_1(y)$, with $\mathfrak{s}_1(x) \neq a_2$, hence $\mathfrak{h}(\mathfrak{s}_1(x)) = \mathfrak{s}_1(y) \Leftrightarrow \mathfrak{h}(\mathfrak{s}'_1(x)) = \mathfrak{s}'_1(y) \Leftrightarrow \mathfrak{h}'(\mathfrak{s}'_1(x)) = \mathfrak{s}'_1(y)$, thus $(\mathfrak{U}, \mathfrak{s}_1, \mathfrak{h}) \models \ell \Leftrightarrow (\mathfrak{U}', \mathfrak{s}'_1, \mathfrak{h}') \models \ell$.
- If $x, y \notin V$ then $\mathfrak{s}'_1(x) = a_o$, $\mathfrak{s}'_1(y) = a_{o'}$, with $\mathfrak{s}_1(x) = a_{o-1}$, $\mathfrak{s}_1(y) = a_{o'-1}$ and $o, o' \geq 3$ thus $\mathfrak{h}'(\mathfrak{s}'_1(x)) = \mathfrak{s}'_1(y) \Leftrightarrow o = o' - 1 \Leftrightarrow \mathfrak{h}(\mathfrak{s}_1(x)) = \mathfrak{s}_1(y)$.
- If $x \in V$ and $y \notin V$, then $\mathfrak{s}'_1(y) = a_o$ with $\mathfrak{s}_1(y) = a_{o-1}$ and $o \in [\![3 \mathinner{..} i]\!]$. We distinguish two cases. If $x \in \{x_1, \ldots, x_n\}$, then $\mathfrak{h}(\mathfrak{s}_1(x)) \notin \{a_1, \ldots, a_l\}$ (by Definition 15 (2)) thus $\mathfrak{h}(\mathfrak{s}_1(x)) = \mathfrak{h}'(\mathfrak{s}'_1(x)) \neq \mathfrak{s}_1(y), \mathfrak{s}'_1(y)$ and $\ell$ is false in both structures. Otherwise, $x = y_j$, for some $j \in [\![1 \mathinner{..} m]\!]$ such that $b'_j \notin \{a_3, \ldots, a_i\}$. If $b'_j = a_1$ then $\mathfrak{h}(\mathfrak{s}_1(x)) = a_2$ and $\mathfrak{h}'(\mathfrak{s}'_1(x)) = a_3$, thus $\mathfrak{h}(\mathfrak{s}_1(x)) = \mathfrak{s}_1(y) \Leftrightarrow a_2 = \mathfrak{s}_1(y) \Leftrightarrow a_2 = a_{o-1} \Leftrightarrow o = 3 \Leftrightarrow a_3 = \mathfrak{s}'_1(y) \Leftrightarrow \mathfrak{h}'(\mathfrak{s}'_1(x)) = \mathfrak{s}'(y)$, hence $\ell$ has the same truth value in $(\mathfrak{U}, \mathfrak{s}_1, \mathfrak{h})$ and $(\mathfrak{U}', \mathfrak{s}'_1, \mathfrak{h}')$. If $b_j \neq a_1$ then $\mathfrak{h}'(\mathfrak{s}_1(x)) = \mathfrak{h}(\mathfrak{s}_1(x))$, and $\mathfrak{s}_1(x) \notin \{a_1, \ldots, a_i\}$, thus $\mathfrak{h}(\mathfrak{s}_1(x) \notin \{a_2, \ldots, a_{i+1}\}$, hence $\ell$ is false in both structures.
- If $y \in V$ and $x \notin V$ then there exists $o \in [\![3 \mathinner{..} i]\!]$ such that $\mathfrak{s}_1(x) = a_{o-1}$ and $\mathfrak{s}'_1(x) = a_o$, with $\mathfrak{s}_1(y) = \mathfrak{s}'_1(y) \notin \{a_2, \ldots, a_i\}$. We have $\mathfrak{h}'(\mathfrak{s}'_1(x)) = a_{o+1}$ and $\mathfrak{h}(\mathfrak{s}_1(x)) = a_o$, thus $\mathfrak{h}'(\mathfrak{s}'_1(x)), \mathfrak{h}(\mathfrak{s}_1(x)) \in \{a_3, \ldots, a_{i+1}\}$. By definition of $i$, $a_{i+1} \notin \{b_1, \ldots, b_m\}$ (since $a_{i+1} \notin \{b'_1, \ldots, b'_m\}$ and $i+1 > i$), moreover $a_{i+1} \notin \mathfrak{s}(\{x_1, \ldots, x_n\})$ by Definition 15 (3). Thus $a_{i+1} \neq \mathfrak{s}_1(y)$. Since $\mathfrak{s}_1(y) \notin \{a_2, \ldots, a_i\}$ we deduce that $\mathfrak{s}_1(y) \notin \{a_3, \ldots, a_{i+1}\}$, thus $\ell$ is false in both structures.

As a consequence, $(\mathfrak{U}', \mathfrak{s}'_1, \mathfrak{h}')$ and $(\mathfrak{U}, \mathfrak{s}_1, \mathfrak{h})$ necessarily coincide on $\phi$, and consequently $(\mathfrak{U}, \mathfrak{s}_1, \mathfrak{h}) \models \neg\phi$, hence $(\mathfrak{U}, \mathfrak{s}, \mathfrak{h}) \not\models \forall y_1, \ldots, y_m \mathrel{.} \phi$ which contradicts our hypothesis. □

**Lemma 17.** *Let $\varphi = \forall y_1, \ldots, y_m \mathrel{.} \phi$ be a prenex formula of $\mathsf{SL}^1$ of free variables $x_1, \ldots, x_n$ (with $n > 0$) where $\phi$ is a boolean combination of universe-independent test formulas. If $(\mathfrak{U}, \mathfrak{s}, \mathfrak{h})$ is a finite minimal model of $\varphi$ then $||\mathfrak{U}|| \leq 2n \cdot (m+3)$.*

*Proof.* Let $\mathcal{I} = (\mathfrak{U}, \mathfrak{s}, \mathfrak{h})$ be a minimal finite model of $\varphi$ and let $a_i = \mathfrak{s}(x_i)$ for $i = 1, \ldots, n$. We inductively define a sequence $l_i$ $(1 \leq i \leq n)$ of natural numbers as follows: $l_i$ is the minimal natural number such that either $\mathfrak{h}^{l_i}(a_i) \notin \mathsf{dom}(\mathfrak{h})$ or $\mathfrak{h}(\mathfrak{h}^{l_i}(a_i)) \in \{a_1, \ldots, a_n\} \cup \{\mathfrak{h}^j(a_i) \mid j \in [\![1 \mathinner{..} l_i - 1]\!]\} \cup \{\mathfrak{h}^j(a_k) \mid k \in [\![1 \mathinner{..} i-1]\!], j \in [\![1 \mathinner{..} l_k]\!]\}$. Because the domain of $\mathfrak{h}$ is finite, the numbers $l_i$ always exist, for all $i = 1, \ldots, n$. Note that by construction, given $i \in [\![1 \mathinner{..} i]\!]$, if $\mathfrak{h}^j(a_i) \neq \mathfrak{h}^k(a_i)$ for all $k < j$ and $\mathfrak{h}^j(a_i) \notin \{\mathfrak{h}^k(a_p) \mid k \in \mathbb{N}\}$ for all $p < i$, then $j \leq l_i$. Hence, since by Proposition 14, we have $\mathfrak{U} = \{\mathfrak{h}^j(\mathfrak{s}(x_i)) \mid i \in [\![1 \mathinner{..} n]\!], j \in \mathbb{N}\}$, we deduce that $\mathfrak{U} = \bigcup_{i=1}^{n} \{\mathfrak{h}^j(a_i) \mid j \in [\![0 \mathinner{..} l_i]\!]\}$. Furthermore, by definition of $l_i$, all locations $\mathfrak{h}^j(a_i)$, for $i \in [\![1 \mathinner{..} n]\!]$ and $j \in [\![1 \mathinner{..} l_i]\!]$, are pairwise distinct.

We define the following subsets of $\mathfrak{U}$: $\mathfrak{U}_1 \stackrel{\mathsf{def}}{=} \{a_i \mid i \in [\![1 \mathinner{..} n]\!]\}$, $\mathfrak{U}_2 \stackrel{\mathsf{def}}{=} \{\mathfrak{h}^{l_i}(a_i) \mid i \in [\![1 \mathinner{..} n]\!], \mathfrak{h}^{l_i}(a_i) \notin \mathsf{dom}(\mathfrak{h})\}$, and $\mathfrak{U}_3 \stackrel{\mathsf{def}}{=} \{\mathfrak{h}(\mathfrak{h}^{l_i}(a_i)) \mid i \in [\![1 \mathinner{..} n]\!], \mathfrak{h}^{l_i}(a_i) \in \mathsf{dom}(\mathfrak{h})\}$. By definition, $\mathfrak{U}_2 \cup \mathfrak{U}_3$ contains at most $n$ elements, thus $||\mathfrak{U}_1 \cup \mathfrak{U}_2 \cup \mathfrak{U}_3|| \leq 2n$. We have that every element $c$ such that there exist $a \neq b$ with $\mathfrak{h}(a) = \mathfrak{h}(b) = c$ is in $\mathfrak{U}_3$. Indeed, assume that there exist two such elements $a, b \in \mathfrak{U}$. Then there exist $i, j \in [\![1 \mathinner{..} n]\!]$, $i' \in [\![0 \mathinner{..} l_i]\!]$, $j' \in [\![0 \mathinner{..} l_j]\!]$ such that $a = \mathfrak{h}^{i'}(a_i)$ and $b = \mathfrak{h}^{j'}(a_j)$. We assume by symmetry that $i \leq j$. Then by definition of $l_j$ we must have $j' = l_j$, so that $\mathfrak{h}(b) = c \in \mathfrak{U}_3$. The reader may refer to Fig. 1 for an illustration. Now, consider a sequence of the form $(\mathfrak{h}^j(a_i), \ldots, \mathfrak{h}^{j'}(a_i))$ (with $j \leq j'$) containing no element in $\mathfrak{U}_1 \cup \mathfrak{U}_2 \cup \mathfrak{U}_3$. By definition, this sequence fulfills Conditions 3 and 1 from Definition 15. If the sequence does not fulfill Condition 2, then there exist $k \in [\![j \mathinner{..} j'-1]\!]$ such that $\mathfrak{h}^{k+1}(a_i)$ is a fork element, hence $\mathfrak{h}^{k+1}(a_i) \in \mathfrak{U}_3$, which contradicts our hypothesis. Consequently, $(\mathfrak{h}^j(a_i), \ldots, \mathfrak{h}^{j'}(a_i))$ is a line for $(\mathcal{I}, \varphi)$. By Proposition 16 such lines cannot be of length greater than $m+2$, therefore $\mathfrak{U} \setminus (\mathfrak{U}_1 \cup \mathfrak{U}_2 \cup \mathfrak{U}_3)$ contains at most $(m+2) \cdot L$ elements, where $L$ is the number of sequences $(\mathfrak{h}^j(a_i), \ldots, \mathfrak{h}^{j'}(a_i))$ of maximal length not containing elements in $\mathfrak{U}_1 \cup \mathfrak{U}_2 \cup \mathfrak{U}_3$. Thus $||\mathfrak{U}|| \leq (m+2) \cdot L + 2n$. By definition, all such sequences necessarily start by some element $\mathfrak{h}(a)$, where $a \in \mathfrak{U}_1 \cup \mathfrak{U}_2 \cup \mathfrak{U}_3$, thus there are at most $||\mathfrak{U}_1 \cup \mathfrak{U}_2 \cup \mathfrak{U}_3|| \leq 2n$ such sequences. Hence $L \leq 2n$ and $||\mathfrak{U}|| \leq 2n \cdot (m+3)$. □

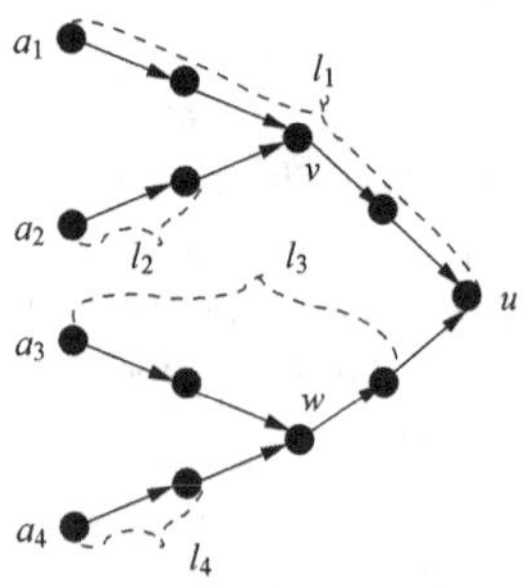


**Fig. 1.** Heap decomposition example. We have $l_1 = 5$, $l_2 = 1$, $l_3 = 3$ and $l_4 = 1$. Moreover, $\mathfrak{U}_1 = \{a_1, a_2, a_3, a_4\}$, $\mathfrak{U}_2 = \{u\}$ and $\mathfrak{U}_3 = \{u, v, w\}$.

**Corollary 18.** *The finite and infinite satisfiability problems for formulas of* $\mathsf{BSR}(\mathsf{SL}^1)$ *are PSPACE-complete.*

*Proof.* PSPACE-hardness follows from the proof that satisfiability of the quantifier free fragment of $\mathsf{SL}^2$ is PSPACE-complete [4, Proposition 5]. Indeed, this proof does not depend on the universe being infinite or the fact that $k = 2$. There remains to show PSPACE-membership for both problems. Observe that this does not directly follow from Lemmas 4 and 17, because (i) the sets $\mu^{inf}(\phi)$ and $\mu^{fin}(\phi)$ are of exponential size hence no efficient algorithm can compute them and, (ii) Lemma 17 only holds for universe-independent formulas. W.l.o.g., we assume that the considered formula contains at least one free variable and is of the form $\forall y_1, \ldots, y_m . \phi$. It is sufficient to focus on the finite satisfiability problem. Indeed, by Lemma 4, $\forall y_1, \ldots, y_m . \phi \equiv^{inf} \bigvee_{M \in \mu^{inf}(\neg\phi)} M$. By Theorem 8, $\forall y_1, \ldots, y_m . \phi$ has an infinite model iff $\forall y_1, \ldots, y_m . \phi \wedge \lambda_{n+m}$ has a finite model, where the size of $\lambda_{n+m}$ is quadratic in $n+m$. Moreover, since $\lambda_{n+m}$ is a $\mathsf{BSR}(\mathsf{SL})$ formula, $\forall y_1, \ldots, y_m . \phi \wedge \lambda_{n+m}$ is also a $\mathsf{BSR}(\mathsf{SL})$ formula. Hence infinite satisfiability can be reduced polynomially to finite satisfiability.

Let $\psi = \bigvee_{M \in \mu^{fin}(\neg\phi)} M$ (note that the size of $\psi$ is exponential w.r.t. that of $\phi$). Let $L$ be the maximal number $l$ such that a test formula $|h| \leq l$ or $|h| \leq |U| - l$ occurs in $\mu^{inf}(\phi)$. By Lemma 4, the number $L$ is polynomial w.r.t. $\mathsf{size}(\phi)$. We guess a structure $\mathcal{I} = (\mathfrak{U}, \mathfrak{s}, \mathfrak{h})$ and check that it is a model of $\varphi$ as follows. We first guess the set $\mathcal{C}$ of literals of the form $|U| \leq i$, $|U| < i$, $|h| \leq i$, $|h| > i$, $|h| \leq |U| - i$, or $|h| > |U| - i$ with $i \in [\![0 \,..\, L]\!]$ that are true in $\mathcal{I}$. It is clear that $\varphi$ is satisfiable iff $\varphi \cup \mathcal{C}$ is satisfiable for some such set $\mathcal{C}$. Up to redundancy, $\mathcal{C}$ contains at most 6 literals (one literal of each kind). With each test formula $\ell \in \mathcal{C}$ we may associate an equivalent formula $\gamma(\ell)$ in $\mathsf{BSR}(\mathsf{SL1})$ built on atoms $x \approx y$ or $\mathsf{alloc}(x)$ using the following equivalence statements:

- $|h| \leq i \;\Leftrightarrow\; \forall x'_1, \ldots, x'_{i+1} . \mathsf{dist}(x'_1, \ldots, x'_{i+1}) \rightarrow \bigvee_{j=1}^{i+1} \neg\mathsf{alloc}(x'_j)$,
- $|h| \leq |U| - i \;\Leftrightarrow\; \exists x'_1, \ldots, x'_i . \mathsf{dist}(x'_1, \ldots, x'_i) \wedge \bigwedge_{j=1}^{i} \neg\mathsf{alloc}(x_j)$,
- $|U| \leq i \;\Leftrightarrow\; \forall x'_1, \ldots, x'_{i+1} \neg\mathsf{dist}(x'_1, \ldots, x'_{i+1})$.

Let $\vartheta$ be the conjunction of all formulas $\gamma(\ell)$ where $\ell \in \mathcal{C}$. Note that $\vartheta$ contains (up to redundancy) at most $3L + 2$ existential variables and $3L + 2$ universal variables. Now consider the formula $\psi'$ obtained from $\psi$ by replacing every test formula such that $\ell \in \mathcal{C}$ (resp. $\overline{\ell} \in \mathcal{C}$) by $\top$ (resp. $\bot$). Let $\varphi'$ be the formula obtained by putting $\forall y_1, \ldots, y_m . \neg\psi' \wedge \vartheta$ in prenex form. It is clear that $\varphi'$ is in $\mathsf{BSR}(\mathsf{SL1})$ and that all test formulas in $\varphi'$ are universe-independent, furthermore $\varphi'$ contains at most $n' = n + (3L + 2)$ free or existential variables and $m' = m + (3L + 2)$ universal variables. Moreover, $\varphi' \equiv \varphi \wedge \vartheta$, hence $\varphi'$ is satisfiable iff $\varphi$ admits a model satisfying $\mathcal{C}$. By Lemma 17, $\varphi'$ is satisfiable iff $\varphi'$ admits a model $(\mathfrak{U}, \mathfrak{s}, \mathfrak{h})$ such that $||\mathfrak{U}|| \leq 2n' \times (m' + 3)$. We may thus check that $\varphi'$ is satisfiable by fixing such a set $\mathfrak{U}$, guessing the value of $\mathfrak{s}(x)$ on each variable $x$ free in $\varphi$, guessing some heap $\mathfrak{h}$ on $\mathfrak{U}$, and checking that $(\mathfrak{U}, \mathfrak{s}, \mathfrak{h}) \models \mathcal{C}$ and that $(\mathfrak{U}, \mathfrak{s}, \mathfrak{h}) \models \varphi$. The former test is easy to perform by counting the number of allocated and nonallocated cells. For the latter test, we check the negation $(\mathfrak{U}, \mathfrak{s}, \mathfrak{h}) \not\models \varphi$, by testing that there exists a store $\mathfrak{s}'$ coinciding with $\mathfrak{s}$ on $x_1, \ldots, x_n$ such that $(\mathfrak{U}, \mathfrak{s}', \mathfrak{h}) \models \neg\phi$, i.e., such that $(\mathfrak{U}, \mathfrak{s}', \mathfrak{h}) \models \bigvee_{M \in \mu^{fin}(\neg\phi)} M$. To this aim, we guess the value of each variable $y_i$ in $\mathfrak{s}'$, guess a minterm $M$, check that $M \in \mu^{fin}(\neg\phi)$ (which can be done in polynomial space by Lemma 4) and check that $(\mathfrak{U}, \mathfrak{s}', \mathfrak{h})$ validates every test formula in $M$ (it is clear that this can be done in polynomial time). □

## 6 Conclusion

We have shown that the prenex fragment of Separation Logic over heaps with one selector, denoted as $\mathsf{SL}^1$, is decidable in time not elementary recursive. Moreover, the Bernays-Schönfinkel-Ramsey $\mathsf{BSR}(\mathsf{SL}^1)$ is PSPACE-complete. These results settle an open question raised in [6] and allow one to draw a precise boundary between decidable and undecidable cases inside $\mathsf{BSR}(\mathsf{SL}^k)$. As far as applications are concerned, the logic $\mathsf{BSR}(\mathsf{SL}^1)$ can be used to reason on singly linked data-structures, where $*$ and $\mathrel{-\!\!*}$ are used to state dynamic transformations of the heap and the quantifiers are useful to state general properties of the considered data-structure (e.g., to check that a loop invariant is preserved). Theorem 8, relating infinite and finite satisfiability, holds for any $k \geq 1$ and we believe that it could pave the way to further decidability results for prenex fragments of $\mathsf{SL}^k$.

**Acknowledgments.** The authors wish to thank Stéphane Demri, Etienne Lozes and Alessio Mansutti for the insightful discussions during the preparation of this paper.

## References

1. Börger, E., Grädel, E., Gurevich, Y.: The Classical Decision Problem. Perspectives in Mathematical Logic. Springer, Heidelberg (1997)
2. Brochenin, R., Demri, S., Lozes, E.: On the almighty wand. Inf. Comput. **211**, 106–137 (2012)
3. Calcagno, C., Distefano, D.: Infer: an automatic program verifier for memory safety of C programs. In: Bobaru, M., Havelund, K., Holzmann, G.J., Joshi, R. (eds.) NFM 2011. LNCS, vol. 6617, pp. 459–465. Springer, Heidelberg (2011). https://doi.org/10.1007/978-3-642-20398-5_33
4. Calcagno, C., Yang, H., O'Hearn, P.W.: Computability and complexity results for a spatial assertion language for data structures. In: Hariharan, R., Vinay, V., Mukund, M. (eds.) FSTTCS 2001. LNCS, vol. 2245, pp. 108–119. Springer, Heidelberg (2001). https://doi.org/10.1007/3-540-45294-X_10
5. Demri, S., Galmiche, D., Larchey-Wendling, D., Méry, D.: Separation logic with one quantified variable. In: Hirsch, E.A., Kuznetsov, S.O., Pin, J.É., Vereshchagin, N.K. (eds.) CSR 2014. LNCS, vol. 8476, pp. 125–138. Springer, Cham (2014). https://doi.org/10.1007/978-3-319-06686-8_10
6. Echenim, M., Iosif, R., Peltier, N.: The Bernays-Schönfinkel-Ramsey class of separation logic on arbitrary domains. In: Bojańczyk, M., Simpson, A. (eds.) FoSSaCS 2019. LNCS, vol. 11425, pp. 242–259. Springer, Cham (2019). https://doi.org/10.1007/978-3-030-17127-8_14

7. Fitting, M.: First-Order Logic and Automated Theorem Proving. Texts and Monographs in Computer Science. Springer, New York (1990). https://doi.org/10.1007/978-1-4684-0357-2
8. Ishtiaq, S.S., O'Hearn, P.W.: Bi as an assertion language for mutable data structures. In: ACM SIGPLAN Notices, vol. 36, pp. 14–26 (2001)
9. Lozes, É.: Expressivité des logiques spatiales. Thèse de doctorat, Laboratoire de l'Informatique du Parallélisme, ENS Lyon, France, November 2004. http://www.lsv.ens-cachan.fr/Publis/PAPERS/PS/PhD-lozes.ps
10. Rabin, M.O.: Decidability of second-order theories and automata on infinite trees. Trans. Am. Math. Soc., 141:1–35 (1969). http://www.jstor.org/stable/1995086
11. Reynolds, J.C.: Separation logic: a logic for shared mutable data structures. In: Proceedings of LICS 2002 (2002)

# Dynamic Doxastic Differential Dynamic Logic for Belief-Aware Cyber-Physical Systems

João G. Martins[1,2](✉), André Platzer[1,3], and João Leite[2]

[1] Computer Science Department, Carnegie Mellon University, Pittsburgh, USA
{jmartins,aplatzer}@cs.cmu.edu
[2] NOVA LINCS, Universidade NOVA de Lisboa, Caparica, Portugal
jleite@fct.unl.pt
[3] Fakultät für Informatik, Technische Universität München, Munich, Germany

**Abstract.** Cyber-physical systems (CPS), such as airplanes, operate based on sensor and communication data, i.e. on potentially noisy or erroneous beliefs about the world. Realistic CPS models must therefore incorporate the notion of beliefs if they are to provide safety guarantees in practice as well as in theory. To fundamentally address this challenge, this paper introduces a first-principles framework for reasoning about CPS models where control decisions are explicitly driven by controller beliefs arrived at through observation and reasoning. We extend the differential dynamic logic $\mathsf{d}\mathcal{L}$ for CPS dynamics with belief modalities, and a learning operator for belief change. This new dynamic doxastic differential dynamic logic $\mathsf{d}^4\mathcal{L}$ does due justice to the challenges of CPS verification by having (1) real arithmetic for describing the world and beliefs about the world; (2) continuous and discrete world change; (3) discrete belief change by means of the learning operator. We develop a sound sequent calculus for $\mathsf{d}^4\mathcal{L}$, which enables us to illustrate the applicability of $\mathsf{d}^4\mathcal{L}$ by proving the safety of a simplified belief-triggered controller for an airplane.

**Keywords:** Differential dynamic logic · Dynamic epistemic logic · Sequent calculus · Hybrid systems · Cyber-physical systems

## 1 Introduction

Cyber-physical systems (CPS) mix discrete cyber change and continuous physical change. Examples of CPS include self-driving cars, airplane autopilots, and industrial machines. With widespread espousal of automation in transportation, it is imperative that we develop methods capable of verifying the safety of the algorithms driving the CPSs on which human lives will increasingly depend.

Supported by the Alexander von Humboldt Foundation, NSF grant CNS-1446712, CMU | Portugal grant SFRH/BD/51886/2012, and PTDC/CCI-COM/30952/2017.


S. Cerrito and A. Popescu (Eds.): TABLEAUX 2019, LNAI 11714, pp. 428–445, 2019.
https://doi.org/10.1007/978-3-030-29026-9_24

However, because CPSs rely on sensors and partial human operation, both of which are imperfect, they face a possible discrepancy between reality, and the perception, understanding and beliefs thereof. Critical system components are engineered to be exceptionally reliable, so safety incidents often originate from just such a discrepancy between what is believed to be true versus what is actually true. This can be highlighted by three (of many) tragedies, some now known to be preventable, e.g., through neutral control inputs [1,5,12]. However, non-critical sensor failures led to erroneous pilot beliefs. These beliefs resulted in the pilots' inability to perform informed, safe control decisions, leading to 574 fatalities in these three incidents alone.

Verification efforts for practical system designs must therefore augment initial analyses which assume perfect information with an awareness of factors such as sensor errors, actuator disturbances, and, crucially, incomplete or incorrect perceptions of the world. Ideally, such factors ought to become an explicit part of the model so that CPS design and verification engineers can confront this challenge of uncertainty head on at design time, before safety violations occur.

We argue that the notion of beliefs (*doxastics*) about the state of the world, which has been extensively studied, can succinctly capture such phenomena. We develop a first-principles language and verification method for reasoning about *changing beliefs in a changing world.* Using this language, CPS designers may create more realistic controllers whose decisions are explicitly driven by their beliefs. The consequences of such decisions are borne out in the continuous-time and continuous-space evolution of these *belief-aware CPS.*

In this new paradigm, control decisions are grounded *only* in what can be observed and reasoned. By providing the tools to develop such *belief-triggered controllers*, we help bridge the gap between the theoretical safety of *CPS models*, and the practical safety of the *CPS vehicles* that will soon be driving and flying us to our destinations.

## 2 Technical Approach

Our approach is to integrate a framework for specifying and verifying real-world CPS with a suitable notion of dynamic beliefs. The result should be a single cohesive framework capable of complex reasoning about changing beliefs in a changing world, as required by belief-aware CPSs.

Work on control-theoretic *robust* solutions for CPS models seem promising, since they entail asymptotical steering towards a desired target domain despite perturbations in the system [11]: sensor and actuator noise could be modeled as perturbations rather than beliefs. However, perturbation analysis does not capture the complex causal relationship from observation, to reasoning, to actuation in an explicit way that can lead to e.g. malfunction checklists or pilot best practices. Accurate analyses for safety incidents such as [1,5,12] require the power to (1) model agents with reasoning capabilities, and (2) leverage complex logical arguments about perception versus fact in the pursuit of safety guarantees.

The differential dynamic logic $\mathsf{d}\mathcal{L}$ [16,17,19] is a successful tool for designing and verifying belief-*unaware* CPS, i.e. a "changing world" in a real-valued domain. Dynamic epistemic logics (DELs), on the other hand, deal with changing knowledge (which is tightly connected to beliefs[1]) in a propositional static world that never changes [3,4,7,10], again through the lens of modal logic. Some previous work exists at the intersection of these two. However, belief-aware CPS requires unobservable world change under the real numbers, which is in conflict with the public propositional world-change in [6]; and a more comprehensive and less restrictive treatment of belief that goes beyond using the underlying dynamic modalities of world-change to emulate noise as in [14].

Since both $d\mathcal{L}$ and DELs are dynamic modal logics, they are prime candidates for inspiration in the pursuit of a unified dynamic modal logic that can reason about changing beliefs in a changing world. We develop the *dynamic doxastic differential dynamic logic* $d^4\mathcal{L}$, as an extension of $d\mathcal{L}$ with (1) belief modalities, and (2) a learning operator for describing belief-change, inspired by DELs.

This new framework requires a fundamental conceptual shift in the design of CPS. Let `ctrl` be a program describing control decisions (e.g. a pilot pressing a button), and `plant` be a program for continuous evolution (e.g. an airplane flying). In the current, belief-unaware $d\mathcal{L}$ paradigm, the primary mode of establishing the safety of CPS is by the validity of a formula $pre \rightarrow [(\texttt{ctrl; plant})^*]\, safe$. It states that, starting from precondition *pre*, every possible execution of the program $(\texttt{ctrl; plant})^*$ ends with the safety property *safe* being true, with the star $^*$ operator repeating `ctrl` followed by `plant` any number of times.

*Example 1.* As a running example, suppose an airplane is controlled by directly setting its vertical velocity to 1 or -1 in thousands of feet per second. The safety goal of the controller is to keep the airplane above ground:

1. $pre \equiv safe \equiv (alt > 0)$, i.e., the airplane is above ground.
2. $\texttt{ctrl} \equiv (?alt > 1; yv := -1) \cup yv := 1$, in which two things may happen, on either side of $\cup$. If the airplane is above 1000 ft ($?alt > 1$), it may descend by setting vertical velocity $yv$ to -1000 ft per second. Alternatively, it can climb with $yv := 1$, which may always happen since this action has no ? test.
3. $\texttt{plant} \equiv t := 0; t' = 1, alt' = yv \,\&\, t \leq 1$ describes, using differential equations, that altitude changes with vertical velocity ($alt' = yv$) for a maximum of 1 unit of time using time counter $t' = 1$. The *evolution domain constraint* $t \leq 1$ bounds how much time may pass before the pilot reassesses this choice.

Intuitively, this CPS is safe because the controller can only decide to descend if it is high enough above ground such that descending for 1 second at a velocity of -1000 ft per second, traveling a total of 1000 ft, keeps it above ground. This condition is based on *ontic* (real world, or factual) truth and does not capture the reality that altitude is read from a noisy altimeter, and that *pilot beliefs* trigger actions, not ontics.

In contrast, in *belief-aware* CPS, control decisions are triggered by some belief $B_a(\phi)$, not ontic truth $\phi$. This minor syntactic change belies the complexity of

[1] Beliefs may be erroneous, knowledge may not.

the underlying paradigm shift. The CPS model must now explicitly describe how an agent learns about the world and acquires such beliefs $B_a(\phi)$. In $\mathsf{d}^4\mathcal{L}$, this process of observation and reasoning is specified by means of a *learning operator.*

A $\mathsf{d}\mathcal{L}$ program $\alpha$, describing ontic change, does not alter beliefs. In contrast, a learning operator program $L_a(\alpha)$ changes *only* agent $a$'s beliefs, with the change described by $\alpha$ becoming doxastic rather than ontic. The pattern $\alpha; L_a(\alpha)$ describes *observed* ontic change, which also affects beliefs. This learning operator may be used in a program `obs` to describe the agent's learning processes of observation and reasoning. This leads to the addition of the belief-changing `obs` to the safety formula $pre \rightarrow [(\texttt{obs; ctrl; plant})^*]\, safe$ used for belief-aware CPS.

*Example 2.* Consider a belief-triggered controller for the airplane of Example 1. The model now incorporates the fact that observation is imperfect, and that the altimeter, while operating properly, has some noise bounded by $\varepsilon > 0$.

1. $\texttt{obs} \equiv L_a(?alt - alt_a < \varepsilon)$. The pilot $a$ learns, by observing the altimeter with known error bounds $\epsilon$, that the *perceived* altitude $alt_a$ can be lower than the *true* altitude $alt$ by at most $\varepsilon$. Thus, the belief $B_a(alt - alt_a < \varepsilon)$ comes to be.
2. $\texttt{ctrl} \equiv (?B_a(alt_a - \varepsilon > 1); yv := -1) \cup yv := 1)$. Climbing, being safe, remains an always acceptable choice. However, the trigger for descending is that the pilot *believes* that the *perceived* altitude with worst-case noise is still high enough for the airplane to descend for one second, i.e. $B_a(alt_a - \varepsilon > 1)$.

We must add $\varepsilon > 0$ to *pre*, but `plant` does not change since beliefs do not directly affect the behavior of the real world: they do so only through agent actions.

More generally, $\mathsf{d}^4\mathcal{L}$ allows for arbitrary combinations of ontic $\mathsf{d}\mathcal{L}$ actions and the learning operator, representing any interleaving of physical and doxastic change, the former potentially unobservable, and the latter potentially imperfect, e.g. through noisy sensors.

## 3 Syntax of $\mathsf{d}^4\mathcal{L}$

In this section, we will describe $\mathsf{d}^4\mathcal{L}$ terms, formulas and programs. As in $\mathsf{d}\mathcal{L}$, real arithmetic is used to accurately model CPSs. Thus, terms are real-valued.

The safety of well-functioning belief-aware CPS is often predicated on beliefs being grounded in reality so that informed decisions can be made, cf. formula $B_a(alt - alt_a < \varepsilon)$ of Example 2, where perceived altitude can underestimate factual altitude by at most $\epsilon$. This relation between belief and truth is at the core of many safety arguments, and should be describable within the logic. We must therefore be able to refer to both ontic (factual) and doxastic (belief) states in the same context, as in $B_a(alt - alt_a < \varepsilon)$.

### 3.1 $d^4\mathcal{L}$ Terms and Formulas

State variables describe ontic truth, e.g. *alt* is the airplane's real altitude. Doxastic variable $alt_a$ is agent $a$'s perception of *alt*. Basic arithmetic is also in the language, e.g. $x-y$. Constants $c \in \mathbb{Q}$ allow for digitally representable numbers in the syntax, e.g. 2.5 but not $\pi$, though the semantics can give variables any value in $\mathbb{R}$. Logical variables $X$ are introduced by quantifiers over $\mathbb{R}$ to e.g. discharge reasoning about continuous time, or to find witnesses for existential modalities.

Let $\mathbb{A}$ be a finite set of agents, $\Sigma$ be a countable set of logical variables, $\mathbb{V}$ be a countable set of state variables, and $\mathbb{V}_a = \{x_a : x \in \mathbb{V}\}$ the set of doxastic variables for agent $a \in \mathbb{A}$. The following definition distinguishes between terms with and without doxastic variables. The distinction is crucial when assigning to state or doxastic variables, as we will see in Definition 3.

**Definition 1.** *The doxastic terms $\theta$ and non-doxastic terms $\zeta$ of $d^4\mathcal{L}$, with $\otimes \in \{+,-,\times,\div\}$, $X \in \Sigma$, $x \in \mathbb{V}, x_a \in \mathbb{V}_a$, $a \in \mathbb{A}$, $c \in \mathbb{Q}$, are given by the grammar:*

$$\begin{aligned} \theta &::= \theta \otimes \theta \mid X \mid c \mid x \mid x_a \\ \zeta &::= \zeta \otimes \zeta \mid X \mid c \mid x \end{aligned}$$

The formulas of $d^4\mathcal{L}$ are a superset of $d\mathcal{L}$'s [17], which are a superset of those of first-order logic for real arithmetic. Alongside logical connectives, we may write propositions such as $\theta_1 \leq \theta_2$ and logical quantifiers $\forall X\ \phi$. To this, $d^4\mathcal{L}$ adds the belief modality $B_a(\phi)$, meaning agent $a$ believes $\phi$. The dynamic modality formula $[\alpha]\,\phi$ (after all executions of program $\alpha$, $\phi$ is true), and its dual $\langle\alpha\rangle\,\phi$ (after some execution of $\alpha$, $\phi$ is true) capture belief-aware CPS behavior. The language of the programs $\alpha$ will be specified later in Definition 3.

Since $d^4\mathcal{L}$ beliefs are *only* about the state of the world, it is useful to distinguish between formulas $\xi$ which may appear inside belief modalities, and those $\phi$ which may not. We still allow doxastic terms $\theta$ in $\phi$, since safety proofs may generate such formulas.

**Definition 2.** *The formulas $\phi, \xi$ of $d^4\mathcal{L}$ are given by the grammar:*

$$\begin{aligned} \phi &::= \phi \vee \phi \mid \neg\phi \mid \theta \leq \theta \mid \forall X\ \phi(X) \mid [\alpha]\,\phi \mid B_a(\xi) \\ \xi &::= \xi \vee \xi \mid \neg\xi \mid \theta \leq \theta \end{aligned}$$

The remaining logical connectives, $\wedge, \rightarrow$ and duals $\langle\alpha\rangle\,\phi$, $\exists X\ \phi(X)$, $P_a(\xi)$ are defined as usual, e.g. $\langle\alpha\rangle\,\phi \equiv \neg\,[\alpha]\,\neg\phi$, and $P_a(\xi) \equiv \neg B_a(\neg\xi)$ when $a$ considers $\xi$ possible. We may now generalize the noisy but accurate sensors of Example 2.

*Example 3 (Noisy sensors).* Sensors often come with known error bounds $\varepsilon$. A pilot reading from the altimeter should thus come to believe the indicated value to be within $\varepsilon$ of the real *alt*, as captured by $B_a\left((alt_a - alt)^2 \leq \varepsilon^2\right)$, with integer exponentiation being definable from multiplication.

Belief modalities with both state and doxastic variables are *meta-properties* of belief, e.g., how far doxastic truth is from ontic truth. Thus, their truth value indeed changes as either the world or beliefs change. Section 6 will show such formulas are part of the core argument for some belief-aware CPS safety proofs. When formulas such as $B_a\left((alt_a - alt)^2 \leq \varepsilon^2\right)$ are not true, it can become impossible for $a$ to make informed decisions. Safety may then instead rely on very conservative actions, e.g. bringing a car to a stop, or flying straight and level.

### 3.2 Doxastic Hybrid Programs

The hybrid programs (HPs) of d$\mathcal{L}$ [17] are able to describe both discrete and continuous ontic change. They are the starting point for the *doxastic hybrid programs* (DHPs) of d$^4\mathcal{L}$. We introduce a learning operator $L_a(\gamma)$ for doxastic change, where $\gamma$ encodes an agent observing the world, reading from a sensor, or suspecting some change to have happened. In this paper, the language of the learned program $\gamma$ is nearly identical to that of hybrid programs, and to the epistemic actions of the epistemic action logic EAL [7].

**Changing Physical State.** Assignment $x := \zeta$ performs instantaneous ontic change, e.g. pushing the autopilot button, $autopilot := 1$, or resetting a time counter with $t := 0$, as in Examples 1 and 2. No doxastic variables are allowed in $\zeta$, since ontic truth is not directly a function of belief!

Differential equations $x' = \zeta \;\&\; \chi$ describe continuous motion over a nondeterministic duration, so long as the evolution domain constraint formula $\chi$ is true throughout. For example, $alt' = yv, t' = 1 \;\&\; t \leq 10$ describes linear change of altitude for up to 10 seconds according to vertical velocity $yv$. Nondeterministic ontic assignment $x := *$ is definable as $x' = 1; x' = -1$, which assigns any value in $\mathbb{R}$ to $x$ by increasing then decreasing it arbitrarily.

The test $?\phi$ transitions if and only if d$^4\mathcal{L}$ formula $\phi$ is true. It was used in Example 1 as an ontic trigger $?(alt > 1)$ determining whether an airplane could descend, and similarly as a belief trigger $?B_a\,(alt_a - \varepsilon > 1)$ in Example 2, where a pilot can only descend if they believe the airplane is safely above 1000 ft while taking worst-case noise into account.

Sequential composition $\alpha; \beta$ is self-explanatory. The choice $\alpha \cup \beta$ nondeterministically executes either $\alpha$ or $\beta$. It may be used to encode multiple possible outcomes or actions, e.g. $(?alt > 1; yv := -1) \cup yv := 1$ from Example 2.

Nondeterministic repetition $\alpha^*$ lets $\alpha$ be iterated arbitrarily many times. It was used in $(\texttt{obs}; \texttt{ctrl}; \texttt{plant})^*$ to ensure the safety proof applies to a system that can run for a long time, not just to a one-time control decision.

**Changing Belief State.** Agent beliefs are updated by means of the *learning operator* $L_a(\gamma)$, where $\gamma$ is a program describing belief change. Notably, to interleave ontic and belief change, the learning operator is a program itself rather than a modality as in [6,8]. Under d$^4\mathcal{L}$'s possible world semantics, each agent $a$ considers multiple worlds possible. The intuitive behavior of $L_a(\gamma)$ is to execute program $\gamma$ at each such world, and consider all outcomes of such executions as possible worlds.

The language of $\gamma$ is a slightly modified subset of that of hybrid programs. Inside a learning operator, ontic assignment $x := \zeta$ becomes doxastic assignment $x_a := \theta$. Since doxastic change (unlike ontic change) may depend on previous beliefs, the assigned term $\theta$ allows doxastic variables. The language also includes test $?\phi$, choice $\gamma_1 \cup \gamma_2$ and sequential composition $\gamma_1; \gamma_2$.

This language of doxastic change captures the bulk of observation and reasoning phenomena found in belief-aware CPS, which tend to occur at distinct and discrete intervals, e.g. looking at a sensor periodically. The literature [6,20] suggests that learned differential equations and repetition pose a very significant additional challenge, which is useful only in more specialized scenarios.

Learned programs may contain nondeterminism, as in $L_a(\gamma_1 \cup \gamma_2)$. Intuitively, this says that agent $a$ is aware that either $\gamma_1$ or $\gamma_2$ happened, but cannot ascertain which: agent $a$ must consider possible all outcomes of $\gamma_1$ and of $\gamma_2$. Thus, in $\mathsf{d}^4\mathcal{L}$, learned nondeterminism is unobservable, and leads to the *indistinguishability of outcomes*, as in action models and epistemic actions [3,7]. This is in contrast to program $L_a(\gamma_1) \cup L_a(\gamma_2)$, in which agent $a$ either learns $\gamma_1$, or learns $\gamma_2$, but in both case knows precisely which one happened.

Learned test $L_a(?\xi)$ eliminates those possible worlds for which $?\xi$ does not succeed, i.e. in which $\xi$ is false. In this way, $[L_a(?\phi)]\,\psi$ is analogous to public announcements and the tests of epistemic actions [7].

So far, the set of possible worlds may contract through learned tests and finitely expand with learned choice. The nondeterministic doxastic assignment $x_a := *$ further enables uncountable expansion of possibilities by assigning any value in $\mathbb{R}$ to $x_a$. To let $x_a$ take *any* value satisfying some property $\phi(x_a)$, the program $L_a(x_a := *; ?\phi(x_a))$ first "resets" the values $x_a$ can take using nondeterministic assignment, and then contracts the set of possible worlds with $?\phi(x_a)$.

The grammar of programs divides programs into two categories. The first, denoted $\alpha$, describes the language of ontic change, or the ontic fact $L_a(\gamma)$ that program $\gamma$ was learned. The second, denoted $\gamma$, describes the language of doxastic change, and, as we have seen, is a subset of the first with minor modifications.

**Definition 3.** *Let $x \in \mathbb{V}$, $a \in \mathbb{A}$, $x_a \in \mathbb{V}_a$ $\phi, \xi$ be formulas per Def. 2, $\theta, \zeta$ be terms per Def. 1. Doxastic hybrid programs (DHP) $\alpha$ and learnable programs $\gamma$ are defined thus:*

$$\begin{array}{lcl} \alpha & ::= & x := \zeta \mid x' = \zeta \,\&\, \chi \mid ?\phi \mid \alpha;\alpha \mid \alpha \cup \alpha \mid \alpha^* \mid L_a(\gamma) \\ \gamma & ::= & x_a := \theta \mid x_a := * \mid ?\xi \mid \gamma;\gamma \mid \gamma \cup \gamma \end{array}$$

With a better understanding of $\mathsf{d}^4\mathcal{L}$ programs, we may now describe exactly how the belief of Example 3, $B_a\left((alt_a - alt)^2 \leq \varepsilon^2\right)$, is acquired.

*Example 4 (Noisy sensors, cont'd).* By observing a *trusted* altimeter, the pilot decides to forget previous beliefs about altitude and trust the current reading. Then, because the altimeter has a known error bound of $\varepsilon$, the pilot must now consider possible all altitude values at most $\varepsilon$ away from the true value of *alt*.

$$L_a\left(alt_a := *;\ ?(alt_a - alt)^2 \leq \varepsilon^2\right)$$

# 4 Semantics of $\mathsf{d}^4\mathcal{L}$

The $\mathsf{d}^4\mathcal{L}$ semantics are designed to allow agents to hold potentially erroneous beliefs (proper belief, not knowledge) about a world which may undergo unobserved change. We are inspired by the modal Kripke semantics, but diverge from it by completely decoupling the valuation describing ontic truth, denoted $r$ in $\mathsf{d}^4\mathcal{L}$, from agent beliefs, since unobservable actions must change ontic truth *only*.

Because beliefs are exclusively about the world and not about other beliefs, different agents' worlds need not interact with one another. Therefore, each agent $a$ has their own set of worlds $W_a$, which they consider possible. Each agent $a$'s valuation $V_a(t)$ function holds the values of all doxastic variables at every world $t \in W_a$, e.g. agent $a$'s perception of altitude at $t \in W_a$ is $V_a(t)(alt_a)$.

In these sets of possible worlds, every world $t_1 \in W_a$ is indistinguishable from any other world $t_2 \in W_a$. Under the usual Kripke semantics, this means that the accessibility relation $\sim_a$, determining indistinguishability between worlds, is an equivalence relation, i.e. an S5 system. Equivalence relations traditionally encode knowledge, and belief is usually obtained by waiving the reflexivity requirement. In such belief systems, a *distinguished world* $s \in W_a$ determines ontic truth, and yet may not be accessible through $\sim_a$.

In $\mathsf{d}^4\mathcal{L}$, we achieve belief by allowing discrepancies between the valuations of the possible worlds, including the distinguished one, and the separate ontic valuation $r$. Thus, a pilot could believe the airplane to be high with $V_a(t)(alt_a) > 1000$ for every $t \in W_a$, while it could be low in reality, with $r(alt) \leq 1000$.

This allows us to omit the accessibility relations entirely. It also simplifies learned program semantics since the learning operator can never inadvertently change ontic truth by altering the valuation of the distinguished world. We keep the distinguished world in Definition 4 as a means by which we may interpret every formula in every context, as we will see in Definitions 5 and 6.

This gives us the models of $\mathsf{d}^4\mathcal{L}$, called physical-doxastic models, or PD-models for short. For simplicity, we consider only one agent $a$ from now on, and we omit the subscript where it can be easily inferred, e.g. $V$ instead of $V_a$.

**Definition 4 (Physical/doxastic model).** *A physical/doxastic model or PD-model $\omega = \langle r, W, V, s\rangle$ consists of (1) $r : \mathbb{V} \to \mathbb{R}$, the state of the physical world; (2) $W$ a set of worlds called the* possible worlds*; (3) $V : W \to (\mathbb{V}_a \to \mathbb{R})$, a valuation function in which $V(t)(x_a)$ returns agent $a$'s perceived value of the doxastic variable $x_a$ at world $t \in W$; and 4) $s \in W$, a distinguished world.*

PD-models are sufficient to give meaning to all terms, formulas and programs. We use $\omega, \nu, \mu$ to denote PD-models, and sub- and super-scripts are applied everywhere, e.g. $\omega' = \langle r', W', V', s'\rangle$. The shortcut $t \in \omega$ means $t \in W$; $\omega(t)(x_a)$ means $V(t)(x_a)$; and $\omega(x)$ means $r(x)$. The distinguished world of $\omega$ is $\mathrm{DW}(\omega)$ and its distinguished valuation $\mathrm{DV}(\omega) = \omega(\mathrm{DW}(\omega)) = \omega(s) = V(s)$. The real world is $\mathrm{R}(\omega) = r$. Finally, let $\langle r, W, V, s\rangle \oplus t = \langle r, W, V, t\rangle$ for any $t \in \omega$.

**Interpretation of Terms, Formulas, and Programs.** The interpretation of terms and formulas is standard, with logical variables $X$ given meaning by a variable assignment $\eta : \Sigma \to \mathbb{R}$, state variables $x$ by the physical state $\mathrm{R}(\omega)$, and doxastic variables $x_a$ by the distinguished valuation $\mathrm{DV}(\omega)$. Terms and formulas such as $alt_a$ and $alt_a > 1000$ may appear outside doxastic modalities during calculus proofs. The distinguished valuation (for the distinguished world) ensures that they have a well-defined meaning and can thus be used as part of the proof.

**Definition 5 (Term interpretation).** *Let $\omega = \langle r, W, V, s\rangle$ be a PD-model, and $\eta : \Sigma \to \mathbb{R}$ be a logical variable assignment. Then, the interpretation of terms is defined inductively as follows: $val_\eta(\omega, x) = r(x)$ for state variable $x$; $val_\eta(\omega, X) = \eta(X)$ for logical variable $X$; $val_\eta(\omega, x_a) = \mathrm{DV}(\omega)(x_a)$ for doxastic variable $x_a$; $val_\eta(\omega, \theta_1 \otimes \theta_2) = val_\eta(\omega, \theta_1) \otimes val_\eta(\omega, \theta_2)$ for $\otimes \in \{+, -, \times, \div\}$.*

Formula interpretation is derived directly from d$\mathcal{L}$, first-order logic for real arithmetic, and simplified Kripke semantics for beliefs. Definitions 6 and 7 are mutually recursive due to the box modality formula $[\alpha]\,\phi$ and test program $?\phi$.

**Definition 6 (Interpretation of formulas).** *Let $\omega = \langle r, W, V, s\rangle$ be a PD-model, $\eta$ be a variable assignment, and $\langle r, W, V, s\rangle \oplus t = \langle r, W, V, t\rangle$. Then, the valuation of a formula $\phi$ as 1 (true) or 0 (false) is defined inductively as follows.*

$$
\begin{array}{lll}
val_\eta(\omega, \theta_1 \le \theta_2) = 1 & \textit{iff} & val_\eta(\omega, \theta_1) \le val_\eta(\omega, \theta_2) \\
val_\eta(\omega, \phi_1 \vee \phi_2) = 1 & \textit{iff} & val_\eta(\omega, \phi_1) = 1 \textit{ or } val_\eta(\omega, \phi_2) = 1 \\
val_\eta(\omega, \neg\phi) = 1 & \textit{iff} & val_\eta(\omega, \phi) = 0 \\
val_\eta(\omega, \forall X\ \phi) = 1 & \textit{iff} & \textit{for all } v \in \mathbb{R}, val_{\eta[X \mapsto v]}(\omega, \phi) = 1 \\
val_\eta(\omega, B_a(\xi)) = 1 & \textit{iff} & \textit{for all } t \in \omega, val_\eta(\omega \oplus t, \xi) = 1 \\
val_\eta(\omega, [\alpha]\,\phi) = 1 & \textit{iff} & \textit{for all } (\omega, \omega') \in \rho_\eta(\alpha), val_\eta(\omega', \phi) = 1
\end{array}
$$

Under these semantics, $B_a(x = 0)$ is equivalent to $x = 0$ since state variable $x$ is independent of the choice of distinguished world, unlike $x_a$. CPS designers have no reason to write such formulas, but when they do appear in calculus proofs, the doxastic modality is eliminated using the equivalence $B_a(x = 0) \leftrightarrow x = 0$.

**Program Semantics.** The program semantics is given as a reachability relation over PD-models, with $(\omega, \omega') \in \rho_\eta(\alpha)$ meaning that PD-model $\omega'$ is reachable from $\omega$ using program $\alpha$. The semantics of DHPs starts with that of d$\mathcal{L}$'s hybrid programs. Most cases are intuitive. Differential equations use their solution $y$ to evolve $\mathrm{R}(\omega)$ for a nondeterministic duration, and ensure the evolution domain constraint $\chi$ is satisfied throughout. For a more in-depth treatment, see [17].

To this we add doxastic assignment, which affects the distinguished valuation $\mathrm{DV}(\omega)$, and the learning operator, which represents the "execute $\gamma$ at each possible world" semantics from DELs, as illustrated in Fig. 1.

In Fig. 1, let $(\omega, \omega') \in \rho_\eta(L_a(\gamma))$. Then, each world $\nu \in \omega'$ *after* learning has an "origin" world $t \in \omega$ from *before* learning, e.g. $t_1$ is the origin world for $\nu_1$

and $\nu_2$. Every PD-model $\nu$ that $\gamma$ can reach from each origin world $t \in \omega$ (i.e. $(\omega \oplus t, \nu) \in \rho_\eta(\gamma)$) becomes a possible world $\nu \in \omega'$ after $L_a(\gamma)$. The valuation $\omega'(\nu)$ reflects the effects of $\gamma$, which can be found in the distinguished valuation of $\nu$, and thus, we let $\omega'(\nu) = \text{DV}(\nu)$.

Finally, the distinguished world of $\omega'$ is chosen as any $t' \in \omega'$ whose origin world is $\text{DW}(\omega)$. This applies the principle of *learned nondeterminism as indistinguishability of outcomes to the distinguished world.*

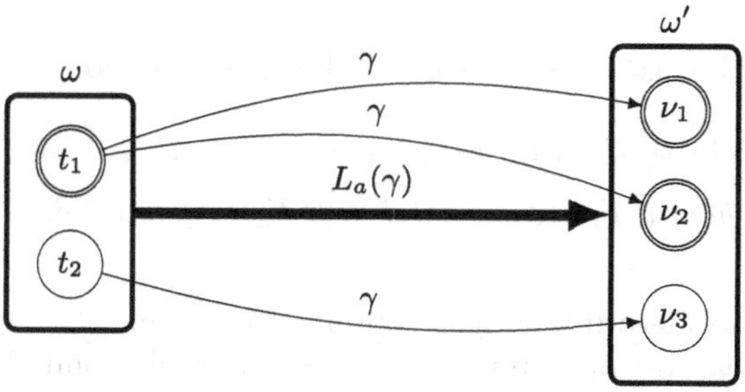


**Fig. 1.** The double-circled $t_1 = \text{DW}(\omega)$ creates, through $\gamma$'s nondeterminism, two post-learning worlds $\nu_1, \nu_2 \in \omega'$ worlds, either of which can be nondeterministically chosen as $\text{DW}(\omega')$. The world $t_2 \in \omega$ leads to $\nu_3 \in \omega'$, which cannot be chosen as $\text{DW}(\omega')$.

**Definition 7 (Transition semantics).** *Let $\omega = \langle r, W, V, s\rangle$ be a PD-model, and $\eta$ be a variable assignment. The transition relation for doxastic dynamic programs is inductively defined by:*

- $(\omega, \omega') \in \rho_\eta(x := \zeta)$ *iff* $\omega' = \omega$ *except* $\text{R}(\omega')(x) = val_\eta(\omega, \zeta)$
- $(\omega, \omega') \in \rho_\eta(x_a := \theta)$ *iff* $\omega' = \omega$ *except* $\text{DV}(\omega')(x_a) = val_\eta(\omega, \theta)$
- $(\omega, \omega') \in \rho_\eta(x_a := *)$ *iff* $\omega' = \omega$ *except* $\text{DV}(\omega')(x_a) = v$ *for some* $v \in \mathbb{R}$
- $(\omega, \omega') \in \rho_\eta(x' = \zeta \mathbin{\&} \chi)$ *iff* $\omega' = \langle r[x \mapsto y(\tau)], W, V, s\rangle$ *for the solution* $y : [0, \text{T}] \to \mathbb{R}$ *of the diff. eq., with* $\tau \in [0, \text{T}]$ *for some* $\text{T} \geq 0$*. Furthermore, for all* $t_i \in [0, \tau]$*, and* $val_\eta(\langle r[x \mapsto y(t_i)], W, V, s\rangle, \chi) = 1$.
- $(\omega, \omega) \in \rho_\eta(?\phi)$ *iff* $val_\eta(\omega, \phi) = 1$
- $\rho_\eta(\alpha; \beta) = \rho_\eta(\alpha) \circ \rho_\eta(\beta)$
$$= \{\omega_3 : \textit{ there is } \omega_2 \textit{ s.t. } (\omega_1, \omega_2) \in \rho_\eta(\alpha) \textit{ and } (\omega_2, \omega_3) \in \rho_\eta(\beta)\}$$
- $\rho_\eta(\alpha \cup \beta) = \rho_\eta(\alpha) \cup \rho_\eta(\beta)$
- $(\omega, \omega') \in \rho_\eta(\alpha^*)$ *iff there is* $n \in \mathbb{N}$ *such that* $(\omega, \omega') \in \rho_\eta(\alpha^n)$*, where* $\alpha^n$ *is* $\alpha$ *sequentially composed* $n$ *times.*
- $(\omega, \omega') \in \rho_\eta(L(\gamma))$ *if:* $r' = r$*,* $W' = \{\nu : \textit{ there is } t \in \omega \textit{ s.t. } (\omega \oplus t, \nu) \in \rho_\eta(\gamma)\}$*,* $\omega'(\nu) = \text{DV}(\nu)$ *for all* $\nu \in \omega'$*, and* $\text{DW}(\text{DW}(\omega')) = \text{DW}(\omega)$.

Figure 1 and Definition 7 show that $\mathsf{d}^4\mathcal{L}$'s learning operator applies the DEL semantics to *any* language of change, so long as it has a transition semantics, as in $(\omega \oplus t, \nu) \in \rho_\eta(\gamma)$. It is possible to extend this operator to traditional multi-agent Kripke structures by letting two after-learning worlds be indistinguishable in $\omega'$ iff their origin worlds were indistinguishable in $\omega$, as is standard in DELs.

## 5 Sound Sequent Calculus

Our main contribution towards the verification of belief-aware CPS is a sound proof calculus for d$^4\mathcal{L}$. The meaning of a sequent $\Gamma \vdash \phi$ with a d$^4\mathcal{L}$ formula $\phi$ and a set of d$^4\mathcal{L}$ formulas $\Gamma$ is captured with the following definition of validity.

**Definition 8 (Validity).** *A sequent $\Gamma \vdash \phi$ is* valid *iff for all $\omega$ and $\eta$,*

$$val_\eta\left(\omega, \bigwedge_{\psi \in \Gamma} \psi \rightarrow \phi\right) = 1$$

For simplicity's sake, we use a single definition of soundness for proof rules.

**Definition 9 (Global Soundness).** *A proof rule PR, as in* $\text{PR}\,\dfrac{\Gamma_1 \vdash \phi_1}{\Gamma_2 \vdash \phi_2}$, *is* globally sound *when, if $\Gamma_1 \vdash \phi_1$ is valid then $\Gamma_2 \vdash \phi_2$ is valid.*

**Overview of the Calculus.** Fig. 2 contains the fragment of the calculus that pertains to the learning operator. The d$\mathcal{L}$ calculus [16] is omitted as it is easily adaptable to d$^4\mathcal{L}$. Single-modality agent rationality axioms can be adopted for belief, i.e. $B_a(\phi_1 \rightarrow \phi_2) \rightarrow (B_a(\phi_1) \rightarrow B_a(\phi_2))$ and, if $\phi$ is valid, then $B_a(\phi)$ is too. The proof for the following theorem can be found in [13].

**Theorem 1** *The proof rules in Fig. 2 are globally sound.*

Sequent contexts $\Gamma$ are partitioned into $\Gamma_R; \Gamma_B; \Gamma_P; \Gamma_O$. The set $\Gamma_R$ is the set of formulas with only state and logical variables and no doxastic modalities, e.g. $alt > 0$. $\Gamma_B$ and $\Gamma_P$ are the sets of belief and possibility formulas respectively, e.g. $B_a\left((alt_a - alt)^2 \leq \varepsilon^2\right)$ and $P_a\left((alt_a - alt)^2 \leq \varepsilon^2\right)$. $\Gamma_O$ is the set of formulas with doxastic variables but no modalities, e.g. $alt_a > 0$. The rules in Fig. 2 are only applicable once this partitioning has been achieved. Finally, let $\Gamma\backslash_{x_a} = \{\phi \in \Gamma : x_a \text{ does not occur in } \phi\}$.

Proof rules for learned programs that change doxastic state, like assignment or test, work by altering the contexts in suitable ways. Each learned program has two rules, for the $\Box$ and $\Diamond$ dynamic modalities, which deal with the nondeterminism in the choice of the distinguished world. The exception is $L_a(\alpha \cup \beta)$, where doxastic and dynamic modalities interact much more subtly.

The proof rules for assignment $L_a(x_a := \theta)$ capture the intuition that, since $x_a$ now has the value of $\theta$ at each possible world, syntactically substituting all occurrences of $x_a$ with $\theta$ ought to mean the same thing.

Since nondeterministic assignment $L_a(x_a := *)$ gives $x_a$ any possible value, then anything previously possible about $x_a$ remains possible. However, beliefs about $x_a$, which must hold for *all* worlds, do not survive the assignment (unless they are tautologies). The proof rules $[L{:=}*]$ and $\langle L{:=}*\rangle$ eliminate the formulas which may no longer hold after assignment from the context.

Formulas describing the distinguished world, i.e. in $\Gamma_O$, are retained or removed, respectively, depending on whether the dynamic modality allows us pick our distinguished world to suit our goals, as with $\Diamond$, or not, as with $\Box$.

$$[L{:=}]\ \frac{\Gamma \vdash \phi(\theta)}{\Gamma \vdash [L_a(x_a := \theta)]\, \phi(x_a)}\ ^1 \qquad \langle L{:=}\rangle\ \frac{\Gamma \vdash \phi(\theta)}{\Gamma \vdash \langle L_a(x_a := \theta)\rangle\, \phi(x_a)}\ ^1$$

$$[L{:=}*]\ \frac{\Gamma_R; \Gamma_B\backslash_{x_a}; \Gamma_P; \Gamma_O\backslash_{x_a} \vdash \phi}{\Gamma \vdash [L_a(x_a := *)]\, \phi} \qquad \langle L{:=}*\rangle\ \frac{\Gamma_R; \Gamma_B\backslash_{x_a}; \Gamma_P; \Gamma_O \vdash \phi}{\Gamma \vdash \langle L_a(x_a := *)\rangle\, \phi}$$

$$[L?]\ \frac{\Gamma_R; \Gamma_B; \emptyset; \Gamma_O \vdash B_a(\xi) \rightarrow \psi}{\Gamma_R; \Gamma_B; \Gamma_P; \Gamma_O \vdash [L_a(?\xi)]\, \psi} \qquad \langle L?\rangle\ \frac{\Gamma_R; \Gamma_B; \emptyset; \Gamma_O \vdash B_a(\xi) \wedge \psi}{\Gamma_R; \Gamma_B; \Gamma_P; \Gamma_O \vdash \langle L_a(?\xi)\rangle\, \psi}$$

$$[L;]\ \frac{\Gamma \vdash [L_a(\gamma_1); L_a(\gamma_2)]\, \phi}{\Gamma \vdash [L_a(\gamma_1; \gamma_2)]\, \phi} \qquad \langle L;\rangle\ \frac{\Gamma \vdash \langle L_a(\gamma_1); L_a(\gamma_2)\rangle\, \phi}{\Gamma \vdash \langle L_a(\gamma_1; \gamma_2)\rangle\, \phi}$$

$$[LB\cup]\ \frac{\Gamma \vdash [L_a(\gamma_1)]\, B_a(\xi) \wedge [L_a(\gamma_2)]\, B_a(\xi)}{\Gamma \vdash [L_a(\gamma_1 \cup \gamma_2)]\, B_a(\xi)} \qquad \langle LB\cup\rangle\ \frac{\Gamma \vdash \langle L_a(\gamma_1)\rangle\, B_a(\xi) \wedge \langle L_a(\gamma_2)\rangle\, B_a(\xi)}{\Gamma \vdash \langle L_a(\gamma_1 \cup \gamma_2)\rangle\, B_a(\xi)}$$

$$[LP\cup]\ \frac{\Gamma \vdash [L_a(\gamma_1)]\, P_a(\xi) \wedge [L_a(\gamma_2)]\, P_a(\xi)}{\Gamma \vdash [L_a(\gamma_1 \cup \gamma_2)]\, P_a(\xi)} \qquad \langle LP\cup\rangle\ \frac{\Gamma \vdash \langle L_a(\gamma_1)\rangle\, P_a(\xi) \vee \langle L_a(\gamma_2)\rangle\, P_a(\xi)}{\Gamma \vdash \langle L_a(\gamma_1 \cup \gamma_2)\rangle\, P_a(\xi)}$$

$$[L\cup]\ \frac{\Gamma \vdash [L_a(\gamma_1)]\, \phi \wedge [L_a(\gamma_2)]\, \phi}{\Gamma \vdash [L_a(\gamma_1 \cup \gamma_2)]\, \phi}\ ^2 \qquad \langle L\cup\rangle\ \frac{\Gamma \vdash \langle L_a(\gamma_1)\rangle\, \phi \vee \langle L_a(\gamma_2)\rangle\, \phi}{\Gamma \vdash \langle L_a(\gamma_1 \cup \gamma_2)\rangle\, \phi}\ ^2$$

[1] **The substitution of $x_a$ by $\theta$ must be admissible in $\phi$, see Doxastic Assignment**
[2] **Formula $\phi$ does not contain doxastic modalities or variables, or learning operators**

**Fig. 2.** Dynamic doxastic fragment of the $\mathsf{d}^4\mathcal{L}$ calculus, with $\Gamma$ being $\Gamma_R; \Gamma_B; \Gamma_P; \Gamma_O$

Learned test results in the belief about the test result, as in public announcements. The test contracts the set of possible worlds, so we must remove the set of possibility formulas from the context, as they may no longer hold. The underlying dynamic modality determines whether this belief is a precondition for $\psi$ or a necessity ($\Diamond$ implies at least one transition, $\Box$ does not).

Learned sequential composition is merely reduced to regular sequential composition. Doxastic assignment and choice deserve further attention below.

**Doxastic Assignment.** The rule for doxastic assignments relies on its syntactic substitution being equivalent to the semantic substitution effected by learned assignment. This nontrival result can be captured succinctly by Lemma 1, whose full proof is found in [13]. This result only holds when the substitution is *admissible* with respect to a given formula $\phi$, i.e. that syntactic conditions are in place ensuring the substitution will not change the meaning of the substituted variables, and therefore, of the formula [13].

**Lemma 1 (Doxastic Substitution Lemma).** *Let $\phi$ be a formula. Let $\sigma$ be an admissible substitution for $\phi$ which replaces only doxastic variable $x_a$. Then, for every $\eta$ and $\omega = \langle r, W, V, s\rangle$, we have $val_\eta(\omega, \sigma(\phi)) = val_\eta(\sigma(\omega), \phi)$, where $\sigma(\phi)$ is syntactic substitution, and $\sigma(\omega)$ is semantic substitution, defined as $\sigma(\omega) = \langle r, W, \sigma(V), s\rangle$, with $\sigma(V)(t)(x_a) = val_\eta(\omega \oplus t, \sigma(x_a))$ and $\sigma(V)(t)(y_a) = V(t)(y_a) = \omega(t)(y_a)$ for $y_a \neq x_a$, for all $t \in \omega$.*

**Nondeterministic Choice.** Learned choice influences *doxastic* modalities, and the choice of distinguished world is influenced by *dynamic* modalities. This makes for some subtlety in the rules for learned choice. Consider the potential rule below, which assumes $L_a(\gamma_1 \cup \gamma_2)$ is equivalent to $L_a(\gamma_1) \cup L_a(\gamma_2)$.

$$\frac{P_a(\neg\xi)\,;\xi \vdash \langle L_a(?\xi)\rangle\, B_a(\xi) \vee \langle L_a(?\mathit{True})\rangle\, B_a(\xi)}{P_a(\neg\xi)\,;\xi \vdash \langle L_a(?\xi \cup ?\mathit{True})\rangle\, B_a(\xi)}$$

The sequent contexts tell us that $\xi$ holds in the distinguished world $\mathrm{DW}(\omega)$, but not in some other $t \in \omega$. The disjunction holds, since $\langle L_a(?\xi)\rangle\, B_a(\xi)$ is trivially true. The program $L_a(?\xi \cup ?\mathit{True})$ preserves all worlds, including $t$, because of $?\mathit{True}$. Since $\xi$ is not true in $t$, agent $a$ cannot therefore believe $\xi$. But if the top is valid and the bottom is not, this rule would be unsound.

This phenomenon occurs because the conclusion of the rule requires us to prove $B_a(\xi)$ for worlds originated through both $?\xi$ and $?\mathit{True}$. However, the premise of the rule implies we need only check those from either $?\xi$ *or* $?\mathit{True}$, as if the $\Diamond$ dynamic modality had control over learned nondeterminism. It does not: outcomes of learned nondeterminism are *always* considered indistinguishable.

Proof rules for $L_a(\gamma_1 \cup \gamma_2)$ must therefore be as conservative as the most conservative of their dynamic and doxastic modalities: the only proof rule that allows disjunction in the premise is $\langle LP\cup\rangle$ since both modalities $\Diamond$ and $P_a(\cdot)$ are existential. This realization informs the soundness proofs for learned choice.

*Proof (Soundness sketch for $\langle LB\cup\rangle$).* Let $\omega$ be an arbitrary PD-model. We must show that $\mathrm{val}_\eta(\omega, \langle L_a(\gamma_1 \cup \gamma_2)\rangle\, B_a(\xi)) = 1$, i.e. that $\xi$ is true at every world $\nu$ reachable by either $(t,\nu) \in \rho_\eta(\gamma_1)$ or $(t,\nu) \in \rho_\eta(\gamma_2)$ for $t \in \omega$.

Let $(t,\nu) \in \rho_\eta(\gamma_1)$. By hypothesis, $\mathrm{val}_\eta(\omega, \langle L_a(\gamma_1)\rangle\, B_a(\xi)) = 1$, i.e. $\xi$ is true at every world reachable by $\gamma_1$, and $\nu$ in particular. The argument is symmetrical for $(t,\nu) \in \rho_\eta(\gamma_2)$, but only because the premise is a conjunction. Thus, for any world $\nu$ created by $L_a(\gamma_1 \cup \gamma_2)$, $\xi$ is true at that world. Therefore, $B_a(\xi)$. □

## 6 Validation and Application

We will now use $\mathsf{d}^4\mathcal{L}$ to illustrate how to prove the safety of a small belief-aware CPS. The scenario is similar to that of Example 2, and it is useful to have a reference for some of the most used $\mathsf{d}\mathcal{L}$ proof rules that $\mathsf{d}^4\mathcal{L}$ inherits [16].

$$[;]\,\frac{\Gamma \vdash [\alpha]\,[\beta]\,\phi}{\Gamma \vdash [\alpha;\beta]\,\phi} \qquad [?]\,\frac{\Gamma \vdash \phi \rightarrow \psi}{\Gamma \vdash [?\phi]\,\psi} \qquad \rightarrow\!R\,\frac{\Gamma, \phi \vdash \psi}{\Gamma \vdash \phi \rightarrow \psi}$$

We let the pilot observe the altimeter with $\mathsf{O} \equiv L_a(alt_a := *; ?Noise)$, with $Noise \equiv (alt_a - alt < \varepsilon)$. The control program $\mathsf{C}$ climbs or descends by setting vertical velocity depending on whether descent is believed to be safe, $\mathsf{CB} \cup \mathsf{CP} \equiv (?B_a(alt_a - T - \varepsilon > 0)\,; yv := -1) \cup (?P_a(alt_a - T - \varepsilon \leq 0)\,; yv := 1)$. The two tests are mutually exclusive, leading to dual belief operators: descending requires the strong condition of belief, whereas the mere possibility of being too low triggers a climb. We use $\mathtt{F} \equiv t := 0; t' = 1, alt' = yv \;\&\; t < T$ as very simplified flight dynamics, and an invariant $inv \equiv (alt > 0 \land T > 0)$ to handle repetition.

We will prove the validity of the formula $alt > 0, T > 0 \vdash [(\mathtt{O}; \mathtt{C}; \mathtt{F})^*]\, alt > 0$ by successively applying sound proof rules from d$\mathcal{L}$ and Fig. 2 to it. The leaves of the proof tree will be formulas that can be easily discharged using only d$\mathcal{L}$ rules or real arithmetic. Once the proof tree is complete, we will know this safety formula is valid, and thus that the modeled system is safe.

$$\text{loop}\;\dfrac{\dfrac{*}{alt > 0, T > 0 \vdash inv} \qquad [;]\,[;]\,\dfrac{inv \vdash [\mathtt{O}]\,[\mathtt{C}]\,[\mathtt{F}]\, inv}{inv \vdash [\mathtt{O}; \mathtt{C}; \mathtt{F}]\, inv} \qquad \dfrac{*}{inv \vdash alt > 0}}{alt > 0, T > 0 \vdash [(\mathtt{O}; \mathtt{C}; \mathtt{F})^*]\, alt > 0}$$

The middle branch continues in:

$$[\cup]\;\dfrac{inv; B_a\,(Noise) \vdash [\mathtt{CB}]\,[\mathtt{F}]\, inv \qquad inv; B_a\,(Noise) \vdash [\mathtt{CP}]\,[\mathtt{F}]\, inv}{inv; B_a\,(Noise) \vdash [\mathtt{C}]\,[\mathtt{F}]\, inv}$$

$$[\mathrm{L}?] \to\! R\;\dfrac{inv; B_a\,(Noise) \vdash [\mathtt{C}]\,[\mathtt{F}]\, inv}{inv \vdash [L_a(?Noise)]\,[\mathtt{C}]\,[\mathtt{F}]\, inv}$$

$$[L{:=}*]\;\dfrac{inv \vdash [L_a(?Noise)]\,[\mathtt{C}]\,[\mathtt{F}]\, inv}{inv \vdash [L_a(alt_a := *)]\,[L_a(?Noise)]\,[\mathtt{C}]\,[\mathtt{F}]\, inv}$$

$$[\mathrm{L};]\,[;]\;\dfrac{inv \vdash [L_a(alt_a := *)]\,[L_a(?Noise)]\,[\mathtt{C}]\,[\mathtt{F}]\, inv}{inv \vdash [L_a(alt_a := *; ?Noise)]\,[\mathtt{C}]\,[\mathtt{F}]\, inv}$$

The branch on the right closes using d$\mathcal{L}$ proof rules and standard d$\mathcal{L}$ reasoning independent of beliefs: if the airplane is above ground and climbs, it remains above ground. The left branch requires some doxastic reasoning.

$$\text{cut}\;\dfrac{inv; B_a\,(Noise)\,, B_a\,(alt_a - T - \varepsilon > 0) \vdash alt > T \qquad inv; alt > T \vdash [\mathtt{F}(-1)]\, inv}{inv; B_a\,(Noise)\,, B_a\,(alt_a - T - \varepsilon > 0) \vdash [\mathtt{F}(-1)]\, inv}$$

$$[:=]\;\dfrac{inv; B_a\,(Noise)\,, B_a\,(alt_a - T - \varepsilon > 0) \vdash [\mathtt{F}(-1)]\, inv}{inv; B_a\,(Noise)\,, B_a\,(alt_a - T - \varepsilon > 0) \vdash [yv := -1]\,[\mathtt{F}(yv)]\, inv}$$

$$[;]\,[?] \to\! R\;\dfrac{inv; B_a\,(Noise)\,, B_a\,(alt_a - T - \varepsilon > 0) \vdash [yv := -1]\,[\mathtt{F}(yv)]\, inv}{inv; B_a\,(Noise) \vdash [?B_a\,(alt_a - T - \varepsilon > 0)\,; yv := -1]\,[\mathtt{F}(yv)]\, inv}$$

The left side of the cut rule must show that $alt > T$, and for that we will use the S5 rationality axioms that allow for reasoning about arithmetic. Thus, the agent may conclude (1) $B_a\,(alt > alt_a - \varepsilon)$ from $B_a\,(Noise)$, and (2) $B_a\,(alt_a > T + \varepsilon)$ from $B_a\,(alt_a - T - \varepsilon > 0)$. But (1) and (2) together lead to $B_a\,(alt > T)$, which no longer contains any doxastic variables. It is therefore equivalent to $alt > T$. We have thus used the belief meta-property (1), relating ontic and doxastic truth, to obtain an important fact about the world which we may now use in the right side of the proof.

This right side is a standard d$\mathcal{L}$ proof without doxastics: the rules for differential equations show that, after evolving for at most $T$ time at a speed of $-1$, the airplane cannot end up below ground, since it started above $T$ altitude.

This completes the sequent proof. It leveraged a mix of ontic, doxastic and meta-doxastic statements in order to make the argument for the safety of this controller. When working with trusted sensors, we also see an intuitive partitioning of the proof: first, doxastic formulas ($B_a\,(alt_a - T - \varepsilon > 0)$) and meta-doxastic formulas ($B_a\,(Noise)$) are used to derive ontic formulas ($alt > T$). Second, such ontic statements form the basis for arguments made in d$\mathcal{L}$-exclusive proof branches that ensure post-control actuation results in safe behavior.

This clear separation of concerns allows CPS engineers to work more intuitively and compositionally during the design and verification stages of belief-aware CPS.

The ways in which agents learn and reason influence the ontic facts that can be deduced, but those facts must in turn be informed by safety requirements of the CPS's physical evolution. Doxastics and ontics clearly play off each and have, in the past, contributed to safety incidents. By making this explicit in the model, $\mathsf{d}^4\mathcal{L}$ ensures adequate attention is given to such dynamics so that hopefully, ontic/doxastic concerns can be identified before they lead to tragedy.

## 7 Related Work

The logic $\mathsf{d}^4\mathcal{L}$ takes heavy inspiration from two bodies of work: one for reasoning about a changing world, and one for reasoning about changing beliefs.

**Changing World.** The logic $\mathsf{d}\mathcal{L}$ for reasoning about the ontic dynamics of CPS [16,17,19] has shown itself to be capable of verifying interesting and relevant real world systems [17,18]. However, it requires manual modeling discipline to express noise [14], rather than having noise or beliefs thereof as built-in primitives.

The example used in this paper is so simple that it can still be converted to $\mathsf{d}\mathcal{L}$ using modeling tricks [14]. The trick is to transform $alt_a$ into a state variable and remove the learning operator from the observation program, i.e. $alt_a := *; ?Noise$ rather than $L_a(alt_a := *; ?Noise)$. The agent's control would then be $(?alt_a - T - \varepsilon > 0; yv := -1) \cup (?alt_a - T - \varepsilon \leq 0; yv := 1)$.

However, this conversion relies fundamentally on the box dynamic modality $[\alpha]\,\phi$, which checks safety for *all* executions of $alt_a := *; ?Noise$. With liveness formulas using the diamond dynamic modality $\langle\alpha\rangle\,\phi$, safety need only be checked for *one* execution. Thus, in liveness formulas, this method would fail to capture the intended behavior of both the learning operator and the belief modality, which should still apply to *all* possible worlds, or, in $\mathsf{d}\mathcal{L}$ terms, all executions.

This conversion can also quickly become complex. A more detailed controller for a pilot trying to remain around or above cruising altitude $A$ could be $(?B_a\,(alt_a - T - \varepsilon > A)\,; yv := -1) \cup (?P_a\,(alt_a - T - \varepsilon > A)\,; yv := -0.5) \cup (?B_a\,(alt_a - T - \varepsilon \leq 0)\,; yv := 1)$. This is similar to previous controllers, but allows for a more gentle descent when the pilot considers the possibility of being close to $A$. The equivalent $\mathsf{d}\mathcal{L}$ controller is $(?alt_a - T - \varepsilon > A; yv := -1) \cup (?alt_a - T + \varepsilon > A; yv := -0.5) \cup (?alt_a - T - \varepsilon \leq 0; yv := 1)$. However, this elimination of doxastic modalities requires a change in the arithmetic itself, e.g. $(?P_a\,(alt_a - T - \varepsilon > A)$ turns into $(?alt_a - T + \varepsilon > A)$. Belief must consider worst case noise, whereas possibility can consider the best case. This can quickly become complex when going beyond simpler interval-based noise scenarios.

Both $\mathsf{d}\mathcal{L}$ and $\mathsf{d}^4\mathcal{L}$ controllers allow tests for deciding which action to take, but represent action triggers in first-order logic or doxastic logic, respectively, e.g. $alt_a - T + \varepsilon > A$ and $P_a\,(alt_a - T - \varepsilon > A)$. Decisions in real CPS are based on belief, and as the conversion from doxastic to non-doxastic action triggers quickly becomes non-trivial, it is best to avoid subtle modeling mistakes by working with belief during design and verification. With $\mathsf{d}^4\mathcal{L}$, safety engineers can rely on doxastic intuitions during verification, rather than having to infer them from formulas such as $alt_a - T + \varepsilon > A$, which does not clearly convey the concept of possibility that is so clear in $P_a\,(alt_a - T - \varepsilon > A)$.

The notion of robustness in hybrid systems control can capture complex notions of sensor and actuator noise [11], but is ultimately restrictive for the purpose of belief-aware CPS, as discussed at the beginning of Sect. 2. Adaptive control, where no *a priori* constraints are known, often depends on neural networks [15], and safety guarantees for systems relying on learning are known to add significant complexity to such efforts [9].

**Changing Belief.** On the other side, we have dynamic epistemic logics (DELs) [3,4,6,7,10], of which a good overview can be found in the literature [8]. They provide several notions of learning for different languages, some similar to our programs [6]. *Public* propositional world-change [6] would make ontic change implicitly observable, which is in direct conflict with the unobservability requirements of belief-aware CPS. Furthermore, relevant DEL axiomatizations rely on creating a conjunction out of properties of each accessible possible world [4,8], which is incompatible with the uncountably many worlds that CPS demand.

Belief revision through the AGM postulates [2] is an axiomatic, declarative approach to belief change. Because it is such a different approach, it presents many challenges in its integration with model-theoretic work such as $\mathsf{d}\mathcal{L}$.

In order to begin addressing safety concerns around ontic/doxastic interactions at design time, CPS engineers and agents must make complex logical arguments from both ontic facts and beliefs, as in Sect. 6. Despite their many successes, the works described in this section do not address this particular challenge directly in a principled way.

## 8 Conclusions

This paper considers interactions between belief and fact, which have significant safety implications. We proposed belief-aware CPSs as a first-principles paradigm under which safety concerns with such ontic/doxastic dynamics are expressly dealt with at design time, before safety violations occur. Our contribution is the logic $\mathsf{d}^4\mathcal{L}$ for modeling and verifying belief-aware CPSs, requiring simultaneous, complex belief- and world-change. Its formulas can describe ontic, doxastic and meta-doxastic statements, and its programs can model belief-aware CPS with belief-triggered controllers that make decisions based only on what they can observe and reason. We proposed a learning operator for belief-change, which is capable of transforming any transition-based semantics of change into a semantics of *belief*-change. We presented a sequent calculus for $\mathsf{d}^4\mathcal{L}$, which is proven to be sound, and used it to show the safety of a simple belief-aware CPS. This is, to the best of our knowledge, the first calculus for a dynamic logic of belief/knowledge change that can handle an uncountable domain, as in CPS.

**Acknowledgment.** We thank the anonymous reviewers for their helpful feedback.

## References

1. Aircraft Accident Investigation Bureau of Ethiopia: Report No. AI-01/19, Aircraft Accident Investigation Preliminary Report, Ethiopian Airlines Group, B737–8 (MAX) Registered ET-AVJ (2019)
2. Alchourrón, C.E., Gärdenfors, P., Makinson, D.: On the logic of theory change: partial meet contraction and revision functions. J. Symb. Log. **50**(2), 510–530 (1985)
3. Baltag, A., Moss, L.S.: Logics for epistemic programs. Synthese **139**(2), 165–224 (2004)
4. Baltag, A., Moss, L.S., Solecki, S.: The logic of public announcements, common knowledge, and private suspicions. In: TARK, pp. 43–56. Morgan Kaufmann Publishers Inc., San Francisco (1998)
5. Bureau d'Enquêtes et d'Analyses (BEA): Final report on the accident on 1st June 2009 to the airbus A330–203 registered F-GZCP operated by Air France flight AF 447 from Rio de Janeiro to Paris (2012)
6. van Ditmarsch, H.P., van der Hoek, W., Kooi, B.P.: Dynamic epistemic logic with assignment. In: AAMAS, pp. 141–148. ACM, New York (2005)
7. van Ditmarsch, H.P.: Descriptions of game actions. J. Logic, Lang. Inf. **11**(3), 349–365 (2002)
8. van Ditmarsch, H., van der Hoek, W., Kooi, B.: Dynamic Epistemic Logic. Springer, Netherlands (2005). https://doi.org/10.1007/978-1-4020-5839-4
9. Fulton, N., Platzer, A.: Verifiably safe off-model reinforcement learning. In: Vojnar, T., Zhang, L. (eds.) TACAS 2019. LNCS, vol. 11427, pp. 413–430. Springer, Cham (2019). https://doi.org/10.1007/978-3-030-17462-0_28
10. Gerbrandy, J., Groeneveld, W.: Reasoning about information change. J. Logic, Lang. Inf. **6**(2), 147–169 (1997)
11. Goebel, R., Hespanha, J.P., Teel, A.R., Cai, C., Sanfelice, R.: Hybrid systems: generalized solutions and robust stability. In: Proceedings of the 6th IFAC Symposium on Nonlinear Control Systems, September 2004
12. Komite Nasional Keselamatan Transportasi: Preliminary Aircraft Accident Investigation Report, PT. Lion Mentari Airlines, Boeing 737–8 (MAX); PK-LQP (2018)
13. Martins, J.G., Platzer, A., Leite, J.: A sound calculus for a logic of belief-aware cyber-physical systems. Tech. Rep. CMU-CS-19-116, School of Computer Science, Carnegie Mellon University, Pittsburgh, PA, July 2019
14. Mitsch, S., Ghorbal, K., Vogelbacher, D., Platzer, A.: Formal verification of obstacle avoidance and navigation of ground robots I. J. Robotics Res. **36**(12), 1312–1340 (2017)
15. Nguyen, N.T., Krishnakumar, K.S., Kaneshige, J.T., Nespeca, P.P.: Flight dynamics and hybrid adaptive control of damaged aircraft. J. Guidance Control and Dyn. **31**(3), 751–764 (2008)
16. Platzer, A.: Differential dynamic logic for hybrid systems. J. Autom. Reas. **41**(2), 143–189 (2008)
17. Platzer, A.: Logics of dynamical systems. In: LICS, pp. 13–24. IEEE (2012)
18. Platzer, A.: Logic & proofs for cyber-physical systems. In: Olivetti, N., Tiwari, A. (eds.) IJCAR 2016. LNCS, vol. 9706, pp. 15–21. Springer, Cham (2016). https://doi.org/10.1007/978-3-319-40229-1_3

19. Platzer, A.: Logical Foundations of Cyber-Physical Systems. Springer, Cham (2018). https://doi.org/10.1007/978-3-319-63588-0
20. Platzer, A., Tan, Y.K.: Differential equation axiomatization: the impressive power of differential ghosts. In: Dawar, A., Grädel, E. (eds.) LICS, pp. 819–828. ACM, New York (2018)

# Operational Semantics and Program Verification Using Many-Sorted Hybrid Modal Logic

Ioana Leuştean, Natalia Moangă(✉), and Traian Florin Şerbănuţă

Faculty of Mathematics and Computer Science, University of Bucharest,
Str. Academiei 14, 010014 Bucharest, Romania
{ioana,traian.serbanuta}@fmi.unibuc.ro, natalia.moanga@drd.unibuc.ro

**Abstract.** We propose a general framework to allow: (a) specifying the operational semantics of a programming language; and (b) stating and proving properties about program correctness. Our framework is based on a many-sorted system of hybrid modal logic, for which we prove its completeness results. We believe that our approach to program verification improves over the existing approaches within modal logic as (1) it is based on operational semantics which enables a more natural description of the execution than Hoare-style weakest precondition used by dynamic logic; (2) since it is multi-sorted, it allows for a clearer encoding of semantics, with a smaller representational distance to its intended meaning.

**Keywords:** Operational semantics · Program verification · Hybrid modal logic · Many sorted logic

## 1 Introduction

Program verification within *modal logic*, as showcased by *dynamic logic* [15], is following the mainstream axiomatic approach proposed by Hoare/Floyd [11,17]. In this paper, we continue our work from [18] in exploring the amenability of dynamic logic in particular, and of modal logic in general, to express operational semantics of languages (as axioms), and to make use of such semantics in program verification. Consequently, we consider the SMC Machine described by Plotkin [21], we derive a dynamic logic set of axioms from its proposed transition semantics, and we argue that this set of axioms can be used to derive Hoare-like assertions regarding functional correctness of programs written in the SMC language.

The main idea is to define a general logical system that is powerful enough to represent both the programs and their semantics in a uniform way. With respect to this, we follow the line of [14] and the recent work from [22].

The logical system that we developed as support for our approach is a *many-sorted hybrid polyadic modal logic*, built upon our general many-sorted polyadic

All authors contributed equally to this work.


S. Cerrito and A. Popescu (Eds.): TABLEAUX 2019, LNAI 11714, pp. 446–476, 2019.
https://doi.org/10.1007/978-3-030-29026-9_25

modal logic defined in [18]. We chose a modal setting since, as argued above, through dynamic logic and Hoare logic, modal logic has a long-standing tradition in program verification (see also [9] for a modal logic approach to separation logic [23]) and it is successfully used in specifying and verifying hybrid systems [20].

In [18] we defined a general many-sorted modal logic, generalizing some of the already existing approaches, e.g. [24,25] (see [18] for more references on many-sorted modal logic). This system allows us to specify a language and its operational semantics and one can use it to certify executions as well. However, both its expressivity and its capability are limited: we were not able to perform symbolic execution and, in particular, we were not able to prove Hoare-style invariant properties for loops. In Remark 1, we point out some theoretical aspects related to these issues.

In the present paper we employ the procedure of *hybridization* on top of our many-sorted modal logic previously defined. We drew our inspiration from [8,22] for practical aspects, and from the extensive research on *hybrid modal logic* [1,7] on the theoretical side. Our aim was to develop a system that is strong enough to perform all the addressed issues (specification, semantics, verification), but also to keep it as simple as possible from a theoretical point of view. To conclude: in our setting we are able to associate a sound and complete many-sorted hybrid modal logic to a given language such that both operational semantics and program verification can be performed through logical inference.

Given a propositional modal logic, a hybrid corresponding system is defined by adding some special atomic symbols (called *nominals*) to name the states of a model. Apart from nominals, some hybrid systems have a special operator $@_j$ (which is interpreted as a jump to the state denoted by the nominal $j$), as well as binders such as $\forall$ and $\exists$. Whenever binders are employed, one also needs *state variables*, special variables that are bind to states (see [1] for details). We have to make a methodological comment: sometimes nominals are presented as another *sort* of atoms (see, e.g. [7]). Our sorts come from a many-sorted signature $(S, \Sigma)$, as in [14], so all the formulas (in particular the propositional variables, the state variables, the nominals) are $S$-sorted sets. When we say that the hybrid logic is *mono-sorted* we use sorted according to our context, i.e. the sets of propositional variables, nominals and state variables are regular sets and not $S$-sets.

The many-sorted polyadic modal logic defined in [18] is briefly presented in Sect. 2. The hybridization is performed in Sect. 3. A concrete language and its operational semantics are defined in Sect. 4; we also show how to perform Hoare-style verification. A section on related and future work concludes our paper. We refer to [19] for more details and full proofs.

## 2 Preliminaries: A Many-Sorted Polyadic Modal Logic

In this section we recall the many-sorted polyadic modal logic defined in [18]. Recall that the most well-known system of modal logic uses only unary modalities (e.g $\Box$ and $\Diamond$), but more general systems using modal operators of arbitrary arities (called *polyadic modalitaties* [6, Section 6]) are also studied.

In the many-sorted setting, the arity of each operator is defined by a many-sorted signature. Consequently, our language is determined by a fixed, but arbitrary, many-sorted signature $\mathbf{\Sigma} = (S, \Sigma)$ and an $S$-sorted set of propositional variables $P = \{P_s\}_{s \in S}$ such that $P_s \neq \emptyset$ for any $s \in S$ and $P_{s_1} \cap P_{s_2} = \emptyset$ for any $s_1 \neq s_2$ in $S$. For any $n \in \mathbb{N}$ and $s, s_1, \ldots, s_n \in S$ we denote $\Sigma_{s_1 \ldots s_n, s} = \{\sigma \in \Sigma \mid \sigma : s_1 \cdots s_n \to s\}$.

The set of formulas is an $S$-indexed family inductively defined by:

$$\phi_s :: = p \,|\, \neg\phi_s \,|\, \phi_s \vee \phi_s \,|\, \sigma(\phi_{s_1}, \ldots, \phi_{s_n})$$

where $s \in S$, $p \in P_s$ and $\sigma \in \Sigma_{s_1 \cdots s_n, s}$.

We use the classical definitions of the derived logical connectors: for any $\sigma \in \Sigma_{s_1 \ldots s_n, s}$ the *dual operation* is $\sigma^{\Box}(\phi_1, \ldots, \phi_n) := \neg\sigma(\neg\phi_1, \ldots, \neg\phi_n)$.

In the sequel, by $\phi_s$ we mean that $\phi$ is a formula of sort $s \in S$. Similarly, $\Gamma_s$ means that $\Gamma$ is a set of formulas of sort $s$. When the context uniquely determines the sort of a state symbol, we shall omit the subscript.

In order to define the semantics we introduce $(S, \Sigma)$-*frames* and $(S, \Sigma)$-*models*. An $(S, \Sigma)$-*frame* is a tuple $\mathcal{F} = (W, (R_\sigma)_{\sigma \in \Sigma})$ such that:

- $W = \{W_s\}_{s \in S}$ is an $S$-sorted set and $W_s \neq \emptyset$ for any $s \in S$ (the elements of $W$ are called *worlds*, *states* or *points*),
- $R_\sigma \subseteq W_s \times W_{s_1} \times \ldots \times W_{s_n}$ for any $\sigma \in \Sigma_{s_1 \cdots s_n, s}$.

An $(S, \Sigma)$-*model based on* $\mathcal{F}$ is a pair $\mathcal{M} = (\mathcal{F}, V)$ where $V = \{V_s\}_{s \in S}$ such that $V_s : P_s \to \mathcal{P}(W_s)$ for any $s \in S$. Note that, for any $\sigma \in \Sigma$, the relation $R_\sigma$ is the interpretation of $\sigma$ in any model based on the frame $\mathcal{F}$. The model $\mathcal{M} = (\mathcal{F}, V)$ will be simply denoted as $\mathcal{M} = (W, (R_\sigma)_{\sigma \in \Sigma}, V)$. For $s \in S$, $w \in W_s$ and $\phi$ a formula of sort $s$, the many-sorted *satisfaction relation* $\mathcal{M}, w \overset{s}{\models} \phi$ is inductively defined as follows:

- $\mathcal{M}, w \overset{s}{\models} p$ iff $w \in V_s(p)$
- $\mathcal{M}, w \overset{s}{\models} \neg\psi$ iff $\mathcal{M}, w \overset{s}{\not\models} \psi$
- $\mathcal{M}, w \overset{s}{\models} \psi_1 \vee \psi_2$ iff $\mathcal{M}, w \overset{s}{\models} \psi_1$ or $\mathcal{M}, w \overset{s}{\models} \psi_2$
- if $\sigma \in \Sigma_{s_1 \ldots s_n, s}$, then $\mathcal{M}, w \overset{s}{\models} \sigma(\phi_1, \ldots, \phi_n)$ iff for any $i \in [n]$ there exist $w_i \in W_{s_i}$ such that $R_\sigma w w_1 \ldots w_n$ and $\mathcal{M}, w_i \overset{s_i}{\models} \phi_i$.

**Definition 1 (Validity and satisfiability).** *Let $s \in S$ and assume $\phi$ is a formula of sort $s$. Then $\phi$ is* satisfiable *if $\mathcal{M}, w \overset{s}{\models} \phi$ for some model $\mathcal{M}$ and some $w \in W_s$. The formula $\phi$ is* valid *in a model $\mathcal{M}$ if $\mathcal{M}, w \overset{s}{\models} \phi$ for any $w \in W_s$; in this case we write $\mathcal{M} \overset{s}{\models} \phi$. The formula $\phi$ is valid in a frame $\mathcal{F}$ if $\phi$ is valid in all the models based on $\mathcal{F}$; in this case we write $\mathcal{F} \overset{s}{\models} \phi$. Finally, the formula $\phi$ is valid if $\phi$ is valid in all frames; in this case we write $\overset{s}{\models} \phi$.*

The deductive system is presented in Fig. 1.

The *set of theorems* of $\mathbf{K}_{\mathbf{\Sigma}}$ is the least set of formulas that contains all the axioms and it is closed under deduction rules. Note that the set of theorems is obviously closed under $S$-*sorted uniform substitution* (i.e. propositional variables

**The system $\mathbf{K}_\Sigma$**

- For any $s \in S$, if $\phi$ is a formula of sort $s$ which is a theorem in propositional logic, then $\phi$ is an axiom.
- Axiom schemes: for any $\sigma \in \Sigma_{s_1 \cdots s_n, s}$ and for any formulas $\phi_1, \ldots, \phi_n, \phi, \chi$ of appropriate sorts, the following formulas are axioms:
  $(K_\sigma)$ $\sigma^\Box(\ldots, \phi_{i-1}, \phi \to \chi, \phi_{i+1}, \ldots) \to$
  $(\sigma^\Box(\ldots, \phi_{i-1}, \phi, \phi_{i+1}, \ldots) \to \sigma^\Box(\ldots, \phi_{i-1}, \chi, \phi_{i+1}, \ldots))$
  $(Dual_\sigma)$ $\sigma(\psi_1, \ldots, \psi_n) \leftrightarrow \neg\sigma^\Box(\neg\psi_1, \ldots, \neg\psi_n)$
- Deduction rules: *Modus Ponens* and *Universal Generalization*

  $(MP)$ if $\stackrel{s}{\vdash} \phi$ and $\stackrel{s}{\vdash} \phi \to \psi$ then $\stackrel{s}{\vdash} \psi$
  $(UG)$ if $\stackrel{s_i}{\vdash} \phi$ then $\stackrel{s}{\vdash} \sigma^\Box(\phi_1, .., \phi, ..\phi_n)$

  where $\stackrel{s}{\vdash} \phi$ means that $\phi$ is a provable formula of sort $s$.

**Fig. 1.** $(S, \Sigma)$ modal logic

of sort $s$ are uniformly replaced by formulas of the same sort). If $\phi$ is a theorem of sort $s$ write $\stackrel{s}{\vdash}_{\mathbf{K}_\Sigma} \phi$. Obviously, $\mathbf{K}_\Sigma$ is a generalization of the modal system $\mathbf{K}$ (see [7] for the mono-sorted version).

In modal logic one can speak about *local* and *global* logical consequence, both from a syntactical and a semantical point of view. Given a set of premises, a formula is a local consequence if, for any model, whenever the premises are satisfied at some state, the formula is also satisfied at the same state (the truth is preserved point-to-point). From the global point of view, the formula is satisfied at any point of a model, whenever the premises are satisfied at any point. We refer to [7, 1.5] for the mono-sorted setting and to [18, Section 3] for the many-sorted one. The distinction between local and global deduction is deepened in the many-sorted approach: *locally*, the conclusion and the hypotheses have the same sort, while *globally*, the set of hypotheses is a many-sorted set. In the sequel we only consider the local setting.

**Definition 2 (Local deduction).** *[18] If $s \in S$ and $\Gamma_s \cup \{\phi\}$ is a set of formulas of sort $s$, then we say that $\phi$ is (*locally*)* provable from *$\Gamma_s$ if there are $\gamma_1, \ldots, \gamma_n \in \Gamma_s$ such that $\stackrel{s}{\vdash}_{\mathbf{K}_\Sigma} (\gamma_1 \wedge \ldots \wedge \gamma_n) \to \phi$. In this case we write $\Gamma_s \stackrel{s}{\vdash}_{\mathbf{K}_\Sigma} \phi$.*

The construction of the canonical model is a straightforward generalization of the mono-sorted setting. For more details, we refer to [18]. The last result we recall is the (strong) completeness theorem with respect to the class of all frames.

**Theorem 1.** *[18] Let $\Gamma_s$ be a set of formulas of set $s$. If $\Gamma_s$ is a consistent set in $\mathbf{K}_\Sigma$ then $\Gamma_s$ has a model. Moreover, if $\phi$ is a formula of sort $s$, then $\Gamma_s \models_{\mathbf{K}_\Sigma} \phi$ iff $\Gamma_s \vdash_{\mathbf{K}_\Sigma} \phi$, where $\Gamma_s \models_{\mathbf{K}_\Sigma} \phi$ denotes the fact that any model of $\Gamma$ is also a model of $\phi$.*

*Remark 1 (Problems).* The many-sorted modal logic allows us to define both the syntax and the semantics of a programming language (see [18] for a complex example). However, there are few issues, both theoretical and operational, that we could not overcome:

(i1) the logic can be used to certify executions, but not to perform symbolic verification; in particular, in order to prove the invariant properties for loops, the existential binder is required;
(i2) the completeness theorem for extensions of $\mathbf{K}_{\mathbf{\Sigma}}$ from [18] only refers to model completeness, but says nothing about frame completeness (see [12] for a general discussion on this distinction);
(i3) the sorts are completely isolated formally, but in our example elements of different sorts have a rich interaction.

These issues will be addressed in the following sections.

## 3 Many-Sorted Hybrid Modal Logic

The hybridization of our many-sorted modal logic is developed using a combination of ideas and techniques from [1,3,4,7,12,13]. We refer to [19] for the full proofs of the results presented in this section.

Hybrid logic is defined on top of modal logic by adding *nominals*, *states variables* and specific operators and binders. Nominals allow us to directly refer the worlds (states) of a model, since they are evaluated to singleton sets in any model. However, a nominal may refer to different worlds in different models. In the sequel we introduce the *constant nominals*, which are evaluated to singletons, but they refer to the same world (state) in all models. Our example for constant nominals are `true` and `false` from Sect. 4.

**Definition 3 (Signature with constant nominals).** *A* signature with constant nominals *is a triple* $(S, \Sigma, N)$ *where* $(S, \Sigma)$ *is a many-sorted signature and* $N = (N_s)_{s \in S}$ *is an* $S$*-sorted set of constant nominal symbols. In the sequel, we denote* $\mathbf{\Sigma} = (S, \Sigma, N)$.

As before, the sorts will be denoted by $s$, $t$, ... and by $\text{PROP} = \{\text{PROP}_s\}_{s \in S}$, $\text{NOM} = \{\text{NOM}_s\}_{s \in S}$ and $\text{SVAR} = \{\text{SVAR}_s\}_{s \in S}$ we will denote some countable $S$-sorted sets. The elements of PROP are ordinary propositional variables and they will be denoted $p, q, \ldots$; the elements of NOM are called *nominals* and they will be denoted by $j$, $k$, ...; the elements of SVAR are called *state variables* and they are denoted $x$, $y$, .... We shall assume that for any distinct sorts $s \neq t \in S$, the corresponding sets of propositional variables, nominals and state variables are distinct. A *state symbol* is a nominal, a constant nominal or a state variable.

As in the mono-sorted case, nominals and state variables will be semantically constrained: they are evaluated to a singleton set, which means they will always refer to a unique world of our model. In addition, the constant nominals will refer to the same world(state) in any evaluation, so they will be defined at the frames' level.

In the mono-sorted setting, starting with a modal logic, the simplest hybrid system is obtained by adding nominals alone. However, the *basic hybrid system* is obtained by adding the *satisfaction modality* $@_j\phi$ (which states that $\phi$ is true at the world denoted by the nominal $j$). The most powerful hybrid systems are obtained by further adding the binders $\forall$ and $\exists$ that bind state variables to worlds, with the expected semantics [1,2,4]. In the sequel we will develop the hybrid modal logic $\mathcal{H}_{\mathbf{\Sigma}}(@,\forall)$ in our many-sorted setting. As mentioned in Remark 3, the system $\mathcal{H}_{\mathbf{\Sigma}}(@)$ can be similarly analyzed (see also [19] for more details).

Note that, whenever the context is clear, we will simply write $\overset{s}{\models}$ instead of $\overset{s}{\models}_{\mathcal{H}_{\mathbf{\Sigma}}(@,\forall)}$, and $\overset{s}{\vdash}$ instead of $\overset{s}{\vdash}_{\mathcal{H}_{\mathbf{\Sigma}}(@,\forall)}$. We will further assume that the sort of a formula (set of formulas) is implied by a concrete context but, whenever necessary, we will use subscripts to fix the sort of a symbol: $x_s$ means that $x$ is a state variable of sort $s$, $\Gamma_s$ means that $\Gamma$ is a set of formulas of sort $s$, etc.

**Definition 4 ($\mathcal{H}_{\mathbf{\Sigma}}(@,\forall)$ formulas).** *For any $s \in S$ we define the formulas of sort $s$:*

$$\phi_s := p \mid j \mid y_s \mid \neg\phi_s \mid \phi_s \vee \phi_s \mid \sigma(\phi_{s_1}, \ldots, \phi_{s_n})_s \mid @_k^s\phi_t \mid \forall x_t\, \phi_s$$

*Here, $p \in \mathrm{PROP}_s$, $j \in \mathrm{NOM}_s \cup N_s$, $t \in S$, $k \in \mathrm{NOM}_t \cup N_t$, $x \in \mathrm{SVAR}_t$, $y \in \mathrm{SVAR}_s$ and $\sigma \in \Sigma_{s_1 \cdots s_n, s}$. For any $\sigma \in \Sigma_{s_1 \ldots s, s}$, the dual formula $\sigma^{\Box}(\phi_1, \ldots, \phi_n)$ is defined as in Sect. 2. We also define the dual binder $\exists$: for any $s, t \in S$, if $\phi$ is a formula of sort $s$ and $x$ is a state variable of sort $t$, then $\exists x\, \phi := \neg\forall x\, \neg\phi$ is a formula of sort $s$. The notions of* free state variables *and* bound state variables *are defined as usual. For any $s \in S$, the set of all formulas of sort $s$ is denoted* $\mathrm{FORM}_s$.

*Remark 2 (Expressivity).* As a departure from our sources of inspiration, we only defined the satisfaction operators $@_j$ for nominals, and not for state variables. Hence, $@_x$ is not a valid formula in our logic. Our reason was to keep the system as "simple" as possible, but strong enough to overcome the problems encountered in the non-hybrid setting (see Remarks 1). More issues concerning expressivity are analyzed in Sect. 5.

One important remark is the definition of the satisfaction modalities: *if $k$ and $\phi$ are a nominal and a formula both of the sort $t \in S$, then we define a family of satisfaction operators $\{@_k^s\phi\}_{s\in S}$* such that $@_k^s\phi$ is a formula of sort $s$ for any $s \in S$. This means that $\phi$ is true at the world denoted by $k$ on the sort $t$ and is acknowledged on any sort $s \in S$. For example, if $j$ and $k$ are nominals of sort $t$ and $s \neq t$ the formula $@_j^s\neg k$ expresses the fact that at any world of sort $s$ we know that the worlds of sort $t$ named by $j$ and $k$ are different. So, our sorted worlds are not isolated any more, both from a syntactic and a semantic point of view.

**Definition 5.** *If $\mathbf{\Sigma} = (S, \Sigma, N)$ then a* $\mathbf{\Sigma}$-frame *is $\mathcal{F} = (W, (R_\sigma)_{\sigma\in\Sigma}, N^{\mathcal{F}})$ where $(W, (R_\sigma)_{\sigma\in\Sigma})$ is an $(S, \Sigma)$-frame and $N^{\mathcal{F}} = (N_s^{\mathcal{F}})_{s\in S}$ and for any $s \in S$*

$N_s^{\mathcal{F}} = (w^c)_{c \in N_s} \subseteq W_s$. *We will further assume that distinct constant nominals have distinct sorts, so we shall simply write* $N^{\mathcal{F}} = (w^c)_{c \in N}$.

Let $\mathbf{\Sigma} = (S, \Sigma, N)$ be a many-sorted signature with nominal constants and let $\mathcal{F}$ be a $\mathbf{\Sigma}$-frame. A *model* (based on $\mathcal{F}$) is a pair $\mathcal{M} = (\mathcal{F}, V)$ such that $V : \mathrm{PROP} \cup \mathrm{NOM} \to \mathcal{P}(W)$ is an $S$-sorted function such that $V_s(j)$ is a singleton set for any $s \in S$ and $j \in \mathrm{NOM}_s$. In order to define the semantics for $\mathcal{H}_{\mathbf{\Sigma}}(@, \forall)$ more is needed. Given a model $\mathcal{M} = (W, (R_\sigma)_{\sigma \in \Sigma}, (w^c)_{c \in N}, V)$, an *assignment* is an $S$-sorted function $g : \mathrm{SVAR} \to W$. If $g$ and $g'$ are assignment functions, $s \in S$ and $x \in \mathrm{SVAR}_s$, then we say that $g'$ is an *$x$-variant* of $g$ (and we write $g' \stackrel{x}{\sim} g$) if $g_t = g'_t$ for $t \neq s \in S$ and $g_s(y) = g'_s(y)$ for any $y \in \mathrm{SVAR}_s$, $y \neq x$. Moreover, if $V$ is an $S$-sorted evaluation, we define $V^N : \mathrm{PROP} \cup \mathrm{NOM} \cup N \to \mathcal{P}(W)$ by $V_s^N(c) = \{w^c\}$ for any $s \in S, c \in N_s$ and $V_s^N(v) = V_s(v)$ otherwise.

**Definition 6 (The satisfaction relation in $\mathcal{H}_{\mathbf{\Sigma}}(@, \forall)$).** *In the sequel*

$$\mathcal{M} = (W, (R_\sigma)_{\sigma \in \Sigma}, (w^c)_{c \in N}, V)$$

*is a model and* $g : \mathrm{SVAR} \to W$ *an $S$-sorted assignment. The satisfaction relation is defined as follows for any sort $s \in S$:*

- $\mathcal{M}, g, w \stackrel{s}{\models} a$, *if and only if* $w \in V_s^N(a)$, *where* $a \in \mathrm{PROP}_s \cup \mathrm{NOM}_s \cup N_s$,
- $\mathcal{M}, g, w \stackrel{s}{\models} x$, *if and only if* $w = g_s(x)$, *where* $x \in \mathrm{SVAR}_s$,
- $\mathcal{M}, g, w \stackrel{s}{\models} \neg\phi$, *if and only if* $\mathcal{M}, g, w \stackrel{s}{\not\models} \phi$ *where* $\phi \in \mathrm{FORM}_s$,
- $\mathcal{M}, g, w \stackrel{s}{\models} \phi \vee \psi$, *if and only if* $\mathcal{M}, g, w \stackrel{s}{\models} \phi$ *or* $\mathcal{M}, g, w \stackrel{s}{\models} \psi$ *where* $\phi, \psi \in \mathrm{FORM}_s$,
- *if* $\sigma \in \Sigma_{s_1 \ldots s_n, s}$ *then* $\mathcal{M}, g, w \stackrel{s}{\models} \sigma(\phi_1, \ldots, \phi_n)$, *if and only if there is* $(w_1, \ldots, w_n) \in W_{s_1} \times \cdots \times W_{s_n}$ *such that* $R_\sigma w w_1 \ldots w_n$ *and* $\mathcal{M}, g, w_i \stackrel{s_i}{\models} \phi_i$ *for any* $i \in [n]$,
- $\mathcal{M}, g, w \stackrel{s}{\models} @_k^s \psi$ *if and only if* $\mathcal{M}, g, u \stackrel{t}{\models} \psi$ *where* $k \in \mathrm{NOM}_t \cup N_t$, $\psi$ *has the sort* $t$ *and* $V_t^N(k) = \{u\}$,
- $\mathcal{M}, g, w \stackrel{s}{\models} \forall x\, \phi$, *if and only if* $\mathcal{M}, g', w \stackrel{s}{\models} \phi$ *for all* $g' \stackrel{x}{\sim} g$ *where* $\phi \in \mathrm{FORM}_s$, $x \in \mathrm{SVAR}_t$ *for some* $t \in S$.
  *Consequently,*
- $\mathcal{M}, g, w \stackrel{s}{\models} \exists x\, \phi$, *if and only if* $\exists g'(g' \stackrel{x}{\sim} g$ *and* $\mathcal{M}, g', w \stackrel{s}{\models} \phi)$.

Following the mono-sorted setting, satisfiability in $\mathcal{H}(@, \forall)$ is defined as follows: a formula $\phi$ of sort $s \in S$ is *satisfiable* if $\mathcal{M}, g, w \stackrel{s}{\models} \phi$ for some model $\mathcal{M}$, some assignment $g$ and some $w \in W_s$. Consequently, the formula $\phi$ is *valid* in a model $\mathcal{M}$ if $\mathcal{M}, g, w \stackrel{s}{\models} \phi$ for any assignment $g$ and any $w \in W_s$. One can speak about validity in a frame as in Sect. 2. In the presence of nominals, we can speak about *named models* and *pure formulas*, as in [7, Section 7.3].

**Definition 7 (Named models and pure formulas).** *A formula is* pure *if it does not contain propositional variables. A* pure instance *of a pure formula is obtained by uniformly substituting nominals for nominals of the same sort. We say that a formula is* $\forall\exists$-pure *if it is* pure *or it has the form*

$\forall x_1 \ldots \forall x_n \exists y_1 \ldots \exists y_m \psi$, *where* $\psi$ *contains no propositional variables and the only state symbols from* $\psi$ *are in* $\{x_1, \ldots, x_n, y_1, \ldots, y_m\}$.

*A model* $\mathcal{M} = (W, (R_\sigma)_{\sigma \in \Sigma}, (w^c)_{c \in N}, V)$ *is* named *if for any sort* $s \in S$ *and world* $w \in W_s$ *there exists* $k \in \mathrm{NOM}_s \cup N_s$ *such that* $\{w\} = V^N_s(k)$.

Note that a model is named if any world (state, point) is named by a nominal or a nominal constant (which means that it can be referred at a syntactic level). As in the mono-sorted case, pure formulas and named models are important since they give rise to strong completeness results with respect to the class of frames they define. Can we prove a similar result for the system $\mathcal{H}_\Sigma(@, \forall)$ when state variables are involved? We give a positive answer to this question, inspired by the discussion on existential saturation rules from [3, Lemma 1]. In order to do this, we define $\forall\exists$-*pure* formulas and we characterize frame satisfiability for such formulas. Consequently, Proposition 1 will lead to completeness results with respect to frame validity.

**Proposition 1 (Pure formulas in $\mathcal{H}_\Sigma(@, \forall)$).** *Let* $\mathcal{M}$ *be a named model where* $\mathcal{M} = (W, (R_\sigma)_{\sigma \in \Sigma}, (w^c)_{c \in N}, V)$, $\mathcal{F} = (W, (R_\sigma)_{\sigma \in \Sigma}, (w^c)_{c \in N})$ *the corresponding frame and* $\phi$ *a* $\forall\exists$*-pure formula of sort* $s$. *Then* $\mathcal{F} \models^s \phi$ *if and only if* $\mathcal{M} \models^s \phi$.

**The system $\mathcal{H}_\Sigma(@, \forall)$**

- The axioms and the deduction rules of $\mathcal{K}_\Sigma$
- Axiom schemes: any formula of the following form is an axiom, where $s, s', t$ are sorts, $\sigma \in \Sigma_{s_1 \cdots s_n, s}$, $\phi, \psi, \phi_1, \ldots, \phi_n$ are formulas (when necessary, their sort is marked as a subscript), $j, k$ are nominals or constant nominals, and $x$, $y$ are state variables:

$(K@)$ $@^s_j(\phi_t \to \psi_t) \to (@^s_j\phi \to @^s_j\psi)$ $\quad$ $(Agree)$ $@^t_k @^{t'}_j \phi_s \leftrightarrow @^t_j \phi_s$

$(SelfDual)$ $@^s_j \phi_t \leftrightarrow \neg @^s_j \neg \phi_t$ $\quad$ $(Intro)$ $j \to (\phi_s \leftrightarrow @^s_j \phi_s)$

$(Back)$ $\sigma(\ldots, \phi_{i-1}, @^{s_i}_j \psi_t, \phi_{i+1}, \ldots)_s \to @^s_j \psi_t$ $\quad$ $(Ref)$ $@^s_j j_t$

$(Q1)$ $\forall x\, (\phi \to \psi) \to (\phi \to \forall x\, \psi)$ where $\phi$ contains no free occurrences of x

$(Q2)$ $\forall x\, \phi \to \phi[y/x]$ where $y$ is substitutable for $x$ in $\phi$

$(Name)$ $\exists x\, x$

$(Barcan)$ $\forall x\, \sigma^\Box(\phi_1, \ldots, \phi_n) \to \sigma^\Box(\phi_1, \ldots, \forall x \phi_i, \ldots, \phi_n)$

$(Barcan@)$ $\forall x\, @_j \phi \to @_j \forall x\, \phi$

$(Nom\, x)$ $@_k x \wedge @_j x \to @_k j$

- Deduction rules:

$(BroadcastS)$ if $\models^s @^s_j \phi_t$ then $\models^{s'} @^{s'}_j \phi_t$

$(Gen@)$ if $\models^{s'} \phi$ then $\models^s @_j \phi$, where $j$ and $\phi$ have the same sort $s'$

$(Name@)$ if $\models^s @_j \phi$ then $\models^{s'} \phi$, where $j$ does not occur in $\phi$

$(Paste)$ if $\models^s @_j \sigma(\ldots, k, \ldots) \wedge @_k \phi \to \psi$ then $\models^s @_j \sigma(\ldots, \phi, \ldots) \to \psi$ where $k$ is distinct from $j$ that does not occur in $\phi$ or $\psi$

$(Gen)$ if $\models^s \phi$ then $\models^s \forall x \phi$ where $\phi \in Form_s$ and $x \in \mathrm{SVAR}_t$ for some $t \in S$.

Here, $j$ and $k$ are nominals or constant nominals having the appropriate sort.

**Fig. 2.** $(S, \Sigma)$ hybrid logic

We are ready now to define the deductive system of our logic, which is presented in Fig. 2.

Theorems and (local) deduction from hypothesis are defined as in Sect. 2. In order to further develop our framework, we need to analyze the *uniform substitutions*. Apart from being $S$-sorted, in the hybrid setting, more restrictions are required: state variables are uniformly replaced by state symbols that are *substitutable* for them (as in the mono-sorted setting [4]).

The system $\mathcal{H}_{\mathbf{\Sigma}}(@,\forall)$ is sound with respect to the intended semantics.

**Proposition 2 (Soundness).** *The deductive system for $\mathcal{H}_{\mathbf{\Sigma}}(@,\forall)$ from Fig. 2 is sound.*

The following lemma generalizes the results from [3], being essentially used in the proof of the completeness theorem.

**Lemma 1.** *1. The following formulas are theorems:*

(*Nom*) $@_k^s j \to (@_k^s \phi \leftrightarrow @_j^s \phi)$
*for any* $t \in S$, $k, j \in \mathrm{NOM}_t \cup N_t$ *and* $\phi$ *a formula of sort* $t$.

(*Sym*) $@_k^s j \to @_j^s k$
*where* $s \in S$ *and* $j, k \in \mathrm{NOM}_t \cup N_t$ *for some* $t \in S$,

(*Bridge*) $\sigma(\ldots \phi_{i_1}, j, \phi_{i+1} \ldots) \wedge @_j^s \phi \to \sigma(\ldots \phi_{i-1}, \phi, \phi_{i+1}, \ldots)$
*if* $\sigma \in \Sigma_{s_1 \ldots s_n, s}$, $j \in \mathrm{NOM}_{s_i} \cup N_{s_i}$ *and* $\phi$ *is a formula of sort* $s_i$.

*2. if* $\overset{s}{\vdash} \phi \to j$ *then* $\overset{t}{\vdash} \sigma(\ldots, \phi, \ldots) \to \sigma(\ldots, j, \ldots) \wedge @_j^t \phi$
*for any* $s, t \in S$, $\sigma \in \Sigma_{t_1 \cdots t_n, t}$, $j \in \mathrm{NOM}_s \cup N_s$ *and* $\phi$ *a formula of sort* $s$.

Let $\perp_s$ denote a formula of sort $s$ that is *nowhere true*. If $s \in S$ and $\Gamma_s$ is a set of formulas of sort $s$, then $\Gamma_s$ is *consistent* if $\Gamma_s \overset{s}{\not\vdash} \perp_s$. An *inconsistent* set of formulas is a set of formulas of the same sort that is not consistent. Maximal consistent sets are defined as usual.

In the rest of the section, we develop the proof of the strong completeness theorem for our hybrid logical systems, possibly extended with additional axioms. If $\Lambda$ is a set of formulas, we denote by $\mathcal{H}(@,\forall) + \Lambda$ the system obtained when the formulas of $\Lambda$ are seen as additional axiom schemes. The main steps are: the extended Lindenbaum Lemma, the construction of the Henkin model and the Truth Lemma (all of them extending the similar results in the mono-sorted case). In order to state our extended Lindenbaum Lemma, we need to define the *named, pasted and @-witnessed* sets of formulas. The following definition is technical, but its purpose is to define a set of conditions that allow us to prove the Truth Lemma (Lemma 4), a central result defining the satisfaction of a formula in the Henkin model (Definition 9), and a main step in the proof of the completeness theorem (Theorem 2).

**Definition 8 (Named, pasted and @-witnessed sets).** *Let* $s \in S$ *and* $\Gamma_s$ *be a set of formulas of sort* $s$ *from* $\mathcal{H}_{\mathbf{\Sigma}}(@,\forall)$. *We say that*

- $\Gamma_s$ *is* named *if one of its elements is a nominal or a constant nominal,*
- $\Gamma_s$ *is* pasted *if, for any* $t \in S$, $\sigma \in \Sigma_{s_1 \cdots s_n, t}$, $k \in \mathrm{NOM}_t \cup N_t$, *and* $\phi$ *a formula of sort* $s_i$, *whenever* $@_k^s \sigma(\ldots, \phi_{i-1}, \phi, \phi_{i+1}, \ldots) \in \Gamma_s$ *there exists a nominal* $j \in \mathrm{NOM}_{s_i}$ *such that* $@_k^s \sigma(\ldots, \phi_{i-1}, j, \phi_{i+1}, \ldots) \in \Gamma_s$ *and* $@_j^s \phi \in \Gamma_s$.

- $\Gamma_s$ *is* @-witnessed *if the following two conditions are satisfied:*
  (-) *for* $s', t \in S$ , $x \in \mathrm{SVAR}_t$, $k \in \mathrm{NOM}_{s'} \cup N_{s'}$ *and any formula* $\phi$ *of sort* $s'$, *whenever* $@_k^s \exists x\, \phi \in \Gamma_s$ *there exists* $j \in \mathrm{NOM}_t$ *such that* $@_k^s \phi[j/x] \in \Gamma_s$,
  (-) *for any* $t \in S$ *and* $x \in \mathrm{SVAR}_t$ *there is* $j_s \in \mathrm{NOM}_t$ *such that* $@_{j_x}^s x \in \Gamma_s$.

**Lemma 2 (Extended Lindenbaum Lemma).** *Let* $\Lambda$ *be a set of formulas in the language of* $\mathcal{H}_{\mathbf{\Sigma}}(@, \forall)$ *and* $s \in S$. *Then any consistent set* $\Gamma_s$ *of formulas of sort* $s$ *from* $\mathcal{H}_{\mathbf{\Sigma}}(@, \forall) + \Lambda$ *can be extended to a named, pasted and @-witnessed maximal consistent set by adding countably many nominals to the language.*

We are now ready to define a Henkin model, see [1,3] for the mono-sorted hybrid modal logic.

**Definition 9 (The Henkin model).** *Let* $s \in S$ *and assume* $\Gamma_s$ *is a maximal consistent set of formulas of sort* $s$ *from* $\mathcal{H}_{\mathbf{\Sigma}}(@, \forall)$. *For any* $t \in S$ *and any* $j \in \mathrm{NOM}_t \cup N_t$, *we define* $|j| = \{k \in \mathrm{NOM}_t \cup N \mid @_j^s k \in \Gamma_s\}$. *The* Henkin model *is* $\mathcal{M}^{\Gamma_s} = (W^\Gamma, (R_\sigma^\Gamma)_{\sigma \in \Sigma}, (|c|)_{c \in N}, V^\Gamma)$ *where*

$$W_t^\Gamma = \{|j| \mid j \in \mathrm{NOM}_t \cup N_t\} \text{ for any } t \in S$$
$$(|j|, |j_1|, \ldots, |j_n|) \in R_\sigma^\Gamma \text{ iff } @_j^s \sigma(j_1, \ldots, j_n) \in \Gamma_s \text{ for any } \sigma \in \Sigma_{t_1 \cdots t_n, t}$$
$$V_t^\Gamma(p) = \{|j| \mid j \in \mathrm{NOM}_t \cup N_t, @_j^s p \in \Gamma_s\} \text{ for any } t \in S \text{ and } p \in \mathrm{PROP}_t$$
$$V_t^\Gamma(j) = \{|j|\} \text{ for any } t \in S \text{ and } j \in \mathrm{NOM}_t.$$

*Under the additional assumption that* $\Gamma_s$ *is @-witnessed, we define the assignment* $g^\Gamma : \mathrm{SVAR} \to W^\Gamma$ *by*

$$g_t^\Gamma(x) = |j| \text{ where } t \in S,\ x \in \mathrm{SVAR}_t \text{ and } j \in \mathrm{NOM}_t \text{ such that } @_j^s x \in \Gamma_s.$$

**Lemma 3.** *The Henkin model from Definition 9 is well-defined.*

**Lemma 4 (Truth Lemma).** *Let* $s \in S$ *and assume* $\Gamma_s$ *is a named, pasted and @-witnessed maximal consistent set of formulas of sort* $s$ *from* $\mathcal{H}_{\mathbf{\Sigma}}(@, \forall)$. *For any sort* $t \in S$, $j \in \mathrm{NOM}_t \cup N_t$ *and for any formula* $\phi$ *of sort* $t$, *we have* $\mathcal{M}^\Gamma, g^\Gamma, |j| \models^t \phi$ *iff* $@_j^s \phi \in \Gamma_s$.

We are ready now to prove the strong completeness theorem for the hybrid logic $\mathcal{H}_{\mathbf{\Sigma}}(@, \forall)$ extended with axioms from $\Lambda$. For a logic $\mathcal{L}$, the relation $\vdash^s_{\mathcal{L}}$ denotes the local deduction, the relation $\models^s_{Mod(\mathcal{L})}$ denotes the semantic entailment w.r.t. models satisfying all the axioms of $\mathcal{L}$, while $\models^s_{\mathcal{L}}$ denotes the semantic entailment w.r.t. frames satisfying all the axioms of $\mathcal{L}$.

**Theorem 2 (Completeness).**

1. Strong model-completeness. *Let* $\Lambda$ *be an* $S$*-sorted set of formulas and assume* $\Gamma_s$ *is a set of formulas of sort* $s$ *for some* $s \in S$. *If* $\Gamma_s$ *is a consistent set in* $\mathcal{L} = \mathcal{H}_{\mathbf{\Sigma}}(@, \forall) + \Lambda$ *then* $\Gamma_s$ *has a model that is also a model of* $\Lambda$. *Consequently, for a formula* $\phi$ *of sort* $s$, $\Gamma_s \models^s_{Mod(\mathcal{L})} \phi$ *iff* $\Gamma_s \vdash^s_{\mathcal{L}} \phi$.

2. Strong frame-completeness for pure extensions. *Let $\Lambda$ be an $S$-sorted set of $\forall\exists$-pure formulas and assume $\Gamma_s$ is a set of formulas of sort $s$ for some $s \in S$. If $\Gamma_s$ is a consistent set in $\mathcal{L} = \mathcal{H}_{\mathbf{\Sigma}}(@, \forall) + \Lambda$ then $\Gamma_s$ has a model based on a frame that validates every formula in $\Lambda$. For a formula $\phi$ of sort $s$, $\Gamma_s \overset{s}{\models}_{\mathcal{L}} \phi$ iff $\Gamma_s \overset{s}{\vdash}_{\mathcal{L}} \phi$.*

The following useful results can be easily proved semantically:

**Proposition 3.** *1. (Nominal Conjunction) For any formulas and any nominals of appropriate sorts, the following hold:*
*(i1)* $\sigma(\ldots, \phi_{i-1}, \phi_i, \phi_{i+1}, \ldots) \wedge @_k(\psi) \leftrightarrow \sigma(\ldots, \phi_{i-1}, \phi_i \wedge @_k(\psi), \phi_{i+1}, \ldots)$
*(i2)* $\sigma^{\Box}(\ldots, \phi_{i-1}, \phi_i, \phi_{i+1}, \ldots) \wedge @_k(\psi) \leftrightarrow$
$\sigma^{\Box}(\ldots, \phi_{i-1}, \phi_i \wedge @_k(\psi), \phi_{i+1}, \ldots) \wedge @_k(\psi)$
*2. If $\phi_1, \ldots \phi_n$ are formulas of appropriate sorts and $x$ is a state variable that does not occur in $\phi_j$ for any $j \neq i$ then:*
*(i3)* $\exists x \sigma^{\Box}(\ldots, \phi_{i-1}, \phi_i, \phi_{i+1}, \ldots) \to \sigma^{\Box}(, \ldots, \phi_{i-1}, \exists x \phi_i, \phi_{i+1}, \ldots)$

In the many-sorted setting one can wonder what happens if we have an $S$-sorted set of deduction hypothesis $\mathbf{\Gamma} = \{\Gamma_s\}_{s \in S}$. The following considerations hold for any of $\mathcal{H}_{\mathbf{\Sigma}}(@)$ and $\mathcal{H}_{\mathbf{\Sigma}}(@, \forall)$. Clearly, a model $\mathcal{M}$ is a model of $\mathbf{\Gamma}$ if $\mathcal{M} \overset{s}{\models} \gamma_s$ for any $s \in S$ and $\gamma_s \in \Gamma_s$ (in this case we write $\mathcal{M} \models \mathbf{\Gamma}$). Using the "broadcasting" properties of the $@_i$ operators, we define another syntactic consequence relation:

$\mathbf{\Gamma} \overset{s}{\mid\!\sim} \phi$ *iff there are* $s_1, \ldots, s_n \in S$, $j_1 \in \mathrm{NOM}_{s_1}, \ldots, j_n \in \mathrm{NOM}_{s_n}$ *and* $\gamma_1 \in \Gamma_{s_1}, \ldots, \gamma_n \in \Gamma_{s_n}$ *such that* $\overset{s}{\vdash} @^s_{j_1} \gamma_1 \wedge \cdots \wedge @^s_{j_n} \gamma_n \to \phi$.

**Proposition 4 ($\overset{s}{\mid\!\sim}$ soundness).** *Let $\mathbf{\Gamma}$ be an $S$-sorted set and $\phi$ a formula of sort $s \in S$. If $\mathbf{\Gamma} \overset{s}{\mid\!\sim} \varphi$ then $\mathcal{M} \models \mathbf{\Gamma}$ implies $\mathcal{M} \overset{s}{\models} \phi$ for any model $\mathcal{M}$.*

*Remark 3* **(The modal logic $\mathcal{H}_{\mathbf{\Sigma}}(@)$).** The formulas of $\mathcal{H}_{\mathbf{\Sigma}}(@)$ are:
$\phi_s := p \mid j \mid \neg\phi_s \mid \phi_s \vee \phi_s \mid \sigma(\phi_{s_1}, \ldots, \phi_{s_n})_s \mid @^s_k \phi_t$
The *deduction system* is defined as follows:

- The axioms and the deduction rules of $\mathcal{K}_{\mathbf{\Sigma}}$
- Axiom schemes: $(K@), (SelfDual), (Back), (Agree), (Intro), (Ref)$
- Deduction rules: $(BroadcastS)$, $(Gen@)$, $(Subst)$, $(Name@)$, $(Paste)$

The system $\mathcal{H}_{\mathbf{\Sigma}}(@)$ is sound and complete. We mention that one can prove strong model-completeness results and strong frame-completeness results for pure extensions as in Theorem 2. See [19] for more details and full proofs.

## 4 A SMC-like Language and a Hoare-Like Logic for It

To showcase the application of our logic into program verification, we have chosen to specify a state-machine, whose expressions have side effects and where Hoare-like semantics are known to be hard to use. Note that handling side effects in Hoare logic does not come naturally, requiring one to define a separate axiomatic semantics for expressions with a modified form of assertions in order to allow for expression results tracking (see e.g. [26, Section 4.3]).

In Fig. 3, we introduce the signature $\mathbf{\Sigma} = (S, \Sigma, N)$ of our logic as a context-free grammar (CFG) in a BNF-like form. We make use of the established equivalence between CFGs and algebraic signatures (see, e.g., [16]), by mapping non-terminals to sorts and CFG productions to operation symbols. Note that, due to non-terminal renamings (e.g., `AExp ::= Nat`), it may seem that our syntax relies on subsorting. However, this is done for readability reasons only. The renaming of non-terminals in syntax can be thought of as syntactic sugar for defining injection functions. For example, `AExp ::= Nat` can be thought of as `AExp ::= nat2Exp(Nat)`, and all occurrences of an integer term in a context, in which an expression is expected, could be wrapped by the `nat2Exp` function.

Our language is inspired by the *SMC machine* [21] which consists of a set of transition rules defined between configurations of the form $\langle S, M, C\rangle$, where $S$ is the *value stack* of intermediate results, $M$ represents the *memory*, mapping program identifiers to values, and $C$ is a *control stack* of commands representing the control flow of the program. Since our target is to extend Propositional Dynamic Logic (PDL) [15], we identify the control stack with the notion of *program* in dynamic logic, and use the ";" operator to denote stack composition. We define our formulas to stand for *configurations* of the form $\langle vs, mem\rangle$ comprising a value stack and a memory. Hence, the sorts *CtrlStack* and *Config* correspond to programs and formulas from PDL, respectively. Inspired by PDL, we use the dual modal operator $[\_]\_ : \mathit{CtrlStack} \times \mathit{Config} \to \mathit{Config}$ to assert that a configuration formula must hold after executing the commands in the control stack. The axioms defining the dynamic logic semantics of the SMC machine are then formulas of the form $cfg \to [\mathit{ctrl}]cfg'$ saying that a configuration satisfying $cfg$ must change to one satisfying $cfg'$ after executing $\mathit{ctrl}$. The usual operations of dynamic logic " ;, $\cup$, $^*$ " are defined accordingly [15, Chapter 5]. We depart from PDL with the definition of ? (test): in our setting, in order to take a decision, we test the top value of the value stack. Consequently, the signature of the test operator is $? : \mathit{Val} \to \mathit{CtrlStack}$.

A deductive system, that allows us to accomplish our goal, is defined in Fig. 3. In this way we define an expansion of $\mathcal{H}(@,\forall)$. Our definition is incomplete (e.g. we do not fully axiomatize the natural numbers), but one can see that, e.g. $N_{Bool} = \{\mathbf{true}, \mathbf{false}\}$. To simplify the presentation, we omit sort annotations in the sequel; these should be easily inferrable from the context.

*Remark 4.* Assume that $\Lambda$ contains all the axioms from Fig. 3 and denote $\mathcal{L} = \mathcal{H}(@,\forall) + \Lambda$. Then $\mathcal{L}$ is a many-valued hybrid modal system associated to our language, and all results from Sect. 3 applies in this case. In particular,

**Domains**

```
Nat ::=  natural numbers
Bool ::= true | false | Nat == Nat | Nat <= Nat
```

**Syntax**

```
Var ::=  program variables
AExp ::= Nat | Var | AExp + AExp
       | ++ Var
BExp ::= AExp <= AExp
Stmt ::= x := AExp
       | if BExp
         then Stmt
         else Stmt
       | while BExp do Stmt
       | skip
       | Stmt ; Stmt
```

**Semantics**

```
     Val ::= Nat | Bool
ValStack ::= nil
           | Val . ValStack
     Mem ::= empty | set(Mem, x, n)
CtrlStack ::= c(AExp)
           | c(BExp)
           | c(Stmt)
           | asgn(x)
           | plus   | leq
           | Val ?
           | c1 ; c2
  Config ::=  < ValStack, Mem >
```

**Domains axioms (incomplete)**

$(B1)$ **true**$\leftrightarrow \neg$ **false** $\qquad (I1)$ $@^{Nat}_{\mathbf{true}}(x == y) \to (x \leftrightarrow y)$

... $\qquad$ ...

**PDL-inspired axioms**

$(A\cup)$ $[\pi \cup \pi']\gamma \leftrightarrow [\pi]\gamma \wedge [\pi']\gamma$ $\qquad (A;)$ $[\pi;\pi']\gamma \leftrightarrow [\pi][\pi']\gamma$

$(A?)$ $\langle v \cdot vs, mem\rangle \to [v?]\,\langle vs, mem\rangle$ $\qquad (A\neg?)$ $\langle v \cdot vs, mem\rangle \wedge @_v(\neg v') \to [v'?]\bot$

$(A^*)$ $[\pi^*]\gamma \leftrightarrow \gamma \wedge [\pi][\pi^*]\gamma$ $\qquad (AInd)$ $\gamma \wedge [\pi^*](\gamma \to [\pi]\gamma) \to [\pi^*]\gamma$

Here, $\pi$, $\pi'$ are formulas of sort *CtrlStack* ("programs"), $\gamma$ is a formula of sort *Config* (the analogue of "formulas" from PDL), $v$ and $v'$ are state variables of sort *Var*, *vs* has the sort *ValStack* and *mem* has the sort *Mem*.

**SMC-inspired axioms**

$(CStmt)$ $c(s1;s2) \leftrightarrow c(s1);c(s2)$

$(Aint)$ $\langle vs, mem\rangle \to [c(n)]\,\langle n \cdot vs, mem\rangle$ where $n$ is an integer

$(Aid)$ $\langle vs, set(mem,x,n)\rangle \to [c(x)]\,\langle n \cdot vs, set(mem,x,n)\rangle$

$(A++)$ $\langle vs, set(mem,x,n)\rangle \to [c(++x)]\,\langle n+1 \cdot vs, set(mem,x,n+1)\rangle$

$(Dplus)$ $c(a1+a2) \leftrightarrow c(a1);c(a2);\mathbf{plus}$

$(Aplus)$ $\langle n2 \cdot n1 \cdot vs, mem\rangle \to [\mathbf{plus}]\,\langle (n1+n2) \cdot vs, mem\rangle$

$(Dleq)$ $c(a1 <= a2) \leftrightarrow c(a2);c(a1);\mathbf{leq}$

$(Aleq)$ $\langle n1 \cdot n2 \cdot vs, mem\rangle \to [\mathbf{leq}]\,\langle (n1 \leq n2) \cdot vs, mem\rangle$

$(Askip)$ $\gamma \to [c(\mathbf{skip})]\gamma$

$(Dasgn)$ $c(x := a) \leftrightarrow c(a);\mathbf{asgn}\ (x)$

$(Aasgn)$ $\langle n \cdot vs, mem\rangle \to [\mathbf{asgn}\ (x)]\,\langle vs, set(mem,x,n)\rangle$

$(Dif)$ $c(\mathbf{if}\, b\, \mathbf{then}\, s1\, \mathbf{else}\, s2) \leftrightarrow c(b);((\mathbf{true}\,?;c(s1)) \cup (\mathbf{false}\,?;c(s2)))$

$(Dwhile)$ $c(\mathbf{while}\ \ b\ do\ s) \leftrightarrow c(b);(\mathbf{true}?;c(s);c(b))^*;\mathbf{false}?$

**Memory consistency axioms**

$(AMem1)$ $set(set(mem,x,n),y,m) \leftrightarrow set(set(mem,y,m),x,n)$
where $x$ and $y$ are distinct

$(AMem2)$ $set(set(mem,x,n),x,m) \to set(mem,x,m)$

**Fig. 3.** Axioms defining an SMC-like programming language

the system enjoys strong model-completeness. Moreover, we can safely assume that $\Lambda$ contains only $\forall\exists$-pure formulas, so $\mathcal{L}$ is strongly complete w.r.t the class of frames satisfying $\Lambda$ by Theorem 2.

We present below several *Hoare-like rules of inference*. Note that they are provable from the PDL and language axioms.

**Proposition 5.** *The following rules are admissible:*

1. **Rules of Consequence**
   *If* $\vdash \phi \to [\alpha]\psi$ *and* $\vdash \psi \to \chi$ *then* $\vdash \phi \to [\alpha]\chi$.
   *If* $\vdash \phi \to [\alpha]\psi$ *and* $\vdash \chi \to \phi$ *then* $\vdash \chi \to [\alpha]\psi$.
2. **Rule of Composition, iterated**
   *If* $\phi_0 \to [\alpha_1]\phi_1, \ldots, \phi_{n-1} \to [\alpha_n]\phi_n$, *then* $\phi_0 \to [\alpha_1; \ldots; \alpha_n]\phi_n$.
3. **Rule of Conditional**
   *If $B$ is a formula of sort Bool, and $vs$, $mem$, $P$ are formulas of appropriate sorts such that*
   *(h1)* $\vdash \phi \to [c(b)](\langle B \cdot vs, mem\rangle \wedge P)$,
   *(h2)* $\vdash \langle vs, mem\rangle \wedge P \wedge @_{true}(B) \to [c(s1)]\chi$
   *(h3)* $\vdash \langle vs, mem\rangle \wedge P \wedge @_{false}(B) \to [c(s2)]\chi$
   *(h4)* $\vdash P \to [\alpha]P$ *for any* $\alpha$ *of sort CtrlStack,*
   *then* $\vdash \phi \to [c(\textbf{\textit{if}}\, b\ \textbf{\textit{then}}\ s1\ \textbf{\textit{else}}\ s2)]\chi$

Note that our Rule of Conditional requires two more hypotheses, (h1) and (h4) than the inspiring rule in Hoare-logic. (h1) is needed because language expressions are no longer identical to formulas and need to be evaluated; in particular this allows for expressions to have side effects. (h4) is useful to carry over extra conditions through the rule; note that (h4) holds for all $@_j\varphi$ formulas.

Similarly, the Rule of Iteration needs to take into account the evaluation steps required for evaluating the condition. Moreover, since assignment is now handled by a forwards-going operational rule, we require existential quantification over the invariant to account for the values of the program variables in the memory, and work with instances of the existentially quantified variables.

As before, one can see [19] for the full proofs of the subsequent results.

**Proposition 6 (Rule of Iteration).** *Let $B$, $vs$, $mem$, and $P$ be formulas with variables over $\mathbf{x}$, where $\mathbf{x}$ is a set of state variables. If there exist substitutions $\mathbf{x_{init}}$ and $\mathbf{x_{body}}$ for the variables of $\mathbf{x}$ such that:*

*(h1)* $\vdash \phi \to [c(b)](\langle B \cdot vs, mem\rangle \wedge P)[\mathbf{x_{init}}/\mathbf{x}]$,
*(h2)* $\vdash \langle vs, mem\rangle \wedge P \wedge @_{true}(B) \to [c(s); c(b)](\langle B \cdot vs, mem\rangle \wedge P)[\mathbf{x_{body}}/\mathbf{x}]$
*(h3)* $\vdash P \to [\alpha]P$ *for any formula* $\alpha$ *of sort CtrlStack*

*then* $\vdash \phi \to [c(\textbf{\textit{while}}\ b\ \textbf{\textit{do}}\ s)]\exists\mathbf{x}\,\langle vs, mem\rangle \wedge P \wedge @_{false}(B)$.

*Proving a Program Correct.* Let us now exhibit proving a program using the operational semantics and the Hoare-like rules above. Consider the program:

```
s := 0; i := 0;
while ++ i <= n do s := s + i;
```

Let *pgm* stand for the entire program. We want to prove that if the initial value of $n$ is any natural number, then the final value of $s$ is the sum of numbers from 1 to $n$. Formally,

$$\langle vs, set(mem, n, vn\rangle \rightarrow [c(pgm)]\ \langle vs, set(set(set(mem, n, vn\rangle\ , s, vn * (vn+1)/2), i, vn+1))$$

Let *Cnd* stand for $++i <= n$ and *Body* stand for $s := s+i$. By applying the axioms above we can decompose *pgm* as

$$c(pgm) \leftrightarrow c(0); \textbf{asgn}\ (s); c(0); \textbf{asgn}\ (i); c(\textbf{while}\ Cnd\ \textbf{do}\ Body)$$

Similarly, $c(Cnd) \leftrightarrow c(++i); c(n); \textbf{leq}$ and $c(Body) \leftrightarrow c(s); c(i); \textbf{plus}; \textbf{asgn}\ (s)$.

We have the following instantiations of the axioms:

$\langle vs, set(mem, n, vn)\rangle \rightarrow [c(0)]\ \langle 0 \cdot vs, set(mem, n, vn)\rangle$ *Aint*

$\langle 0 \cdot vs, set(mem, n, vn)\rangle \rightarrow [asgn(s)]\ \langle vs, set(set(mem, n, vn), s, 0)\rangle$ *Aasgn*

$\langle vs, set(set(mem, n, vn), s, 0)\rangle \rightarrow [c(0)]\ \langle 0 \cdot vs, set(set(mem, n, vn), s, 0)\rangle$ *Aint*

$\langle 0 \cdot vs, set(set(mem, n, vn\rangle\ , s, 0))$
$\rightarrow [asgn(i)]\ \langle vs, set(set(set(mem, n, vn), s, 0), i, 0)\rangle$ *Aasgn*

And by applying the Rule of Composition we obtain:

(1) $\langle vs, set(mem, n, vn)\rangle$
$\rightarrow [c(0); \textbf{asgn}\ (s); c(0); \textbf{asgn}\ (i)]\ \langle vs, set(set(set(mem, n, vn), s, 0), i, 0)\rangle$

We now want to apply the Rule of Iteration. First let us handle the condition. Similarly to the "stepping" sequence above, we can use instances of (A++), (Aid), (Aleq), and the Rule of Composition to chain them to obtain:

$\langle vs, set(set(set(mem, n, vn), s, 0), i, 0)\rangle$
$\rightarrow [c(Body)]\ \langle (1 \leq vn) \cdot vs, set(set(set(mem, s, 0), i, 1), n, vn)\rangle$

Let $\mathbf{x} = vi$, $B = vi \leq vn$, $vs = vs$, $mem = set(set(set(mem, s, (vi-1) * vi/2), i, vi), n, vn)$, $P = @_{\mathbf{true}}(vi \leq vn+1)$. For $\mathbf{x_{init}} = 1$ we have that $B[1/vi] = 1 \leq vn$, $mem[1/vi] = set(set(set(mem, s, (1-1) * 1/2), i, 1), n, vn)$, $P[1/vi] = @_{\mathbf{true}}(1 \leq vn+1)$. Using that $(1-1) * 1/2 \leftrightarrow 0$ and $1 \leq vn+1$ we obtain

(2) $\langle vs, set(set(set(mem, n, vn), s, 0), i, 0)\rangle \rightarrow [c(Cnd)](\langle B \cdot vs, mem\rangle \wedge P)[1/vi]$

Now, we can again use instances of (Aid), (Aid), (Aplus), (Aasgn), (AMem), (A++), (AId), (Aleq), and the Rule of Composition to derive

$\langle vs, set(set(set(mem, i, vi), n, vn), s, (vi-1) * vi/2)\rangle \rightarrow [c(Body); c(Cnd)]$
$\langle (vi+1 \leq vn) \cdot vs, set(set(set(mem, s, vi * (vi+1)/2, i, vi+1), n, vn)\rangle$

By applying equivalences between formulas on naturals, the above leads to

$\langle vs, set(set(set(mem, i, vi), n, vn), s, (vi-1) * vi/2)\rangle$
$\rightarrow [c(Body); c(Cnd)]\ \langle B \cdot vs, mem\rangle\ [vi+1/vi]$

Using Proposition 3 ($i2$) and the fact that $vi \leq vn \leftrightarrow vi+1 \leq vn+1$, we obtain

(3)$\langle B \cdot vs, mem\rangle \wedge P \wedge @_{\mathbf{true}}(B)$
$\rightarrow [c(Body); c(Cnd)](\langle B \cdot vs, mem\rangle \wedge P)[vi+1/vi]$

Now using the Rule of Iteration with (2) and (3) we derive that
$\langle vs, set(set(set(mem, n, vn), s, 0), i, 0)\rangle$
$\rightarrow [c(\textbf{while}\ Cnd\ \textbf{do}\ Body)]\exists vi.\ \langle B \cdot vs, mem\rangle \wedge P \wedge @_{\textbf{false}}(B)$
By arithmetic reasoning, $\vdash$ $(\textbf{false} \rightarrow vi \leq vn) \leftrightarrow (\textbf{true} \rightarrow vn + 1 \leq vi)$, hence $\vdash @_{\textbf{false}}(vi \leq vn) \leftrightarrow @_{\textbf{true}}(vn + 1 \leq vi)$. Moreover, $@_{\textbf{true}}(vn + 1 \leq vi) \wedge @_{\textbf{true}}(vi \leq vn+1) \leftrightarrow @_{\textbf{true}}(vn+1 \leq vi \wedge vi \leq vn+1)$ which by arithmetic reasoning is equivalent to $@_{\textbf{true}}(vi =_{Nat} vn + 1)$, which by (I1) is equivalent to $vi \leftrightarrow vn + 1$ which allows us to substitute $vi$ by $vn + 1$ and eliminate the quantification, leading to

$$\exists vi.\ \langle vs, mem\rangle \wedge P \wedge @_{\textbf{false}}(B) \leftrightarrow \langle vs, mem\rangle\, [vn + 1/vi],\ \text{hence,}$$

(4) $\langle vs, mem'\rangle \rightarrow [c(\textbf{while}\ Cnd\ \textbf{do}\ Body)]\, \langle vs, mem''\rangle$
where $mem'' = set(set(set(mem, s, vn * (vn + 1)/2), i, vn + 1), n, vn)$,
$mem' = set(set(set(mem, n, vn), s, 0), i, 0)$.
Using the Rule of Composition on (1) and (4) we obtain our goal.

## 5 Conclusions and Related Work

We defined a general many-sorted hybrid polyadic modal logic that is sound and complete with respect to the usual modal semantics. From a theoretical point of view, we introduced nominal constants and we restricted the application of the satisfaction operators to nominals alone. We proved that the system is sound and complete and we also investigated the completeness of its pure axiomatic expansions. Given a concrete language with a concrete SMC-inspired operational semantics, we showed how to define a corresponding (sound and complete) logical system and we also proved (rather general) results that allow us to perform Hoare-style verification. Our approach was to define the weakest system that allows us to reach our goals.

There is an abundance of research literature on hybrid modal logic, we refer to [1] for a comprehensive overview. Our work was mostly inspired by [3,5,12,13], where a variety of hybrid modal logics are studied in a mono-sorted setting. We need to make a comment on our system's expressivity: the strongest hybrid language employs both the existential binder and satisfaction operator for state variables (i.e. $@_x$ with $x \in$ SVAR). Our systems seems to be weaker, but the exact relation will be analyzed elsewhere.

Concerning hybrid modal systems in many-sorted setting, we refer to [8,10]. The system from [8] is built upon differential dynamic logic, while the one from [10] is equationally developed, does not have nominals and satisfaction operators, the strong completeness being obtained in the presence of a stronger operator called *definedness* (which is the modal global operator). Note that, when the satisfaction operator is defined on state variables, the global modality is definable in the presence of the universal binder. However, we only have the satisfaction operator defined on nominals, so, again, our system seems to be weaker.

There are many problems to be addressed in the future, both from theoretical and practical point of view. We should definitely analyze the standard translation

[6, Section 2.2] and clarify the issues concerning expressivity; we should study the Fischer-Ladner closure [15, Section 6.1] and analyze completeness w.r.t. standard models from the point of view of dynamic logic; of course we should analyze more practical examples and even employ automatic techniques.

To conclude, the analysis of hybrid modal logic in a many-sorted setting leads us to a general system, that is theoretically solid and practically flexible enough for our purpose. We were able to specify a programming language, to define its operational semantics and to perform Hoare-style verification, all within the same deductive system. Modal logic proved to be, once more, the right framework and in the future we hope to take full advantage of its massive development.

**Acknowledgement.** The authors wish to thank the anonymous reviewers whose comments and suggestions have led to an improved version of our work.

## A Proofs from Sect. 3

**Proposition 1 (Pure formulas in $\mathcal{H}_{\mathbf{\Sigma}}(@,\forall)$).** Let $\mathcal{M}$ be a named model where $\mathcal{M} = (W, (R_\sigma)_{\sigma\in\Sigma}, (w^c)_{c\in N}, V)$, $\mathcal{F} = (W, (R_\sigma)_{\sigma\in\Sigma}, (w^c)_{c\in N})$ the corresponding frame and $\phi$ a $\forall\exists$-pure formula of sort $s$. Then $\mathcal{F} \models^s \phi$ if and only if $\mathcal{M} \models^s \phi$.

*Proof* Let $\phi$ be a pure formula of sort $s$ and suppose $\mathcal{F} \not\models^s \phi$. Then there exist a valuation $V'$ and some state $w \in W_s$ in the model $\mathcal{M}' = (\mathcal{F}, V')$ such that $\mathcal{M}', w \not\models^s \phi$.

On each sort $s \in S$ we will notate $j_1^s, \dots, j_t^s$ all the nominals occurring in $\phi$. But because we are working in a named model, $V$ labels every state of any sort in $\mathcal{F}$ with a nominal of the same sort. Hence, on each sort $s \in S$ there exist $k_1^s, \dots, k_t^s$ nominals such that $V_s^N(j_1^s) = V_s'(k_1^s)$, … ,$V_s^N(j_t^s) = V_s'(k_t^s)$. Therefore, if $\mathcal{M}', w \not\models^s \phi$ and $\psi$ is obtained by substituting on each sort each nominal $j_i^s$ with the corresponding one $k_i^s$, then $\mathcal{M}, w \not\models^s \psi$.

But $\phi$ is a pure formula, and by substituting the nominals contained in the formula with other nominals of the same sort, the new instance it is also a pure formulas like $\psi$. Therefore, by hypothesis, we have $\mathcal{M}, v \models^s \psi$ for any $v \in W_s$. But also $w \in W_s$, hence $\mathcal{M}, w \models^s \psi$, and we have a contradiction.

Next, suppose $\mathcal{M} \models^s \forall x_1 \dots \forall x_n \exists y_1 \dots \exists y_n \phi$ where $y_1, \dots, y_n$ do not occur in $\phi$. Hence, for any $g$ and any $w$ of sort $s$, $\mathcal{M}, g, w \models^s \forall x_1 \dots \forall x_n \exists y_1 \dots \exists y_n \phi$ where $y_1, \dots, y_n$ do not occur in $\phi$. So, for any assignment $g' \overset{x_1,\dots,x_n}{\sim} g$ exists an assignment $g'' \overset{y_1,\dots,y_n}{\sim} g'$ such that $\mathcal{M}, g'', w \models^s \phi(x_1, \dots, x_n, y_1, \dots, y_n)$. Let $g'(x_i) = \{w_i\}$ and $g''(y_i) = \{w_i'\}$ for any $i \in [n]$. Because we work with named model, there exist nominals $k_i$ and $j_i$ such that $V_s^N(k_i) = \{w_i\}$ and $V_s^N(j_i) = \{w_i'\}$ for any $i \in [n]$. Therefore, we get for any $k_1, \dots, k_n$ exist $j_1, \dots, j_n$ such that $\mathcal{M}, g'', w \models^s \phi[k_1/x_1, \dots, k_n/x_n, j_1/y_1, \dots, j_n/y_n]$. But now we have a pure formula and the assignment function will not affect the satisfiability of the formula. Therefore, for any $k_1, \dots, k_n$ exist $j_1, \dots, j_n$ such that $\mathcal{F} \models^s \phi[k_1/x_1, \dots, k_n/x_n, j_1/y_1, \dots, j_n/y_n]$. Therefore, for any assignment $g$ and any $w$ of sort $s$ we have that for any $k_1, \ldots, k_n$ there exist $j_1, \ldots, j_n$ such that $\mathcal{M}', g, w \models^s \phi[k_1/x_1, \ldots, k_n/x_n, j_1/y_1, \ldots, j_n/y_n]$. We use the contrapositive of $(Q2)$ axiom to get that $\mathcal{M}', g, w \models^s \exists y_1 \ldots \exists y_n \phi[k_1/x_1, \ldots, k_n/x_n]$ and by Lemma 7 we get that for any assignment $g$ and any $w$ of sort $s$ we have that $\mathcal{M}', g, w \models^s \forall x_1, \ldots, \forall x_n \exists y_1 \ldots \exists y_n \phi$ if and only if $\mathcal{F} \models^s \forall x_1, \ldots, \forall x_n \exists y_1 \ldots \exists y_n \phi$. □

Before proceeding with the next results from Sect. 3, we need to prove some lemmas that are generalization of [4].

Nominals and constant nominals are always substitutable for state variables of the same sort. If $x$ and $z$ are state variables of the sort $s$, then we define:

- if $\phi \in \mathrm{PROP}_s \cup \mathrm{SVAR}_s \cup \mathrm{NOM}_s \cup N_s$, then $z$ is substitutable for $x$ in $\phi$,
- $z$ is substitutable for $x$ in $\neg\phi$ iff $z$ is substitutable for $x$ in $\phi$,
- $z$ is substitutable for $x$ in $\phi \vee \psi$ iff $z$ is substitutable for $x$ in $\phi$ and $\psi$,
- $z$ is substitutable for $x$ in $\sigma(\phi_1, \ldots, \phi_n)$ iff $z$ is substitutable for $x$ in $\phi_i$ for all $i \in [n]$,
- $z$ is substitutable for $x$ in $@_j^s \phi$ iff $z$ is substitutable for $x$ in $\phi$,
- $z$ is substitutable for $x$ in $\forall y\, \phi$ iff $x$ does not occur free in $\phi$, or $y \neq z$ and $z$ is substitutable for $x$ in $\phi$.

In the sequel, we will say that a substitution is *legal* if it perform only allowed replacements. If $\phi$ is a formula and $x$ is a state variable we denote by $\phi[z/x]$ the formula obtained by substituting $z$ for all free occurrences of $x$ in $\phi$ ($z$ must be a nominal, a constant nominal or a state variable substitutable for $x$).

**Lemma 5 (Agreement Lemma).** *Let $\mathcal{M}$ be a standard model. For all standard $\mathcal{M}$-assignments $g$ and $h$, all states $w$ in $\mathcal{M}$ and all formulas $\phi$ of sort $s \in S$, if $g$ and $h$ agree on all state variables occurring freely in $\phi$, then:*

$$\mathcal{M}, g, w \models^s \phi \textit{ iff } \mathcal{M}, h, w \models^s \phi$$

*Proof.* We suppose that $g$ and $h$ agree on all state variables occurring freely in $\phi$ on each sort. We prove this lemma by induction on the complexity of $\phi$:

- $\mathcal{M}, g, w \models^s a$ iff $a \in \mathrm{PROP_s} \cup \mathrm{NOM_s} \cup N_s$ we have $w \in V_s^N(a)$ iff $\mathcal{M}, h, w \models^s a$.
- $\mathcal{M}, g, w \models^s x$ iff $x \in \mathrm{SVAR_s}$ we have $w = g_s(x)$, but $g_s(x) = h_s(x)$, therefore $\mathcal{M}, h, w \models^s x$.
- $\mathcal{M}, g, w \models^s \neg\phi$ iff $\mathcal{M}, g, w \not\models^s \phi$. But, if $g$ and $h$ agree on all state variables occurring freely in $\neg\phi$, then same for $\phi$. Therefore, from the induction hypothesis, $\mathcal{M}, g, w \models^s \phi$ iff $\mathcal{M}, h, w \models^s \phi$. Then $\mathcal{M}, g, w \not\models^s \phi$ iff $\mathcal{M}, h, w \not\models^s \phi$. Then $\mathcal{M}, h, w \models^s \neg\phi$.
- $\mathcal{M}, g, w \models^s \phi \vee \psi$, iff $\mathcal{M}, g, w \models^s \phi$ or $\mathcal{M}, g, w \models^s \psi$. But, $g$ and $h$ agree on all state variables occurring freely in $\phi$ or $\psi$, then from induction hypothesis, we have ($\mathcal{M}, g, w \models^s \phi$ iff $\mathcal{M}, h, w \models^s \phi$) or ($\mathcal{M}, g, w \models^s \psi$ iff $\mathcal{M}, h, w \models^s \psi$). Then, ($\mathcal{M}, h, w \models^s \psi$ or $\mathcal{M}, h, w \models^s \psi$) iff $\mathcal{M}, h, w \models^s \phi \vee \psi$.

- $\mathcal{M}, g, w \models^s \sigma(\phi_1, \ldots, \phi_n)$ iff there is $(w_1, \ldots, w_n) \in W_{s_1} \times \cdots \times W_{s_n}$ such that $R_\sigma w w_1 \ldots w_n$ and $\mathcal{M}, g, w_i \models^{s_i} \phi_i$ for each $i \in [n]$, then, by induction hypothesis $\mathcal{M}, h, w_i \models^{s_i} \phi_i$ for each $i \in [n]$. Hence, we have that there is $(w_1, \ldots, w_n) \in W_{s_1} \times \cdots \times W_{s_n}$ such that $R_\sigma w w_1 \ldots w_n$ and $\mathcal{M}, h, w_i \models^{s_i} \phi_i$ for each $i \in [n]$ iff $\mathcal{M}, h, w \models^s \sigma(\phi_1, \ldots, \phi_n)$.
- $\mathcal{M}, g, w \models^s @^s_j \phi$ iff $\mathcal{M}, g, v \models^{s'} \phi$ where $V^N_{s'}(j) = \{v\}$ iff $\mathcal{M}, h, v \models^{s'} \phi$ where $V^N_{s'}(j) = \{v\}$ (induction hypothesis) iff $\mathcal{M}, h, w \models^s @^s_j \phi$.
- $\mathcal{M}, g, w \models^s \forall x \phi$ iff $\forall g'(g' \stackrel{x}{\sim} g$ implies $\mathcal{M}, g', w \models^s \phi)$. But $g$ and $h$ agree on all state variables occurring freely in $\forall x \phi$ and because $x$ is bounded, then $h_s(y) = g_s(y)$ for any $y \neq x$. Therefore, $\forall g'(g'_s(y) = g_s(y) = h_s(y)$ for any $y \neq x$ implies $\mathcal{M}, g', w \models^s \phi)$ equivalent with $\forall g'(g' \stackrel{x}{\sim} h$ implies $\mathcal{M}, h', w \models^s \phi)$ iff $\mathcal{M}, h, w \models^s \forall x \phi$. □

**Lemma 6 (Substitution Lemma).** *Let $\mathcal{M}$ be a standard model. For all standard $\mathcal{M}$-assignments $g$, all states $w$ in $\mathcal{M}$ and all formulas $\phi$, if $y$ is a state variable that is substitutable for $x$ in $\phi$ and $j$ is a nominal then:*

- $\mathcal{M}, g, w \models^s \phi[y/x]$ *iff* $\mathcal{M}, g', w \models^s \phi$ *where* $g' \stackrel{x}{\sim} g$ *and* $g'_s(x) = g_s(y)$
- $\mathcal{M}, g, w \models^s \phi[j/x]$ *iff* $\mathcal{M}, g', w \models^s \phi$ *where* $g' \stackrel{x}{\sim} g$ *and* $g'_s(x) = V^N_s(j)$

*Proof.* By induction on the complexity of $\phi$.

- $\phi = a$, $a \in \mathrm{PROP}_s \cup \mathrm{NOM}_s \cup N_s$. Then $a[y/x] = a$ and $\mathcal{M}, g, w \models^s a[y/x]$ if and only if $\mathcal{M}, g, w \models^s a$ if and only if $w \in V^N_s(a)$. But $g' \stackrel{x}{\sim} g$ and by Agreement Lemma $\mathcal{M}, g', w \models^s a$.
- $\phi = z$, where $z \in \mathrm{SVAR}_s$. We have two cases:
  1. If $z \neq x$, then $\mathcal{M}, g, w \models^s z[y/x]$ if and only if $\mathcal{M}, g, w \models^s z$ if and only if $\mathcal{M}, g', w \models^s z$ (Agreement Lemma).
  2. If $z = x$, then $\mathcal{M}, g, w \models^s z[y/x]$ if and only if $\mathcal{M}, g, w \models^s y$ if and only if $w \in g_s(y)$ if and only if $w \in g'_s(x)$ if and only if $w \in g'_s(z)$ if and only if $\mathcal{M}, g', w \models^s z$.
- $\phi = \neg\phi$, then $\mathcal{M}, g, w \models^s \neg\phi$ if and only if $\mathcal{M}, g, w \not\models^s \phi$ if and only if $\mathcal{M}, g', w \not\models^s \phi$ (inductive hypothesis) if and only if $\mathcal{M}, g', w \models^s \neg\phi$.
- $\phi = \phi \vee \psi$, then $\mathcal{M}, g, w \models^s (\phi \vee \psi)[y/x]$ if and only if $\mathcal{M}, g, w \models^s \phi[y/x]$ or $\mathcal{M}, g, w \models^s \psi[y/x]$ if and only if $\mathcal{M}, g', w \models^s \phi$ or $\mathcal{M}, g', w \models^s \psi$ (inductive hypothesis) if and only if $\mathcal{M}, g', w \models^s \phi \vee \psi$.
- $\phi = \sigma(\phi_1, \ldots, \phi_n)$, then $\mathcal{M}, g, w \models^s \sigma(\phi_1, \ldots, \phi_n)[y/x]$ if and only if $\mathcal{M}, g, w \models^s \sigma(\phi_1[y/x], \ldots, \phi_n[y/x])$ if and only if exists $(u_1, \ldots, u_n) \in W_{s_1} \times \ldots \times W_{s_n}$ such that $R_\sigma w u_1 \ldots u_n$ and $\mathcal{M}, g, u_i \models^{s_i} \phi_i[y/x]$ for any $i \in [n]$ if and only if there exists $(u_1, \ldots, u_n) \in W_{s_1} \times \ldots \times W_{s_n}$ such that $R_\sigma w u_1 \ldots u_n$ and $\mathcal{M}, g', u_i \models^{s_i} \phi_i$ for any $i \in [n]$ (inductive hypothesis) if and only if $\mathcal{M}, g', w \models^s \sigma(\phi_1, \ldots, \phi_n)$.
- $\phi = @^s_j \phi$, then $\mathcal{M}, g, w \models^s @^s_j \phi[y/x]$ if and only if $\mathcal{M}, g, v \models^s \phi[y/x]$ where $V^N_{s'}(j) = \{v\}$ if and only if $\mathcal{M}, g', v \models^{s'} \phi$ where $V^N_{s'}(j) = \{v\}$ (inductive hypothesis) if and only if $\mathcal{M}, g', w \models^s @^s_j \phi$.

- $\phi = \forall x\phi$, then $\mathcal{M}, g, w \models^{s} (\forall x\phi)[y/z]$ if and only if $\mathcal{M}, g, w \models^{s} (\forall x\phi)[y/z]$ if and only if $\mathcal{M}, g, w \models^{s} \forall x\phi$ if and only if $\mathcal{M}, g', w \models^{s} \forall x\phi$ (Agreement Lemma).

For the next case we will use the notation $g^{x\leftarrow y}$ to specify that $x$ is substituted by $y$, therefore, if $x$ if free in a formula, after substitution we will not have any more $x$.

*Claim 1 (1).* The following two statements are equivalent:

- For all $g'$, if $g' \overset{z}{\sim} g$ then $\mathcal{M}, g'^{x\leftarrow y}, w \models^{s} \phi$.
- For all $g'$, if $g' \overset{z}{\sim} g^{x\leftarrow y}$ then $\mathcal{M}, g', w \models^{s} \phi$.

*Proof.* Suppose for all $g'$, if $g' \overset{z}{\sim} g$ then $\mathcal{M}, g'^{x\leftarrow y}, w \models^{s} \phi$ and $g' \overset{z}{\sim} g^{x\leftarrow y}$. Since $g'_s(o) = g_s^{x\leftarrow y}(o)$ for any $o \neq z$ and $x \neq z$, then $g'_s(x) = g_s^{x\leftarrow y}(x) = g_s(y)$. Therefore, $g'_s = {g'_s}^{x\leftarrow y}$ and $g' = g'^{x\leftarrow y}$. Hence, $\mathcal{M}, g', w \models^{s} \phi$. Next, suppose for all $g'$, if $g' \overset{z}{\sim} g^{x\leftarrow y}$ then $\mathcal{M}, g', w \models^{s} \phi$ and $g' \overset{z}{\sim} g$. Therefore, $g_s'^{x\leftarrow y} \overset{z}{\sim} g_s^{x\leftarrow y}$, so $g'^{x\leftarrow y} \overset{z}{\sim} g^{x\leftarrow y}$. From second case, we have that $\mathcal{M}, g'^{x\leftarrow y}, w \models^{s} \phi$.

- $\phi = \forall z\phi$, where $z \neq x$. Suppose $\mathcal{M}, g, w \models^{s} (\forall z\phi)[y/x]$ iff $\mathcal{M}, g, w \models^{s} \forall z(\phi[y/x])$ iff for all $g'$, if $g' \overset{z}{\sim} g$ then $\mathcal{M}, g', w \models^{s} \phi[y/x]$ iff for all $g'$, if $g' \overset{z}{\sim} g$ then $\mathcal{M}, g'^{x\leftarrow y}, w \models^{s} \phi$ (induction hypothesis) iff or all $g'$, if $g' \overset{z}{\sim} g^{x\leftarrow y}$ then $\mathcal{M}, g', w \models^{s} \phi$ (Claim 1) iff $\mathcal{M}, g^{x\leftarrow y}, w \models^{s} \forall z\phi$ where $g'_s(x) = g(y)$ and $g' \overset{z}{\sim} g$ iff $\mathcal{M}, g', w \models^{s} \forall z\phi$ where $g'_s(x) = g_s(y)$ and $g' \overset{z}{\sim} g$ (Agreement Lemma).

For the second case, when substituting with a nominal, the proof is similar. □

**Lemma 7 (Generalization on nominals).** *Assume* $\mid\!\overset{s}{-}\phi[i/x]$ *where* $i \in \mathrm{NOM}_t$ *and* $x \in \mathrm{SVAR}_t$ *for some* $t \in S$. *Then there is a state variable* $y \in \mathrm{SVAR}_t$ *that does not appear in* $\phi$ *such that* $\mid\!\overset{s}{-}\phi[y/x]$

*Proof.* There are two cases. First, let us suppose that $x$ does not occur free in $\phi$, therefore $\phi[j/x]$ is identical to $\phi[y/x]$, hence as $\phi[j/x]$ is provable, so is $\forall y\phi[y/x]$ for any choice of $y$.

Secondly, suppose that $x$ occur free in $\phi$. Suppose $\phi[j/x]$. Hence we have a proof of $\phi[j/x]$ and we choose any variable $y$ that does not occur in the proof, or in $\phi$. We replace every occurrence of $j$ in the proof of $\phi[j/x]$ with $y$. It follows by induction on the length of proofs that this new sequence is a proof of $\phi[y/x]$. By generalization we extend the proof with $\forall y(\phi[y/x])$ and we can conclude that $\forall y(\phi[y/x])$ is provable. □

We are ready now to proceed with the proves from Sect. 3.

**Proposition 2 (Soundness).** The deductive systems for $\mathcal{H}_{\mathbf{\Sigma}}(@, \forall)$ from Fig. 2 is sound.

*Proof.* Let $\mathcal{M}$ be an arbitrary model and $w$ any state of sort $s$.

($K_@$) Suppose $\mathcal{M}, g, w \models^{s} @_j^s(\phi_t \to \psi_t)$ if and only if $\mathcal{M}, g, v \models^{t} \phi_t \to \psi_t$ where $V_t^N(j) = \{v\}$ iff $\mathcal{M}, g, v \models^{t} \phi_t$ implies $\mathcal{M}, g, v \models^{t} \psi_t$ where $V_t^N(j) = \{v\}$. Suppose

$\mathcal{M}, g, w \stackrel{s}{\models} @_j^s \phi_t$ and $V_t^N(j) = \{v\}$. Then $\mathcal{M}, g, v \stackrel{t}{\models} \phi_t$ where $V_t^N(j) = \{v\}$ , but this implies that $\mathcal{M}, g, v \stackrel{t}{\models} \psi_t$ where $V_t^N(j) = \{v\}$ iff $\mathcal{M}, g, w \stackrel{s}{\models} @_j^s \psi_t$.

(*Agree*) Suppose $\mathcal{M}, g, w \stackrel{t'}{\models} @_k^{t'} @_j^t \phi_s$ iff $\mathcal{M}, g, v \stackrel{t}{\models} @_j^t \phi_s$ where $V_t^N(k) = \{v\}$ iff $\mathcal{M}, g, u \stackrel{s}{\models} \phi_s$ where $V_t^N(k) = \{v\}$ and $V_s^N(j) = \{u\}$. Then $\mathcal{M}, g, u \stackrel{s}{\models} \phi_s$ where $V_s^N(j) = \{u\}$ which implies that $\mathcal{M}, g, w \stackrel{t'}{\models} @_j^{t'} \phi_s$.

(*SelfDual*) Suppose $\mathcal{M}, g, w \stackrel{s}{\models} \neg @_j^s \neg \phi_t$ iff $\mathcal{M}, g, w \stackrel{s}{\not\models} @_j^s \neg \phi_t$ iff $\mathcal{M}, g, v \stackrel{t}{\not\models} \neg \phi_t$ where $V_t^N(j) = \{v\}$ iff $\mathcal{M}, g, v \stackrel{t}{\models} \phi_t$ where $V_t^N(j) = \{v\}$ iff $\mathcal{M}, g, w \stackrel{s}{\models} @_j^s \phi_t$.

(*Back*) Suppose $\mathcal{M}, g, w \stackrel{s}{\models} \sigma(\ldots, \phi_{i-1}, @_j^{s_i} \psi_t, \phi_{i+1}, \ldots)_s$ if and only if there is $(w_1, \ldots, w_n) \in W_{s_1} \times \cdots \times W_{s_n}$ such that $R_\sigma w w_1 \ldots w_n$ and $\mathcal{M}, g, w_i \stackrel{s_i}{\models} \phi_i$ for any $i \in [n]$. This implies that there is $w_i \in W_{s_i}$ such that $\mathcal{M}, g, w_i \stackrel{s_i}{\models} @_j^{s_i} \psi_t$, then $\mathcal{M}, g, v \stackrel{t}{\models} \psi_t$ where $V_t^N(j) = \{v\}$. Hence, $\mathcal{M}, g, w \stackrel{s}{\models} @_j^s \psi_t$

(*Ref*) Suppose $\mathcal{M}, g, w \stackrel{s}{\not\models} @_j^s j_t$. Then $\mathcal{M}, g, v \stackrel{t}{\not\models} j$ where $V_t^N(j) = \{v\}$, contradiction.

(*Intro*) Suppose $\mathcal{M}, g, w \stackrel{s}{\models} j$ and $\mathcal{M}, g, w \stackrel{s}{\models} \phi_s$. Then $V_s^N(j) = \{w\}$ and $\mathcal{M}, g, w \stackrel{s}{\models} \phi_s$ implies that $\mathcal{M}, g, w \stackrel{s}{\models} @_j^s \phi_s$. Now, suppose $\mathcal{M}, g, w \stackrel{s}{\models} j$ and $\mathcal{M}, g, w \stackrel{s}{\models} @_j^s \phi_s$. Because, from the first assumption, we have $V_s^N(j) = \{w\}$, then, form the second one, we can conclude that $\mathcal{M}, g, w \stackrel{s}{\models} \phi_s$.

($Q1$) Suppose that $\mathcal{M}, g, w \stackrel{s}{\models} \forall x(\phi \to \psi)$ iff $\mathcal{M}, g', w \stackrel{s}{\models} \phi \to \psi$ for all $g' \stackrel{x}{\sim} g$. Results that for all $g' \stackrel{x}{\sim} g$ we have $\mathcal{M}, g', w \stackrel{s}{\models} \phi$ implies $\mathcal{M}, g', w \stackrel{s}{\models} \psi$. But $\phi$ contains no free occurrences of $x$, then for all $g' \stackrel{x}{\sim} g$ we have ($\mathcal{M}, g, w \stackrel{s}{\models} \phi$ implies $\mathcal{M}, g', w \stackrel{s}{\models} \psi$). Hence, $\mathcal{M}, g, w \stackrel{s}{\models} \phi$ implies that, for all $g' \stackrel{x}{\sim} g$, $\mathcal{M}, g', w \stackrel{s}{\models} \psi$. Then, $\mathcal{M}, g, w \stackrel{s}{\models} \phi$ implies that $\mathcal{M}, g, w \stackrel{s}{\models} \forall \psi$ iff $\mathcal{M}, g, w \stackrel{s}{\models} \phi \to \forall x \psi$.

($Q2$) Suppose that $\mathcal{M}, g, w \stackrel{s}{\models} \forall x \phi$. We need to prove that $\mathcal{M}, g', w \stackrel{s}{\models} \phi[y/x]$. But this is equivalent, by Substitution Lemma, with proving that $\mathcal{M}, g', w \stackrel{s}{\models} \phi$ where $g' \stackrel{x}{\sim} g$ and $g'_s(x) = g_s(y)$. But $\mathcal{M}, g, w \stackrel{s}{\models} \forall x \phi$ iff $\mathcal{M}, g', w \stackrel{s}{\models} \phi$ for all $g' \stackrel{x}{\sim} g$. Let $g'_s(z) = g_s(y)$, if $z = x$, and $g'_s(z) = g_s(z)$, otherwise. Therefore, we have $g' \stackrel{x}{\sim} g$ , $g'_s(x) = g_s(y)$ and $\mathcal{M}, g', w \stackrel{s}{\models} \phi$. For the case of substituting with a nominal is similar. We define $g'_s(x) = V_s^N(j)$, if $z = x$, and $g'_s(z) = g_s(z)$, otherwise.

(*Name*) Suppose that $\mathcal{M}, g, w \stackrel{s}{\models} \exists x x$ iff exists $g' \stackrel{x}{\sim} g$ and $\mathcal{M}, g', w \stackrel{s}{\models} x$. We choose $g'$ an $x$-variant of $g$ such that $g'_s(x) = \{w\}$.

(*Barcan*) Suppose $\mathcal{M}, g, w \stackrel{s}{\models} \forall x\, \sigma^\square(\phi_1, \ldots, \phi_n)$ then for all $g' \stackrel{x}{\sim} g$, and for all $w_i \in W_{s_i}$, $i \in [n]$, $R_\sigma w w_1 \ldots w_n$ implies $\mathcal{M}, g', w_i \stackrel{s_i}{\models} \phi_i$ for all $i \in [n]$. But $g$ and $g'$ agree on all state variables occurring freely. Therefore, for all $w_i \in W_{s_i}$, $i \in [n]$, $R_\sigma w w_1 \ldots w_n$ and all $g' \stackrel{x}{\sim} g$ , we have $\mathcal{M}, g, w_i \stackrel{s_i}{\models} \phi_i$ for all $i \in [n]$ and $i \neq l$ and $\mathcal{M}, g', w_l \stackrel{s_l}{\models} \phi_l$. Hence, for the $l$-th argument, we have $\mathcal{M}, g, w_l \stackrel{s_l}{\models} \forall x \phi_l$. So, $\mathcal{M}, g, w \stackrel{s}{\models} \sigma^\square(\phi_1, \ldots, \forall x \phi_l \ldots \phi_n)$.

(*Barcan@*) Suppose $\mathcal{M}, g, w \stackrel{s}{\models} \forall x @_j^s \phi$ iff $\mathcal{M}, g', w \stackrel{s}{\models} @_j^s \phi$ for all $g' \stackrel{x}{\sim} g$. Then, $\mathcal{M}, g', v \stackrel{t}{\models} \phi$ for all $g' \stackrel{x}{\sim} g$ where $V_t^N(j) = \{v\}$ and so $\mathcal{M}, g, v \stackrel{t}{\models} \forall x \phi$ where $V_t^N(j) = \{v\}$. Hence, $\mathcal{M}, g, w \stackrel{s}{\models} @_j^s \forall x \phi$.

($Nom\ x$) Suppose $\mathcal{M}, g, w \models^{s} @_j^s x$ and $\mathcal{M}, g, w \models^{s} @_k^s x$. Then $\mathcal{M}, g, v \models^{t} x$ where $V_t^N(j) = \{v\}$ and $\mathcal{M}, g, u \models^{t} x$ where $V_t^N(k) = \{u\}$. This implies that $u = v$, so $V_t^N(j) = V_t^N(k)$. Then $\mathcal{M}, g, w \models^{s} @_j^s k$ for any model $\mathcal{M}$ and any world $w$.

($BroadcastS$) Suppose $\mathcal{M}, g, w \models^{s} @_j^s \phi_t$ if and only if $\mathcal{M}, g, v \models^{t} \phi_t$ where $V_t^N(j) = \{v\}$. Hence, for any $s' \in S$ we have $\mathcal{M}, g, w \models^{s'} @_j^{s'} \phi_t$.

Now, let $\mathcal{M}$ be an arbitrary named model.

($Name@$) Suppose $\mathcal{M}, g, w \models^{s} @_j^s \phi$ iff $\mathcal{M}, g, v \models^{s'} \phi$ where $V_{s'}^N(j) = \{v\}$, but we work in named models, therefore, in any model $\mathcal{M}$ there exist $v$ and $j$ where $V_{s'}^N(j) = \{v\}$ and this implies $\mathcal{M}, g, v \models^{s'} \phi$.

($Paste$) Suppose $\mathcal{M}, g, w \models^{s} @_j^s \sigma(\psi_1, \ldots, \psi_{i-1}, k, \psi_{i+1}, \ldots, \psi_n) \wedge @_k^s \phi \to \psi$ iff $\mathcal{M}, g, w \models^{s} @_j^s \sigma(\psi_1, \ldots, \psi_{i-1}, k, \psi_{i+1}, \ldots, \psi_n)$ and $\mathcal{M}, g, w \models^{s} @_k^s \phi$ implies $\mathcal{M}, g, w \models^{s} \psi$. Hence, $\mathcal{M}, g, v \models^{s'} \sigma(\psi_1, \ldots, \psi_{i-1}, k, \psi_{i+1}, \ldots, \psi_n)$ where $V_{s'}^N(j) = \{v\}$ iff exists $(v_1, \ldots, v_n) \in W_{s_1} \times \ldots \times W_{s_n}$ such that $R_\sigma v v_1 \ldots v_i \ldots v_n$ where $V_{s'}^N(j) = \{v\}$ and $\mathcal{M}, g, v_e \models^{s'} \psi_e$ for any $e \in [n], e \neq i$ and $\mathcal{M}, g, v_i \models^{s_i} k$ iff $V_{s_i}^N(k) = \{v_i\}$. If $\mathcal{M}, g, w \models^{s} @_k^s$ and $V_{s_i}^N(k) = \{v_i\}$, then $\mathcal{M}, g, v_i \models^{s_i} \phi$.

Then, if there exists $(v_1, \ldots, v_n) \in W_{s_1} \times \ldots \times W_{s_n}$ such that $R_\sigma v v_1 \ldots v_i \ldots v_n$ where $V_{s'}^N(j) = \{v\}$ and $\mathcal{M}, g, v_e \models^{s'} \psi_e$ for any $e \in [n], e \neq i$ and $\mathcal{M}, g, v_i \models^{s_i} \phi$, these imply $\mathcal{M}, g, w \models^{s} \psi$. So, $\mathcal{M}, g, v \models^{s'} \sigma(\psi_1, \ldots, \psi_{i-1}, \phi, \psi_{i+1}, \ldots, \psi_n)$ where $V_{s'}^N(j) = \{v\}$ implies $\mathcal{M}, g, w \models^{s} \psi$.

In conclusion, $\mathcal{M}, g, w \models^{s'} @_j^s \sigma(\psi_1, \ldots, \psi_{i-1}, \phi, \psi_{i+1}, \ldots, \psi_n) \to \psi$. □

In the sequel, by PL we mean classical propositional logic and by ML we mean the basic modal logic.

**Lemma 1.**

1. The following formulas are theorems:
   ($Nom$) $@_k^s j \to (@_k^s \phi \leftrightarrow @_j^s \phi)$
   for any $t \in S$, $k, j \in \mathrm{NOM}_t \cup N_t$ and $\phi$ a formula of sort $t$.
   ($Sym$) $@_k^s j \to @_j^s k$
   where $s \in S$ and $j, k \in \mathrm{NOM}_t \cup N_t$ for some $t \in S$,
   ($Bridge$) $\sigma(\ldots \phi_{i_1}, j, \phi_{i+1} \ldots) \wedge @_j^s \phi \to \sigma(\ldots \phi_{i-1}, \phi, \phi_{i+1}, \ldots)$
   if $\sigma \in \Sigma_{s_1 \ldots s_n, s}$, $j \in \mathrm{NOM}_{s_i} \cup N_{s_i}$ and $\phi$ is a formula of sort $s_i$.
2. if $\vdash^{s} \phi \to j$ then $\vdash^{t} \sigma(\ldots, \phi, \ldots) \to \sigma(\ldots, j, \ldots) \wedge @_j^t \phi$
   for any $s, t \in S$, $\sigma \in \Sigma_{t_1 \cdots t_n, t}$, $j \in \mathrm{NOM}_s \cup N_s$ and $\phi$ a formula of sort $s$.

*Proof.* 1. ($Nom$)

(1) $\vdash^{t} j \to (\phi \leftrightarrow @_j^t \phi)$ ($Intro$)
(2) $\vdash^{s} @_k^s (j \to (\phi \leftrightarrow @_j^t \phi))$ ($Gen@$)
(3) $\vdash^{s} @_k^s (j \to (\phi \leftrightarrow @_j^t \phi)) \to (@_k^s j \to @_k^s (\phi \leftrightarrow @_j^t \phi))$ ($K@$)
(4) $\vdash^{s} @_k^s j \to @_k^s (\phi \leftrightarrow @_j^t \phi)$ ($MP$) : (2), (3)
(5) $\vdash^{s} @_k^s (\phi \leftrightarrow @_j^t \phi) \leftrightarrow (@_k^s \phi \leftrightarrow @_k^s @_j^t \phi)$ ML

(6) $\vdash^{s} @_k^s j \to (@_k^s \phi \leftrightarrow @_k^s @_j^t \phi)$ PL:(4), (5)
(7) $\vdash^{s} @_k^s @_j^t \phi \leftrightarrow @_j^s \phi$ $(Agree)$
(8) $\vdash^{s} @_k^s j \to (@_k^s \phi \leftrightarrow @_j^s \phi)$ PL:(6), (7)

$(Sym)$
(1) $\vdash^{s} @_k^s j \wedge @_j^s k \to @_j^s k$ $Taut$
(2) $\vdash^{s} @_k^s j \wedge @_j^s k \to @_j^s k) \to (@_k^s j \to (@_j^s k \to @_j^s k))$ $Taut$
(3) $\vdash^{s} @_k^s j \to (@_j^s k \to @_j^s k)$ $(MP)$ : (1), (2)
(4) $\vdash^{s} (@_j^s k \to @_j^s k) \to @_j^s k$ PL
(5) $\vdash^{s} @_k^s j \to @_j^s k$ PL
(6) $\vdash^{s} @_j^s k \to @_k^s j$ Analogue
(7) $\vdash^{s} @_j^s k \leftrightarrow @_k^s j$ PL:(5), (6)

$(Bridge)$
(1) $\vdash^{s} \sigma(\ldots \phi_{i_1}, j, \phi_{i+1} \ldots) \wedge \sigma^{\Box}(\ldots, \neg\phi_{i-1}, \phi, \neg\phi_{i+1}, \ldots) \to$
$\sigma(\ldots \phi_{i-1}, j \wedge \phi, \phi_{i+1}, \ldots)$ ML
(2) $\vdash^{s} j \wedge \phi \to @_j^s \phi$ $(Intro)$
(3) $\vdash^{s} \sigma(\ldots \phi_{i-1}, j \wedge \phi, \phi_{i+1}, \ldots) \to \sigma(\ldots \phi_{i-1}, @_j^s \phi, \phi_{i+1}, \ldots)$ ML
(4) $\vdash^{s} \sigma(\ldots \phi_{i-1}, @_j^s \phi, \phi_{i+1}, \ldots) \to @_j^s \phi$ $(Back)$
(5) $\vdash^{s} \sigma(\ldots \phi_{i-1}, j, \phi_{i+1} \ldots) \wedge \sigma^{\Box}(\ldots, \neg\phi_{i-1}, \phi, \neg\phi_{i+1}, \ldots) \to @_j^s \phi$ PL
(6) $\vdash^{s} \sigma(\ldots \phi_{i-1}, j, \phi_{i+1} \ldots) \wedge \sigma^{\Box}(\ldots, \neg\phi_{i-1}, \neg\phi, \neg\phi_{i+1}, \ldots) \to @_j^s \neg\phi$ (5)
(7) $\vdash^{s} \neg @_j^s \neg\phi \to \neg(\sigma(\ldots \phi_{i-1}, j, \phi_{i+1} \ldots) \wedge \sigma^{\Box}(\ldots, \neg\phi_{i-1}, \neg\phi, \neg\phi_{i+1}, \ldots))$ PL
(8) $\vdash^{s} @_j^s \phi \to (\neg\sigma(\ldots \phi_{i-1}, j, \phi_{i+1} \ldots) \vee \neg\sigma^{\Box}(\ldots, \neg\phi_{i-1}, \neg\phi, \neg\phi_{i+1}, \ldots))$ PL
(9) $\vdash^{s} @_j^s \phi \to (\neg\sigma(\ldots \phi_{i-1}, j, \phi_{i+1} \ldots) \vee \sigma(\ldots, \phi_{i-1}, \phi, \phi_{i+1}, \ldots))$ $(Dual)$
(9) $\vdash^{s} @_j^s \phi \to (\sigma(\ldots \phi_{i-1}, j, \phi_{i+1} \ldots) \to \sigma(\ldots, \phi_{i-1}, \phi, \phi_{i+1}, \ldots))$ PL
(10) $\vdash^{s} @_j^s \phi \wedge \sigma(\ldots \phi_{i-1}, j, \phi_{i+1} \ldots) \to \sigma(\ldots, \phi_{i-1}, \phi, \phi_{i+1}, \ldots)$ PL

2.
(1) $\vdash^{s} j \to (\neg\phi \leftrightarrow @_j^s \neg\phi)$ $(Intro)$
(2) $\vdash^{s} j \to (\neg\phi \leftrightarrow @_j^s \neg\phi) \to (j \to (@_j^s \neg\phi \to \neg\phi))$ PL
(3) $\vdash^{s} j \to (@_j^s \neg\phi \to \neg\phi)$ $(MP)$ : (1), (2)
(4) $\vdash^{s} (j \to (@_j^s \neg\phi \to \neg\phi)) \to (j \wedge @_j^s \neg\phi \to \neg\phi)$ PL
(5) $\vdash^{s} j \wedge @_j^s \neg\phi \to \neg\phi$ $(MP)$ : (3), (4)
(6) $\vdash^{s} \phi \to (\neg j \vee @_j^s \phi)$ PL,$(SelfDual)$
(7) $\vdash^{s} \phi \to j$ hypothesis
(8) $\vdash^{s} \phi \to (\neg j \vee @_j^s \phi) \wedge j$ PL
(9) $\vdash^{s} \phi \to @_j^s \phi \wedge j$ PL
(10) $\vdash^{s} (\phi \to @_j^s \phi) \wedge (\phi \to j)$ PL
(11) $\vdash^{s} \phi \to @_j^s \phi$ PL

Therefore, if $\vdash^{s} \phi \to j$ then $\vdash^{s} \phi \to @_j^s \phi$.
(1) $\vdash^{s} \phi \to j$ hypothesis
(2) $\vdash^{t} \sigma(\ldots, \psi_{i-1}, \phi, \psi_{i+1}, \ldots) \to \sigma(\ldots, \psi_{i-1}, j, \psi_{i+1}, \ldots)$ ML(1)
(3) $\vdash^{s} \phi \to @_j^s \phi$ (1)
(4) $\vdash^{t} \sigma(\ldots, \psi_{i-1}, \phi, \psi_{i+1}, \ldots) \to \sigma(\ldots, \psi_{i-1}, @_j^s \phi, \psi_{i+1}, \ldots)$ ML(3)
(5) $\vdash^{t} \sigma(\ldots, \psi_{i-1}, \phi, \psi_{i+1}, \ldots) \to @_j^t \phi$ $(Back)$,PL(4)
(6) $\vdash^{t} \sigma(\ldots, \psi_{i-1}, \phi, \psi_{i+1}, \ldots) \to (\sigma(\ldots, \psi_{i-1}, j, \psi_{i+1}, \ldots) \wedge @_j^t \phi)$ PL:(2), (5)
Therefore, if $\vdash^{s} \phi \to j$ then $\vdash^{t} \sigma(\ldots, \phi, \ldots) \to \sigma(\ldots, j, \ldots) \wedge @_j^t \phi$. □

**Lemma 2 (Extended Lindenbaum Lemma).**
Let $\Lambda$ be a set of formulas in the language of $\mathcal{H}_{\mathbf{\Sigma}}(@,\forall)$ and $s \in S$. Then any consistent set $\Gamma_s$ of formulas of sort $s$ from $\mathcal{H}_{\mathbf{\Sigma}}(@,\forall) + \Lambda$ can be extended to a named, pasted and @-witnessed maximal consistent set by adding countably many nominals to the language.

*Proof.* The proof generalizes to the $S$-sorted setting well-known proofs for the mono-sorted hybrid logic, see [7, Lemma 7.25], [3, Lemma 3, Lemma 4], [4, Lemma 3.9].

For each sort $s \in S$, we add a set of new nominals and enumerate this set. Given a set of formulas $\Gamma_s$, define $\Gamma_s^k$ to be $\Gamma_s \cup \{k_s\} \cup \{@^s_{j_x} x \mid x \in \mathrm{SVAR_s}\}$, where $k_s$ is the first new nominal of sort $s$ in our enumeration and $j_x$ are such that if $x$ and $y$ are different state variables of sort $s$ then also $j_x$ and $j_y$ are different nominals of same sort $s$. Now that we know we are working on the sort $s$, we will write $k$ instead of $k_s$.

Suppose $\Gamma_s^k$ is not consistent. Then there exists some conjunction of formulas $\theta \in \Gamma_s$ such that $\overset{s}{\vdash} k \to \neg\theta$. We use the $(Gen@)$ rule and the $(K@)$ axiom to prove that $\overset{s}{\vdash} @^s_k k \to @^s_k \neg\theta$. From the $(Ref)$ axiom and the $(MP)$ rule it follows $\overset{s}{\vdash} @^s_k \neg\theta$. Remember that $k$ is a new nominal, so it does not occur in $\theta$ and we use $(Name@)$ rule to get that $\overset{s}{\vdash} \neg\theta \Rightarrow \neg\theta \in \Gamma_s$. But this contradicts the consistency of $\Gamma_s$. Now, we prove the case for the additional $@^s_{j_x} x$ formulas. Suppose $\overset{s}{\vdash} \theta \to \neg @^s_{j_x} x$. We use the $(SelfDual)$ axiom to get $\overset{s}{\vdash} \neg\theta \vee @^s_{j_x} \neg x$. If $\overset{s}{\vdash} \neg\theta$, this contradicts the consistency of $\Gamma_s$. If $\overset{s}{\vdash} @^s_{j_x} \neg x$, then $\overset{s}{\models} @^s_{j_x} \neg x$. Hence, for any model $\mathcal{M}$, any assignment function $g$ and any world $w \in W_s$, we have $\mathcal{M}, g, w \overset{s}{\models} @^s_{j_x} \neg x$ if and only if $\mathcal{M}, g, v \overset{s}{\models} \neg x$ where $V^N_s(j_x) = \{v\}$. Then for any model $\mathcal{M}$ and any assignment $g$, $g_s(x) \neq V^N_s(j_x)$, contradiction.

Now we enumerate on each sort $s \in S$ all the formulas of the new language obtained by adding the set of new nominals and define $\Gamma^0 := \Gamma_s^k$. Suppose we have defined $\Gamma^m$, where $m \geq 0$. Let $\phi_{m+1}$ be the $m+1-th$ formula of sort $s$ in the previous enumeration. We define $\Gamma^{m+1}$ as follows. If $\Gamma^m \cup \{\phi_{m+1}\}$ is inconsistent, then $\Gamma^{m+1} = \Gamma^m$. Otherwise:

(i) $\Gamma^{m+1} = \Gamma^m \cup \{\phi_{m+1}\}$, if $\phi_{m+1}$ is neither of the form $@_j\sigma(\ldots,\varphi,\ldots)$, nor of the form $@_j\exists x\varphi(x)$, where $j$ is any nominal of sort $s''$, $\varphi$ a formula of sort $s''$ and $x \in \mathrm{SVAR}_{s''}$.
(ii) $\Gamma^{m+1} = \Gamma^m \cup \{\phi_{m+1}\} \cup \{@_j\sigma(\ldots,k,\ldots) \wedge @_k\varphi\}$, if $\phi_{m+1}$ is of the form $@_j\sigma(\ldots,\varphi,\ldots)$.
(iii) $\Gamma^{m+1} = \Gamma^m \cup \{\phi_{m+1}\} \cup \{@_j\varphi[k/x]\}$, where $\phi_{m+1}$ is of the form $@_j\exists x\varphi(x)$.

In clauses $(ii)$ and $(iii)$, $k$ is the first new nominal in the enumeration that does not occur neither in $\Gamma^i$ for all $i \leq m$, nor in $@_j\sigma(\ldots,\varphi,\ldots)$.

Let $\Gamma^+ = \bigcup_{n\geq 0}\Gamma^n$. Because $k \in \Gamma^0 \subseteq \Gamma^+$, this set in named, maximal, pasted and @-witnessed by construction. We will check if it is consistent for the expansion made in the second and third items.

Suppose $\Gamma^{m+1} = \Gamma^m \cup \{\phi_{m+1}\} \cup \{@_j\sigma(\ldots,k,\ldots) \wedge @_k\varphi\}$ is an inconsistent set, where $\phi_{m+1}$ is $@_j\sigma(\ldots,\varphi,\ldots)$. Then there is a conjunction of formulas $\chi \in \Gamma^m \cup \{\phi_{m+1}\}$ such that $\overset{s}{\vdash} \chi \to \neg(@_j\sigma(\ldots,k,\ldots) \wedge @_k\varphi)$ and so

$\overset{s}{\vdash} @_j\sigma(\ldots,k,\ldots) \wedge @_k\varphi \to \neg\chi$. But $k$ is the first new nominal in the enumeration that does not occur neither in $\Gamma^m$, nor in $@_j\sigma(\ldots,\varphi,\ldots)$ and by Paste rule we get $\overset{s}{\vdash} @_j\sigma(\ldots,\varphi,\ldots) \to \neg\chi \Rightarrow \overset{s}{\vdash} \chi \to \neg @_j\sigma(\ldots,\varphi,\ldots)$, which contradicts the consistency of $\Gamma^m \cup \{\phi_{m+1}\}$.

Suppose $\Gamma^{m+1} = \Gamma^m \cup \{\phi_{m+1}\} \cup \{@_j\varphi[k/x]\}$ is inconsistent, where $\phi_{m+1}$ is $@_j\exists x\varphi(x)$. Then there is a conjunction of formulas $\chi \in \Gamma^m \cup \{\phi_{m+1}\}$ such that $\overset{s}{\vdash} \chi \to \neg @_j\varphi[k/x]$, where $k$ is the new nominal. By generalization on nominals (Lemma 7) we can prove $\overset{s}{\vdash} \forall y(\chi \to \neg @_j\varphi[y/x])$, where $y$ is a state variable that does not occur in $\chi \to \neg @_j\varphi[k/x]$. Using $(Q1)$ axiom, we get $\overset{s}{\vdash} \chi \to \forall y \neg @_j\varphi[y/x]$ and by $(SelfDual)$ $\overset{s}{\vdash} \chi \to \forall y @_j \neg\varphi[y/x]$. Next, we use $(Barcan@)$ to get $\overset{s}{\vdash} \chi \to @_j\forall y \neg\varphi[y/x])$. Because $x$ has no free occurrences in $\varphi[y/x]$, we can prove that $@_j\forall y\neg\varphi[y/x]) \leftrightarrow @_j\forall x\neg\varphi$. Therefore, $\overset{s}{\vdash} \chi \to @_j\forall x\neg\varphi$ , so $\overset{s}{\vdash} \chi \to @_j\neg\exists x\varphi$. Use once again $(SelfDual)$ and we have $\overset{s}{\vdash} \chi \to \neg @_j\exists x\varphi$. Then $\neg @_j\exists x\varphi \in \Gamma^m \cup \{\phi_{m+1}\}$, but this contradicts the consistency of $\Gamma^m \cup \{\phi_{m+1}\}$. □

**Lemma 3.** The Henkin model from Definition 9 is well-defined.

*Proof.* Let $s \in S$ and assume that $\Gamma_s$ is a set of formulas of sort $s$. Note that $R^\Gamma_\sigma$ is well-defined by $(Nom)$ and $(Bridge)$ from Lemma 1. For $t \in S$ and $j \in \mathrm{NOM}_t$, $V^\Gamma(j)$ is well-defined by axiom $(Ref)$. For the system $\mathcal{H}_\Sigma(@,\forall)$, we further that $\Gamma_s$ is also @-witnessed so, for any $t \in S$ and $x \in \mathrm{SVAR}_t$, there is a nominal $j \in \mathrm{NOM}_t$ such that $@^s_j x \in \Gamma$. The fact that $g^\Gamma$ is well-defined follows by $(Nom\, x)$. □

**Lemma 4 (Truth Lemma).** Let $s \in S$ and assume $\Gamma_s$ is a named, pasted and @-witnessed maximal consistent set of formulas of sort $s$ from $\mathcal{H}_\Sigma(@,\forall)$. For any sort $s' \in S$, $j \in \mathrm{NOM}_{s'} \cup N_{s'}$ and for any formula $\phi$ of sort $s'$ we have $\mathcal{M}^\Gamma, g^\Gamma, |j| \overset{s'}{\models} \phi$ iff $@^s_j\phi \in \Gamma_s$.

*Proof.* We make the proof by structural induction on $\phi$.

- $\mathcal{M}^\Gamma, g^\Gamma, |j| \overset{s'}{\models} a$, where $a \in \mathrm{PROP}_{s'} \cup \mathrm{NOM}_{s'} \cup N_{s'}$ iff $|j| \in V^N_{s'}(a)$ iff $@^s_j a \in \Gamma_s$.
- $\mathcal{M}^\Gamma, g^\Gamma, |j| \overset{s'}{\models} x$, where $x \in \mathrm{SVAR}_{s'}$ iff $g^\Gamma_{s'}(x) = |j|$ iff $@^s_j x \in \Gamma_s$.
- $\mathcal{M}^\Gamma, g^\Gamma, |j| \overset{s'}{\models} \neg\phi$ iff $\mathcal{M}^\Gamma, g^\Gamma, |j| \overset{s'}{\not\models} \phi$ iff $@^s_j\phi \notin \Gamma_s$, but we work with consistent sets, therefore $@^s_j\phi \notin \Gamma_s$ iff $\neg @^s_j\phi \in \Gamma_s$ iff $@^s_j\neg\phi \in \Gamma_s$ $(SelfDual)$.
- $\mathcal{M}^\Gamma, g^\Gamma, |j| \overset{s'}{\models} \phi \vee \varphi$ iff $\mathcal{M}^\Gamma, g^\Gamma, |j| \overset{s'}{\models} \phi$ or $\mathcal{M}^\Gamma, g^\Gamma, |j| \overset{s'}{\models} \varphi$ iff (inductive hypothesis) $@^s_j\phi \in \Gamma_s$ or $@^s_j\varphi \in \Gamma_s$ iff $@^s_j\phi \vee @^s_j\varphi \in \Gamma_s$ iff $@^s_j(\phi \vee \varphi) \in \Gamma_s$.
- $\mathcal{M}^\Gamma, g^\Gamma, |j| \overset{s'}{\models} \sigma(\phi_1,\ldots,\phi_n)$ iff exists $|k_i| \in W_{s_i}$ such that $R|j||k_1|\ldots|k_n|$ and $\mathcal{M}^\Gamma, g^\Gamma, |k_i| \overset{s_i}{\models} \phi_i$ for any $i \in [n]$. Using the induction hypothesis, we get $@^s_{k_i}\phi_i \in \Gamma_s$. But $R|j||k_1|\ldots|k_n|$ iff $@^s_j\sigma(k_1,\ldots,k_n) \in \Gamma_s$. Use the Bridge axiom to prove $@^s_j\sigma(k_1,\ldots,k_n) \wedge @^s_{k_1}\phi_1 \wedge \ldots \wedge @^s_{k_n}\phi_n \to @^s_j\sigma(\phi_1,\ldots,\phi_n)$, so $@^s_j\sigma(\phi_1,\ldots,\phi_n) \in \Gamma_s$. Now, suppose $@^s_j\sigma(\phi_1,\ldots,\phi_n) \in \Gamma_s$. We work with

pasted models, so there are some nominals $k_i$ such that $@_j^s\sigma(k_1,\ldots,k_n) \in \Gamma_s$ and $@_{k_i}^s\phi_i \in \Gamma_s$ for any $i \in [n]$. Therefore, exists $k_i$ such that $R|j||k_1|\ldots|k_n|$ and, by induction hypothesis, $\mathcal{M}^\Gamma, g^\Gamma, |k_i| \overset{s_i}{\models} \phi_i$ for any $i \in [n]$ if and only if $\mathcal{M}^\Gamma, g^\Gamma, |j| \overset{s'}{\models} \sigma(\phi_1,\ldots,\phi_n)$.

- $\mathcal{M}^\Gamma, g^\Gamma, |j| \overset{s'}{\models} @_k^{s'}\phi$ iff $\mathcal{M}^\Gamma, g^\Gamma, |k| \overset{s''}{\models} \phi$, but from induction hypothesis $@_k^s\phi \in \Gamma_s$ and by applying $(Agree)$ we get $@_j^s @_k^s\phi \in \Gamma_s$.
- $@_j^s\exists x\phi \in \Gamma_s$, then there exists $l \in \mathrm{NOM}_{s'}$ such that $@_j^s\phi[l/x] \in \Gamma_s$. Let $g' \overset{x}{\sim} g^\Gamma$ such that $g'_{s'}(x) = \{|l|\}$. Therefore, there exists $l \in \mathrm{NOM}_{s'}$ such that $g'_{s'}(x) = \{|l|\}$, $g' \overset{x}{\sim} g^\Gamma$ and $\mathcal{M}^\Gamma, g', |j| \overset{s'}{\models} \phi$ iff $\mathcal{M}^\Gamma, g^\Gamma, |j| \overset{s'}{\models} \exists x\phi$.
- $\mathcal{M}^\Gamma, g^\Gamma, |j| \overset{s'}{\models} \exists x\phi$ iff exists $g' \overset{x}{\sim} g^\Gamma$ and $\mathcal{M}^\Gamma, g', |j| \overset{s'}{\models} \phi$. Let $g'_{s'}(x) = \{|l|\}$. Hence, there exists $l \in \mathrm{NOM}_{s'}$ such that $g'_{s'}(x) = \{|l|\}$, $g' \overset{x}{\sim} g^\Gamma$ and $\mathcal{M}^\Gamma, g', |j| \overset{s'}{\models} \phi$ iff $\mathcal{M}^\Gamma, g, |j| \overset{s'}{\models} \phi[l/x]$ and from inductive hypothesis $@_j^s\phi[l/x] \in \Gamma_s$. Use the contrapositive of the $(Q2)$ axiom, $\overset{s'}{\vdash} \phi[l/x] \to \exists x\phi$ and the $(Gen@)$ and $(K@)$ rules to obtain $@_j^s\phi[l/x] \to @_j^s\exists x\phi \in \Gamma_s$. Therefore, $@_j^s\exists x\phi \in \Gamma_s$.

□

**Theorem 2 (Completeness).**

1. Strong model-completeness. Let $\Lambda$ be a set of formulas in the language of $\mathcal{H}_\Sigma(@,\forall)$ and $s \in S$ and assume $\Gamma_s$ is a set of formulas of sort $s$. If $\Gamma_s$ is a consistent set in $\mathcal{L} = \mathcal{H}_\Sigma(@,\forall) + \Lambda$ then $\Gamma_s$ has a model that is also a model of $\Lambda$. Consequently, for a formula $\phi$ of sort $s$, $\Gamma_s \overset{s}{\models}_{Mod(\mathcal{L})} \phi$ iff $\Gamma_s \overset{s}{\vdash}_{\mathcal{L}} \phi$.
2. Strong frame-completeness for pure extensions. Let $\Lambda$ be a set of pure formulas in the language of $\forall\exists$-pure formulas in the language of $\mathcal{H}_\Sigma(@,\forall)$ and $s \in S$ and assume $\Gamma_s$ is a set of formulas of sort $s$. If $\Gamma_s$ is a consistent set in $\mathcal{L} = \mathcal{H}_\Sigma(@,\forall) + \Lambda$ then $\Gamma_s$ has a model based on a frame that validates every formula in $\Lambda$. For a formula $\phi$ of sort $s$, $\Gamma_s \overset{s}{\models}_{Mod(\mathcal{L})} \phi$ iff $\Gamma_s \overset{s}{\vdash}_{\mathcal{L}} \phi$.

*Proof.* Since 1. is obvious, we only prove 2. If $\Gamma_s$ is a consistent set in $\mathcal{H}_\Sigma(@,\forall) + \Lambda$ then, applying the Extended Lindenbaum Lemma, then $\Gamma_s \subseteq \Theta_s$, where $\Theta_s$ is a maximal consistent named, pasted and @-witnessed set (in an extended language $\mathcal{L}'$). If $\mathcal{M}^\Theta$ is the Henkin model and $g^\Theta$ is the assignment from Definition 9 then, by Truth Lemma, $\mathcal{M}^\Theta, g^\Theta, |j| \overset{s}{\models} \Gamma_s$ for any $t \in S$ and $j \in \mathrm{NOM}_t \cup N_t$. Moreover, $\mathcal{M}^\Theta$ is a named model (in the extended language) that is also a model of $\Lambda$. By Proposition 1, the underlying frame of $\mathcal{M}^\Theta$ satisfies the $\forall\exists$-pure formulas from $\Lambda$. Hence the logic $\mathcal{H}_\Sigma(@,\forall) + \Lambda$ is strongly complete w.r.t to the class of frames satisfying $\Lambda$. Assume that $\Gamma_s \overset{s}{\models}_\Lambda \phi$ and suppose that $\Gamma_s \overset{s}{\nvdash} \phi$. It follows that $\Gamma_s \cup \{\neg\phi\}$ is inconsistent, so there exists a model of $\Gamma_s$ based on a frame satisfying $\Lambda$ that is not a model of $\phi$. We get a contradiction, so the intended completeness result is proved. □

**Proposition 3.**

1. (Nominal Conjunction) For any formulas and any nominals of appropriate sorts, the following hold:
   (i1) $\sigma(\ldots,\phi_{i-1},\phi_i,\phi_{i+1},\ldots) \wedge @_k(\psi) \leftrightarrow \sigma(\ldots,\phi_{i-1},\phi_i \wedge @_k(\psi),\phi_{i+1},\ldots)$
   (i2) $\sigma^{\Box}(\ldots,\phi_{i-1},\phi_i,\phi_{i+1},\ldots) \wedge @_k(\psi) \leftrightarrow$
   $\sigma^{\Box}(\ldots,\phi_{i-1},\phi_i \wedge @_k(\psi),\phi_{i+1},\ldots) \wedge @_k(\psi)$
2. If $\phi_1,\ldots\phi_n$ are formulas of appropriate sorts and $x$ is a state variable that does not occur in $\phi_j$ for any $j \neq i$ then:
   (i3) $\exists x\sigma^{\Box}(\ldots,\phi_{i-1},\phi_i,\phi_{i+1},\ldots) \to \sigma^{\Box}(,\ldots,\phi_{i-1},\exists x\phi_i,\phi_{i+1},\ldots)$

*Proof.* 1. (Nominal Conjunction)

(i1) $\mathcal{M},g,w \overset{s}{\models} \sigma(\ldots,\phi_{i-1},\phi_i,\phi_{i+1},\ldots) \wedge @_k(\psi)$ iff $\mathcal{M},g,w \overset{s}{\models} @_k(\psi)$ and $\mathcal{M},g,w \overset{s}{\models} \sigma(\ldots,\phi_{i-1},\phi_i,\phi_{i+1},\ldots)$ iff $\mathcal{M},g,v \overset{s'}{\models} \psi$ where $V_{s'}^N = \{v\}$ and there exist $w_1 \in W_{s_1},\ldots,w_n \in W_{s_n}$ such that $R_\sigma ww_1\cdots w_n$ and $\mathcal{M},g,w_j \overset{s_j}{\models} \phi_j$ for all $1 \leq j \leq n$ iff there exist $w_1 \in W_{s_1},\ldots,w_n \in W_{s_n}$ such that $R_\sigma ww_1\cdots w_n$ and $\mathcal{M},g,w_j \overset{s_j}{\models} \phi_j$ for all $1 \leq j \leq n$, $j \neq i$, and $\mathcal{M},g,w_i \overset{s_i}{\models} \phi_i \wedge @_k(\psi)$ iff $\mathcal{M},g,w \models \sigma(\ldots,\phi_{i-1},\phi_i \wedge @_k(\psi),\phi_{i+1},\ldots)$.

(i2) $\mathcal{M},g,w \overset{s}{\models} \sigma^{\Box}(\ldots,\phi_{i-1},\phi_i,\phi_{i+1},\ldots) \wedge @_k(\psi)$ iff $\mathcal{M},g,w \overset{s}{\models} @_k(\psi)$ and $\mathcal{M},g,w \overset{s}{\models} \neg\sigma(\ldots,\neg\phi_{i-1},\neg\phi_i,\neg\phi_{i+1},\ldots)$ iff $\mathcal{M},g,v \overset{s'}{\models} \psi$ where $V_{s'}^N = \{v\}$ and for all $w_1 \in W_{s_1},\ldots,w_n \in W_{s_n}$ for which $R_\sigma ww_1\cdots w_n$, there exists $1 \leq j \leq n$ such that $\mathcal{M},g,w_j \overset{s_j}{\models} \phi_j$ iff $\mathcal{M},g,v \overset{s'}{\models} \psi$ where $V_{s'}^N = \{v\}$ and for all $w_1 \in W_{s_1},\ldots,w_n \in W_{s_n}$ for which $R_\sigma ww_1\cdots w_n$, there exists $1 \leq j \leq n$, $j \neq i$ such that $\mathcal{M},g,w_j \overset{s_j}{\models} \phi_j$ or $\mathcal{M},g,w_i \overset{s_i}{\models} \phi_i$ iff $\mathcal{M},g,v \overset{s'}{\models} \psi$ and for all $w_1 \in W_{s_1},\ldots,w_n \in W_{s_n}$ for which $R_\sigma ww_1\cdots w_n$, there exists $1 \leq j \leq n$, $j \neq i$ such that $\mathcal{M},g,w_j \overset{s_j}{\models} \phi_j$ or $\mathcal{M},g,w_i \overset{s_i}{\models} \phi_i \wedge @_k(\psi)$ iff $\mathcal{M},g,w \overset{s}{\models} \sigma^{\Box}(\ldots,\phi_{i-1},\phi_i \wedge @_k(\psi),\phi_{i+1},\ldots) \wedge @_k(\psi)$.

2.

(i3) $\mathcal{M},g,w \overset{s}{\models} \exists x\sigma^{\Box}(\phi_1,\ldots,\phi_{i-1},\phi_i,\phi_{i+1},\ldots,\phi_n)$ iff exists $g' \overset{x}{\sim} g$ such that $\mathcal{M},g',w \overset{s}{\models} \sigma^{\Box}(\phi_1,\ldots,\phi_{i-1},\phi_i,\phi_{i+1},\ldots,\phi_n)$ iff exists $g' \overset{x}{\sim} g$ such that for all $(v_1,\ldots,v_n) \in W_{s_1} \times \ldots \times W_{s_n}$, $R_\sigma wv_1\ldots v_n$ implies $\mathcal{M},g',v_j \overset{s_j}{\models} \phi_j$ for some $j \in [n]$. Then, for all $(v_1,\ldots,v_n) \in W_{s_1} \times \ldots \times W_{s_n}$, $R_\sigma wv_1\ldots v_n$ implies there exists $g' \overset{x}{\sim} g$ such that $\mathcal{M},g',v_j \overset{s_j}{\models} \phi_j$ for some $j \in [n]$. But $x$ does not occur in $\phi_j$ for any $j \in [n]$ and $j \neq i$, so for all $(v_1,\ldots,v_i,\ldots,v_n) \in W_{s_1} \times \ldots \times W_{s_i} \times \ldots \times W_{s_n}$, $R_\sigma wv_1\ldots v_i\ldots v_n$ implies $\mathcal{M},g',v_j \overset{s_j}{\models} \phi_j$ and there exists $g' \overset{x}{\sim} g$ such that $\mathcal{M},g',v_i \overset{s_i}{\models} \phi_i$ for some $i,j \in [n]$ and $j \neq i$. We use Agreement Lemma, then for all $(v_1,\ldots,v_i,\ldots,v_n) \in W_{s_1} \times \ldots \times W_{s_i} \times \ldots \times W_{s_n}$, $R_\sigma wv_1\ldots v_i\ldots v_n$ implies $\mathcal{M},g,v_j \overset{s_j}{\models} \phi_j$ and $\mathcal{M},g,v_i \overset{s_i}{\models} \exists x\phi_i$ for some $i,j \in [n]$ and $j \neq i$. Therefore, $\mathcal{M},g,w \overset{s}{\models} \sigma^{\Box}(\phi_1,\ldots,\phi_{i-1},\exists x\phi_i,\phi_{i+1},\ldots,\phi_n)$.

□

**Proposition** 4 **($\mid\!\stackrel{s}{\sim}$ soundness).** Let $\mathbf{\Gamma}$ be an $S$-sorted set and $\phi$ a formula of sort $s \in S$. If $\mathbf{\Gamma} \mid\!\stackrel{s}{\sim} \varphi$ then $\mathcal{M} \models \mathbf{\Gamma}$ implies $\mathcal{M} \stackrel{s}{\models} \phi$ for any model $\mathcal{M}$.

*Proof.* Let $\mathcal{M}$ be a model and assume $\stackrel{s}{\vdash} @^s_{j_1}\gamma_1 \wedge \cdots \wedge @^s_{j_n}\gamma_n \to \phi$ as above. If $\mathcal{M} \models \mathbf{\Gamma}$ then, by $(Gen@)$, $\mathcal{M} \stackrel{s}{\models} \Gamma_s \cup \{@^s_{j_1}\gamma_1, \ldots, @^s_{j_n}\gamma_n\}$. Using the soundness of the local deduction, we get the desired conclusion. □

# B Proofs from Sect. 4

**Proposition** 5 **(Hoare-like Admissible Rules).** The following rules are admissible:

1. **Rules of Consequence**
   If $\vdash \phi \to [\alpha]\psi$ and $\vdash \psi \to \chi$ then $\vdash \phi \to [\alpha]\chi$.
   If $\vdash \phi \to [\alpha]\psi$ and $\vdash \chi \to \phi$ then $\vdash \chi \to [\alpha]\psi$.
2. **Rule of Composition, iterated**
   If $\phi_0 \to [\alpha_1]\phi_1, \ldots, \phi_{n-1} \to [\alpha_n]\phi_n$, then $\phi_0 \to [\alpha_1; \ldots; \alpha_n]\phi_n$.
3. **Rule of Conditional**
   If $B$ is a formula of sort $Bool$, and $vs$, $mem$, $P$ are formulas of appropriate sorts such that
   (h1) $\vdash \phi \to [c(b)](\langle B \cdot vs, mem\rangle \wedge P)$,
   (h2) $\vdash \langle vs, mem\rangle \wedge P \wedge @_{\mathbf{true}}(B) \to [c(s1)]\chi$
   (h3) $\vdash \langle vs, mem\rangle \wedge P \wedge @_{\mathbf{false}}(B) \to [c(s2)]\chi$
   (h4) $\vdash P \to [\alpha]P$ for any $\alpha$ of sort $CtrlStack$,
   then $\vdash \phi \to [c(\mathbf{if}\ b\ \mathbf{then}\ s1\ \mathbf{else}\ s2)]\chi$

*Proof.* In the sequel we shall mention the sort of a formula only when it is necessary.

1. Rule of Consequence follows easily by $(UG)$.
2. Rule of Composition follows easily by $(UG)$ and $(CStmt)$.
3. Rule of Conditional. Since $B$ is a formula of sort $Bool$, using the axiom $(B1)$ and the completeness theorem, one can easily infer that

   $\vdash B \leftrightarrow (\mathbf{true} \wedge @_{\mathbf{true}}B) \vee (\mathbf{false} \wedge @_{\mathbf{false}}B)$
   Using the fact that any operator $\sigma \in \Sigma$ commutes with disjunctions, Proposition 3 we get
   $(*) \vdash \langle B \cdot vs, mem\rangle \to ((\langle \mathbf{true} \cdot vs, mem\rangle \wedge @_{\mathbf{true}}B) \vee$
   $(\langle \mathbf{false} \cdot vs, mem\rangle \wedge @_{\mathbf{false}}B)$
   Now we prove that

   $\vdash \langle \mathbf{true} \cdot vs, mem\rangle \wedge @_{\mathbf{true}}B \to [(\mathbf{true}?; c(s1)) \cup (\mathbf{false}; c(s2))]\chi$.
   Note that $\vdash @_{\mathbf{true}}(\neg\, \mathbf{false})$, so we use $(A?)$ and $(A\neg?)$ as follows:
   $\vdash \langle \mathbf{true} \cdot vs, mem\rangle \wedge @_{\mathbf{true}}B \to \langle \mathbf{true} \cdot vs, mem\rangle \wedge @_{\mathbf{true}}B \wedge @_{\mathbf{true}}(\neg\, \mathbf{false})$
   $\vdash \langle \mathbf{true} \cdot vs, mem\rangle \to [\mathbf{true}?]\, \langle vs, mem\rangle$
   $\vdash \langle \mathbf{true} \cdot vs, mem\rangle \wedge @_{\mathbf{true}}(\neg\, \mathbf{false}) \to [\mathbf{false}?]\bot$

Next we prove that
$(@[]) \vdash @_k\varphi \to [\alpha]@_k\varphi$
for any formulas $\alpha$, $\varphi$ and nominal $k$ of appropriate sorts. Note that $\vdash [\alpha]\top$ so, using Proposition 3.3, we have the following chain of inferences:
$\vdash @_k\varphi \to @_k\varphi \wedge [\alpha]\top$
$\vdash @_k\varphi \wedge [\alpha]\top \to [\alpha]@_k\varphi$
and $(@[])$ easily follows.
Consequently, $\vdash @_{\mathbf{true}}B \to [\mathbf{true}?]@_{\mathbf{true}}B$
Since dual operators $\sigma^\Box$ for $\sigma \in \Sigma$ commutes with conjunctions, using also (h4) we get
$\vdash \langle \mathbf{true} \cdot vs, mem\rangle \wedge P \wedge @_{\mathbf{true}}B \to ([\mathbf{true}?](\langle vs, mem\rangle \wedge P \wedge @_{\mathbf{true}}B))\wedge [\mathbf{false}?]\bot$
By (h2) and $(K)$ it follows that
$\vdash \langle \mathbf{true} \cdot vs, mem\rangle \wedge \wedge P @_{\mathbf{true}}B \to [\mathbf{true}?; c(s1)]\chi \wedge [\mathbf{false}?]\bot$
Since $\bot \to [c(s2)]\chi$, and using $(A\cup)$ we proved
$\vdash \langle \mathbf{true} \cdot vs, mem\rangle \wedge P \wedge @_{\mathbf{true}}B \to [(\mathbf{true}?; c(s1)) \cup (\mathbf{false}?; c(s2))]\chi$.
In a similar way, we get
$\vdash \langle \mathbf{false} \cdot vs, mem\rangle \wedge P \wedge @_{\mathbf{false}}B \to [(\mathbf{true}?; c(s1)) \cup (\mathbf{false}?; c(s2))]\chi$.
By $(*)$ we infer
$\vdash \langle B \cdot vs, mem\rangle \to [(\mathbf{true}?; c(s1)) \cup (\mathbf{false}?; c(s2))]\chi$
Using $(K)$ and $(Dif)$ we get the conclusion. □

**Proposition 6 (Rule of Iteration).** Let $B$, $vs$, $mem$, and $P$ be formulas with variables over $\mathbf{x}$, where $\mathbf{x}$ is a set of state variables. If there exist substitutions $\mathbf{x_{init}}$ and $\mathbf{x_{body}}$ for the variables of $\mathbf{x}$ such that:

(h1) $\vdash \phi \to [c(b)](\langle B \cdot vs, mem\rangle \wedge P)[\mathbf{x_{init}}/\mathbf{x}]$,
(h2) $\vdash \langle vs, mem\rangle \wedge P \wedge @_{\mathbf{true}}(B) \to [c(s); c(b)](\langle B \cdot vs, mem\rangle \wedge P)[\mathbf{x_{body}}/\mathbf{x}]$
(h3) $\vdash P \to [\alpha]P$ for any formula $\alpha$ of sort $CtrlStack$

then $\vdash \phi \to [c(\mathbf{while}\ b\ \mathbf{do}\ s)]\exists\mathbf{x}\, \langle vs, mem\rangle \wedge P \wedge @_{\mathbf{false}}(B)$.

*Proof.* Denote $\theta := \langle B \cdot vs, mem\rangle \wedge P$ and $\theta_I := \exists\mathbf{x}\theta$. We think of $\theta_I$ as being the invariant of **while** $b$ **do** $s$. Note that, using the contraposition of $(Q2)$ and (h1) we infer that
(c1) $\vdash \phi \to [c(b)]\theta_I$
In the following we firstly prove that
(c2) $\vdash \theta_I \to [\alpha]\theta_I$,
where $\alpha =\mathbf{true}?; c(s); c(b)$. Since
$\vdash B \leftrightarrow (\mathbf{true}\ \wedge @_{\mathbf{true}}B) \vee (\mathbf{false}\ \wedge @_{\mathbf{false}}B)$
it follows that
$\vdash \theta \to (\langle \mathbf{true} \cdot vs, mem\rangle \wedge P \wedge @_{\mathbf{true}}B) \vee (\langle \mathbf{false} \cdot vs, mem\rangle \wedge P \wedge @_{\mathbf{false}}B)$
By $(A?)$, (h3) and $(@[])$ (from the proof of Proposition 5) we infer
$\vdash \langle \mathbf{true} \cdot vs, mem\rangle \wedge P \wedge @_{\mathbf{true}}B \to [true?](\langle vs, mem\rangle \wedge P \wedge @_{\mathbf{true}}B)$
and, by (h2)
$\vdash \langle \mathbf{true} \cdot vs, mem\rangle \wedge P \wedge @_{\mathbf{true}}B) \to [\alpha]\theta[\mathbf{x_{body}}/\mathbf{x}]$

Since $\vdash @_{\mathbf{false}}(\neg\, \mathbf{true})$, by $(A\neg?)$ we get

$\vdash \langle \mathbf{false} \cdot vs, mem\rangle \wedge @_{\mathbf{false}}(\neg\, \mathbf{true}) \to [\mathbf{true}?]\bot$, so

$\vdash \langle \mathbf{false} \cdot vs, mem\rangle \wedge P \wedge @_{\mathbf{false}}B) \to [\alpha]\theta[\mathbf{x_{body}}/\mathbf{x}]$

As consequence $\vdash \theta \to [\alpha]\theta[\mathbf{x_{body}}/\mathbf{x}]$ and, using the contraposition of $Q_2$, we infer that $\theta \to [\alpha]\theta_I$. We use now the fact that

$\vdash \forall x(\varphi(x) \to \psi) \to (\exists x\varphi(x) \to \psi)$ if $x$ does not appear in $\psi$,

which leads us to $\vdash \theta_I \to [\alpha]\theta_I$. Using $(UG)$ we get $\vdash [c(b);\alpha^*](\theta_I \to [\alpha]\theta_I)$.

By (c1) it follows that

$\vdash \phi \to ([c(b)]\theta_I \wedge ([c(b);\alpha^*](\theta_I \to [\alpha]\theta_I))$

Using the induction axiom, $(UG)$, $(K)$ and the fact that the dual operators commutes with conjunctions, we get

$\vdash ([c(b)]\theta_I \wedge ([c(b);\alpha^*](\theta_I \to [\alpha]\theta_I)) \to [c(b);\alpha^*]\theta_I$

So $\vdash \phi \to [c(b);\alpha^*]\theta_I$, which proves the invariant property of **while** $b$ **do** $s$.

To conclude, so far we proved

$\vdash \phi \to [c(b);\alpha^*]\exists\mathbf{x}\theta$

We can safely assume that the state variables from $\mathbf{x}$ do not appear in $\phi$, $b$

Note that $c(\mathbf{while}\ b\ \mathbf{do}\ s) \leftrightarrow c(b);\alpha^*;\mathbf{false}?$

As before,

$\vdash \theta \to (\langle \mathbf{true} \cdot vs, mem\rangle \wedge P \wedge @_{\mathbf{true}}B) \vee (\langle \mathbf{false} \cdot vs, mem\rangle \wedge P \wedge @_{\mathbf{false}}B)$

Using again $(A?)$ and $(A\neg?)$ we have that

$\vdash \langle \mathbf{false} \cdot vs, mem\rangle \to [\mathbf{false}?]\, \langle vs, mem\rangle$

$\vdash \langle \mathbf{true} \cdot vs, mem\rangle \wedge @_{\mathbf{true}}(\neg\, \mathbf{false}) \to [\mathbf{false}?]\bot$

It follows that

$\vdash \theta \to [\mathbf{false}?](< vs, mem > \wedge P \wedge @_{\mathbf{false}}B)$ so, using the properties of the existential binder

$\vdash \exists\mathbf{x}\theta \to \exists\mathbf{x}[\mathbf{false}?](< vs, mem > \wedge P \wedge @_{\mathbf{false}}B)$

Since the state variables from $\mathbf{x}$ do not appear in **false**?, by Proposition 3 it follows that

$$\vdash \exists\mathbf{x}[\mathbf{false}?](< vs, mem > \wedge P \wedge @_{\mathbf{false}}B) \to [\mathbf{false}?]\exists\mathbf{x}(< vs, mem > \wedge P \wedge @_{\mathbf{false}}B)$$

We can finally obtain the intended result:

$\vdash \phi \to [c(b);\alpha^*;\mathbf{false}?]\exists\mathbf{x}(< vs, mem > \wedge P \wedge @_{\mathbf{false}}B)$ □

## References

1. Areces, C., ten Cate, B.: Hybrid logics. In: Blackburn, P., et al. (eds.) Handbook of Modal Logic, vol. 3, pp. 822–868. Elsevier, Amsterdam (2007)
2. Blackburn, P., Seligman, J.: Hybrid Languages. J. Log. Lang. Inf. **4**, 251–272 (1995)
3. Blackburn, P., ten Cate, B.: Pure extensions, proof rules, and hybrid axiomatics. Stud. Log. **84**(2), 277–322 (2006)
4. Blackburn, P., Tzakova, M.: Hybrid completeness. Log. J. IGPL **4**, 625–650 (1998)
5. Blackburn, P., Tzakova, M.: Hybrid languages and temporal logic. Log. J. IGPL **7**, 27–54 (1999)
6. Blackburn, P, van Benthem, J.: Modal logic: a semantic perspective. In: Blackburn, P., et al. (eds.) Handbook of Modal Logic, vol. 3, pp. 1–84 (2007)

7. Blackburn, P., Venema, Y., de Rijke, M.: Modal Logic. Cambridge University Press, Cambridge (2002)
8. Bohrer, B., Platzer, A.: A hybrid, dynamic logic for hybrid-dynamic information flow. In: LICS 2018 Proceedings of the 33rd Annual ACM/IEEE Symposium on Logic in Computer Science, pp. 115–124 (2018)
9. Calcagno, C., Gardner, P., Zarfaty, U.: Context logic as modal logic: completeness and parametric inexpressivity. In: POPL 2007 Proceedings of the 34th Annual ACM SIGPLAN-SIGACT Symposium on Principles of Programming Languages, pp. 123–134 (2007)
10. Chen, X., Roşu, G.: Matching mu-Logic. In: LICS 2019. Technical report http://hdl.handle.net/2142/102281 (2019, to appear)
11. Floyd, R.W.: Assigning meanings to programs. Proc. Am. Math. Soc. Symp. Appl. Math. **19**, 19–31 (1967)
12. Gargov, G., Goranko, V.: Modal logic with names. J. Philos. Log. **22**, 607–636 (1993)
13. Goranko, V., Vakarelov, D.: Sahlqvist formulas in hybrid polyadic modal logics. J. Log. Comput. **11**, 737–754 (2001)
14. Goguen, J., Malcolm, G.: Algebraic Semantics of Imperative Programs. MIT Press, Cambridge (1996)
15. Harel, D., Tiuryn, J., Kozen, D.: Dynamic Logic. MIT Press Cambridge, Cambridge (2000)
16. Heering, J., Hendriks, P.R.H., Klint, P., Rekers, J.: The syntax definition formalism SDF —reference manual—. ACM Sigplan Not. **24**(11), 43–75 (1989)
17. Hoare, C.A.R.: An axiomatic basis for computer programming. Commun. ACM **12**(10), 576–580 (1969)
18. Leuştean, I., Moangă, N., Şerbănuţă, T.F.: A many-sorted polyadic modal logic. arXiv:1803.09709 (2018, submitted)
19. Leuştean, I., Moangă, N., Şerbănuţă, T.F.: Operational semantics using many-sorted hybrid modal logic. arXiv:1905.05036 (2019)
20. Platzer, A.: Logical Foundations of Cyber-Physical Systems. Springer, Cham (2018). https://doi.org/10.1007/978-3-319-63588-0
21. Plotkin, G.D.: A structural approach to operational semantics (1981) Technical report DAIMI FN-19, Computer Science Department, Aarhus University, Aarhus, Denmark. (Reprinted with corrections in J. Log. Algebr. Program) 60–61, 17–139 (2004)
22. Roşu, G.: Matching logic. Log. Methods Comput. Sci. **13**(4), 1–61 (2017)
23. Reynolds, J.C.: Separation logic: a logic for shared mutable data structures. In: Proceedings 17th Annual IEEE Symposium on Logic in Computer Science (2002)
24. Schröder, L., Pattinson, D.: Modular algorithms for heterogeneous modal logics via multi-sorted coalgebra. Math. Struct. Comput. Sci. **21**(2), 235–266 (2011)
25. Venema, Y.: Points, lines and diamonds: a two-sorted modal logic for projective planes. J. Log. Comput. **9**, 601–621 (1999)
26. von Oheimb, D.: Hoare logic for Java in Isabelle/HOL. Concurr. Comput. Pract. Exp. **13**(13), 1173–1214 (2001)

# Author Index

Printed in the United States
By Bookmasters